MANAGEMENT OF SITE INVESTIGATIONS FOR RADIOACTIVE WASTE DISPOSAL FACILITIES

The following States are Members of the International Atomic Energy Agency:

AFGHANISTAN
ALBANIA
ALGERIA
ANGOLA
ANTIGUA AND BARBUDA
ARGENTINA
ARMENIA
AUSTRALIA
AUSTRIA
AZERBAIJAN
BAHAMAS
BAHRAIN
BANGLADESH
BARBADOS
BELARUS
BELGIUM
BELIZE
BENIN
BOLIVIA, PLURINATIONAL STATE OF
BOSNIA AND HERZEGOVINA
BOTSWANA
BRAZIL
BRUNEI DARUSSALAM
BULGARIA
BURKINA FASO
BURUNDI
CABO VERDE
CAMBODIA
CAMEROON
CANADA
CENTRAL AFRICAN REPUBLIC
CHAD
CHILE
CHINA
COLOMBIA
COMOROS
CONGO
COSTA RICA
CÔTE D'IVOIRE
CROATIA
CUBA
CYPRUS
CZECH REPUBLIC
DEMOCRATIC REPUBLIC OF THE CONGO
DENMARK
DJIBOUTI
DOMINICA
DOMINICAN REPUBLIC
ECUADOR
EGYPT
EL SALVADOR
ERITREA
ESTONIA
ESWATINI
ETHIOPIA
FIJI
FINLAND
FRANCE
GABON
GAMBIA
GEORGIA
GERMANY
GHANA
GREECE
GRENADA
GUATEMALA
GUINEA
GUYANA
HAITI
HOLY SEE
HONDURAS
HUNGARY
ICELAND
INDIA
INDONESIA
IRAN, ISLAMIC REPUBLIC OF
IRAQ
IRELAND
ISRAEL
ITALY
JAMAICA
JAPAN
JORDAN
KAZAKHSTAN
KENYA
KOREA, REPUBLIC OF
KUWAIT
KYRGYZSTAN
LAO PEOPLE'S DEMOCRATIC REPUBLIC
LATVIA
LEBANON
LESOTHO
LIBERIA
LIBYA
LIECHTENSTEIN
LITHUANIA
LUXEMBOURG
MADAGASCAR
MALAWI
MALAYSIA
MALI
MALTA
MARSHALL ISLANDS
MAURITANIA
MAURITIUS
MEXICO
MONACO
MONGOLIA
MONTENEGRO
MOROCCO
MOZAMBIQUE
MYANMAR
NAMIBIA
NEPAL
NETHERLANDS, KINGDOM OF THE
NEW ZEALAND
NICARAGUA
NIGER
NIGERIA
NORTH MACEDONIA
NORWAY
OMAN
PAKISTAN
PALAU
PANAMA
PAPUA NEW GUINEA
PARAGUAY
PERU
PHILIPPINES
POLAND
PORTUGAL
QATAR
REPUBLIC OF MOLDOVA
ROMANIA
RUSSIAN FEDERATION
RWANDA
SAINT KITTS AND NEVIS
SAINT LUCIA
SAINT VINCENT AND THE GRENADINES
SAMOA
SAN MARINO
SAUDI ARABIA
SENEGAL
SERBIA
SEYCHELLES
SIERRA LEONE
SINGAPORE
SLOVAKIA
SLOVENIA
SOUTH AFRICA
SPAIN
SRI LANKA
SUDAN
SWEDEN
SWITZERLAND
SYRIAN ARAB REPUBLIC
TAJIKISTAN
THAILAND
TOGO
TONGA
TRINIDAD AND TOBAGO
TUNISIA
TÜRKİYE
TURKMENISTAN
UGANDA
UKRAINE
UNITED ARAB EMIRATES
UNITED KINGDOM OF GREAT BRITAIN AND NORTHERN IRELAND
UNITED REPUBLIC OF TANZANIA
UNITED STATES OF AMERICA
URUGUAY
UZBEKISTAN
VANUATU
VENEZUELA, BOLIVARIAN REPUBLIC OF
VIET NAM
YEMEN
ZAMBIA
ZIMBABWE

The Agency's Statute was approved on 23 October 1956 by the Conference on the Statute of the IAEA held at United Nations Headquarters, New York; it entered into force on 29 July 1957. The Headquarters of the Agency are situated in Vienna. Its principal objective is "to accelerate and enlarge the contribution of atomic energy to peace, health and prosperity throughout the world".

IAEA NUCLEAR ENERGY SERIES No. NW-T-1.40

MANAGEMENT OF SITE INVESTIGATIONS FOR RADIOACTIVE WASTE DISPOSAL FACILITIES

INTERNATIONAL ATOMIC ENERGY AGENCY
VIENNA, 2024

Printed and bound by CPI Group (UK) Ltd, Croydon, CR0 4YY
September 2024
STI/PUB/2093
https://doi.org/10.61092/iaea.jy3v-m5p4

IAEA Library Cataloguing in Publication Data

Names: International Atomic Energy Agency.
Title: Management of site investigations for radioactive waste disposal facilities / International Atomic Energy Agency.
Description: Vienna : International Atomic Energy Agency, 2024. | Series: IAEA nuclear energy series, ISSN 1995-7807 ; no. NW-T-1.40 | Includes bibliographical references.
Identifiers: IAEAL 24-01697 | ISBN 978-92-0-121424-9 (paperback : alk. paper) | ISBN 978-92-0-121524-6 (pdf) | 978-92-0-121624-3 (epub)
Subjects: LCSH: Radioactive waste disposal. | Radioactive waste disposal — Management. | Radioactive waste sites. | Radioactive waste disposal in the ground.
Classification: UDC 621.039.7 | STI/PUB/2093

FOREWORD

The IAEA's statutory role is to "seek to accelerate and enlarge the contribution of atomic energy to peace, health and prosperity throughout the world". Among other functions, the IAEA is authorized to "foster the exchange of scientific and technical information on peaceful uses of atomic energy". One way this is achieved is through a range of technical publications including the IAEA Nuclear Energy Series.

The IAEA Nuclear Energy Series comprises publications designed to further the use of nuclear technologies in support of sustainable development, to advance nuclear science and technology, catalyse innovation and build capacity to support the existing and expanded use of nuclear power and nuclear science applications. The publications include information covering all policy, technological and management aspects of the definition and implementation of activities involving the peaceful use of nuclear technology. While the guidance provided in IAEA Nuclear Energy Series publications does not constitute Member States' consensus, it has undergone internal peer review and been made available to Member States for comment prior to publication.

The IAEA safety standards establish fundamental principles, requirements and recommendations to ensure nuclear safety and serve as a global reference for protecting people and the environment from harmful effects of ionizing radiation.

When IAEA Nuclear Energy Series publications address safety, it is ensured that the IAEA safety standards are referred to as the current boundary conditions for the application of nuclear technology.

This publication is intended to support the sharing of knowledge and provide a platform for initiating well planned and focused site investigations in the large number of Member States with a need for disposal facilities for radioactive waste. Site investigation is a series of coordinated activities undertaken to understand and describe the nature of a site. In the context of this publication, it is a critical component of a siting process intended to discriminate between potential sites, with an ultimate goal to identify a suitable site for a repository and acquire the necessary authorizations for construction and disposal operations.

A particular objective of the publication is to ensure that site investigation efforts are founded on a clear understanding of data and information requirements. The data and information requirements are derived from many sources, including international obligations, national policy, legislation and regulations, stakeholder expectations and the nature of the proposed disposal concept. They also stem from the specific needs of specialists within a radioactive waste management organization, particularly those involved with safety assessment, repository design and environmental impact studies.

In addition the publication provides practical guidance on strategic and operational management for planning and implementing a site investigation project, including criteria that might be used to conclude a site investigation project. The publication also includes examples of established tools and techniques that might be used to acquire data and provide interpretations relevant to the production of an integrated understanding of a site for a prospective disposal facility. These examples are derived from several disciplines, with a focus particularly on geology, hydrogeology and hydrogeochemistry.

The IAEA is grateful to the experts listed at the end of the publication for their contributions to the drafting and review of this publication, in particular P. Degnan and R. Chaplow (United Kingdom) for substantially developing the initial draft as well as providing iterative updates based on the input of international experts.

The IAEA officers responsible for this publication were S. Mayer and H. Jung of the Division of Nuclear Fuel Cycle and Waste Technology.

CONTENTS

1. INTRODUCTION

1.1. BACKGROUND

Radioactive waste is produced during all stages of the nuclear fuel cycle, as well as during the manufacture and use of a wide range of nuclear materials and as a result of the decommissioning of nuclear facilities. At the time of writing, there are 31 Member States operating approximately 440 nuclear power reactors for electricity generation. These reactors and associated nuclear fuel cycle activities produce a wide range of radioactive waste materials, including spent nuclear fuel (SNF) where this is declared as waste. Three countries have nuclear power plants under construction for the first time and more newcomers are expected over the coming decade. In addition, several countries have inventories of radioactive waste derived from the previous use of nuclear power in electricity generation programmes. Other countries use nuclear energy for applications in research, industry, medicine and agriculture. Where recycling, reuse, exemptions, export, and decay storage are not feasible end-of-life management options, the IAEA encourages all Member States that generate radioactive waste to have in place policies, strategies, infrastructure and resources to safely and securely dispose of all of the waste generated within the country as this is a fundamental national responsibility.

There are many alternative disposal concepts that might be employed to deal with radioactive waste, and the choice of a particular option will depend on a number of factors. But in all cases the overriding objective with respect to radiological safety is to protect people and the environment from the harmful effects of ionizing radiation [1]. That said, the hazard posed by radioactive waste varies greatly depending on its characteristics. Consequently, where waste is segregated and to be disposed of in dedicated facilities for a particular category or group of categories of waste, safety measures may not need to be applied with the same stringency for every type of repository. For example, the way in which safety measures are applied for the disposal of very low level waste (VLLW) in a near surface repository with a very low associated radiological hazard to facility staff, the public and the environment, would be very different from the way in which measures would be applied for disposal in a deep geological disposal facility for high level waste (HLW). A self-evident corollary of this fact is that the degree of effort required to apply and demonstrate safety during the planning, construction, operation and closure of a disposal facility would vary and this should be commensurate with the level of hazard posed by the waste, both now and in the future. This position is termed a graded approach; it is important to appreciate its significance when planning and implementing site investigations for disposal facilities.

Where disposal is selected as an endpoint in the management of radioactive waste, it will be necessary to ensure the inventory of concern is well characterized to determine a suitable disposal concept. However, knowledge of an inventory alone is not sufficient when deciding on the design of a disposal facility as it is also imperative that detailed site investigations be undertaken to provide a comprehensive understanding of the environment in which the facility will be situated.

Site investigation is used in this report as a general term for an activity or series of activities undertaken to understand the nature of a site, and site characterization means the investigative process used to more precisely describe the attributes of a site [2].[1] In this sense, while both terms involve data acquisition and interpretation activities, site characterization is considered a particular aspect of site investigations that is focused specifically on describing the detailed nature of a site and its physical

[1] *Site investigations* and *site characterization* are terms that are often used synonymously in the literature. Site characterization alone has a clearly defined meaning in the IAEA Safety Glossary [2]: "Detailed surface and subsurface investigations and activities at a site to determine the radiological conditions at the site or to evaluate candidate disposal sites to obtain information to determine the suitability of the site for a repository and to evaluate the long term performance of a repository at the site."

processes, at the present time and in the relevant past. Typically, the characterization of a site is sufficient to address a particular purpose. For a disposal programme, this is ultimately to confirm whether or not the site is suitable for constructing and operating a repository and to support an evaluation of the long term performance of a disposal facility at the site.

Previous IAEA publications have separately addressed the issue of site investigations for near surface [3] and geological disposal facilities [4–6] and have also provided information about the use of nuclear geophysical techniques [7] and hydrogeological methods for site investigations [8]; this includes the most recent IAEA report on the subject, dated 2001 [9]. Since then, there have been significant developments in the science and technology associated with site investigation data acquisition and interpretation, as well as a deepening of management experiences in Member States. Of particular note, several radioactive waste management organizations (RWMOs) have now designed and undertaken extensive multidisciplinary and strategically focused site investigation projects for geological disposal facilities, for example: Ontario Power Generation and Nuclear Waste Management Organization in Canada; Posiva in Finland; Andra in France; the Federal Company for Nuclear Waste Management in Germany; Nuclear Waste Management Organization of Japan (NUMO); the Swedish Nuclear Fuel and Waste Management Company (SKB) in Sweden; Nagra in Switzerland; Nuclear Waste Services (NWS) in the United Kingdom (UK) (formerly Radioactive Waste Management, or RWM); and the US Department of Energy and Sandia National Laboratories in the United States of America (USA). The challenges and successes that arose as a result of these experiences are highly informative for others who are yet to undertake their own investigations. It is therefore timely for the IAEA to provide this compilation of lessons learned on the management of site investigations as part of an overall repository development process.

1.2. OBJECTIVE

This publication is intended to support the sharing of knowledge between Member States and provide a platform for initiating well planned and tightly focused site specific investigations in the numerous Member States with a need for disposal facilities for radioactive waste.

In advance of any decisions about the use of particular investigation technologies and techniques, it is essential that Member States considering the planning and implementation of site investigations go about the task in a strategically oriented manner, with a clear understanding of the reasons why site data and information need to be collected and how they will be used. Without this clear understanding and without sufficiently detailed planning, resource use (human effort, equipment, funding and time) is unlikely to be optimized, the rationale for decisions will be unclear and confidence in the implementing organization may be significantly diminished. There may also be safety implications. Thus the principle aim for this publication is to provide guidance to Member States on the need for an approach to the planning and implementation of site investigations that is requirements driven and to highlight the significance of key management issues to be addressed in designing and undertaking site investigation projects that are required to characterize sites for radioactive waste disposal facilities.

A second objective is to provide interested readers with relatively high level information on a range of data acquisition and interpretation technologies and techniques that are currently potentially available to support site investigations. It is emphasized that decisions concerning the choice of particular technologies and techniques will depend on specific national and local circumstances such as the data and understanding required (including the levels of accuracy and precision needed), the availability of equipment and personnel, the costs involved and the degree to which alternative and equally viable methods may exist. In all cases, suitably qualified experts should be used in the planning and implementation of the data acquisition and interpretation activities.

Guidance and recommendations provided here in relation to identified good practices represent experts' opinions but are not made on the basis of a consensus of all Member States.

1.3. SCOPE

This publication is primarily concerned with the provision of guidance concerning the strategic and operational management of site investigations for radioactive waste disposal facilities. As such, the discussion relates to a range of issues and circumstances, including:

(a) Disposal concepts: The management of site investigation activities described in this publication is particularly oriented towards applications for the underground disposal of radioactive waste in a geological repository. The rationale for this is that the site investigations required for near surface repositories can generally be considered as a subset of the investigations required for a mined geological repository. This reflects the fact that many of the activities undertaken to characterize the surface environment and the type of information required will be similar whether for near surface or geological disposal concepts. But for a near surface repository concept, the subsurface investigation needs will be much less than those necessary for a geological disposal facility in most cases. Furthermore, although the scope of site investigation technologies and techniques covered in this publication are principally focused on addressing requirements associated with geological disposal concepts, the strategic and wider management considerations that will be necessary in the planning and implementation of site investigation projects for geological and near surface repositories may be very similar. Therefore, the guidance is intended to be widely applicable and the reader is encouraged to consider the relevance of the information provided here for one's own national situation, which might only be considering a near surface disposal concept at the current time.
(b) Characterization of the natural environment: Site investigations as part of a disposal facility siting process are necessary to describe the natural environment at a site (e.g. the topography, the nature of groundwater flow, the rock types and structures present, their geometrical relationships, etc.) and the socioeconomic features of a site (e.g. land use, local transport infrastructure, human habits, etc.). Furthermore, and depending on the stage and detail of the investigations, the concentrations and distribution of natural and artificial radioisotopes present at the site may also be required for establishing baseline radiological conditions. This publication pertains solely to the management of site investigations relating to the study of the natural environment. Consequently, site investigations related to socioeconomic factors and radiological monitoring are outside of the scope of the publication, as are issues concerning stakeholder engagement and societal consent (whilst strongly acknowledging the importance of this topic for effective strategic planning and for ensuring the successful implementation of site investigations [10]).
(c) Site investigations as part of the siting process: In this publication the distinction is made between site investigation activities to be undertaken as part of the siting process prior to an authorization for repository construction and any continued environmental monitoring and other site investigation activities that might continue once repository construction commences (i.e. that might be required to contribute to ongoing modelling efforts and enhanced confidence building for emplacement operations, closure and beyond). Post-siting environmental monitoring and other site investigation activities are therefore excluded from the scope of this publication.
(d) Investigations at coastal sites: In some Member States, disposal programmes are being developed where a repository may be located at a coastal site, either under the land or under the seabed (but importantly, conceptual designs need to include underground access to a sub-seabed repository from surface works on land to ensure conformance with international agreements and conventions). In such situations, extensive characterization of the marine environment and the sub-seabed may be required. It is anticipated that much of the guidance provided in this publication will be applicable for repository programmes focused on inland and coastal settings. However, it is appreciated that additional specialist guidance beyond that available here may be required in relation to management issues and the tools and techniques to be applied in coastal settings where seabed and sub-seabed characterization is anticipated.

(e) Investigations in challenging environments: It is also recognized that site investigations may be required in particularly challenging environments. A challenging environment may be technically demanding and could include regions displaying a relatively high degree of geological complexity or a tectonically active terrain characterized by neotectonic faulting and/or volcanism. Whilst these situations are acknowledged and there are active site investigation projects currently under way in these types of environments, the challenges and solutions associated with them are not specifically addressed in detail here as each case should be appraised individually. Challenging environments may also reflect difficult operational climatic conditions, distance from labour markets and sparse transport infrastructure, etc. The choices and the consequences of logistical decisions to be made in such situations are not addressed here.

(f) Repository depths: Geological disposal is considered to take place in intact rock formations at depths greater than several tens of metres, with mined deep geological repositories (DGRs) for HLW typically considered to be located at depths between a few hundred metres to a thousand metres. The tools and techniques for use in site investigations for deeper repository concepts beyond 1000 m, including a deep borehole disposal concept with a disposal zone that may extend 5 km below the surface, are not specifically addressed in this publication. However, much of the management guidance provided is expected to be equally applicable.

(g) Emphasis on total system understanding: The principle scientific disciplines employed during site investigations for disposal facilities are related to the Earth sciences. More specifically, they include geology, hydrogeology, hydrogeochemistry, geophysics, rock mechanics, climatology and environmental studies. Information from all of these disciplines needs to be integrated to provide a holistic 3-D and dynamic understanding of a site and its evolution, as well as an appreciation of the potential for natural disturbances to impact on the performance of engineered or natural barriers that may be part of a future repository system. This integrated system understanding is the basis for site specific radionuclide transport modelling that will ultimately determine the calculated dose and risk associated with a specific waste inventory and disposal concept.

(h) End user requirements: The data and understanding derived from site investigations are key inputs for developing a safety assessment and for confidence in the wider safety case. Site specific data and information are also critical for developing the detailed design of the engineering components of a disposal facility. The guidance on the management and use of data and information from a site investigation project in this publication is oriented especially towards satisfying the requirements for both of these areas. Although partially addressed, it does not focus in such great detail on site investigation data needs for supporting environmental impact assessments (EIAs). This is because such assessments may or may not be required in different Member States and, where they are required, the information needs are potentially highly variable. Furthermore, there are commonalities in the data and information needed to satisfy an EIA and for use in characterizing the biosphere to support a safety assessment.

(i) Management vs technical focus: This publication is primarily concerned with the management of site investigations for supporting safety assessment and repository design studies, with respect to strategic planning and for designing and controlling implementation activities. However, certain technical aspects of a site investigation project cannot be ignored as RWMO staff responsible for investigations need to be sufficiently knowledgeable about the availability and use of technologies and techniques so they can make informed decisions. There are often several alternative methods available to determine and describe physical properties at the surface or underground such as remote sensing, surface mapping, and subsurface in situ measurements on various scales and from measurements on samples in a laboratory. All such measurements provide data that have variable accuracy and precision and they are representative of natural conditions on varying length scales. Examples of some of the principle tools used in site investigations to date are provided, but the listing is not intended to be definitive or exhaustive and the choice of a particular method is expected to be selected on a site specific basis to reflect data requirements and resource availability.

(j) Specialist topics: Lastly, concerning the scope of the report, it is recognized that the technical topics to be addressed during site investigations are very wide ranging and in a publication such as this it will not be possible to include equal consideration of every facet of a major site investigation project (e.g. one that might be developed to address the disposal needs of HLW and SNF from a nuclear power generating programme). Consequently, specialist planning and operational management of activities involving site specific research into colloidal transport, gas migration and the presence and impact of microbes, to name but a few topics, are not explicitly mentioned. It is, however, expected that most of the management guidance in this report is provided at a sufficiently high level that it would be broadly applicable in many technical areas.

1.4. STRUCTURE

After this introductory section, Section 2 provides an overview of site investigations for repository development, focusing on the evolution of understanding associated with management issues and their place in the overall waste disposal process.

Section 3 highlights a requirements driven approach as a key imperative for directing the planning and implementation of site investigations. Data and information are required from site investigations to support different aspects of a repository programme, such as the operational and post-closure safety assessments, repository engineering design studies and EIAs. Each of these will place various requirements on the data and information to be collected from a site investigation project and these requirements are addressed here. Section 3 also clearly sets out high level objectives and the broad scope of the various site investigation disciplines associated with a site investigation project, in relation to the geology, hydrogeology, hydrogeochemistry, rock mechanics, and the biosphere research topics.

An outline of the strategic planning process, as well as the various management planning activities and considerations that will be necessary to achieve a successful site investigation project are set out in Section 4.

In Section 5, information is provided on alternative ways to organize and structure a site investigation project to facilitate planning and the communication of activities.

The following two sections deal with data acquisition and processing (6) and with data analysis, interpretation and integration (7). Examples of tools, techniques, management approaches, issues, challenges and solutions are provided, as well as guidance on the development of conceptual and descriptive models.

Prior to Section 9, which provides a series of conclusions, Section 8 briefly suggests the basis for a decision to conclude site investigations undertaken as part of the siting phase of a repository development programme.

Finally, several case studies are provided in the Annexes as examples of where various issues and challenges associated with the management of site investigations have been recognized and dealt with in Member States with experience of mature site investigation projects. These experiences are complementary to the main body of the report and provide essential lessons learned for others.

1.5. USERS

There is a wide spectrum of capabilities and experience in organizations responsible for radioactive waste disposal across the globe. The level of maturity of a disposal programme generally reflects the nature of the nuclear infrastructure present within a country and the length of time a nuclear programme has been in place, as well as the availability of resources and the degree of government support. Different radioactive waste inventories will require different disposal solutions to be available at different times, typically reflecting the hazard posed by the characteristics of the waste and its volume. This publication is intended to have value in Member States with developed and less well-developed programmes and for all types of radioactive waste and for near surface and geological disposal concepts (although primarily oriented towards site investigations for geological disposal, as noted above).

Regardless of the nature of the inventory for disposal and any preference for a particular disposal concept, this publication will be of interest to senior decision makers, scientists and engineers working within organizations charged with the planning and implementation of site investigation projects intended to lead to the disposal of radioactive waste (i.e. working within RWMOs). Other organizations may be charged with responsibility for initial planning in advance of a dedicated RWMO being established (e.g. waste producers or scientific and nuclear research institutes) and this publication is also intended for their use.

In addition to aiding staff within RWMOs responsible for the planning and implementation of site investigations, professionals in technical support organizations and those working within regulatory authorities will also benefit from accessing the report. The report is also expected to have relevance for decision makers in national and local governments as well as academics and other stakeholder groups including students and interested members of the public.

Although the publication deals with many specialist technical areas that involve the use of sometimes complex data acquisition technology and interpretation methods, it is intended to be readily understandable by anyone with a basic interest in site investigations and geoscience.

2. SITE INVESTIGATION: AN OVERVIEW

2.1. INTRODUCTION

The IAEA, through its Safety Standards Series, sets out fundamental safety principles [1], specific safety requirements [11] and related guidance relevant to the disposal of radioactive waste in dedicated facilities [12–14]. These Safety Standards are complemented by various IAEA technical publications including the Nuclear Energy Series which, at the highest level, present the nuclear energy principles [15] and radioactive waste management objectives [16]. In addition to these references, which provide a basic framework for developing any radioactive waste disposal programme, the IAEA provides a generalized radioactive waste classification scheme that relates the activity and half-life of the radionuclides in a waste inventory to broad disposal options [17]. This defines six classes of waste: exempt waste (known as EW), very short lived waste (known as VSLW), VLLW, low level waste (LLW), intermediate level waste (ILW), and HLW.

Radioactive waste arises from the generation of electricity in nuclear power plants, from nuclear fuel cycle operations including the reprocessing of SNF, from other activities in which radioactive material is produced or used (e.g. sealed radioactive sources), from the decommissioning of nuclear facilities and from the remediation of contaminated sites.

Depending on its origin, radioactive waste occurs in a variety of physical and chemical forms with different characteristics including radioactivity content and the rate of decay of the radionuclides present. A common characteristic of all radioactive waste is its potential to present a hazard to people and the environment. It is imperative, therefore, that radioactive waste is to be safely managed to reduce any associated risks to acceptable levels. Effective radioactive waste management considers the complete chain of activities from waste generation to ultimate waste disposal. These activities successively cover the processing and handling of each waste stream to produce stable and solid waste forms, reduced in volume and immobilized as far as practicable and placed in suitable casks, drums or other containers to facilitate their storage, transport and ultimate disposal.

The preferred strategy for the management of all radioactive waste destined for disposal is to contain (i.e. to confine radionuclides within the waste matrix, the packaging and the disposal facility) and isolate it from the accessible biosphere. However, it is important to note there are a number of possible endpoints in the management of radioactive waste, of which disposal is only one. These end points also include decay storage, exemption and clearance, recycling and reuse, as well as the possible export of waste under certain circumstances. It is also important to stress that storage of radioactive material not suitable for clearance is not a permanent management option, because it can only ever be an interim solution and not an endpoint in itself.

Different disposal options exist for radioactive waste, which may broadly be classified into:

— Near surface disposal suitable for VLLW and LLW;
— Geological disposal suitable for ILW, HLW and SNF declared as waste (note however that to reflect national preferences, VLLW and LLW may also be disposed of in geological disposal facilities).

While there is no precise numerical value that allows to group a disposal facility as either a near surface disposal or a geological disposal, the depth of disposal relative to the surface is an obvious indicator. Near surface disposal facilities are typically sited on or above the surface or in trenches excavated into the ground or located underground in natural caverns or engineered facilities constructed at depths of up to several tens of metres from the surface. Consequently, a definition of geological disposal would be waste emplacement at depths in excess of about 30 to 50 m from the surface in consolidated rock environments. Geological repositories can be subdivided into facilities that are situated at intermediate or deep depths. The subset of geological disposal facilities termed DGRs are typically constructed at several hundred metres deep (generally considered between 200 to 300 m to a maximum of 1000 m, although this could, in principle, be deeper if required, depending on the nature of the rock environment and the engineered system). Consequently, intermediate depth disposal facilities would be located at depths of between approximately 30 to 50 m and 200 to 300 m. It is important to note, however, that depth alone is not sufficient to demonstrate the adequate containment and isolation of a specified radioactive waste inventory intended for disposal. Rather, it is the overall performance of the disposal system that is important, reflecting the collective behaviour of engineered and natural barriers, as well as distance from the surface.

VLLW and LLW inventories contain low activity and relatively short lived waste, although they may contain small quantities of longer lived waste dispersed over a large repository volume. They present a potential hazard for durations generally not exceeding a few centuries and so can safely be contained within a suitable near surface facility. Approximately 140 near surface disposal facilities have been successfully sited worldwide and are in operation or already closed. ILW, HLW and SNF declared as waste present a greater potential hazard than found in waste acceptable for disposal in near surface repositories (i.e. LLW) due to higher activity levels and the presence of radionuclides with longer half-lives. They therefore require disposal in a geological environment capable of ensuring long term safety without human intervention (i.e. demonstrating passive safety) for thousands or even hundreds of thousands of years. Several geological disposal facilities for LLW and ILW are in operation worldwide (e.g. Germany, Republic of Korea, and USA) and a few countries (Sweden, Finland, and France) are well advanced in the development and licensing of deep geological disposal facilities for HLW and SNF.

Despite these successes, and despite scientific consensus that indicates we have the technology and understanding required to safely site, construct, operate and close a geological disposal facility, the implementation of geological disposal strategies for HLW and SNF declared as waste remains one of the greatest ongoing challenges in the management of radioactive waste in many Member States. There are several explanations for the difficulty and delay in implementing geological disposal solutions for these waste types at a suitable site and in a timely manner. In many instances the key reasons are generally accepted to relate to sociopolitical concerns, especially societal acceptance and political will, but the fact that the costs and infrastructure requirements for developing a mined geological disposal facility are appreciable is also an important factor hindering progress in many Member States, especially those with relatively small volumes of high activity and long lived radioactive waste.

Figure 1 illustrates the relationships between the key factors that influence the choice of a potentially suitable disposal concept and site. It is to be recognized that the waste inventory characteristics are the starting point for considerations of a suitable disposal concept and may have a significant bearing on the nature of the conditions required from a site, as well as impacting on the repository design. At the same time, a preference for a specific type of repository design may be an important consideration in selecting appropriate biosphere and geosphere conditions for siting a repository, while the site itself would dictate some of the detailed design elements of an engineered facility, such as the locations of buildings and roads on the surface and the layout of disposal arrangements. Disposal concept choices that take into account these three factors (inventory, site conditions and repository design) will also be affected by wider sociopolitical considerations and expectations, as well as regulatory requirements that would stem from national policy choices and safety and security considerations.

Various rock types are potentially suitable for hosting a repository, and safety cases have been developed for disposal in crystalline (e.g. granite and metamorphic rocks in Sweden and in Finland) and sedimentary (e.g. indurated clay rocks in France and Switzerland and salt rock in the USA and Germany) environments. Table 1 presents illustrative rock types[2] and some of the properties associated with them (relevant for investigating, constructing, operating and closing a geological facility and for limiting radionuclide transport), noting that there may be a wide variation in some of these properties and characteristics even within a single rock type and over time.

It should be noted the rock categories listed in Table 1 are general and may contain a high degree of variation in terms of properties, especially clays (from very plastic to indurated claystone) and sandstones. For example, the degree of consolidation, cementation, deformation history and weathering in a sandstone

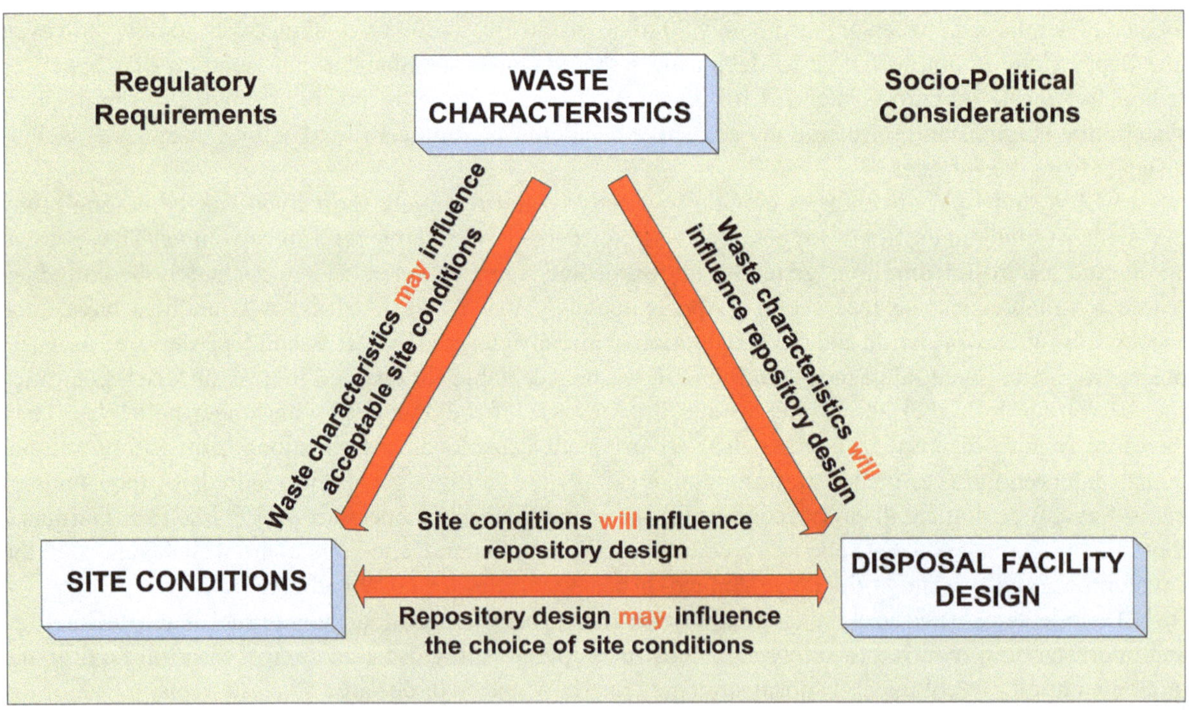

FIG. 1. The interrelationships between key factors influencing the choice of a disposal concept.

[2] Other rock types have also been considered for hosting a repository, such as basalt, volcaniclastic and various metamorphic rocks.

TABLE 1. ILLUSTRATIVE PROPERTIES AND CHARACTERISTICS OF ROCK TYPES

	Sedimentary			Crystalline
	Clay	Salt	Sandstone	Granite
Hydraulic conductivity (including dominant flow mechanism to be expected and degree of uniformity)	Very low to negligible (matrix); Homogeneous at DGR scale	Very low to negligible (through intercalated sediments only); Homogeneous at DGR scale	Moderate to very high (matrix); Heterogeneous at DGR scale	Very low to negligible (matrix); Very low to very high (fracture); Heterogeneous at DGR scale
Dominant solute transport mechanism	Molecular diffusion	Molecular diffusion	Advection–dispersion (matrix flow)	Advection–dispersion (fracture flow)
Dissolution potential	Very low	High	Very low	Very low
Geochemical stability at depth & likely redox conditions	Stable; reducing	Stable; reducing	Stable; reducing	Stable; reducing
Sorption capacity	Very high	Low	Low to moderate	Low to moderate
Thermal conductivity	Low	High	Moderate	Moderate
Rock strength	Low to high	Low	Moderate to high	High to very high
Deformation behaviour	Plastic to brittle	Visco-plastic	Brittle	Brittle
Construction experience at DGR depths	Low to moderate	High	Low	High
Post-closure safety dependence on engineered barriers	Low	Low	Depends on host rock	Low to high

Note: Clay, salt and granite are examples of rock types typically considered for hosting a geological disposal facility. Sandstone properties are also provided as an example of a rock type that might be found above or surrounding a host formation.

may significantly alter rock strength, flow mechanism and hydraulic conductivity (e.g. fractured sandstone may display matrix flow and fracture flow).

It is also noted that all rock types can have less than ideal individual attributes (properties) associated with them, but the suitability of a rock to host a repository does not generally depend only on a single property or even a group of characteristics. Rather, it is essential to consider the performance of the disposal system in its entirety, including the nature of any surrounding rocks, as any deficiencies in one or several individual properties associated with a particular type of rock may be more than adequately compensated for by other physical attributes or by engineered solutions.

It is emphasized that it will never be possible to find 'the best' or 'the safest' site on the basis of rock properties, because it will be impossible to investigate and determine the detailed nature of every possible site, as would be necessary to allow such an absolute comparison. Instead, the key will be to find a site that provides assurance that the required level of system safety and performance can be met

FIG. 2. The relationships among the four principle areas in which data and information from site investigations are to be used (according to end users).

(e.g. by meeting regulatory safety requirements) and is acceptable to decision makers and the majority of stakeholders.

Lastly in this introduction, four principal areas where data and information derived from site investigations are to be used within a disposal programme are introduced, namely: (i) in safety assessments; (ii) in repository design and engineering studies; (iii) for use in EIAs; and (iv) more widely, to demonstrate scientific understanding and promote confidence in the safety case for a disposal facility and for supporting dialogue with stakeholders (Fig. 2). The application of data and information in each of these areas will reflect specific needs, but much of the data and information to be collected from a site investigation programme will have multiple uses (shown by the overlapping domains in Fig. 2). Section 3 presents in more detail the information requirements established by the end users of the information.

2.2. EVOLUTION IN THE MANAGEMENT OF SITE INVESTIGATION PROJECTS

This publication emphasizes a requirements driven approach to site investigations, reflecting data collection and interpretation for a specific need that advances a repository development programme. This approach has been gradually gaining momentum since the 1990s as research requirements to support a demonstration of safety have become better understood and the need for the efficient and effective use of scarce resources has become even more pronounced.

During the initial development of nuclear power programmes and the growing use of nuclear applications in medicine, research and industry during the 1950s and 1960s, considerations about what to do with radioactive waste were typically secondary to ensuring that the benefits of nuclear energy were being made available. Waste management, and in particular disposal, were often afterthoughts to be dealt with later. Consequently, numerous problems arose such as:

- Radioactive waste was not properly characterized as it was generated;
- Waste generation and tracking were not controlled;
- Solid and liquid waste accumulated without adequate conditioning;
- Loss of control (ownership) of disused sealed radioactive sources;
- Waste stored on an ad hoc basis under unsuitable conditions;
- Waste processed without consideration of waste acceptance criteria for disposal;
- Unlicensed or unsafe disposal;
- Unclear links and responsibilities among the institutions involved in waste generation and management;
- Financial liabilities not defined and insufficient allocation of funds for waste management;
- Political ambiguity and lack of commitment;
- Difficulty to obtain social consent for the construction of new nuclear power plants and other facilities because no suitable endpoint solutions defined for existing waste.

In relation to planning for disposal facilities, the earliest advisory committees charged with studying disposal options recognized that finding suitable geological conditions would be required to contain, delay or retard any radionuclides that might potentially migrate from disposal facilities in harmful concentrations to people and the environment. However, details in terms of relating engineered and natural barrier safety functions and performance with other system components were generally poorly defined. Furthermore, regulatory guidance and oversight was often sparse or missing. Consequently, in early site investigations there was a tendency to collect all the natural system data that could be collected, in anticipation of it having some use in the future.

Experience highlights another feature of early radioactive waste disposal programmes. It is now well appreciated that a RWMO responsible for undertaking site investigations for a repository development programme will need to have an appropriate level of competence and capability in the earth and physical sciences available to be able to act as an intelligent client for geoscientific and wider research services. In the past, in cases where the detailed planning and implementation of site investigations had to be undertaken by, or outsourced to, academic institutes and research oriented technical support organizations (sometimes called TSOs), the entity responsible for disposal sometimes did not have sufficient internal competence or experience in the production of technical specifications nor for managing field operations and reviewing results. Consequently, there was a danger that the work would not be sufficiently focused on meeting the direct requirements of the disposal programme. As a result, data and information of marginal value was sometimes collected, resulting in an inefficient use of resources.

The key to limiting the potential for unfocused research during site investigations and to avoid the collection of data for its own sake, is to ensure the following:

- Site investigation projects are well planned and driven by specific requirements;
- The RWMO has a sufficient body of competent and experienced staff to enable it to act as an intelligent client for outsourced services (i.e. sufficient to define the work that needs to be done, discriminate between alternative approaches, and to critically manage and review the work as it proceeds).

Another major development in the evolution of radioactive waste disposal programmes, especially since the 1980s, is the recognition of the need for an iterative and intimate relationship between the planning, acquisition and interpretation of site data on the one hand, and their use in safety assessments and repository design studies on the other. As an example, in the past it was sometimes the case that geoscientists would presume what data and information would be needed to complete a safety exercise that examined the consequences of postulated radionuclide migration through the geosphere, based on their experience of what data was available, what could feasibly be acquired, what was geologically of interest and what was relevant. Consequently, they would decide what data to collect in the field, often without regard for parameter uncertainty evaluations or quality control (QC) and then undertake interpretations as needed. Conversely, radiological consequence modellers would develop safety assessment methods with incomplete knowledge about the type and quality of data they required, what was feasible or the nature of geological processes that could impact on risk and their interactions. Therefore, on the one hand requests made by safety assessment modellers for information from geoscientists were often incomplete or otherwise unsatisfactory and, on the other hand, data provided by the geoscientists were sometimes delivered, but not wanted or of limited value, or sometimes not delivered at all (Fig. 3). It is apparent that these dangers still exist with regards to immature disposal programmes.

Today, data are typically acquired during site investigations by specialists working in the Earth sciences. That data are then processed and interpreted to provide secondary data and information. Process models are often employed to address specific questions and evaluate the significance of observations and data. At some stage, the data and information derived from a site needs to be provided to safety assessment specialists who further analyse the significance and meaning of the site data and information with a view to evaluate the dose or risk of harm arising from the release of radionuclides from a repository. A simplified illustration of the links from data collection through data processing and interpretation to their use in the derivation in dose and risk estimates via risk assessments is provided in Fig. 4. However, the simplified process shown in Fig. 4 is not a one time or one-way activity and Fig. 5 better illustrates the actual process for data acquisition and interpretation in more detail, as observed in mature repository development programmes today.

The starting point for a site investigation programme at a specific site is an initial hypothesis regarding site conditions, generally based on a literature survey or expert judgement, and the definition of information requirements derived from regulatory requirements together with any additional needs relating to safety assessment, environmental impact studies and repository design studies. Based on these defined needs and an initial hypothesis, a gap analysis is undertaken, comparing the required site

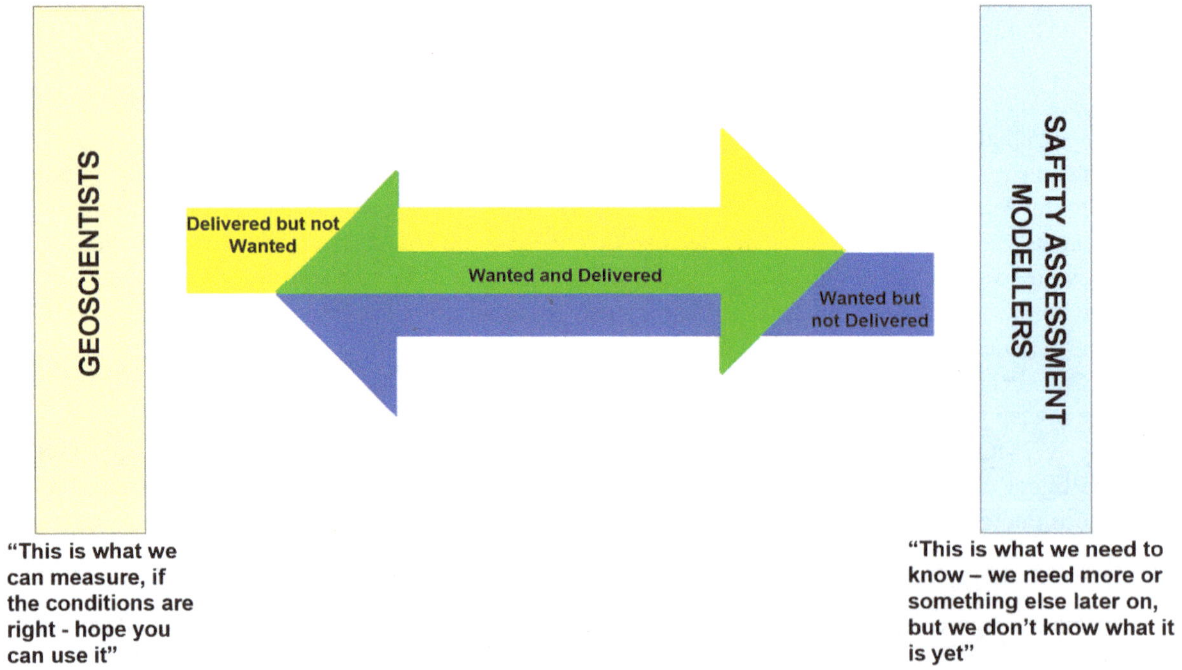

FIG. 3. Data provision and requirements specification between geoscientists and safety assessment modellers in immature repository development programmes (adapted from information provided by Nagra).

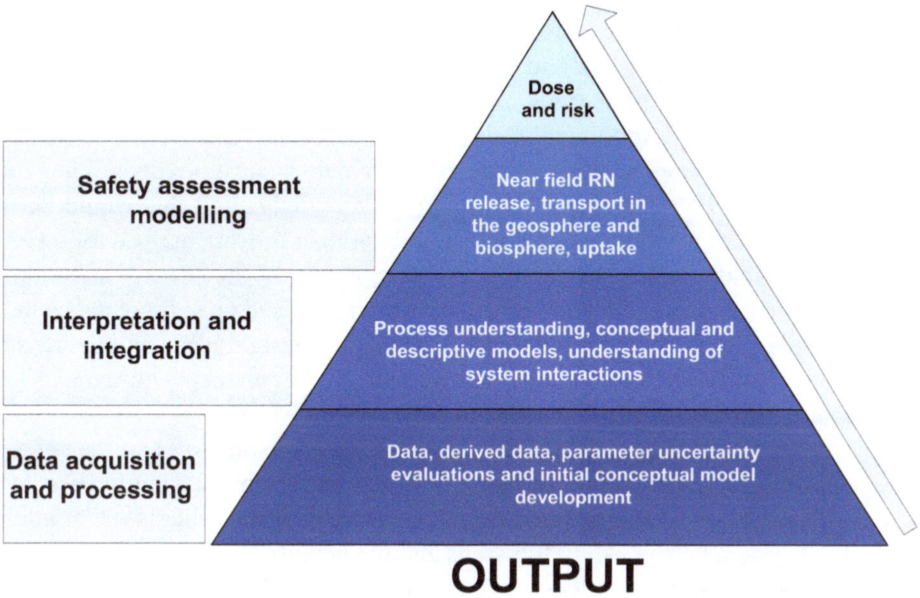

FIG. 4. Simplified illustration of the links between data acquisition during site investigations and their use in the derivation of dose and risk as end points in safety assessment.

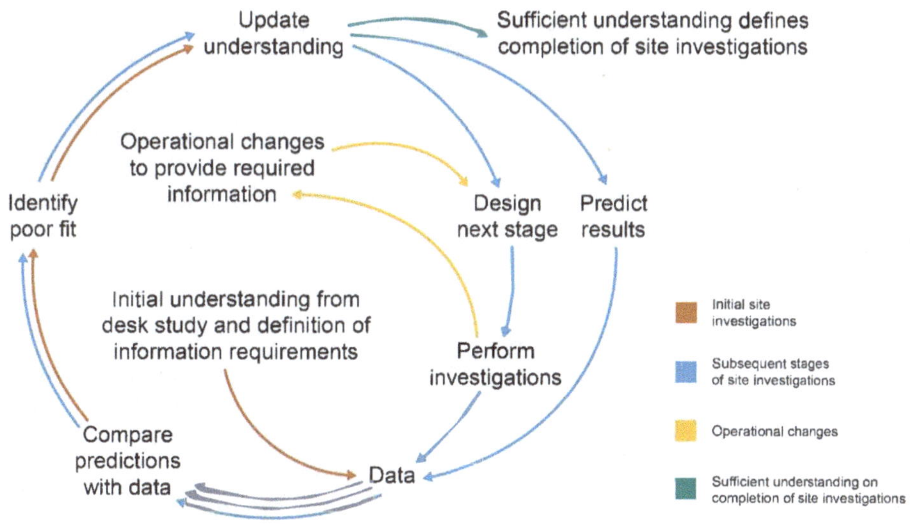

FIG. 5. The life cycle of a site investigation project showing the interactions among components of the iterative loops that need to be established as part of a site investigation project for the provision of data and understanding and the identification of remaining uncertainties and knowledge gaps [18] (reproduced courtesy of NDA, UK).

data and understanding to the current state of knowledge. This results in a data acquisition plan that in turn leads to the implementation of the various site investigation surveys and other activities. Relevant data are acquired, processed and interpreted and the results are compared against the initial hypothesis and any predictions established prior to the start of investigations. It is essential that the reasons for the identification of any poor fits between data and predictions are explained in terms of parameter uncertainty and/or a deficiency in process understanding. Depending on the strength of correspondence between data and predictions, and the reasons for a mismatch, the initial hypothesis and an assessment of site understanding are updated. Additionally, supporting process models would be revised to support further predictions of system behaviour. These comparisons of data against predictions and model updates would generate a recognition of remaining or new gaps in knowledge and this would be used to drive another round of data acquisition activity. This circular and iterative process may be followed any number of times until there is sufficient understanding to allow completion of the site investigations. A decision about when it would be possible to exit the iterative loop would be based on sufficient data being available and satisfactory site understanding existing to address repository design needs and demonstrate whether or not site construction and performance would conform with any regulatory requirements (see Section 8 for further discussion of criteria that might be used to define completion).

There should be close ongoing collaboration between site investigation and the safety assessment personnel throughout the investigations to transfer data and agree on the significance of outstanding uncertainties,[3] as well as the nature of, and interactions between, physical processes that might impact the engineered and natural system barrier safety functions and the potential migration, retardation and uptake of radionuclides. Safety assessment modellers would also be closely involved with site investigation personnel to agree and refine conceptual models of site behaviour and evolution. After the periodic and formal transfer of data and information, safety assessment calculations should be updated in parallel with the site investigation activities to assess the impacts of the enhanced understanding on dose and risk. As part of this effort, the site investigation and safety assessment personnel should collaboratively establish where missing and uncertain data may have to be obtained from further site investigation efforts, by seeking expert judgement or from research undertaken at other sites (e.g. field experiments or the use of natural analogue studies).

It is emphasized that it is not sufficient for site investigation personnel to simply transfer data without context, nor for safety assessment personnel to request additional site data without explaining the need. A close working relationship is also necessary between site investigation staff and those involved in developing the wider safety case and, more specifically, with repository design personnel. Some RWMOs have developed dedicated processes and permanent coordination teams to ensure close collaboration among the various groups.

In relation to the development of scenarios for the future evolution of a repository system and the potential impact of dynamic biosphere and geosphere conditions on engineered and natural barriers, and on potential radionuclide migration pathways, the importance of demonstrating geosphere stability and identifying possible system perturbations (natural and human) has become much more apparent since the 1970s and 1980s. It has long been an established geological axiom that the past is the key to the present; past and present conditions at a site are also a key to understanding possible future conditions. The practice of predicting geological trends on the basis of observations and interpretations of the past is critical for the development of time varying conceptual and long term numerical models based on alternative, but plausible, site evolution scenarios required for post-closure safety assessments. There has been tremendous progress in this area in recent decades as understanding of the natural environment has grown and computational power has multiplied. A summary of some of the natural long term processes and events that might impact scenario development for a DGR system in the UK is provided in Ref. [19].

The use of underground research laboratories (URLs, or underground research facilities, URFs) is another area that has grown as their potential to support repository development programmes has become apparent. Ever since early URLs were constructed, such as Whiteshell in Canada (operational in a crystalline rock environment between 1985 and 2010), Stripa in Sweden (constructed in crystalline rock, operational from 1976 to 1992), and Tono in Japan (sedimentary rock, 1986 to 2004), URLs have been

[3] There will likely be many uncertainties recognized during site investigations, but not all of them will be significant and hence lead to further site investigation work. For example, if a large uncertainty has a negligible effect on the performance of the system in terms of safety, it may not be necessary to gather more data.

used not only to investigate the 3-D in situ characteristics of a rock mass, but also to identify and improve suitable equipment and demonstrate the development of novel techniques and operational processes. Over time, experiments and demonstrations have become more sophisticated as the tools for data collection and the use of interpretation methods have advanced in line with understanding of natural safety relevant processes. These have not only increased confidence in the safety of repository systems but have also in many cases enhanced cost effectiveness. A further development has been the recognition that URLs may have an important role to play in generating wider stakeholder confidence and acceptance, well exemplified by the success of the Äspö visitor centre in Sweden and the Bure visitor centre (at the Meuse & Haute-Marne Centre: CMHM) in France.

The last significant evolutionary factor to mention has been the recognition of the need for more detailed and strategically focused planning and management of site investigations regardless of the magnitude of the task, while remaining aware of the need for a graded approach. This development has been driven by a greater emphasis on the demonstration of the safety of a repository system, but it is also a response to the need to ensure the optimal use of scarce resources and to reflect the growth in stakeholder requirements and expectations, especially those originating from the public, regulators and funding bodies.

2.3. SITE INVESTIGATIONS WITHIN A RADIOACTIVE WASTE DISPOSAL PROGRAMME

A programme for radioactive waste disposal begins as soon as a decision making entity (e.g. the national government) formally commits as a matter of policy to dispose of radioactive waste as a management end point. However, a repository programme can be considered to truly begin only once an organization has been established and charged with the authority and responsibility to dispose of the radioactive waste. In addition to having a suitable mandate, sufficient resources are essential to enable the organization to carry out its mission.

At its simplest level, a repository programme comprises three stages (Fig. 6): (i) the preoperational period prior to waste emplacement; (ii) the operational period which begins as soon as first waste emplacement commences; and (iii) the post-closure phase, the period after a repository has been backfilled and sealed.

Siting is the process of selecting a suitable site for a disposal facility and it begins as soon as a decision is taken to investigate potential regions or sites to host a repository. Siting terminates when the

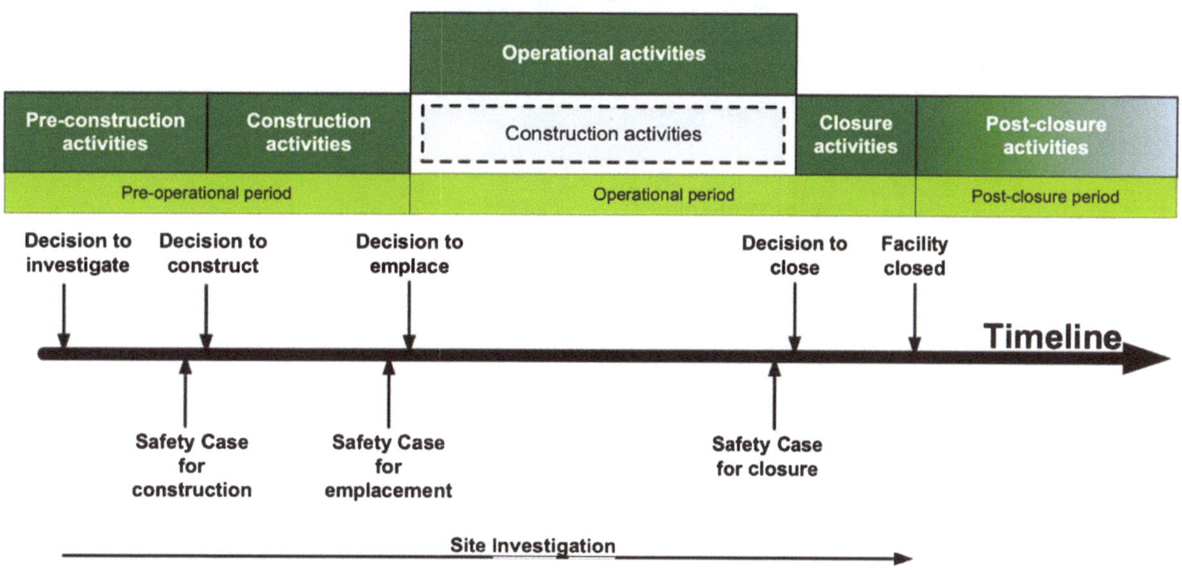

FIG. 6. Timeline in the development of a repository programme [12].

suitability of a site has been confirmed, an authorization has been provided by a regulatory body for the construction of a repository and a final decision is taken to construct the facility at the chosen site. The goal of siting is therefore to select a suitable site for hosting a disposal facility and to obtain the necessary authorizations to construct a disposal facility. This requires the provision of evidence to demonstrate the suitability of a site in the following three areas.

(1) Safety and environmental suitability:
- Demonstrate, through the application of the concepts of passive safety and defence in depth, that the site has characteristics which, when combined with the facility design and the nature of the waste inventory, adequately isolate and contain the waste or otherwise limit the occurrence in the biosphere of any radionuclides derived from the repository that might be released in potentially hazardous concentrations, for a period of time that is appropriate to that posed by the waste hazard due to its toxicity and half-life;
- Demonstrate that the complexity and stability of the geology and hydrogeology of a site would not adversely affect confidence in interpretations;
- Demonstrate that the proposed development would have acceptably low environmental impacts.

(2) Construction and operational suitability:
- Demonstrate an appropriate level of resource availability for the duration of the repository development programme (human, financial and equipment);
- Demonstrate that there is or would be satisfactory access to a site (in terms of land ownership and transport infrastructure) and adequate space for operations at the surface;
- Demonstrate that the nature of the site would not present unreasonable demands on construction;
- Demonstrate that the site would not present unreasonable demands on emplacement operations;
- Demonstrate that the repository could be successfully sealed at the end of emplacement operations.

(3) Societal and political acceptability:
- Demonstrate local community acceptance;
- Demonstrate national approval and acceptance.

The main tool used to present evidence of site suitability and to demonstrate confidence in the safety of the disposal concept is the safety case [20]. A safety case may be supported by a complementary environmental impact statement or the two might be combined into one overall environmental safety case for presentation to regulators.

The starting point for siting can be anywhere on a continuum ranging from the full national map to one or a short list of a priori preferred sites. For further reference, these two end members for the siting strategy are labelled as:

- Site screening starting from a national map;
- Preferred site (or list of sites) with a priori favourable characteristics.

The first approach (Fig. 7) comprises initial desk based studies that consider the potential for large regions (possibly countrywide) to host a repository. Using a range of predetermined and agreed exclusionary criteria and possibly also using avoidance and suitability criteria, prospective search areas within those regions would be identified and subject to further studies.[4] After a comparison against discriminating criteria, search areas would be successively screened out until a relatively small number of areas containing prospective sites are identified. In turn, these sites would then be considered for their potential suitability to host a repository based on preliminary ground based investigations. Once an

[4] These further studies would likely comprise more detailed desk based studies and possibly limited field surveys and other data acquisition methods.

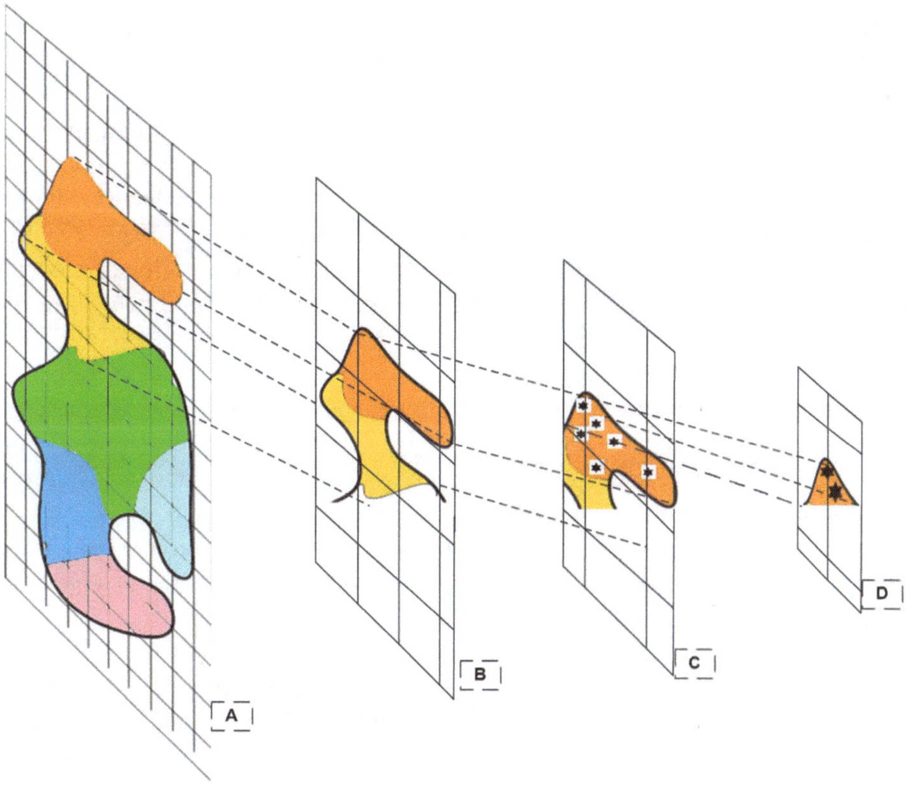

FIG. 7. A screening strategy involving desk studies of large regions to identify search areas of promise for hosting a repository (A), then focusing in on these areas (B) and, in turn, identifying several sites for preliminary investigation (C). Finally, one or a very small number of sites are selected for detailed site characterization (D).

acceptable level of data and site understanding has been established for each site,[5] the criteria, possibly supported by preliminary safety assessment modelling, which is recommended, would be used to screen out more sites from further consideration.

The number of sites selected at each stage and the number of rounds of screening depend on national circumstances and decisions, not least of which would be a siting effort proportionate to the hazard posed by the waste (i.e. application of a graded approach). This would lead to a choice to characterize in detail only one or at most a few of those remaining sites until the final choice of disposal facility location is made through site confirmation.

The other siting approach begins with one or a short list of sites with a priori favourable characteristics such as volunteerism, land ownership, the presence of existing nuclear facilities (potentially strengthened by proximity to storage of a major part of the national waste inventory), or takes advantage of a site that was previously investigated well. The preferred site is then characterized in detail and evaluated to ensure it meets safety and other requirements for disposal. Application of this approach allows one to move straight from the conceptual planning stage to detailed site characterization at the preferred site. It is important to bear in mind that the field investigations in this case would be progressively advanced, as described elsewhere in the publication (i.e. prioritized and carried out in phases). In particular, any indicators that suggest the site may not be suitable should be investigated early on.

For many large scale geological disposal programmes the siting approach iteratively focusing on smaller areas from an initial countrywide or regional search area has often been used), while especially for

---

[5] Section 8 provides some discussion on when sufficient data and understanding might exist for site screening. During a site comparison process, it is generally accepted that the level of understanding between sites should be comparable prior to an assessment of potential suitability. However, it is recognized that under some circumstances the selection process may allow for deviation from this assumption.

near surface repository programmes, the targeted approach identifying a preferred site has been followed. It is noteworthy that there are several variants on the two approaches illustrated above and the choice as to which should be followed would be based on national decisions.

In the second half of the twentieth century, site selection for a radioactive waste disposal facility was typically part of a decision making process that did not necessarily involve the public or other key stakeholders until very late in the process, if at all. Sites were selected without widespread consultation, the decision on a site was declared (or in some cases kept secret) and if and when concern was voiced, the siting decision was strongly defended with little opportunity offered for finding an alternative. This strategy has been termed the decide–announce–defend approach. It has been found to be counterproductive in a number of Member States. As a result, a more open and inclusive approach evolved whereby, on the basis of preliminary studies and technical criteria, regions or large areas were identified as being potentially suitable to host a repository. Then, prospective communities within those regions were approached to enquire whether or not they would be willing to consider hosting a facility, should suitable site conditions be established. This approach, termed the techno–socio approach to siting, has been successful in some Member States, but has not resulted in the approval of sites in other States where it has been tried. Consequently, a further step has been taken in some Member States whereby all communities in a country are invited to consider their willingness to host a repository, regardless of any initial assessment of technical suitability. On the basis of a community expressing an interest, a full discussion of the implications of hosting a repository and a review of the suitability of the local site characteristics ensue, leading to an informed decision by the community as to whether or not to move forward in collaboration with the proponent. The host community is commonly provided with resources to enable independent technical advice to be provided and has designated veto rights. This siting approach has been termed full volunteerism and, under the right conditions, it can result in a comprehensive partnership between the RWMO and a host community. However, there is no guarantee that a 'suitable' host community might come forward. Furthermore, at some point, relinquishing veto rights is inevitable.

The above presentation of siting approaches and the range of considerations is a gross simplification of the actual process adopted in Member States that have carried out such studies. Many examples of successful and unsuccessful siting approaches can be found in the literature. The Nuclear Decommissioning Authority (NDA) in the UK has published a useful report comparing the various siting approaches adopted in a number of countries [21].

Regardless of the siting strategy and the decision making process adopted, a structured approach that has widespread support should be agreed on in advance. The various countries that have undertaken siting exercises have tended to follow a phased approach involving stages separated by key decision points. Figure 8 illustrates the four main stages in a siting process using a screening approach [12]. After initial concept development and planning, one or more large regions of interest are selected to identify smaller candidate areas with potentially many sites (area survey stage). These sites are in turn screened on the basis of predefined criteria, mainly through desk studies and possibly some limited field surveys, to leave a manageable number of sites that could be investigated more thoroughly by surface based investigations and possibly through limited borehole investigations (site investigation stage). After further screening, more detailed surface based and subsurface studies are then carried out on a single site or a small number of sites (typically up to three sites, but no maximum is prescribed) during the detailed site characterization stage until a single site is chosen and an application is made to authorize construction of a repository. The following subsections presents more information about the stages that could be used in such a phased screening approach to siting.

2.3.1. Conceptual planning stage

The first phase of any siting programme is the conceptual planning stage, shown in Fig. 8. This involves the identification of potentially suitable disposal concepts, in line with defined boundary conditions such as the nature of the waste inventory and national policy, and also high level strategic planning for the site investigations. It is also during this phase that criteria should be established to assess

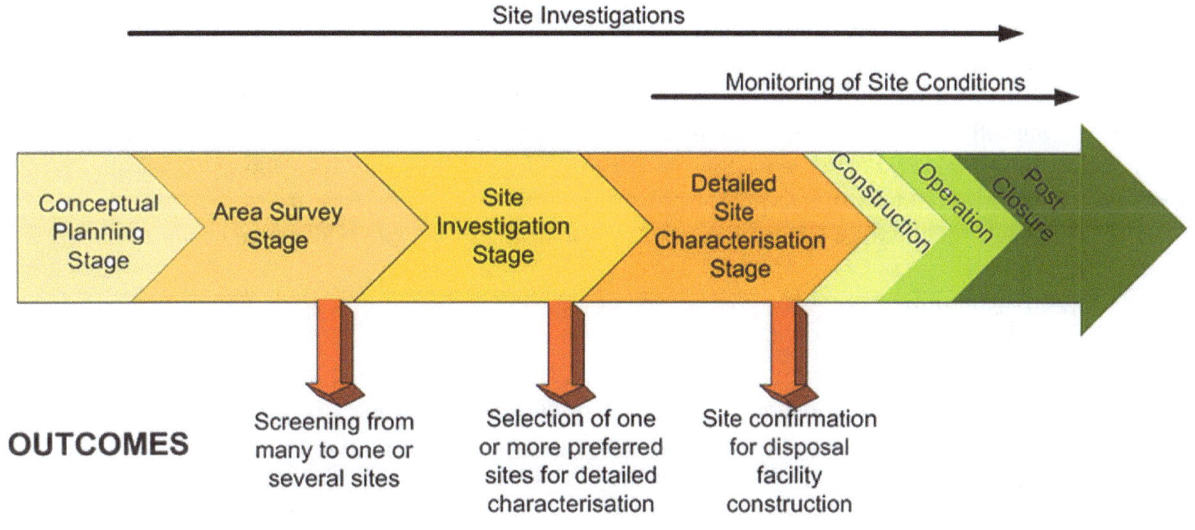

FIG. 8. The four main stages in a siting process [12].

the potential for a site to host a repository. The organization charged with defining the site selection criteria might be the RWMO or might be an independent body selected by government.

In a progressive screening approach to siting, the initial selection criteria would typically include regional considerations that would exclude an area or a site from receiving further attention at an early stage, such as one with a risk of major seismic activity compromising activities during construction or emplacement operations or impacting the long term containment function of natural and engineered barriers. Other factors that could be of early concern include potential volcanic activity or the presence of other natural hazards such as flooding, tsunami and rock fall, or the presence of exploitable resources that might prevent a site from hosting a geological disposal facility. Socioeconomic criteria may also be used for high level screening, such as proximity to population centres, the lack of existing transport infrastructure and an embargo on investigating areas of special cultural, scientific or ecological interest. In addition to deciding on exclusionary criteria, 'preferential' criteria might also be developed during the conceptual planning phase of a siting effort, relating to favourable geology, hydrogeology and topography. General siting criteria for a geological disposal facility are provided in [12].

The site selection criteria that are chosen should be developed in accordance with national regulatory requirements and take into account wider stakeholder interests and needs. The choice of criteria is generally best undertaken in an open and transparent manner and it may be appropriate, depending on the national context and the particular circumstances of the disposal programme, to include a wide range of stakeholders in the decision making process. It is also to be noted that exclusionary criteria based on quantitative threshold values for single parameters (e.g. hydraulic conductivity, fracture frequency, etc.) are generally best avoided as the safety of a site never depends on the value of a single property, rather it reflects the behaviour of the total integrated engineered and natural repository system.

2.3.2. Area survey stage

For a siting strategy that focuses in on smaller areas, from large regions to specific sites, the next phase would comprise a review of one or more areas within a country. In this context, an area might be an administrative unit or it might be defined by natural characteristics. Consequently, the size of an area is not defined, but it would generally be large enough to contain several prospective sites. These areas would be determined largely on the basis of desk studies involving a review of existing information. The information collected at this stage would have to be sufficient to allow: (i) a fair comparison between the areas; and (ii) the screening out (exclusion) of some areas plus the identification of smaller areas and specific sites that show promise and that would be suitable for more detailed investigations.

At some point, literature surveys and a review of archived information will have to be supplemented by additional data and understanding that can only be collected in the field at a specific site. This can formally be considered the start of operational site investigations and it may occur during the area survey stage if existing information is scarce. As a result of work during this stage, some regions will be screened out as being unsuitable to host a repository, while a smaller number of areas and specific sites within those defined areas would remain as prospective candidates to be taken forward to the next stage. Such sites might all be located within the same general region or might be widely dispersed across a country.

2.3.3. Site investigation stage

The number of sites remaining at the start of this stage is not defined, but might typically number three to ten. This is when ground operations begin in earnest. If not already obtained, remote sensing data are acquired and preliminary surface based geophysical and mapping surveys are planned in detail and carried out. A limited number of boreholes are also likely to be drilled during this phase to collect subsurface information that does not already exist as a result of previous studies.

The objective of the continued screening undertaken now is to reduce the number of prospective sites to one, two or possibly three sites, each of which would be characterized in more detail during the next and final stage to demonstrate site suitability to host a repository and to allow a final choice of site to be made.

It is useful to note here that the nature of the waste, preferences concerning the engineered system and the characteristics of the geological and surface environment collectively make up what is termed the disposal system. Therefore, it can be appreciated that a more detailed description of the disposal system will become evident during site investigations and repository design and safety assessment studies would be able to advance significantly.

2.3.4. Detailed site characterization stage

The final stage in an idealized siting strategy that focuses in from large regions of interest is termed the detailed site characterization stage. During this phase of siting, a small number of candidate sites (typically one to three) are intensively investigated through further surface based surveys and underground through the use of boreholes.[6] If required, a site specific URL might also be constructed now to more fully understand the nature of the 3-D characteristics of a rock mass at a site. Sections 6 and 7 describe in detail the types of site investigation activities that could be undertaken at a site during this stage or earlier (data acquisition and its interpretation and integration).

On the basis of the knowledge acquired during site investigations, as well as results from the safety assessment studies and detailed repository design efforts that would occur in parallel, this stage concludes when a single site is chosen by the proponent to host a disposal facility, when all of the siting studies have been integrated into a comprehensive safety case (and an environmental impact statement has been prepared if required), when an application for authorization to construct a repository has been submitted, and when the licence has been provided by the designated authority.

---

[6] If a preferred site approach has been used, the detailed site characterization efforts could start immediately once appropriate plans and approvals have been made (i.e. the area survey and site investigation stages described above could be omitted).

3. A REQUIREMENTS DRIVEN APPROACH TO SITE INVESTIGATIONS

3.1. INTRODUCTION

For the purposes of safety, security, environmental responsibility, efficiency of resource use and for focus, a site investigation project for a radioactive waste disposal facility should be founded on a clear understanding of data and information requirements. This is termed a requirements driven approach. Without a clear understanding of what data and information needs to be obtained and why, there is no rational basis for designing or implementing the site investigations. Furthermore, it cannot be known when and how to optimally collect the data and information, nor in what level of detail.

This requirements driven approach to the planning and management of site investigations has been developed and refined by several RWMOs since the 2000s (e.g. SKB [22, 23], Ontario Power Generation [24], Andra [25], NUMO [26], and Nagra [27]). As a result, a general requirements framework has been created for site investigation projects (Fig. 9).

As can be seen from Fig. 9, there is a hierarchy of needs led, at the highest level, by international obligations and commitments, followed by national policy and strategy choices that in turn determine national legislation and regulations.[7] National policy decisions based on the radioactive waste inventory to be disposed of will also influence the range of disposal concepts that might be considered. As soon as decisions have been made about viable disposal concepts, generic engineering studies and generic safety assessments can commence. An early start to these studies, even in the absence of knowledge from specific sites, builds up internal competencies and capabilities in these key areas and begins to focus in on organizational strengths and weaknesses, including technical limitations and knowledge gaps that will have to be addressed in due course. Note that there is an iterative loop shown in Fig. 9 by red arrows and also note that international obligations, national policy and regulatory requirements may change at any time and these may alter the disposal system's research, development and demonstration (RD&D) requirements (shown by dotted arrows). Section 8 discusses criteria to establish exit conditions.

Based on a preliminary disposal concept and regulatory requirements, which will evolve in parallel with a repository programme, knowledge gaps can be identified. These will define a framework for RD&D activities. Safety assessment, preliminary repository engineering design studies and investigations to support environmental impact studies can all be considered forms of RD&D and they will all be informed by data and information to be collected at a site. In many disposal programmes there is also a generic RD&D portfolio of activities in place. This is for collecting and analysing information that may not be readily available from a prospective site, to contribute to safety assessment and repository design efforts and to enhance confidence in the repository concept. An example of such generic studies is research into natural analogues of repository-related processes that are not be feasible to investigate at a site (also see Section 3.5).

Leading on from international obligations, national policy decisions and supporting legislation and regulations, the RD&D programme, in turn, identifies specific data and information requirements for use in safety case development, comprising safety assessment, repository design studies and EIAs. This also contributes to more general confidence building. The process is then taken further to relate these requirements to the natural system parameters that need to be measured and the geosphere and biosphere processes that need to be understood. For each data set identified for a particular use, it will be important

---

[7] It is noted that national policy decisions should logically dictate the underpinning legislation and regulatory requirements. However, the reality is that in many cases legislation and regulation may already exist (reflecting previous policy choices) and these may in fact significantly influence subsequent policy considerations and choices.

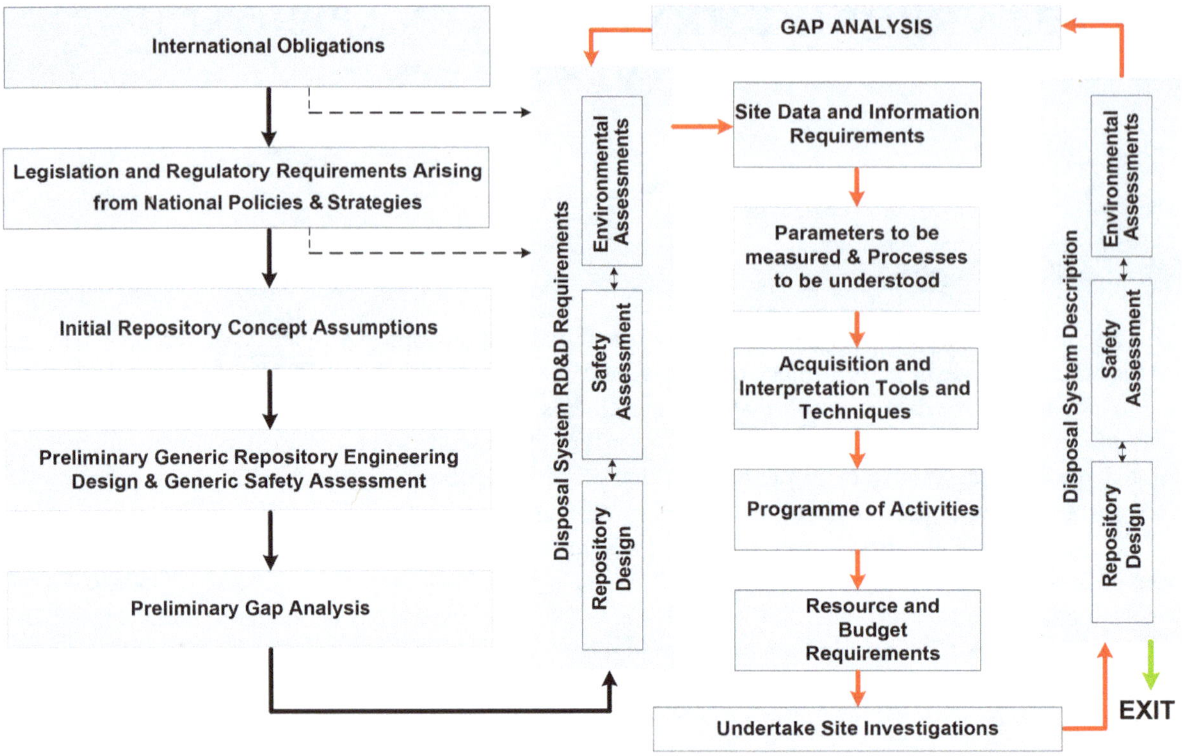

FIG. 9. The chain of requirements and how they control the specification of a site investigation programme. The red arrows show an iterative loop.

to specify the accuracy, precision and representativeness required from parameter measurements to reduce or manage uncertainty and deal with spatial variability. On this basis, the suitability of various data acquisition tools and techniques can be assessed for their prospective use in site investigation surveys, as can appropriate interpretation processes and modelling tools. Dependent on which surveys are required and which tools and techniques are considered appropriate, a detailed work programme can then be developed which can be used as the basis for cost estimates and requests for budgets and other resources.

Figure 9 shows that by defining initial requirements and exploring how they might be addressed as a result of knowledge concerning the inventory for disposal and site conditions, a suitable site specific disposal concept can be proposed. This will, in turn, define more detailed data and information requirements — to be collected through another iteration within the site investigation project to further refine and strengthen engineering, environmental and safety assessment studies.

It is to be emphasized that due to the project timescales typical for site investigations, which have been measured in decades in some major national disposal programmes, the requirements underpinning data acquisition and interpretation efforts are considered dynamic and likely to evolve. Consequently, it is essential to closely monitor and amend requirements as necessary.

Based on these considerations, it can be appreciated that as part of the planning process for any site investigation project, it will be important to document as fully as possible the requirements and other information needs and preferences associated with each of the following:

- *International obligations:* Responsibilities and requirements stemming from, for example, commitments associated with the Joint Convention on the Safety of Spent Fuel Management and on the Safety of Radioactive Waste Management or other international conventions, treaties and agreements.
- *Legislation and regulatory requirements arising from national policies and strategies:* A listing should be established of specific requirements relating to national policy and acts of law, both

national and local, if appropriate. These might include environmental, nuclear, radiation protection, building, mining and transport acts and ordinances. In addition, regulatory authorities may issue additional regulatory requirements stemming from legislation in other areas and these should be well appreciated. The RWMO may have to respond to the needs of one or several regulatory bodies and there should be regular communications and reviews to ensure appropriate guidance is provided in interpretation of the regulations, whilst maintaining a strict division of roles and responsibilities between the implementer and regulator(s).

— *Disposal system requirements:* Laws and regulations cannot easily be used directly to define the detailed requirements necessary to be demonstrated by the disposal system and its various parts. Consequently, it is best for the RWMO to translate regulatory requirements into functional system requirements and preferences, to be demonstrated by the repository and its surrounding environment on the basis of (i) established safety principles and preferences, such as the need for containment and isolation of the waste, passive safety, the use of multiple barriers, etc. and (ii) any requirements related to the construction and operation of a disposal facility, such as the availability of a satisfactory volume of rock with characteristics amenable to construction.

— *Site data and information requirements:* Laws, ordinances and regulations cannot readily be used to stipulate the detailed requirements associated with components of the natural system, such as rock properties and groundwater characteristics. However, by carefully defining the functional requirements associated with barriers and understanding the need to study in detail the nature of pathways from the source of a hazard to the potential receptors in the biosphere, through the use of performance and safety assessment approaches, more specific requirements and preferences can be generated to inform the implementer about the parameters that will be required to be measured and the ranges of acceptability associated with those parameters, recognizing that there will be process feedbacks involved and deficiencies in one parameter may be compensated by another (i.e. need to consider total system performance and safety).

— *Data and interpretation tools and techniques:* On the basis of the identification of requirements and functional analysis associated with the repository system, that in turn establishes the data and information to be collected in the geosphere and biosphere, the next stage is to establish the range of equipment and other tools and techniques that can be used to obtain the information required. Sections 6 and 0 provide more details in this regard.

— *Programme of activities:* The site investigations should be developed step by step, with clear programming of activities based on the appropriate allocation of time frames for surveys, reviews, stakeholder interactions, etc. and separation of the various phases by milestones and decision points. A complex site investigation project will require sophisticated programming to be undertaken on several levels, following a hierarchical approach to reflect the needs of different users (e.g. using a Gantt chart, programme evaluation reviews, and other tools). As much as is possible, activities should be carried out in parallel. Critical paths in the programme should be established through the identification of potential challenges and critical activities that can impact other areas. Clearly, the specification of a comprehensive, logically and optimally structured, and defensible programme is a major endeavour that calls upon the need for established project management skills.

— *Resource and budget requirements:* Only once the range of tools and techniques to be considered for use is decided on and a programme of activities developed can reliable cost estimates be established to allocate budgets and evaluate workforce needs. There are many alternative approaches to cost estimation; specialist guidance will be required in this area as it is not dealt with further in this document, although it is addressed further in other IAEA reports [28].

In following a requirements driven approach, it should be possible to clearly demonstrate that only information needed to support a disposal programme has been collected, ensuring the effective use of resources and establishing an appropriate level of confidence and quality associated with the data and information, for example in terms of sufficiency, fitness for purpose, accuracy, precision and how representative it is. The first case study (Annex I) illustrates the approach adopted by NWS in the UK for identifying information requirements associated with a site investigation project.

The remainder of the report expands on several of the above issues, but first general data and information requirements in relation to their end use within a RWMO are considered. In this context, end use relates to post-closure safety and environmental assessments and repository engineering and design (Sections 3.2 to 3.4). There are also implicit requirements that relate to addressing external stakeholders' expectations and promoting confidence in the safety and viability of the disposal concept (Section 3.5). Then key objectives and requirements in terms of disciplines are looked at in more detail in Section 3.6. It is important to recognize in this discussion that the data and information required by end users specifies and is directly reflected in the data and information requirements to be delivered by the discipline oriented activities. It is also reiterated, by reference to Fig. 2, that there are overlaps in the uses to be made of the data and information by each of the end user groups. Thus, for example, some of the geological information to be provided responds to end user needs in all three areas of safety assessment, repository design studies and environmental impact studies, as well as providing additional confidence for external stakeholders.

3.2. REQUIREMENTS FOR POST-CLOSURE SAFETY ASSESSMENT

The fundamental safety objective set out in the IAEA Safety Standards [1] is to protect people and the environment from the harmful effects of ionizing radiation. In the context of radioactive waste disposal, this is to be achieved primarily through the containment and isolation of the waste. Post-closure safety assessment is the principal tool employed to illustrate the expected performance of a sealed repository and its safety, in terms of quantified potential radiological consequences arising from the hazard posed by radioactive waste (i.e. radiation dose and risks).

Experience and skills in designing and carrying out safety assessments within a RWMO should be initiated early on in the disposal programme and should ideally start before or at least during the concept design stage, at the same time as the site investigation effort is being planned. From generic first steps in the absence of site specific data, the capability to develop robust safety assessment models should increase significantly as data and information from a site become available. Regular updating of a safety assessment should occur over the lifetime of the siting studies (and beyond into operations) and, as shown in Section 2, there needs to be close and iterative collaboration between the site investigation team and the safety assessment modellers to specify requirements, identify outstanding uncertainties, develop and test predictions, and design further stages of the investigation efforts.

During a post-closure safety assessment exercise there are essentially five key questions to be answered: (i) What are the conditions under which radionuclides might be released from a repository into the wider environment? (ii) How will the engineered and natural barriers perform under a range of plausible future conditions and what pathways could be exploited? (iii) Over what timescales might radionuclide migration to the biosphere occur. (iv) In what form and concentrations might potentially hazardous radionuclides return to the biosphere? and (v) What would be the consequences for humans and the environment? In addressing these questions, it can be appreciated that the broad natural system information needed to prepare a site safety assessment includes the following:

- Geological framework of the site including the distribution of geological units and structural features that might act as barriers to radionuclide migration or as fast pathways;
- Hydrogeological framework including the nature of groundwater flow in terms of flow mechanisms and solute transport retardation processes, potential radionuclide migration pathways, flow velocities and hydraulic gradients;
- The hydrogeochemical framework, describing the nature and distribution of groundwater bodies, their evolution over time and their potential to affect barrier safety functions and potential pathways in the future;

— Nature of the driving forces that influence the potential migration of fluids through the geosphere (groundwater, gas) such as topography and groundwater density;
— Descriptive models of the site in relation to geology, hydrogeology, hydrogeochemistry and geotechnical properties that define the 3-D geometry of units and the parameters and parameter values that describe the properties of those units and their impacts in terms of, for example, solute migration and retardation processes, potential pathways, flow velocities, among others;
— An understanding of the regional stress and thermal regimes;
— Understanding of the effects of repository construction and design on the nature of the zone surrounding a repository (i.e. the volume of rock that is damaged or disturbed during excavation and drilling (termed an excavation damaged zone or an excavation disturbed zone (sometimes referred to as an EDZ))).

In the context of safety assessment studies, site investigation experts should place special emphasis not just on collecting robust measures of parameters and quantifying their uncertainty but also on understanding, in an integrated manner, the evolution of the site and its potential for change in the future, as a result of natural phenomena and also anthropogenic influences.

For geological disposal facilities intended to contain long lived waste, processes that might occur in or at the Earth's surface due to the impacts of climate change, volcanism, tectonics and seismicity have to be considered over long times scales (certainly tens of thousands of years for deep geological disposal concepts involving long lived waste) due to the potential for these processes to affect radionuclide release and migration in the geosphere and at the geosphere–biosphere interface (through sea level rise and fall, permafrost development, glaciation, faulting and land uplift, among other processes). Conversely, the timescales of concern for a near surface disposal facility would be measured in terms of centuries or a few thousand years.

Projections exploring the effects of climate variability have been developed over a wide range of timescales. They take into account natural Milankovitch (astronomical) forcing and consideration of our impact on climate [19, 29]. Confidence in near-term climate projections is greater than for those made at timescales measured in tens or hundreds of thousands of years, but justified scenarios involving the effects of climate change and associated exogenic processes can be developed and the range of uncertainty associated with the potential impacts on a repository system can be evaluated.

In some cases, data and information required for populating safety assessment models cannot always be readily obtained from site investigations. In these circumstances generic information may suffice in the absence of reliable site data and in other cases, especially those concerning processes that operate over long timescales, the use of natural analogues may be required [30]. Furthermore, by applying a graded approach to site investigations, other generic data may be suitable for use on a site specific basis where the effort and costs involved would not be commensurate with the hazard posed, being aware at all times that safety considerations are paramount and that cost alone is never the prime consideration.

3.3. REQUIREMENTS FOR REPOSITORY DESIGN

The mechanical properties of the prospective rock mass being considered for a disposal facility influence the containment and isolation characteristics of the repository, but also strongly affect the detailed design of the repository, including its dimensions, the disposition of shafts, tunnels and disposal vaults and the construction methods to be employed. In addition to the geotechnical attributes of the rock mass (e.g. in terms of rock quality, in situ stress and other properties), it is critical to evaluate information derived from other disciplines in the context of the engineering design — especially geology and hydrogeology — but also including the hydrochemistry where groundwater composition may have a bearing on the long term integrity of engineered materials and barriers that are part of the repository design.

There are two complementary approaches commonly used to support engineering design and construction: the use of empirical rock quality indices and numerical modelling. The data and information to be obtained from a site investigation project should support the data needs for both approaches as they relate to the following site aspects:

- — Topography;
- — Geological succession;
  - Rock units;
  - Depths to boundaries.
- — Geological structures;
  - Faults and fractures;
  - Other discrete geological features and their disposition (e.g. folds, dykes, sills).
- — Rock mass quality;
  - Weathering and alteration;
  - Geotechnical units;
  - Rock quality ratings (e.g. Q, RMR).
- — Engineering properties of soils;
- — Engineering properties of rocks;
- — Temperatures;
- — In situ stresses;
- — Groundwater conditions;
  - Predicted inflows (mechanisms, fluxes and locations);
  - Predicted pressures.
- — Hydrochemistry;
  - Dissolution potential;
  - Precipitation potential;
  - Mineralogical alteration.
- — Construction materials;
- — Presence and nature of gases;
- — Rock deterioration (e.g. spalling or creep).

Seismic hazard analysis will also need to be taken into consideration as earthquake activity may impact on some of the above properties and features. During site investigations, information from the above areas are used to support decisions concerning the design of the engineering components of the disposal facility, for example to define the following activities and features of the facility and inform the proposed method of construction:

- — Surface layout of works, including access to underground (drifts and shafts);
- — Design of surface works;
- — Repository volume and layout;
- — Repository footprint (plan area at the surface reflecting the disposition of the underground components of the repository);
- — Depth of waste emplacement vaults;
- — Dimensions of waste vaults and access ramps, tunnels, shafts, etc.;
- — Excavation methods for underground openings;
- — Support requirements for underground openings;
- — Identification of geotechnical and environmental hazards (such as gases, water, in situ stresses and creep, etc.) and other risks;
- — Identification of surface hazards (e.g. landslides, alluvial flood risk and tsunami, pyroclastic and lava flow, etc. with the potential to affect construction, existing buildings and operations);
- — Risk management strategy for minimizing risk and dealing with identified hazards;
- — Interactions between engineering structures and groundwater (concrete, bentonite, steel, etc.);
- — Materials for construction use;

— Strategy for dealing with excavation and construction wastes.

Although there are a number of aspects of the information requirements that appear similar to those required for safety assessment purposes, it is important to recognize that the methods of satisfying the information needs may vary depending on the use that will be made of the data. Thus, for example, construction engineers may require different types of information on groundwater flow (for example to evaluate likely inflows into underground openings and to design appropriate ground treatment to reduce inflows should that be necessary) than would be required by the hydrogeological modelling team seeking to evaluate the long term transport of radionuclides through the geosphere. The length scales and spatial distribution of information related to engineering design and construction is also likely to be somewhat different from that required for other purposes, namely geoscientific understanding and safety assessment.

Much of the information necessary to assist in the early repository design effort can be obtained from surface based investigations (e.g. surveys of surface outcrop and investigations on core from boreholes, as well as wireline logging in boreholes). However, for geological repositories, it is generally the case that more accurate and representative data will need to be obtained from underground investigations in a generic or site specific URL or from surveys undertaken as part of the underground exploratory phase of a siting programme where, subject to regulatory approvals, an underground tunnel or vault might later become part of the repository development, as demonstrated for example from Posiva's repository programme in Finland.

3.4. REQUIREMENTS FOR ENVIRONMENTAL IMPACT ASSESSMENT

Where required by national legislation and regulations, a consequence analysis is to be undertaken to establish the likely effects of a proposed development on the environment. This exercise is commonly called an EIA and its main drivers are to:

— Avoid serious and irreversible damage to the environment through the identification of potential impacts and the application of measures for avoiding or mitigating them;
— Ensure efficient resource use;
— Enhance social aspects of a development;
— Inform decision making and establish conditions for approval.

The type of information required for an EIA comprises not only a description of the proposed disposal facility, including the construction methods and operational activities to be undertaken, but also consideration of possible impacts in terms of, for example, the following:

— Air quality;
— Noise;
— Surface water bodies;
— Groundwater;
— Land use;
— Flora and fauna;
— Socioeconomic factors.

An important aspect of an EIA is that when there are alternative options for facility design and construction, operation and closure methods, they are to be reviewed to assess the cost–benefit in terms of environmental degradation or improvement. Where damaging environmental impacts are unavoidable, mitigation measures are to be established.

An EIA for a radioactive waste disposal facility would not only have to consider conventional impacts arising from typical development activities (e.g. excessive noise during construction of a building), but also the potential for harmful radiological impacts on people and the natural environment during operational and post-closure periods.

An assessment of possible environmental impacts would be underpinned by field surveys, monitoring and other methods, largely relating to investigations of the surface environment. As much of the data and information required to support EIAs is essentially similar to that needed to inform biosphere studies for use in safety assessment and repository design purposes, it can be readily appreciated that there would be significant overlaps between the two approaches. Consequently, site investigations should be optimized by ensuring that activities address data and information acquisition needs in all three areas without unnecessary duplication of effort.

A particular area of overlap in the use of site data for EIA and safety assessments relates to the establishment of baseline conditions. Baseline can be defined as a description of the condition of a system prior to some form of disturbance. Due to the natural temporal variability associated with some key parameters (e.g. groundwater pressure), baseline conditions are generally not established at a particular moment in time but over a specified period. By understanding the environmental situation at a site prior to repository construction, any significant impacts arising as a result of that construction and the later operation of a facility should be readily recognizable, provided that a suitable monitoring programme is in place. Consequently, to establish the baseline condition at a site, a series of measurements are required prior to construction, for comparison with a series of measurements undertaken during or after construction.

Confidence in observations and their meaning are dependent upon ensuring data and information that is sufficient quality assured are available to adequately describe the undisturbed conditions at a site and to understand the potential for any subsequent disturbance. Such data would be used to develop hypotheses and to populate and test models of the site so that cause and effect can be confidently distinguished. Monitoring data to be compared against baseline data would therefore need to reflect periodic measurements of specified parameters, with sampling frequency and resolution established on the basis of a precise definition of requirements and scientific reasoning. The parameters to be monitored might be defined by the regulatory authority, but in the absence of any such requirements, monitoring related to the baseline would, at a minimum, typically require periodic measurements of groundwater pressures and hydrochemistry, surface water flows and water chemistry, meteorological conditions and environmental concentrations of selected radioisotopes. Other parameters relevant to the baseline might also be required by the regulator.

An example of one of the main environmental impacts that might be expected to result from repository construction is a local change in groundwater levels and a wider groundwater pressure response. This could have relevance for a situation where, for example, groundwater is being abstracted in proximity to a potential repository for drinking or the irrigation of crops, with the potential impacts being either groundwater pollution or a reduction in groundwater yield. In order to distinguish the effects of repository construction on groundwater levels and pressures, it would be necessary to ensure spatial and temporal pressure fluctuations due to any other human activities in the area have been adequately identified and that the occurrence and impact of natural processes are sufficiently understood. For example, natural groundwater pressure fluctuations are observed due to episodic variations in recharge and due to minute crustal deformations arising from changes in atmospheric pressure and the gravitational effects of the Moon and Sun (Earth tides). They can also result from a seismic event.

Numerical modelling is typically used to evaluate the likelihood of impacts and their magnitude. To support the construction, testing and use of numerical models, site data[8] may be used in one of the following three ways:

(i) For the initial population of models;

[8] For some models it may also be appropriate to use generic data; for example, where it is not possible to obtain site data of sufficient quality.

(ii) To improve the match between model results and observations by adjusting input parameters to reflect data or conceptual uncertainties (calibration);
(iii) By comparing model output with measurements and observations that have not previously been used to populate and calibrate the models (validation).

In this manner, data and models are used to increase confidence in understanding the processes that might affect radionuclide transport at a site.

Any environmental monitoring system put in place for establishing baseline conditions or otherwise supporting an EIA would be implemented during the siting phase, but more specifically would have to be put in place during the detailed site characterization stage or earlier. Typically at least a year is needed to establish baseline seasonal meteorological effects, but several years of data might be expected for a major development reliant on populating and calibrating advanced numerical groundwater flow and transport models where there might be a large degree of seasonal variability.

The case study in Annex V is based on the experiences of Andra in France and contains a summary of the approach adopted and lessons learned concerning the design and implementation of an environmental monitoring system to support site investigations and the use of site data in safety assessment, EIA, and repository design and operation.

3.5. ADDRESSING ADDITIONAL STAKEHOLDER NEEDS

This section is concerned with identifying and responding to additional site investigation needs, especially for establishing and strengthening stakeholder confidence and acceptance of a safety case. In this discussion, the term confidence [31][9] is used in its everyday meaning and not in relation to statistical concepts of a confidence interval or confidence level. Consequently, confidence, as used here, is entirely subjective.

A safety case includes a collection of arguments and evidence in support of the safety and viability of a disposal facility [2, 18]. It is also an important vehicle for engaging in dialogue with regulatory authorities and other stakeholders. It evolves in parallel with advances in other areas of the repository development programme and at any stage in this process the 'current' safety case is intended to support decisions during and at the end of that stage.

Safety assessment is an integral part of a safety case as one of its key roles is to inform, guide and support decisions that have a radiological or wider safety consequence. Therefore, it is important to recognize that a safety case is much broader than the safety assessment alone. As such, safety case documentation can present supplementary data and information beyond that strictly needed for safety assessment calculations leading to dose and risk. Taken together with safety assessment results and the specifics of a repository design, this additional data and information not only justifies the feasibility of a proposal to a regulator, but is also intended to enhance wider stakeholder confidence in the RWMO and its work in implementing a disposal concept.

It is obvious that to promote wider confidence and, in turn, acceptance of a repository development programme, it is important the RWMO demonstrates that the disposal concept defined by the site conditions and repository design is capable of providing a degree of safety and security commensurate with the hazard posed by the waste inventory. This is evidenced primarily by the safety assessment results, the repository design and a security plan. However, it is also important to recognize that reliance on broad acceptance of the results of safety assessment and engineering calculations is not necessarily sufficient to generate confidence in an overall safety case, as a viable disposal concept needs to be demonstrably feasible (e.g. in terms of construction, operation and closure) and also demonstrate an efficient and effective use of resources in comparison to other potentially available solutions. Furthermore, another

---

[9] The OECD/NEA has defined confidence as "to have reached a positive judgement that a given set of conclusions are well-supported" [31].

part of the confidence building effort requires a convincing and defensible demonstration of the overall understanding of a site and knowledge about the processes operating at a site, beyond that strictly necessary for the safety assessment or engineering purposes. Consequently, in addition to robust site specific data intended to inform models and designs, additional data and information might be fortuitously or deliberately collected and used to support confidence in the disposal concept and the wider safety case. These data and information obtained from the proposed repository site may be supplemented by research that draws on wider data and information derived away from the proposed disposal site. When such data and information are to be used, it is important to clearly establish that they are applicable in the context of the site specific repository proposal and, hence, appropriate for use in a safety case.

Studies providing potentially useful data and information obtained from sites other than where the site investigations for a disposal facility are ongoing are termed generic. Generic studies include field observations and measurements at sites that could be considered representative of repository conditions under alternative boundary conditions (i.e. natural and/or anthropogenic analogues) or where in situ experiments have already been carried out, providing generically applicable data, or in circumstances where samples obtained from elsewhere have been used to obtain laboratory measurements that might be proxies for repository conditions. Such generically derived data and information are typically obtained when they cannot readily (i.e. safely, easily, economically, reliably, etc.) be acquired from the specific site considered for hosting a repository or because the temporal and spatial scales associated with certain processes are so long that any observations or experiments at the investigation site(s) would be impractical or unreliable for testing hypotheses and models. In selecting generic data and information from a field or laboratory setting, it will be important to establish the degree to which boundary conditions can be identified and the degree to which the data are representative of the field conditions at a site and hence transferable (Table 2).

Generic studies are potentially useful not just for providing data and information relevant for use in safety assessment and repository design studies, but also because they demonstrate the wider applicability of key processes and the widespread existence of important features. Because many generic studies might have already been undertaken in advance of site specific investigations, often by academics or organizations not associated with a repository development programme, data and conclusions would likely be viewed favourably as impartial, but there is a strong obligation on the RWMO to ensure data used or conclusions derived from such data are reliable and actually applicable to the site being investigated.

To make any data and information from site specific and generic studies readily available in a comprehensive, understandable and timely manner, the RWMO will need to provide reports at different levels of detail to address the different needs of different audiences. This will require the development

TABLE 2. TYPES OF GENERIC STUDIES THAT COULD BE USED TO SUPPORT A SAFETY CASE

Type of generic study	Boundary condition characteristics	Timescale of relevance for process understanding	Length scale of relevance for process understanding
Natural	Uncertain initial and evolving boundary conditions, real environment	Long term (range of 10 000 years and beyond)	Variable, from microscopic to regional
Field observations and in situ experiments	Complex boundary conditions, realistic environment	Short to medium term (up to tens of years)	Variable, typically from centimetre to regional
Laboratory experiments	Well defined imposed boundary conditions, potentially poorly realistic environment	Short term (up to several years)	Variable, typically from microscopic to metre

of an appropriate publications strategy that itself should sit within a broader stakeholder engagement and communications strategy [10]. The resulting plan for publications would likely identify the need for the production of a number of different types of reports of variable complexity and detail, ranging from high level summary reports written in non-technical language and targeted at the public, to very detailed presentations of technical data and their analysis and targeted at specialists. Further guidance on reporting is provided in Section 4.2.8.

While stakeholder needs and expectations for confidence building may be difficult to predict with a high degree of certainty, they can be broadly anticipated. Consequently, confidence relies not just on an acceptance of the robustness of the proposed disposal concept, underpinned by data and information, but in large part confidence also relates to trust in the RWMO and trust in the overall decision making process (i.e. the process that also includes decision making activities outside of the immediate control of the RWMO, being dominated especially by the actions and behaviours of the regulatory authority and the role of the public and of local and national government bodies).

Frequently cited aspects relating to the culture and behaviour of an RWMO that contribute to stakeholder trust and confidence concern issues of openness and transparency. The degree to which a culture of openness and transparency is practiced will be very specific to an organization and may reflect wider norms within a Member State. Openness typically refers to a willingness on the part of an RWMO to engage with stakeholders and involve them in an internal decision making process, whilst transparency is concerned with active information sharing to promote understanding. It can be appreciated that openness and transparency go together and, in this regard, stakeholders need to be informed and helped to understand not just the issues but also the nature of any underpinning evidence. Stakeholders should also be provided with unbiased information to allow them to come to their own informed conclusions. Consequently, wherever possible, site relevant data and information including their context and meaning should be made available to stakeholders, together with any other influences underpinning key decisions. Multiple lines of independent evidence used to illustrate conclusions and justify decisions are especially powerful in this regard.

To have confidence in an RWMO and its associated disposal concept, stakeholders also need to be convinced of the credibility, integrity and reliability of the organization. Confidence in an organization reflects positive attitudes built on a knowledge of past and current actions, the nature of the people within the RWMO (comprising the leadership, the experts and supporting staff), the internal governance, the documented procedures and other internal processes that make up a management system [32]. Of equal importance for establishing confidence are the tools employed to acquire, interpret, manage, store and present data, information and knowledge (comprising the physical equipment, the data management infrastructure and computational models, particularly relating to the validation of such models) and the reliability and robustness of the data generated and the information used (Fig. 10). These factors that all contribute towards stakeholder confidence are directly under the control of an RWMO (shown by the boxes linked by arrows in Fig. 10), but there are also external factors that are largely outside of the control of the RWMO, such as the broader national decision making process, regulatory oversight and best international practice.

Most of the factors shown in Fig. 10 that contribute towards promoting stakeholder confidence are evaluated on a subjective basis; therefore, their relationship to site investigation as specific requirements cannot easily be made with precision and certainty. However, an RWMO should recognize the factors that influence confidence building in their own society and ensure those factors within its control are acknowledged and managed in an appropriate manner. This will enhance stakeholder confidence in the site investigations and in the wider safety case. A more comprehensive discussion of factors affecting stakeholder acceptance for the implementation of geological disposal can be obtained in Ref. [33].

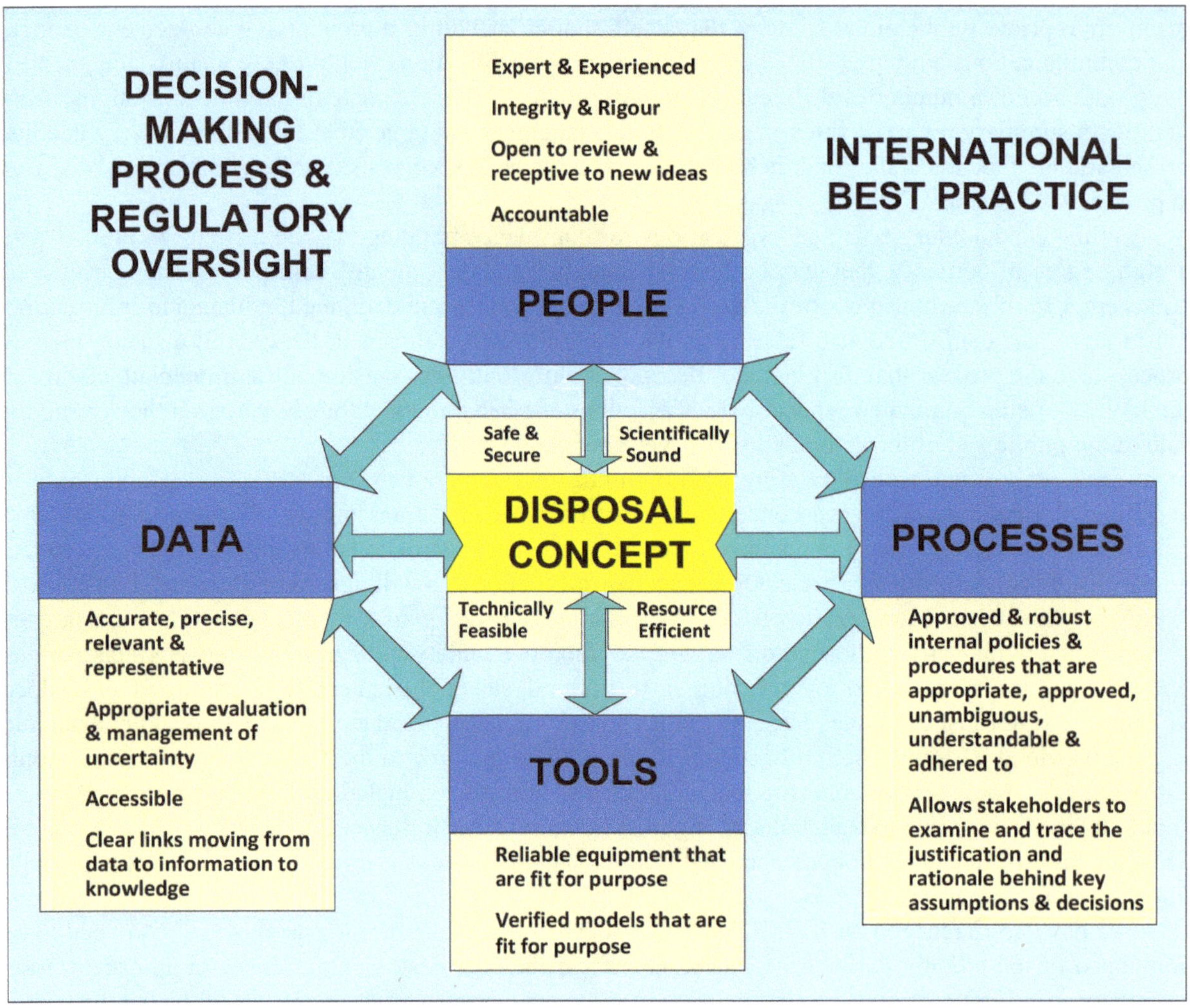

FIG. 10. Key aspects contributing to stakeholder confidence in an RWMO, its site investigation project and the wider safety case for a disposal concept.

3.6. OBJECTIVES AND SCOPE OF INFORMATION REQUIREMENTS IN RELATION TO DISCIPLINES

Requirement 15 in the IAEA Specific Safety Requirements document SSR-5 (Disposal of Radioactive Waste [11]) relates to site characterization for a disposal facility. It specifies:

> "The site for a disposal facility shall be characterized at a level of detail sufficient to support a general understanding of both the characteristics of the site and how the site will evolve over time. This shall include its present condition, its probable natural evolution and possible natural events, and also human plans and actions in the vicinity that may affect the safety of the facility over the period of interest. It shall also include a specific understanding of the impact on safety of features, events and processes associated with the site and the facility."

Reflecting the need to respect the fundamental safety requirement, some underpinning discipline oriented work programme objectives and requirements can be developed to structure and guide the planning and management of a site investigation project. The five major disciplines typically comprising a site investigation project relate to:

— Geology;
— Hydrogeology;
— Hydrochemistry;
— Geotechnics;
— Biosphere studies.

This high level division of requirements according to discipline contains many subdisciplines. Subdiscipline expertise in one field can often be readily applied in other disciplines (for example data derived from structural geology has applications in geology, hydrogeology and geotechnical areas). It can be readily appreciated that any requirements based on disciplines will likely contain a degree of complexity and overlap. Nevertheless, for each of the major disciplines, illustrative high level objectives and work programme requirements are provided in Sections 3.6.1 to 3.6.5.

3.6.1. Geological work programme

Objective: To provide data and information on the geological features and processes operating at a site, as well as understanding concerning past and future events that may impact on the geology of a site, which in turn could degrade or bypass disposal system barrier functions in the future. Such data and information are needed to provide a framework for the preparation of models to support safety assessments, to inform repository design studies and EIA, and to provide confidence in the understanding of the site. In particular, it will be necessary to demonstrate that the geological complexity of the site can be understood with confidence and that geosphere conditions would remain sufficiently stable over the timescale of concern (i.e. when radionuclides would remain a potential hazard), such that any disturbance to the system would not unreasonably impact on the performance of engineered and natural barriers and so pose an unreasonable risk to people and the environment.

The geological information required from site investigations comprises:

— The regional geological setting of a site, to provide context and understanding for more local (site scale) studies;
— The topographic form of the site and its surrounding region to provide the framework within which the investigations and the disposal facility are located;
— The geomorphology of the site as an aid to interpreting the history of the development of the site, the nature of potential hazards and the surficial distribution of the geological materials present at the site;
— The nature, distribution and properties of the soils and any superficial sediments at the site (alluvium, regolith, etc.);
— The nature, distribution and properties of any cover rocks overlying the host formation at and around the site;
— The nature, distribution and properties of the host rock formation within which it is proposed to construct the disposal facility;
— The nature, statistics and characteristics of the structural geological features present at the site including folds, faults, bedding planes, joints, etc.;
— The nature, extent, distribution and history of exploitation of mineral deposits and other natural resources within the area (as well as any indicators of potential future exploitation);
— An assessment and description of the spatial heterogeneity of the geological units present at the site;
— The geological evolution of the area including the genesis, relative age and nature of fracture filling materials and studies of displacements and movements along discontinuity surfaces;
— The nature of volcanic activity at the site, from a knowledge of Quaternary and earlier volcanism;

— The nature of tectonic activity of the site from knowledge of geologically recent fault movements and the aid of seismological monitoring;
— The nature of geodynamic processes at a site (e.g. erosion, uplift and subsidence).

3.6.2. Hydrogeological work programme

Objective: To provide data and information on the hydrogeology of a site, including relevant features, events and processes (FEPs) at a regional scale and on a local scale. It is essential data and information be sufficient to delimit and define properties and boundary conditions for groundwater flow models and to achieve an appropriate hydrogeological understanding (i.e. one that justifies the hydrogeological descriptions for use in process modelling and safety assessments) to inform repository design and EIA, to establish baseline conditions in advance of any disturbances created by repository construction and operations, and to provide confidence in the understanding of the site.

The hydrogeological information required from site investigations comprises:

— The extent of the flow system affecting the site and the nature of its boundaries;
— The extent of the wider area judged to be needed for groundwater flow modelling;
— The surface water hydrology including the location and extent of water bodies, the temporal variation in flow and the impact of weather and climate variations (precipitation and temperature);
— The nature of the near surface hydrology in the form of recharge and discharge areas and establishment of a water balance;
— The geometry of the different hydrogeological units within the overall flow system and any relation to observable geological features;
— The hydraulic and transport properties associated with the hydrogeological units (e.g. transmissivity, hydraulic conductivity, porosity, storativity, flowing fracture frequency, dispersion lengths) to describe flow and solute migration within and between them;
— The nature of flow within and between the different hydrogeological units (e.g. contribution from matrix, role of fractures and joints, etc.) and the influence of scale-length on the flow properties;
— Patterns and trends in the flow properties within and between the hydrogeological units (e.g. changes with depth and other forms of spatial heterogeneity);
— The nature of the driving forces (e.g. gravity, density, temperature, osmosis, etc.);
— The relative contribution to overall flux provided by the different hydrogeological units;
— The observable spatial and temporal variations in groundwater pressures within the groundwater flow system;
— The nature of or the potential for any geodynamic processes acting on the groundwater flow system (e.g. seismic pumping) and of the overall evolution of the system for a period equivalent to the timescale of interest for safety assessment;
— The hydrogeological evolution of the area, including the nature and degree of mixing over time between different groundwater bodies (to be derived through palaeohydrogeological studies incorporating hydrochemical measurements).

3.6.3. Hydrochemical work programme

Objective: To provide data and information on the hydrogeochemistry of the site to (a) characterize the baseline (undisturbed) hydrogeochemical conditions at the site; (b) describe the origin and flux of the water in the rock, vertically and horizontally; and (c) obtain specific data on parameters that are of importance for safety assessment and repository design, such as pH, Eh, sulphates, colloids and chlorides, as well as isotopes for investigating residence times.

The hydrogeochemical information required from site investigations comprises:

— The presence of human influences on groundwater composition (may indicate fast pathways);

— The presence and identity of different groundwater bodies, their distribution and the extent to which mixing is occurring or has occurred (may indicate connectivity across the site);
— The compositions of the groundwater endmembers from which any identified groundwater bodies have been derived by mixing and/or reaction;
— The characterization of rock/water interactions influencing the groundwater compositions now and in the past and that may have relevance for radionuclide retardation;
— The identity of chemical and isotopic constituents in the groundwater to provide evidence of the genesis of groundwater components (e.g. meteorological, palaeomarine, deep-seated crustal origin, etc.) and past/current flow directions;
— The isotopic composition and other data that indicate the ages or 'residence times' of water and solutes in the groundwater system, from which travel times or groundwater body stability can be inferred;
— The presence of structural or stratigraphic controls on groundwater composition;
— The presence of deleterious materials or groundwater compositions that could influence the design of the engineering works;
— The chemical and microbiological conditions affecting long term performance of engineered barriers (waste form, waste packaging, buffer and backfill);
— The chemical conditions affecting the engineering of an excavation, the possibility of any chemical or natural radiological hazards, and the short term durability of introduced materials;
— The composition of groundwater at disposal depth that will re-saturate the backfill and any buffer material, and eventually influence the dissolution, speciation, and mobility of radionuclides when the primary containment (i.e. waste containers) is breached;
— The hydrogeochemical interpretation (i.e. interpretation of the chemical compositions of groundwater bodies and interactions with the rock mass) that contributes to hydrogeological understanding of present day groundwater movements including recharge and discharge;
— The hydrogeochemical data that can be used to populate, calibrate and test models of how groundwater movement and solute transport have evolved over a long period of time in the past ('palaeohydrogeological modelling');
— The compositions of groundwater bodies in the regional volume of rock around a facility (above, below, and laterally) that might influence how chemical conditions evolve in the future;
— The distributions of trace elements, of natural radionuclides and of related stable nuclides that are analogues for the geochemical behaviour of waste derived radionuclides.

3.6.4. Geotechnical work programme

Objective: To provide data and information on the geotechnical characteristics of the soils and rocks at the site that are sufficient to inform the engineering design and construction of the disposal facility and support the development of models in which geomechanically relevant processes and effects are important and may relate to coupling with other processes relevant for demonstrating safety.

The geotechnical information required from site investigations comprises:

— The vertical and lateral extent of the rock mass considered potentially suitable to determine whether the selected site area or volume is large enough to accommodate the facility;
— The mechanical properties of the rocks and soils that could influence the design and construction of the facility and provide rock quality ratings such as Q, RMR, GSI;
— The characteristics and mechanical properties of discontinuities and of the rock mass at all depths of interest;
— The characteristics of the rock mass and the ways in which these could be changed as a result of construction activities (e.g. rock deformations, deterioration, etc.);
— The characteristics of the construction materials that could be used for construction purposes;

— The magnitude and orientation of in situ stress and determination of how stress/strain has changed in the past and might evolve, if this is relevant to safety (e.g. for controlling or influencing aspects of groundwater flow).

3.6.5. Biosphere work programme

Objective: To provide data and information on the biosphere characteristics of the site, needed to: (i) establish baseline biosphere conditions in advance of any disturbances created by repository construction and operations; (ii) support an understanding of potential radionuclide migration pathways and compartments in the biosphere (Fig. 11); (iii) facilitate the development of process models and safety assessment models through the use of current biosphere characteristics (as an analogue for future conditions that might exist at the site); (iv) inform repository design (surface works); (v) support EIA; and (vi) to provide confidence in the overall understanding of the site as part of the safety case and for dialogue with stakeholders.

The biosphere information required from site investigations comprises the following:

- Geographic location, land use and topography (and bathymetry if relevant);
- A description of the geomorphological features of the site, including a description of potential hazards associated with those features (e.g. flooding, tsunami, avalanche, rock fall and land slips);
- Measurements of, and seasonal variability in, local weather conditions including precipitation, wind strength and direction, solar insolation and temperature, actual and potential evaporation, etc. including a description of potential hazards associated with extreme weather conditions;
- Long term variability in climate (to reflect the timescales of concern for the inventory to be disposed of);
- The disposition and characteristics of surface water bodies and wetlands (hydrology) including the extent of surface catchments and river/stream flow rates and fluxes, seasonal fluctuations in flow, relationship to weather conditions and interactions with groundwater;
- The characteristics of the near surface geology relevant for biosphere interpretations, including locations and extent of surface outcrop and overburden, as well as a description of the Quaternary geological history of the site (reflecting the geomorphology of the site and possible land uplift or subsidence and/or eustatic sea level rise and fall);
- The near surface hydrogeology relevant for biosphere interpretations, including percolation (unsaturated zone hydraulic conductivity), matric potential and interactions with surface water and deeper groundwaters at the geosphere–biosphere interface;
- The hydrochemical composition of surface waters and groundwaters relevant for biosphere interpretations and the effects of interactions with rocks, soils and organic matter;
- A description of the soils across the site including their composition, disposition and depths and how they have likely evolved over the Quaternary period in response to climate change;
- The extent of terrestrial vegetation (species) across the site, including cultivated arable crops;
- A survey of terrestrial fauna including habitats, estimated numbers and species, as part of baseline studies and for possible inclusion as representative biota in food chain studies;
- A description of freshwater, estuarine and marine biota (fauna and flora) that might be present at the site including species surveys;
- Human habits, behaviours and characteristics at the site, possibly including anthropological studies of former site occupiers, to support definitions of future representative groups in safety assessments.

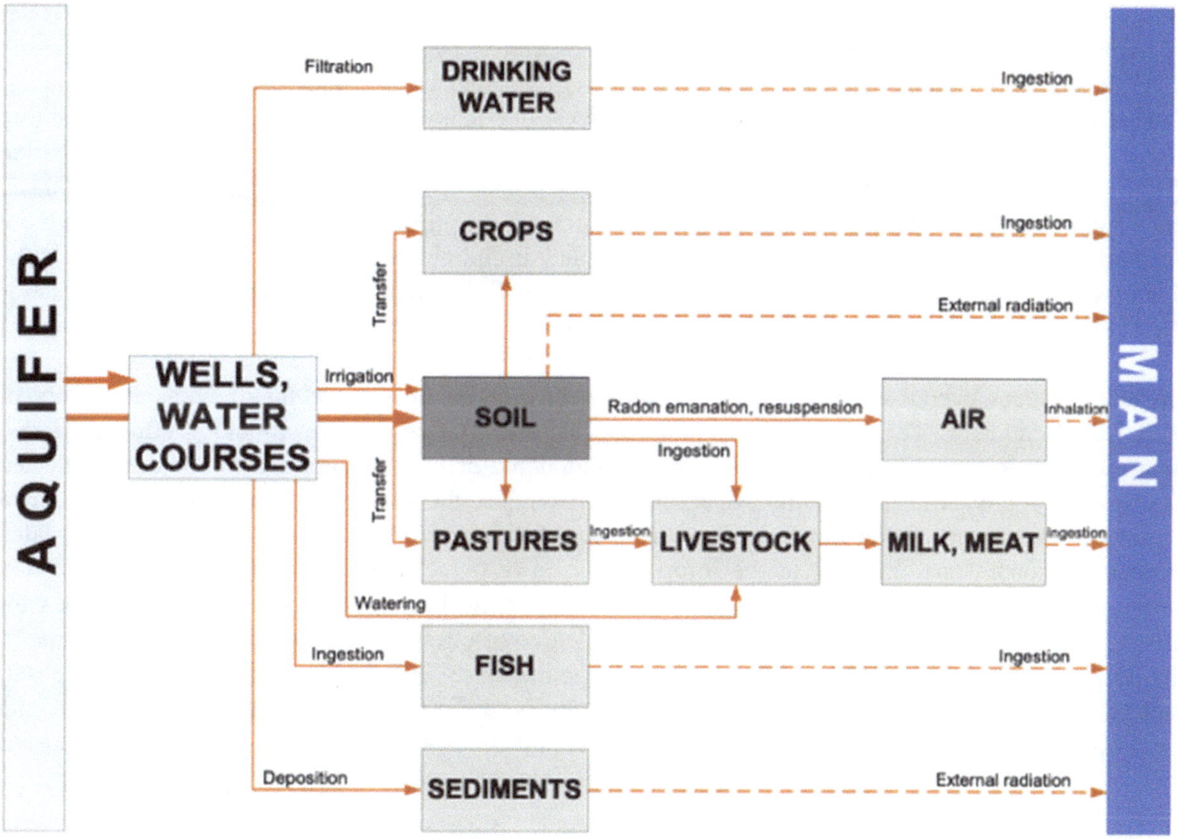

FIG. 11. A simple biosphere compartment model requiring site information to run safety assessment calculations (reproduced courtesy of ONDRAF/NIRAS).

3.7. CONCLUDING COMMENTS

Physical investigations of the natural environment at a site require the acquisition and interpretation of a large volume of data and information in response to recognized requirements. As investigations proceed, further data and information will be gathered which is likely to strengthen or challenge the understanding of the site and its evolution. Because of the nature of the iterative site investigation cycles described in Section 2.2, site description and conceptual models should be periodically updated to reflect a growing appreciation of the characteristics of a site based on further rounds of data acquisition and interpretation. It is possible that new areas of investigation will be required to assess the significance of newly recognized features and processes that could impact on safety or the design of the engineered system. This means it will not be feasible to fully define all the data and information requirements for site investigations at the outset of the project. Consequently, flexibility in operations is needed to iteratively adapt both the detailed requirements as well as further site characterization activities as the investigation proceeds.

The discipline based acquisition of site data and information, as described above, is intended to permit a suitably comprehensive 3-D description of the natural system that might exist for any disposal project for radioactive wastes. It is, however, important to recognize that the requirements of each discipline are presented above at a rather high level. Although they are sufficient to structure and inform the strategic planning needs for a site investigation project, they are not detailed enough to develop scopes of work needed to specify the parameters that would need to be collected or the natural processes to be understood for end use in safety assessment, environmental impact and repository design studies. Furthermore, they do not say anything about the levels of accuracy, precision and the representability required to describe a system adequately. In order to inform the specification of site investigation activities, it can be appreciated that the high level requirements presented in this publication would need to be augmented by deeper level requirements. Examples of a requirements hierarchy, including the more detailed requirements associated with end user needs and discipline oriented activities are provided in the first case study presented in Annex I and in NDA Refs [34–37]. It is noted that dedicated software systems exist to support requirements management, including off-the-shelf software offered by several vendors.

Lastly, it is important to remember that site investigation data and information will be an important element within an overall safety case for a repository construction licence and that the regulator is one of the ultimate end users. The specification of site information requirements, the amount of effort expended in meeting those requirements and the rigour required for demonstrating safety should be reflected in a graded approach to the site investigations. For example, the resources, objectives and associated information requirements for borehole disposal of a small volume of category 4 and 5 disused, sealed radioactive sources are different to the information and resource needs for a near surface repository designed for LLW disposal. Those are, in turn, different than the information and resource needs for investigating a site for a DGR for SNF. Hence, the degree of effort to be expended addressing some of the objectives above (developed for this publication particularly in the context of geological disposal) will vary depending on the nature of the waste inventory, the preferred disposal concept and the site conditions. Indeed, for some simple waste inventories and concepts, some of the information discussed may not be required to present a compelling and defensible safety case.

4. PLANNING SITE INVESTIGATIONS

4.1. INTRODUCTION TO THE STRATEGIC PLANNING PROCESS

The ultimate goal of site investigations is to locate a suitable site for a repository and to acquire the necessary authorizations for construction, disposal operations and closure. In order to achieve this goal, a wide range of scientific, technical, commercial, and managerial expertise will need to be applied to the planning and implementation of a site investigation project. During any site investigation project there will be a range of external considerations and restrictions that influence the internal management functions and the internal decision making process. Figure 12 illustrates some of the wide range of external considerations and the main internal management responsibilities that reside within a multidisciplinary and integrated site investigation project.

Strategic planning is an especially important management activity that reflects external considerations and influences all aspects of internal management responsibilities. It should be carried out at an early stage and with significant diligence. The strategic planning process requires systematic analysis of a particular challenge or issue to identify and plan for a defined long term goal, taking into account the range of decision options available and their associated risks and benefits as well as present and future driving forces (internal and external to the organization). The deliverable from a strategic planning exercise is an implementation plan that establishes a series of coordinated actions (i.e. the durations and sequencing of specific activities and their interactions, as well as the identification of decision points (milestones), and the teams or individuals responsible for implementation). Furthermore, a comprehensive plan should include contingency measures to

FIG. 12. Indicative external considerations that establish key requirements, inform project design and focus engagement efforts with stakeholders (outer circles) as well as indicative internal management areas necessary in a site investigation project (inner circle) (reproduced courtesy of NWS).

address situations where identified risks might materialize and set out the necessary allocation of appropriate human resources and finances required to realize the preferred plan of action. At its simplest level, this process can be illustrated as in Fig. 13 or shown as a series of steps, as in Fig. 14.

Strategic planning is carried out in all industrial, research and government sectors and there are various approaches that may be employed for carrying out a strategic planning exercise. Consequently, there are a plethora of frameworks, methods and models available to support detailed strategic planning for site investigations. The approach presented in Figs 13 and 14 is only one description of a planning process among many, but it is considered sufficiently general and reliable to be broadly applicable. Figure 15 is a simple flow chart illustrating how strategic analysis leading to an associated implementation plan then underpins the entire site investigation effort.

Requirements, objectives, a well developed strategy and a corresponding implementation plan are prerequisites to initial data acquisition, and all are to be updated and revised as necessary in response to new information and changing circumstances derived from new requirements, knowledge gaps and uncertainty

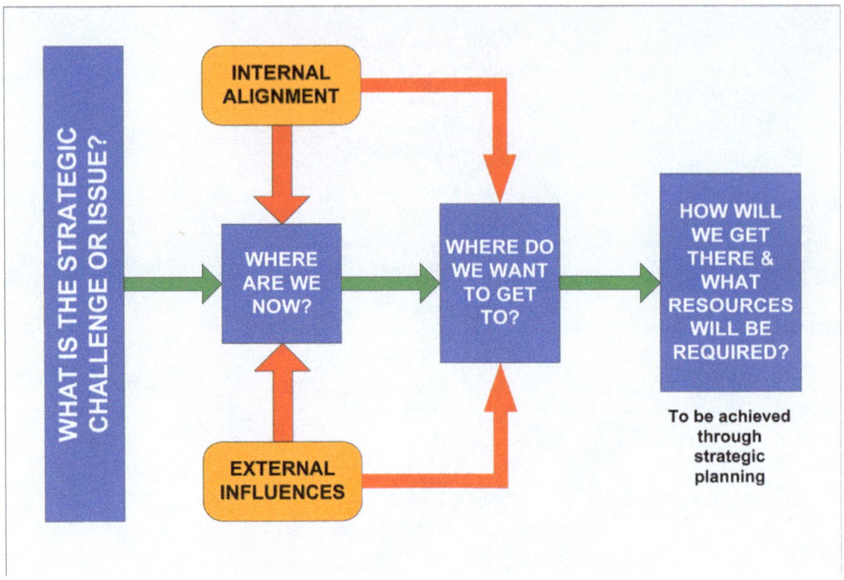

FIG. 13. A generalized approach to strategic planning.

analysis. Exit is achieved when all essential requirements have been addressed and there are no outstanding knowledge gaps or uncertainties that might lead to unacceptable dose and risk results or otherwise undermine confidence in implementation of the concept (see Section 8 for further discussion).

A strategic plan identifies actions and sets out the path for optimizing the allocation of scarce resources to safely and efficiently meet the site investigation goal and objectives in a timely manner. In order to develop such a plan, it is of paramount importance that an organization be fully aware of the current situation in which it operates. This includes the national context and an appreciation of external and internal driving forces, as well as other circumstances that do or could impact on the organization itself and the environment that it operates within. This introduces the important concept of project risk identification and management, a key activity that should be undertaken as an integral part of the management system for site investigations.

A strategically focused site investigation implementation plan should be developed and regularly updated on the basis of a careful analysis of the driving forces and influencing factors that could affect operations in the future. Thus, constantly monitoring the implementation of the plan is required and contingency measures need to be established that can be adopted should circumstances change and conditions require it. Such strategic forethought permits the RWMO to comprehensively understand its situational context, now and in the future, and supports rapid and efficient decision making concerning choices to be made from among many alternatives.

The benefits of robust strategic analysis and planning for site investigations should be self-evident, but a well constructed strategically oriented site investigation plan will provide an organization with:

- — Clarity and focus regarding strategic imperatives, goals and objectives;
- — Recognition of internal strengths and weaknesses;
- — Recognition of present and potential external influences and entities that do or could drive change;
- — A basis for deriving and evaluating the merits and weaknesses associated with strategic options;
- — The definition of activities required to achieve objectives and how they relate to each other;
- — A framework and pathway for optimizing resource use and the management of technical activities;
- — A timeline within which to structure activities (start, duration and finish);
- — A baseline against which actual progress can be measured;
- — A means to communicate with stakeholders, demonstrating how objectives will be and have been achieved in accordance with responsibilities and expectations;
- — Identification of current and future risks and opportunities and establishing a basis for countering or exploiting them.

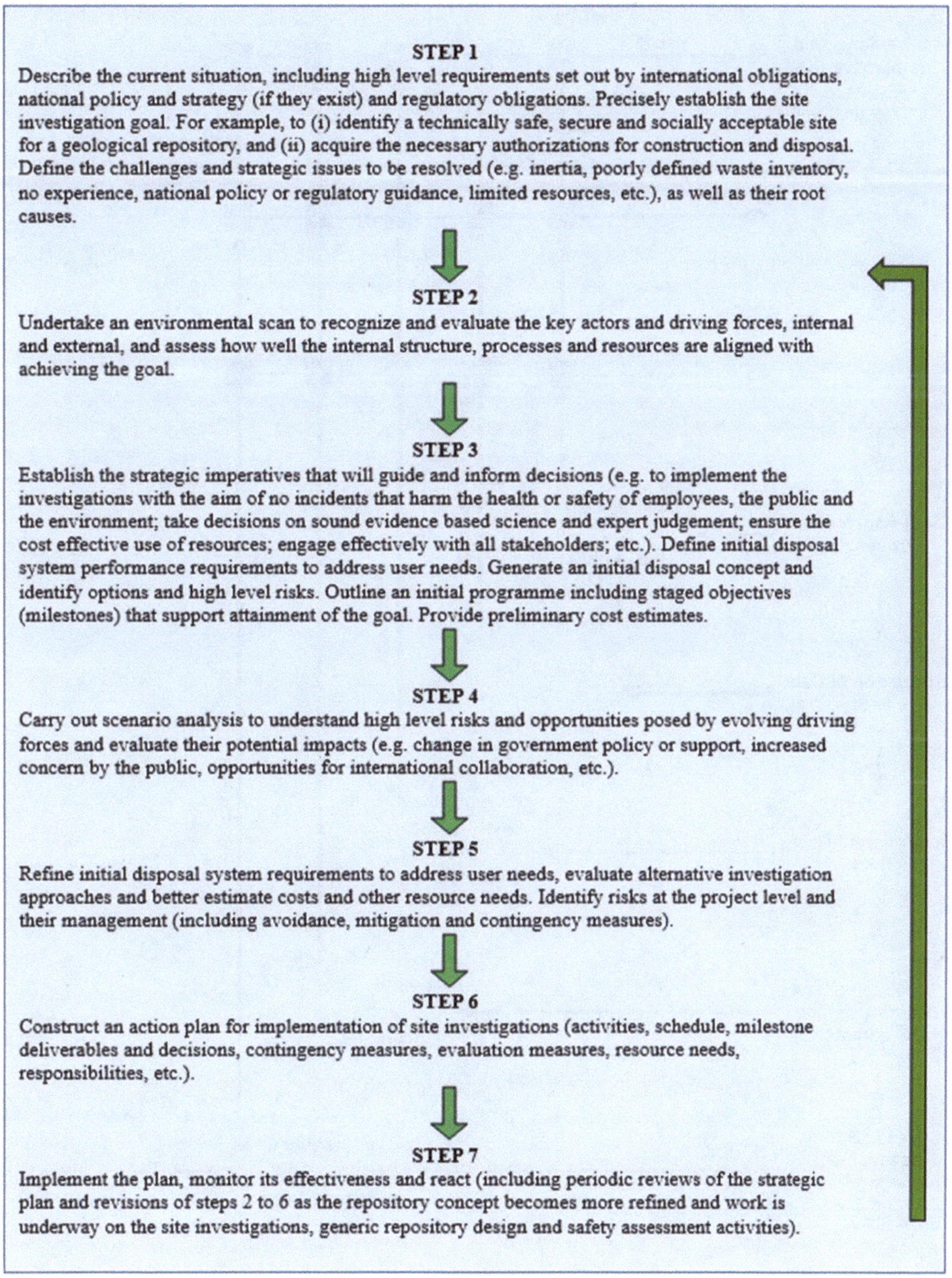

FIG. 14. A stepwise approach to the strategic planning process.

Conversely, without a strategic plan an organization responsible for site investigations is essentially committing itself to let circumstances dictate its future direction, with all the risks that entails.

As described in several places in this publication, the site investigations at a particular site will evolve as more knowledge is gained and the extent and relevance of uncertainties are recognized. It can be therefore be appreciated that strategic planning has to be dynamic, as the implementation of site investigations needs to be updated in response to feedbacks from the wider programme and changing circumstances, including external influences.

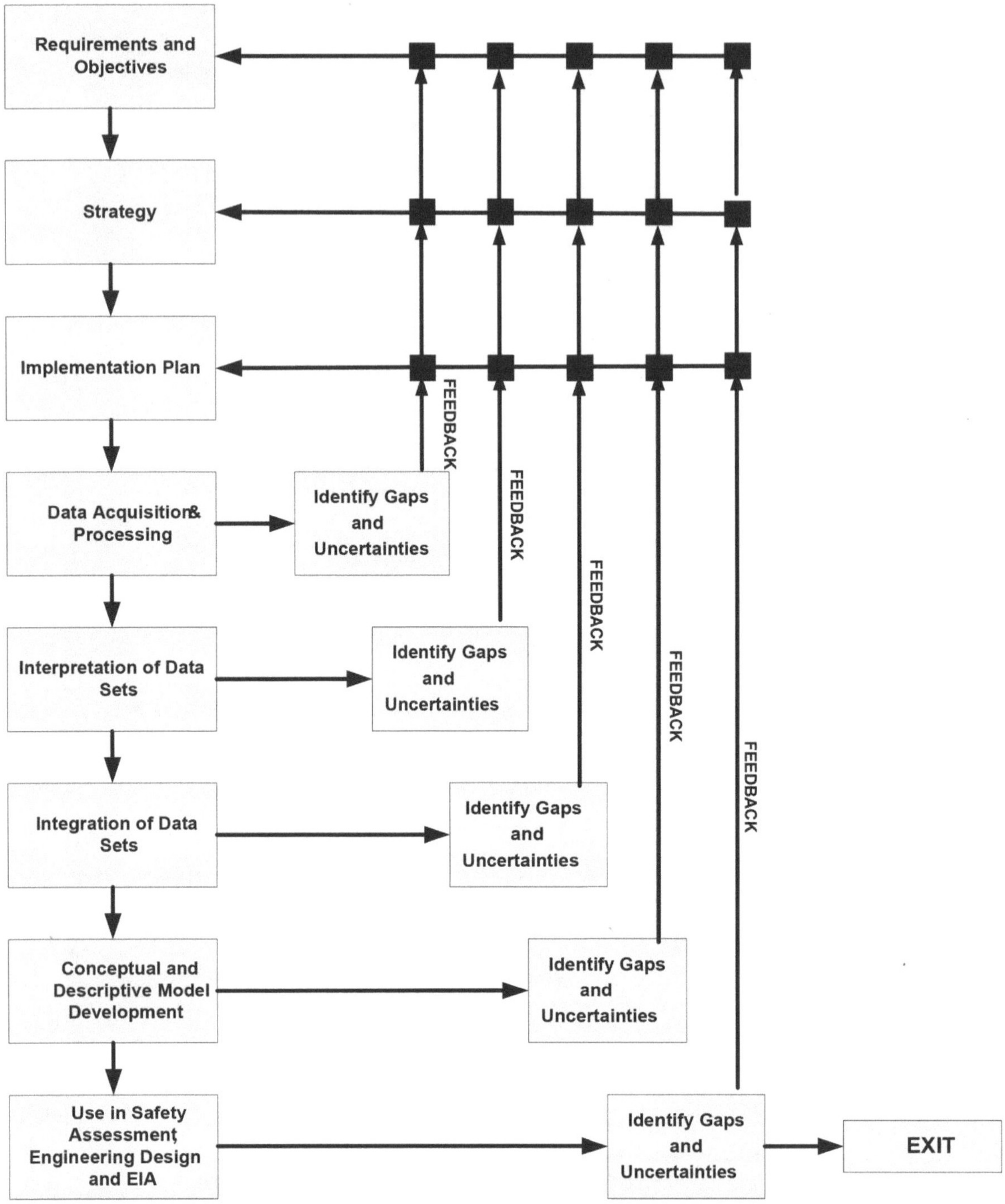

FIG. 15. Key stages and feedback loops involved in a site investigation project.

The strategic planning of site investigations necessarily involves forecasting the allocation of scarce resources in an environment where health and safety considerations are paramount for workers and the public, and where impacts on the environment are to be minimized to a level acceptable to stakeholders and the organization itself. One of the key factors providing stakeholder confidence in the RWMO carrying out site investigations (and hence ultimately confidence in the safety of the repository system during site investigations, but also during operations and for the post-closure period) is the QC and quality assurance (QA) process built into the planning and implementation approach. The definition of quality targets and the manner and degree to which they are to be met are one of four key constraints that can be varied during project planning and while site investigations are under way. These constraints collectively dictate when and how a project might be implemented. The other variable key constraints alongside quality considerations are: the range of and detail in the activities to be undertaken in the site investigations (i.e. the scope of work); the expenditures associated with the site investigation project (costs); and the duration and scheduling of the site investigation activities (time). Each of these constraints can be varied to reach a specified goal but, within an environment of limited resource availability, an increase or decrease in one area has to generally be compensated for by a corresponding change in another. For example, an increase in the scope of work, while maintaining levels of quality, would typically result in an increase in costs and/or time. Figure 16 illustrates this planning and management dilemma in terms of a 'project management pyramid' model, balancing constraints and finite resource availability.

It can be appreciated that because confidence in an organization is essential and because preoperational, operational and post-closure safety is of paramount importance, the use of robust quality controls and systematic checks on all activities have to be consistently applied. It is therefore the other factors of the project management pyramid that constrain a project (scope of work, time allocation and costs) that are to be adapted as and when needed. This requires that a comprehensive and strong management system be in place and that there is an appropriate organizational culture to ensure strategic decisions regarding priorities are understood and followed [32, 38]. Risk identification and analysis during the strategic planning process should be used to anticipate circumstances that might lead to potential conflicts between the four factors during site investigation implementation, with avoidance and mitigation measures established to determine how they should be addressed (see bullet point seven in Section 4.2.1).

As illustrated in Fig. 12, in addition to a high level of competence in strategic planning there are other key management activities that will necessarily be applied by an RWMO during the planning and implementation of a site investigation project, such as human resource management, financial and commercial management, etc. It is emphasized that the development and application of best practices in all management areas should be in place and functioning by the time of site investigation implementation.

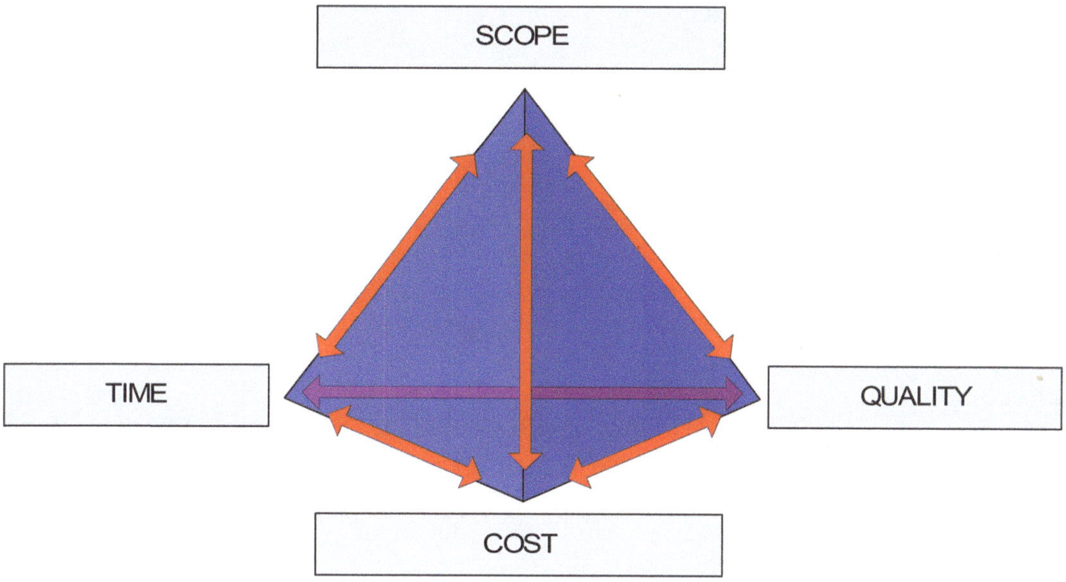

FIG. 16. The project management pyramid.

4.2. STRATEGIC PLANNING CONSIDERATIONS FOR SITE INVESTIGATIONS

4.2.1. Site investigation principles

Based on a review of best practice, it is recommended that strategic planning for a site investigation project should incorporate the following principles:

— Ensure effective lines of communication are established at an early stage with waste producers, regulators and all other external stakeholder groups to be involved in reviews, consultation and decision making. Monitor effectiveness and ensure these lines of communication are maintained.
— Strategically plan and manage the site investigations as an integrated project and ensure there are effective interfaces with other technical areas of the repository development programme, especially safety assessment, repository design, EIA and generic research.
— Ensure effective interfaces with dedicated management areas within the organization such as procurement, legal support, human resource management and public relations.
— Carry out the site investigations using suitably qualified and experienced personnel and ensure expertise is maintained by implementation of an appropriate programme of personal development and training, as well as the design and implementation of an appropriate knowledge management system.
— Identify site investigation requirements and design the investigations to gather the necessary data and information in line with a graded approach, avoiding the collection of data and information that has no value in relation to the defined objectives.
— Carry out the site investigations in a manner devised to gather the required data and information in a way that ensures the health and safety of workers and the public, minimizes environmental impact, avoids unnecessary geosphere disturbance[10] and demonstrates timeliness and value for money.
— In situations where there is a potential conflict between constraints and resource availability, the order of priority for effective project management should be established. It is suggested, that unless there are compelling reasons otherwise, the order for setting priorities should be:
  (i) A consideration of safety and quality in all planning and implementation activities;
  (ii) The minimization of environmental impact and geosphere disturbance;
  (iii) Ensuring the entire scope of work is covered, as needed for addressing end user requirements;
  (iv) Undertaking the project components in a timely manner;
  (v) Any other cost considerations.

The main methods to be used to ensure these are achieved is the rigorous application of a suitable management system contains clear policies and procedures derived from best practice and strategic analysis.

— During investigations intended to discriminate between alternate sites for their suitability to host a repository, collect easily accessible and less costly data first, target especially the data that can be used to confirm whether there are reasons to exclude a particular site from further consideration.
— Use data acquisition tools and techniques of demonstrable reliability, accuracy and precision to gather the required information.
— Ensure critical parameters are determined by more than one technique where possible and develop methodologies to evaluate the uncertainty associated with the measurement of these parameters.
— Ensure all data and information generated by the site investigation project are managed in a manner that is traceable, maintains integrity and makes them readily available to all.
— Identify project risks and establish strategies for avoiding or mitigating them.

---

[10] Site investigations should be undertaken with the intention to minimize disturbance to a site, both to ensure reduced environmental impacts and so that the geological barrier function is not unnecessarily affected by activities.

— Ensure there is sufficient delegated responsibility and flexibility in the management of the site investigation project to adjust to new requirements and circumstances as they arise.

This list is not in any order, nor is it exhaustive as national or local circumstances may dictate that additional or alternative principles be defined. The remainder of this Section discusses important management aspects to be dealt with as part of the site investigation planning process.

4.2.2. Evaluating the scale of site investigation effort

A fundamental safety principle is that the protection of people and the environment be optimized to provide the highest level of safety that can reasonably be achieved (Principle 5 [1]). This statement should be qualified by the statement that "The resources devoted to safety… and the scope and stringency of regulations and their application, have to be commensurate with the magnitude of the radiation risks and their amenability to control" [1]. This reasoning was already introduced in Section 1 and has been termed a graded approach to the management of hazards and risks. It is reiterated here, as it is the primary basis for evaluating the scale of effort required to undertake site investigations.

When considering a national inventory of radioactive waste, different categories of radioactive waste are likely to have different disposal routes for reasons of safety, logistics or economics. While the IAEA classification scheme for radioactive waste [17] does not provide quantitative waste characterization boundaries for activity and half-life limits, it does nevertheless provide guidance on suitable disposal options for the different classes of waste that are defined in the scheme. Consequently, it is essential that national choices and regulations define the basis for different classes of waste and determine appropriate disposal options for them. For example, while in many Member States LLW is generally considered suitable for near surface disposal involving an appropriately engineered repository, Germany and Switzerland have stipulated that such waste be disposed of underground in geological repositories.

Clearly, the starting point before embarking on a strategic planning process for site investigations is to be aware of existing national policy, legislation and any existing regulations and to thoroughly characterize and understand the nature of the radioactive waste that the RWMO is responsible for (now and in the future), including its origin, volume, location and the chemical and radiological behaviour of conditioned waste streams. Only on the basis of thoroughly understanding the nature of the waste can informed decisions about end point disposal options be made. Thus, accurate and comprehensive knowledge about the waste streams will naturally have a significant impact on the scale of the site investigation project and on the entire disposal facility programme.

The preferred siting strategy will also impact on the extent and type of site investigations to be adopted. If a strategy of regional investigations involving a large number of sites followed by more targeted investigations at an ever decreasing number of sites is adopted (i.e. a site screening approach), then clearly the overall labour and equipment requirements, the time necessary and the costs will be significantly more than a situation where a single or small number of potential sites have been preselected (i.e. a preferred site approach — see Section 2.3 on site selection strategies).

4.2.3. Programming and scheduling

As described in Section 2 and Fig. 5, the main phases in a site screening approach to siting, an approach that focuses on smaller areas and fewer sites can be described by four stages; these can, in turn, be used to structure high level technical activities and management actions (Table 3 provides some examples). As planning progresses, high level descriptions of technical and management activities would be progressively refined to generate more reliable cost estimates and a more detailed programme.[11] This would be one of the key outputs arising from a comprehensive planning exercise.

[11] The term *programme* is used here in the sense of a plan comprising a list of activities and events organized on the basis of time (i.e. a schedule), but also including additional elements such as responsibilities and deliverables.

TABLE 3. ILLUSTRATIVE ACTIVITIES AND KEY MANAGEMENT ACTIONS ASSOCIATED WITH SITING STAGES

Siting stage	Main activity	Key management actions
Conceptual & planning stage	Planning for investigations	Establish high level requirements; Decide on initial repository concept; Establish site data and information requirements; Set investigation objectives and responsibilities; Define site selection methodology and criteria; Define generic specifications and scopes of work; Develop costs and programme; Establish management systems.
Area survey, site investigation & detailed site characterization stages (implementation of site investigations)	Preparation for each stage	Obtain licences and authorizations; Identify suitable contractors for any commissioned work; Call for bids and evaluate offers; Choose contractors and place orders; Agree on detailed planning with contractors.
	Data acquisition & processing during each stage	Undertake desk studies and initial screening; Manage external relationships at site(s); Supervise site data acquisition; Review data during its generation and ensure the management system is functioning.
	Interpretation & integration during each stage	Manage data flow and interpretation activities; Coordinate interpretation activities; Ensure interdisciplinary collaboration and synthesis.
	Reviews & reporting during and after each stage	Manage reviews; Approve reports and make them available.

Implementation activities should proceed in a progressive manner according to a well defined schedule, recognizing that they will be carried out in a series of steps separated by three key decision points, namely:

— The selection of one or a small number of sites for detailed site characterization;
— A decision to 'go underground' (i.e. the construction of site specific underground research infrastructure to investigate the subsurface at one or more sites);
— The choice of a preferred site for hosting a repository.

Figure 17 illustrates the system adopted by SKB during its site investigations for a spent fuel repository in Sweden. It shows the various implementation steps (0, 1.1, 2.1, etc.) leading to a preliminary judgement about site suitability for constructing a geological disposal facility. In the Swedish case, an authorization to construct a repository will be provided prior to in situ underground investigations during the construction of tunnels and shafts (i.e. the underground investigations of the rock mass will be carried out as part of the initial repository development). However, a final authorization to emplace waste depends on the results of a safety case that incorporates the additional understanding generated by the underground phase of investigations. This approach, which is also currently part of the planning approach in the UK, compares with the situation where major underground in situ investigations are carried out prior to an application for repository construction, in which case it is important that the investigation facility be carefully designed and constructed so it can be incorporated into the final disposal facility (as in Finland) or can be sited adjacent to the disposal facility proper (as in France).

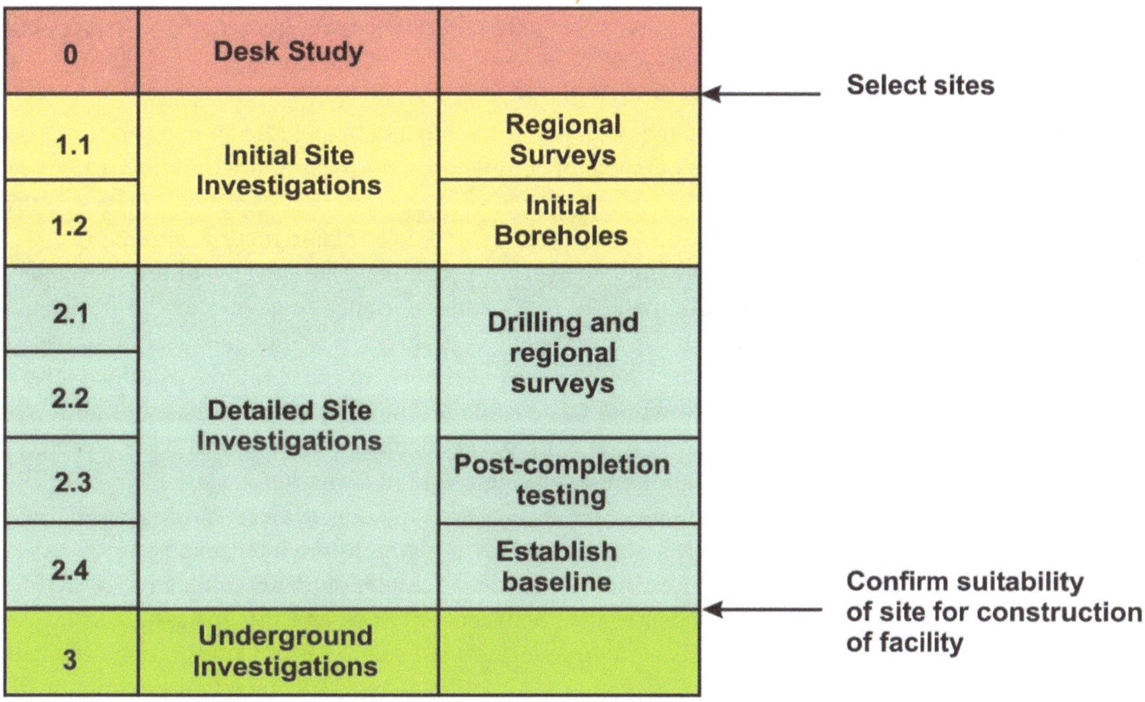

FIG. 17. Main steps of a site investigation programme (reproduced courtesy of NWS).

For a siting strategy that involves the preselection of a preferred site, rather than by screening out a large number of sites and focusing down the scale of interest from regions and large areas to one or more specific sites, there are two key decision points and even these may be combined in appropriate situations. These are:

— A decision to go underground for in situ investigation of the subsurface;
— Confirmation of the suitability of the preferred site for construction of the waste disposal facility (either at the end of the surface based investigations, in which case underground investigations would proceed as part of the initial facility development, or after all investigations have been completed, surface based and underground).

In either siting strategy, the decision as to whether investigation tunnels, shafts and an underground laboratory are constructed and operated prior to or after a repository construction licence is provided will depend on several factors which may include:

— Legislation and regulatory requirements;
— Complexity of the rock mass and degree of uncertainty and confidence in data and models;
— Need to demonstrate and test equipment and methods;
— Wider stakeholder expectations and requirements, including issues of acceptance.

During the planning phase, a decision may be made about which approach should be adopted, especially if there is a clear regulatory requirement, but the final choice may have to wait for interim results from surface based investigations.

If a decision is made to employ a site specific underground laboratory for rock mass investigations and characterization, a choice will have to be made as to whether the voids constructed for accessing the rock mass and carrying out measurements, demonstrations and experiments would eventually be used as part of the disposal facility, should a construction licence be provided. Alternatively, it may be decided to keep these separate from the disposal facility. It is likely that this type of choice will only be possible during the detailed site characterization phase.

The level of detail by which the site investigation activities are defined and scheduled will vary considerably depending on the particular stage or step of implementation and also in line with the graded approach. It is expected that those activities planned for completion in the near future (possibly 1 to 2 years) could be defined in a considerable amount of detail. However, those activities expected to be undertaken sometime further into the future (say, 3 years or more) may only be defined in more general terms.

It is important to emphasize that an implementation programme is anyway likely to change in response to various events internally within the RWMO and externally (e.g. due to the influence of government, regulators, etc.). Hence, as the site investigations progress, it will be necessary to update the implementation programme as more information becomes available. In order to do this effectively, and to support its risk management function, it would be sensible during the planning phase for an RWMO to initially define a baseline scenario incorporating anticipated conditions and, on this basis, develop the corresponding programme as well as a small number of viable variant scenarios.

Typically, an implementation programme for a site investigation project is illustrated using a Gantt chart, such as that shown in Fig. 18. This comprises a hierarchy of interlinked activities based on a work breakdown structure, each with estimated start and completion dates (indicative of duration) and with key milestones shown.[12] The dependencies between activities and the critical activities upon which others depend can also be illustrated using other techniques, such as PERT (programme evaluation review technique). However, the Gantt or PERT charts do not alone make up an actual programme, they are simply tools to support and illustrate an implementation plan. A programme and plan should at a minimum also specify objectives, responsibilities, resource requirements, deliverables and outcomes.

It is expected that for, practical purposes and economic reasons, there will be a continuous programme of investigations in which the steps at a site would proceed without significant breaks. However, the regulator or other stakeholders may require a series of interim decision points to evaluate the project status as the focus of the investigations will change from step to step.

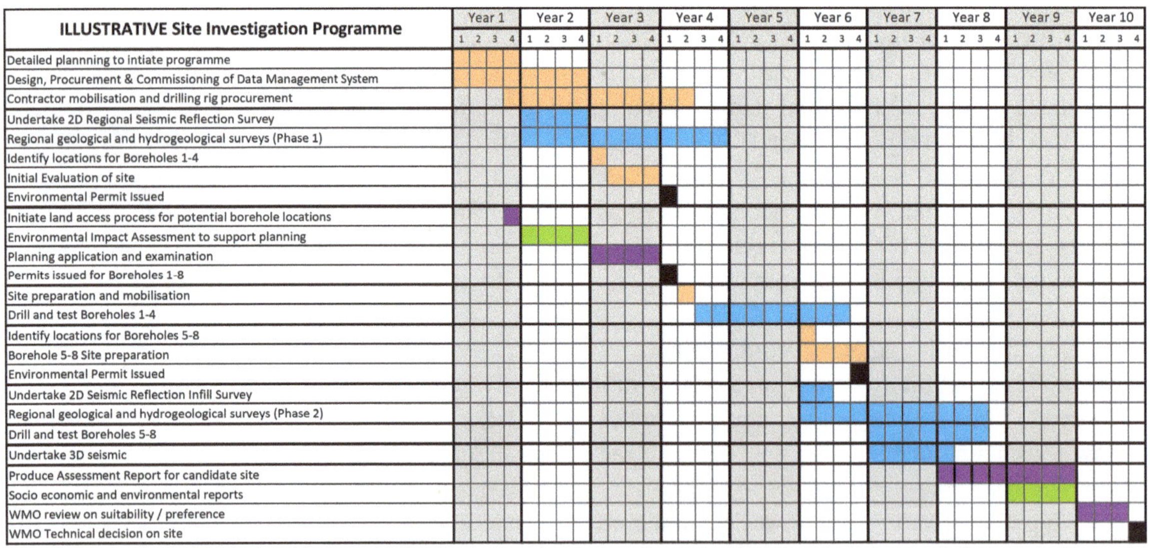

FIG. 18. The duration bars of processes displayed in a Gantt chart for a site investigation project. Process descriptions fill the left column and dates head the other columns.

[12] A *milestone* is an event (i.e. it has zero duration in a programme) which may represent completion of a task, attainment of a key objective, or a key decision point.

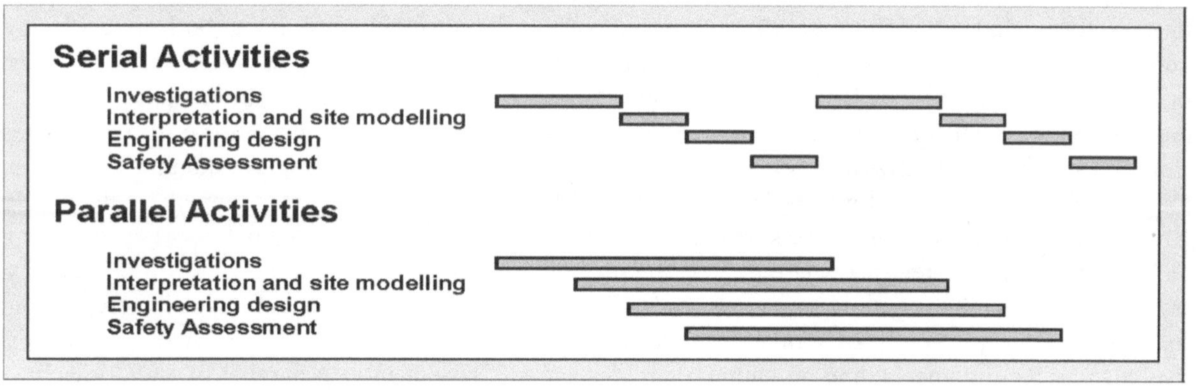

FIG. 19. Illustration of the timing of parallel vs serial site investigation activities displayed in a Gantt chart (reproduced courtesy of NWS).

The site investigation activities undertaken at each step shown in Fig. 17 involve the acquisition of data from various investigation activities and interpretation of the data. The resultant data and information would then be provided to the end users involved in repository engineering design and safety and environmental assessments. In order to optimize the time available and the use of resources, these activities should be logically interrelated and carried out in parallel as much as is possible, rather than in a serial or consecutive manner (Fig. 19).

At discrete, predetermined times during the investigations there should be a series of planned 'data freezes' when a package of data and information is defined by its availability at a specific date, frozen in the data management system (DMS) and handed on to end users so that a new round of engineering design and safety assessment updates can commence. During the time these updates are under way, further site investigations and generic research proceed, and more data are collected so that at the following data freeze captures any new or updated data and information that can, again, be given to the design and safety assessment teams so that they can maintain a rolling programme of work.

In order for there to be adequate opportunities for feedback into the design of the ongoing investigations from the interpretation efforts and the repository design and assessment studies, it is essential that the rate at which data are acquired is not so fast that excessive amounts of new data and information are received before the previous freeze has been interpreted, assessed and documented. Irrespective of the actual scheduling of activities, it is inevitable that the data streams and interpretation streams will be somewhat out of phase and that some inefficiencies will occur in terms of the data to be gathered and their processing and interpretation. However, experience shows these inefficiencies are less than those that would occur due to the loss of momentum and the dissipation of resources associated with running a serial programme of activities.

An essential element of this parallel approach to programming is ensuring that the interpretation, repository design and assessment personnel are able to send feedback to the site investigation team concerning the additional information and understanding that they require. Timely feedback is needed so that investigations can be adjusted to respond effectively to the developing understanding of the ground conditions at the site (see Section 2.2 and Fig. 3). It is, therefore, essential that wider disposal programme planning ensures those undertaking the various technical activities are integrated into a coherent project team that shares information and experiences, and that this team coordinates effectively.

4.2.4. Authorizations and permissions

Here, it is simply emphasized that the relevant authorities and other key stakeholders should be consulted early in the siting process to ensure requirements concerning authorizations in the form of licences and permissions are recognized and so that applications can be prepared appropriately and in a timely manner. Although some Member States may have stringent regulatory requirements in relation to repository siting, generally speaking, site investigations do not require special authorizations from a nuclear or radiological licensing authority, as typically there would not be any environmental release of radionuclides. Nevertheless, it would be sensible to ensure a continuous information exchange with a nuclear facility regulator during the site investigations. Planning permission may, however, be required from a local government authority to drill boreholes or to site a meteorological monitoring station, for example (see Fig. 12). Special licences may also be required from an environmental authority in advance of hydrological stream measurements or disturbing groundwater during hydrogeological testing and possibly for other reasons where there is the potential to affect the environment. It will be essential that the RWMO and its contractors comply with national and local legislation and regulations regarding, for example, the protection of groundwater resources and the wider environment. Other permissions may also be required depending on national circumstances. Consequently, a number of authorities may need to be engaged.

In order to achieve a successful investigation project, it will be essential to develop and maintain a good working relationship with regulators and planning authorities, as well as any other national or local government bodies from whom authorizations and permissions will be required. Good relationships should of course also be maintained with the communities affected by the investigations. This type of interaction should be prepared for during the planning phase through the production of a strategic engagement plan [10, 39][13]. Whether community engagement also includes a role for local residents in decision making is a strategic decision and possibly a legal or regulatory one. In any event, the views of affected communities regarding concerns and expectations may provide opportunities to strengthen the implementation of the site investigations but, if ignored, could have seriously adverse consequences.

Specifically in relation to implementer–regulator interactions, it will be important to ensure that the professional separation of responsibilities and decisions are maintained at all times, and any actions that might lead to the conflict of interests, or any perception of a conflict, are avoided.

4.2.5. Quality control and assurance

A management system is the complete set of interacting policies, procedures, processes, structures and resources used to initiate and direct organizational activities including designs, procurement, documentation, inspections, tests, model validation and verification, equipment calibration, audits and continuous improvement (more fully described in Refs [32, 38]). As part of a repository development programme, it is essential that a site investigation project be managed, performed, assessed and controlled through the application of a suitable management system to provide confidence that safety, technical, environmental, quality and economic requirements and objectives are being met. Consequently, an appropriate management system should be agreed and in place in advance of the site investigations, and it should be continuously assessed and optimized as the project progresses to ensure it is robust and remains appropriate.

The range of management system components to be developed specifically in relation to safety concerns can be appreciated by reference to IAEA safety requirements and the corresponding guide [38, 40]. A management system contributes to ensuring the existence and continued enhancement of a pragmatic and strong safety culture and provides for the use of good practices during all activities relevant to disposal, including activities specific to site investigations. Demonstrating the design and application of a robust management system would form an important component of a safety case and provide confidence in the RWMO and its results. This is an area that would be closely scrutinized by regulatory authorities as part of a licensing process.

A key aspect of a management system is the provision and application of QC measures (the system for controlling the quality of services and outputs by verifying personnel and equipment correspond to predetermined requirements) and quality assurance procedures (QA: the system used to ensure the quality

[13] The development of such a strategy is outside of the scope of the current publication, but examples and guidance on communication and stakeholder involvement are provided in Refs [10, 39].

of services and products and provide confidence that specific requirements are being fulfilled), but it is emphasized that a management system is much more than these aspects alone [2, 38].

As part of its management approach, it will be essential that an RWMO prepare a quality plan to define the procedures to be followed during each of the investigation's activities. The plan and its procedures would include references to the various national and international standards that apply to the investigations (e.g. ISO 9001:2015 or equivalent [41]). Such a quality plan would include (not exhaustive):

— Scope of the work to be undertaken within the project and covered by QA procedures including how activities will be managed and their durations;
— Relevant internal policies and procedures;
— Organizational structure, roles and responsibilities;
— Outputs (deliverables) and expected outcomes;
— QA and QC activities ensuring:
  - Specifications and methods statements include measurable quality requirements (standards and targets) using defined evaluation criteria and including statements demonstrating how conformance with QA requirements will be provided and cleared;
  - Methods for evaluating performance through inspections and frequency of reviews;
  - Procedures for identifying and managing project risks;
  - Records of qualifications and experience of RWMO personnel undertaking QA audits (and for counterpart QA officers in contracting organizations as appropriate);
  - Records of observations and notification of non-compliance;
  - Monitoring of implementation of countermeasures and their effectiveness.
— A strategy for continuous improvement.

Similarly, any contractor undertaking work on behalf of the RWMO should also produce their own quality plan and submit it to the RWMO for approval before the commencement of investigations. The RWMO should reserve the right to undertake on-site surveillance audits of the consultant's quality system during the implementation of the contract, as well as any post hoc checks.

4.2.6. Composition and structure of a site investigation team

The composition and structural organization of the RWMO team to be responsible for implementing site investigations is an important component of a management system. In general, staff numbers and composition will depend on:

— The organization's responsibilities (in terms of waste types to be dealt with, their volume and the size of programme);
— Availability of funding;
— Availability of expertise;
— Nature of the disposal concept and site conditions;
— The strategy to be adopted concerning contractor service procurement and the implementation of technical activities.

In relation to the last bullet point, a prerequisite in advance of investigations will be a senior management decision about whether:

(a) The RWMO is to undertake all or the majority of the site investigations based on an in-house capability and capacity to carry out the work, in which case a relatively large workforce of specialist employees would be required;
(b) Or a small team of in-house technical managers who would be used to procure the services of external contractors and consultants commissioned to carry out the work required.

These two options, of course, represent endmember alternatives and a decision could be made during the planning phase to use a more equitable mixture of internal and external human resources instead. The strategy for managing human resources may also evolve as the project progresses. In deciding on which approach to adopt, there are a number of considerations. These can be exemplified by firstly considering the endmember options in terms of the benefits (pros) and disbenefits (cons).

Some of the benefits of the in-house implementation approach include:

— Strong control over the planning, coordination and review of all site acquisition and interpretation activities;
— Continuous staff availability and the flexibility to reassign personnel as required;
— Control over employment conditions (negotiated pay and conditions such as permanent employees or the use of time limited contracts) and a detailed understanding of individual skills and expertise;
— Detailed understanding of the tools and processes used to acquire and turn primary data into information and knowledge;
— The ability to ensure all site investigation personnel operate as a single integrated team, sharing information and experiences and producing synergies;
— Development of a strong culture of responsibility and ownership and engendering of the importance of safety and environmental responsibility in all activities.

Some of the disbenefits of this approach might include the following:

— It may not be possible to easily employ all of the required specialists, either because prospective employees recognize that the site investigation project would have a finite lifetime or because of a reluctance to work in what might be perceived by some prospective employees to be a rather narrow or problematic field. Also, it may be the case that in some Member States the expertise and experience required is not available domestically, in which case foreign applicants would have to be sought.
— Budgetary constraints, especially in later years when conflicting pressures may become apparent, could jeopardize the viability of the project.
— A large and guaranteed multi-year budget for labour and equipment would be required.
— Administrative and logistical issues relating to the employment of a relatively large workforce (pensions, career development, health care, etc.).
— The employment of the vast majority of site investigation personnel would likely be terminated once a repository construction licence was provided (except in the case where underground investigations via tunnels and shafts are carried out as the first phase of construction, in which case the termination of the majority of the team would be stretched out in time up to the provision of a licence for waste emplacement). Redundancies would have to be carefully managed and would be sensitive.

Some of the benefits of the outsourcing approach would include:

— The ability to appoint a single management contractor for a turnkey contract (either a consortium or a large technical service provider who would subcontract out specialist work as needed);
— Devolved responsibility to the contractor(s) for detailed planning and undertaking the investigations, based on the specification of the data and information required, but not how to acquire them;
— The use of well-trained experts with established positions in the employ of contractors (in a case where such experts would be unlikely to consider direct employment with an RWMO, if offered);
— No RWMO responsibility for the external site investigation team in terms of employment conditions, meaning that contractors can be engaged as required to undertake specific tasks without the need to pay or administer work benefits such as sickness and holiday leave, career development, pensions, etc.;
— No RWMO requirement to purchase or maintain specialist equipment.

Some of the disbenefits of the outsourcing approach might include:

— A possible lack of control in specifying and managing the site investigations.
— Contractors may work in small teams operating in isolation from other teams, meaning the likelihood of cross-fertilization of ideas and knowledge building may be reduced without strong guidance from the RWMO.
— A reliance on external procedures and QC, including the use of suitably qualified and experienced personnel. Health and safety in particular might be an issue. Furthermore, the importance of carrying out work with due regard to post-closure safety considerations might not be well appreciated.
— The possibility of higher overall costs (although costs may actually be lower in some circumstances).
— Contractors may not be available when needed, for example if the schedule is not well planned or if opportunities arose to reschedule work and bring tasks forward or due to conflicts with the needs of other clients.
— Penalties would be likely for contract variations, which are more likely when there is a lack of specialist in-house expertise and where there are a large number of contracts. This may be likely for complex projects, given the research focus of site investigations.
— A relatively high administrative burden especially in terms of procurement contracts and audits.

Clearly the pros and cons noted above would have to be closely evaluated by each RWMO in the context of their own circumstances and any disbenefits could be largely or entirely mitigated with proper planning. In reality, most large and advanced site investigation projects have recognized labour requirements that will vary over time, and the project has a finite lifetime. These organizations have therefore tended to use a staffing model whereby there is a strong in-house capability to manage and contribute to some of the site investigation work, but with significant support from external contractors and consultants. In all cases though, there is a requirement to have an appropriate number of suitably qualified and experienced people available in-house to act as an educated customer for specialist services in order to proactively manage and implement the site investigations. Some site investigation teams, such as that built up by SKB in Sweden, used a mixture of permanent in-house employees working side-by-side for the duration of the project with an equal number of contractors who were essentially seconded. In this way, a common and strong organizational culture and safety ethos was fostered, which made information and knowledge more readily and effectively disseminated across the team.

In the following discussion, it is assumed that the RWMO has adopted a model whereby there is initially a limited number of in-house specialists comprising senior experienced personnel with experience in all of the disciplines that will be required as part of the investigations (a high level site investigation team or 'core team'). It should be anticipated that this core team would maintain knowledge and continuity across the entirety of the project as far as possible and would be responsible for oversight and coordination of all activities. They would also perform an external facing role with stakeholders. In this way, the RWMO would be able to function as an 'educated client' for the services to be provided by external contractors. In addition, other staff would be required to appropriately manage contractors.

The expertise to be covered by the core team would largely reflect geoscience needs, but also knowledge about the physics and chemistry underpinning understanding of the process and ecosystems. Given the range and complexity of the disciplines that would need to be covered in a site investigation programme, it may or may not be possible to find a single person who would have highly detailed knowledge in all areas, but it should be possible to employ someone as the overall site investigation director who would be sufficiently experienced and knowledgeable.[14] This person could be supported by other senior managers responsible for planning and implementing the investigations in line with their own experience and based around the disciplines of geology, geophysics, hydrogeology, hydrogeochemistry, rock mechanics and biosphere research. Each of these disciplines comprises a multitude of

[14] Sufficiency in this context is subjective but would require someone who can define high level needs and define specifications, manage activities appropriately and evaluate the meaning and significance of results.

subdisciplines and each could even be further subdivided on the basis of them being data acquisition or interpretation activities.

In addition to technical competencies, experience will be needed regarding conventional management functions directly relevant to the successful planning and implementation of the site investigations. These areas will include commercial management (procurement), data and information management, QC, the design and operation of other management systems, public relations and communications management, infrastructure management (to obtain, for example, local government and environmental agency planning permissions for works), project management, and health and safety management (Fig. 12). Some of these functions may be shared across the organization, especially during the planning phase, or may be dedicated solely to support the site investigations to ensure there are no issues resulting from conflicting demands and priorities in other areas of the organization.

The core team could be kept relatively small, supported by other organizational resources as necessary. The core team should ideally be established in advance of the siting process or at least built up and operating during the conceptual planning stage, so that the strategic elements of the siting programme can be decided and so that the detailed planning for the site investigation project can be established as comprehensively as possible in advance of ground surveys, drilling and sampling. The core team would also act as the coordination or steering group responsible for planning and implementation throughout the siting phase. Figure 20 provides two alternative organization charts for the top-level core team and supporting functions, to be established prior to or during the planning phase. Note that the hypothetical structure shown by chart A in Fig. 20 reflects planning responsibilities assigned on the basis of disciplines, while B reflects responsibilities relating to operational functions (data acquisition, interpretation and data management), each covering multiple disciplines.

It is assumed that as the siting programme advances (e.g. from Step 0 into Step 1 and Step 2 in Fig. 17), the in-house capability would be augmented by additional specialists, as required to coordinate an increasing workload. Those additional personnel would work especially on the specification of tasks, the daily management of contractors, reviewing the output of the acquisition activities, ensuring the application of robust interpretation methods and checking results, integrating the data and information and making it available to end users. No assumption is made as to whether the enhanced in-house team would comprise permanent RWMO employees or contractors on limited time contracts or possibly seconded from technical support organizations.

Because there may be multiple sites being investigated at the same time during the area survey and site investigation stages, several teams may be working in parallel. It is noteworthy that during the investigations, local field offices might be established at each site, primarily to have a local coordination function for the site investigation personnel, but also to ensure good relations with local communities and their representatives. It will be necessary for the core team to provide oversight to ensure investigation activities are being carried out in such a manner that fair and reasonable comparisons between sites can be made. An expanded organization chart for when site investigations are under way (Step 1.1 onward) might look like Fig. 21.

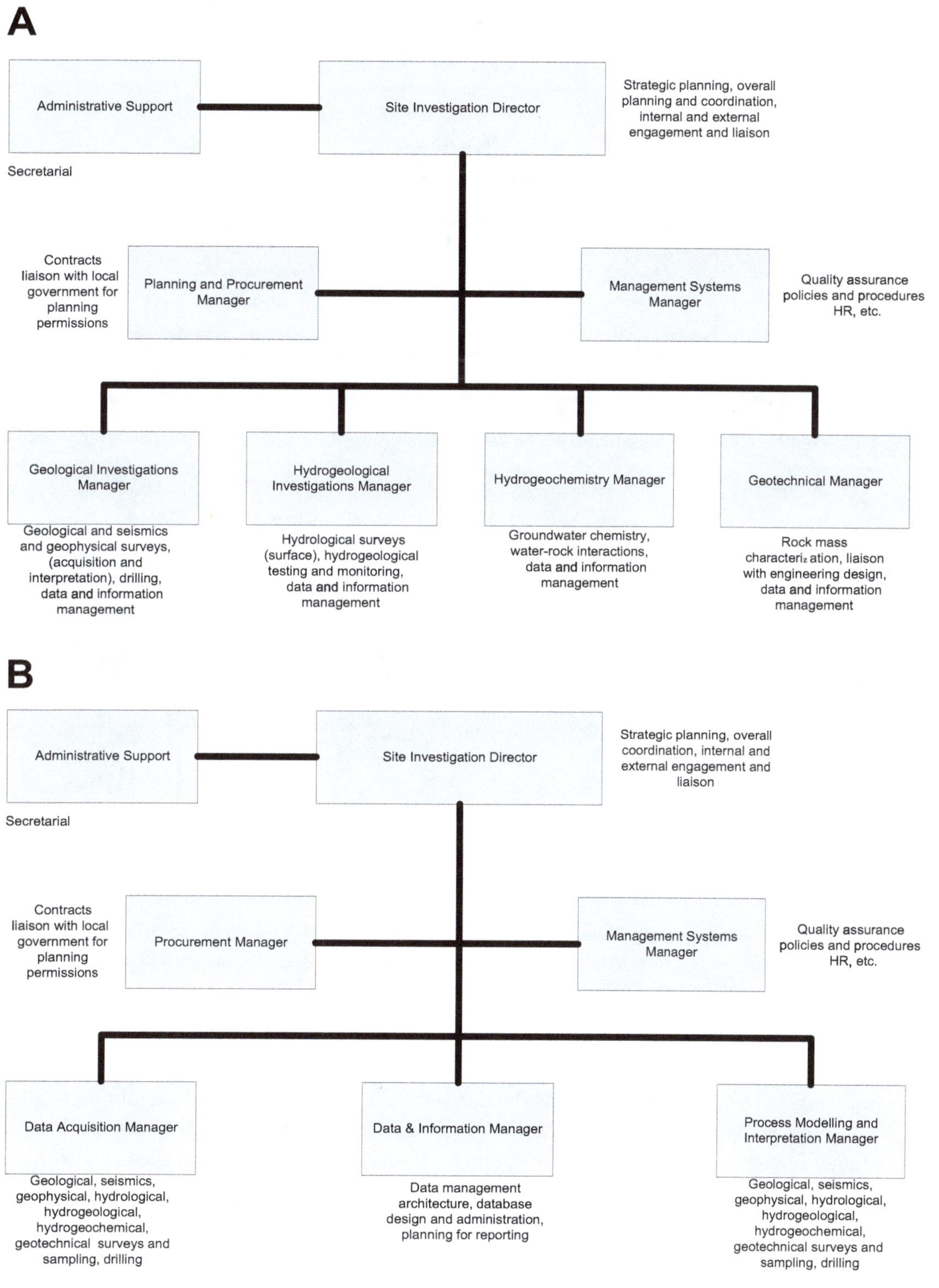

FIG. 20. Alternative organization chart illustrating the high level structure of a site investigation team prior to the conceptual planning stage (Fig. 8) and during Step 0 (Fig. 17).

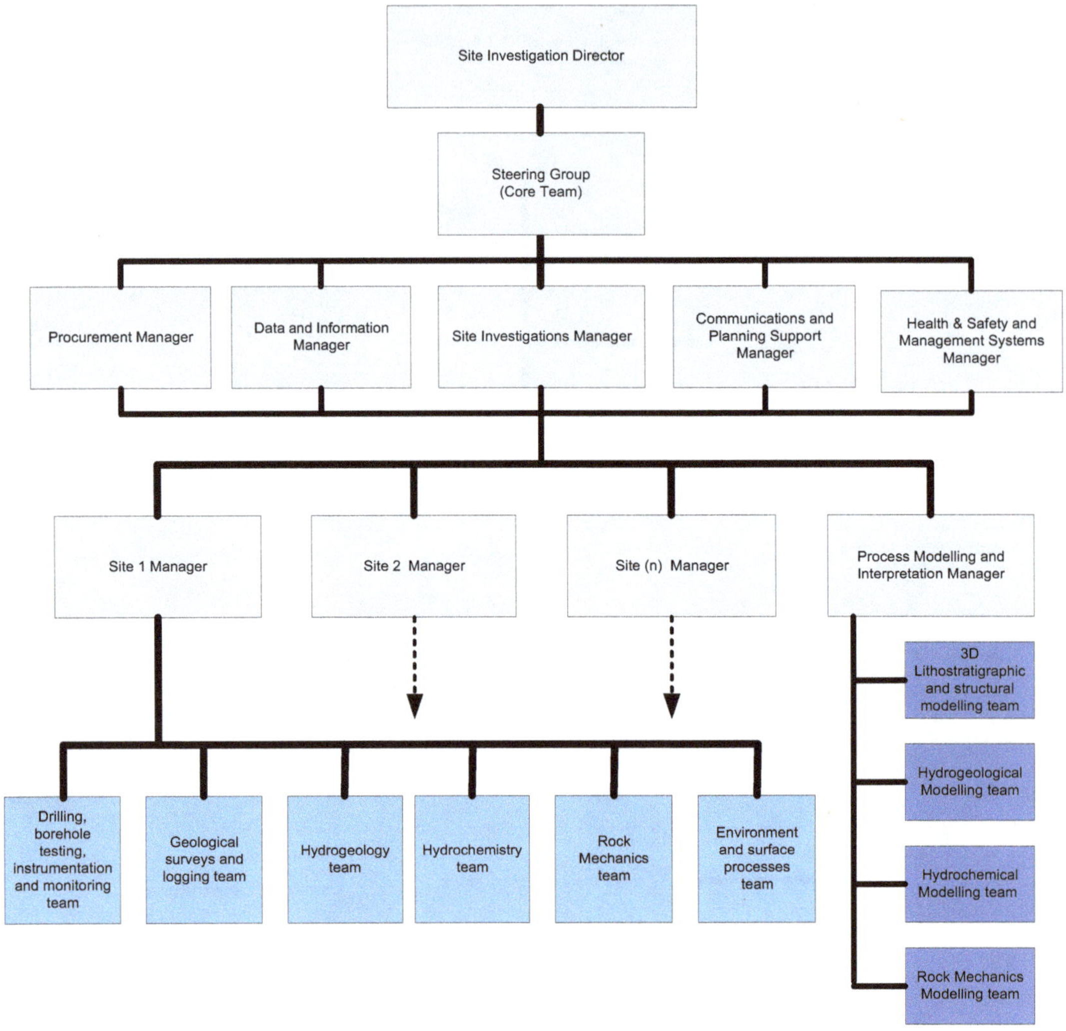

FIG. 21. Organization chart for the structure of site investigation personnel responsible for investigating several sites during Steps 1.1 to 2 (see Fig. 17). Note that the the steering group would likely comprise individual senior personnel with other responsibilities elsewhere in this chart or from elsewhere in the organization.

Depending on the waste inventory to be disposed of, the site investigations may require detailed consideration of the thermal properties of the various rock units present (such as thermal conductivity, specific heat and heat transfer coefficients) to be derived from the geological investigations. Furthermore, combined geological, hydrogeological and hydrogeochemical information would be used in radionuclide transport modelling. Responsibility for investigations in these specialist topical areas are not explicitly represented in the organization chart, but any specialist functions should be accommodated in the organizational structure. It is therefore apparent that Figs 20 and 21 are illustrative only, as are the position titles, and it is emphasized there are a great many structural variants that might be considered, including managing the site investigation team within a matrix. By the time site investigations have progressed to Step 3 (underground investigations accessed by tunnels and shafts) the composition and structure of the site investigation team will strongly reflect a multitude of organizational preferences and so no illustrative organizational chart is shown.

4.2.7. Data and information management

Data and information management is another aspect of a comprehensive management system that should be planned for prior to site investigations getting under way. As site investigations progress, an ever-increasing amount of data and processed information will be generated. Some data will reflect spatially defined one time measurements, while others will reflect time series sampling. Many of the data sets will be very extensive, reflecting high spatial or temporal resolution and/or large investigation volumes, such as those derived from seismic surveys. Data will be collected using a wide range of techniques to reflect investigating a wide range of features and processes. The accuracy, precision and the representativeness of the parameters measured will vary greatly, depending on the technique used and the feature or process being considered. Technical and descriptive metadata, providing context for underlying data, will also need to be captured.

Raw data collected in the field or in the laboratory may need to be processed to provide derived data (e.g. field derived parameter values reflecting water table fluctuation or groundwater pressure measurements are necessary for producing 'hydraulic conductivity'). Raw and derived data and information will, in turn, be analysed and their meaning assessed, including through the use of process models. Bearing in mind that a prospective site for a radioactive waste disposal facility has to be considered as an holistic system containing features that can be considered on different scales, and that it contains many important interactions and feedback processes, disparate data sets and information on individual system components will need to be combined and interpretations reassessed in a wider context. Based on an appreciation of the interactions and dependencies between processes, the holistic site understanding can then be described in terms of a total system conceptual model. The outputs from these integrated interpretations will then need to be presented in ways amenable for use by different end users and a wider audience of interested stakeholders. This will require numeric tabulations, the drafting of maps, graphs, diagrams and illustrations and the use of descriptive text, as well as the provision of contextual information such as processing and interpretation methods, assumptions and bias. Such data can be stored in spreadsheets or relational database formats and reporting can be accomplished by linking to visualization software as necessary. Computerized 2-D (surface) and 3-D (rock volume) geographical information systems (GIS) are now commonplace and can provide significant advantages in data management, such as easy access to records and reporting, consistency in scales and symbolic meaning and the use of spatially layered data.

End users (primarily the safety assessment modellers and repository design engineers) will take data and information derived at the site, as well as any elicited information and wider research data (e.g. from natural analogue studies or generally applicable information derived from elsewhere), and use them further for the parameterization or calibration of models and for developing engineering designs. These models will in turn produce further data, information and understanding that can be used to identify areas where there is a need to investigate a site further to provide the additional data required to reduce key uncertainties and improve confidence in interpretations. Ultimately the results of safety analysis and engineering design studies, as well as a wider appreciation of a site, are for use in the safety case to support regulatory authorizations and provide information for awarding construction contracts. The process of data and information acquisition and flow to end users can be illustrated as in Fig. 22. This diagram provides a simplified illustration of data and information flows through various stages of data acquisition, processing to provide derived data, analysis leading to interpretation, integration leading to conceptualization and site understanding, and eventual application by end users before synthesis and presentation in a safety case or an 'environmental safety case', depending on regulatory requirements (see Section 7 for further details).

Given the volume of data typically generated during site investigations, the long chain of acquisition and processing activities, the potential for misunderstanding (through, for example, the use of inconsistent or unexplained terminology, measurement units and context) and the multiple objectives and users involved, it can be readily appreciated that data management is of critical importance and should be planned for in advance of the actual site investigations. This requires the design and construction of a DMS comprising hardware and a structured software platform that interacts with data providers and users as well as other applications, to capture, store and present data. A DMS plan would also identify selected people and organizations authorized to enter data or modify structures and would have in place safeguards

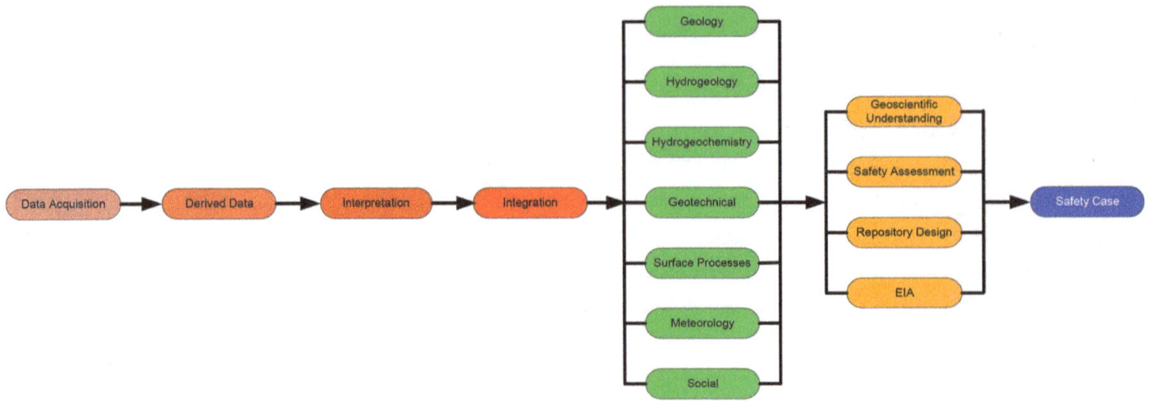

FIG. 22. Simplified site data and information flows. Note that the information domains shown in the tallest group (Geology, Hydrogeology, etc.) represent the information underpinning discipline specific descriptive models that, when amalgamated, provide a unifying descriptive model of a site.

to ensure access is provided to specified groups of users, as required. The types of data providers and users and brief descriptions of their roles are provided in the systems context diagram of Fig. 23.

The DMS should be designed as an integrated system to allow links between hardware components and remote access for multiple users who are geographically dispersed. The various components might be part of a universal custom design especially commissioned for the project or could be a series of linked off-the-shelf software and hardware for databases and presentation. If the latter is envisaged, interoperability considerations will of course be important.

The design and governance[15] of the DMS architecture should ensure:

— Coherent and consistent management through the application of scientific standards and the use of an agreed on and unambiguous terminology, and through the use of established conventions (e.g. measurement units) and QA procedures;
— Data are captured, stored and presented using off-the-shelf or custom system components that are certified or otherwise approved to meet the needs of the site investigation project;
— A clear audit trail to data sources;
— Recording of the personnel involved in entering, adapting or moving data and, if required, those accessing data;
— An environment that permits full version control (for phased data freezes) and maintenance of archives;
— Accessibility as and when required and with the ability to provide data and reports efficiently and in agreed formats;
— Maintenance of a secure and controlled environment, with multiple copies of data and information held off site in case of data corruption, hardware failure, disaster management, etc.

While a major aspect of the DMS is a database function, it is emphasized that a comprehensive data management strategy and resulting DMS is more than that alone. Advanced features that could be incorporated into the design of the DMS include the following:

— Automated data collection and transmission systems through the use of telemetry during investigations at borehole sites and at other locations (e.g. seismic and meteorological stations) so electronic data collected at the sites can be transmitted to network servers for storage;

[15] Governance is required to improve and maintain the quality of the data, to define data ownership and set accountability for the quality of the data, and to define or select standards, policies and procedures as well as monitor their implementation.

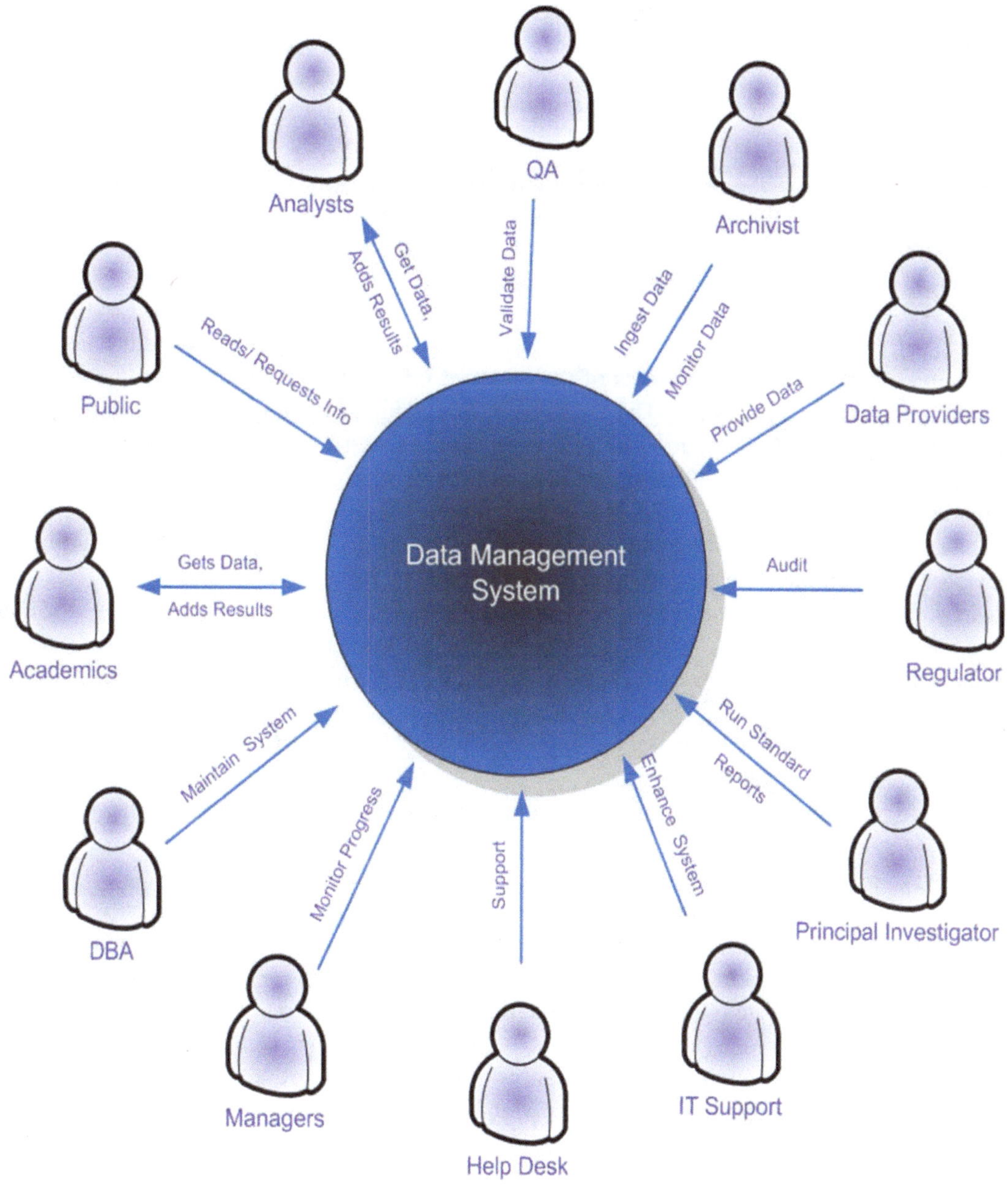

FIG. 23. An example of a system context diagram for a DMS. DBA = database administrator; QA = quality assurance and management systems operator (reproduced courtesy of RWM).

— Storage of data in suitable formats on secure network servers so users can access data as and when required simply by logging onto the network;

— Links to GIS and other software systems to facilitate the interrogation of the database to obtain information related to specific requirements, for example, to identify information that is available within a specified distance in any direction from a given point in space or to locate information related to a particular geoscientific unit;

— An ability to interface with a range of presentation and visualization tools to allow information to be presented in different formats to facilitate understanding and communication.

DMS planning should include consideration of resource requirements and development lead times. The planning and the DMS implementation should be undertaken within a strict regime of QC and according to approved procedures and standards, many of which exist internationally, for example, ISO 15836 (The Dublin Core Metadata standard), ISO 19115:2003 (Metadata), and ISO 19119 (Open Geospatial Consortium Services Architecture).

Even over the lifetime of a siting programme, one that might be measured in terms of a decade or more, hardware and software will change to take advantage of increased performance, functionality and to improve the user's experience. In this regard, a DMS will need to evolve and it should be designed as a modular system to take advantage of hardware and software upgrades, ensuring previous versions of the database can remain accessible and intact without corruption or loss of data.

Proper regard should also be taken to ensure robust digital preservation and archiving (see also Section 4.2.8) for timescales that might extend beyond the lifetime of a repository development programme. It is good practice to ensure the various databases making up the site investigation DMS are backed up regularly and that duplicates of the data and information are provided to an archive away from the where the central servers are. The archived material should employ media and formats that aim to ensure preservation and long term accessibility. While the need for a policy on archiving data may seem a long way off for a nascent RWMO, it is an essential component of a comprehensive and robust management system.

An example of a data management approach and system, as developed by Andra as part of the French disposal programme for a geological repository, is found in Ref. [42].

Clearly, the planning and design of a comprehensive DMS for site investigations is a complex endeavour and specialist guidance should be sought for further information and support. However, based on experience and the foregoing discussion the, following general guidance can be offered:

— Typically it is planned in advance of the data acquisition activities and is to be made ready to accept data as soon as investigations commence;
— One should not combine a technical DMS with a project management system or other administrative database system (i.e. to produce a 'universal' organization-wide DMS) but rather the DMS should be dedicated to manage relevant technical site data and information only;
— Digital data inputs, editing, transfers and access need to be secure within the DMS;
— Verified data should be entered into the DMS following agreed QA procedures and as soon as is possible, along with supporting contextual information (metadata);
— It needs to be accessible to multiple users with different skills, different needs, and different security clearance levels;
— It should have user friendly interfaces that allow options for data presentation;
— Version control is important, as is automatic logging of user activity and data traceability;
— No actions affecting data should be anonymous;
— It needs to handle time series and geographically located data, as well as other data and information sets that may not have a spatial context;
— As much as possible, use monitoring equipment should allow electronic recording and the automatic transfer of data to a DMS, to save effort and minimize the potential for human errors;
— An agreed taxonomy and vocabulary are imperative (consistent terms and meanings, measurement units, etc.) and are applied across disciplines;
— Open digital formats should be used as much as is possible to ensure data remain readily readable and transferable;
— Digital preservation needs to be considered at an early stage and appropriate decisions and actions need to be taken (e.g. records management policy, data recording formats, storage media and use of multiple off-site locations for storage and archiving);
— An approved process will be needed to capture the source of any external information and for recording it in a DMS (e.g. screenshot and attribution) as an RWMO is unlikely to have control over the quality of this, and recording generic data from web sites for integration into a DMS can be problematic as records may change over time or become inaccessible;
— Controls should be in place to ensure there is no editing of archived data and information;

— A risk based approach should be taken to data management and an RWMO has to be constantly vigilant to anticipate issues that might arise so they can be avoided or mitigated.

It is noteworthy that data, information and knowledge are three different but related aspects that support understanding. Knowledge management is outside of the scope of this report but will be, in many respects, as important as data and information management in achieving a successful disposal programme.

Site investigation data flow

As described above and illustrated by Fig. 22, there will be a voluminous flow of site investigation data and information arising from the acquisition of raw data through to its end use in a safety case. Planning for this site investigation data flow is a critical part of a DMS strategy. A particularly useful tool has been developed to not only plan, but also to manage and record key data sets and data flows: a 'site investigation flow diagram' (SIFD). This tool was initially developed by Nagra (Switzerland) and was further refined and expanded by Japan Atomic Energy Agency and NUMO (Japan), Nirex (UK) and others for application in various geological settings. Notably, NUMO (Japan) makes extensive use of SIFDs during site investigations. An SIFD uses a network framework where individual data and derived information can be traced through the various steps of processing, data analysis and interpretation, integration, use in subsystem process models and ultimately an end use. As data can be clearly traced, an SIFD can be used for planning site investigation activities, as a project management tool during implementation, for presentational purposes and as a record of the use made of data in any post hoc auditing.

Different organizations have developed slightly different approaches to the design of an SIFD although the framework typically follows the stages presented in Fig. 22. Generally, the flow of data and information proceeds from initial data acquisition (in relation to a particular acquisition tool or technique) and any necessary processing to generate derived data. Data and information then flow through to interpretation, integration and the development of site descriptive models (SDMs) and finally to the various uses to be made of the data and information. Each stage in this chain of flow typically includes a large number of individual activities that are structured as a column within the SIFD. Each activity at each stage has a precise role and is essentially a work programme element within a site investigation project. Each activity produces discrete outputs comprising data and information, including contextual information. In the SIFD, all data and information have an origin and an end use and they are connected by flows between activities at each of the intermediate stages. The flow of these 'packages' of data and information are shown by links to precursor or successor activities where appropriate. In many cases there are multiple links between activities. The links demonstrate the derivation of supporting data and information (when considered looking backward) and how the data and information will be used in the next stage (when looking forward). An example of an applied SIFD is shown in Fig. 24 (geological data flow in Switzerland).

The construction of a comprehensive SIFD can be rather labour intensive and takes time. In addition, the number and types of activities, as well as links between them, will likely change as the project progresses, to take account of feedback from end users and new information. Hence it can readily be foreseen that there will be discrete SIFDs developed for each of the steps in the investigation project (for example, as shown in Fig. 17). Therefore, an SIFD should be specified for a particular stage or moment in time. To support the construction of an SIFD and its upkeep, significant benefit can be taken from using software application.

An SIFD can be developed for discipline specific areas (e.g. considering geotechnical aspects of site investigations alone) or activities can be aggregated to provide a full description of the data flow associated with a complete site investigation project. In addition, an SIFD can be produced or presented in different levels of detail to satisfy a range of objectives and audiences. An SIFD can be constructed largely in a top-down manner, by end users specifying data requirements (in line with the principle requirements driven theme of this publication), but it can also be generated in a bottom-up manner, by using experts with a detailed understanding of the applicability of a range of tools and techniques for data

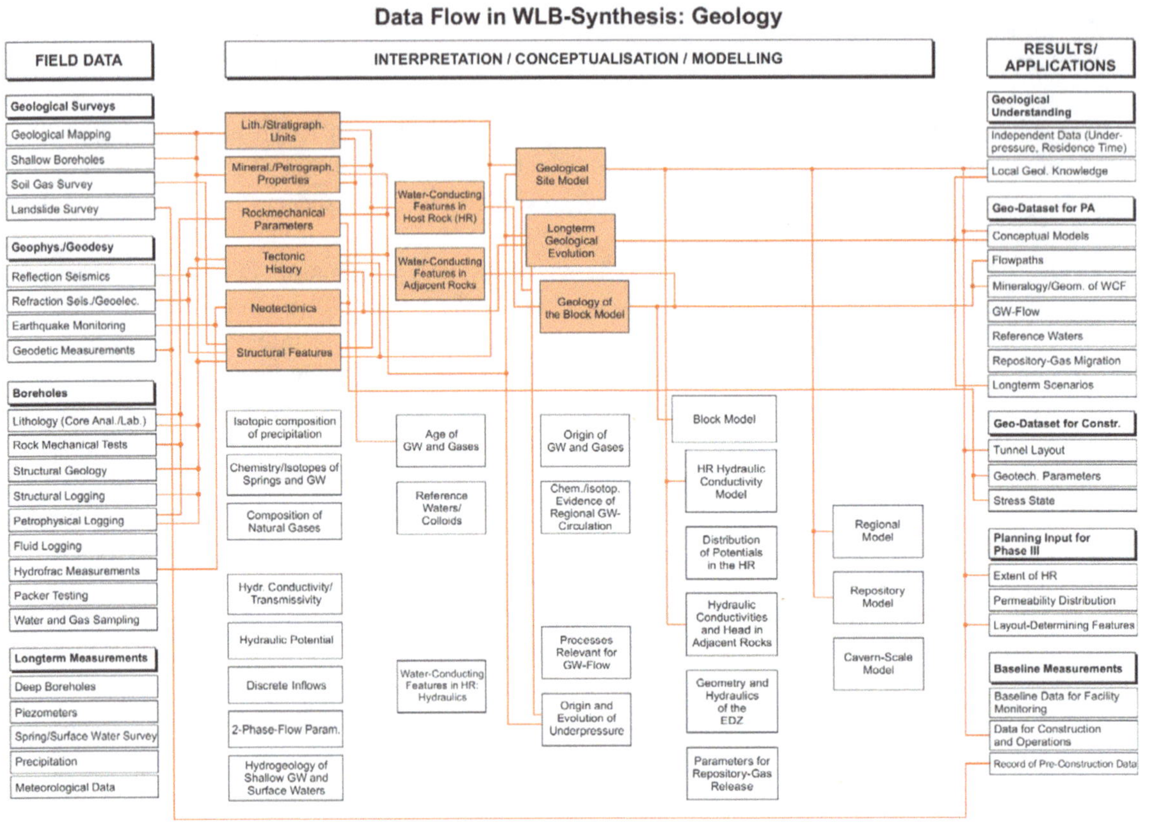

FIG. 24. SIFD for geology showing flow from left to right, Wellenberg (WLB) investigations, Switzerland [43] (reproduced courtesy of Nagra).

acquisition and interpretation. The best approach in practice has been found to be a mixture of top-down and bottom-up approaches, ensuring effective communication between data users and data providers.

If the SIFD approach is to be adopted to support efficient data flow and management, it should be planned in advance of the site investigations to fulfil its planning and project management potential. However, it should be remembered that once produced, an SIFD should not be used dogmatically as an inflexible framework to justify activities. Instead, the SIFD should be adjusted to reflect anticipated and actual changes in data flow.

In summary, an SIFD can be used to support a site investigation project at different times and for different purposes, as follows:

— *Planning*: Construction of an SIFD provides clear and logical links between end user data needs (what information will be required, why and when) and how those needs will be addressed through identified data acquisition tools and methods, data processing, interpretation and integration activities. An SIFD therefore supports programme development and provides ready justification for the site investigation activities to be commissioned. Constructing an SIFD can also be used to help establish resource allocations through the identification of specific tools, techniques and work activities.
— *During investigations*: An SIFD can be used as a project management support tool to reflect changes in needs and priorities over time and to ensure data flows are functioning as planned. It also facilitates communication and understanding between the site investigation team and the safety assessment, repository design and EIA personnel.
— *Post-investigation*: An SIFD supports auditing and review of the project by providing a clear demonstration of how the acquired data have been collected and used, and which packages of

acquired data contribute to each element of end use in safety assessment, engineering design and construction, EIA, general scientific understanding and ultimately for a safety case.

— *Presentation*: An SIFD can be simplified and presented at a high level for general audiences or in a high degree of detail, as required for technical audiences. It can also readily demonstrate changes over time in data and information generation and flow. An SIFD is therefore potentially very powerful for supporting a safety case and for wider communication efforts.

4.2.8. Reporting and archiving

Over the course of a repository siting programme, with a duration that may well take several years or more, there will be a great many reports produced covering the planning of the investigations, the specification of requirements, objectives and activities, the methods used for acquiring and interpreting data, recording the data collected and reporting the results of interpretations, as well as summaries. According to the siting strategy, site investigation reports may need to be produced for a number of different sites. A formal organizational reporting procedure and structure should be prepared, and a detailed schedule of the major reports anticipated should be produced during the planning phase as part of the design of the project. Table 4 presents a scheme that might be adopted for the structured production of reports arising from site investigations.

Although certain reports are likely to be 'one time only' reports, in that they satisfactorily define and achieve a set of objectives, many other reports would be periodically updated, such as the site investigation master plan document, regular compilations of requirements, results from monitoring programmes and an annual summary report describing project progress and the current status of site understanding.

All of the reports shown in Table 4 would be 'owned' by the RWMO, in other words, responsibility for their production and content would reside entirely with the commissioning organization, even if they are actually produced by contractors (most factual and interpretation reports would typically be produced by the contractors charged with undertaking the work). An early decision should be made about whether contractors should use formats defined by the RWMO in the reports they are responsible for, or whether they are free to use their own reporting formats. At a minimum, contractors should be made aware of the need to ensure consistency across the project, for example in the use of naming conventions (such as for 'bedding dip and strike'), terminology and measurement units, etc. The RWMO should be responsible for developing the information and documents necessary to ensure such consistency (e.g. the codes for use when contractors log core). Similarly, it is up to the RWMO to decide or agree on the content expected of a contractor's report. One way to agree on detailed report content is for the RWMO to draft an outline of the report (e.g. table of contents) in collaboration with the contractor on the basis of the specifications and a contractor proposal. The size and complexity of the ensuing report will of course depend on the scale of the survey or the complexity of the topic.

For data acquisition reports, the overall content should be specified to at least include the following:

— An explanation of what information is to be obtained and how the information will be used (i.e. justification for why the information is required — responding to requirements provided by the RWMO end users);

— QC and QA measures;

— A description of the data acquisition methods and activities to be undertaken to provide the required information, including the reasons why particular methods were selected;

— A summary of the programme for acquiring and reporting the information and an assessment of how well it was met;

— Data reporting (i.e. the data and information that were collected and processed, along with relevant contextual information);

— An assessment of the environmental impact of the investigations and the measures taken to minimize any adverse environmental impacts;

— Conclusions and an assessment of lessons learned.

To the maximum extent practicable, all completed and approved reports should be prepared as hard copy (in limited numbers) and in electronic format to facilitate broad distribution, as required. As a general principle, an RWMO should seek to make data and information in reports as widely available as possible as quickly as possible without compromising on quality. However, it is recognized that national or local circumstances may place constraints on these intentions due to possible legal or regulatory obligations and also possible commercial considerations. With regards to commercial considerations, it is noted that the ownership of intellectual property rights should be established at the outset. They would normally reside with the RWMO as the commissioning organization, but in the interests of openness, transparency and confidence building, contractors and collaborators could be encouraged to use and publish data and information in peer reviewed journals, subject to approval.

TABLE 4. ILLUSTRATION OF A TYPICAL REPORTING STRUCTURE (TECHNICAL REPORTS)

Type of report		Examples & content	Intended audience
Planning and design reports		Project Plan (project definition, costs and programme). Other reports to specify objectives and requirements, describe the design of equipment, data systems and major surveys, laboratory testing requirements and procedures, programmes of testing and calibration, requirements and design of monitoring networks, etc.	Internal use, contractors, regulators
Factual reports	Data acquisition methods	Accounts of the work undertaken (data acquisition and processing methods, a log of activities, types of measurements made, who did what and when, site conditions, factors that might affect data quality, etc.).	Internal use, regulators, academics, other interested stakeholders
	Data reports	Documenting the raw data obtained (time series, sampling, mapping, one-off measurements and observations etc.) parameter uncertainty evaluations and any processing (reasons, method and results). Would include compilation of data used for establishing site baseline conditions.	Internal use, regulators, academics, other interested stakeholders
Interpretation reports	Topical reports	Typically, single discipline interpretation reports (e.g. geology, hydrogeology, etc.) or reports oriented towards narrow topics. Contents would include objectives, data sources, analytical methods, assumptions, possible bias, conceptual model development, process models and their description, results, sensitivity evaluations and alternative interpretations, and identification of areas of conceptual and modelling uncertainty.	Internal use, regulators, international organizations, academics, other interested stakeholders
Integrated accounts		Multidisciplinary integrated understanding of site wide and holistic system conceptual and descriptive models.	As for topical reports
Summary reports		High level syntheses providing accounts of selected activities and their results or compilations of information concerning the project as a whole.	Decision makers and other interested stakeholders, including the public

It is recommended that an internal policy and associated procedures for report production and reviews be included as part of the management system, to promote confidence in the RWMO and its documents. The management system should also specify requirements for indexing and storing reports and for tracking their progress through to approval.

Internal reviews should be undertaken by one or more suitably qualified senior managers for every report produced. However, the length of time to review a draft report and then have it revised and published can be considerable, and it will be important to ensure that the staff responsible for reviewing and authorizing reports do not cause unnecessary delays in publication. Internal report review objectives would be defined as needed, but at a minimum would seek to ensure:

— Work was carried out in line with specifications and expectations;
— Methods used were sound and appropriate;
— Sources of data are documented and the data are correct;
— Results are unambiguous and adequately documented;
— Conclusions presented are logical, on the basis of evidence provided;
— There are no unanswered questions;
— The structure and report content are clear and attributable.

In addition to internal reviews, external reviews should be used selectively for certain classes of report or for particularly difficult or sensitive subject matter. The terms of reference for external reviews should always be specified and it would be good practice to ensure review comments are provided using a common template so that the revision process can be optimized. Several RWMOs have established international advisory groups of experts to support the planning and management of site investigation programmes. Terms of reference include the identification of challenges and provision of guidance leading to solutions. An additional role for advisory group members is to review key publications under preparation.

Once reporting has been successfully completed and a specified phase of site investigations has been formally concluded, it will be necessary to archive and preserve the data and information acquired during the site investigations, including reports. At the outset of a major site investigation project, it is uncertain how long the project might last and for how long access to data and information records might be needed. Many regulatory authorities now stipulate minimum periods for which data and information is to be readily accessible and readable. This may impose requirements on records preservation well beyond the lifetime of the site investigations and indeed beyond the operational period of the repository.

The challenges involved in data preservation, both as hardcopy and in digital form, are considerable. Archiving paper records over timescales measured in hundreds of years is costly given the required storage space, time, durability of the physical storage media and efforts to ensure records preservation. It is anticipated that all site investigation reports and records will also be preserved on electronic media, with the associated requirements of verifying their availability and readability within a rapidly evolving IT landscape. In particular, the continued accessibility of now obsolete electronic data formats, many of which were based on proprietary database systems, is a challenge often encountered today and cannot be left untreated for the future – all data have to remain accessible and readable with widely available digital archival systems.

The development and adoption of good practices for digital preservation and archiving is a common issue for all RWMOs. The OECD/NEA Working Party on Information, Data and Knowledge Management (known as the WP-IDKM) has an Expert Group on a Data and Information Management Strategy for the Safety Case (called the EGSSC).[16] Prior to the formation of that working party, there was an initiative on the Preservation of Records, Knowledge and Memory (known as an RK&M) across generations. This was launched to minimize the risk of losing records, knowledge and memory, with a focus on the period

[16] See: https://www.oecd-nea.org/jcms/pl_25233/working-party-on-information-data-and-knowledge-management-wp-idkm

of time after repository closure. Also of use, NASA, through its Consultative Committee for Space Data Systems has developed a reference model and conceptual framework for an archival system dedicated to preserving and maintaining access to digital information over the long term. This is called the Open Archival Information System (shortened to OAIS), published as an ISO reference model [44].

4.3. CONCLUDING COMMENTS

The above guidance should not be considered exhaustive as there are many other management responsibilities and considerations that should be planned for and put in place prior to the commencement of site investigation activities (see Fig. 12 for a wider set of the management skills necessary for a site investigation project). However, the foregoing discussion is intended to cover many of those issues that would be especially important in the context of planning a site investigation project. It is emphasized that the strategic elements presented above should be continuously monitored for effectiveness and revised as necessary.

5. STRUCTURING A SITE INVESTIGATION PROJECT

5.1. INTRODUCTION

The process of site investigation comprises a large number of distinct but interrelated activities that lead to the understanding of a site and its evolution to date. The manner in which the work activities are to be planned, managed and reported will depend on, among other things, the application of various technical and scientific areas of expertise, the goal and end uses to be made of the understanding derived from the site, the definition of the physical investigation areas and consideration of the project time frame, including the concept of a staged approach to investigations. This section relates the site investigation project to a series of structured frameworks so it can be adequately described. The frameworks presented in this section comprise: (i) a technical activity based approach; (ii) a discipline based approach; (iii) a physical area based approach; and (iv) a time based approach. It is noted that a definition of end user requirements and how site data and information can fulfil those needs has already been presented in Section 3.

5.2. SITE INVESTIGATIONS IN RELATION TO TECHNICAL ACTIVITIES

This framework considers technical activities that would typically be structured as discrete packages of work when developing a plan and its associated schedule. Work packages based on activities are necessary to ensure an appropriate project programme is developed and so that reliable cost estimates can be established during the planning phase. For this discussion, a high level hierarchy is presented, with most technical activities relating to one of the four major project work areas described as follows:

(i) Planning for site investigations — comprising preparatory activities taking into account high level objectives and requirements, a preferred or reference repository concept and a knowledge of resource availability and other actual or potential constraints;

(ii) Data acquisition — activities undertaken to obtain site specific measurements using a range of tools and techniques categorized as follows:
   (a) Desk based studies involving the compilation and analysis of existing information;
   (b) Non-intrusive methods that do not require any significant disturbance at the surface:
      — Remote sensing and imagery (satellite based systems);
      — Airborne geophysics and imagery;
      — Surface based geophysics, including seismic and resistivity surveys and other specialist techniques;
      — Surface based surveys including geological mapping, surface water mapping and fluxes (hydrology), sampling and monitoring.
   (c) Invasive methods that require excavation and drilling to provide direct access to the subsurface for sampling, in situ testing and the installation of monitoring equipment using:
      — Boreholes;
         • Downhole geophysical logging and other survey methods;
         • Hydrogeological testing (hydraulic and transport properties);
         • Production logging;
         • Groundwater and rock sampling;
         • Instrumentation for monitoring groundwater pressure and chemistry.
      — Trial pits and trenches;
      — URL experiments, tests and demonstrations.

(iii) Data compilation, processing, analysis, interpretation and integration — a series of linked activities to:
   (a) Assemble and transform measurement data (primary and derived) into information on the properties and evolution of the site;
   (b) Understand the features and processes of the site (e.g. groundwater flow, geology, environmental processes, etc.);
   (c) Develop a self-consistent understanding of the site as a whole, reflecting all the individual aspects (features and processes, as well as any significant events that have affected the site) in an integrated manner.

(iv) Reviews and reporting — a formal and critical examination of the content and quality of the output arising from the site investigations and the communication of data and understanding to personnel involved in safety assessment, repository design and EIAs, as well as to the public and other and key stakeholders.

The high level scope and relationships between the listed activities are illustrated in Fig. 25.

It can be readily appreciated that in line with the project management concept of a work breakdown structure, the high level activities presented in Fig. 25 can be further broken down into their constituent activities and these in turn may be broken down yet further. For RWMO planning and high level management purposes, a three layer decomposition of site investigation activities is generally sufficient (as shown for example in a project management Gantt chart). Individual contractors may, of course, further refine the hierarchy of activities to maintain their own control of resource usage and scheduling.

5.3. SITE INVESTIGATIONS IN RELATION TO DISCIPLINES

The idea of structuring data and information requirements into five specific disciplines was introduced in Section 3.6, as follows:

— Geology;
— Hydrogeology;
— Hydrochemistry;
— Geotechnics;
— Biosphere studies.

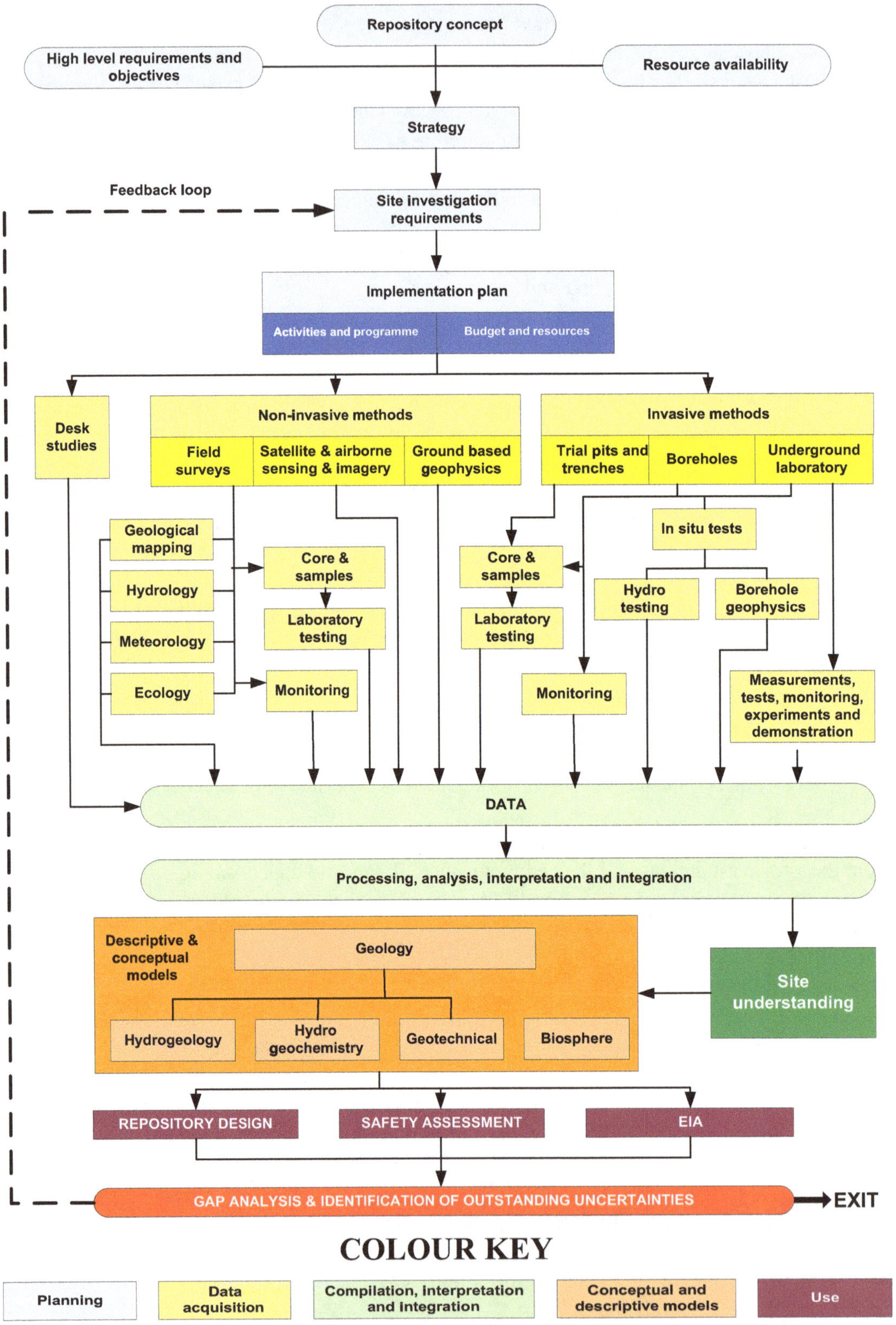

FIG. 25. Generalized scope of site investigation activities and their relationships.

The five part discipline based structure is a useful framework to not only organize end user requirements, but also to support the definition and organization of work activities and to package data and information that is to be transferred to end users specializing in safety assessment, repository design and EIA. The five areas also usefully relate to a series of discipline based SDM domains — a concept originated by SKB in Sweden [45] and adopted by Posiva (Finland), NUMO (Japan), NWS (UK), and others. The case study in Annex III is based on the experiences of SKB in Sweden and it contains a summary of the approach adopted and lessons learned concerning the development and implementation of the concept of SDMs to aid site investigations (see also Section 6.13.2 for further information on the derivation and use of SDMs).

It is emphasized that considering a site investigation project as a series of discipline oriented technical activities should not be a reason to accept the development of artificial barriers between the individual experts and teams engaged on the project. Site investigation requires an interdisciplinary approach. While it can sometimes be true that terminology adopted in one discipline may not be used in or may have a slightly different meaning in another domain, and the methods and concepts employed by one group to support understanding can also vary from one discipline to another, there is an absolute need for effective interaction to share understanding and knowledge across disciplines and to try to achieve synergies that are mutually beneficial. Even within a discipline, there might be communication challenges between the teams tasked with acquiring data and those seeking to analyse and make sense of the data. Experience has shown that these concerns are very real and significant during a site investigation project. The communication challenge is compounded when site investigation personnel (mainly geoscientists) have to translate the geoscientific understanding into methods, algorithms or boundary conditions that would be then used by the safety assessment mathematicians, physicists, chemists and engineers involved in repository design studies. For these reasons, effective senior managers establish and actively manage mechanisms to ensure opportunities for collaboration and knowledge exchange are exploited, so work is not undertaken within discipline oriented silos, and so there is a common basis for understanding across the entire project domain.

5.4. SITE INVESTIGATIONS IN RELATION TO PHYSICAL AREAS

The size of an area to be investigated for hosting a disposal facility will depend on many factors, including the following:

- The volume and hazard posed by the waste inventory for disposal (adopting a graded approach);
- The regional setting of the site;
- The nature of the surface conditions at a site, in particular the topography and the presence of any natural hazards considered relevant for operational safety;
- The nature of the geological conditions at a site, in particular the spatial variability of key parameters and other properties considered relevant for post-closure safety (i.e. those impacting on containment and isolation functions);
- Regulatory requirements, logistical restrictions and stakeholder expectations.

These factors will also influence the degree of detail considered necessary for adequately characterizing a site, bearing in mind that the greatest level of detailed understanding is likely to be required in the immediate vicinity of the disposal volume, and that less detailed knowledge may be sufficient at increasing distances away from the site. This is the main reason why adopting a ‘nested approach’ to site investigations is generally a suitable strategy. Such an approach recognizes that understanding the regional setting of a site is fundament, as it provides context and hence supports one’s understanding of the nature of the immediate site of waste disposal.

In addition to providing a deeper conceptual understanding of the evolution and present day nature of a site, the regional context also provides important data and information needed to populate and calibrate a wide range of models that provide information about key processes operating at the site and that are potentially of use in developing projections of possible future disposal system behaviour. This includes the specification of boundary conditions for subregional scale numerical models of the repository and its setting. This nested investigation and modelling approach is applied within all major site investigation programmes, where a focus on establishing the regional geological context for a prospective disposal site is a necessary prerequisite for site confirmation. The only strategic difference between programmes is whether the major part of any regional investigations should be initiated at an early stage, prior to detailed site scale studies (e.g. as might be done during a site screening siting strategy), or whether a large amount of detailed site scale investigations precede significant resource expenditure on studies at a regional scale (e.g. as might be expected when adopting a preferred siting approach). In reality, for a major characterization programme, the site scale and regional investigations are likely to be carried out largely in parallel.

Understanding the regional context for a site may be especially important for disposal programmes located in relatively dynamic geological settings. For example, Japan is situated in proximity to four major tectonic plates (the Pacific, North American, Eurasian and Filipino plates). These plates are constantly in motion and they interact in a relatively complex manner, which is reflected in a high incidence of seismicity and volcanic activity within a complex stress field. It is imperative that information concerning plate movements is well understood (e.g. rates and directions of movement and the nature of the interactions between the plates) as these will have a direct bearing on hydrogeological, hydrochemical and geomechanically relevant FEPs that could affect safety functions at a site in Japan, once one is selected. Although the geodynamic context in some countries may be relatively complicated compared to tectonically quiescent settings, and processes might operate over a variety of spatial scales and timescales, they may be sufficiently well understood through careful palaeohydrogeological and geophysical investigations undertaken over a range of scales, as well as from natural analogue studies undertaken elsewhere.

Different RWMOs have adopted different terms to describe the different site investigation areas associated with their particular disposal concept. In this publication, for the purpose of designing a framework for a site investigation project, the following domains are defined, noting that they may be applied to more than one site until a choice is made for a single preferred site:

The '**regional area of interest**'. This corresponds to a relatively large area, potentially of several hundred square kilometres surrounding one or more prospective sites. The boundaries of a 'region' may be prescribed by an administrative unit, by topographic features (e.g. mountains, rivers or the sea), by a relatively consistent set of lithostratigraphic units at or near the surface, or be based on the identification of other features such as a water catchment. It is likely that early stage investigations at a regional scale would take advantage of data and information from pre-existing surveys and boreholes where possible, rather than sinking a number of new investigation boreholes for the project (although this might not be the case where there is no existing subsurface information or where there is a potential site within a regional area of interest that shows a high prospect for being suitable). The regional site investigations would therefore be largely based on desk studies during early stage siting studies. However, once specific sites are selected, and to adequately characterize a site to determine its suitability to host a geological disposal facility, it will be necessary to understand a relatively wide area around the site in more detail as well as the nature of the site itself. As noted above, this is necessary to more precisely determine the geological context and framework for a site and to establish appropriate boundary conditions for use in nested site scale groundwater flow and radionuclide transport models. Consequently, although a regional area of interest would be designated prior to the selection of a specific site in the case of a site screening strategy, certain regional scale investigations would likely continue in tandem with local site scale studies. For a siting strategy using a preferred site approach (see Section 2.3), the regional area of interest might only be determined after the selection of a site.

The '**site area**' is defined here as the location of a prospective geological disposal facility and the domain within which the majority of detailed investigation activities would be carried out, using intrusive (i.e. drilling) and non-intrusive methods. It is essential a site area boundary be clearly defined by reference to national coordinates. The site area is the surface expression of the subsurface investigation domain within the rock mass that is expected to undergo relatively detailed characterization. It includes the area in which it is intended to excavate and operate a disposal facility as well as its surrounding environs. It would also include the area where surface infrastructure would eventually be located to support waste emplacement operations. For large scale investigations relating to an extensive underground repository, the site area might be anticipated to represent an area of between 25 km^2 to approximately 50 km^2. A general site area will likely be identified during the area survey stage of the siting process (Section 2.3 and Fig. 3) unless a site is predefined by ownership or it exists as an administrative unit. Consequently, the size and boundaries of the site area might not be precisely defined at the start of ground investigations. However, the margins of a site area should be defined as soon as possible to support the detailed operational planning and for communication with stakeholders. Alternatively, a site location might be predefined in advance of detailed characterization studies in a siting strategy without the need for the area survey and site investigation stages, as discussed in Section 2.3.

The '**target area**' is defined here as the surface expression of the total expected repository domain at depth (i.e. the footprint) and its immediate vicinity. It is an area that would undergo detailed characterization. The target area would be nested within the site area boundaries. The target area might be of the order of 1 km^2 to approximately 10 km^2 dependant on the size of the expected waste inventory, the disposition of underground tunnels, shafts and vaults, and the nature of the rock mass within which the waste disposal vaults and associated underground facilities are to be constructed. For a homogeneous rock mass extending from surface to repository depths and laterally, it may be possible to identify a suitable target area and the disposal horizon at depth on the basis of non-intrusive investigations such as surface geological mapping and by use of geophysical and remote sensing techniques before any drilling is carried out at the site. However, for locations where the potential host formation is hidden beneath several hundreds of metres of sedimentary cover rocks or in a structurally complex domain, it is unlikely to be possible to identify suitable target zones using non-intrusive methods alone. For such locations, an alternative approach is required whereby the target area is to be defined after a phase of regional surveys and deep drilling (i.e. after Stage 1.2 of the investigations, see below). If a site specific URL is to be constructed, it may or may not be considered as part of the target area, dependant on local decisions about whether the laboratory would or would not eventually form a part of the repository infrastructure. Note that some repository concepts have considered waste emplacement at multiple depths within the same site to reduce the size of the repository footprint (e.g. Nagra in Switzerland).

5.5. SITE INVESTIGATIONS IN RELATION TO TIME

The strategic importance of carefully understanding data and information requirements in relation to a phased approach to site investigations was introduced in Section 4.2.3. There, information is provided on the range of data acquisition activities that will be required to be undertaken in relation to a number of steps, as presented in Fig. 17 and Table 3. This information is augmented by Table 5 that more precisely sets out the main activities expected during each step and expands the decision points that define the end of each step.

Regardless of whether a screening or a preferred site approach is to be employed for siting a radioactive waste disposal facility (Section 2.3), a phased approach to site investigations is of benefit for planning and managing activities as the steps describe durations to be used in programming and decision making. Thus, the steps can also be used to bound activities in time, as discrete subprojects with defined objectives separated by clearly defined milestones that could be associated with key decisions about whether to proceed to a new stage or not. As a site investigation project progresses, the complexity of the site investigations will increase in terms of the number of activities being undertaken, their links and dependencies, as well as their sophistication.

A case study presented by NUMO in Annex II is based on the extensive experience in Japan. It contains a summary of the development and lessons learned concerning the adoption of a stepwise approach to the planning of site investigations.

5.6. CONCLUDING COMMENTS

Definition of a site investigation project requires that it be structured in an appropriate manner. This can be accomplished in a number of ways as presented above on the basis of activities, disciplines, physical areas and in relation to time through the use of stages. It is best that any framework adopted be selected to aid a specific management objective, which would likely relate to one of planning, implementation or communication, but could combine all three.

It is emphasized that these frameworks are not in any way mutually exclusive and all of the site investigation activities, together with the data and information they produce, can be described within any of them. It is also emphasized that a framework definition (e.g. in terms of the activities or the description of specialist disciplines) can be readily broken down further and refined as needed. Various well-known tools can be used within these structured approaches and may link across them, such as Gantt and PERT charts, GIS, influence diagrams and decision trees.

Lastly, it is noted that these frameworks should not be considered as exhaustive; they are simply those most commonly employed to structure a site investigation project. Other frameworks may be devised by an RWMO as needed, for example in relation to addressing the needs of regulators or other defined stakeholder groups.

6. DATA ACQUISITION AND PROCESSING

6.1. INTRODUCTION

This section describes management considerations relating to the planning and implementation of data acquisition and processing activities. As part of the discussion, information concerning the types of equipment and techniques that can be used during site investigations for a geological radioactive waste disposal programme are provided, based on the high level division of activities presented in Section 5.2. This is not intended to comprehensively list all of the available tools and techniques that might be used, but rather to indicate those most commonly used in repository site investigations and the types of data sets that they provide. Important definitions and concepts are introduced first.

Data obtained from site investigations will tend to be dominated by measurements of parameters expressed by numbers, but data can also be descriptive. Data should be factual (i.e. based on actual measurements and critical observations) but it is recognized that in some circumstances site data may be based on deductions or subjective description. Raw data are unprocessed data as originally measured and recorded in the field or laboratory. Derived parameters (i.e. secondary data or 'developed parameters') are processed data that have undergone screening or manipulation to remove or modify the raw data into a form amenable for further use. Data are meaningless without context; consequently it will be important to ensure metadata are associated with the site data, such as whether the data are deduced or observed, the spatial location of measurements, the time of a measurement and/or the equipment and methods used.

6.2. PLANNING CONSIDERATIONS FOR DATA ACQUISITION ACTIVITIES

During the planning stage for site investigations, it will be essential to have a good understanding of the data and information that will be needed to address end user needs (Section 3). The degree of effort to be expended during the planning and implementation of data acquisition activities should reflect

TABLE 5. ILLUSTRATIVE TECHNICAL ACTIVITIES AND DECISION POINTS IN RELATION TO INVESTIGATION STEPS

Step number	Main activity	Component activities	Decision point at end of stage
0	Desk based study	Planning and preparation for site investigations. Initial data collection from broad areas and sites (desk studies).	Select prospective regions (and possibly some specific sites) for non-intrusive preliminary investigations.
1.1	Regional surveys	Comprehensive compilation and analysis of existing data and information. Undertake additional regional surveys, as required.	Select locations of potential sites within the regions of interest.
1.2	Initial site	Site area scale exploratory geological and hydrogeological surveys, including airborne and ground based geophysical surveys and interpretation of new remote sensing data (e.g. aerial photographs, satellite imagery). Surface based surveys including geological mapping and sampling.	Screen sites to a small number suitable for detailed site investigations. Define the size of site areas that need to be investigated at each prospective site and estimate the number of deep boreholes likely to be required to characterize the site. Confirm locations and requirements for deep boreholes to be drilled in initial drilling campaign. Define extent to which monitoring network for regional hydrogeology studies needs to be supplemented (borehole monitoring, river gauges, meteorological stations, etc.).
2.1	Initial detailed site	Site wide 2-D and 3-D geophysical surveys focused especially on preliminary target area. Establish seismic and meteorological monitoring network. Drill first multipurpose boreholes, test and install instrumentation (possibly 1 to 3 deep boreholes). Install supplementary components of monitoring network for regional hydrogeological studies. Start site area scale mapping and surveys.	Confirm locations and requirements for further deep boreholes to be drilled in secondary campaign and define extent of more detailed subsurface investigations.
2.2	Detailed site	Drill further boreholes, carry out specialist single hole hydrogeological testing (e.g. fracture network, small interval and tracer testing). Install instrumentation for monitoring. Continue with regional and local hydrogeological monitoring.	Confirm (or otherwise) that there are sufficient deep boreholes to have adequately characterized the target area and wider site.
2.3	Post-completion testing	Carry out cross-hole hydrogeological testing, multihole tracer testing and multiborehole pumping test(s). Cross-hole tomography surveys.	Confirm (or otherwise) that, subject to subsequent establishment of baseline conditions, there is sufficient information to have characterized the site.
2.4	Establish baseline conditions	Sampling, measurement and monitoring of key parameters.	Confirm that baseline conditions have been established and that in situ investigations are now required underground.
3	Underground	Predictions, testing, sampling, measurements, experiments, demonstrations and monitoring.	Confirm (or otherwise) that site understanding is sufficient to enable application for regulatory approval to construct a repository.

the nature of the radioactive waste to be disposed of (graded approach) and the nature of the disposal concept, comprising site conditions, the repository conceptual design, the barrier functions planned and the timescales associated with containment, isolation and potential radionuclide migration. However, the data requirements and the level to which they are to be met will also vary in time, depending on whether the investigations are intended to discriminate between alternative sites that are potentially suitable as a location for a repository (area and site investigation stages in a screening process) or whether a site is to be investigated in detail to allow for site confirmation and an application for a repository construction licence (detailed site characterization stage).

Site conditions and their geological setting are highly variable across the globe, even though there may be commonalities (in rock type for example). Some countries are considering or implementing disposal in what might be considered dynamic terrains characterized by the presence of natural disruptive events and processes, such as relatively rapid uplift and erosion, seismic activity and volcanism. Others are located in intracratonic areas characterized by crustal stability and a lack of geological complexity. The nature of the geological environment will of course influence site requirements and the tools and techniques to be employed to adequately characterize site conditions in response to those requirements. In general, the more geologically dynamic or complex a region is, the greater the effort required will be to adequately characterize site conditions and confidently assess the presence, frequency and impact of natural disruptive events (especially early on in the siting process) and the geological evolution of the site to date.

Only on the basis of fully understanding requirements can a site investigation planner rationally evaluate which data acquisition equipment and techniques would be suitable in a particular environment of interest (i.e. taking into account the rock types present and any possible logistical or practical constraints, such as limitations on access). In a situation where there are multiple tools and techniques available, knowledge of requirements and the investigation environment will influence choices about which ones should be favoured and on what basis.

As far as possible, tools and techniques should be used where they can address data needs in several areas. For example, where it is appropriate to do so, drilling a multipurpose borehole would typically be far more efficient than drilling several boreholes with each borehole dedicated to a single purpose.

It is very important to recognize that site understanding is significantly enhanced by the acquisition and analysis of multiple data sets and these are best interpreted holistically, rather than in isolation. Therefore, integration of expert knowledge is required during the planning phase, as well as during implementation of site investigations, to ensure the combination of tools and techniques used is complementary and optimizes the value of data collected.

If no available investigation technique readily exists to address a defined data need at a site, the planner will have to investigate the potential significance of the value of the parameter(s) to be measured (e.g. in terms of safety or confidence building) and then evaluate whether a new tool or method can be developed or whether data not specific to the site or elicited data can be provided instead.

An important consideration when managing site investigations in relatively dynamic regions characterized by active tectonics is that the occurrence and potential impact of natural disruptive events may exclude prospective sites. Therefore, particular effort should be expended during the early stages of a siting process to identify and exclude clearly unsuitable areas (in line with established site selection criteria). Most of the tools and techniques described in this section would be generally usable in almost any terrain. However, investigation of sites in tectonically active areas may require the use of specialist equipment or methods to investigate particular events, features and processes (for example to more accurately assess uplift rates in tectonically active mountainous areas).

In selecting investigation tools and techniques, the basic principles presented in Section 4.2.1 should be remembered. In addition, it is noted that preferences for using particular site characterization techniques can be selected on the basis of the following criteria (based in part on information contained in Ref. [24]):

— *History of safe operation:* The method should have measures in place to protect the operator and others, comprising established and well defined procedures, as well as physical components designed to ensure safe operations. There should be a history of safe use.
— *Environmentally benign:* Where possible, the method should be non-intrusive or minimally intrusive but, in all cases, choices should be made taking into account potential environmental impacts such as contamination, noise, derogation of land and water, etc.
— *Practicality:* The application of a selected method should be feasible for a defined set of conditions, and it would have a high probability of success in acquiring the required data using available technology within the time available.
— *Demonstrated effectiveness:* The method needs to have been used successfully under conditions similar to those anticipated at the site to be investigated. For equipment and instrumentation this means they should be reliable, durable and suitable.
— *Repeatability:* The method is expected to provide consistent results under identical operational conditions.
— *Accuracy and precision:* The method needs to be able to collect representative data with an accuracy and precision that is sufficient for its intended use.
— *Compatibility:* The method should be compatible with and not limit other data collection methods.
— *QA:* The method needs to comply with authorized and well documented procedures, data control/management, monitoring of results, etc.
— *Cost-effectiveness:* The method should be cost effective relative to other data collection methods, taking into consideration all of the above factors.

Equipped with knowledge about which tools and techniques are available and could be used in the investigations (on the basis of understanding pros and cons in relation to the data required and the environment of interest), discussions can ensue with suitable contractors. Note that the choice of a good contractor will be vitally important to the success of an activity and cost should not be the only factor to be taken into consideration. Rather, a project risk based approach should be adopted during contractor selection, based on the likelihood of objectives being achieved with no adverse impacts on safety or the reputation of the RWMO.

It is well appreciated that project risks are greatest at the start of the project, and they reduce as time progresses because more in-house knowledge is gained about managing the project (Fig. 26 (a), shows the expected reduction in project risk through time as the RWMO becomes more knowledgeable). To reduce risks as much and as early as possible, the RWMO should be sufficiently informed to be able to act as an educated client in dealings with contractors. An educated client is one who has defined the objectives precisely and is aware of the data acquisition tools, techniques and services available, their uses and limitations and their associated market costs.

Typically, especially in the early stages of a project, a sensible approach would be for an RWMO to develop specifications focused as much as possible on identifying what data and information is expected to be delivered by a contractor along with quality criteria, rather than specifying the methods to be used. However, if the RWMO employs highly experienced personnel or if the necessary expertise is gained over the life of the project, it may wish to stipulate the use of certain tools and techniques for the collection of data to ensure adequate control of costs and activities and the achievement of objectives. Note that in the absence of available in-house expertise, there is a danger that if an RWMO attempts to over-specify the detail in a task or relies heavily on advice from contractors to develop very detailed specifications (which may be beneficial for the contractor, but less so for the RWMO), this also has an impact on project risk. Figure 26 (b) shows the relationship between the risk of not meeting requirements adequately and the detail provided in contract specifications for site investigation activities for situations where the RWMO

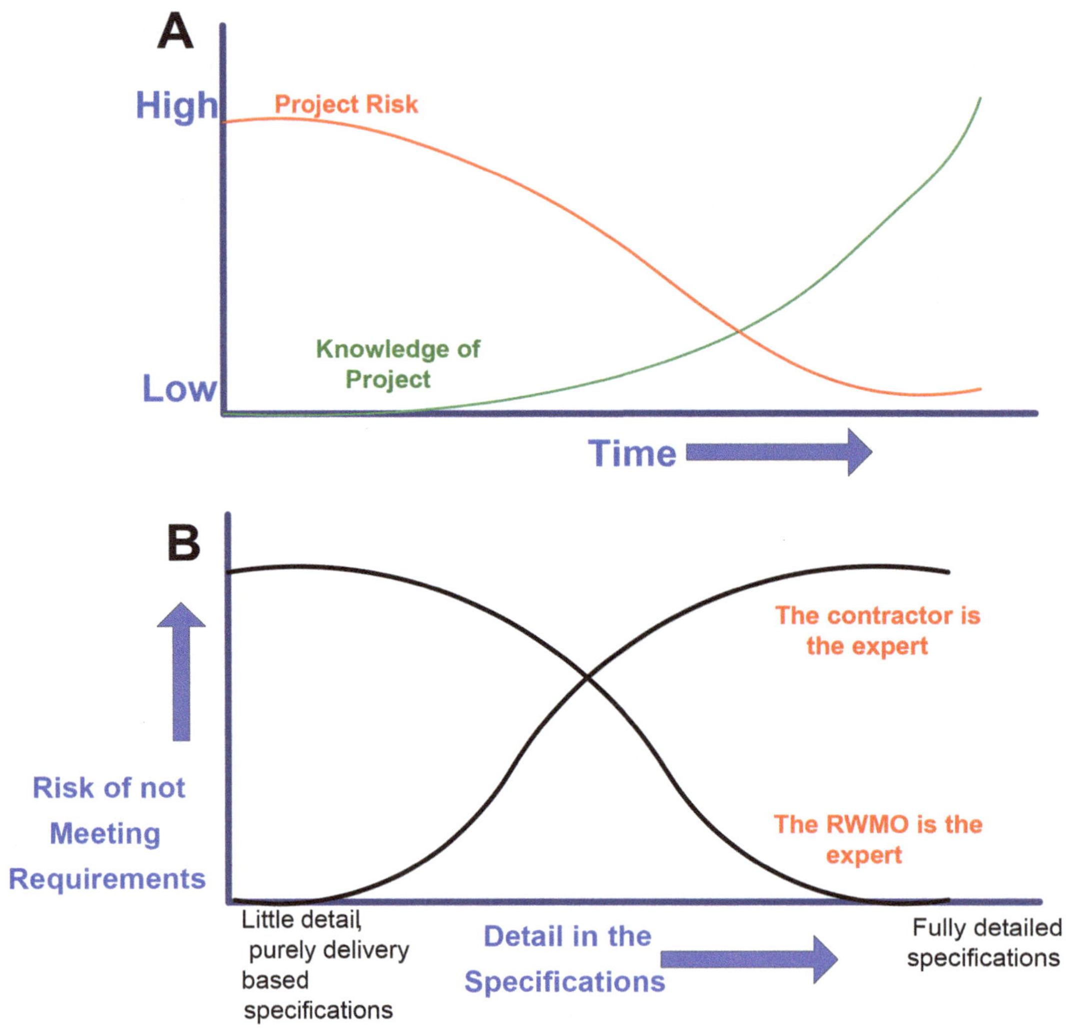

FIG. 26. Generalized project risk plots.

has staff that are, or are not, sufficiently expert to specify in detail methods, tools and approaches to be adopted, as well as expectations concerning results.

Given the timescales involved in a major site investigation project and the paramount importance of safety in all its facets, as well as the significant costs involved, it would be prudent for an RWMO to establish a close and trusting working relationship with contractors in the early stages of a site investigation project while at the same time seeking to develop enhanced in-house competencies to be able to act as an educated client for site investigation services.

Note that where they exist, international and national standards or codes of practice should be used for designing and operating equipment and for the provision of services to guide data acquisition activities by promoting safety, quality and efficiency. For example, relevant standards are published by the British Standards Institution (known as the BSI), American Society for Testing and Materials (known as the ASTM), and the German Institute for Standardization (known as the DIN), among others.

6.3. DATA ACQUISITION TOOLS AND TECHNIQUES — NON-INTRUSIVE METHODS

Non-intrusive site investigation methods are those that do not require any significant disturbance at the ground surface. They comprise the following three broad categories:

— Remote sensing (satellite and airborne based imagery);
— Surface based geophysics including terrain mapping, seismic and resistivity surveys and other geophysical techniques;
— Surface based walk-over surveys for onetime measurements, sampling and monitoring including surveys of topography and geomorphology, meteorology, geological and hydrogeological mapping, surface water mapping (hydrology) and surveys of flora and fauna.

These non-intrusive investigation methods are typically initiated early in a siting process and may be used as soon as specific prospective sites are identified. Some activities may continue through to site confirmation and beyond (e.g. monitoring).

6.3.1. Remote sensing

Remote sensing is the acquisition of data and information about the Earth or any other object, without making physical contact with that object. Hence, the term is most commonly applied in relation to investigations of the Earth's surface or subsurface using satellite and aircraft based systems. It encompasses a very wide range of techniques that are of most use especially during the early reconnaissance phases of a siting programme. This is because the techniques are able to effectively and rapidly measure and map properties uniformly over large areas. Consequently, they are cost effective when compared to ground based methods that could be used to obtain the same information. They can, however, be used at any stage in an investigation.

Remote sensing using satellite based systems produces images mostly by means of capturing electromagnetic energy reflected or emitted from the Earth's surface. The various techniques depend on observed spectral differences in the energy captured by sensors; essentially the 'brightness' or radiance of objects and features (although this applies not just in the optical spectrum, but ranges from the imaging of gamma rays with a wavelength limit of <0.03 nm to radio waves at >30 cm). Airborne remote sensing techniques include gravity and magnetic methods, as well as electromagnetic spectrometry. These can often be employed simultaneously using helicopter or fixed wing aircraft.

The main techniques associated with satellite based systems map the land in different parts of the electromagnetic spectrum. Multispectral data have been commercially available since Landsat came into operation in the early 1970s and other satellites provide similar services (e.g. SPOT Image Corp.). These systems take images of the Earth in multiple wavelengths of electromagnetic radiation (multispectral). Landsat used seven channels, each of which formed a separate image that emphasized different aspects of the Earth's features, including vegetation and rock type. Images from the non-visible range of the spectrum are converted into false colour images as it is the midinfrared channel (2.08 to 2.35 m) that is particularly used to discriminate between rock types.

Unlike multispectral imaging that employs relatively few and broad wavelength bands to produce images, hyperspectral sensing uses imaging spectrometers to simultaneously capture the electromagnetic energy from solar radiation reflected off the Earth's surface in hundreds of narrow, adjacent spectral bands, creating a continuous spectrum for each pixel. The images produced therefore contain a mass of data that by reference to a spectral library is diagnostic of differences in vegetation, rocks and mineral composition.

Interferometric synthetic aperture radar (known as InSAR) is a highly sensitive geodetic remote sensing technique using narrow beam microwave frequencies originating from satellite or airborne systems and reflected from the surface. It generates detailed maps of millimetre scale (or larger) deformation and movement in the Earth's surface over time. Consequently, the technique is especially useful for recognizing and monitoring fault activity at the surface, subsidence due to mining and movement associated with volcanic activity. Differential global navigation satellite systems (known as DGNSS) and lidar (light detecting and ranging) can also be used to monitor ground movement. Lidar projects a pulsed laser light at the Earth's surface and records the difference in the return time and wavelength. The data allows the accurate measure of altitude of the land to produce high resolution maps of surface topography. The technique can also be used to map the seabed (bathymetry).

Gravity and magnetic techniques use highly sensitive gravity meters and magnetometers to measure the natural spatial variability of the Earth's gravity and magnetic fields to generate contour maps. These maps reflect contrasts in rock density and their magnetic properties and take account of a rock volume and not just surface properties. The interpretations of these anomalies are relatively large scale and, whilst they reflect different properties, magnetic and gravitational characteristics can often be correlated. The processed information can also be combined with other data sets to indicate gross rock characteristics and to identify the presence of large scale structures.

For airborne electromagnetic methods, the depth of investigation is rather limited compared to ground based systems (up to a few hundred metres), nevertheless, such surveys can be carried out rapidly and over large areas.

6.3.2. Surface based geophysics

As part of a staged approach to siting a disposal facility, a range of surface based geophysical methods could be employed at different times for deriving data on various properties. Generally speaking, surface based geophysical methods provide data and information on the 2-D or 3-D structure of the rock mass and the spatial distribution of various groundwater and rock properties. The types of technique and survey that would be commonly used during site investigations include seismic refraction and reflection (including tomographic methods for 3-D imaging); ground penetrating radar, magnetic, electromagnetic and magnetotelluric methods, measurements of electrical resistivity and of gravity. A brief description of the physical basis for the techniques and their use is provided in Table 6.

Because some geophysical methods sample relatively large volumes of rock relatively quickly, as with remote sensing and airborne methods, they are used to provide information that is particularly useful during the reconnaissance stages of investigations, to provide a general appreciation of the nature of the subsurface, regionally and more locally. Resistivity surveys are one such technique.

Some other techniques, such as 2-D and 3-D seismic surveys, are today also considered indispensable for characterizing subsurface conditions at a site as part of detailed investigations for a major disposal programme. Basic 2-D surveys may be used early on in a site investigation project, relying on readily obtainable equipment to characterize the subsurface to moderate depths with coarse resolution. For more detailed investigations undertaken as part of a major investigation project, 3-D tomographic and even 4-D (time-lapse) seismic surveys may be employed. For full coverage of a relatively large volume of rock (cubic kilometre scale) requiring detailed resolution, high energy sources (typically trucks fitted with vibroseis equipment) would be used in conjunction with a large array of geophones and significant computer processing power would be necessary for interpretations. Although not strictly a surface based geophysical investigation technique, vertical seismic profiling (VSP) takes advantage of seismic monitoring equipment placed in suitable boreholes within the survey area to augment surface based equipment. It entails positioning closely spaced geophones in a vertical borehole to record measurements of acoustic energy signals generated at the surface or within a borehole. Two or more boreholes may be instrumented, and various source and detector array configurations may be used. Downhole responses to the source signals are correlated with seismic response measurements collected in other boreholes or at the surface. A range of data processing techniques can be used to reduce noise and further optimize the value of the data collected. The result of VSP is highly accurate time-depth and seismic velocity information resulting in significantly higher resolution subsurface images that can be used to discriminate small scale structural and lithological features for use in populating models or for planning underground works (e.g. tunnelling).

TABLE 6. NON-INTRUSIVE SURFACE BASED GEOPHYSICAL TOOLS & TECHNIQUES COMMONLY USED FOR DATA ACQUISITION DURING SITE INVESTIGATIONS

Non-intrusive tools & techniques				
Name of technique	Description	Data acquired	Stage	Comments
Ground penetrating radar	Measures travel time of an electromagnetic pulse through the ground. Structures and contacts mapped by virtue of reflectivity, thus an electromagnetic analogue of seismic.	Lithological contacts and geological structures, including induced damage zones.	DSCS	Depth of penetration is very limited except in highly resistive environments. Coverage is fast over smooth ground, by hand or towed on skids or trailers.
Seismic reflection	Acoustic waves are generated at 'shot' points located at or near ground surface using explosives or mechanical sources. The signal is transmitted into the subsurface and reflected back towards the surface at rock layering and structures. High density units are strong reflectors. Acoustic travel times are recorded using an array of receivers (geophones). In 2-D, geophones are laid out in a line and recorded responses are presumed to be reflected signals from contacts and structures vertically below the survey line. In 3-D, the shot points and geophones are laid out in an area array.	Lithological contacts and geological structures. Higher resolution and cross-hole interpretation possible from VSP. Through acoustic impedance, which controls reflection strength, correlations with additional borehole data possibly gives insights into hydraulic heterogeneity.	SIS and DSCS	Offers greatest resolution over greatest depth range of all geophysical techniques. Works best in sedimentary environments, but now also used successfully in hard rock terrains. Very high level of development due to use in oil industry. Relatively high costs, especially for 3-D and 4-D applications and requires intensive processing. Correlations between seismic responses and other parameters measured in boreholes shows promise for interpolated 3-D visualization of other properties.
Seismic refraction	As in seismic reflection, acoustic waves are generated at an energy source using explosives or mechanical devices and travel times are recorded using an array of geophones. However, analysis is based on the refraction of P-waves at interfaces between rocks with contrasting density using analysis of first arrival times.	Lithological contacts and geological structures, as well as depth to groundwater in some circumstances and calculation of elastic moduli.	SIS and DSCS	Seismic refraction is especially useful for very shallow investigations. Technique is relatively inexpensive, and data are simple to process; however, surveys provide coarse resolution and require subhorizontal layering and velocity increasing consistently with depth.

TABLE 6. NON-INTRUSIVE SURFACE BASED GEOPHYSICAL TOOLS & TECHNIQUES COMMONLY USED FOR DATA ACQUISITION DURING SITE INVESTIGATIONS (cont.)

Non-intrusive tools & techniques				
Name of technique	Description	Data acquired	Stage	Comments
Resistivity	Maps differences in electrical conductivity using electrodes transmitting a current injected into the ground and recording the voltage at multiple locations. Electrodes connected by cable, so ground coverage can be rapid. Ground resistivity inferred according to Ohm's law. Resolution dictated by electrode spacing.	Maps higher and lower resistance zones in 2-D or 3-D (tomography through use in combination with cross-hole resistivity), differentiates clays and saturated-conditions.	SIS and DSCS	Ground coverage with large electrode spacing (100 m or more) is relatively slow. Technique requires electrical contact with the ground. Depth of exploration is reduced as conductivity increases but can be several hundred metres. Resolution possible at ~5 m scale. Can also be used in boreholes.
Transient electromagnetics (TEM)	Electromagnetic method in which currents are induced in the ground by discontinuous pulses of electric current in coils or loops.	Maps conductivity variations with potential to delineate stratigraphy, cover thickness and depth of an oxidation zone.	DSCS	Can be recorded on the ground or from air. Highly variable loop size and hence penetration.
Frequency domain electromagnetics	Magnetotellurics (MT) and higher frequency audio magnetotellurics (AMT) exploit natural geomagnetic fluctuations and are 'natural source' techniques. In controlled source audio magnetotellurics (CSAMT) an alternating current is injected into a grounded electric dipole.	Maps conductivity anomalies and CSAMT, with interpretation, was found very useful for mapping chloride concentrations at Sellafield.	DSCS	MT and AMT allow greater depth of penetration than CSAMT, but resolution is poorer. Airborne and ground based systems.
Microseismics	A passive seismic monitoring technique that measures the intensity, location and timing of spontaneous underground movements (dislocations) in the rock mass. A monitoring system comprises geophones, accelerometers and force-balance accelerometers in uniaxial, biaxial and triaxial combinations (triaxial provides most accurate estimate of source parameters).	Can be used as part of a seismic monitoring network. Maps the location of faults and fractures dislocating in response to seismic events, as well as indicating failure mechanisms from focal plane solutions.	DSCS	Can be located in boreholes for monitoring purposes, as well as for identifying individual structures.

TABLE 6. NON-INTRUSIVE SURFACE BASED GEOPHYSICAL TOOLS & TECHNIQUES COMMONLY USED FOR DATA ACQUISITION DURING SITE INVESTIGATIONS (cont.)

Non-intrusive tools & techniques				
Name of technique	Description	Data acquired	Stage	Comments
Nuclear magnetic resonance (NMR)	Radio frequency resonance results from the application of an applied static magnetic field. The resonance (usually of protons) is subsequently excited at radio frequency by coils or antennas to provide a quantitative measurement of proton relaxation times in large bulk samples.	Pore size and, by calibration, hydraulic conductivity.	DSCS	Can be used during ground surveys or during borehole logging. Strength of correlations with hydraulic conductivity depend on the rock types being assessed.

Note: SIS = site investigation stage, DSCS = detailed site characterization stage.

Another important development is the growing use of seismic attributes[17] using derived data such as acoustic impedance.[18] In addition to use for VSP applications, borehole seismic logging information can be used to condition these 2-D and 3-D seismic attributes derived from seismic surveys. Where the borehole values for the attributes can be correlated with other geophysically derived properties, such as porosity from wireline geophysical logging or from core analysis (which might in turn be correlated with permeability), it becomes possible to image hydraulically significant rock mass heterogeneities throughout a seismically characterized volume. However, the confidence associated with such extrapolations is highly variable.

Figure 27 provides an example of a synthetic image of a rock volume based on seismic attributes. It is derived from the results of an analysis in the UK investigating the empirically observed relationship between acoustic impedance, *AI*, (the product of rock density and seismic velocity), core derived porosity, φ (%), and hydraulic conductivity, K (m/s). The relationship between permeability and porosity at the site is $\varphi = 2.6 \log(K) + 34$ and the empirical relationship observed between core derived porosity, φ, and acoustic impedance is $\varphi = -2.6\ AI + 40$. This results in the derived relationship: $\log(K) = -AI + 2.31$, permitting acoustic impedance to be converted across the seismically imaged volume into a 3-D pseudomeasure of hydraulic conductivity. This example undertaken by Nirex in the UK was one of the earliest studies to employ seismic attributes for characterizing a repository site.

Due to the costs involved, major seismic surveys would probably only be undertaken during the detailed site characterization stage of a siting process, although in some cases data may already have been collected by others (e.g. as part of seismic investigations for resource investigations). Seismic surveys are very powerful for imaging the subsurface in sedimentary environments where there is a strong and orderly contrast in rock types and their use in crystalline rock terrains for imaging structures and heterogeneities in the rock mass is becoming more commonplace.

Other more specialist non-intrusive geophysical techniques shown in Table 6, such as CSAMT, would tend to be used only during the detailed site characterization phase at a limited number of sites due to availability and costs and the fact that prospective sites investigated at an early stage in the siting process may not need such information for screening purposes.

6.3.3. Surface mapping and walk-over surveys

Walk-over surveys involve the use of specialist personnel to collect measurements, observations and samples in the field to better understand the characteristics of a site. Mapping is a specific type of field survey that places measurements, observations and samples into a spatial framework. Depending on the type of map to be generated, mapping can be accomplished remotely (e.g. on the basis of satellite and airborne systems) or by direct investigations on the ground, or a combination of both. Ground-truthing of remotely obtained data is sometimes required to verify and enhance the resolution of maps generated remotely, although with today's technology (e.g. lidar and high resolution imagery) this is becoming less prevalent.

Various types of map will be required to provide a framework for site investigation interpretations, as well as to provide measurements for use in repository design studies and EIA and for populating and calibrating safety assessment models. The types of map expected to be generated at a site includes:

— *Geographical maps:* For general usage, good quality geographical maps will be required which show administrative divisions, built up areas, roads, railways, powerlines and named localities. Coordinates should be provided using accepted national conventions and these should be applied to all other types of spatial mapping. The length scale and an orientation marker (e.g. true north) should also always be shown, as well as any necessary key to the use of symbols.

[17] Seismic attributes are measurements and parameters measured or derived from seismic reflection data.

[18] Acoustic impedance is the product of compressional (P-wave) velocity and rock density.

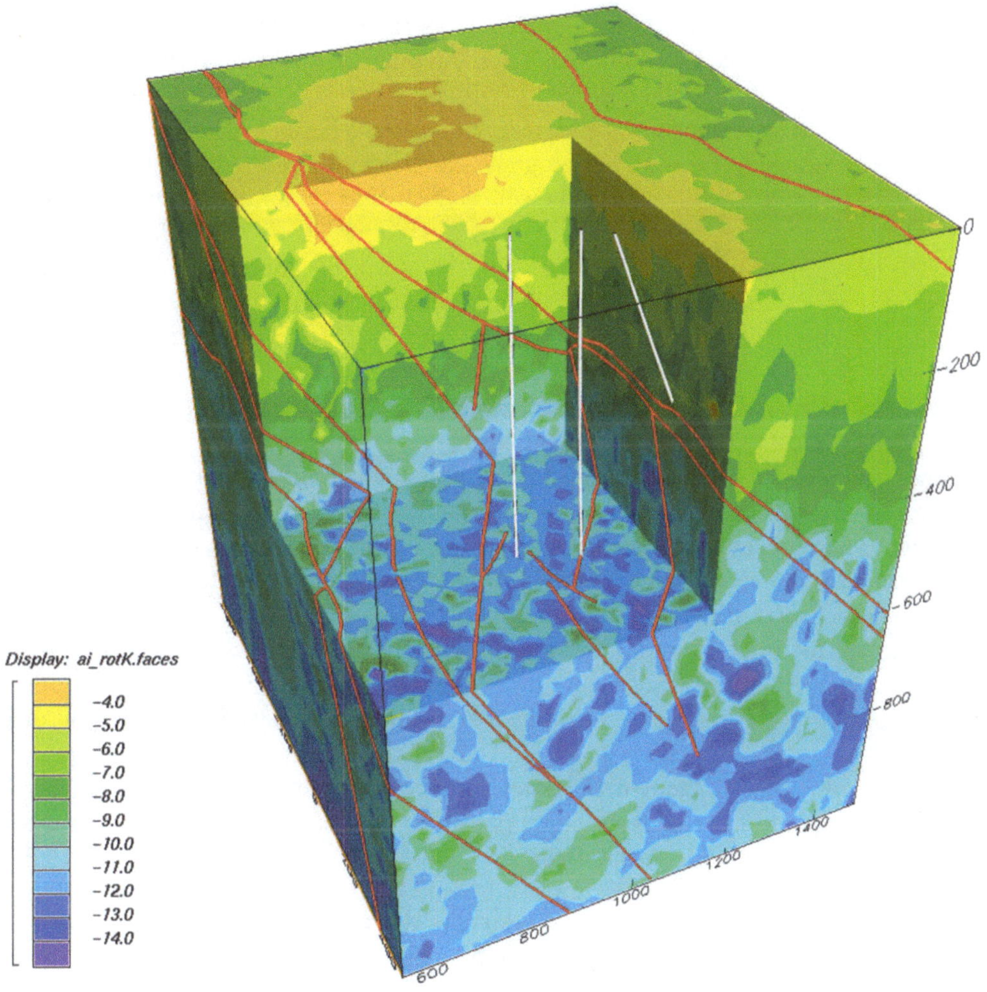

FIG. 27. Synthetic image of a rock volume at Sellafield, UK, based on seismic attributes correlated with borehole porosity from core logging, in turn correlated to hydraulic conductivity. The colour coding represents an index that highlights an inferred hydraulic heterogeneity existing between and within two major rock types at the site: the Borrowdale Volcanic Group, a metamorphic rock and the overlying Sherwood Sandstone Group, a layered sedimentary sequence. The red lines are faults and the white lines are boreholes. Vertical and horizontal scale in metres (reproduced courtesy of NWS).

— *Topographic and bathymetric maps:* Topographic maps are likely to be based on existing surveys in many Member States but may require specialist mapping services to produce new high resolution digital terrain maps for use in computerized systems. These may be generated by laser scanning from airborne mounted platforms, as well as from aerial photographs and ground surveys. For coastal sites or sites that incorporate deep lakes, it would be appropriate to obtain or produce bathymetric maps.

— *Geological maps:* Several map types in this category can be used to illustrate the spatial distribution of different rock types, the type, location and orientation of structures and the nature of the hydrogeology in terms of groundwater units and piezometric maps of groundwater head and flow directions. Specialist maps may also be used that indicate the inferred distribution of coal, oil, gas and ore resources. Geological maps include:

- *Lithological maps:* Although remote sensing technologies may be employed to generate geological maps showing the distribution of different rock units, they are primarily based on ground surveys, supplemented by subsurface information where available. They are especially intended to illustrate the extent and character of superficial sediments (e.g. regolith and alluvium) and geological units exposed at the surface (outcrop) or inferred with high confidence to exist at shallow depths (subcrop). The inferred spatial extent of deep geological units can also be shown on geological maps, with stratigraphic contours representing the upper surface topography of a unit in the subsurface and isopach maps showing the variation in thickness of a rock unit. The measured strike and dip of a rock unit is also typically shown and, for metamorphic rocks, any foliation trend might also be represented. Surface based geological mapping of lithology might be augmented by aerial photographs and satellite and airborne geophysics, including multi- and hyperspectral imagery, as well as the results of laboratory studies (e.g. thin section microscopy: scanning electron microscopy (known as SEM), X ray fluorescence (known as XRF) or X ray diffraction (known as XRD)). It will be important to understand the spatial relationships of different geological units to derive a reliable lithostratigraphic column and an understanding of how different geological units are juxtaposed at depth due to depositional or structural complexity. An understanding of the disposition of lithostratigraphic units and structural features is the primary framework used in all other aspects of the hydrogeological and geotechnical subsurface investigations.

- *Structural maps:* In addition to geological units, structural features present at the surface or inferred at depth can also be shown, such as faults and folds. It is noted that in hard rock terrains, fracture mapping as a specific type of structural map would be very important, whereby the characteristics of faults and fractures are evaluated to indicate their potential to act in the subsurface as fast pathways or barriers for groundwater flow and hence suggest how they might facilitate potential radionuclide transport. Discrete fracture network and process models require statistics on fracture orientations, their frequency, spacing, aperture and trace lengths and the nature of any mineralogical infilling materials. Microstructural mapping and analysis might also be used to understand the relative timing of fracture generating events that might be important for recognizing different subsets of fractures and their orientations. Surface based geological mapping of structural lineaments might be augmented by aerial photographs and satellite and airborne geophysics.
- *Hydrogeological maps:* Subsurface hydrogeological maps include information on hydrogeological units, borehole locations and discharge areas, with measured groundwater levels and pressures interpolated to provide contours and general flow directions based on the interpreted hydraulic gradient. The framework for such maps is based on the geological mapping but may be refined by facies analysis of sedimentary rocks and fracture networks to support interpretations of groundwater flow patterns in different formations.

— *Soil mapping:* Depending on the nature of the site investigations and the actual data and information requirements, soil mapping as part of the biosphere investigations may be required in great detail or may not be required at all. The distribution and nature of different types of soil would be based primarily on ground mapping and laboratory analysis of samples. Satellite imagery and aerial photographs might also be used.

— *Land use, ecosystem and vegetation maps:* Likewise, the need for these maps would depend on the nature of the site investigations and data and information requirements. Although land use, ecosystem and vegetation maps might possibly be considered more relevant for near surface disposal concepts, taking account of the time frames in which the conditions observed today could be used to support relatively near future scenario development, it should be appreciated that they might also be used as proxies for future biosphere conditions for a DGR concept in a scenario where there are similar climate boundary conditions. Furthermore, it will be important to have these types of maps available to support the planning and construction of surface infrastructure and to support EIAs. The surveys required to generate these maps could initially be based on ground surveys and aerial reconnaissance if they do not already exist, but satellite imagery might also be used (e.g. multispectral mapping).

— *Surface hydrology map:* Survey information to be indicated on these maps would include the location of surface water bodies (lakes, rivers, streams, etc.) and the flows associated with them, as well as the boundaries of catchment areas. Surface water sampling and stream and river gauging locations would also be indicated. Meteorological stations would provide supporting information and these locations would also be included if sited in the domain of interest. Hydrological maps would primarily be used in the assessment of recharge and discharge zones and for deriving water balances. They could also be used in prospective radionuclide migration studies if relevant to the disposal concept.

— *Natural hazard maps:* These types of maps and additional information obtained through the associated surveys would be generated especially to support site selection, but in the case of preferred candidate sites where the risks associated with hazards are considered to be present, but acceptable, they might also be used for planning repository infrastructure layout and operations. They would be especially based on geomorphological observations to highlight, for example, areas of flood risk or proximity to slopes displaying risk of rockfall and avalanches. They might also include tsunami hazard. In seismically and volcanically active areas, hazard maps might include the surface locations of geologically recently active fault zones and volcanoes, as well as the extent of potential lava and pyroclastic flows. Hazard assessment maps might also be used to support the generation of scenarios for use in safety assessment.

— *Other types of map:* These might be produced as required and could include, for example, maps of marine or fluvial terraces to investigate sea level changes, or the locations of monitoring networks and maps indicating differential crustal movements over time.

6.4. DATA ACQUISITION TOOLS AND TECHNIQUES — INTRUSIVE METHODS

Intrusive data acquisition requires the surface environment be disturbed through the use of excavations or rock drilling to provide direct access to the subsurface environment. Boreholes (also termed drill holes), trial pits, trenches, tunnels and shafts can all be used to provide this access. The type of access and the method selected to provide the access will depend on the data and information required at a particular stage/step of the investigations. While trial pits and a small number of boreholes might be required to discriminate between alternate sites during site investigations (area survey and site investigation stages), their use would increase as the investigation effort progresses at a potential repository site that displays good promise (detailed site characterization stage). Due to cost considerations, mined access (shafts and tunnels) would likely only occur at a site when necessary and as part of the detailed characterization effort at a potential repository site. In some countries, there may be a decision to construct a generic underground research facility, or to convert an existing underground structure into such a facility. Such a generic underground rock facility would be at a location other than the prospective repository site and it would act primarily as an in situ laboratory focused on providing generically applicable samples, data and information, but might also be used to provide demonstrations of technology and methods or to inform and encourage communication with stakeholders. An underground research facility might also be constructed at a prospective repository site, to provide critical information on the subsurface in the immediate environs of a repository in advance of site confirmation. Such a facility might be constructed in advance of a repository construction licence application or as part of the first phase of repository development. Because of the focus of this report, Section 6.6 provides a limited further discussion of the use of underground facilities as part of site investigations.

Trial pits and trenches are useful as they allow direct observations of soil, superficial sediments and intact rock. They are particularly useful during geological mapping and for sampling. Trenches are relatively linear and allow horizontal profiles to be made and, depending on their width, generate what might be considered 3-D information in the very near subsurface. Pits are generally subcircular excavations and, like trenches, they clear away surface vegetation and superficial deposits to allow observations to be made of the underlying rock. Pits and trenches can be especially useful for acquiring structural and lithological data where outcrop is limited. In comparison, boreholes provide vertical to subvertical profiles and provide what is essentially 1-D information.

Lysimeters are instruments that measure vegetative evapotranspiration and assess the amount of precipitation that can infiltrate through soil. Strictly speaking, they require ground disturbance for installation, although the impact is intended to be minimal. Measurements and observations supplement meteorological and groundwater data and are used in biosphere studies.

Boreholes provide some of the most useful data and information needed to characterize subsurface conditions at a site and they are used extensively for siting disposal facilities, although for near surface disposal concepts the need for deep boreholes would be very limited.

Despite an apparent 1-D limitation, when two boreholes are in relatively close proximity, and for certain rock mass characteristics, it is possible to construct with some confidence a 2-D vertical section between the boreholes through the use of cross-hole testing data or interpolated measurements of rock properties. With geoscientific justification, such interpretations might extend out beyond the boreholes. With three or more boreholes in relatively close proximity, any interpretations of the rock mass conditions between the boreholes would represent a 3-D volume and the orientations of any regular planar features that cross the volume and intersect all three of the boreholes, such as the water table or faults, might be derived with some confidence. Some structural data sets intercepted by a vertical to subvertical borehole, such as high angle fault zones, are likely to be undersampled in comparison to subhorizontal features.

6.4.1. Borehole drilling

Given the focus of this report and its intended audience (given in Sections 1.3 and 1.5), the following discussion concentrates on management considerations associated with the drilling and siting of boreholes, rather than focusing on drilling technology per se.

General considerations

In the absence of other forms of access to the subsurface (e.g. a mine or pre-existing boreholes) the drilling of new boreholes at one or more sites will be required during a siting programme for a geological disposal facility. Drilling intended to obtain high quality data for scientific and engineering purposes is technically challenging and one of the most costly parts of a site investigation project. It is also the activity that would likely have the most significant environmental impact. Therefore, not only should the extent and quality of data that are to be obtained from boreholes be well understood, but well thought out borehole design and operational plans should be developed to ensure the boreholes are drilled to reflect optimum value for money, efficient use of time and to minimize environmental impacts.

Depending on national conditions, the availability of drill rigs and experienced personnel may be limited. Furthermore, the procedure and time involved in obtaining permits and permissions will vary between Member States, and this can be a complicated and time consuming process. Any unforeseen delays due to a failure to have drill rigs on site or to obtain necessary licences could have significant knock-on effects elsewhere in the site investigation programme.

There are significant project and safety risks associated with drilling, and the activity should only be carried out by experienced drilling contractors using technically appropriate equipment that meets national safety and environmental standards. For example, if deep borehole drilling is planned, there may be a risk of encountering gas pockets in certain geological environments.

An important consideration is that boreholes drilled for site investigation purposes will have to be sealed at some point in the future, regardless as to whether or not the site is eventually selected to host a repository. Some boreholes may be sealed at the conclusion of the site investigations undertaken to support a repository construction licence application, while others may remain open through the repository construction and waste emplacement phases, to allow for continued monitoring, sampling or testing. There may be a regulatory requirement to develop sealing plans as a condition of planning approval for drilling a borehole, but in any event it is good practice to design an appropriate sealing methodology at the outset, one that identifies suitable sealing and backfilling materials, as well as the sealing procedure to be followed. Further discussion on sealing is provided below.

Borehole objectives

Boreholes are a tool. Individual boreholes may be used to obtain a wide range of information (multipurpose boreholes) or may be designed to obtain a particular type of information (single-use boreholes). They can be constructed for a range of purposes, including the following:

- To obtain physical samples of the rock, typically as rock chippings or core, to be used in laboratory testing for geological, hydrogeological, geochemical and geotechnical characterization;
- To provide access for downhole geophysical logging to determine certain characteristics of the groundwater in the vicinity of the borehole and of the rock forming the walls of the borehole and adjacent to them;
- To provide access for hydrogeological testing to determine how well fluids are retained in the rock (storage characteristics), how they move through the rock (flow mechanisms) and the ease of flow through the rock (hydraulic conductivity and transmissivity);
- To provide access for production logging to identify the location and velocity of groundwater inflows within a borehole and under appropriate conditions allowing for measurements of flow direction;
- To undertake tracer tests to establish the nature of the properties that impact on the transport of solutes through the rock mass;
- To access samples of groundwater for hydrochemical and isotopic analysis;
- To provide access for measurements of the in situ stress state within the rocks;
- To provide access to measure the ambient rock temperatures and geothermal gradient;
- To provide access for the installation of instrumentation for measuring long term groundwater pressure and water table fluctuations;
- To provide access for monitoring variations in groundwater chemistry over time;
- To provide access for other forms of testing and installation of instrumentation (e.g. microseismic monitoring);
- To allow preliminary identification of the target formation to host disposal vaults and tunnels or, in the case of borehole disposal, the disposition zones.

It is noteworthy that in addition to the listed purposes, data derived from boreholes are also used to calibrate and validate surface based geophysical modelling methods. The value of combining this information and understanding with those derived from surface based geophysical surveys (e.g. VSP and correlations with seismic attributes) is alluded to in Section 6.3. Such methods provide 2-D and 3-D imaging of the subsurface over relatively large areas, however, they are highly dependent on the use of models and associated assumptions. When such interpretations are augmented by subsurface data derived from boreholes, confidence in the inferred characteristics of the rock mass is significantly strengthened.

Drilling techniques

The choice of drilling technique adopted at a site will depend on the equipment available, access and ground conditions, the rock type(s) to be drilled and the borehole design, reflecting the dimensions of the planned borehole and the type of data and information that is required to be obtained from it.

The main types of rock drilling techniques used for site investigations are based on the application of percussive and rotary forces, with some drilling techniques employing both forces simultaneously. The various types of drilling methods in common use include cable drilling, top-hammer and down-the-hole air hammer drilling (mainly percussive) and direct rotary and reverse circulation mud drilling (rotary). Other variants include sonic drilling, a technique employing high-frequency vibration forces with rotation, a technique that is able to quickly penetrate a range of rock types (but only to relatively limited depths of a few hundred metres), wireline diamond core drilling which uses hollow, diamond impregnated bits and, as its name suggests, is particularly useful for obtaining intact high quality core, and coiled-tube drilling that uses a downhole motor and a continuous reel of tubing to eliminate manual handling of drill rods. All drilling methods have their pros and cons relating to the size of the rig, design of the drill bit, the speed of drilling, cost, stability of the borehole, the depth of penetration, the borehole diameter, the quality of core recovery, their ability to be 'steered' (directional drilling) and impacts on the environment. It will therefore be important to select the most appropriate drilling method for the particular circumstances being considered in consultation with drilling experts.

Drilling fluids

In addition to a range of drilling rigs which may be available, choices will be required concerning the use of drilling fluid. Drilling fluids are used for a number of reasons, including to stabilize a rock formation during drilling, to control fluid losses, to increase drilling efficiency through lubrication and cooling of the drill bit, and to facilitate the return of rock cuttings to the surface. The types of drilling fluid used are determined by the drilling technique selected, the rock environment, borehole construction and the purpose expected of the borehole. Fluids can range in composition from air, to water, to high density muds (e.g. bentonite) and polymers.

The choice of a particular drilling fluid should never be left solely to a driller to decide as it is imperative that an appropriate fluid is used when drilling, especially if uncontaminated groundwater sampling is required. Drilling fluids may enter a formation and significantly alter measurements of groundwater chemistry.

Drilling fluid losses should be recorded in driller's daily records as they may indicate discrete features or zones of higher permeability than background conditions. Losses are most likely to occur where the pressure in the borehole exceeds that in the formation. Consequently, to minimize formation invasion, it is desirable to reduce the pressure in the borehole so that it is always a little less than the formation pressure (this is termed underbalanced drilling). To limit pressure in the borehole and hence minimize contamination of the groundwater with drilling fluid, heavy drilling muds should not be used where possible. However, it is understood that in some cases they may be necessary, as higher density fluids, including those containing viscosifiers and additives to control thixotropy, have good lifting characteristics to bring rock cuttings to the surface and therefore the borehole design and the nature of the rock mass may dictate their use. The lowest density drilling fluid (i.e. air) is likely to penetrate far more into a formation than a high density fluid, and air may significantly alter redox conditions in the local formation. Consequently, the need for airlifting should also be carefully evaluated prior to use. Generally, locally derived freshwater can be a reasonable compromise when drilling a borehole for geochemical sampling purposes, but even here, different groundwater chemistry at different depths can complicate later interpretations. It is also advantageous to produce fine cuttings, if possible, so they are able to be lifted easily. This can be achieved using a coring drilling method, for example, which of course also lessens the volume of rock cuttings to be lifted and can facilitate direct sampling in the laboratory of any fluids trapped within the core and not exposed to the atmosphere during handling.

In situations where non-reactive drilling fluids cannot readily be used, the fluid may be dosed with a tracer such as uranine, fluorescein or iodide. One or even two inert artificial tracers might be used, but good control over their concentration is essential. Tracers are added to the drilling fluid to understand the impact of the fluid on in situ groundwater sampling quality as the use of a tracer of known initial concentration allows the degree of drilling fluid contamination to be identified by understanding the change in concentration during mixing. This mixing can be corrected for through simple regression analysis.

It is important to recognize that lubricants are used extensively on drilling tool and drill string connections and these typically contain many reactive chemical constituents that could also affect hydrochemical measurements and analysis. As a general principle, testing and characterizing the lubricants in a laboratory prior to use, and selecting them on the basis of their potential impact on chemical groundwater constituents and measurement quality, is an essential prerequisite to drilling a borehole for groundwater sampling.

Higher density drilling fluids that invade the borehole wall, which can potentially affect hydrogeochemistry, can also induce 'skin' effects. These effects may modify permeability measurements near the borehole, as they can cause blockages along hydraulic pathways. The driller should ensure skin effects are reduced as much as possible through the use of approved drilling fluids to ensure high quality groundwater pressure data are collected from a borehole.

Drilling costs

Drilling is expensive and rigs come in a range of sizes that are rated to drill under different ground conditions to various depths. The power necessary to drill down to a specified depth with a prescribed borehole diameter will vary in relation to the rock type being drilled and this would be reflected in the size and complexity of rig needed and hence in the cost.

For hard rock environments, such as are found in Sweden, Finland and large parts of Canada, quite small and highly mobile rigs may be used to drill relatively narrow diameter investigation boreholes to depths of several hundred metres. For example, in the granitic rocks investigated by SKB in Sweden, a 76 mm drill hole provides a core diameter of approximately 50 mm and this was considered to be adequate for the majority of geological logging needs and for obtaining core samples for laboratory tests (Fig. 28 (a)). Conversely, for investigations in weak or friable sedimentary environments prone to instability, such as those encountered in cover rocks at Sellafield in the UK, 160 mm diameter boreholes were required for obtaining 95 mm core. These boreholes were drilled down to 2000 m depths and expensive oilfield type drilling technology was required using large static rigs placed on specially constructed pads (Fig. 28 (b)). Wider diameter boreholes were also needed to enable the use of downhole testing, monitoring and geophysical logging equipment that would not be suitable in slim-line drill holes.

Borehole depth and diameter

For a large site investigation project, it would be difficult to specify in advance the depths, orientations (vertical, inclined or deviated), diameter and total number of boreholes that might be required at a site. Instead, such judgements would depend on an ongoing assessment of the site conditions and periodic revisions to data requirements in line with the developing site understanding. However, for a major site investigation project undertaken in a relatively simple geological environment,[19] it might be expected that the number of boreholes required would be in the order of ten to a few tens and ranging in depth from shallow (<100 m) to deep (1000 m depth or more). Despite the difficulty of anticipating precisely what drilling effort will be required, some assumptions will have to be made during the planning phase about the number and depths of boreholes to plan a programme and establish provisional cost estimates.

A borehole design should be determined by the objectives set for the borehole and knowledge about expected subsurface conditions, and possibly by regulatory constraints. On this basis, one can specify the design characteristics to reflect the types of sampling and testing that are to be carried out in the boreholes and the types of instrumentation that are to be installed. In a situation where there is no knowledge of the subsurface at or beyond proposed repository depths, at least one borehole should be targeted with completion deeper than the repository. More deep boreholes might be necessary where potentially significant features or conditions are recognized from geophysical imaging or interpretations (e.g. major faults or possible up-coning of saline groundwater).

When specifying a borehole, a critical design dimension is the diameter at the base of the borehole. This diameter is established to ensure the required sampling and testing can be carried out down to the base of the borehole, that the required instrumentation can be installed and, once the borehole is no longer required, that it can be effectively sealed and abandoned. Hence, the need to run geophysical logging tools, undertake particular tests (e.g. stress measurements by over-coring), obtain samples of a defined

[19] The definition of a relatively simple geological environment is relative and highly subjective but might include, for example, a regionally extensive granite batholith or a horizontally layered sedimentary formation containing thick and laterally extensive clay horizons.

FIG. 28. Drilling rigs used during (a) the Forsmark site investigations in Sweden, and (b) the Sellafield site investigations, UK (reproduced courtesy of SKB and NWS).

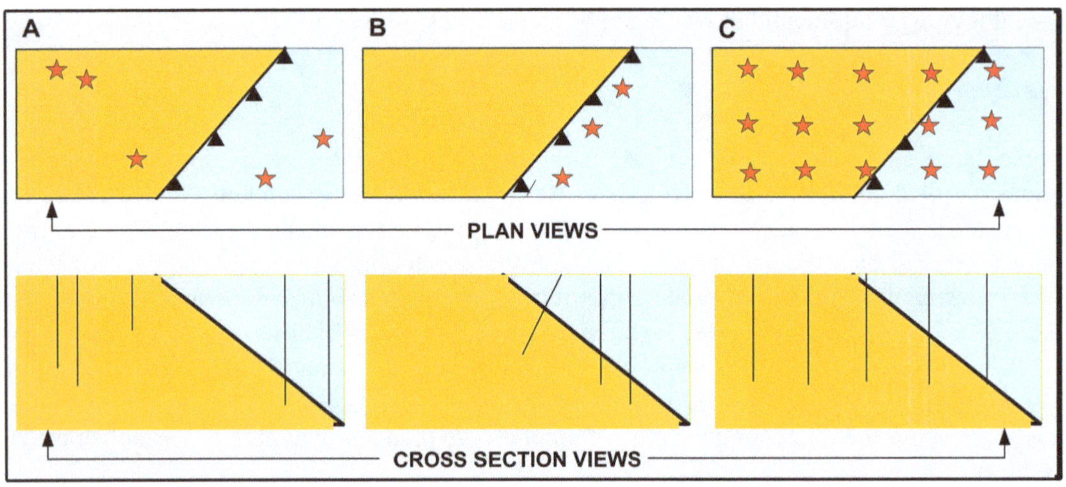

FIG. 29. Plan and cross-section views of random (A), targeted (B) and structured (C) borehole locations.

size, or install particular instrumentation will generally constrain the design of the borehole with regard to its diameter.

Drilling rig use

Depending on the number of drill rigs needed to complete an investigation, their commercial availability (for hire or purchase) and the amount of funding available, a plan should be established for rig use. For a large scale investigation, the high level strategic alternatives to be considered might include those shown in Table 7. Clearly, for such a major cost element, the strategy to be employed regarding the use and procurement of drilling rig services should be very carefully thought out, especially because the required equipment might be scarce in the marketplace and because of long lead times. Most repository site investigation projects undertaken to date have tended to employ a strategy of hiring drill rigs and sometimes several at the same time, depending of the scale of investigations. It will be imperative that the RWMO maintain control over the drilling operations by being an educated customer and by having staff on hand to provide oversight and make immediate decisions as required. This will of course mean that staff have to be delegated with sufficient authority to make such decisions.

Borehole locations

There are essentially three alternative approaches that may be used to determine the location of boreholes to be drilled as part of site investigations: random, targeted and structured (Fig. 29).

In Fig. 29 we see a plan view (top set) and a cross-section (bottom set) of an area with geological units shown in yellow and blue. A reverse fault is indicated in the plan view (tooth marks on hanging wall) and borehole locations are shown as stars in the plan views and as vertical and subvertical lines in the cross-section views. The line of cross-section is towards the bottom of the top set of illustrations, although borehole locations have been migrated to coincide with the line of section. The different approaches, or strategies, to locate boreholes can be described as follows:

- *Random:* While not necessarily absolutely random, the approach to selecting drilling locations within a site area shown in Fig. 29 (a) is typically based on little or no a priori subsurface information and is employed without the use of any guiding principles or without objectively discriminating between potential drilling locations. In this manner, typical or atypical rock types and structures have an equal chance of being identified, although, unless the methodology for selecting a location is truly random, there will always be a degree of bias.
- *Targeted:* The approach shown in Fig. 29 (b) ensures boreholes are located to investigate particular features and/or rock conditions that are recognized but not yet fully characterized. The type of features that might be targeted include faults (to investigate their impact on groundwater flow) while rock conditions to be investigated might include anomalous or typical domains within an investigation area, as identified from resistivity or seismic studies (typical domains are those considered likely to be representative of a repository location). This is an efficient approach but it depends on relevant and reliable a priori knowledge.
- *Structured:* In the approach shown in Fig. 29 (c), the prospective locations of boreholes are set out in a defined array (pattern) to methodically investigate rock, mineral and/or groundwater characteristics using geostatistical or other methods intended to investigate spatial variability in a systematic manner.[20] The approach is typically used in ore resource estimation when the spatial correlations of rock properties need to be understood with a high degree of confidence. The strength of the approach is that it limits bias during the selection of borehole locations, but the major weaknesses are cost and potential inefficiency (variability might be established with fewer boreholes or the variability of important properties might reflect smaller scale characteristics than can be captured by the borehole spacing that is used).

Each of the three strategies has potential benefits and disadvantages, as shown in Table 8. The approach selected for identifying borehole locations should be based on a clear understanding of technical objectives and logistical constraints, such as planning permission, funding and ease of access. However, it is important to recognize that the approach will likely be modified as an investigation proceeds and increased site understanding is obtained. Thus, in the absence of any pre-existing data, a random approach might be initially employed, but a more targeted approach then typically develops. Site investigations for an underground repository have rarely used a structured approach requiring a dense array of deep boreholes, mainly due to cost considerations but more importantly because safety assessments can usually be undertaken satisfactorily using the data available from targeted approaches. Where multiple boreholes are required, the order of drilling should be prioritized on a rational basis to reflect data needs and the availability of drilling equipment.

[20] Geostatistics is a special form of spatial analysis that differs from classical descriptive statistics.

TABLE 7. ALTERNATIVE STRATEGIES FOR DRILLING RIG USE AND OWNERSHIP

	Use strategy	Pros	Cons
A	RWMO to purchase one rig to drill boreholes as needed. Drilling staff could be directly employed by RWMO or contracted.	Rig would be available at all times and could respond rapidly to revised priorities, giving complete flexibility to integrate drilling needs into other aspects of the programme. Relatively high confidence in achieving a defined schedule. Close association of rig staff with the RWMO means they would appreciate significance of QA and potential impacts on safety.	High capital costs, depreciation and running costs, including need for specialist staffing. Idle staffing during borehole testing and turnaround times. Drilling of one borehole could take several months, therefore for a large drilling programme requiring many boreholes the duration at one drill site may become unreasonable. Potential issues regarding staff redundancies (if employed) and rig disposal after the drilling programme. Breakdowns could cause a major bottleneck. Different types of rig may be needed for different ground conditions that might exist at the site.
B	Contract one rig and crew from a on a long term basis to drill boreholes continuously until a set number of boreholes have been completed or for a set period of time (campaign).	Rig would be available at all times during a campaign and could respond rapidly to revised priorities. Some confidence in achieving a defined schedule. No responsibility for maintenance.	Mobilization and demobilization costs for each campaign. Drilling one borehole could take several months. Contract variation and additional costs may be needed to respond to revised needs. Different rigs may be needed for different ground conditions that might exist at the site. Drill rig teams may change, and each would have to learn significance of quality and potential impacts on safety.
C	Purchase several rigs to drill concurrently and move on to new drill sites as needed.	As in option A. Also, different rig capacities would optimize effort when different types of borehole are required. More rapid completion of drilling campaigns. If one rig becomes unavailable (breakdown), adjustments can be made to reschedule with lower impact than when using one rig alone.	As in option A's cons, except that there would be a more rapid campaign due to parallel drilling. More drilling staff required and very high capital and operational expenditure.
D	Hire several rigs to drill several boreholes in parallel, then move the rigs to new borehole sites as needed. Rigs would be operated continuously until all boreholes were completed, then demobilized at the end of a campaign.	Rigs would be available at all times during a campaign and could respond rapidly to revised priorities. Reasonable to high confidence in achieving a defined schedule. Different rig capacities would optimize effort when different types of borehole are required. No responsibility for maintenance.	Mobilization and demobilization costs for each campaign (a continuous drilling programme is unlikely to be possible). Contract variation and additional costs may be needed to respond to revised needs. Requires good coordination and possibly greater on-site presence of RWMO staff to reduce potential for lack of direct control of operations (with resultant impacts of health and safety, quality, post-closure safety, etc.).

TABLE 8. ADVANTAGES AND DISADVANTAGES ASSOCIATED WITH ALTERNATIVE STRATEGIES FOR LOCATING INVESTIGATION BOREHOLES

Approach	Pros	Cons	When to use
Random	Used when there is little pre-existing information at a site or in the absence of any information at all. A randomly located borehole may fortuitously intercept an unknown feature or zone of interest.	Wasteful use of resources when data could be analysed to identify features to target. Without other information to support interpretations, there could be a low level of confidence in the degree to which data are representative of wider conditions.	Reconnaissance stage or later on to specifically investigate for the presence of potentially significant features or a 'typical' rock volume that might be considered as an average geosphere for repository conditions.
Targeted	Allows focused and efficient investigation of specific features or conditions.	None early on in an investigation, except that it requires a priori information. Later on, targeted boreholes may provide diminishing returns.	As soon as potentially relevant targets are recognized and their impact on post-closure safety or repository design has to be evaluated.
Structured	Allows geostatistical evaluations that may provide a very good insight into the spatial variability of rock properties.	Expensive in time and cost. Borehole spacing may not sufficiently capture the variability of important properties. Extensive drilling into the geosphere could compromise the barrier function.	Generally not recommended for siting deep boreholes as part of repository investigations.

Borehole sealing

On completion of the preconstruction phase of a site investigation project, it is prudent to seal boreholes at sites not selected to host a repository very soon after a decision to discard the site, in accordance with the guidance published by the relevant regulatory authority. Boreholes located on a site selected to host a repository may also be sealed at an early stage after site confirmation, although some or all boreholes may be left accessible for long term monitoring purposes.

After they have fulfilled their purpose, boreholes need to be sealed to ensure they do not provide a preferential pathway for surface water to quickly reach groundwater bodies or for non-potable (e.g. saline or brackish) groundwater, if present, to move upward from depth and enter a freshwater aquifer or reach the surface under artesian conditions. Inadequately sealed boreholes at a disposal facility site may also potentially act as a fast pathway for the transport of any radionuclides that might be released from a repository far in the future. Consequently, major efforts should be made to ensure effective sealing [46–49]. The functions required from the seals should be specified early on and measures established so conformance can be demonstrated. Once the borehole has been drilled, these plans should be revisited and, with knowledge of the disposition of rock formations and structural features, the plans can be advanced or revised, particularly regarding the locations of seals (compared to borehole sections to be occupied by other backfilling materials).

Whether at a repository site or one previously screened out for further consideration, experience of sealing and abandoning deep boreholes has clearly indicated that the removal of instrumentation from redundant boreholes can present significant practical difficulties. Casing that has been in place for some time or at depth can also become difficult to remove for a number of reasons. When investigation boreholes are to be sealed at a site selected to host a repository, it will very likely be a requirement that casing is removed, at least over the zones designated to be sealed tightly. This is because carbon steel and cement will corrode and degrade, and fast pathways could be formed in the voids that result. Consequently, sealing materials (such as bentonite) should be set to expand tight against the rock wall face. In addition, in some Member States regulatory guidance demands that even away from a repository site, any borehole casing that has significantly corroded is required be removed from the borehole prior to sealing if such corrosion could potentially allow for the contamination of freshwater aquifers. In designing boreholes, it is therefore essential to have due regard to the manner in which those boreholes could subsequently be sealed and abandoned, should the need arise.

6.4.2. Downhole geophysical logging

Downhole geophysical logging uses a variety of probes and sensors lowered into a borehole by a wireline cable from the surface to make in situ measurements of formation properties at high spatial resolution. Many proprietary logging tools were originally developed for use in the oil industry, requiring use in wide diameter boreholes and many were specially engineered to withstand the aggressive environments commonly encountered under oilfield conditions. Consequently, they were relatively difficult to obtain and expensive to use. Today, wireline geophysical logging is commonplace in mining, construction, geothermal investigations and in the water resource sector. Consequently, the costs associated with investigations using geophysical logging tools are relatively reasonable, compared to in the past. Given this factor, it would be advisable for a site investigation manager to plan for geophysical logging in most of the boreholes to be drilled at a site, especially where it has not been possible to obtain rock core.

Many of the geophysical logging instruments typically applied in site investigations for a potential repository are suitable for use in a wide range of borehole diameters. They are robust, fast and relatively easy to use, allowing rapid and comprehensive rock mass characterization to variable depths beyond the borehole rock face into the formation, from several centimetres to many metres depending on the tool. Some instruments can only be used in a column of water and some tools require an uncased borehole to allow measurements to be taken.

Depending on the technique and equipment employed, tools may make continuous measurements of rock mass properties or may provide measurements at point locations. Some equipment employs dedicated sensors for measuring only one type of parameter, while other tools comprise multiple sensors that can be operated in combination, allowing the possibility of recording many different geophysical parameters during a single logging run. Overlaying the results of multiple geophysical methods may be of great value, providing synthetic attributes that confirm, reinforce and advance interpretations of rock mass properties (see for example, Fig. 30).

Table 9 lists the main geophysical logging tools that might be used in a site investigation project and the data that might be acquired. A more extensive overview of downhole geophysical techniques and their interpretation is provided in Ref. [50]. Geophysical logging would generally be undertaken in most if not all boreholes that are drilled at a site, as the information they provide is significant and can be used to not only characterize the immediate rock mass but, by correlating properties (e.g. derived from core) or through calibrating other methods that image the rock mass such as seismic resistivity and attributes (see Fig. 27), they provide for interpolations and extrapolations between and beyond boreholes.

Several tools are of particular relevance to site investigations in rock environments dominated by fracture flow, as they allow discrete fractures to be imaged and hydraulic conductivity to be estimated. These include full-bore FMI and acoustic borehole televiewer logs for fracture identification (Fig. 31) and full waveform acoustic logging to facilitate Stoneley wave analysis allowing the estimation of formation permeability as well as the presence of fractures. In combination with the results of hydraulic testing and in some cases augmented by mineralogical studies from core, the subset of the total fracture population that may be responsible for transmitting water might be derived. The important fracture characteristics to be captured would include orientation, aperture, frequency, length, connectivity and transmissivity. Then, based on these parameter values, statistical models of discrete fractures can be developed for use in advective flow and radionuclide transport codes.

NMR logging is a geophysical logging technique. Although currently largely confined to oilfield exploration, it is finding wider application in site investigations for radioactive waste repositories. The

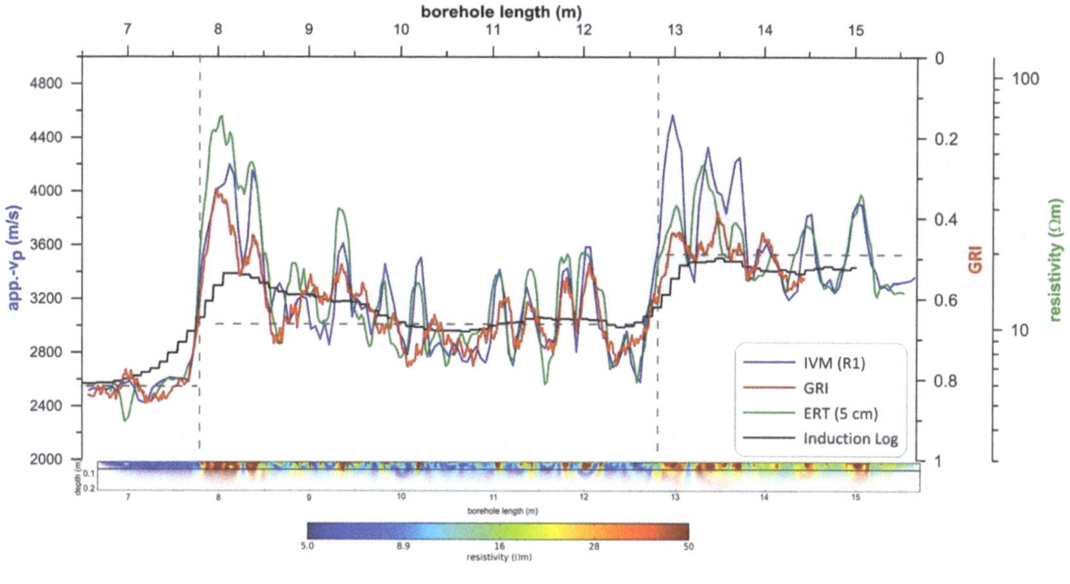

FIG. 30. Comparison of interval velocity measurements (seismic/ultrasonic P- and S-wave velocities) providing information on dynamic elastic parameters (e.g. Poisson's ratio, Young's modulus, shear modulus), ERT (2-D electrical resistivity tomography), GRI- and conventional data. These attributes can be used for characterization of the borehole excavation damaged zone and other features such as fractures to be used in anisotropy studies, flow modelling and identifying major and minor stress orientations in local rock (reproduced courtesy of the Federal Institute for Geosciences and Natural Resources, Germany).

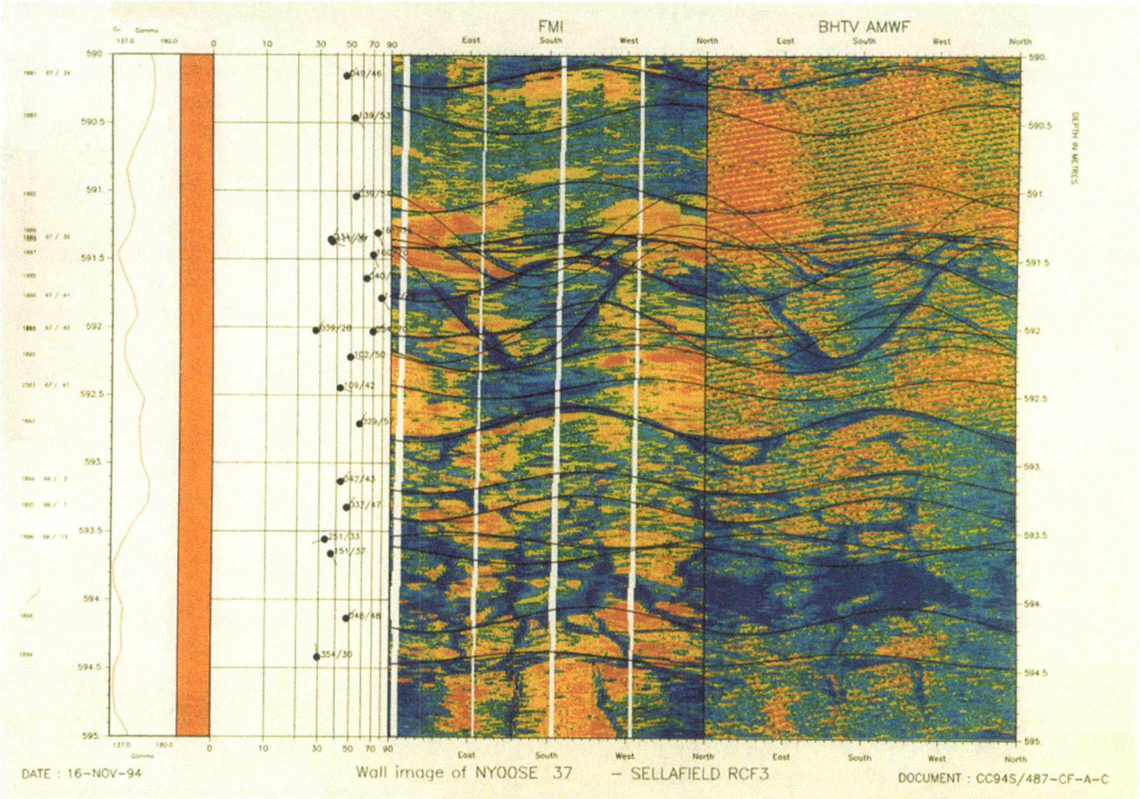

FIG. 31. An example of borehole fracture images derived from geophysical logging using FMI and borehole televiewer tools. Planar features intercepting the borehole are shown as sinusoidal waves due to the unfolding of the cylindrical borehole wall to produce a 2-D image (reproduced courtesy of NWS).

technique is based on the transmission of an electromagnetic signal of a specific frequency in the presence of a strong magnetic field and recording the corresponding resonance of hydrogen protons in groundwater as they become excited and relax. Inversion of the measurements enables direct inference of the bulk hydraulic conductivity based Xon estimated water content and an interpretation of the mean pore throat size within a matrix dominated (interstitial) flow system.

TABLE 9. SUMMARY OF TYPICAL INVASIVE GEOPHYSICAL LOGGING TOOLS AND TECHNIQUES USED FOR DATA ACQUISITION DURING SITE INVESTIGATIONS

Name of technique	Description	Data acquired	Stage	Comments
Acoustic borehole televiewer (known as BHTV)	Generates continuous images of borehole wall using an ultrasonic beam and rotating mirror. Images are oriented by means of accelerometers, magnetometers or gyros permitting dip and strike measurements of features (structures and bedding planes). Both amplitude and arrival times are measured, providing two images.	Efficient collection of downhole structural data especially fractures. Can be viewed in 3-D as 'virtual core.' Provides measures of borehole shape for geotechnical studies.	DSC	Operates in water filled boreholes only. Quality of image influenced by roughness of the borehole wall.
Borehole camera	Generates continuous optical images of the borehole wall using a high resolution camera.	Continuous measurements. Provides pictures of the borehole shape and texture.	DSC	Operates in air or clear fluids.
Spectral gamma ray sensor	The natural gamma ray activity of ^{40}P, ^{238}U and ^{232}Th is measured. The distribution of the activity is characteristic for different rock types.	Continuous measurements. Counting rate.	DSC	Operates in air, fluids or casings. Bulk radioactive emissions may also be recorded using a simple gamma ray logger.
Density log	A gamma ray source is emitting radiation into the rock. A detector, shielded from direct radiation, is measuring the scattered gamma radiation using the Compton effect which depends on the density of the rock.	Continuous measurements. Counting rate.	DSC	The depth of investigation is in the range of 1 dm.
Electrical resistivity log or conductivity log	An electric current is introduced into the rock using electrodes. Other electrodes at various distances to each other measure the received voltage which is proportional to resistivity or conductivity in that distance range. Includes high definition formation microimaging (FMI).	Continuous measurements of electrical conductivity and resistivity.	DSC	Conductive fluids or casings have influence to the result.
Neutron–neutron probe	With a neutron generator or ^{252}Cf as source, a detector is measuring the neutron radiation on collision with atoms of the rock. Measurement of the radiation resulting from neutron capture.	Counting rate is inversely proportional to porosity. Element concentration for each element. Hydrogen sensitive.	DSC	Calibration necessary. Resolution in dm range.

TABLE 9. SUMMARY OF TYPICAL INVASIVE GEOPHYSICAL LOGGING TOOLS AND TECHNIQUES USED FOR DATA ACQUISITION DURING SITE INVESTIGATIONS (cont.)

Name of technique	Description	Data acquired	Stage	Comments
Sonic log	An acoustic source generates a longitudinal wave along the borehole wall. The travel time over a defined distance is recorded and is proportional to rock porosity.	Compressional and shear wave interval velocities can be determined.	DSC	Sensitive to borehole diameter changes.
Vertical seismic profile	A seismic source generates acoustic waves at the surface and geophones recording amplitude and travel time along the borehole.	Stepwise measurement in discrete depths intervals.	DSC	—
Caliper log	A wireline tool that uses up to six independent articulated mechanical arms that run along the borehole wall as the tool is moved up or down or acoustic (ultrasonic) methods.	Continuous measurements of the diameter of the borehole, its shape with depth and deviation from vertical.	DSC	—
Self-potential log	The natural electrical potential is measured between an electrode at ground surface and an electrode in the borehole drilling fluid.	Continuous measurements provide indications of shale volume and freshwater horizons.	DSC	Operates in fluid filled boreholes, but not with oil based drilling mud. Sensitive to electrical noise.
Magnetic susceptibility log	A transmitter coil generates an alternating magnetic field around the borehole, which induces electrical eddy currents that are proportional to the conductivity of the rock. The in-phase part is proportional to susceptibility.	Continuous measurements.	DSC	Resolution in m range. Calibration with known rock samples.
Dual laterlog tool	Analogue to resistivity log. Additional bucking electrodes focus the current to right angles to the logging tool. The vertical resolution is high.	Continuous measurements.	DSC	Thin layers can be detected.

Note: DSC = detailed site characterization (stage).

6.4.3. Hydrogeological testing

Hydrogeological testing is a general term for various types of test that provide insights into the present day nature of groundwater flow and solute transport within a rock mass, including the magnitude of flow and its orientation, among other properties. These tests include pumping tests (hydraulic tests) and flow logging, both of which are carried out in boreholes; tracer testing, which can be carried out in the field or in a laboratory; and other specialist testing undertaken on rock and groundwater samples in a laboratory. There are numerous references providing information on hydrogeological testing principles and methods, as well as the physical constants and properties needed for flow and transport calculations [51–55]. Table 10 presents some of the key physical properties that can be determined by hydrogeological testing in the field.

This section summarizes some of the basic principles concerning field based hydraulic testing, tracer testing and flow logging methods. It also introduces some important management considerations. Laboratory based methods using rock and groundwater samples follow in Section 6.5. It is acknowledged this section focuses primarily on investigations relating to the saturated (phreatic) zone at a site, in which groundwater at greater than atmospheric pressure permanently resides within accessible pore space. Research methods involving data acquisition and process understanding in the unsaturated (vadose) zone above the groundwater table at a site are not covered in detail. However, it will be important to investigate flow and transport processes in the unsaturated zone, especially for near surface disposal concepts or geological disposal concepts with waste emplacement in the unsaturated zone (as previously considered at Yucca Mountain, USA [56]), but also for understanding the nature of the present day geosphere–biosphere interface, demonstrating the robustness of numerical flow and transport models and as an analogue for deriving exposure scenarios under future conditions.

During the early stages of an investigation, some of the parameters described in Table 10 are typically considered as constant across a formation or a hydrogeological unit, but in reality they are all likely to vary in space and may also change as a function of time in response to changing boundary conditions. Consequently as investigations proceed, it may be appropriate to ensure spatial variability is addressed by taking several measurements at different locations, as well as ensuring uncertainty estimates are carried out as part of the testing activities, depending on the requirements of the investigation.

The objectives and scope of information required from a hydrogeological work programme are provided in Section 3.6.2, but it is emphasized that a key goal when collecting hydrogeological information is to provide insights into the potential migration of radionuclides that might be released from a repository system at some time far in the future. Therefore, a good understanding of water flow and solute transport mechanisms at a site will be required and this should be set within a framework of the geology and hydrogeological units, populated by appropriate physical properties and parameter values and defined by appropriate boundary conditions. The physical scale of the hydrogeological investigation should reflect the scale over which local hydrogeological processes operate and over which radionuclides might migrate. In undertaking hydrogeological investigations, it is important to recognize that a holistic view of the system is required, and this means drawing on understanding from other disciplines, with a special emphasis on hydrogeochemistry and water–rock interaction studies. Consequently, a multidisciplinary and multiscalar approach is needed, with regional scale studies supporting an understanding of the nature and extent of the local groundwater system and its controls, possibly focusing down to studies at the molecular scale at a site. Furthermore, as the predominant concern is with the potential migration of long lived mobile radionuclides from a future repository, scenarios for the evolution of the present day groundwater system have to be postulated and tested, including assessments of the potential for radionuclide movement, retardation, dispersion, dilution and the nature of future potential pathways (types of pathway, path length, timescales for transport and discharge locations). These are to be based not only on an understanding of the present-day groundwater system, but also on how the system has evolved to its current condition, which is the focus of palaeohydrogeology, a topic studied in several international disposal programmes [57].

Hydraulic tests

Hydraulic tests are used for understanding the nature of groundwater flow in the geosphere and for determining the significance of properties that impact groundwater storage and movement, such as storativity, transmissivity and connectivity. The values assigned to these parameters reflect the nature of the rock itself (i.e. the composition of the rock and its structure in terms of active fluid pathways) and of the groundwater, density and viscosity in particular. The presence and nature of any hydraulic boundaries may also be deduced, for example barrier or recharge boundaries. This information is of critical value in the context of a repository siting project because it supports the development of robust conceptual models of the flow system, incorporating process understanding over the spatial domain of interest, and provides data for the parameterization and calibration of groundwater flow and transport models. Hydraulic data and information are also critical inputs during the design and construction phases of underground workings, to provide

TABLE 10. EXAMPLES OF ROCK PROPERTIES DETERMINED BY HYDROGEOLOGICAL TESTING IN THE FIELD

Property	Description	Notation	Dimensions
Porosity	Voids in a rock may be connected or isolated and may contain water or a gas. Total porosity reflects the proportion of a volume of rock made up of pores or other types of void space and it is calculated as the volume of total void space divided by the total volume of the bulk solid. This is distinct from effective porosity, which is the volume of interconnected void space available for flow divided by the volume of the bulk solid.	n (n_{Σ} & n_e)	Dimensionless (usually shown as a percentage or decimal fraction)
Intrinsic permeability	A property of the rock alone that describes the ease with which a fluid is able to pass through a saturated medium. It is related to hydraulic conductivity by $k = K\frac{\mu}{pg}$ where μ is dynamic viscosity, ρ is fluid density and g is gravitational acceleration.	k	L^2
Hydraulic conductivity	A property of the rock and the groundwater that describes the ease with which groundwater is able to flow. It is defined as the volume of water that is moving through a porous medium in unit time under a unit hydraulic gradient, through a unit area measured at 90° to the direction of flow. The hydraulic conductivity of a fractured hard rock reflects the aperture of the flowing fractures, their connectivity and density.	K	L/T
Transmissivity	The product of hydraulic conductivity and the saturated thickness of a permeable rock mass.	T	L^2/T
Specific storage	The volume of water in a unit volume of a confined saturated permeable rock that is released from storage under a unit decline in hydraulic head. It reflects the compressibility of the rock mass (i.e. the unit change in the volume of rock and its contained water per unit of pressure variation applied under constant temperature).	S_s	L–1
Storativity	In a confined volume of permeable rock, the amount of water released from storage per unit surface area of the aquifer, or aquitard per unit decline in hydraulic head. It is the product of specific storage and the saturated thickness of the confined rock under consideration ($S = S_s b$). In an unconfined aquifer, storativity is the product of specific yield and specific storage ($S = S_y \times S_s b$), but the value is very close to the value for specific yield because there is relatively little aquifer compression and water expansion under unconfined conditions.	S	Dimensionless

TABLE 10. EXAMPLES OF ROCK PROPERTIES DETERMINED BY HYDROGEOLOGICAL TESTING IN THE FIELD (cont.)

Property	Description	Notation	Dimensions
Specific yield	The volume of water that a unit area of unconfined aquifer is able to release from storage per unit decline in the water table elevation due to gravity drainage. It reflects the proportion of void space able to yield water and therefore approximates effective porosity.	S_y	Dimensionless
Hydraulic diffusivity	The ratio of the transmissivity and the storativity of a saturated rock mass (also the ratio of hydraulic conductivity and specific storage). It reflects the rate at which groundwater pressure changes are able to propagate through the rock mass.	D	L^2/T
Hydraulic gradient	Groundwater moves from high hydraulic head to an area with lower hydraulic head. This loss in head (hL) describes a vector gradient along a flow path of unit distance (L): $hL/L = i$	i	Dimensionless (usually shown as a decimal)
Hydraulic head	Also called piezometric elevation, the hydraulic head is a point measurement of groundwater potential, expressed as an elevation in relation to a reference datum. It reflects the height pressurized water would attain in a piezometer and, as such, hydraulic head can also be considered as the total mechanical energy available to a unit weight of fluid. The equation for hydraulic head comprises three components within a simplified expression of the Bernoulli equation that contains terms for pressure, potential energy and kinetic energy. In laminar groundwater flow, kinetic energy can be ignored, so: $h = P / \rho g + z$ (where P is fluid pressure and z is the elevation). Note that hydraulic head varies with the density of water in a water column and this is to be taken into account in any calculations involving variable density flow.	h	L
Dispersivity	A property of the rock and groundwater that varies with the scale of measurement and describes the degree to which a solute spreads laterally away from the principle direction of groundwater flow, while mixing along the advancing solute front in the flow path. It is caused by differences in groundwater velocity due to channel friction and connected void aperture size variation, and alternate solute flow paths due to tortuosity (together termed *mechanical dispersion*). Dispersivity is a function of a coefficient of mechanical dispersion and of average linear velocity. It is evaluated parallel to flow (longitudinal dispersivity) and perpendicular to flow (transverse dispersivity).	α (αL & αT)	L

TABLE 10. EXAMPLES OF ROCK PROPERTIES DETERMINED BY HYDROGEOLOGICAL TESTING IN THE FIELD (cont.)

Property	Description	Notation	Dimensions
Hydrodynamic dispersion	A property that describes the spread of a tracer due to mechanical dispersion and molecular diffusion. It is represented in transverse and longitudinal directions in relation to the main flow direction.	D (D_L & D_T)	L^2/T

advance information about expected ground conditions and to ensure safe operations. Finally, knowledge about the hydraulic nature of a site allows the RWMO to ensure potential environmental impacts are avoided or minimized due to interactions at the geosphere–biosphere interface, which may otherwise result in the derogation of water supplies and water quality in the ground or on the surface.

Many hydraulic testing and interpretation methods were originally designed to investigate water resource potential in aquifers, permeable domains within a rock mass displaying relatively high groundwater flow rates and the potential for releasing large volumes of water from storage [51]. Other testing methods have been designed for application in geotechnical investigations and for oil and gas exploration. More recently, specialist tests and interpretation methods have been developed to investigate fractured low permeability rocks that might be considered for hosting radioactive waste repositories. Table 11 provides a summary of selected hydraulic testing methods.

Hydraulic parameters measured in the field are derived from tests using controlled pressure stimuli imposed on a rock mass to induce stress in the groundwater system, either by pumping groundwater out of a borehole (abstraction) or by pumping in water or a gas under pressure (injection). Injection tests are especially applicable for testing low transmissivity formations and can be impractical for testing higher transmissivity rocks. Testing can be undertaken in an open uncased borehole or may be targeted on hydraulically isolated sections within a borehole using one or more inflatable packers.[21] A pump may be situated on the surface or a down-the-hole pump with sufficient power can be used. Testing may involve several hours, days or even weeks of pumping and monitoring. Careful, detailed measurements of the water table movement and/or groundwater pressure responses to abstraction or injection rates are made in the control borehole being pumped and, if present, in any surrounding observation boreholes set at different distances and orientations away from the pumped site. Under tight financial, logistical or time constraints, or where multiple boreholes are not necessary, a single borehole may be used for pumping and monitoring groundwater responses. However, the volume of the rock mass able to be investigated and the confidence in the hydraulic data obtained is more limited in this situation than where observation boreholes are also employed.

Groundwater pressure changes are most commonly achieved by pumping water out of a borehole at an abstraction (discharge) rate that is set to produce a pronounced drawdown in the water level in the borehole. Water extraction may be applied almost instantaneously as a ‘slug’ or over time, either at a constant rate or at a variable rate intended to maintain a constant head. The abstraction rate can also be applied as a stepwise series of varying constant discharge rates (a step test). In an extraction test the perturbed groundwater level or groundwater pressure at a point reduces with time in comparison to ambient conditions, either as a smooth function or with a break in slope reflecting a pressure response to a change in formation conditions away from the immediate vicinity of the borehole (including ‘skin’ effects), or as a series of steps in relation to the pumping pattern. It is advantageous to monitor the change in the free water level (water table) or the groundwater pressure at a point in a control borehole and in any observation boreholes that may be available

[21] Sections of the borehole can be compartmentalized by inflating one or more ‘packers’ to create watertight seals. One packer is sufficient to isolate the basal section of a borehole, whereas a minimum of two are needed to adequately isolate a midborehole section (additional packers may be used to create guard zones either side of the interval of interest to ensure hydraulic isolation). Hydraulic testing and geochemical sampling can then carried out within the isolated interval of the borehole. Intervals can vary in length from less than a metre to hundreds of metres.

TABLE 11. SUMMARY OF SELECTED INVASIVE TOOLS & TECHNIQUES FOR HYDRAULIC TESTING

Name of technique	Description	Discharge (q) & hydraulic head (h) in pumped borehole	Data acquired	Comments
Constant rate	Groundwater is extracted at a constant rate from an open or isolated column of water in a borehole (A). The discharge rate is monitored over time in relation to the drawdown in the groundwater level (or pressure) observed in the pumped borehole and any observation boreholes (B). Test duration should reflect time required to reach a quasi-steady drawdown state. After pumping ceases, the rise in groundwater level or pressure back towards ambient conditions is recorded.	A 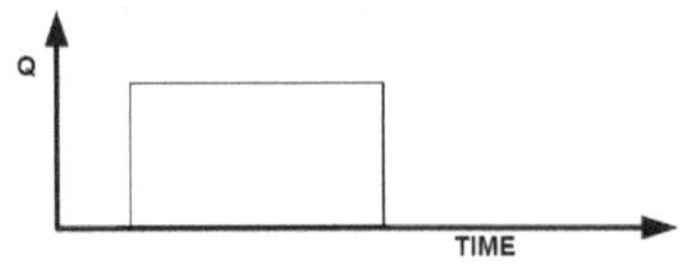 B 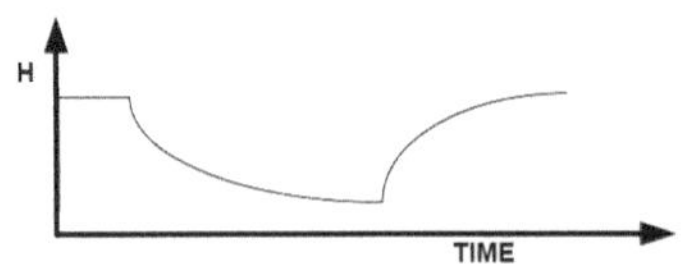	Hydraulic conductivity, transmissivity, storativity, specific yield, specific storage, hydraulic diffusivity and nature of boundary conditions. Hydraulic aperture in fracture flow systems.	Requires a good appreciation of likely conditions to plan the test appropriately. May be of long duration (several days to weeks). Achieving a constant discharge in a very low permeability environment may be difficult. Requires monitoring in observation boreholes for storage parameters. Dual permeability affects can be recognized from analysis. The accuracy of measurements reflects the resolution of the equipment or manual monitoring.
Constant head	Groundwater is extracted at an initially high rate from an open or isolated column of water in a borehole. The drawdown or hydraulic head (groundwater pressure) is observed in the pumped borehole and, once a predetermined level is reached (B), the discharge rate is reduced to maintain a constant drawdown level in the borehole (A). After pumping ceases, the rise in groundwater level back towards ambient conditions is recorded.	A 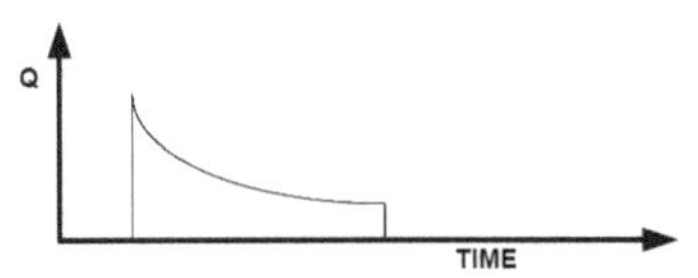 B 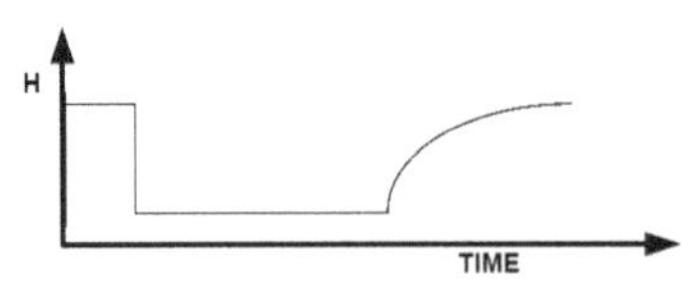	Hydraulic conductivity, transmissivity, storativity, specific yield, specific storage, hydraulic diffusivity and nature of boundary conditions. Hydraulic aperture in fracture flow systems.	Requires a good appreciation of likely conditions to plan the test appropriately. Requires monitoring in observation boreholes for storage parameters. Dual permeability affects can be recognized from analysis. The accuracy of measurements reflects the resolution of the equipment or manual monitoring.

TABLE 11. SUMMARY OF SELECTED INVASIVE TOOLS & TECHNIQUES FOR HYDRAULIC TESTING (cont.)

Name of technique	Description	Discharge (q) & hydraulic head (h) in pumped borehole	Data acquired	Comments
Step test	The borehole is pumped (extraction) at an initially low discharge rate until drawdown (pressure) stabilizes. The pumping rate is then increased until drawdown again stabilizes or for a period of time equal to the first step. This progressive equalization of flow and groundwater level or pressure is carried out several times until maximum drawdown is established (A). The pump is then turned off and the groundwater rise is monitored (B).	A 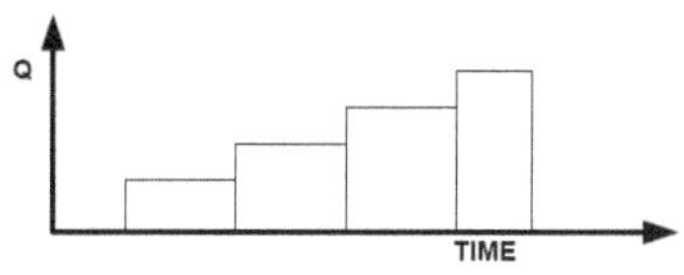 B 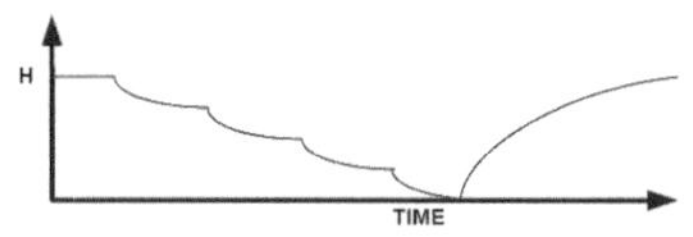	Hydraulic conductivity, transmissivity and well loss parameters. Hydraulic aperture in fracture flow systems.	This is a single borehole test that does not require monitoring in observation boreholes. Discharge at different rates should be for approximately equal periods of time, sufficient to approach pressure or groundwater level stabilization. A minimum of three steps is required. A variation used in fractured rock formations is a Lugeon test in which increasing constant-rate injection pressure is applied over a number of steps.
Slug test (rising head & falling head)	A slug test is a rapid injection or withdrawal test typically carried out in an isolated borehole section (A). A rising head test requires the rapid extraction of a large slug of groundwater to quickly lower the groundwater level in the interval being tested	A 	Hydraulic conductivity, transmissivity, storativity, specific yield, specific storage and boundary conditions. Hydraulic aperture in fracture flow systems.	During injection in a puckered interval (falling head test), the pressure stimulus should not be so great that it might induce hydraulic fracturing. Multiple rising head and falling head tests may be carried out in the same borehole interval. Results should be the same; any differences may reflect skin effects, fracture deformation or non-Darcy flow conditions.

TABLE 11. SUMMARY OF SELECTED INVASIVE TOOLS & TECHNIQUES FOR HYDRAULIC TESTING (cont.)

Name of technique	Description	Discharge (q) & hydraulic head (h) in pumped borehole	Data acquired	Comments
Slug test (rising head & falling head) (cont.)	(B), while a falling head test requires the addition of groundwater or a gas — typically nitrogen or air	B 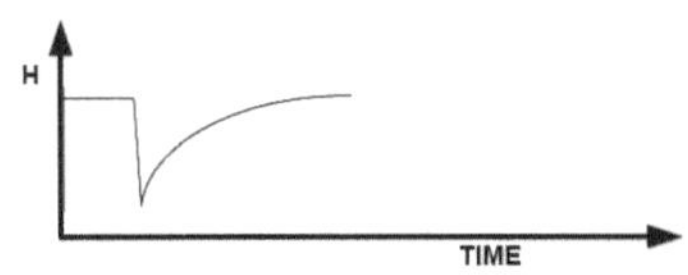		
	(C). The speed of the groundwater response will depend on the transmissivity of the formation and this also affects the duration of the test.	C 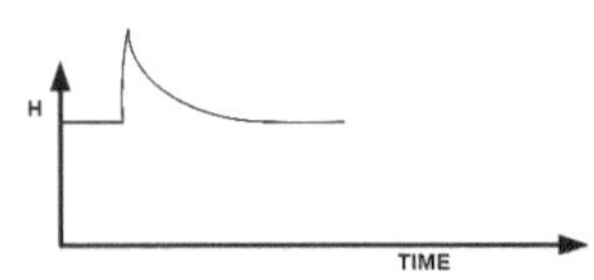		

during the application of the pressure stimulus and also during the subsequent pressure re-equilibration once the pressure signal is relaxed. This is because the groundwater level or pressure recovery can be separately analysed to augment test results achieved during the imposition of the pressure stimulus (groundwater extraction or fluid injection).

As water is drawn into the borehole from the surrounding formation, an artificial hydraulic gradient is induced. In an unconfined aquifer with a porous medium, the pumping creates a temporarily unsaturated zone termed a cone of depression that migrates outward until a quasi-steady state is achieved. Figure 32 shows an idealized field configuration for a drawdown test undertaken in a permeable rock mass with an unconfined aquifer.

When pumping in an aquifer confined by a porous medium, water would typically be drawn into a borehole from the adjacent connected network of pore space, with a low-pressure zone extending outward with time until there is pressure equilibrium or a hydraulic barrier is met. A layered sedimentary sequence is often anisotropic (horizontal flow greater than vertical flow) and therefore radial flow towards the borehole might be expected. In a fractured hard rock where pumping is carried out from an isolated section of a borehole, the nature of the induced flow towards the borehole would be dictated by the connectivity of the fracture network. This may theoretically be described as 1-D flow where groundwater flow is concentrated along a single channel, 2-D reflecting radial flow towards the borehole (typically horizontal), or 3-D whereby the volume of rock providing the induced flowing groundwater is spherical around the pumping interval (Fig. 33). In reality, flow in a fractured medium would often reflect an intermediate situation between 1-D and 2-D or between 2-D and 3-D. The flow dimension in a rock can be ascertained by appropriate testing and analysis [58].

As an alternative to reducing the hydraulic head locally during groundwater extraction, over-pressure can be applied to a hydraulically isolated zone in a borehole by injecting water or a gas into the zone, either as a single slug or in a series of steps (e.g. during a Lugeon test). Other variant tests commonly used in site investigations for underground disposal facilities include pressure pulse tests and drill stem tests. Such tests are especially useful for investigating fracture flow transmissivity in hard rocks where flow rates are low. Under suitable flow conditions, they can be carried out during drilling operations to test contiguous sections of a borehole immediately after they are drilled. In this situation active drilling is punctuated by periods when the drill string is withdrawn and testing equipment, comprising a downhole pump, pressure monitoring transducers and a single packer, is lowered on a tool string. The packer isolates a newly drilled interval from the temporary base of the borehole up to a known distance above the floor. The packer may be relocated within the borehole several times during a testing interval, to investigate the impact of the interval length on results. In order to minimize disruption to the costly drilling process, such testing has to be well planned and carried out as rapidly as possible, taking into account the investigation objectives and the site conditions.

A typical post-drilling test programme in low permeability rocks characterized by fracture flow might consist of a multipacker tool string positioned within the borehole at a series of depths, with pumping or injection undertaken from a central zone and monitoring of pressure responses carried out in zones above and below the pressure source interval. The objective of such testing is to investigate the hydraulic conductivity of individual fractures or groups of fractures and also the connectivity between sections of the borehole. Where pressure responses are also recorded in isolated intervals in one or more

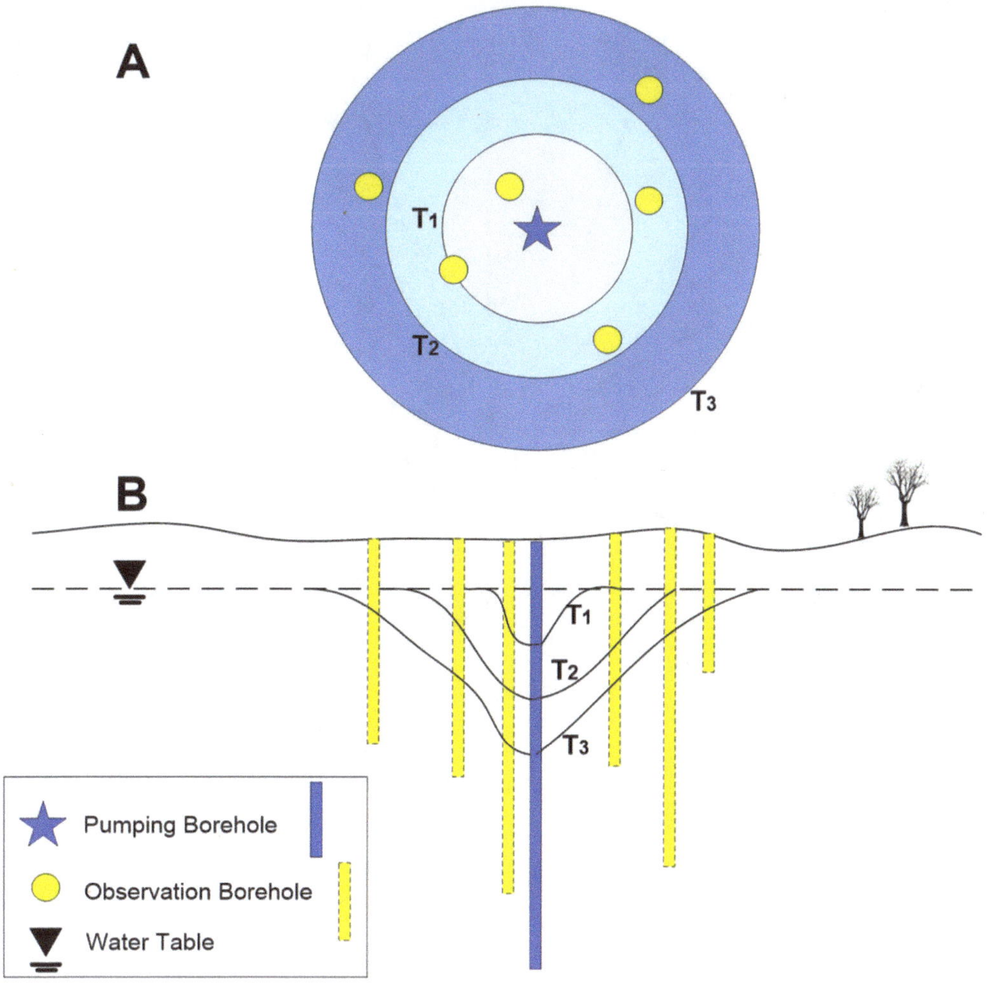

FIG. 32. Schematic (a) plan view and (b) cross-section showing the type of arrangement expected when conducting a long term pumping test in a porous medium. The cone of depression at times T1 to T3 is caused by the lowering of the water table in the vicinity of the pumped borehole.

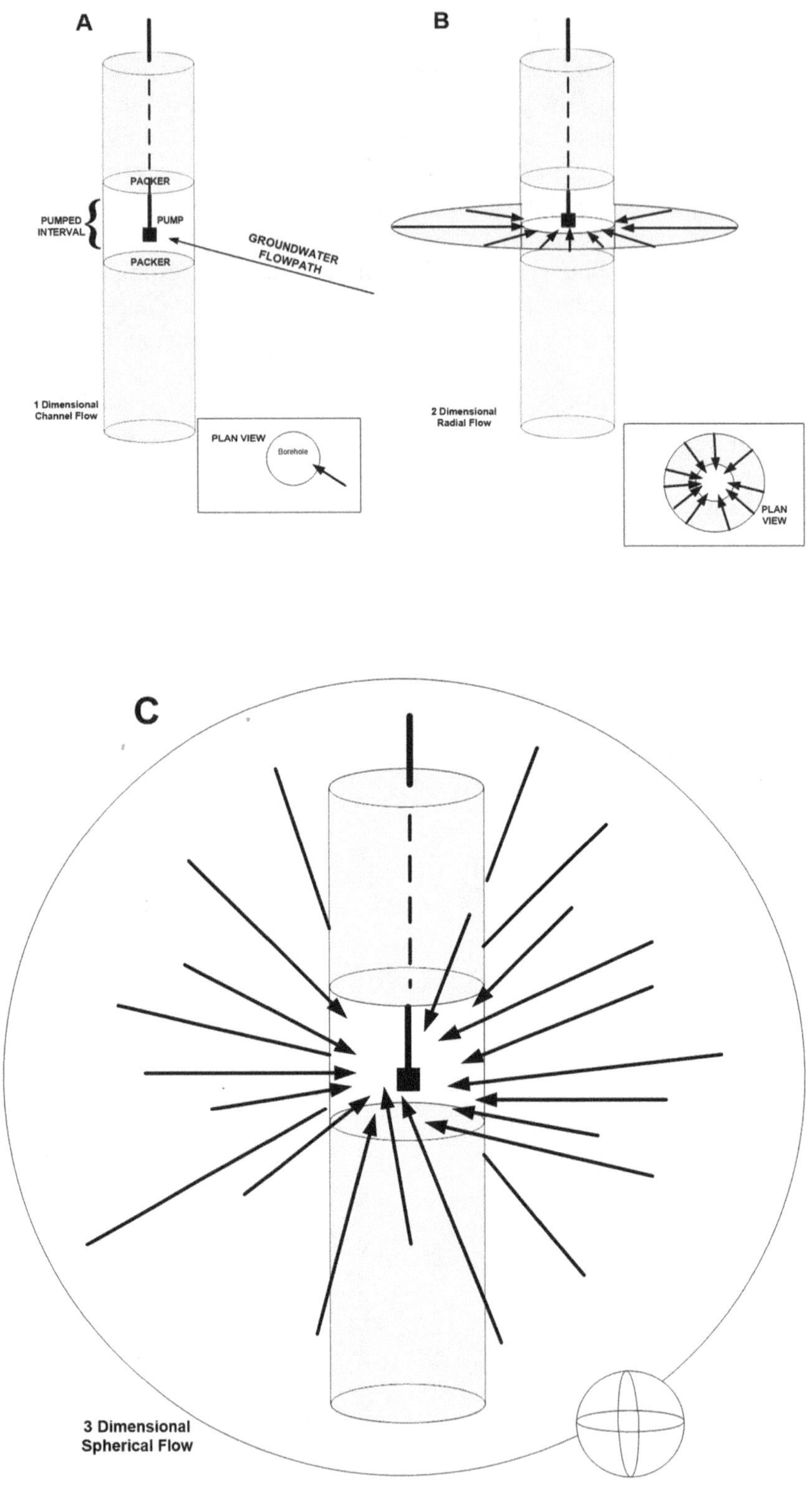

FIG. 33. Examples of (A) 1-D channel flow, (B) 2-D radial flow and (C) 3-D spherical flow patterns in response to induced flow towards a pumped borehole interval.

observation boreholes, not just at the control borehole, the tests provide information on larger scale cross-hole connections due to the presence of discrete flowing fractures within the rock mass.

When designing a pumping test, it is necessary to have at least an approximate appreciation of the conditions that are likely to be encountered. This is necessary so that the correct equipment can be obtained for optimal test performance, so that the location of the equipment placement (pumps, packers and monitoring points) can be selected appropriately and so that the timescale allocated for a test and the pumping rate(s) applied will produce pressure responses that can be properly analysed. Clearly, specialist hydrogeologists should be employed to design and carry out the tests. If contractors are to be used, they should be clear about the uses that will be made of the data and the need for diligent planning, testing and records keeping. It is a best practice for hydrogeologists to be present during preparation and testing and the RWMO needs to ensure there is clear and good communication with the drilling team when testing and drilling are carried out at approximately the same time.

With sufficient time, hydraulic testing in the field can yield results representative of relatively large volumes of rock. As such, they are the primary basis for deriving hydraulic and transport parameters for use in modelling. However, geosphere understanding can be significantly enhanced by small scale measurements of permeability and other properties made in the laboratory on selected samples of rock (see Section 6.5.1). The interpretation of these laboratory tests complements field testing, as they are able to provide many, but not all, of the same hydraulic parameters as measured in the field. Differences occur because measurements reflect different investigation lengths and the fact that rock sampled for testing in a laboratory is a generally a biased sample that may not be representative of in situ conditions. Nevertheless, with an appropriate understanding of potential sampling and measurement errors, a hydraulic testing programme based on field and laboratory measurements may allow a high degree of confidence to be assigned to derived hydraulic parameter values. It also provides a better understanding of scale effects, including the relative contributions of matrix flow compared to fracture flow, in a situation where both mechanisms might be operating at the same time, for example in a weathered hard rock or fractured sandstone.

Tracer tests

Hydrogeological tests employing natural (environmental) and artificial (applied) tracers are used to investigate solute transport with flowing groundwater in a rock mass. The mechanisms contributing to solute transport are:

— Advective transport, in which solutes are carried downstream with the bulk groundwater flow;
— Diffusive transport, in which the flux of a solute is proportional to its concentration gradient;
— Dispersive transport, arising from variations in fluid velocities within the domain which, in turn, are caused by heterogeneity in the hydrogeological and rock properties.

Testing provides insights into the rates of groundwater flow, recharge and discharge locations, the degree and nature of connectivity in the subsurface and also the degree to which solute concentrations reduce over distance and time due to irreversible sorption, retardation, spreading (dispersion) and bulk dilution through mixing. Tracers can also be employed to investigate solute movement in the absence of groundwater flow, through the process of molecular diffusion along a concentration gradient. Understanding and quantifying all of these processes is critical for populating and calibrating models of radionuclide migration in the geosphere and for enhancing confidence in dose and risk assessments. For this reason (and despite some limitations[22]) wherever possible, tracer tracing should be undertaken to supplement information on advective flow obtained from hydraulic testing.

[22] Tracer tests are usually carried out where it is known that there is good connectivity in the rock mass. This may not be representative of the larger volume of significance for radionuclide transport away from a repository. Furthermore, the timescale of a tracer test is very limited in comparison to the timescale of interest in a post-closure safety assessment and this may be insufficient to permit all relevant reactions to occur, especially those concerning rock matrix diffusion in fractures.

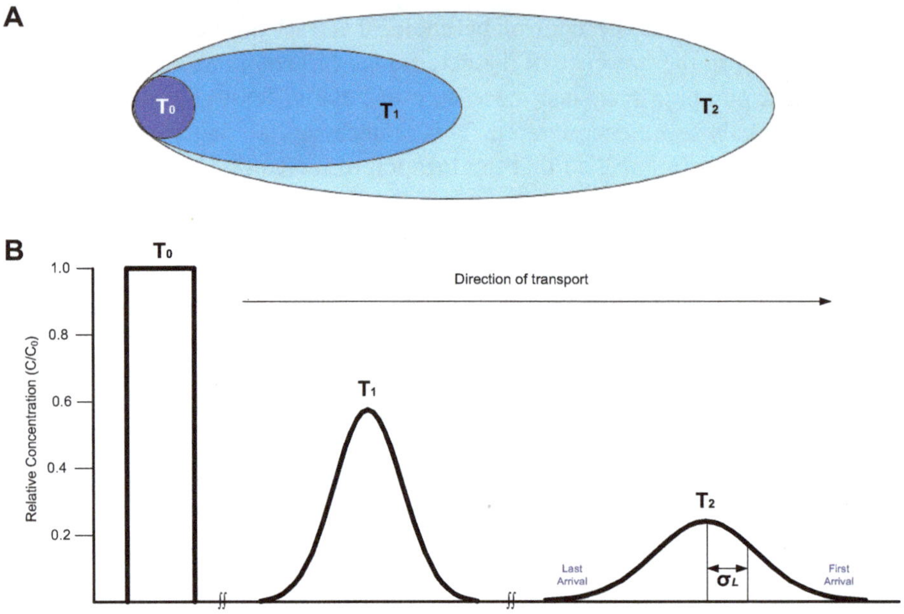

FIG. 34. (A) Plan view showing the effects of longitudinal and transverse dispersion of a package of groundwater containing a slug of a tracer of known concentration; (B) plot showing the relative change in tracer concentration with time and distance due to longitudinal dispersion (adapted from Fetter [59]). σ_L is the standard deviation of the tracer concentration in a longitudinally spreading plume.

The basic principle involved in tracer testing is to use the known natural chemical or isotopic composition of a groundwater body, or to introduce a change in the natural groundwater composition by adding a defined concentration of one or more new constituents. By knowing the composition and nature of a labelled and discrete package of water at a known time ($t = 0$) at the starting point of a test (e.g. an injection borehole or recharge area), one can compare the concentrations measured at the start of the test to those observed over time at another location (e.g. an observation borehole), with differences in concentration assumed to reflect the differential movement of the tracer through the rock mass under a natural or induced hydraulic gradient, assuming of course the water package originating at the starting point is flowing towards the monitoring point and can be measured there. The concentration observed at a monitoring point will vary with time due to hydrodynamic dispersion from initially low concentrations at the first arrival time up to a peak concentration, and will then decline. This pattern describes a breakthrough curve, which is usually modelled as Gaussian (Fig. 34). Where artificial tracers are used (added constituents), it will be necessary to be assured the flow rates have the potential to record the arrival of tracer over the experimental time frame.

By reference to Fig. 34, the longitudinal component of hydrodynamic dispersion (D_L) is calculated from:

$$D_L = \sigma_L^2 / 2\,t$$

where

σ_L^2 is the variance;

and t is the mean arrival time.

Transverse hydrodynamic dispersion is calculated in the same manner. Empirically this has always been found to be less than the longitudinal value for dispersion in porous media (approximately by a factor of 1/5 to 1/20). However, the derivation of longitudinal and transverse length is much more complex in fracture networks [60].

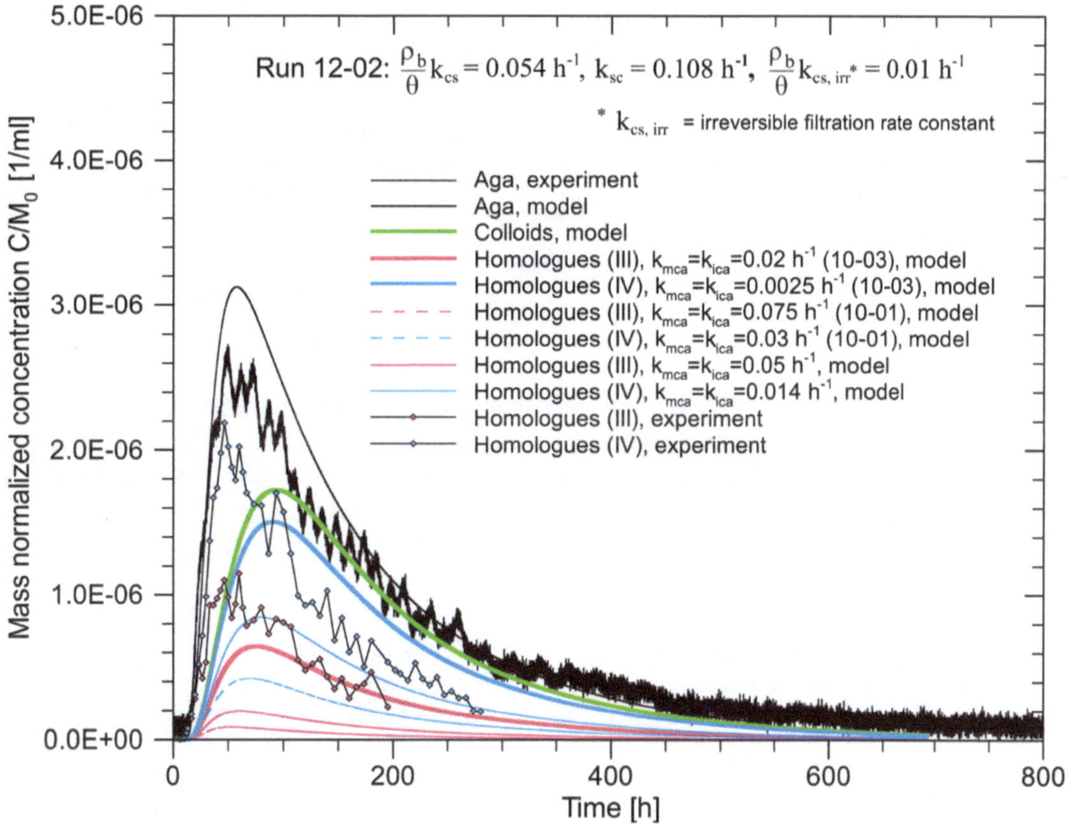

FIG. 35. Simulated and experimental breakthrough curves for colloids and radionuclides from a tracer test carried out as part of the colloid formation and migration project in the Grimsel Underground Laboratory, Switzerland [62] (reproduced courtesy of Nagra).

Various types of tracer were already introduced in Section 6.4.1 in relation to the dosing of drilling fluids to understand the effects of drilling fluid contamination on groundwater compositions. Uranine, fluorescein and iodide were noted as relatively inert tracers that might be suitable for this task. They can also be used for dedicated tracer testing, along with many other inorganic and organic solutes, microbes and particulates such as colloids and spores. In general, tracers should be non-toxic, have no potential to modify hydraulic conductivity or react chemically with ambient rock and groundwater, be relatively inexpensive and easy to detect in low concentrations. It is essential that the choice of tracer be aligned with the site conditions and test objectives to better understand potential radionuclide migration. For example, a conservative stable chloride tracer might be used as an analogue (e.g. ^{36}Cl in a waste inventory) or a more reactive tracer might be used to ensure rock matrix diffusion involving sorption is incorporated into any results. When considering using a tracer containing ions that are naturally present in the groundwater, it is important to first obtain naturally occurring groundwater concentrations and then make relevant ion concentrations in the tracer significantly above this level to take account for expected dispersion, rock matrix diffusion and dilution, although it is important that the concentration not be so much that groundwater density is significantly altered. In some countries, short half-life radioactive tracers may be used in low concentrations with approval from the appropriate regulatory authorities (e.g. tracer testing undertaken at the Grimsel Underground Laboratory, Switzerland [61, 62]). Figure 35 shows modelled and empirical breakthrough curves for various tracers, including radionuclides, as deduced from Nagra tracer testing studies.

Tracer testing in the field is best accomplished when the injection point is separate from the monitoring point(s), although single borehole tracer testing methods are commonly used, for example:

(i) *Single borehole pulse method (Huff-Puff) test:* This method involves the injection and abstraction of a tracer over a given time. It requires the instant injection of a known concentration of tracer followed by mixing in the water column. Locally derived formation water is then injected into the borehole to force water into the surrounding formation. The water is allowed to migrate under the natural gradient for a period of time and then the borehole is pumped to reverse the flow back towards the borehole. The tracer concentration is measured as a function of the pumped volume of groundwater and time. There are several necessary assumptions, but this technique allows the estimation of the distance the tracer has travelled, its average velocity, and a measure of longitudinal dispersivity.

(ii) *Point dilution test:* A tracer of known concentration is rapidly added to the standing water column or to a borehole interval isolated by packers. After mixing, periodic monitoring of the concentration of the tracer in the borehole should show a decrease over time, which is recorded. The results are used to derive the magnitude of tracer flow and hydraulic conductivity, but analysis depends on several assumptions about the borehole and surrounding formation. There are several correction factors that can be applied to compensate for known or assumed conditions.

More typically, where rock mass and logistical conditions allow, two or more boreholes would be used to calculate hydrodynamic dispersion and porosity. There are several multiborehole tracer testing methods that can be employed, and they may use either a natural or induced hydraulic gradient:

(i) In order to characterize solute transport under a natural gradient, a tracer (or cocktail of tracers) of known concentration is injected as a slug into a control borehole and left to mix and disperse in the water column and then into the formation, with advective movement towards one or more observation boreholes situated downstream. Monitoring might be undertaken in an observation borehole over an open water column or from isolated intervals separated by packers. Monitoring would record the first arrival time and the concentration of tracer over time. The groundwater velocity and the rock mass conditions, the volume of tracer used, the time expected for breakthrough and the number of observation points could all impact significantly on the arrival time and concentration distribution, or if anything is detected at all. Analytical and numerical methods can be used to match the field results to calculated results using the advection–dispersion equation.

(ii) For investigating tracer transport under diverging radial flow conditions, locally derived water is injected into a borehole for a given time, accompanied by a slug or the continuous addition of a tracer (or cocktail of tracers). In order to establish a reasonable gradient, injection is often undertaken in an isolated borehole interval. A major assumption is that, in a porous formation, the water injection sets up a radial flow field under a constant artificial gradient in the surrounding formation. Monitoring points in one or more observation boreholes await the tracer's arrival (i.e. they are not using pumps to induce flow in their direction).

(iii) A similar type of tracer test can be used when there are only two boreholes available, except the test uses a converging radial flow field instead of diverging. In this case, the tracer is rapidly injected as a slug into a borehole and pumping in a monitoring borehole creates an artificial hydraulic gradient that induces flow and the tracer towards it.

Many methods can be used to detect and measure the concentration of a tracer or a mix of tracers in a groundwater body. These include the use of electrical conductivity meters, liquid chromatography, specific ion electrode analysis, titration, photoluminescence techniques, mass spectrometry, gamma emission counting and colorimetric techniques. Several of these can be accomplished in the field, while others require groundwater sampling and laboratory analysis.

To understand recharge–discharge relationships, groundwater residence times and the mixing of groundwater bodies, isotopic tracers that occur naturally in groundwater may be used. A number of radionuclides are created naturally in the atmosphere including ^{3}H, ^{14}C, ^{36}Cl and ^{85}Kr; these can fall with rain and then percolate into the groundwater system. Other manufactured radionuclides resulting from atomic bomb testing in the atmosphere and dispersion from nuclear facilities can also be used to provide information on residence times. Groundwater sampling and the use of these tracers is discussed further in Section 6.5.2.

Tracer tests can be expensive, especially if observation boreholes have to be specially drilled. However, in a site investigation project it may be possible to use boreholes already drilled for other purposes. In some situations, tracer testing may not be feasible (e.g. due to rock conditions, lack of equipment, or personnel limitations and time constraints) or there might be high uncertainty about whether a tracer might be detected at a monitoring point. Obtaining results from tracer testing will be especially difficult in an area characterized by a sparsely connected rock mass. Even in areas of extensive fracturing, the pathway to be followed by a tracer originating from a control interval (source) within a borehole may not correspond to the direction implied by the local hydraulic gradient. Furthermore, within any given volume of rock more than one interconnected set of fractures may coexist in isolation from others. These issues tend to be amplified with depth and with both smaller source and monitored intervals within a borehole. Prior to any tracer testing, note should be made especially of instances where cross-hole hydraulic signals have been recorded during hydraulic testing, as these may be promising targets for testing. These considerations indicate that despite the obvious value of tracer testing, their use should be carefully considered and, as with hydraulic testing, specialist hydrogeologists should be consulted to ensure proper planning and implementation, as well as analysis and interpretation.

If tracer testing is to be undertaken, the RWMO site investigation manager should ensure testing objectives are clearly understood by the hydrogeologists. A high regard for QC and the accurate recording of activities and results is also essential. The locations of a control borehole and dedicated observation boreholes should be based on at least an elementary understanding of the local hydrogeology. There may be various pros and cons associated with the use of different tracers, the use of a particular test and the use of a natural or induced hydraulic gradient. These questions and others will challenge the decision making capacity of those responsible for field investigations.

Flow logging

Flow logging (also termed production logging) is carried out in fluid filled boreholes to determine the location of groundwater inflow points or zones in a borehole, to target hydrochemical sampling and to understand and potentially correlate flow zones between boreholes. Groundwater velocity and flow direction may also be evaluated.

Flow logging can be undertaken using geophysical methods to identify sudden changes in electrical conductivity and differential temperatures in the water column, through to the use of mechanical impellors (spinner flow meter) or by measuring the movement of fine particles using a colloidal borescope (Table 12). Other methods have also been used to investigate groundwater flow zones in a borehole and these may be of special use in formations with low permeability [63] (e.g. heat-pulse flow meter, laser Doppler velocimeters and borehole dilution methods).

A standard flow logging string comprises one or more logging tools that may be run alone or in combination. The tools measure a range of parameters which indicate the nature and movement of fluids within and in close vicinity to the borehole. Logs of the measured parameter value against depth are typically acquired while the borehole is being gently pumped and the measurement tool is being moved vertically, so that groundwater is actively flowing into the borehole under an artificial gradient, but without inducing turbulent flow conditions. The logging tools are lowered to the base of the borehole or the interval to be measured, and slowly raised so as not to unduly disturb the water column. The variations in temperature, electrical conductivity and physical flow are used to identify zones or points where groundwater is entering the borehole (flowing features or flowing zones) and for selecting locations for further hydrogeological testing, groundwater sampling or the installation of monitoring zones. Derivatives of the measurements of temperature and conductivity can be calculated to make identification

TABLE 12. SUMMARY OF INVASIVE FLOW LOGGING TOOLS AND TECHNIQUES FOR DATA ACQUISITION DURING THE DETAILED SITE CHARACTERIZATION STAGE OF SITE INVESTIGATIONS

Name of technique	Description	Data acquired	Comments
Temperature log	One or more temperature sensors are encased in a housing, often with a pressure transducer to record the depth of the probe in the water column. The probe is attached to the surface via a wireline on a winch. With one or more thermometers in the probe, absolute and relative temperatures in the borehole can be measured as the probe is slowly moved up the borehole.	Continuous measurements of water temperature and fluctuations. Anomalies in temperature may indicate locations where there is fluid flow from the surrounding formation.	Operates in a fluid filled undisturbed environment. May be run in tandem with other logs.
Electrical conductivity log	An electric current is introduced into the rock using electrodes. Other electrodes at various distances to each other measure the received voltage, which is proportional to the resistivity or conductivity in that distance range. Equipment for measuring conductivity include the high definition FMI tool.	Continuous measurements of electrical conductivity and resistivity. Anomalies in conductance may indicate locations where there is fluid flow from the surrounding formation.	Conductive fluids or casings influence the result. May be run in tandem with other logs.
Spinner flowmeter	A small free-wheeling impellor (propeller) is lowered into the water column. The rate of impellor rotations is proportional to the flow velocity.	An increase in water velocity may be significant in zones of inflow.	Requires inflow rates of approximately 1 L/s or greater and therefore alternative methods are potentially required in very tight formations (e.g. heat-pulse flowmeter).
Colloidal borescope	The tool comprises magnifying lenses, high definition camera, an illumination source and other equipment housed in a stainless steel probe lowered into the borehole. The optical equipment captures the movement of individual suspended particles (to 10 μ).	The tool measures groundwater velocity, flow direction and colloidal particle size with depth. Variations may indicate inflow zones.	Measures colloidal movement under laminar flow conditions, but correction factors may be required.

of small inflows or subtle changes in gradient easier to recognize. Additionally, several logging runs may be undertaken at different pumping rates, with higher rates inducing more flow into the borehole and increasing the likelihood of observing discrete responses in tight formations. It is noted that if a test section is dominated by one or more large inflow points, the signatures of very small inflows in proximity may well be swamped and unrecognizable.

6.4.4. Groundwater pressure monitoring

After groundwater sampling and hydrogeological testing, typically carried out during or shortly after drilling, some boreholes might be selected to become part of a network of monitoring points aimed at acquiring water table and groundwater pressure time series data. A groundwater pressure monitoring network would be used to firstly establish baseline hydraulic heads and vertical and horizontal groundwater pressure gradients across a site, and subsequently to record changes in groundwater pressure within a borehole over a period of time. Such information can be used to identify and assess the occurrence, impact and nature of cross-borehole responses to transient pressure signals and these, in turn, provide insights into connectivity and large scale hydraulic properties in the rock mass.

The precise objectives for a monitoring network and for individual borehole installations should be established on the basis of site investigation requirements and an existing knowledge of site conditions. Only then should a system be designed and procured. A network may be monitored manually or using automated measurement and recording equipment (transducers and data loggers). For small scale investigation projects, it may be acceptable to undertake periodic manual water level monitoring in boreholes, in line with a graded approach. The frequency of monitoring would reflect investigation aims, but to capture meteorological events and seasonal variability, measurement frequency would ideally be at least weekly or, preferably, daily. If a manual monitoring option is selected, it will be very important to ensure the personnel carrying out the task are well trained and that they carefully follow documented procedures at all times. It is also best for the data used at the borehole to be clearly known and used consistently. Suitable equipment is typically used to record water levels (e.g. an audible dipper lowered on a graduated tape). The borehole should be designed to ensure it is weatherproof against rainfall or flooding so there would be no contamination of the water column. In some Member States, boreholes in relatively isolated and unguarded settings may appear attractive for people intent on stealing or damaging monitoring equipment. Where possible, transducers and the surface expression of any equipment situated down a borehole should be kept in locked boxes affixed to the ground or in the borehole itself, secured with a lockable protective cover.

Establishing a comprehensive automated groundwater pressure monitoring system can be relatively expensive but is considered essential for any large scale site investigation project. Where such a system is judged necessary, groundwater pressure monitoring equipment appropriate for the conditions should be installed at multiple points within each borehole where possible (i.e. within hydraulically isolated intervals). Long term monitoring systems need to be reliable, durable and retrievable, and appropriate measurement, processing and analysis methods are of utmost importance. There are several types of commercially available automated monitoring system that allow pressure sampling at multiple locations within a borehole. Bespoke monitoring systems can also be designed, manufactured and installed in a borehole to be able to monitor pressures and temperature fluctuations continuously or at discrete intervals with the aid of retrievable and reconfigurable pressure sensors. In addition to automated pressure recording, the data obtained from a borehole monitoring network might be sent in real time to a central processing site using cables or an automated wireless communication process (telemetry).

For an automated groundwater monitoring system using pressure sensors and data loggers, sampling times might be essentially continuous or set for specified recording intervals. The recording interval and pressure resolution limit should be established to readily allow monitoring of anticipated pressure signals resulting from various sources.

Sampled data may record the occurrence of the following:

— Naturally occurring groundwater pressure transients caused by earthquakes and discrete meteorological events;
— Daily, seasonal and annual variations in the water table or groundwater pressure due to cyclically and non-cyclically occurring natural conditions (e.g. barometric pressure fluctuations, Earth tides[23] and seasonal changes in recharge);
— Groundwater pressure variations in response to loading by ocean/sea tides, due to hydraulic connection within the formation and to a tidal regime or by direct loading of overlying rocks;
— Artificial pressure perturbations due to drilling or construction.[24]

[23] Earth tides are due to lunar and solar gravitational forces that cause cyclical small scale movements of the Earth's surface.

[24] Periodic groundwater sampling might also be undertaken in monitoring boreholes. Repeat or new forms of hydrogeological testing might also be carried out at various times, in order to ascertain whether hydraulic or transport properties are changing with time.

In some configurations, an individual borehole pressure monitoring system is based on a single string suspended in the borehole, with multiple transducers located at port couplings in a number of isolated borehole intervals separated by hydraulic packers or by permanent grout or bentonite seals (discrete multilevel monitoring). Other configurations use multiple pipes of various lengths within a single borehole (nested piezometer configuration), with the base of each pipe open within a zone that is carefully sealed using permanent seals or removable packers to isolate it from adjacent zones. The pipes are left open to the surface so that the water level reflecting the static groundwater pressure in each zone can be continuously measured by individual transducers or floats. In the first instance, 30 or more pressure sampling ports might be available in a deep borehole for monitoring. In the second example, the number of pressure transducers or water level monitoring points is limited by the diameter of the borehole and the number of individual pipes that it can accommodated. Other monitoring systems also exist.

Prior to the installation of monitoring equipment, zones of potential interest within a borehole should be targeted on the basis of geological, hydrogeological and geophysical interpretations. After installation of a monitoring system, groundwater pressures in the surrounding formation need to settle back to equilibrium. The time this takes reflects the nature of the formation and, in a repository host rock, this might be anything from several days to several months. In a highly disturbed, very tight formation, full equilibration may even take years.

Groundwater pressure monitoring equipment is likely to exhibit drift over time, as shown by the pressure recordings. Equipment may also fail completely; therefore, it is essential that the system be well maintained and that transducers are periodically calibrated. Experience indicates a lot of effort will be expended on these tasks and dedicated staff should be on hand to implement a fixed schedule of inspection and calibration as well as to retrieve and repair equipment when it breaks down. There may be relatively long periods when individual pressure monitoring probes and transducers will be unavailable due to failure and the need for off-site repairs. Consequently, it would be advisable to have a backup stock of transducers and probes available as replacements.

It has been frequently observed that groundwater pressure in a borehole responds to barometric (atmospheric) pressure [64]. This pressure is spatially and temporally variable with no discernible pattern in detail. It is assumed the groundwater pressure response to barometric pressure is due to the variation in load on the rocks overlying an aquifer, resulting in differential fluid pressures that cause flow into or away from a borehole. To ensure accurate and detailed pressure data from formations for comparison over time and space, barometric pressure effects should be removed from groundwater pressure data.

Where the groundwater at a site is of variable density due to temperature and pressure changes with depth in deep boreholes or due to differences in the concentrations of total dissolved solids, calculated head values should take account of these density variations. This is accomplished through the processing of groundwater pressure data that takes into consideration groundwater sample density measurements to calculate freshwater head and environmental head [65]. Knowledge about the distribution of freshwater head across a site allows one to derive the horizontal component of groundwater flow in an equivalent porous medium containing variable density groundwater, while knowledge of the distribution of environmental head allows one to derive the vertical component of groundwater flow at a particular location.

Although specific groundwater monitoring locations, equipment and procedures should be established as early as possible, for a large and complex project it may well be necessary to adapt the system by adding to it; for example, by taking advantage of monitoring any new investigation boreholes that might be drilled to fill gaps in understanding.

It was already noted that at some stage, monitoring installations will have to be retrieved; prior to borehole sealing, for example. This should be planned for. In particular there should be contingency plans in place for situations where the monitoring string in a borehole becomes stuck.

6.5. ROCK, WATER AND GAS SAMPLING, AND LABORATORY TESTING

This section discusses the components of a sampling plan and management considerations associated with collection and laboratory testing of rock, water and gas samples. As with many techniques covered in this report, the actual analytical methods and techniques used during laboratory testing are frequently complex, and beyond the scope of the publication. Consequently, specialist guidance and advice should be sought as needed to ensure sufficiently detailed requirements and scopes of work are produced, and for the activities to be carried out and reported appropriately.

Before commencing any sampling activity, a quality plan should be established setting out the objectives in detail, as well as requirements and procedures to be followed during sampling. The plan would typically include, at a minimum, the following elements:

— Sampling objectives setting out the data and information to be collected;
— Nature of the sample and how much is needed (sample volumes);
— Sample quality criteria;
— Sample locations (using defined, random or structured approaches);
— Number of samples to be collected at each location and, if required, the frequency with which samples are to be collected (for monitoring purposes);
— The priority attached to the collection of the various samples;
— Sampling tools and methods (e.g. what equipment is to be used and under what operating conditions; what procedures are to be followed, including how samples will be collected, the types of sample containers to be used, labelling requirements and whether any sample is to be split in the field or subsamples taken);
— Sample collection, preservation, handling and ownership responsibilities;
— Analytical methods;
— Data management and reporting requirements;
— QC and QA policies and procedures.

Laboratories capable of carrying out the required tests and analysis should be identified well in advance of sample collection in the field. In some circumstances (e.g. where analysis of environmental isotopes will be required), specialist laboratory testing and analysis may not be possible using nationally available resources due to a lack of experience or equipment. Instead, overseas laboratories may need to be contracted. Even where relevant capabilities and equipment exist within a country, the number of samples and volume of data expected may be significant and could overwhelm local capacity. Therefore, when potential contractors have been identified, site investigation managers should also take into account the lead time for establishing contracts and any constraints a laboratory might have to deal with the estimated volume of the various samples to be collected, and their testing and analysis requirements over time.

Even with the best of planning, bottlenecks may sometimes occur, especially in under-resourced laboratories meaning a possible delay in the provision of results. Some activities (e.g. sorption experiments) may take several months or even years to complete. The RWMO should, therefore, ensure it uses reliable laboratories with appropriate staff and certification, and that they follow established national and international standards.

Some contractors may use proprietary methods for analysis that are obscure or difficult to replicate. It will therefore be essential that the basis for experiments and testing is shared with the RWMO, so the proposed approach is acceptable and so that results can be replicated as required to provide evaluations of analytical uncertainty and to facilitate openness and confidence in results.

Laboratory contractors may also use proprietary software to carry out analysis and to store records and results (e.g. a dedicated relational database). This not only makes the sharing of data and information with others more difficult, but it can mean that when files are retrieved from an archive in the future, they are unreadable. Clearly, the software and the media used by contractors to store data and information should be acceptable to the RWMO (see Section 4.2.8 for further discussion).

6.5.1. Rock sampling and testing

Rock samples may be obtained from surface based surveys, borehole drilling or pit and trench excavations. Samples may also be obtained from local mines and from dedicated URLs constructed as part of the investigations. Samples can range in size from rock chippings to cylindrical drill cores with diameters in the tens of centimetres, up to boulder size samples weighing several tons. Size is only limited by sample availability, the investigation objectives and the logistics and cost of transporting and analysing a sample.

Rock samples provide information on the chemical composition, mineralogy and texture of a rock, as well as a large number of physical properties. Rock samples can also provide radiometric and other age determinations to provide information on fault activity, uplift and erosion rates, time since burial for sediments and the relative and absolute dating of mineral formation.

Rock samples naturally represent differ types of rock. These may be described in several different ways, for example, those based on the following:

— The dominant mineralogical composition and texture (granites, sandstones, halites, etc.);
— The mode of genesis (igneous, metamorphic, sedimentary);
— The age of formation;
— A formal hierarchical lithostratigraphic description (groups, formations, members, etc.);
— For sedimentary rocks, the depositional environment and characteristics of the sedimentary architecture (sedimentary facies);
— For metamorphic rocks, the protolith (i.e. the precursor rock type that existed prior to metamorphism) or the fabric.

Any or all of these categorization schemes may be employed. Because these schemes require a relatively elementary geological knowledge, for the most part based on readily observable diagnostic characteristics, the rock type is generally recorded directly in the field or during the logging of core obtained from a borehole (Section 6.5.1). However, the need to differentiate unfamiliar or subtly different rock types may also require laboratory analysis. Standards and conventions for rock-type categorization may exist in a country, having been set by the national geological survey, for example. These schemes should be adopted wherever possible. If they do not already exist in an appropriate form, a suitable scheme should be devised under the direction of the RWMO to facilitate understanding and communication with wider stakeholders.

There is an assumption that the rock samples will be minimally disturbed (i.e. that they preserve within their fine scale structure the conditions that prevailed in the larger rock mass while they were in situ and are, therefore, representative of the local field conditions).[25] It should be noted that core obtained from a formation without extensive fracturing (such as a massive sandstone) may be considered reasonably representative of the rock mass from which it was sampled, but intact core obtained from a fractured medium will of course only be representative of the locally sampled non-fractured portion of the rock mass and not of a larger volume that contains fractures (i.e. it is clearly a biased sample).

In addition to providing descriptive information on composition, mineralogy and texture (obtained from thin sections, X ray diffraction, X ray fluorescence, etc.), the parameters most commonly derived from core analysis are intrinsic permeability, effective porosity, experimental sorption distribution coefficient (K_d), compressive and shear strength, Young's modulus and Poisson's ratio. Information derived from rock samples is used primarily in petrophysical and hydrogeochemical studies for understanding past and present groundwater flow and potential radionuclide migration, as well as in geotechnical studies associated with the design, construction and closure of a disposal facility.

[25] To some degree, even with special care during extraction and with laboratory controls designed to mimic in situ conditions, this assumption is usually not valid as the sample extraction process will cause changes that are often irreversible due to mechanical damage, pressure release and chemical changes.

FIG. 36. Three contiguous ca. 3 m long intact cores of granite ready for cutting, packaging and transport at the Xin Chang site investigation in the Beishan region of NW China (reproduced courtesy of Beijing Research Institute of Uranium Geology).

Rock samples collected at the surface may be highly weathered and/or their position in the stratigraphy may be difficult to ascertain, especially in remote or highly vegetated settings. Consequently, the majority of rock samples are usually derived from drilling. Sample collection techniques range from simple grab methods (e.g. for obtaining rock chippings) to rock coring that may result in intact cores without breaks over many metres in length (Fig. 36). Such cores can be oriented through correlations with geophysical logging of diagnostic features, meaning individual features can also be oriented.

Grab samples are generated during drilling as cuttings returned with drilling fluid and are readily available at a drill site. The person responsible for collecting grab samples typically pays little attention to the exact depth of grab samples and they are impossible to orient. These rock chips are used to identify the general lithology between an estimated range of depths, and any macroscopic structure is generally destroyed during collection.

In the near surface environment, soil and rock samples may be collected using hollow-stem auger or direct-push sample extraction methods rather than drilling. These generally provide relatively undisturbed samples that preserve at least some of the sample structure. Use of these methods is limited to soils and very soft formations (e.g. plastic clays and highly weathered rocks).

Drill cores are very often collected from a borehole for analysis during a repository investigation project. They are obtained by cutting into the rock using a circular diamond impregnated drill bit attached to the end of a hollow core barrel. This technique allows a core of uncut rock to advance into the barrel as drilling progresses. When the barrel is filled, it is pulled back to the surface where the drilled core is then retrieved and examined. In situations where a high percentage of low-strength rock needs to be obtained intact, or under difficult drilling circumstances, some coring systems use an inner barrel that holds the core and fits inside the drill pipe and outer core barrel. These triple-core barrel systems have a wireline device that can be sent down the interior of the drill pipe, where it latches onto the inner barrel filled with rock core and then pulls it back to the surface.

The orientation and spacing of fractures, bedding planes and other planar structures, as well as the depth to other geologic features can be determined from cores if the orientation and depth of the core barrel is known and maintained during the drilling operation. Core can also be oriented later when correlations are made with downhole geophysical logs. Coring devices using oilfield technology are also available allowing targeted collection of sidewall core samples called plugs.

Once retrieved, core samples are labelled with unique identifiers indicating location and depth, then are cut, boxed and photographed. Preservation techniques may be required to stabilize a highly weathered or mineralized rock sample and keep it from degrading after collection. Hold times for samples should be agreed to reflect the preservation technique used and observed to ensure sample integrity. This is especially relevant for samples that are to be analysed for their chemical composition. Special handling procedures may also be required to prevent a sample from drying out and crumbling. This is facilitated through the use of suitable trays for holding and transporting intact rock core and any soft sediment encased in liners.

Sample selection is intended to ensure representative examples of the main lithologies present at a site are collected and analysed, including rocks containing secondary mineralization associated with pore-filling events and veins or mineralized coatings on fracture surfaces that are indicative of past fluid flow through a connected, hydraulically active porous medium or fracture network. Table 13 provides some of the more important rock attributes that should be recorded in relation to core sampling.

TABLE 13. CORE CHARACTERISTICS TO BE RECORDED PRIOR TO AND DURING SAMPLE ANALYSIS

Core characteristic	Description
Depth	Depth in borehole from surface datum.
Lithofacies	Rock type and presence of diagnostic indicators of the facies.
Rock unit	Stratigraphic identifier — group, formation (and member if known).
Texture	Description of the size, shape and arrangement of discrete grains or particles (sedimentary); the relationship between individual crystals and any fabric and foliation (metamorphic and igneous).
Mineralogy	Mineral constituents size and form.
Colour	Dominant colour (according to specified reference colours [e.g. Munsell rock colour chart]).
Alteration	Degree of bulk rock weathering or hydrothermal alteration described according to a reference: — Fresh = unaffected by alteration; — Weak = rock fabric completely preserved, stained/bleached to some extent, or unaltered volume over 50%; — Moderate = original rock texture recognizable, some unaltered volume; — Strong = minerals and fragments altered to clay minerals, original rock texture no longer recognizable.
Fracture density	Number of observable natural fractures in core over a given distance (e.g. 1 m).
Location and dip of fracture	Midpoint of fracture trace in the core, apparent dip direction in relation to core axis and its maximum inclination with respect to horizontal (i.e. perpendicular to core axis). Note: True dip and dip direction is usually obtained by reference to correlated wireline logs.
Fracture aperture	Average width of aperture observed in the core, with maximum and minimum observable values.
Shape of fracture	Overall shape of fracture (rough, wavy, smooth planar), nature of any vugs or other voids, presence of striations and/or slickensides.
Genesis of fracture	Drilling induced, natural or unknown. Shear or tension fracture, or unknown.
Alteration of wall rock along fracture	Degree of local alteration on fracture wall according to reference descriptions.
Width of alteration on fracture	Width into intact rock of any alteration along the fracture (e.g. oxide staining).

TABLE 13. CORE CHARACTERISTICS TO BE RECORDED PRIOR TO AND DURING SAMPLE ANALYSIS (cont.)

Core characteristic	Description
Fracture filling materials	Description of fracture filling mineral (e.g. quartz, carbonate, clay, zeolite, haematite, limonite, manganese oxide, pyrite, gypsum, epidote, chlorite, etc.)
Width of filling materials in fracture	Average width of mineral surface on fracture wall or complete fracture filling width.
Other observations	Evidence for reactivation of fractures (superimposed slickensides or striations); several generations of infilling mineralization and the relationship, evidence for mineral transformation, etc.

Information on past mineralization episodes (there may be several events filling a single vein or veins with different mineralization may be cross-cutting), together with the structural characteristics of the fracture/vein systems (e.g. frequency, orientation, cross-cutting relationships) and the orientation and nature of present day flowing features, can provide important information for understanding the evolution of the groundwater flow system to the present day. Many such studies have been carried out in mature investigation programmes to characterize the mineralization history at a site. For example, at Sellafield in the UK, a well defined framework of fracture movement and mineralization events was established involving nine temporally discrete 'mineralization episodes' [66]. Many aspects of the fracture mineralization were then related to specific patterns of mineralization in the region. Work also focused directly on identifying the subset of recently flowing features from all fractures present and on deriving petrophysical properties important for flow. The samples were collected from boreholes and the information upscaled to provide insights into past and current flow conditions at Sellafield [67].

Chain of custody procedures should be used to document the transfer of samples from the original collector to subsequent custodians, eventually ending only when the sample is disposed of by transfer to another owner or destroyed. Finally, archiving of samples is desirable and may even be necessary, as stipulated by regulations. This may require construction of a core library, designed to preserve and house samples in an ordered and retrievable manner so further work may be undertaken at a later date if needed. The library may be owned or managed by the RWMO or by another party (e.g. regional or national geological survey).

Although there are many analytical methods that can be used on rock samples, exhaustive descriptions are beyond the scope of this report. The International Society for Rock Mechanics provides information on several standard tests for deriving physical properties from rock samples [68]. Table 14 provides information on some of the standard analytical methods used to ascertain key properties and parameters for studying samples.

It should be noted there is a continuous advance in the use of innovative techniques for the analysis of core samples, and techniques that were experimental at the time of writing this report may have since become mainstream.

6.5.2. Groundwater and surface water sampling and testing

Obtaining representative and high quality hydrochemical data from water samples is one of the most important activities to be undertaken during site investigations. However, obtaining this data can be difficult and time consuming, especially in boreholes characterized by 'tight' (i.e. largely impermeable)

TABLE 14. EXAMPLES OF ANALYTICAL TECHNIQUES AVAILABLE FOR THE STUDY OF ROCK SAMPLES

Property	Technique	Description	Limitations
Porosity (total)	Water saturation	The volume of a rock sample is determined precisely by machining and then saturated with water under vacuum. The mass is determined (M_{sat}). Sample is then dried in an oven and desiccator, and its mass measured (M_s). Total porosity (n_Σ) is derived from: $Vv = (M_{sat} - M_s)/\rho w$ $n_\Sigma = 100\ Vv/V$ where Vv is total pore volume, V is bulk volume of the sample and ρw is density of fresh water.	Requires competent samples that (i) can be machined to a geometry suitable for experimentation and that (ii) will not change volume on heating and drying. Presumes all porosity is accessible.
Effective porosity, pore geometry, and pore dimensions	Optical fluorescence microscopy	Sample is impregnated under vacuum with an epoxy resin containing fluorescein. Sample is then thin sectioned and examined with an optical or electron microscope. Connected voids impregnated by the resin (pores and fractures) will contrast with the background matrix.	No standard procedures; resolution limited by the power of the microscope. The resin will not have same viscosity as water so may not invade all pore space.
Effective porosity and pore size distribution	Mercury injection	Sample is dried, then placed in a vacuum sealed vessel containing a known volume of mercury. The sample is impregnated with mercury using increasing pressure. The mercury accesses progressively smaller pores.	Sample should limit creation of new void space during injection. Requires a conceptual model of porosity represented as a series of interconnected cylindrical tubes.
Intrinsic permeability (higher permeability rocks [e.g. 10^{-4} to 10^{-11} m/s])	Triaxial flow cell	The basis for deriving permeability in a laboratory uses the Darcy equation. A rock sample is placed in a triaxial cell separating two reservoirs of water with known properties held at a confining pressure. The upstream reservoir is injected with additional water so it becomes differentially pressurized at a known rate to a constant value. The outflow pressure in the downstream cell is at atmospheric pressure (monitored) to ensure a steady state flow condition. Intrinsic permeability (k) is derived from: $k = \frac{\eta Q L}{A \Delta P}$ Where η is the water viscosity, Q is volumetric flow rate, L is the sample length, A is its cross-sectional area perpendicular to flow and ΔP is the imposed hydraulic gradient. Note intrinsic permeability is related to hydraulic conductivity (K) by the relationship: $K = k\rho g/\eta$ where ρ is the density of the fluid and g is acceleration due to gravity.	The intrinsic permeability may vary depending on the orientation of the core (reflecting inherent anisotropy). A tight seal is essential. In high permeability rocks, pressure should be low enough to ensure laminar flow.

TABLE 14. EXAMPLES OF ANALYTICAL TECHNIQUES AVAILABLE FOR THE STUDY OF ROCK SAMPLES (cont.)

Property	Technique	Description	Limitations
Intrinsic permeability (very low permeability rocks [e.g. 10^{-11} to 10^{-20} m/s])	Pressure pulse decay	For low permeability samples, a gas should be used instead of water as it has a much lower viscosity and thus reduces the experimental time considerably. The rock sample is sealed tightly in a gas permeameter cell. Gas is rapidly introduced and pressurized on one side of the sample. The gas flows through the sample to the downstream cell. The transient pressure signal and downstream response are analysed to match the observed results to porosity and permeability values expected from alternative conceptual and numerical models (inverse modelling). The Klinkenberg factor is used to correct from gas permeability to a liquid permeability.	The intrinsic permeability may vary depending on the orientation of the core (reflecting inherent anisotropy). A tight seal is essential. Microcracks may form in response to the pressure variations placed on the sample. There are many assumptions inherent in the model used (various factors and geometries). Relatively large uncertainty at lower permeabilities.
Intrinsic diffusion coefficient and the experimental sorption distribution coefficient (K_d)	Through-diffusion cell experiment	A rock sample is machined into a disc and sealed in place in a horizontal diffusion cell. Two water bodies are introduced either side, one cell containing a tracer of known concentration and the other, the measurement cell, devoid of tracer. The tracer diffuses into and through the rock sample and is detected in the measurement cell. Measurements continue until the rate of concentration increase is steady. Experiment may be undertaken under uniaxial or triaxial compression to simulate in situ stress. K_d can be estimated if diffusion accessible porosity is known from other tests.	It is important to maintain a perfect seal between the rock and the diffusion cell. The water should conform to the composition of in situ groundwater. Test requires weakly or non-sorbing tracers. The experiment may take a long time to reach steady state (years).
Experimental sorption distribution coefficient (K_d)	Batch sorption	A rock sample of known mass is disaggregated to a very fine grain size by crushing and placed in a vessel containing a synthetic groundwater of known volume, itself containing a known volume and concentration of a tracer. The tracer begins to sorb onto and into the rock and the experiment continues until the concentration in the water reaches an equilibrium concentration. The disaggregated sample is then separated from the liquid by centrifuge or filtration and the final concentration of tracer measured in solution. The experimental sorption distribution coefficient (K_d) is calculated: $K_d = (C_0 - C_t)V / C_tM$ where C_0 is the initial concentration of tracer in solution, C_t is the final concentration of tracer in solution, M is the rock sample mass and V is the volume of solution.	No standard protocols agreed. Crushing likely increases the surface area available for sorption compared to the effective porosity of the intact sample. The tracer may sorb to the vessel wall. The tracer is to be carefully selected to ensure there is no chemical reaction and precipitation. The experiment can take a long time (months to years) and the standard deviation derived from repeated experiments increases with more highly sorbing tracers.

TABLE 14. EXAMPLES OF ANALYTICAL TECHNIQUES AVAILABLE FOR THE STUDY OF ROCK SAMPLES (cont.)

Property	Technique	Description	Limitations
Compressive strength	Uniaxial test	The rock sample is placed in a servo-controlled stiff loading frame where it is unconfined except for loading from one direction only. Rock deformation (strain) is monitored by data loggers as loading increases. A stress-strain curve results and the compressive strength of the rock is determined by the peak stress leading to failure (divide the maximum load by the initial cross-sectional area of the specimen).	There is some sensitivity in results to the rate of loading and water content when unsaturated. A regular cylindrical geometry is required (length to diameter ratio of 2.5 to 3). Good preparation required (smooth, ends perpendicular to main axis).
Shear strength and discontinuity friction angle	Shear strength test	To evaluate the strength of planar discontinuities in the rock sample, the core is placed in a cell comprising two holding rings, and a high strength material is used to surround it (e.g. resin or gypsum cement). A constant shear force is applied perpendicular to the plane being tested by moving one of the rings. Load cells monitor the shear stress. Displacements along the plane and in the normal direction to the applied force are monitored as the encased core is deformed. After failure, the two sections are repositioned and the test is re-run several times to derive the friction angle.	Many discontinuities are non-planar (rough or stepped), so the results can be very inconsistent from sample to sample and when testing the same sample several times as asperities are broken off and the plane is gradually smoothed. Premature tensile failure may occur if the sample is not mounted correctly.

horizons with little groundwater inflow. Furthermore, some important hydrochemical parameters are more difficult to ascertain with confidence than others. For these reasons, it is suggested that significant effort be expended in the planning and the execution of groundwater and surface water sampling and testing programmes.

Water samples can be obtained from surface water bodies (sea, rivers, lakes, springs, etc.) and from the subsurface via boreholes, mines and underground research facilities. Rainwater may also be hydrochemically analysed, to support recharge studies.

Surface waters should be sampled to investigate geosphere–biosphere interactions and to support EIA requirements. These will dictate sampling locations, the density of sampling and the frequency of monitoring of water quality. Surface water sampling can also support the identification of groundwater discharge zones. In tectonically active zones, groundwater sampling and analysis (and associated gases) can provide information on the nature of volcanic activity and the origin of deep-seated fluids, including the potential for faults to act as fast pathways for fluid migration.

Groundwater investigations may use multipurpose boreholes or single purpose boreholes for sampling. Typically, higher quality sampling might be expected from a borehole dedicated to a single task, but this may not be necessary depending on specific objectives and local circumstances. Sampling from boreholes may be undertaken periodically as part of a monitoring regime or may represent a one time only activity with a sample taken from an open water column or over a defined interval. Sampling and analytical methods should conform with established best practice and be in accordance with national and international standards to ensure quality and confidence in results. Where appropriate, a sufficient quantity of water should be collected from several locations to allow for statistically valid descriptions of spatial variability in determinants. Large volumes of groundwater may also be required from individual samples intended for certain types of analysis (e.g. age dating using ^{81}Kr [69]), although new techniques requiring less groundwater are becoming more commonly available.

It is important to appreciate that prior to drilling, there may be natural variations in groundwater composition close to a borehole location. These variations are typically subhorizontal stratifications representing differential flow and residence times in various hydrogeological units or contrasts in fluid density. Groundwater flowing into a low-pressure sink represented by an open hole allows mixing in the water column, and sampling under these conditions is unlikely to provide robust results indicative of any compositional stratification that might be present. For this reason, where stratification is suspected, groundwater sampling should be carried out during drilling when possible and the longer a section of a borehole remains untested, the greater the potential for disturbance of groundwater compositions adjacent to the borehole. It is advantageous to ensure planning for groundwater sampling incorporates provision for conducting tests on the bottom section of the borehole during breaks in drilling. Where sampling while drilling is not possible and for monitoring purposes, the borehole may be isolated into discrete intervals by hydraulic packers and then pumped prior to sampling.[26] The packers should be set either a single packer to isolate the bottom of the borehole or using a straddle packer configuration to isolate an interval away from the base. The sampling interval may be either a predetermined length applied over the whole borehole or it may be decided on by targeting groundwater samples on recognized features, such as inflow zones. When a straddle packer is used, it is good practice to ensure additional packers are also used to provide guard sections around the sampling interval. This can help prevent the possibility of packer leakage and crossflow that might result in contamination of groundwater being sampled by fluid from elsewhere in the borehole. The guard zones and test section would ideally have pressure monitoring in place to identify packer leakage and to monitor formation pressure during shut-in periods.

As noted in Section 6.4.1, the drilling of boreholes is accompanied by the use of a drilling fluid which will enter the formation as the drilling proceeds. When these drilling fluids intrude into the formation, a number of processes might occur that lead to a change in groundwater composition. In addition, a groundwater sample may become contaminated by contact with the drill string and test or sampling equipment. It will therefore be important to recognize when a groundwater sample has been potentially affected by a drilling fluid or contact with equipment (e.g. through the use of tracers) and to correct for this contamination. This can be achieved using a statistical regression approach, whereby a known concentration of a well characterised inert tracer is added to the drilling fluid. Upon sampling, the analyst can mathematically subtract the known contribution made by the contamination.

In addition to bulk water sampling, small volumes of pore water may be obtained for chemical analysis from individual saturated rock samples. Depending on the nature of the material, such water samples are obtained from the rock matrix using high-velocity centrifuge methods, crushing (disaggregation) or by compression (squeezing). In all cases, the principal aim is to obtain groundwater samples representative of the in situ (pre-investigation) conditions at the sampling locations.[27]

The precise suite of chemical determinands and radioisotopes required from analysis will vary depending on the needs of the repository programme and corresponding site investigation objectives. Nevertheless, there are some parameters that will be needed in most if not all programmes, including Eh (redox conditions), pH, sulphide and salinity. Table 15 provides illustrative parameters that might be expected to be collected during site investigations for a major radioactive waste disposal programme.

Section 3.6.3 provides high level objectives and the scope of work expected from hydrogeochemical investigations. Hydrochemical sampling and analysis supports safety assessment modelling and repository engineering through the provision of a database of parameter values which when integrated with other studies is expected to provide a detailed conceptual model of the hydrogeology of a site. This would be based on an understanding of the present day variation in groundwater compositions (chemical and

---

[26] It should be recognized, however, that re-establishing hydrochemical equilibrium in a tight system may take a very long time (years) and there may anyway be hydraulic connections via the drilling damage zone immediately surrounding a borehole wall.

[27] Note where pore water is obtained from a core sample, the chemical and isotopic composition may represent stagnant groundwater, and this may be distinctly different from any water flowing in surrounding fractures that make up an interconnected network formed around intact rock sampled as core.

TABLE 15. ILLUSTRATIVE CHEMICAL AND ISOTOPIC DATA TO BE OBTAINED FROM SAMPLING AND ANALYSIS OF GROUNDWATERS

Parameter	Reason for sampling & analysis
pH	Corrosion, radionuclide solubilities, speciation
HCO_3	Carbonate equilibria controlling pH, complexation
Salinity, TDS	Buffer/backfill stability, competitive sorption
DO	Corrosion, redox conditions and speciation
Na, K, Ca, Mg	Competitive sorption, stability of bentonite
Eh, Fe^{2+}, HS/SO_4	Redox conditions and speciation, proxy indicators for dissolved oxygen, sulphide corrosion of copper
DOC, colloids	Redox conditions, colloid formation and complexation
NH_4	Corrosion
Rn, U, (Ra)	Radiological hazard of excavation, redox indicator (U only), analogue radionuclides
Cl, Br	Total salinity, sources of salinity
Si, Al	Geochemical models for bentonite stability
$^{18}O/^{16}O$, $^{2}H/^{1}H$	Water origins
^{13}C, ^{14}C(TIC)	Carbon sources, biological activity, water residence times
^{3}H	Evidence of recent infiltration
^{4}He	Presence of old groundwater, flow heterogeneity and mixing
U & Th series nuclides	Isotopic disequilibrium in parent–daughter decay series, indicating water–rock reactions
^{36}Cl, $^{37}Cl/^{35}Cl$	Salinity origins and residence time, analogue of source term radionuclide
I, Cs, Se, Sn, Zr	Analogues of radionuclides
REEs	Analogues of radionuclides and diagnostic of water–rock reactions
F, HPO_4, HCO_3	Complex formation with radionuclides
^{14}C(TOC), ^{3}C(TOC)	Groundwater age as transit time for soil-derived dissolved organic material
$^{34}S/^{32}S$ in HS and SO_4, $^{18}O/^{16}O$ in SO_4	Sulphur geochemistry especially with respect to redox transformation and microbial activity
Dissolved $CH_4 + H_2$	Redox conditions, biological activity
$^{129}I/^{127}I$, ^{81}Kr	Groundwater residence times

TABLE 15. ILLUSTRATIVE CHEMICAL AND ISOTOPIC DATA TO BE OBTAINED FROM SAMPLING AND ANALYSIS OF GROUNDWATERS (cont.)

Parameter	Reason for sampling & analysis
Ne, Ar, Kr, Xe	Palaeorecharge temperatures from solubilities; support for palaeoclimate interpretation of stable O and H isotopes
^{40}Ar	Qualitative support for old groundwater ages
Colloid sizes, counts and identification	Potentially mobile particles, colloids capable of sorbing radionuclides
Microbial assemblage	Principal mediators for redox reactions and organic transformation, sources of biomass
$^{11}B/^{10}B$	Sources of saline groundwater
As and F	Species that may pose an environmental hazard

Note: TDS = total dissolved solids; TIC = total inorganic carbon; TOC = total organic carbon (courtesy of A. Bath, Intellisci).

isotopic) across a site and with depth, groundwater residence times and the site's palaeohydrogeological evolution to the present day (at least over a timescale commensurate with that applicable for post-closure safety assessment modelling). This understanding is important for developing scenarios in which there is groundwater ingress into a repository, corrosion, engineered barrier failure and radioisotope solubility, in turn leading to radionuclide releases. Hydrochemical studies also provide data and information for parameterizing and calibrating radionuclide migration models that include advection, dispersion, rock matrix diffusion and reactive transport processes. Furthermore, periodic sampling as part of a monitoring regime would be important for defining baseline hydrochemical conditions and recognizing changes resulting from continued investigations or repository construction and operation, as required by EIAs.

Site investigations will, by their very nature, tend to perturb the natural system and hence the groundwater sample composition, to some extent. Investigations should seek to minimize the opportunity for changes to the groundwater composition, but where they do occur, ensure any compositional modifications made during sampling and analysis can be recognized and quantified appropriately. This is especially important for those determinands that are more susceptible than others to changes resulting from degassing and depressurization during sampling or chemical reactions during retrieval and storage (e.g. pH, Eh, dissolved oxygen and redox-sensitive species). Degassing of samples during recovery from deep boreholes typically results in the loss of CO_2, which causes alkalinity concentrations to decrease and the pH of samples to increase. In turn, this affects the chemical equilibrium of other redox or pH sensitive species resulting in changes to their measured concentration. These unstable determinands are referred to as labile and they should be analysed in situ where possible, using downhole instruments or specialist handling and containment equipment (e.g. the use of inert gas for lifting to the surface or pressurized sample vessels[28]) followed by analysis at an on-site laboratory as quickly as possible. The costs–benefits of using on-site analytical facilities should be weighed, taking into account the investigation objectives and quality requirements in line with the need for a graded approach. While much hydrochemical data will be obtained from the analysis of groundwater samples pumped or 'grabbed' from a borehole, more use is now being made of wireline sensors and probes that are able to directly measure many in situ hydrochemical properties and determinands. Samples intended for isotopic determinations

[28] Downhole samplers should be flushed with inert gas prior to use and should be engineered to control entry of water at formation pressure so that outgassing is prevented.

and gas analysis, together with any samples taken for other specialist investigations, require appropriate preparation and preservation prior to analysis at specialist off-site laboratories.

6.5.3. Gas sampling and testing

Gas samples may be obtained from the release of dissolved gases within water samples as they are depressurized [70] or they may be collected in situ underground from local mines or in a dedicated underground laboratory [71]. Investigating the nature of dissolved gases collected underground is important because of their role in inorganic geochemical modelling used to investigate the stability of engineered barriers through the potential for accelerated corrosion (e.g. due to dissolved oxygen and sulphides) and the dissolution and migration of radionuclides from a repository in response to local groundwater chemistry and redox conditions (e.g. from CO_2 and CH_4 concentrations). Furthermore, understanding the content and composition of dissolved gases in groundwater samples may be important because they provide valuable information on microbiological activity and groundwater origins and evolution, including recharge locations and pathways (e.g. from evidence of anthropogenic influences obtained from dissolved nitrogen and chlorofluorocarbons), palaeotemperature and groundwater residence times (e.g. from noble gas isotope analysis). Radiogenic helium and radon have also been used in seismically active areas to investigate the nature of fault movements.

The sampling tools to be used will depend on the sampling environment, the availability of specialist equipment and the data requirements. For downhole sampling and in underground investigations carried out in mines and laboratories, pressurized sampling vessels and flow through cell samplers have been successfully used in several programmes. Pressurized water samples should be transferred to a laboratory as quickly as possible where gas in the water can then be boiled off under vacuum at water vapour pressure. Analysis of bulk gas samples can be undertaken using gas chromatography while dedicated extraction and analysis techniques are required for investigating gas isotopic ratios. In all cases it is vitally important to avoid air contamination. Because the sampling and analysis of dissolved gases is a highly specialist area that cannot be adequately covered in detail in this publication, further information should be sought from knowledgeable and experienced experts.

6.6. UNDERGROUND RESEARCH LABORATORIES

URLs are constructed and operated to provide in situ data and information on the subsurface environment on a significantly larger scale than can be obtained from an individual borehole. Site specific understanding is obtained from sampling and direct measurements of rock and groundwater, as well as by monitoring. Data are also obtained through the design and application of experiments designed to elucidate physical processes important for radionuclide transport. In addition, URLs permit realistic demonstrations of the application of equipment, tools and methods. Furthermore, they can be used for engaging with stakeholders. The OECD/NEA published a report describing the role of URLs in nuclear waste disposal programmes and their value to build confidence in national programmes, as well as aspects to be considered when planning a URL during stepwise repository development [72]. In the same year, the IAEA produced a technical document illustrating the use of scientific and technical results from URL investigations [73]. More recently, under the auspices of the IAEA, Sandia National Laboratories has produced a report describing the role of URLs in the context of a general timeline for repository development [74].

A URL is a long term and expensive commitment, requiring several years of planning to design the facility and to develop a research programme [75]. A site specific URL would likely only be considered at a prospective repository site after several years of surface based investigations and after significant effort in relation to satisfying planning conditions and obtaining regulatory approvals. Where a URL might become part of a disposal facility, it is imperative to ensure the construction methods, the layout and any experiments or other operations undertaken in the laboratory would not affect the integrity and safety of the completed disposal system. Given its focus, this report does not address the full range of management concerns associated with the planning, constructing and operating an underground laboratory; however, the IAEA is able to provide dedicated guidance and advice in this area through its Underground Research Facility Network.

6.7. MANAGING PARAMETER UNCERTAINTY AND SPATIAL VARIABILITY

There are different types of uncertainty associated with site investigations and site modelling as part of a repository development programme which include:

- — The future evolution of a site, involving the impact of processes and events on the performance of the natural and engineered barriers and pathways at a site and also relating to human habits and behaviours and ecosystem development;
- — The conceptual, mathematical and numerical models used to simulate the behaviour of the disposal system at a site;
- — The parameter values used as inputs into modelling and decision making among which the parameters of concern typically include variables such as material properties or physical dimensions.

Variability is a concept that is closely related to parameter uncertainty. It reflects the heterogeneity of a system described by the random or structured change in a parameter value associated with a property, as observed at different spatial locations or at different times. There is uncertainty because in any complex natural system it may be impossible to define a property's true state at all places and at all scales within a given volume and at all times.

Describing scenarios for the future evolution of a site and evaluating the uncertainty associated with them is typically undertaken in the context of post-closure safety assessment studies and is therefore not addressed in this publication. However, it is noted that a natural system scenario development exercise should always be supported by experts experienced in site investigation. Conceptual and numerical model uncertainty are important aspects that support site understanding; they are addressed in Section 7. Here, this section concentrates on the identification and management of parameter uncertainty and spatial variability in relation to data acquisition and processing. It is noted that the identification and quantitative evaluation of parameter uncertainty are relevant to a wide range of site features and properties that may be investigated in the field, in a laboratory setting or derived from elicited expert opinion.

An individual parameter value expressed as a number may be a constant or a variable, while a parameter comprises one or a series of parameter values that can be described as a frequency distribution or a range. Parameter uncertainty reflects the degree to which a sampled parameter value or the sum of sampled parameter values deviates from the true value of the parameter, consistent with the knowledge available at that time. It is caused by (i) measurement, physical sampling and data processing errors[29] or (ii) imperfect knowledge such as might arise from limited data. Ideally, an uncertainty evaluation provides a range of values associated with a specified parameter, together with the likelihood of each value occurring under a set of specified conditions.

Formally identifying, carefully evaluating and, as necessary, reducing parameter uncertainty is important in the context of radioactive waste disposal programmes for two main reasons. Firstly, because there will be little confidence in any calculations or designs impacting on repository performance if they are based on data that has been acquired without due consideration of the magnitude of errors and omissions. Secondly, because parametric deviations from reality propagate through the process of data acquisition to interpretation and integration and then on to their use in safety assessment models and engineering design. During this process, the cumulative effect of incorporating the uncertainty associated

[29] For ease of use, these three factors (measurement, physical sampling and data processing errors) are collectively termed *sampling errors* in this publication.

with individual parameter values can expand overall uncertainty in a non-linear fashion, potentially affecting key decisions relating to site acceptance and repository approval.

It is important to recognize that knowledge about the uncertainty associated with each set of parameter measurements is not equally significant. For example, safety assessment calculations of dose and risk or repository design calculations may be insensitive to large uncertainty ranges for a number of site parameters. Consequently, as part of the requirements driven approach to site investigations, a comprehensive and systematic uncertainty management plan should be established for identifying those key parameters for which uncertainty evaluations are required and documenting how they are to be dealt with. The following discussion briefly expands on the two types of parameter uncertainty identified above and how they might be managed.

Sampling errors: This type of error generally has specific causes that may be avoided, well characterized or otherwise accounted for, as the potential for them to occur can often be anticipated. For example, equipment limitations can impact the accuracy of measurements while procedural decisions made during measurements and physical sampling may introduce inaccuracy due to censorship, truncation or other forms of bias leading to a systematic over- or under-estimation of a true value.

In some cases, the very act of attempting a measurement may introduce an error. For example, taking a core sample from depth in a borehole causes a local stress change in the rock resulting in a physical response that modifies the framework and fluid pathways associated with the core. Consequently, it is recognized that any laboratory based measurement of hydraulic conductivity obtained from a core sample will not necessarily be representative of the in situ hydraulic conductivity. However, the magnitude of the difference between the in situ and as measured state is uncertain, and this has to be accounted for in any use of the data.

The design and calibration of equipment is another source of sampling error. The equipment used in a site investigation can range from a simple ruler to a highly complex instrument in a laboratory. In relation to uncertainty evaluations associated with the use of any such equipment, there are two independent concepts to be considered: accuracy and precision.

— *Accuracy* indicates the degree to which a measurement reflects a 'true' value;
— *Precision* is a description of the statistical variability associated with the repeated measurement of a parameter value, reflecting the level of consistency in the reproducibility of a measurement.

A highly accurate instrument is by definition precise, but a highly precise instrument is not necessarily accurate. Measurement accuracy and precision are affected by the reliability of the equipment and by its resolution.

— *Reliability* is a measure of performance under anticipated operational conditions and it reflects the quality associated with the design, manufacture and maintenance of equipment;
— *Resolution* can be defined as the smallest change in the parameter being measured that produces a response in the measurement of the parameter value.

Unlike accuracy, high resolution does not necessarily result in high precision (due to the impact of signal noise and calibration drift for example). However, the degree of resolution associated with an instrument does certainly determine the theoretical limit of precision and its potential accuracy. Wherever possible, contractors that undertake analysis and measurements on key data sets requiring rigorous uncertainty evaluations should ensure reporting is accompanied by metadata providing the statistical accuracy and precision associated with the equipment used to make the measurements, as well as commentary on their reliability and resolution.

In order to ascertain the accuracy and precision of measurements made under laboratory conditions (e.g. for hydrochemistry), uncertainty evaluations may be informed by comparisons of interlaboratory analytical results using split samples, the use of internal or international reference standards for instrument calibration, or through the use of manufactured or well studied samples with precisely and accurately known properties for comparison with sample results.

In addition to constraints on the accuracy and precision of measurements caused by field equipment and laboratory instrument limitations, sampling errors can also result from a human failure to adhere to measurement and sampling protocols and procedures or from the introduction of observational decisions and mistakes. For example, a resolution threshold may be introduced by an investigator as a conscious decision to establish the minimum cut-off length for fracture trace measurements made in the field. This may well be a practical action reflecting what can be observed by eye, but nevertheless, the result will be the exclusion of all fracture data smaller than the threshold and this censoring results in a biased data set.

Another source of parameter uncertainty arises from variations in derived (secondary) parameter values caused by the different methods that might be used for the screening and processing of raw data. For example, a set of groundwater pressure data measured from hydraulic testing in the field can be used to derive the hydraulic conductivity of a formation. But the process for translating the raw pressure data into the more valuable hydraulic conductivity parameter requires several assumptions to be made (e.g. laminar groundwater flow, known fluid density, aquifer geometry, etc.). Furthermore, there are several alternative methods that can be used to undertake the necessary data processing and analysis. The value for hydraulic conductivity arising from one set of assumptions and the use of a particular technique will likely be close to, but not exactly the same as, the result arising from a different set of assumptions and the use of another technique. Consequently, where it is not possible to assign a higher degree of confidence to a set of underpinning assumptions or a particular technique, there will be an uncertainty associated with any derived parameter values that incorporate the same raw data, with differences resulting from the use of different assumptions or processing techniques.

Imperfect knowledge: The sampling errors may be considered distinct from parameter uncertainties which arise from an imperfect or incomplete knowledge of a system. In many cases, such knowledge deficiency reflects the temporal or spatial variability of parameter values associated with a physical property, a factor inherent in every naturally heterogeneous system.

Some parameters are highly spatially variable, whereas others may be relatively constant across a given spatial domain. Any such variability might be random or might be organized in a structurally simple or more complex manner and on a range of scales. In simple terms, this means in a relatively homogeneous system, any single measurement of a parameter is likely to be reasonably representative of the total population of measurements. Conversely a highly heterogeneous system, especially one lacking an identifiable trend in the pattern of differences between parameter values, is more likely to be subject to increased uncertainties. A frequency distribution of representative parameter values, for an appropriate sample size, will indicate the degree of heterogeneity present for a particular property. Naturally, a heterogeneous system may require a greater investigation effort than a more homogeneous system, to enable it to be adequately characterized.

Spatial heterogeneity may be present at all scales and consequently the value of a measured property may depend on the scale of the length over which a measurement is made. For example, hydrodynamic dispersion is considered to be influenced by heterogeneities at all scales, and values are therefore scale dependent. For certain features and processes, any differences in measurements due to heterogeneity at a scale much smaller than the length of interest may be effectively ignored, in which case an average or an otherwise representative value may provide an adequate description. Where this is possible, a representative elementary volume for a property can be defined. This is a concept often used to support geomechanical and radionuclide transport modelling. The averaging process applied may be relatively straightforward (i.e. selecting a measurement procedure that ensures that one samples a sufficiently large rock volume) or a suitably representative value can be derived more rigorously by upscaling multiple measurements made at a small scale into an single effective value at the scale of interest. Appropriate treatment of spatial heterogeneity as part of uncertainty management depends on the property under consideration, the scale of interest and the nature of any relevant process that may impact the measurement of parameter values.

Compared to a classical statistical approach, for example using a frequency distribution to derive an average and a measure of dispersion, geostatistics is an alternative method for estimating and describing spatial variability. Geostatistics has a long history of use [76], especially in the mining sector, but it has been used more recently in a wide range of fields, including in site investigations for radioactive waste disposal facilities [77, 78]. It is a specialist area of applied statistics that seeks to characterize the spatial distribution of a correlated set of random variables which exhibit spatial continuity from point to point, but whose value is not known everywhere. Geostatistics relies on modelling and therefore typically requires subjective choices and assumptions. Consequently, the results of a geostatistical analysis, usually in the form of a variogram, are not unique and should themselves be treated to some form of uncertainty analysis.

Classical statistical and geostatistical estimation methods both provide a powerful set of tools for analysing and describing many types of spatially distributed data, but they may be inadequate for other properties and processes at a site, especially when the sample size is zero or limited (i.e. there is little confidence that sample values available are representative of the full population of the variable under consideration). In these circumstances it may be impossible to confidently define a parameter's spatial distribution through a classical statistical or geostatistical description.[30] As an alternative, data elicitation may be used to supplement sparse field data. Typically, such elicitation can provide bounding and best estimate values of a parameter, thus describing a uniform or triangular distribution that is intended to subjectively capture all parameter uncertainty including spatial variability. However, it can be appreciated that demonstrating that this is the case may be problematic. Consequently, data elicitation should be undertaken using a structured, documented and rigorously fair methodology involving suitably qualified experts.

The management of uncertainty: It is important to recognize that uncertainty associated with a parameter always exists in a natural system. However, it is best to adopt a graded approach for evaluating and managing that uncertainty, whereby the effort and rigour involved is commensurate with the hazard posed by the inventory of radioactive waste to be disposed of. In relation to this, it is also important to recognize that parameter values are not all equally significant in terms of their potential impact on safety or engineering design. Therefore, the greatest effort to evaluate and manage uncertainty should be

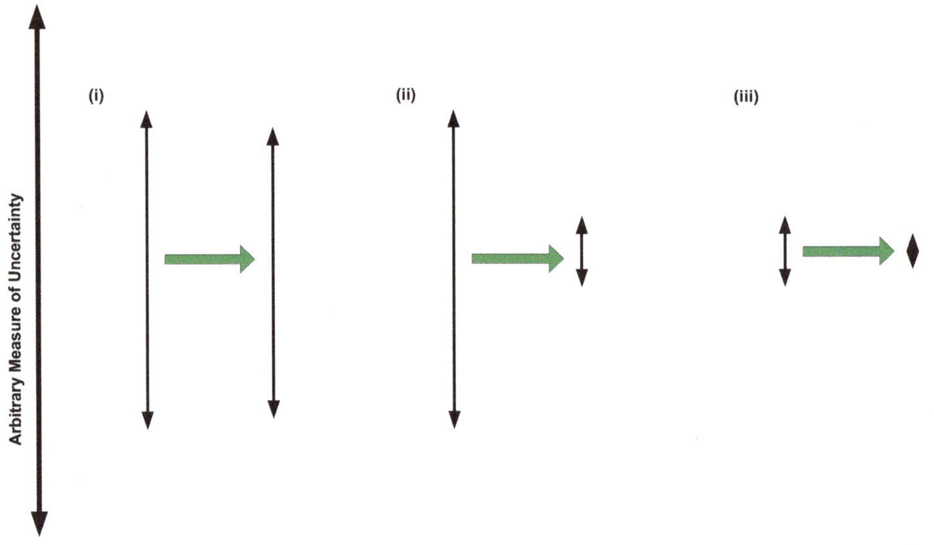

FIG. 37. Schematic illustration of three hypothetical scenarios: Initial parameter uncertainty is further reduced through further site investigation effort, with each scenario illustrating a different relative benefit for overall site understanding.

[30] Nevertheless, there are many other techniques that might be used to find relationships between and within disparate and sparse data sets, such as self-organized mapping (known as SOM) and the use of neural networks.

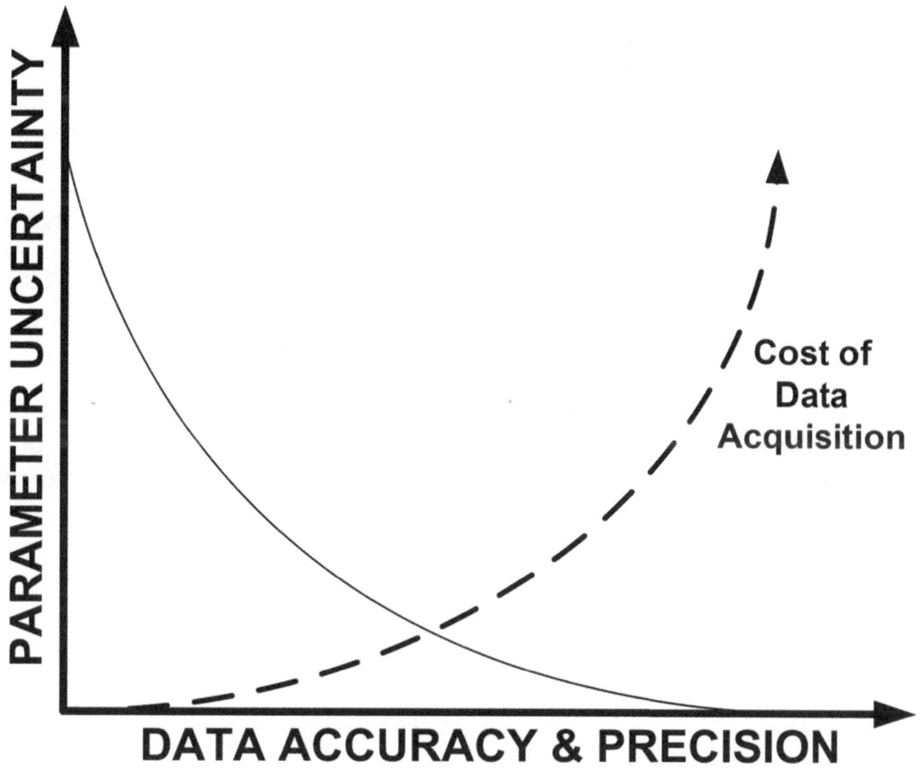

FIG. 38. Correlation between parameter uncertainty and data accuracy/precision, and the relationship to data acquisition costs.

applied to those parameters that have the potential to make a significant impact on dose and risk in safety assessment sensitivity studies or on engineering design choices (Fig. 37).

Each vertical line in Fig. 37 represents a nominally scaled but arbitrary value of uncertainty. Each green arrow represents the same degree of effort to be employed to reduce uncertainty in a parameter. In situation (i), additional effort results in a minimal reduction in uncertainty from an initially large range while the same degree of effort results in a much larger reduction in uncertainty for (ii). And finally in (iii) this same effort reduces an initially very small range in parameter uncertainty to an even smaller range. Other things being equal and where it is possible to estimate the degree to which uncertainty is likely to be reduced, investigation effort might be prioritized to reduce uncertainty in case (ii), as this reduction would appear to be far more significant than in cases (i) and (iii). However, this conclusion may be misleading because decisions should be based on safety assessment results and significance should reflect parameter sensitivity to dose and risk or impact on engineering design and hence inform the value of acquiring additional data for improving confidence and reducing uncertainty. Sensitivity analysis and scoping studies should be used to assess this potential significance.

It is noted there is a direct inverse correlation between the accuracy and precision associated with data measurements and with parameter uncertainty. Furthermore, the degree to which specified accuracy and precision is required can have a major impact on costs (Fig. 38). Therefore, the financial costs associated with reducing parameter uncertainty cannot be ignored; this is especially relevant when considering uncertainty management decisions for what might be considered marginally significant parameter data sets.

To support a cost–benefit analysis of parameter uncertainty evaluations and the use of uncertainty reduction methods, taking into account aspects of accuracy, precision and representativeness (and hence data density), it will be essential to ensure parameters are clearly associated with the uses that will be made of them (Section 4), including an understanding of the impact that individual parameters may have on repository design studies and dose/risk assessments. This conclusion stresses the importance of ensuring that the repository design and safety assessment development studies proceed in parallel with site investigation efforts and that sensitivity analysis is a key component of a safety assessment methodology that needs to be shared with site investigation personnel.

Where rigorous uncertainty management is required, parameter uncertainty resulting from sampling errors and the processing of raw data may be avoided, reduced to an acceptable level or otherwise managed through establishing and implementing a robust and comprehensive management system incorporating approved QC policies and procedures (see Section 4.2.5). This should include measures for managing the potential for introducing bias. It is important to recognize that no management system seeking to address sampling errors is perfect and inadvertent uncertainty may be introduced which is only recognized at a later date. In such a situation, the impact of such errors on parameter measurements would have to be evaluated and potentially rectified on a case by case basis.

The use of multiple data measurement and processing techniques should be considered where they exist and when no single technique has a clear advantage in terms of providing more accurate and precise parameter values. The range of primary data values and any secondary parameter values derived from the use of alternative processing methods but originating from the use of the same raw data is a measure of uncertainty that should be carefully documented and evaluated as necessary.

Depending on the degree of heterogeneity in a system, individual measurements of a parameter may be more or less representative of the true system property being investigated. The sample size is important in this regard as the degree to which the sample statistics (e.g. mean, standard deviation) are an accurate reflection of the whole population is less likely for a small sample size than would be derived from a larger number to samples (Fig. 39). Note, however, the complete population does not generally need to be sampled to provide adequately representative statistics, as the actual number of samples required depends on the nature of the property and the issue to be addressed, as well as the inherent spatial variability. The challenge is to determine what a sufficient number of samples is and how the sampling should be undertaken. The spatial density of measurements and their locations should be appropriate to allow for an adequate uncertainty evaluation leading to the reasonable conceptualization of the system and the parameterization of analytical or numerical calculations that address safety assessment and repository design requirements over the domain of interest.

The observed frequency of spatially distributed variable parameter values can be represented by a frequency distribution. When considered in combination with other sources of uncertainty, the range of values can be described statistically using one of several mathematical functions that show the likelihood of a parameter having a specific value. For example, a Gaussian probability distribution is one in which the full range of sampled measurement results are included, together with interpolations and, depending on the property being considered, possibly also including rational extrapolations (e.g. from regression analysis). This range in values can then be summarized as a mean together with the standard deviation

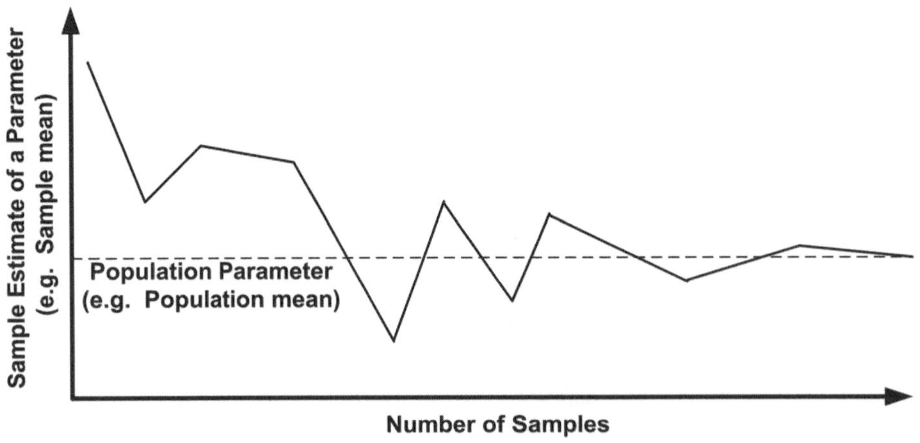

FIG. 39. The effect of increasing sample size on the accuracy of the sample estimate.

as a measure of statistical dispersion. Such an approach is probably the most common formalism for quantifying variability and uncertainty.

The uncertainty associated with certain parameters may be statistically or functionally dependent on the uncertainty of other parameters. Where this is the case, the correlation of the dependence may be used to reduce the uncertainty in a particular data set. Consequently, in developing an uncertainty management plan, it may be advantageous to ensure that there is scope for investigating any correlations that might exist between key parameter data sets.

The magnitude of spatial variability and parameter uncertainties, as well as their effects, will vary for different properties. As noted, spatial variability and parameter uncertainty may be assessed and represented using statistical and geostatistical methods, but a statistical description does not necessarily reduce parameter uncertainty unless it is used to guide further data acquisition[31] or is used for investigating dependencies and correlations. However, the statistical representation of the spatial variability combined with other components of uncertainty as a probability density function, permits their use in Monte Carlo simulations that incorporate and propagate parameter uncertainty in models. This methodology permits an evaluation of the uncertainty in the modelled outcome. Wider discussion concerning the derivation of probability distribution functions and their forms can be obtained from [79] while Fig. 40 illustrates a flow chart to support the production of a probability distribution function (sometimes called a PDF) for a specified parameter, when multiple measurements of that parameter are possible. Note that data elicitation might be employed using suitable experts and methods to augment data measured at a site. This additional information could be derived from expert opinion or from generic data (e.g. natural analogues) in order to provide a continuous range of parameter values using an appropriate distribution function (e.g. a triangular distribution representing three elicited data points).

In addition to the probabilistic approach for uncertainty management and modelling using probability distribution functions, another approach is to accept that uncertainty exists, but to use expert judgement in the event of unreliable evaluation methods or severely limited data to derive a deterministic best estimate or 'conservative' bounding parameter values[32] instead of measures of statistical distributions. Such an approach may usefully be used to address 'What if?' questions involving simple models. However, aside from the potential bias that may be introduced by eliciting data based on judgements, it may be difficult to justify a best estimate value or adequately define what a conservative value is, with a temptation to use a 'worst case' value instead. While the use of what may be an overly pessimistic parameter value might be in line with a precautionary approach, it can be highly misleading and may significantly overestimate risks and impacts, especially taking into account the non-linear propagation of multiple uncertainties.

Any attempt at parameter uncertainty evaluation and management is meaningless if it is predicated on poor quality data. This may not be apparent, especially when quantitative uncertainty evaluations are carried out, because the use of an approved methodology together with the use of numbers may give rise to a false sense of security. To illustrate this point, a groundwater sample is taken from a formation to derive a representative groundwater composition. Chemical analysis of the groundwater sample is used to obtain the concentrations of a range of chemical determinands and, by reference to calibration standards and interlaboratory comparisons on a split sample, an uncertainty range is established for the concentration of each constituent. The methodology for deriving an uncertainty estimate is robust and the quality controls associated with the analysis are rigorous and have been followed. However, the major assumption is that the groundwater sample is representative of the formation. So, unless it can be demonstrated the sample has been obtained from within the target formation and it is uncontaminated by drilling fluids, there would be a low level of confidence associated with any interpretations arising from the results of the analysis and associated uncertainty evaluation. As described in Section 3.5 and Fig. 10, the provision of confidence requires consideration of many dimensions, characterized by the

[31] Uncertainty analysis and the need to reduce the uncertainty associated with key data sets is one of the main drivers of an iterative site investigation project (see Section 2.2 and Fig. 5).

[32] A conservative value is one that is particularly onerous or demanding in the context of its use (i.e. a value that is plausible, but towards or at the limit of what might be feasible under given circumstances).

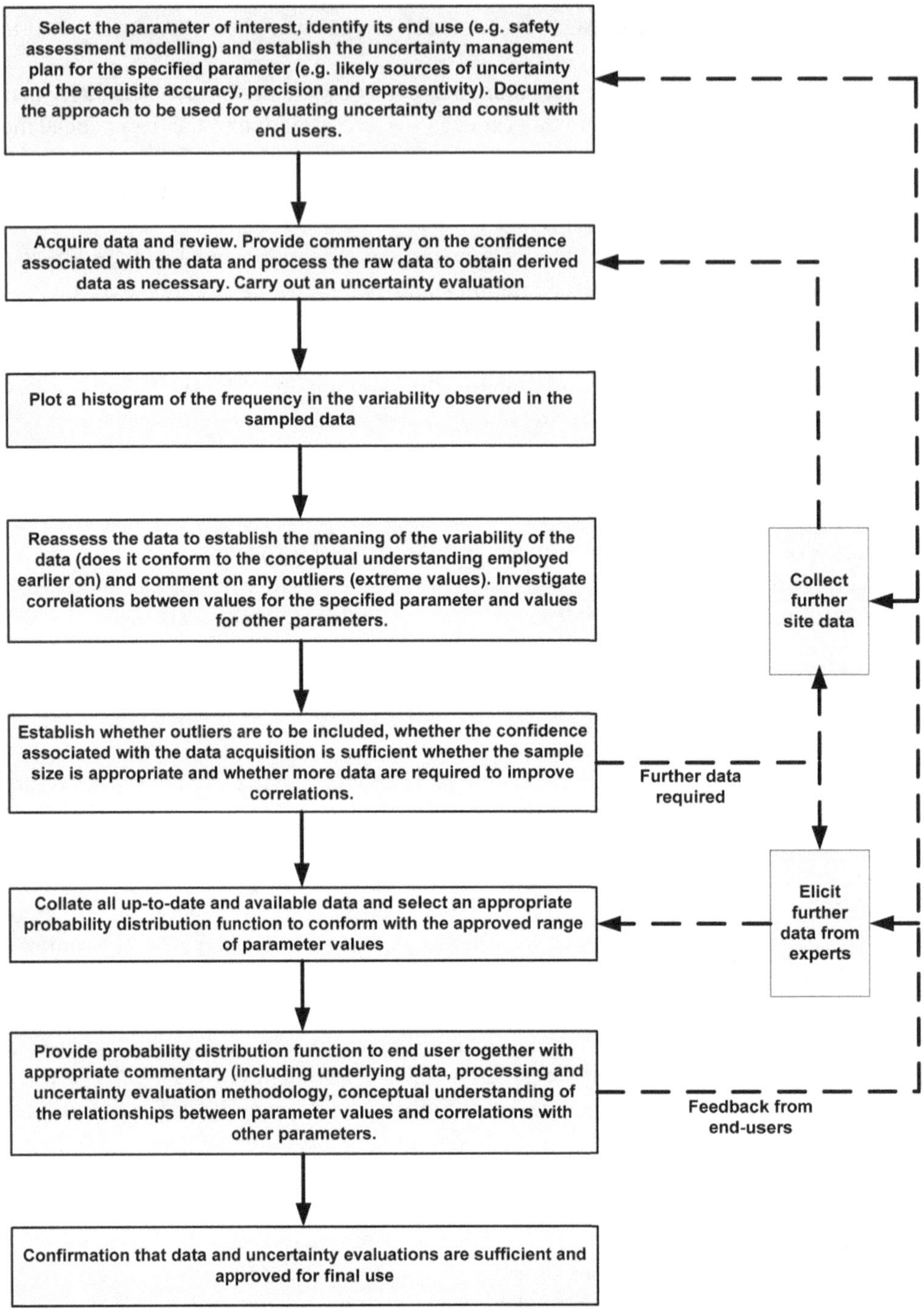

FIG. 40. Process flow chart to support the production of a probability distribution function for a specified parameter.

people, processes, equipment and robustness of data. It is the responsibility of site investigation personnel to ensure that data are appropriate for the end uses to be made of them and that commentary on the confidence associated with data and uncertainty evaluations be provided as part of standard reporting, as they are a key component of a safety case.

In summary, the selection of an appropriate approach to parameter uncertainty management will depend on several issues. Data should be acquired by suitably qualified people, using appropriate equipment and following defined and agreed procedures for measuring parameters and for the systematic identification and evaluation of parameter uncertainty. An understanding of the significance of the potential impact of a parameter data set, reflecting the end uses to be made of the data, allows one to develop an appropriate uncertainty management plan and to prioritize and design future site investigation activities. As part of that plan, it will be important to specify the accuracy, precision and representativeness required from parameter measurements to reduce or manage uncertainty and deal with spatial variability.

6.8. CONCLUDING COMMENTS

The preceding discussion focused on the planning and management of data acquisition activities, as well as some of the tools and techniques that might be used in a repository site investigation project and the uses to be made of the data. The data and information collected are required not just to characterize present day conditions at a site, but also to provide information about the geological evolution of the site to the present day over a timescale that is relevant. This timescale should reflect the future over which the radioactive waste to be disposed of is likely to remain a hazard. For the geological disposal of an inventory containing long lived waste this may be hundreds of thousands of years or longer.

In order to describe past and present site conditions, a large number of data sets will typically be generated during a site investigation project (Fig. 41). These data sets would form a comprehensive database of site wide and regional primary and processed parameter values together with their associated uncertainty limits and contextual confidence assessments. This is a key deliverable from the site investigations, to be fed directly to end users and regulatory authorities and also to site investigation personnel involved in interpretation and integration activities. In addition to quantitative and descriptive data and information, the database would likely also include visual representations of data and information, such as a photographic library of borehole core images, lithological and petrophysical logs (derived from core and downhole geophysical logging), plots of spatially oriented data (e.g. fractures and other structural information or wind direction), plots of hydrochemical information and time series data (e.g. from monitoring of groundwater levels and head in boreholes). The database might also include 2-D and 3-D images of the surface and subsurface estimates derived from ground penetrating radar, resistivity and seismic surveys, or extrapolations and interpolations of discrete features to show their spatial relationships.

It is recognized that not all of the key data sets that might be required have been addressed in this publication, for example discussion is lacking concerning data acquisition techniques for measuring in situ stress, properties and techniques relating to gas migration and rock thermal properties. Nor is Fig. 41 exhaustive. However, a key message is that in order to achieve suitably robust results, a requirements driven approach to data acquisition is required and also that in detail, requirements will vary from site to site.

An important aspect of data acquisition and processing is the identification and management of parameter uncertainty and spatial variability. Although it may be difficult to quantify uncertainty estimates in many situations, attempts should nevertheless always be made whenever possible to identify the existence and the components of parameter uncertainty and evaluate them wherever they occur. This is the basis for then managing that uncertainty appropriately and informing later stages of site investigation. The approach used to manage parameter uncertainty will have a direct impact on building confidence in the disposal concept. In this regard, it will be important to recognize that a regulator may have firm expectations concerning the identification of parameter uncertainty and its treatment in safety assessments and repository design studies. Therefore, it is recommended there should be focused discussion in this area to avoid potential misunderstandings.

It is imperative that all site investigation activities are well planned and justified. Furthermore, data acquisition from the field and the laboratory should proceed in a stepwise fashion, using specified milestones and an agreed decision making process to establish objectives, data freeze points and when to progress on to a new phase of work.

Once objectives have been established in each area of investigation, on the basis of clearly defined requirements, quality plans, schedules, cost estimates, responsibilities, scopes of work, contracts, etc. can

KEY DATASETS DERIVEDFROM DATA ACQUISITION ACTIVITIES		
Digital Terrain Model (Topography & Bathymetry)	Geomorphological map	Hydraulic properties from field tests
Ecosystem (habitat) maps	Soils and regolith map	Fracture network analysis
Lithological Maps & Sections	Meso-scale fracture mapping and rock sampling sites	Hydraulic diffusivity of major units
Faults, lineaments, & other major structures	Hydrological sampling and monitoring sites	Petrology/mineralogy data for rock samples
Database of earthquakes	Surface water sample hydrochemical results	Seismic parameters from boreholes
Gravity maps	Surface hydrology data	Geophysical Formation logs
Magnetic maps	Locations of boreholes, pits and trenches	Core and chipping Logs
Radiometric maps	Engineering test data for soils	In situ stress data
CSAMT maps	Meteorological monitoring sites	Thermal testing data
Seismic reflection and refraction sections and data	Meteorological data	Properties impacting solute transport including dispersion, diffusion, and sorption
Seismic attributes	Groundwater Maps	Rock quality data
Resistivity survey maps and sections	Borehole locations, dimensions, and orientations (inclined or directional)	Groundwater pressures and heads from monitoring
Ground penetrating radar sections	Fluid density profiles	Hydrochemical data from borehole samples (e.g. redox states, major ions, trace elements, dissolved gases, environmental isotopes)
Regional hazards and resources	Borehole temperature profiles	Locations and characteristics of flowing features in boreholes

Legend

Airborne and Surface-based Geophysics	(green)	Surface Mapping and Surveys	(yellow)	Borehole Tests and Measurements, including Geophysics	(orange)

FIG. 41. Typically collected datasets to establish a comprehensive database of past and present site conditions.

then be developed. Frequently, a variety of tools and techniques can be used to obtain particular data sets. While the selection of a preferred set of such tools and techniques tends to reflect local circumstances and conditions, the primary driver always needs to be based on safety considerations. These include conventional operational safety considerations as well as post-closure safety considerations, for example by ensuring that measurements will not compromise barrier safety functions and ensuring that data quality is adequate for the assessment of long term safety.

In conclusion, the following is a summary of some of the key management considerations associated with the planning and implementation of data acquisition activities:

- The RWMO is ultimately responsible for all aspects of the site investigations and should be an educated client for contracting and consultancy services obtained from technical support organizations.
- Site selection criteria should be established in advance of field investigations with a priority given to obtaining information on key aspects that might disqualify a site from further consideration.
- Easily collected information should be obtained early on to rapidly screen alternative sites, including obtaining and compiling data that resides elsewhere (e.g. at universities, national surveys, private companies, etc.).

— Close and continuous collaboration is ensured between site investigation personnel, especially those responsible for data acquisition and interpretation (if different groups are involved).
— Close and continuous and close collaboration is ensured between site investigation personnel, safety assessment specialists and repository design engineers.
— Regulator(s) and local administrative authorities are involved at an early stage and engage them in regular dialogue to facilitate permissions and to ensure expectations can be met.
— A robust and comprehensive management system is essential, with thorough documentation of the planning justifications, as well as the manner in which data acquisition activities will be managed, including written procedures that are developed, applied and monitored to ensure consistency in QA.
— Where they exist, and unless there is good reason to do otherwise, existing international or national standards should be used to establish the data acquisition procedures, design equipment and manage site investigation activities.
— A DMS and its management strategy is ideally in place prior to acquiring the first results of the site investigations, including established standard definitions, formats and units, and apply them across the project.
— A structured approach to risk identification and management should be adopted to ensure not only that physical and technical hazards are identified, assessed, avoided or mitigated, but also that other project hazards relating to traditional management areas are covered.
— The presence of uncertainties involved in the acquisition and processing of data are acknowledged and the significance for their end use in safety assessment and repository engineering studies is evaluated, using this knowledge together with the involvement of end users to establish an uncertainty management plan for key parameters, to design further investigations and acquire additional data, where necessary, to reduce parameter uncertainties where they potentially have a significant impact on dose and risk or engineering decisions.
— Data for developing a true 3-D understanding of the site is sought, incorporating all the relevant features and processes that might impact safety (an integrated and holistic conceptual understanding).
— Data for understanding the evolution of the site to the present day is sought over a timescale commensurate with the hazard posed by the waste inventory.
— Planned phases (or campaigns) are undertaken to ensure that the results feed into parallel safety assessments and engineering design studies and so that their results and revised requirements feed back to the site characterisation activities.
— Laboratory analysis of rock and water samples is planned well in advance because they can take a long time and may require specialised resources.
— The main deliverable from data acquisition activities will be a database of features, properties, events and processes that characterize a site and its evolution to date, including parameter values, uncertainty evaluations, metadata and commentary concerning the confidence associated with data sets. Supplementary material will be included in a series of factual reports describing the site conditions and the tools and techniques employed.
— An effective decision making process has to be in place and key personnel have to be aware of their responsibilities so they are able to effectively respond to any operational challenges that might arise, because they will.

7. DATA ANALYSIS, INTERPRETATION AND INTEGRATION

7.1. INTRODUCTION

The key deliverable from data acquisition activities is a comprehensive database comprising measured parameter values, physical samples and observations concerning present day features and their properties, processes and other characteristics of a site. Each factual data set should be associated with its broader contextual setting, represented by metadata. Together these would have direct application in support of safety assessments, repository design, engineering studies and EIAs. However, a scientific understanding of the characteristics of a potential repository site, especially when one takes into account the evolution of that site, cannot only rely on the acquisition and processing of factual data sets. Rather, site data also needs to be transformed and merged into useful information and understanding that provides wider meaning and deeper knowledge. This information is required by the RWMO's internal end users. It also needs to be communicated more widely, to regulators for discussion and in fulfilment of requirements for authorizations, and to other stakeholders to engender openness and confidence in the RWMO and trust in its decisions. The transformation of raw and processed data into information and knowledge requires significant analysis and the interpretation of individual data sets. Furthermore, to ensure a holistic understanding of a site and its evolution to date, individual data sets need to be integrated, taking due account of all relevant interactions and feedbacks that are recognized. The results are then to be captured in the form of an internally consistent and comprehensive conceptual model (or series of conceptual models) and an appropriately detailed site description, ready for presentation and for use by end users.

Data acquisition, interpretation and integration are, by necessity, iterative activities and would be ongoing throughout the lifetime of investigations at a site until a satisfactory level of knowledge has been achieved. Figures 22 and 25 in Sections 4 and 5 schematically introduced the process leading from data acquisition to the production of information and understanding through interpretation and integration activities. The process flow is also illustrated in Fig. 42 but, as introduced earlier in the report, this diagram emphasizes the possibility to also use additional data and information from outside of the site to aid site specific interpretation and integration. Figure 42 shows inputs from site specific and generic areas and their use in modelling studies to support interpretation and integration activities that result in conceptual and descriptive models for use by internal end users. This also includes the possibility to use generic process modelling. Although it might not be appreciated from Fig. 42, it is important to note that the use of numerical modelling is an aspect of interpretation and integration and not a separate and isolated activity in its own right, as is the generation of conceptual and descriptive models of the site.

Any interpretation and integration exercise will necessarily include a number of assumptions, recognizing also that the confidence one has in the quality of the data sets derived from data acquisition activities may be variable and the degree to which they represent the true situation at a site may be uncertain. Due to the fact that data sets generally represent a subset of a total population and, because of the assumptions, uncertainties and potential bias are inherent in the data, any interpretation may not be unique. Therefore, to capture the range of system possibilities that might exist in terms of combined FEPs and their evolution, alternative but equally valid conceptualizations of the site should be generated. As necessary, data gaps, quality concerns and uncertainties associated with the data and the resulting conceptualizations should be assessed and, if judged important, reduced through progressive phases of further data acquisition and interpretation.

At the end of each stage in a site investigation project, and also at its conclusion, the main deliverables from interpretation and integration activities are an enhanced database of data and information and a series of justified conceptual models and SDMs that collectively summarize the characteristics of a site at a particular moment or interval in time. Important management considerations leading to the generation of these deliverables are the subject of the remainder of this section.

7.2. DATA ANALYSIS AND INTERPRETATION

The process of interpretation has four essential elements.

(1) Firstly, the specification of objectives to articulate a purpose, to establish appropriate approaches and methods for carrying out analysis and to set limits on the interpretation effort that will be needed. These objectives stem from the information requirements and associated questions setting out what the site investigation needs to answer in order to address the needs of internal end users and other stakeholders.

(2) Interpretation then requires the collation and organization of all relevant and available data and observations into formats suitable for analysis, together with any contextual site data and supporting information (e.g. from generic studies).

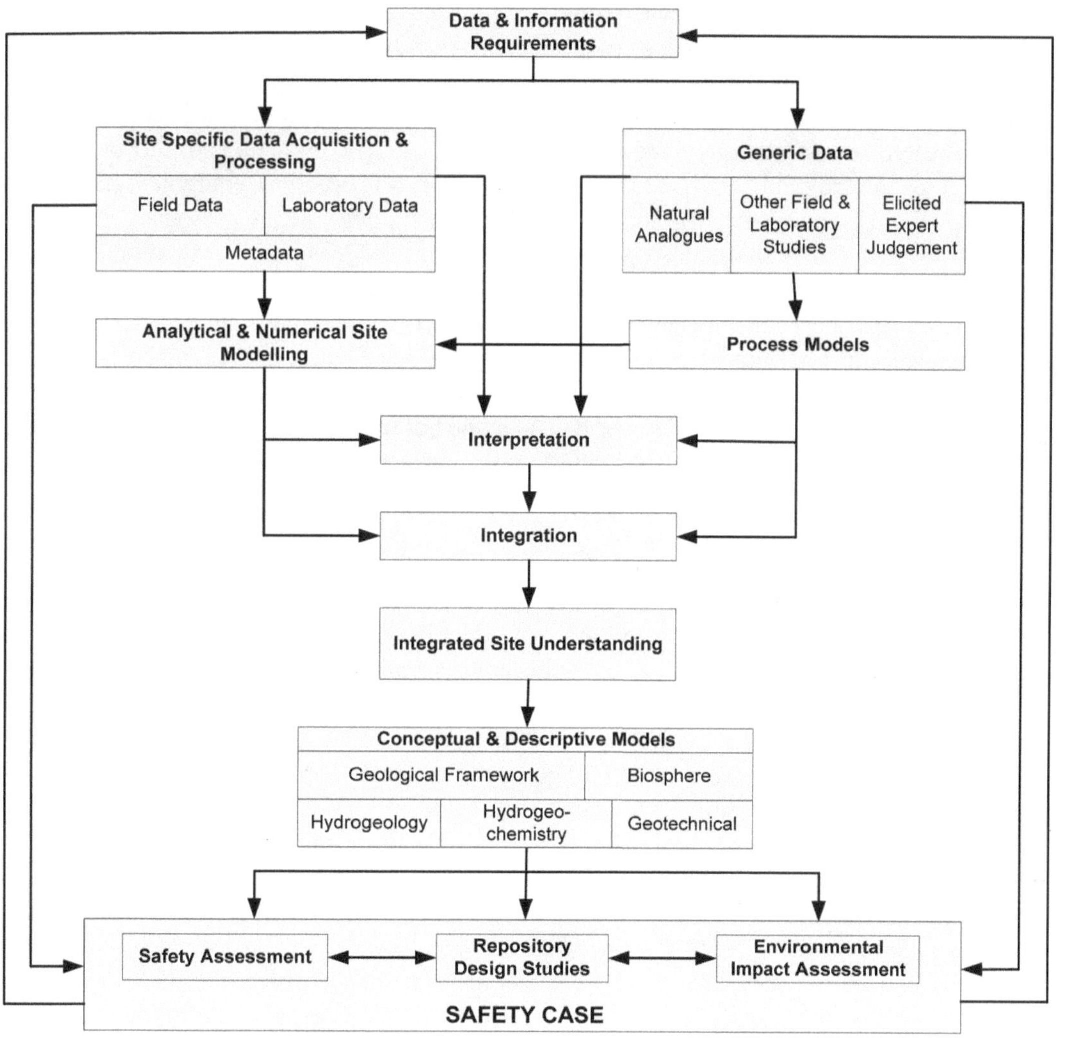

FIG. 42. Data acquisition, interpretation and integration process flow diagram showing that the process is iterative and the discovery of key data gaps and uncertainties at any stage may continually update further data acquisition and information requirements.

(3) Next, careful data analysis is required. This aspect involves searching for relationships and explanations to account for the data and observations by developing hypotheses and testing them. In addition to deductive reasoning, this analysis may include the use of mathematical models to address specific questions. This is a stage during interpretation when further data acquisition may be requested to reduce parameter uncertainties and fill data gaps. This process may occur in several cycles prior to any firm conclusions being drawn.

(4) Finally, conclusions are made about the meaning of the data and observations and these are then documented and presented for review and ultimately for use.

Interpretation involves the application of knowledge about the subject matter, but also calls on the wider technical experience of an expert and, to a degree, their intuition. As such, interpretation may contain an inherent bias. It should be appreciated that conclusions drawn by site investigation personnel may therefore be challenged by end users, both internal and external (e.g. the regulator), or as a result of external reviews. Where such challenges have merit, this may also lead to further loops of iterative site investigation (see Fig. 5).

Due to the subjective inputs required of many interpretations, the activity can be considered both an 'art' and a 'science,' although the emphasis is to be clearly based on logical and defensible stepwise reasoning, together with a clear, comprehensive and auditable chain of actions and documents leading from the raw data through to conclusions, such as might be provided by an SIFD, for example (see Section 4).

Depending on organizational arrangements, the personnel involved in site investigations would typically comprise RWMO technical staff and contractors from support organizations. The composition and structure of the various teams involved in collecting and interpreting data and information would in large part reflect the information needs, the availability of financial resources, equipment and experts, and also organizational preferences (see also Section 4.2.6).Wherever possible the personnel who collect data in the field or in the laboratory should also be involved in their analysis and interpretation. This would be to ensure the circumstances surrounding the data acquisition and processing activities (e.g. the reliability of equipment and their operation), as well as key aspects of the site conditions and field relationships, are well appreciated and not omitted or miscommunicated. However, it is recognized in several areas of investigation that this arrangement is not always feasible. For example, specialist geophysical survey companies may be employed to collect seismic data at a site but may not be involved in the interpretation effort. Also, laboratories commissioned to provide groundwater analyses or to measure geotechnical properties of rock core may not have been involved in the sample collection and may not be involved in later interpretation efforts.

Given the wide range of data collected during a site investigation for a prospective disposal facility, various teams of specialist experts are likely to be involved in the interpretation and integration efforts. For a large and complex disposal programme, it will also be necessary to use generalists who are knowledgeable about particular fields, but who have an added ability to draw together data and understanding from elsewhere and look at the 'big-picture' issues that cut across individual disciplines. This reflects one of the key roles of the RWMO in managing site investigation personnel, which is to ensure that a holistic understanding of the site is made. This requires integration across geosphere and biosphere domains and across individual disciplines. Because experts deeply involved in particular disciplines may sometimes be parochial in outlook, the potential for poor communication or miscommunication about important facts and ideas can be an issue when information is to be shared and discussed outside of a field of normal application. Furthermore, the technical language used in different disciplines may vary somewhat even when discussing ostensibly the same subject. Early on in a site investigation project this may cause communication issues between the data acquisition and interpretation personnel involved in various areas and also with the modellers and engineers who are the internal end users of the information. Special attention should therefore be focused on ensuring that the meaning of the data and information is properly communicated and that there is a common understanding across the disposal programme, through the use of glossaries and reference materials that establish standard nomenclature, terminology, conventions and measurement units.

One solution to address these internal communication issues is to establish and maintain a close and continuous working relationship between all site investigation personnel, as far as possible given the size and complexity of the site investigation. Where this is not completely attainable, effort should be made to at least integrate the senior and experienced site investigation specialists. This can be accomplished through the use of regular meetings in a formal setting to share data and understanding from each subject area and seek to find relationships across disciplines. Because of the crucial role that such a collaborative team plays in the site investigation and the wider repository programme effort, key staff from safety assessment and repository engineering studies should also be involved in these regular meetings.

As a general comment, and of relevance to interpretations in all of the disciplines to be described below, it is important to acknowledge that it is often difficult to measure properties of interest directly at the scale of their application (for use in numerical process models or safety assessment models for example). This can be particularly true for low permeability environments where the length and timescale of interest may be very large, compared to the measurement scale. This generally requires that measurements which have been taken over small scales in the field or laboratory undergo analysis and interpretation to infer corresponding effective properties that are applicable at larger scales. This process is termed upscaling. The difficulty in deriving upscaling relationships generally arises from the presence of small scale heterogeneities in the geological properties, because it may not be possible to explicitly represent these features in the much larger scale models of the system. Where this is the case, an appropriate effective property needs to be derived which accurately represents the behaviour of interest over the larger scales. The difficulty in doing this can be amplified because a local measurement may not be representative of the total population of parameter values and thus may not correspond to the effective behaviour at larger scales. Therefore, considerable interpretive effort is required to ensure any upscaling methodology takes into account parameter uncertainty and is appropriate for the subsystem component under consideration in order to maintain small scale relationships, including feedbacks.

Detailed commentary regarding the vast array of interpretation methods used for the analysis of individual data sets associated with, for example, borehole geophysical logging, hydrogeological test analysis and geomechanical interpretations, is beyond the scope of this report and the reader is advised to consult with specialist experts for more detailed information. Instead, in subsequent sections, the report is restricted to relatively high level considerations regarding the management of interpretation activities and some of the key outputs in relation to the main discipline domains.

7.2.1. Geological interpretations

The geology of a site intended to host a disposal facility and its wider regional setting provides the fundamental framework within which all other geoscientific information typically resides. Consequently, it is vitally important to demonstrate that the geological description reflects reality and that the scope and depth of understanding associated with this description is appropriate. In this regard, the degree to which an interpretation of the geology of a site is complete and correct (in other words the degree to which it reflects reality), should be judged on the basis of the use that is to be made of the representation, following a graded approach. The scope of the information and understanding potentially required from a geological investigation is provided in Section 3.6.1.

Field mapping results in a geological description that at its most basic level comprises a map of part of the Earth's surface showing the spatial distribution of the different rock types that are exposed and their contacts with other units, as well as the nature and spatial distribution of discrete structural features that can be mapped with confidence, such as folds, faults and major fracture zones. Mapping typically requires a degree of interpretation in the field, to identify a specific rock type and associate it with a particular lithological unit and to take account of the nature of the relationships between juxtaposed lithological units and the presence of structural features. In arid or mountainous regions, the geology may be readily observable, while in other areas outcrops may be extremely limited and geomorphological features or changes in vegetation may be required to infer the very near surface geology. With an understanding of the surface geology (whether directly observable or inferred) supported by data derived from geophysics and boreholes, if available, the relationship of geological units with depth and the 3-D extent of structural features may be inferred in the subsurface. Consequently, geologists are able to provide spatial interpretations of the rock mass that take the form of 3-D models, 2-D areal maps and cross sections, or 1-D borehole logs and field transects.

Typically, 1-D logs and transects will be continuous, as observed in a borehole or in the field, and therefore largely complete, albeit reflecting a limited distance with a lower limit of resolution and with the potential for sampling bias due to the orientation of features. Conversely, most 2-D and 3-D representations of geology rely on incomplete physical measurements and observations. Data points may be very sparse, especially in the subsurface. As such, area and volume filling may require significant data interpolation and extrapolation. These require the critical analysis of the acquired data and the use of experience to construct preferred and alternative geologically viable representations.

Remote sensing, surface based geophysics and borehole geophysical investigations can provide data-rich representations of surface and subsurface properties, but all require specialist expertise and software to acquire data and provide interpretations. The data and understanding resulting from these activities can be used to populate and condition basic geological maps as described above, as well as 3-D interpretations relating to the relationship between different rock types and the structural features present. Generally speaking, a large degree of expert judgement is required to create a geological map or 3-D representation of a site. Today, however, 3-D geological modelling software exists that can take uncertain or sparse input data and generate alternative geologically valid models incorporating lithological units and structural features. These modelling tools employ geostatistical and other methods. Where such software assistance is used, there is still an onus on ensuring that appropriate data are input and that checks are provided by an experienced geologist to ensure an uncritical 'black box' approach is not adopted.

GIS are another powerful tool used for overlaying and interrogating multiple types of spatial data and many open source codes, as well as proprietary software, are available. Although conventionally used to represent surface area, there are now add-ons provided by some GIS vendors that permit 3-D visualizations and interrogation.

Typically, the geology of a region and a site would be described at least at the formation-level, but for a detailed understanding of some processes requiring the use of detailed models concerned with advective groundwater flow and solute transport, smaller scale interpretations taking account of 3-D geometry may also need to be developed. Therefore, to refine and augment interpretations concerning lithological units and their relationship to hydrogeological units, rock samples may be collected from outcrops in the field and from trenches, pits, boreholes and underground workings. They can be examined in situ in rock faces or as hand specimens. Samples can also be taken from drill cores and drill chippings. Samples may be analysed essentially intact in a laboratory or machined into precise shapes for testing in specialist equipment or made into thin sections for use in microscopic analysis.

Rock types and the nature of any minerals or secondary rocks that infill discontinuities or pore space in a rock should be categorized using nationally accepted schemes and standards where they exist or a dedicated classification system designed for the purpose, if necessary. Discontinuities such as bedding planes, faults, fractures, joints, veins, dykes, sills, etc. should be recorded using geophysics in the field or in a core, and also categorized using a national systematic scheme or otherwise [80]. Mapping such features can be undertaken at a number of scales, from very long length scales involving lineament mapping derived from satellite images, down to a microscopic scale used for investigating microfractures. Understanding the spatial scaling relationships between fracture sets and their relative timing of formation can be especially beneficial where discrete fracture network models are needed for investigating fluid flow and solute transport. This is because of their potential importance for radionuclide transport at sites involving crystalline rock, although they may also be present in partially consolidated to highly competent sedimentary rock types that do not display self-healing. Where fractures are likely to be important for radionuclide migration studies, geological analysis and interpretations need to be carried out not only at the level of individual fractures but also at a field scale involving fracture networks.

Fractures are brittle features that form in response to rock stress and they reflect a discrete mechanical break in a rock, cutting across or parallel to any pre-existing features and fabrics. They may be generated in successive tectonic regimes in which stress orientations have varied and this can be investigated if considered relevant, through observations of cross-cutting relationships and the grouping of fractures with common orientations. Although the frequency of fracture occurrence varies widely depending on the geological setting and rock type, they are pervasive throughout a rock mass. Fractures may be directly associated with faults (rock mass dislocations involving lateral displacement across a fault face) as fault damage zones or they may exist in the absence of nearby faulting with their formation due to, for example, stress relief during uplift and the erosion of overlying rocks. Joints can also form during the cooling of a volcanic rock or one that has undergone spontaneous combustion (e.g. organic-rich limestones at Maqarin in Jordan, studied as a natural analogue for a cementitious repository [81, 82]). Faults, fractures and other discontinuities can act as either conduits for flow or as a barrier. They may also be 'transparent' in flow terms.

Fractures may be planar or they may comprise irregular and anastomosing surfaces. They are characterized by different spacing intervals and their size may follow scaling laws, as may features on the fracture surface. Because they may be sealed or open at any point and hence may or may not form continuous pathways, not all fractures are hydraulically active. For hydraulically active fractures, variations in aperture can lead to flow channelling and highly variable transmissivity from location to location. Connected fractures may form a continuous network facilitating regional groundwater flow or they may be restricted to isolated compartments in a rock mass. Within a given volume, multiple fracture sets may be present, each with their own degree of connectivity but without any significant hydraulic connection to adjacent fracture sets. For non-flowing networks, they may be filled with stagnant (immobile) groundwater or may be barren due to poor connectivity and no connection to a groundwater source. Flow in fracture systems is dominated by the dimensions and connectivity of the fractures, while solute transport and retardation due to rock matrix diffusion may also be influenced by the nature and extent of fracture infill phases and the extent of solute interactions with the wall rock (flow-wetted surface area) and 'static' pore waters, and also through mixing at fracture intersections. The degree to which rock matrix diffusion occurs at a site is an important aspect to be investigated in a low permeability fractured rock. All of the discussed fracture and fracture network attributes would need to be investigated for a site where potential radionuclide migration in fractures is a consideration.

Because of the complexity of the geometry and physical diversity of fractures, as well as the flow and transport of solutes in a fracture dominated system, there is no possibility to incorporate all the flowing fractures that might exist at a site into deterministic geological and hydrogeological models. Consequently, a crystalline rock mass is typically interpreted as an effectively impermeable background matrix with a superimposed stochastic fracture network that at a minimum takes into account frequency distributions illustrating a range of fracture length, orientation, spacing and aperture. In this approach the fractures may be the only pathways for flow and transport; therefore, the characteristics of the bulk of the rock mass matrix may be ignored. Various software exists that can analyse field data to generate alternative realizations of fracture networks on the basis of underlying models [83], and there are numerous examples of discrete fracture network model applications for repository site investigations or otherwise in the literature [84, 85]. Fractured crystalline rock environments have been or are being investigated extensively in Canada, China, Finland, Japan, Sweden and Switzerland, as well as in other countries. In all of these programmes, detailed fracture characterization has been necessary to understand the potential for radionuclide transport and to stimulate such transport appropriately.

Hybrid numerical models also exist that simulate flow and transport within fracture networks and in interconnected pore space within the bulk of the rock mass. They therefore combine discrete fracture network and continuous porous medium flow and transport concepts. Such models may be appropriate for investigating weathered crystalline rock terranes and sedimentary rock environments characterized by competent, but highly fractured lithologies, such as some sandstones.

Sedimentary rocks are a focus for investigation in some programmes, either because they are the intended host rock for a disposal facility or because the disposal concept includes sedimentary units as a surrounding part of the geosphere, for example in a basement under sedimentary cover disposal concept (known as BUSC). In the past, some site investigations in such environments have tended to concentrate their majority of effort towards characterization of the basement rocks, with less resources applied to acquiring data from the overlying sedimentary rocks. Whilst this may be acceptable if there needs to be early confidence in the integrity of the host rock, there should be acknowledgement that the sedimentary cover sequence will also need to be well characterized at an appropriate level of detail, and adequate resources should be established to do this. Where the host rock is prioritized to the detriment of an overlying sedimentary sequence, which might be due to limited resources and a poorly thought out characterization strategy, it would likely be counterproductive as the cover rocks may play a significant role in isolating and containing the waste and reducing the effects of any future radionuclide transport, for example through retardation, dispersion and dilution.

Sedimentary rocks that have been investigated for hosting a repository include salt (e.g. at the Waste Isolation Pilot Plant in the USA and at Gorleben in Germany), clays (e.g. Bure in France and Mol in Belgium) and volcaniclastic rocks (e.g. Yucca Mountain in the USA). Interpretations of the field observations at these sites have made extensive use of data from modern depositional settings and analogue outcrop studies to develop models of the 'facies architecture' that could be important for controlling flow and transport.

The British Geological Survey has produced a useful report that provides reviews of geological interpretation methods and tools, as applied during site investigations for geological disposal facilities in the UK and elsewhere [86]. The information in that report augments the brief information provided here.

7.2.2. Hydrogeological interpretations

Knowledge about the hydrogeology of a site is required by internal end users of data and information, principally for use in safety assessment calculations leading to estimates of dose and risk, but also to support repository design and construction studies (see Section 3.6.2 for the full range of high level requirements). Focusing first on needs for safety assessment models, the key hydrogeological factors to be considered are the:

(i) Timing and flux of groundwater flow through a postulated repository, taking into account re-saturation and the dissolution of radionuclides leading to their release over time;
(ii) Length and direction of the paths that radionuclides could travel from the repository to the biosphere, taking into account radionuclide retardation, dilution and dispersion;
(iii) Time taken for radionuclides to travel from a repository to the biosphere, taking into account radioactive decay.

The key deliverables expected from hydrogeological investigations, therefore, are not just reliable parameter values acquired at the scale of measurement, but also interpretations providing effective properties with which to populate and calibrate radionuclide transport scenarios and models. These are required to cover very long post-closure time periods and may involve long scales. In addition to directly using measured parameter values, an in-depth understanding of groundwater flow and solute transport mechanisms and constraints is required from the site investigations. The domain of interest includes all potential pathways intersecting a prospective repository, from recharge to discharge. The nature of groundwater driving forces and how the dynamics of the groundwater system have evolved through time should also be evaluated through hydrogeological interpretations.

For repository design and construction purposes, knowledge of the hydrogeology at a site is necessary to plan the layout of surface infrastructure and the subsurface position of shafts, tunnels and vaults. Predictions will need to be made concerning estimated groundwater inflows into these openings during construction and additional protections may need to be put in place where groundwater ingress may be problematic, necessitating continuous pumping or otherwise potentially causing flooding and enhanced corrosion of construction materials. Repository materials will need to be durable for at least the operational period under ambient conditions that could include significant heating in the near field for HLW disposal concepts.

Hydrogeological interpretations rely on groundwater pressure fluctuations in boreholes and recharge/discharge interactions at the surface, as well as the careful analysis of pumping test and tracer test data. Hydrogeological properties can also be directly measured from drill cores and plugs[33] obtained from boreholes. In detail, there may be complex feedbacks between a variety of processes relating to the hydrogeological aspects of a system, as well as with thermal, mechanical and chemical processes. Therefore, to understand the hydrogeology of a site, clearly there also needs to be significant overlap with research within hydrogeochemical, geotechnical and geological domains to take account of the influence of water–rock interactions and impacts on flow and transport relating to differences in various rock mass properties.

Topographic maps and knowledge of the locations and fluxes associated with springs and gaining and losing surface water bodies[34] allows interpretations to be made about likely groundwater recharge and discharge areas. Meteorological data and the monitoring of springs and surface water flows provides information on fluxes into and out of the groundwater system and, in combination with knowledge of the subsurface aquifer or aquitard, allows the construction of a water balance. Hypotheses associated with these interpretations can be investigated using tracer testing and groundwater flow modelling.

In combination with water table and hydraulic head data obtained from boreholes, groundwater level and groundwater pressure contour maps can be produced to provide an indication of the hydraulic gradient and flow directions. Groundwater flow is, for the most part, gravity driven, but flow paths may not necessarily follow a topographic gradient. Therefore, interpretations of flow directions should take account of factors such as heterogeneity and anisotropy in the rock mass and variations in groundwater density. Where groundwater displays variable density in the water column, it is necessary to take into account the temperature profile in a borehole (that is, the geothermal gradient) and knowledge of the chemical constituents in representative samples to derive accurate vertical and lateral components of flow.

There are numerous methods that can be used to derive hydraulic properties based on the analysis of data obtained from hydrogeological testing. Such test analysis also permits interpretations involving the presence and impacts of boundaries, flow mechanisms, connectivity, fluxes and groundwater velocities, for example. Field data can be analysed 'by hand', by plotting pressure and groundwater level data against time on log-log or log-normal graphs and analysing the slope and intercepts. Computerized techniques using automated regression techniques are also commonplace today. The programmes iteratively compare observed time–pressure responses, including pressure derivatives, against a series of programme generated responses until a close match is obtained. In addition to providing derived best fit parameter values, the use of appropriate software allows sensitivity analysis to be readily undertaken. It is important to note, however, that agreement between the observed and modelled data may not be unique and thus any such solution should, at least initially, be taken to indicate only that a qualitative estimate has been found. Consequently, although computer assisted curve matching is a powerful tool, the role of the expert analyst remains important.

All methods of hydraulic testing analysis have their own simplifications and assumptions. As a consequence, there is usually no single definitive result from a borehole test in terms of the derived parameters such as transmissivity and storativity. Because of this, a range of methods of analysis should be applied so that the most appropriate values can be selected, together with a measure of uncertainty. When the use of several analysis methods provides results that converge on a solution, there is likely to be more confidence in an interpretation, presuming of course that the underlying data and assumptions are sound.

Monitoring of groundwater levels and head in boreholes displays periodic fluctuations. These may show annual cycles, representing seasonal changes in precipitation and recharge, or regular and irregular changes that occur on shorter timescales. Irregular changes in groundwater levels that occur throughout

---

[33] Plugs are typically subsamples taken from a core. A key reason to do this is so that three mutually perpendicular plugs are taken at each sampling point and, in this way, the three principle orthogonal vectors for permeability and other properties can be obtained.

[34] 'Gaining' water bodies are fed by groundwater due to the river/stream/lake intersecting a water table, while 'losing' water bodies lose water to the groundwater system as the water table is below their base.

a day or week are known to result from matrix constriction and dilation induced in the rock mass due to transient effects such as barometric pressure changes or seismicity. Individual heavy rainfall events may also impact on groundwater levels in an unconfined aquifer. Regular changes in groundwater heads with monthly, diurnal and semidiurnal frequencies are typically due to the influence of Earth and ocean tides. In a confined or semiconfined aquifer, once the effects of atmospheric pressure have been taken into account, various authors have used spectral analysis methods to estimate effective hydraulic properties from transient groundwater pressure responses. For example, hydraulic diffusivity can be derived by passing fluctuating pressure data into the frequency domain and applying a fast Fourier transform followed by an appropriate filter, before returning the data to the time domain for hydrogeological analysis [87]. Such methods have been developed for deriving regional hydraulic properties for homogeneous porous media and fractured aquifers.

The significance of upscaling was introduced at the start of this section. For large scale models of a site or region, when much of the detailed hydrogeologically relevant structure would not be represented explicitly, simplifications would be used instead. This does not imply that detailed knowledge of the small scale structure is not important, but simply that it is not necessarily directly used for some applications where approximations may be sufficient to address a specific problem. As an example, a continuous porous medium might be characterized in a regional groundwater flow model by a single effective permeability, to represent the groundwater flux through a highly heterogeneous porous medium. It is known that such a model will not predict the distribution of fluxes on a small scale, but if the value of effective permeability is chosen correctly, then the model would adequately predict the bulk flux through the system. The effective permeability can be determined in various ways, such as by simulating the detailed structure over a larger volume and looking at the bulk behaviour of that model. The results of calculations using the detailed model are then used in calculations assuming a simplified homogeneous structure. This is but one aspect of upscaling to derive effective properties. Further examples can be found in Refs [66, 83] in relation to the upscaling of fracture network properties.

It can be argued that the complexity of upscaling and the level of uncertainty associated with some rock types will be significantly lower than that for others, For example, upscaling from rock material to the rock mass at a mudstone dominated site should be more straightforward than for a crystalline site, due to the absence of joints, fractures and other hydraulically relevant heterogeneities. Furthermore, in contrast to a crystalline site, a mudstone site is generally controlled by diffusive pathways that are not scale dependent. Different upscaling approaches are therefore likely to be required for different rock types.

7.2.3. Hydrogeochemical interpretations

Knowledge about the hydrogeochemistry at a site is as important, if not more so, than knowledge of many other conditions, given the potential that hydrogeochemistry has to degrade barriers and control radionuclide dissolution and migration.

Since many of the chemical characteristics of groundwater relate to the time it has been in contact with rock, and the nature of that rock, and since the chemical characteristics of the minerals that fill voids (i.e. fractures and pores) in the rock relate to interactions with the groundwater, the composition of groundwater and minerals sampled at different locations can be interpreted to understand groundwater origins, its movement from recharge to discharge areas and its compositional evolution. With this understanding, one is able to develop and constrain conceptual and numerical models of groundwater flow and solute transport at various scales, also to support interpretations of the evolution of the groundwater system at a site, so that safety assessment studies can more effectively model potential future radionuclide dissolution and transport.

Section 3.6.2 sets out the hydrogeochemical data and information requirements. Note that an understanding of the hydrogeochemistry at a site cannot be divorced from an understanding of the hydrogeology. In order to support the development of hydrogeological and palaeohydrogeological conceptual and numerical models, high quality surface water and groundwater compositional data needs to be acquired through sampling across a site and with depth. Where they are present, minerals coating and infilling voids may also be of value. The analysis of these data and their interpretation provides insights regarding, for example, the mixing relationships between end member water bodies, the location of horizontal and vertical chemical transitions and their hydrogeological significance, as well as the origins of any significant solutes (e.g. salinity) that might affect density and groundwater driving forces, residence times, recharge conditions and past and present flow directions at a site.

Aspects of hydrogeochemical understanding may be required for supporting repository design studies because the ability of groundwater to corrode and degrade engineered barriers and other structures in a repository and thus release radionuclides[35] will, in large part, be dependent on the chemical composition of the groundwater entering the repository. In turn, it is also important to recognize that this interaction with engineered barriers and other repository structures may affect the composition of groundwater, for example interaction with cementitious materials will lead to increasing groundwater alkalinity in the vicinity of these materials and to an alkaline plume further downstream. Such modified groundwater composition and rock interactions along downstream flow paths away from a repository will also influence radionuclide transport. For these reasons, and others, understanding the chemical composition of groundwater at a site, its potential to change composition and interact with repository materials and rock, as well as its potential to mix with other groundwater bodies, is therefore highly significant for post-closure safety studies and repository design and construction.

At a potential repository site, any mixing of compositionally discrete water bodies would reflect advective groundwater flow or diffusive solute transport, or a combination of both. Groundwater mixing relationships can be evaluated using variations in the stable isotope compositions of groundwaters and variations in the concentrations of any constituents assumed to be chemically non-reactive in the groundwater/rock system of interest. Examples of non-reactive solutes that are most useful for this purpose are chloride, bromide, iodide and boron in groundwater systems displaying variable salinity. Once the compositional data are obtained across a number of locations, multivariate statistical analysis of the groundwater compositions can be used for deriving components and interpreting mixing relationships.

Careful examination of mineral infilling relationships in fractures was shown to be particularly powerful for investigating spatial and temporal variations in groundwater salinity at a site for a prospective geological repository in the UK [66]. Nine characteristic and consecutive mineralization episodes (referred to as ME1 to ME9) were distinguished. Of significance for understanding groundwater flow changes over the timescale relevant for repository studies, the latest episode of mineralization, characterized by ME9 calcite crystals, was observed in 'potentially flowing features.' These features were connected fractures that are either flowing at the present time, or were flowing in the geologically recent past, based on the dating of ME9 minerals. Calcite crystals displayed a systematic morphological variation with location (an example is shown in Fig. 43) whereby ME9 calcite morphology was constant or varied progressively from rhombohedral-forms ('nailhead') through intermediate equant forms to elongate scalenohedral forms. The variation in ME9 calcite morphology closely follows the present day variation in groundwater salinity across the site and with depth, where it was established that nailhead forms of calcite typify ME9 precipitates in the freshwater zone, equant crystals tend to occur near the top of the saline transition zone, and scalenohedral forms are characteristic of saline groundwaters. Consequently, in situations where the morphology in an individual crystal has transformed from one form to another, this can be inferred to demonstrate there has been a fluctuation in groundwater salinity at that point. Where the morphological form has remained constant, it suggests that the groundwater salinity has remained constant over the timescale of interest. Because salinity influences groundwater flow and impacts on repository aspects such as the integrity of steel and bentonite barriers, detailed investigations have significant relevance for establishing a robust safety case.

Groundwater age, or more correctly the mean groundwater residence time, generally refers to the average transit time between recharge and the sampling of a groundwater body. Radiogenic and stable isotopes and various natural and anthropogenic tracers can be used to 'date' groundwaters [69] and, therefore, to support inferences concerning pathways (path length, flow directions and time of travel) and compositional evolution. For dynamic near surface groundwater systems in which groundwater bodies are

[35] Solubility and speciation of radionuclides are particularly sensitive to variations in pH, Eh, and pCO_2.

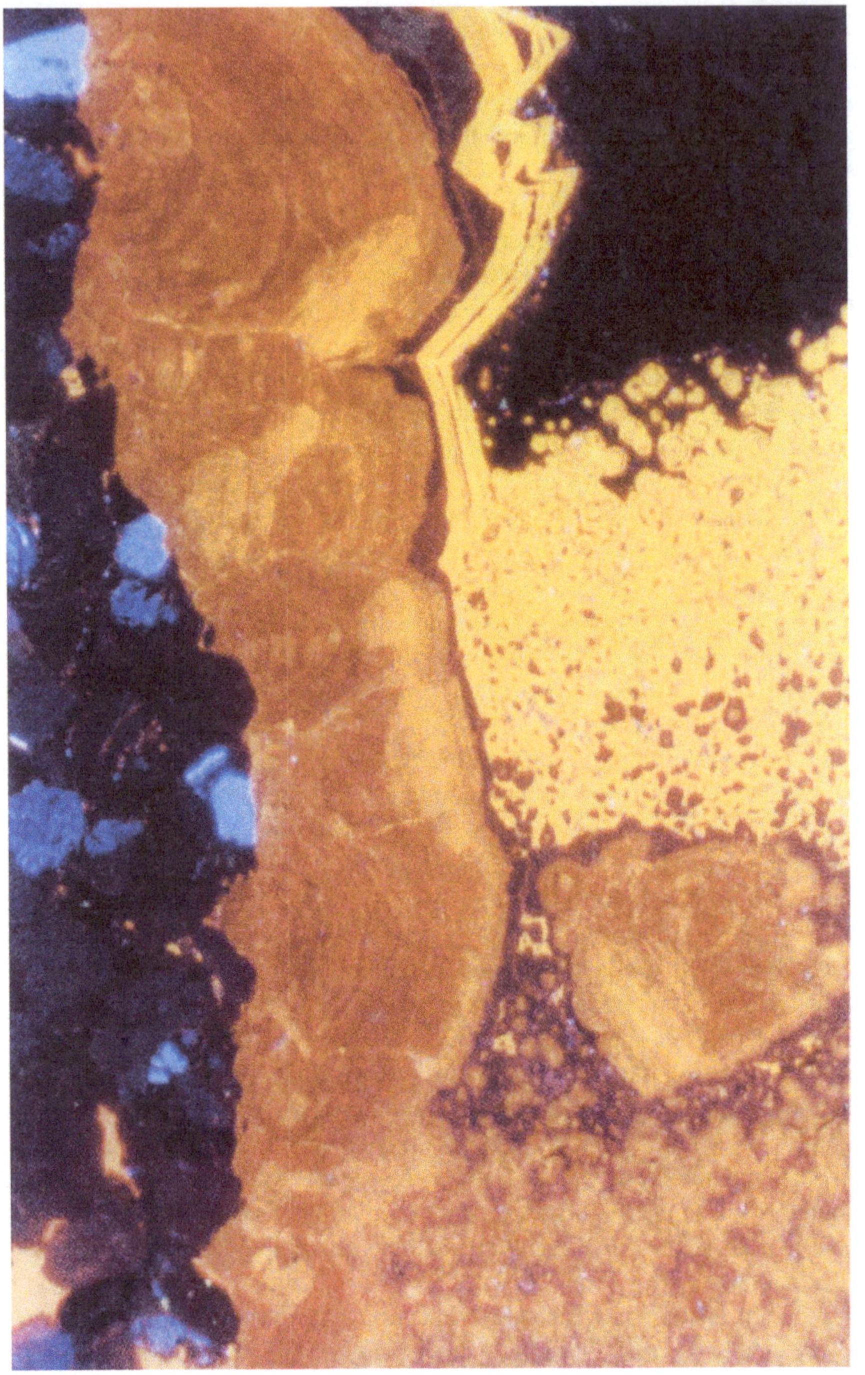

FIG. 43. Cathodoluminescence photomicrograph of a fracture wall (left) lined by ME9 calcite. Generation of an early stage dull orange equant morphology calcite was superseded by precipitation of a finely zoned, bright luminescent scalenohedral ME9 calcite, indicating a transition from relatively brackish to saline groundwater flow locally. Field of width is 3 mm (reproduced courtesy of NWS).

likely to be relatively young, tritium (^{3}H) and krypton-85 (^{85}Kr) can be used to investigate groundwater age and movement. Both are formed naturally by the interaction of cosmic radiation with natural gases in the atmosphere but, since the 1950s, concentrations have been swamped by anthropogenic isotope production. Chlorofluorocarbons (known as CFCs) can also be used to date groundwaters younger than 60 to 70 years.

In deep and/or less dynamic environments, groundwater with ages greater than the timescale of anthropogenic influences can only be inferred hydrochemically from the analysis of natural tracers; usually isotopes. These include natural isotopes of hydrogen and oxygen, which are part of the water molecule, and they can also be used to infer recharge areas (based on altitude or distance from the coast) and climatic conditions at recharge. Other radioisotopes (e.g. ^{14}C and ^{39}Ar) fill the gap between young residence time indicators and radioisotopes used for very old age determinations (i.e. residence times from 60 years and up to approximately 40 000 years before present). For the very long timescales associated with ancient groundwater that are particularly of interest in geological repository studies, ^{81}Kr, ^{129}I, and ^{36}Cl may be used to assess groundwater residence times, potentially up to approximately 1 million years. Naturally occurring radioisotopes may also be used to make inferences concerning the presence of transmissive faults and fractures, for example based on concentrations of radiogenic helium (e.g. ^{4}He derived from the radiogenic decay of U-Th series elements that occur naturally underground). In summary, to interpret the residence times associated with groundwater bodies in a complex groundwater system, a combination of diagnostic age indicators will need to be studied, and this is not a trivial endeavour.

7.2.4. Geotechnical interpretations

Geotechnical studies apply principles and techniques derived from geology, hydrogeology, rock mechanics and soil mechanics. They are used to investigate the nature of soils and rock exposed at the surface to determine their physical and mechanical properties and to assess, monitor and manage hazards posed by natural and artificial slopes and other site conditions. With this knowledge, geotechnical investigations undertaken at the surface are used to support the design and construction of earthworks and structural foundations. Geotechnical studies are also focused on investigating subsurface conditions and materials. In this regard, the underground relationship between different rock types, as well as the discontinuities within and between them, are to be mapped and their physical and geomechanical properties evaluated. This information is used to plan and excavate underground openings such as shafts, tunnels and vaults. The stability of these openings and the hazards posed by subsurface conditions are to be predicted during the design stage and continuously monitored and managed during construction and operations so risks can be avoided and remedial actions put in place as necessary. The geotechnical studies also provide vital information for designing and implementing sealing operations at the conclusion of waste emplacement in a repository.

Given the scope of geotechnical studies to be undertaken at a prospective disposal facility site, it can be appreciated that various factors associated with the acquisition of geotechnical data and their interpretation will directly impact on the engineering design and safety of a repository. In addition to the relative position of the various rock types and their geomechanical properties, they include the local in situ stress field (magnitude and the orientations of the principal stresses), the presence and characteristics of faults and fractures, the impact of weathering and alteration, and the general rock quality.

As well as supporting engineering studies for use during repository design, construction and sealing, geotechnical investigations also have relevance for post-closure safety assessment studies. For example, scenarios used in safety assessments need to take into account interactions whereby geomechanical properties affect and are affected by the hydrogeological and hydrogeochemical conditions of a site. This may be significant in relation to an excavation damage zone, for example. Also, as a conceptual repository design evolves into a detailed design, advanced safety assessment calculations (and the wider safety case) take into account the detailed layout of a repository, which can only be determined with the support of geotechnical knowledge. The scope of data and information requirements relating to the mechanical behaviour of the rock mass at a site is provided in Section 3.6.4.

Insights into quantitative rock mass geomechanical properties are primarily obtained from the laboratory analysis of drill core samples (static and dynamic methods), but can also be derived from surface based geophysical investigations and wireline geophysical logging in boreholes (dynamic methods). Routine geomechanical measurements made on core samples are typically bulk, dry and grain density, uniaxial compressive strength, tensile strength, triaxial strength and triaxial deformability. The important indices of mechanical deformation, Young's modulus, shear modulus and Poisson's ratio are also routinely calculated from stress/strain relationships. Matrix thermal properties (thermal conductivity, specific heat and linear thermal expansion) can be determined if required for use as input parameters in thermal scoping calculations. The variation in rock mechanical effects due to heating are of interest for disposal concepts involving HLW, where coupled thermal-hydro-mechanical processes may be important. Furthermore, a relatively wide range of geophysical measurements can be performed on core samples to assist in the calibration of wireline geophysical logs and to resolve specific issues concerning geotechnical interpretations. These include resistivity, compressional wave (P-wave) ultrasonic velocity for application under unconfined conditions and compressional and shear wave (P-wave and S-wave) ultrasonic velocity for application under confined conditions.

Data on fractures and other discontinuities from field surveys, core and geophysical logging in boreholes can be used in association with measured geomechanical parameter values and with in situ stress data to establish rock quality ratings for the different rock types that might be present across a site. Various classification systems have been devised to assess rock quality and strength, such as the Norwegian Geotechnical Institute Q-system [88, 89], the rock mass rating system [90] and the geological strength index [91]. Essentially, they permit a qualitative assessment of the strength and behaviour of the rock mass, and they are particularly of value for assessing the potential for slip and collapse in the vicinity of underground openings along discontinuities such as fractures and bedding planes. Such effects would be in response to excavation and other disturbances but may also be generated over time due to creep and gradual rock deformation. The rating schemes therefore inform decisions concerning the location and geometry of excavations and the physical support measures needed for ensuring the stability of these manufactured openings underground.

Parameter values obtained from geomechanical studies can be plotted in the form of depth profiles. These may be interpolated between boreholes using appropriate assumptions and methods, such as 3-D geological modelling. Such interpolations are strengthened by correlations with data obtained from the use of geophysical techniques that measure related properties across the 2-D area between boreholes or a 3-D volume, such as would be covered by seismic reflection surveys.

Various computer codes can be used to assist in the design of underground openings in which the deformability of the jointing and the rock can be simulated. The models would be informed by site data and observations, and predictions would be made prior to excavation. In this manner, site data can be used not only to populate numerical models, but also to improve them through calibration and validation. Adjustments to the engineering designs for layout and rock support may then be made as excavations progress on the basis of the results from these numerical models.

7.2.5. Biosphere interpretations

Biosphere data and understanding have multiple uses. For example, in post-closure safety assessment studies, present day biosphere conditions may be used as an analogue for future conditions when a site is under similar climatic boundary conditions. By collecting and interpreting evidence in the present day biosphere, one can interpret past climate and landscape conditions at a site and how they have evolved to the present day. Such interpretations can be obtained from the analysis of data on fauna and flora successions, the contents of lake sediments and radioisotope dating of ice cores and cave deposits, for example, together with knowledge about the likely drivers for landscape changes. In this manner, scenarios can be established to inform safety assessment studies about the timing and nature of future conditions resulting from climate change and landscape evolution [92]. Biosphere data also contribute towards establishing baseline conditions that are used as a starting point for predicting and monitoring the impacts of repository construction. Biosphere data and information would also inform repository design and engineering studies, by constraining the location of surface facilities and the need for drainage, for example. Finally, biosphere data and information have a significant and obvious use in support of the development of environmental impact strategies and assessments.

Given the scope of potential uses for biosphere information, contrasts in environmental conditions and differences in national circumstances (e.g. whether or not an EIA would be necessary), present day biosphere data and information requirements are highly variable between sites. Nevertheless, illustrative data requirements for biosphere studies are provided in Section 3.6.5. In essence, site specific requirements would reflect regulatory needs and expectations, the nature of the waste inventory, the repository concept and the site conditions. As a result, the range of biosphere subdisciplines to be used during biosphere investigations at a site and the focus and prioritization of those studies may be somewhat variable when different programmes are compared, although there will also be similarities.

In general, it could be expected that for a major disposal programme involving relatively large volumes of long lived radioactive waste, comprehensive biosphere investigation studies would be required using experts with knowledge of climatology and other sciences relevant to understanding the climatic and landscape evolution of a site (e.g. palynology, palaeolimnology, speleology, as well as meteorology, geomorphology, hydrology, soil science, ecosystem science (with a focus on soil–plant–animal radionuclide transfers) and biology). Depending on the site and disposal concept, additional inputs might be required in relation to volcanology, seismic impacts, ice sheet and glacier dynamics, oceanography and marine science.

Many of these disciplines will require data obtained directly from investigations at prospective repository sites to develop robust and relevant interpretations and conclusions. It may also be necessary to use proxies from elsewhere or synthetic inputs. The scope of data analysis and interpretation methods in all of these areas is potentially immense and describing them is beyond the scope of this report, but it is important for the RWMO to be aware of their significance for supporting safety assessment and repository engineering studies, as well as in EIAs. Clearly, there is significant overlap with interests in the geological, hydrogeological, hydrochemical and geomechanical domains, especially concerning the nature and potential significance for radionuclide migration across the geosphere–biosphere interface zone (known as the GBIZ).

The IAEA has instigated several major projects of relevance to biosphere studies, including the Biosphere Modelling and Assessment project known as BIOMASS (concerning the development of reference biospheres for post-closure safety assessment), MODARIA I and II (relating to Modelling and Data for Radiological Impact Assessments) and EMRAS I and II (concerned with developing and testing Environmental Modelling for Radiation Safety). Whilst these projects are generic in nature, they have used real site data in many instances and therefore provide a framework for future biosphere site data acquisition and interpretations. Information on all of these projects can be obtained from the IAEA web site. Other collaborative projects of relevance have been carried out by RWMOs and others. For example, BIOCLIM was a project established under the auspices of EURATOM. It aimed at providing a scientific basis and practical methodology for assessing the possible long term impacts on the safety of radioactive waste repositories in deep formations due to climate and environmental change [29]. To achieve this the project developed scenarios for future potential biosphere systems based on knowledge obtained from a range of studies in northeast France, central England, central Spain, northern Germany and central Czech Republic. A series of research reports and publications are available on the Andra website.[36]

Also of note, BIOPROTA[37] is a forum established by several RWMOs and national authorities from around the globe. Its goal is to address key uncertainties in long term assessments of contaminant releases into the environment arising from radioactive waste disposal. The group has undertaken significant research in a number of fields of common interest on behalf of its members. Their work supplements national initiatives and they make a range of interpretive research reports available on their website.

[36] See: https://andra.fr/mini-sites/bioclim/

[37] See: https://bioprota.org/

7.3. USE OF NUMERICAL MODELS FOR PROCESS/SYSTEM UNDERSTANDING

Numerical modelling informs site understanding as part of the interpretation process, as shown in Fig. 42. Models can be used to investigate specific processes operating at a site or to provide a holistic understanding of a larger part of the system. This may be undertaken at any stage of the project due to its iterative nature. Such models would be populated with data from site acquisition activities, but they can also be informed by generic studies undertaken elsewhere (providing raw data and/or modelling conclusions derived from analysis, interpretation and integration). In this publication, a distinction is made between site modelling, used specifically to aid process understanding and inform further site investigation activities, and safety assessment modelling which may use very similar tools and input parameters but which is particularly focused on simulating aspects of the site to derive future prospective dose and risk estimates. Before considering the type of information expected from numerical models, some contextual information is provided next.

Numerical models are one of several types of models that use mathematical concepts to translate aspects of the real world into simplified frameworks so they can be used to provide solutions and insights for the generation of knowledge and to facilitate decision making. They are essential tools to be used in a site investigation project, although they can be time consuming and costly to construct. One major group of mathematical models termed analytical models uses formulations that may provide exact solutions to a problem, whilst numerical formulations approximate underlying equations and thus provide approximate solutions. Because of the complexity associated with many real world site investigation problems, analytical methods may be of limited use, being restricted to an exploration of highly simplified cases. Numerical methods, on the other hand, are often the tool of choice for providing insights into cases of interest in the investigation of a site and its evolution. Other mathematical modelling approaches have also been used to inform problems associated with site investigations, such as statistical modelling, artificial neural networks and fuzzy logic. However, these and other mathematical approaches are not considered further in this publication.

Regardless of the approach, there are seven key steps in a modelling process, as follows:

- Precisely specify the problem to be solved and identify the objectives that need to be addressed by the model.
- Establish a sufficiently detailed conceptual model of the physical system, as needed to address the problem.
- Select an appropriate modelling method together with suitable hardware and software.
- Collate and review all of the available data and other relevant information with which to parameterize and calibrate the model.
- Set up the model framework, including the size of model domain, the processes to be modelled, dimensionality and the discretization. Parameterize the model, including calibration as necessary, and run the calculations.
- Analyse the results to ensure they are reasonable and understood. Undertake further calculations, including sensitivity analysis, as required.
- Present the results and associated interpretations.

These steps may be followed by model validation, if such validation is possible (see below).

It is important to note that the level of detail to be included in any modelling activity should be determined by the complexity of the problem to be addressed and the availability of data to support a suitably detailed representation of the system, as well as the ability of the modelling tool to deal with the challenge. In essence, a model should be as simple as possible, but no simpler.

In the field of numerical modelling, various methods exist, such as finite difference, finite element, and boundary element methods, as well as hybrid methods that combine approaches. All have advantages and limitations and may therefore be appropriate for use in certain applications, but not in others. The approximations inherent in numerical model calculations may be vanishingly small and therefore not of any meaningful concern. However, when dealing with highly complex challenges involving significant time and length scales, and where computational resources do not permit high levels of temporal or spatial resolution, approximations may more significantly affect results and inferences. These approximations stem in large part from discretization issues in which small scale heterogeneities are not adequately represented, which leads to truncation or rounding errors in the mathematical representation of the underlying physics. They are but one source of numerical modelling uncertainty. Aside from contributions from parameter and conceptual model uncertainty, other sources of numerical model uncertainty include necessary simplifications and inaccuracies during model design and set-up, as there will always be differences between a model and reality.

In general when dealing with a relatively complex issue, it is advisable to start the modelling process by creating coarse or highly simplified representations of the problem domain. This may initially be something as basic as very quick and informal analytical calculations. Such models not only provide a very rough estimate of the results one should expect from more detailed models, but they can also help to identify the critical aspects of the system that would need to be rigorously addressed when setting up a more detailed model, compared to those aspects that would have less impact on results (essentially an initial sensitivity study). In some cases, the use of a numerical model may not be justified, due to the lack of data or conceptual understanding. Here, any use of a numerical model would likely give ambiguous (at best) or misleading results. In other cases, the results from numerical models may be uncritically considered as accurate, when they may simply reflect the maxim 'rubbish in, rubbish out,' which means that the quality of the output depends on the quality of the input. Numerical instability can occur in a model when inappropriate time steps are used or the grid spacing is out of balance. There is also always the challenge that a software tool has been used by inexperienced practitioners without any technical knowledge of the application. In every modelling study, the personnel charged with the task of developing a model should clearly understand the underpinning physics and chemistry that direct simulations, the essential mathematical formulations used, and the importance of ensuring that boundary conditions, initialization and other aspects of parameterization are carried out appropriately and are documented. RWMO managers and users of information derived from numerical modelling studies should be aware of these potential issues and take them into account when assessing confidence in interpretations derived from numerical models and how they are to be used in decision making.

Despite such concerns, when properly constructed and used appropriately, numerical models find valuable applications in a wide range of science and engineering areas. This is due to their ability to deal with vast amounts of input data, complex geometries, variable boundary conditions and significant heterogeneities within a system, as well as many non-linearities that are present in coupled process interactions. In site investigations, numerical models are used extensively to support interpretations associated with geology, geophysics, hydrogeology, hydrochemistry and rock mechanics. In each field of application there are a wide range of commercial software programmes available, as well as numerous open source codes. Models are used to simulate groundwater and multiphase flow, radionuclide transport, rock deformation, long term climate variability, land uplift and erosion, glacier dynamics and ecosystem evolution, to name but a few areas relevant to site interpretations. The use of numerical models for simulations and optimization in these and other areas allows an RWMO to perform the following:

(i) Reproduce the observed state of the natural system and improve the understanding of current conditions at a site, including for situations where there are expected to be complex coupled processes of interest in the subsurface;
(ii) Design laboratory and field experiments and analyse the resulting data;
(iii) Make projections based on alternative scenarios about the future performance of a proposed repository system over extended time periods;
(iv) Assess parametric and conceptual uncertainties in these projections (the above points are modified from information in Ref. [93]);

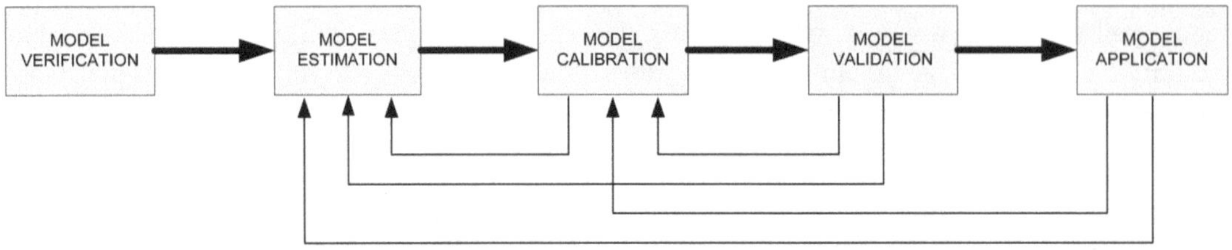

FIG. 44. Relationships among the terminology associated with the development of numerical models and their feedbacks for testing and improvements, as defined in the glossary and body text.

(v) Support decision making and aid the development of safety assessment models by identify key processes relevant to radionuclide migration, establishing boundary conditions and providing derived parameter values.

As an example of how the above aims may be achieved, consider the use of hydrogeological numerical models in support of site interpretation. Hydrogeological models are concerned with groundwater flow, and hydrogeological modelling is an area of application common to almost all site investigation projects. Interpretations typically progress from generic models based on very limited data, to relatively complex models of a system supported by a considerable database of site specific information. Given an appropriate modelling tool and input parameters, it is possible to construct 1-D, 2-D and 3-D models for steady-state and transient conditions. Many groundwater flow modelling codes are available, and several are able to incorporate solute transport processes.

Groundwater flow and solute transport models need to be able to simulate groundwater movement based on the equations for the flow of water through a porous or fractured medium, or a combination of the two. Groundwater velocity in an impermeable potential repository host rock is sufficiently low so that flow can be described by Darcy's law, where the bulk flow is proportional to the gradient of the driving force. This force arises from a combination of pressure gradients and density differences. Salt and other dissolved solids, as well as temperature, affect the density of a fluid and, where present, are to be taken into account in setting up the model. Further equations govern the transport of dissolved solids and heat. These transport-relevant equations are coupled to the equation for groundwater flow through the fluid density. This is appropriate because, as noted, density differences give rise to flows, which in turn lead to a redistribution of the concentration or temperature fields. Dispersion is often represented by a velocity-dependent diffusion coefficient. The transport of solutes (or heat) in groundwater is thus described by an advection–dispersion equation (see Ref. [58] for its derivation).

In a fractured rock, the properties of individual fractures and their connectivity control the overall properties of the flow system. Several algorithms have been developed to calculate groundwater flow and solute transport through the fracture network. It is not generally possible to specify the detail of each fracture with regard to its length and width, location and orientation, so a stochastic approach is typically used in which realizations of the fracture network are generated, each of which exhibits the same statistical properties as the measured fracture network. Typically, the fractures will be categorized into a number of distinct fracture sets, each of which has a characteristic frequency distribution of orientation, fracture length and hydraulic properties. These properties are then described using simple mathematical distributions which are randomly sampled to generate simulations of the fracture network. Often, in the absence of more detailed data, the fractures are uniformly distributed in space. Enough fractures are generated so that the simulated fracture network has the same fracture frequency along appropriately oriented sample lines as the physical fracture frequency (i.e. as seen in a field setting, geophysical wireline images, physical core logs or otherwise inferred). More sophisticated fracture generation schemes can produce clustered fracture distributions or other structures where these are needed. Modelling studies generally assume that the same fracture statistics apply across large volumes of the fractured rock, but this need not be the case. Once the fracture network has been generated, fracture intersections are calculated, and the system is discretized, ready for use in flow and transport calculations (see also Section 7.2.1).

It is important for hydrogeological models to be able to deal with different types of boundary condition, including conditions of specified groundwater pressure and concentration (e.g. to represent onshore and seabed boundaries), no-flow boundaries, and boundaries crossed by a specified flux of water or solute. Selecting an appropriate set of boundary conditions for a fracture network model can be particularly difficult. One solution is to simulate a larger volume than the volume of direct interest and by doing so impose boundary conditions sufficiently far away from the volume of interest internal to the model. Alternatively, the larger volume model (e.g. regional scale) can be used to generate representative boundary conditions for a smaller scale nested model.

The above discussion summarized a general modelling approach in the context of hydrogeological investigations and interpretations. Similar descriptions could be provided for numerical modelling applied to hydrochemistry, rock mechanics or elsewhere. However, in the interest of brevity the reader is advised to consult with experts or the literature if such information is required.

Once a numerical model has been set up to address a well defined problem with specific objectives, it can be used to generate initial solutions (estimates). These may then be improved by comparing preliminary outputs with measured site parameters that have not been previously used in the model's construction. For example, a simple time varying hydrogeological model might produce estimates of groundwater head fluctuations over a year at a particular location where there is a monitoring instrument in place. The modelled time series estimates can then be compared against the monitored data. Where there is a mismatch, the modeller should first seek to understand possible reasons for the difference. Then, and because measurement observations depend strongly on the specific properties local to the measurement locations, the modeller should fine-tune model parameters that relate to those local properties so that updated model estimates better match the measured time series observations from the field. This history matching is a brief and simplified explanation of the process of model calibration, which is a component of model development, rather than an end point in which results are produced for a specific application (Fig. 44). There may be several stages involved in model calibration, allowing for models to be progressively refined until there is sufficient confidence in the ability of the model to represent adequately either the system or some part of it. In this manner, the use of calibrated models not only significantly improves confidence in the understanding of a site, but also provides confidence in any model output that might be used in the form of future projections for which the term 'predictions' are inappropriate, because they can never be proven.[38]

Some models may be tested through the process of validation. Unlike calibration, where model estimates are compared to parameter values unrelated to the population of the model in order to make corrections prior to final outputs, validation requires predictions to be made (i.e. they are a result with which there is some associated confidence). These predictions are then compared with independent measurements that have not been used to set up the initial model or for calibration, and the goodness of fit says something about the accuracy of the model. To aid model validation, specially commissioned tests and experiments might be used to provide additional data in the form of a disturbance independent from those parameters used to initially populate a model. Ideally, validation would involve blind testing, whereby the modeller has no early access to the data with which the model will be validated [94]. The predictions would then be compared with observations. In order to achieve satisfactory model validation, the predictions should fall within an envelope of acceptable results established before checking the results against observations.

As an example, a model that generates predictions of groundwater inflow locations and fluxes might be made in advance of tunnelling associated with the construction of an underground disposal facility. For this type of situation, proposals regarding which models are to be tested, the actual predictions and the evaluation criteria that are to be used for testing the models should all be published in advance.

---

[38] It is recommended that the term prediction not be used when there is no possibility to observe and analyse outcomes. In such situations, where future outcomes may never be known due to long timescales, for example, the terms projection or expectation would be more appropriate (also see footnote 39).

Because validation requires predictions to be made and checked, this can only be achieved where it will be possible to observe actual outcomes.[39] Therefore, for models that are generated to inform about possible outcomes under hypothetical conditions in the far future, as expected from safety assessment models, for example, a projection based on a scenario comprising a specified set of input conditions and assumptions is made, but validation is not strictly possible. However, the same modelling environment might be validated where a prediction relates to a condition that can be observed today and where the starting point for the model was at some time in the past. This was the approach used by Nirex for simulating the dynamics of density-driven saline groundwater movement at the Sellafield site in the UK [95]. It does not validate a model for its application in regards to a future scenario, but if one presupposes the natural conditions that pertained in the past will continue into the future, with any necessary adjustments for human impacts and altered natural conditions, then this would suggest a degree of confidence in the ability of the model to simulate reality.

One final aspect to be aware of in relation to numerical modelling concerns the role of sensitivity studies. These studies are undertaken to investigate which parameters used to populate a model impact significantly on model results, why and to what degree. As such, they may be used to better calibrate a model or to investigate the nature of key uncertainties associated with data and conceptualization. This is accomplished by varying input parameter values in a systematic and structured manner and rerunning the simulations, so the impact of varying each set of parameters can be assessed. The change in a parameter value should be within the range of realistic values that exist for that parameter, such as derived from a frequency distribution of field measurements. These variations would not be chosen randomly, but would be selected methodically, typically using fixed increments. Manual sensitivity analysis can be extremely laborious, but many software programmes are now able to undertake an automated investigation. The results of a sensitivity study should be analysed to consider whether they are consistent with expectations with respect to the underlying conceptual model. Where this is not the case, the modeller should investigate possible reasons for any discrepancies. Where a particular property (or set of properties) is found to have a large impact on results that have some significance (e.g. for safety or engineering designs), but there is a high degree of uncertainty associated with parameter values, this could indicate where additional effort would be beneficial to reduce the uncertainties associated with those data sets to have confidence in the results of the modelling. If, however, the uncertainty is small or irreducible, the sensitivity study could suggest that planning for remedial or mitigating actions might be appropriate.

[39] A *prediction* is defined as a statement about future events, quantities or circumstances that are expected to happen. Following this definition, the term *prediction* can only be applied when one is very confident about a result such that it is probable with a high degree of certainty. Furthermore, a prediction requires that it can later be checked for accuracy against the outcome, otherwise a prediction becomes no better than any other statement about the future. Predictions made as part of the numerical modelling of a system or system component at a site may be entirely appropriate, for example when evaluating anticipated groundwater inflow conditions into a tunnel still to be excavated. Compare this to a situation in which an outcome is considered possible, as opposed to probable, such as for an outcome in relation to long term climate change or radionuclide transport from a geological repository to the biosphere. In such circumstances the confidence in the likelihood of occurrence is less than for a prediction and therefore the term *projection* is preferable. Confidence in realizing an outcome is less than that associated with a prediction because the uncertainty associated with scenarios expands non-linearly into the future. Therefore, it is accepted that any future conditions used to populate a model or generated during its simulations are unlikely to match the actual conditions that will exist at a site over the simulation timescale. Confidence reflects many factors, for example that certain assumptions may not be correct. Consequently, it is accepted that the results of climate models and a post-closure radionuclide migration safety assessment model are unlikely to be entirely accurate and, in any case, they will never be able to be proven during our lifetime. A projection is only one among many possible estimates of a plausible outcome rather than an exact prediction. A 'preferred' or 'best estimate' projection may be specified, supported by justification. Alternatively, the results of several simulations involving different scenarios may be combined to be presented as an envelope of plausible outcomes within which there is high confidence that the modelled event, quantity or circumstance would fall.

7.4. INTEGRATION AND GEOSYNTHESIS

Integration is defined here as the process of synthesizing separate and distinct data sets and information, together with any understanding derived from their interpretation made prior to integration so that, in combination, they can be further analysed and interpreted in the context of the complete disposal system at a site. Thus, integration requires total system thinking, rather than considering data sets as simply discipline specific aspects of a site description or as discrete and separate component parts of a larger system. Integration is here considered, therefore, to combine all available data and information, within specific disciplines and also across them. The expected outcome is a documented self-consistent, comprehensive and holistic understanding of the site and its evolution to be presented as a series of conceptual or SDMs, or as a total system descriptive model (Section 6.13). Note the integration of site data and information may potentially include inputs from generic research studies as required to confirm or augment site specific understanding.

Integration is termed 'geosynthesis' in some programmes (e.g. by Nagra in Switzerland, Andra in France and the Nuclear Waste Management Organization in Canada). The Nuclear Waste Management Organization has formally defined geosynthesis as "a geoscientific explanation of the overall understanding of site characteristics, attributes and evolution (past and future) as they relate to demonstrating long-term deep geological repository performance and safety" [96]. This definition emphasizes that holistic interpretation and integration are not just concerned with the natural evolution of the site to the present day, but also that geosynthesis incorporates an assessment of potential disturbances induced by repository construction, operation and closure, including a geologically rational assessment of the long term future evolution of the site that includes these and other anthropogenic disturbances, as well as the impacts of natural driving forces.

Integration is a vitally important aspect of the interpretation of a site, because the natural system is interconnected on multiple scales and through the interactions of multiple processes. Consequently, it is essential to allow for the potential for coupled processes to influence decisions when considering the post-closure safety of a site, or the optimization of repository design elements.

It can be appreciated that personnel charged with the task of integration need to consider the interpretive effort associated with individual activities and disciplines, for which being familiar with the underlying data acquired in the field and its contextual setting is important. The mechanism for integration is best accomplished by setting up a specially convened group of leading experts from the range of disciplines represented within a site investigation project. This group should meet regularly to provide updates on developments in their own fields and to undertake the task of integration.

An interaction matrix has been one tool that has been used very successfully to illustrate potential feedbacks and relationships at a conceptual level. Originally developed for use as part of a rock engineering systems approach [97], this tool links FEPs to assess the significance of interactions, either qualitatively or through the use of process models. The resulting insight can then be used to generate plausible scenarios and help populate safety assessment models [98]. Because the tool is able to discriminate at some level between safety relevant and/or repository design relevant FEPs, as opposed to those that have little impact, the tool is highly useful for focusing effort. An interaction matrix comprises a 'leading diagonal' of variables or subjects that are interlinked by off-diagonal processes or events and these populate the rest of the matrix (Fig. 45). The subjects of the leading diagonal might be physical domains (e.g. the near field, the geosphere and the biosphere) or attributes of a system that drive process interactions (e.g. rock stress, groundwater flow and fracture network properties).

In addition to generating interaction matrices, the group tasked with integration should be involved with generating conceptual models and SDMs that are approved for use. These are discussed in Section 6.13.

Figure 45 shows how the interaction matrix is used to support integrated interpretations and conceptual model development and for focusing future site investigation activities on aspects of a site system that have a potentially significant impact on safety or repository design. Note the off-diagonal

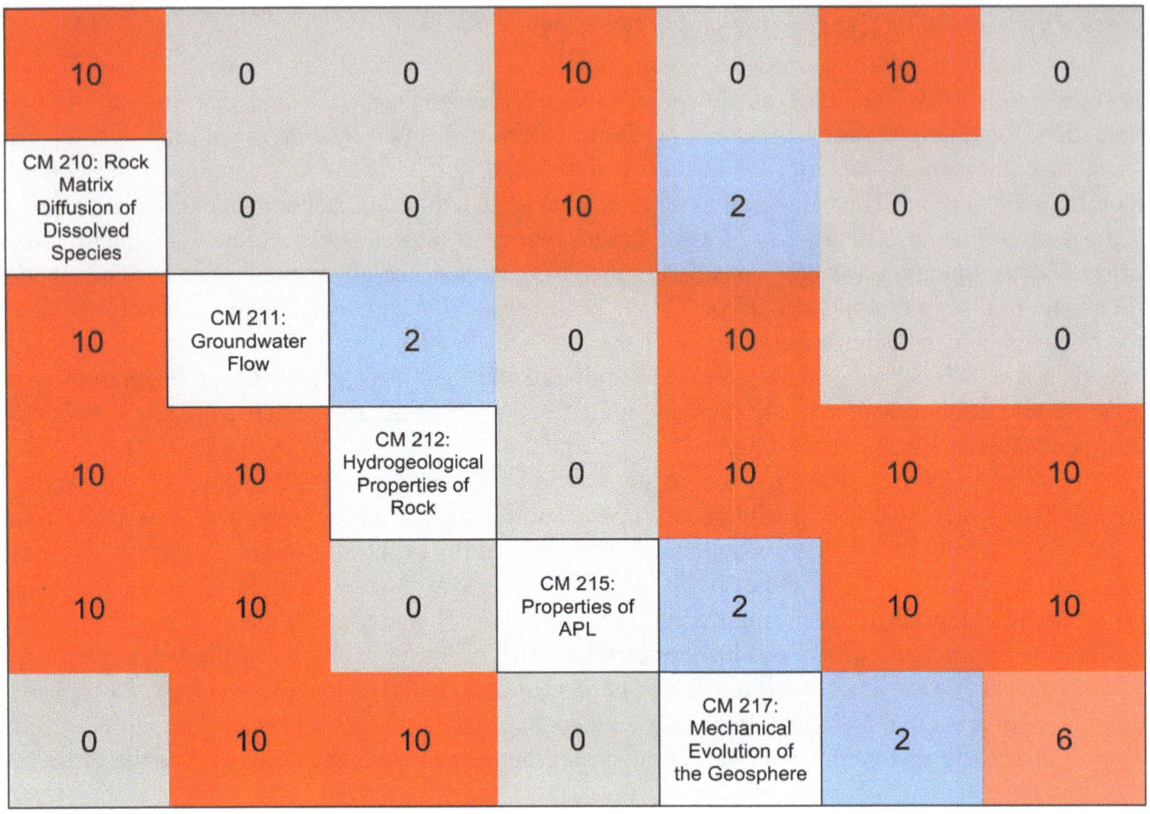

FIG. 45. Part of a sample interaction matrix (reproduced courtesy NWS).

boxes link leading diagonal boxes using a clockwise convention and these record the inferred strength of interactions.

7.5. CONCEPTUAL MODEL AND SITE DESCRIPTIVE MODEL DEVELOPMENT

Conceptual models and SDMs are two related but different instruments intended to succinctly integrate, capture and communicate site understanding. Although they are related to each other, they are considered distinct but complementary, as evidenced by their use by leading RWMOs.

7.5.1. Conceptual modelling

A conceptual model is defined as a brief, clear and unambiguous description of the key features and processes operating within a specified system. It provides a simplified, synoptic representation of reality intended to aid in the development of mathematical models and communicate an understanding of key concepts. It is essentially a framework based on justified simplifications of reality that summarize the understanding of the nature of a site (or other domain) at a particular time, in the sense of when the conceptual model was created and also in relation to the period within which the model applies: either the present day, the past or in a hypothetical future. A conceptual model should define the volume of interest. It would seek to capture relevant and significant FEPs within that volume at various length scales. It would also specify the nature of the boundaries encompassing the volume. The scope of a conceptual model and the level of complexity associated with it should be determined by the use to be made of that model, which itself should reflect a graded approach.

A conceptual model should be based upon observations, data and expert judgement and it may comprise words, illustrations, quantitative data and symbols. Any diagrammatic representation of a

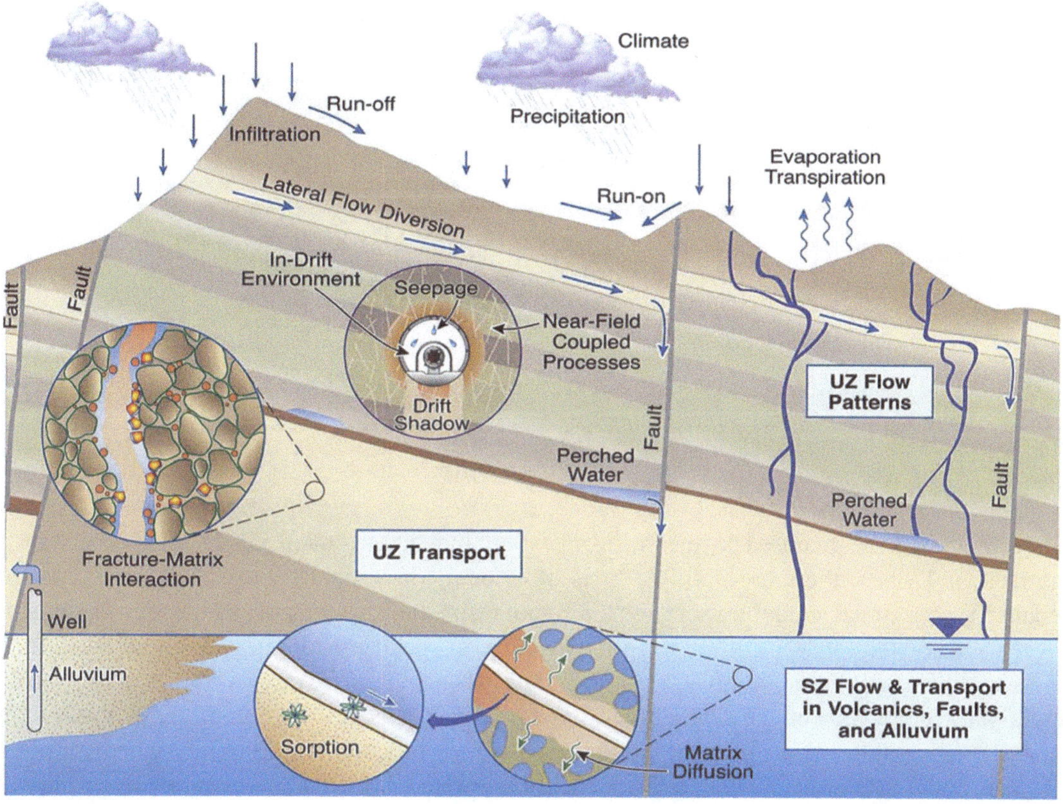

FIG. 46. A conceptual structural model of Yucca Mountain [99] (reproduced courtesy of the US Department of Energy).

conceptual model should be accompanied by text that clearly states the assumptions that underpin the model, as well as the provenance of the data and information used in its generation. Figure 46 is an example of an illustration associated with the conceptual model of the Yucca Mountain site in the USA, developed during the course of those investigations [99].

While some conceptual models may attempt to describe the nature of all relevant features and processes in a given system, others may be used to focus on particular aspects relating to specific processes (or groups of processes defined by a discipline) or for spatially discrete subdomains within a system. Such models can be used to describe, for example, geological, hydrogeological or hydrochemical conditions, and behaviours at a site (Fig. 47). Detailed conceptual models may also be generated to especially illustrate specific features and processes associated with radionuclide transport and the impacts of thermal effects.

A conceptual model is typically generated as a result of the consensus opinion of an expert group, taking into account all of the available information and using all of the experience of the group members. In establishing a conceptual model, this group should take into account the uncertainties associated with the values of the parameters that are used to inform the conceptual model, as well as the uncertainty in effective properties that might have been derived during upscaling, if these were used to inform conceptual model development. These effective properties will themselves be uncertain because of the uncertainty in the underlying parameters and simplifications inherent in the upscaling method. Because such uncertainties will always exist, there may be more than one concept that aligns with the available information. Where this occurs, and where differences between alternative models have a potentially significant impact on post-closure safety or repository design, equally plausible alternative concepts should be carried forward in parallel until more data become available to differentiate between them and in turn exclude them from further consideration.

In addition to parameter uncertainty, it is important to recognize that the development of any conceptual model will itself generate further uncertainty. This is because a considerable amount of

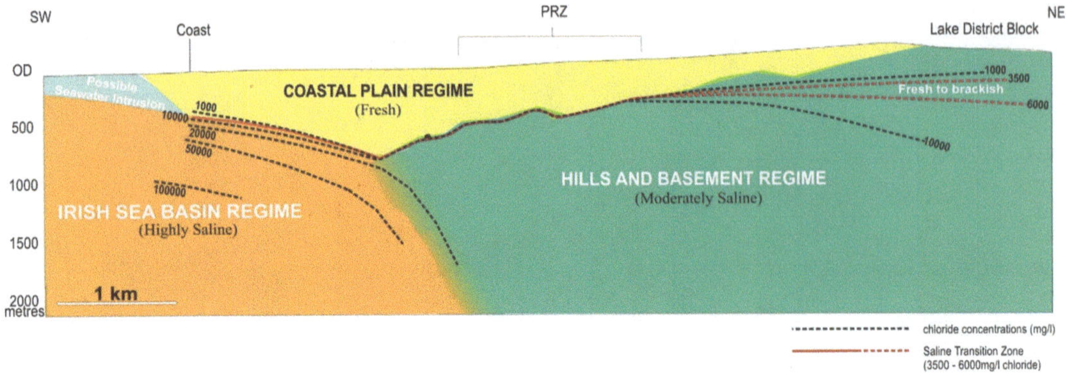

FIG. 47. Illustration supporting a hydrochemical conceptual model of the Sellafield site in the UK [100] (reproduced courtesy of NWS).

subjective input will be required in the form of expert judgement, even when there is a high degree of consensus and where there are multiple lines of evidence in support of an inference. Consequently, and because a conceptual model only reflects a hypothesis about the nature of the site (something that can never be absolutely proven), there is an opportunity to specify parameters and processes for which further data acquisition should be given priority in later stages of site investigation. It is noteworthy that conceptual uncertainty may be the single biggest factor impacting an evaluation of the safety of a site. For this reason, considerable effort should be expended to reduce conceptual uncertainty wherever possible.

As additional data are collected, analysed and interpreted, site understanding evolves and not only is the range of valid conceptual models expected to decrease, but the confidence in remaining conceptual models should become greater. Note, however, that it is to be expected that during site investigations, unforeseen results will arise that may cause periodic setbacks in understanding. These results are then to be investigated and resolved if they potentially impact safety.

Although there may be several equally justified conceptual models of a site at any one time, it is reasonable for experts to select a preferred model, especially during the early stages of an investigation. However, there is always a risk that such a preferred model would become the de facto accepted model, to the exclusion of consideration of all other models.

Once a conceptual model (or series of conceptual models) has been established and approved for use, it can be used as the basis for constructing numerical models of the site and for informing safety assessment studies, as well as for communicating site system understanding to stakeholders. Conceptual models are a prerequisite for numerical modelling, while enhanced conceptual models may be an outcome of numerical models that generate additional understanding.

7.5.2. Site descriptive modelling

Another representation of site understanding is an SDM. According to SKB, the originators of the idea [101], an SDM is an integrated multidisciplinary description of a site and its regional environment with respect to the current state and naturally ongoing processes. The modelling strategy to develop an SDM involves an evaluation of primary data that has undergone interpretation and integration, as well as quantitative modelling in 3-D and an overall confidence evaluation. Such numerical modelling does not include projections (simulations describing possible future conditions) or predictions, thus an SDM applies only to site conditions and behaviour on the present day. For SKB, the disciplines that contribute to the system-wide SDM are:

— Thermal;
— Rock mechanics;
— Hydrology and hydrogeology;

— Hydrogeochemistry;
— Transport properties;
— Ecosystems.

The selection of parameters for use in SDMs and the geometrical framework adopted are based on an underlying conceptual model, one that is populated by primary data. This data and any derived data contributing to the SDM are first evaluated within each discipline and then the evaluations are cross-checked between disciplines to ensure consistency and completeness. It is the individual discipline based models that, when combined, define the total system SDM. The use by SKB of an SDM can thus be taken to represent the total site system based on integrated interpretations that describes the geometry and properties of the bedrock, groundwater system and surface environment, and the interacting processes and mechanisms that are relevant for understanding the evolution of the site to the present day. All individual contributing models and the overall model undergo uncertainty evaluations and confidence assessments and the entire process is documented. The six discipline based descriptive models and the total system SDM are used as an input to the preparation of a performance assessment and engineering designs for the facility at the site.

In SKB's use, SDMs are distinct from conceptual models mainly as their production involves the explicit generation and calibration of supporting numerical models that aid description. Also, conceptual models may incorporate future representations of FEPs and their interactions, whereas an SDM is confined entirely to a description of site conditions today.

There are several components to be included in the preparation of a descriptive model, including:

— Definition of the volume to be represented;
— Geometric units defined so that spatial variability can be adequately represented;
— Parameter values and/or statistical distributions assigned to each of the geometric units.

An example of the strategy used for developing a geological SDM is provided by SKB in Ref. [102] and an illustration of an SDM developed by NUMO in Japan is provided in Fig. 48. Note that in Japan, host rock or site has been selected to date, but a generic safety case methodology adopting an SDM based approach has been developed [103].

7.6. CONCLUDING COMMENTS

The preceding discussion focused on data interpretation and integration activities and the uses that can be made of the resulting information. These activities use various methods to better understand present day conditions at a site and also to provide information about its geological evolution over a timescale that is relevant for safety assessments. Annex IV provides a case study from Sandia National Laboratories (USA) illustrating the relationship between site characterization data and their use in performance assessment modelling.

Interpretation and integration would be ongoing throughout site investigation until its conclusion (which would typically extend beyond siting). Numerical models are an important tool to aid interpretations and are likely to be used to simulate site conditions, as well as to resolve outstanding issues of relevance and importance to site understanding. Conceptual models are a prerequisite for numerical modelling and also one of the results of enhanced understanding derived from the interpretation of such models. Other results from interpretation efforts would inform the planning of further data acquisition activities, including those involving tests and experiments. At various stages, there would be planned data freezes established within the programme so that site data and integrated interpretations in the form of conceptual models and SDMs could be fed as a package to the end users.

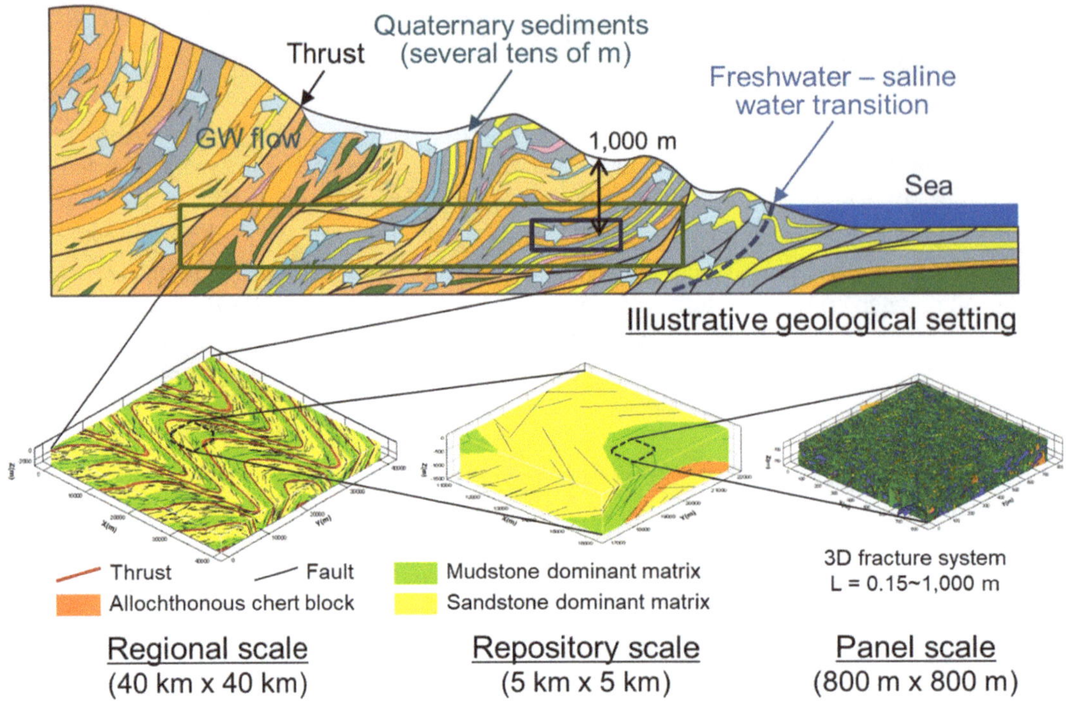

FIG. 48. Sample nested SDM of pre-Neogene sedimentary rocks and associated structures in Japan (reproduced courtesy of NUMO).

The main data sets of primary and processed site data to be provided directly from data acquisition activities were summarized at the conclusion of Section 6. The key deliverables arising from interpretation and integration activities are specified as follows:

— A comprehensive database of secondary (derived) site wide and regional parameter values and information obtained during various stages of interpretation, including from interpretive numerical modelling. These data and information would be accompanied by commentary regarding their derivation and the confidence associated with them. This would take account of uncertain input parameters, numerical modelling uncertainties and alternative conceptualizations of local processes, including the impact of coupled processes and the system as a whole. Supplementary contextual information, such as reports describing the interpretation process and the tools employed, as well as results and conclusions, would also be provided. Results from numerical models might also include a temporal dimension to illustrate the evolution of the site and its environs to the present time.

— A series of conceptual models presented at various length scales and possibly for different time periods, intended to briefly but clearly illustrate and document the interpreted relationship between the key FEPs over the domain of interest and over the timescale of interest. These might be equally plausible alternative models, or they might be a series of discrete conceptual models representing particular aspects of a site, such as the groundwater flow system, the hydrogeochemical distribution of groundwater bodies or the geological framework. A unifying total system conceptual model might also be provided to assist understanding. The conceptual model would comprise text supported by illustrations and numbers as necessary.

— A series of SDMs provided as a series of discipline based site descriptions, all of which would be synthesized into an integrated SDM for the whole site. The descriptive model would also include information concerning the site's regional setting and how this impacts local FEPs. The geology would provide the framework for all other models. The descriptions associated with an SDM would include the geometry of system components, populated by properties. These would comprehensively describe the site at the present day and they would include the results of any numerical modelling undertaken to better understand the site and its evolution to date.

8. WHEN TO CONCLUDE SITE INVESTIGATIONS

8.1. INTRODUCTION

Throughout the site investigation project for radioactive waste disposal, a key challenge is to address all requirements in an efficient manner, iteratively informing decisions concerning site selection and site suitability as a programme progresses towards the implementation of a safe, effective and accepted disposal facility — noting that complexity and the number of requirements may increase as the project evolves and as it calls for a more robust basis to assess safety and feasibility. A closely linked and equally challenging aspect of a site investigation project is knowing when it is acceptable to conclude data acquisition and interpretation activities at the various stages of a siting process. In other words, what criteria can be used to establish when there are enough data and information available to justify a decision to either (i) conclude a stage of investigations undertaken as part of a screening process moving from many sites to one or a few; (ii) move from surface based investigations at a site to a phase of detailed underground characterization; or (iii) complete field and laboratory studies in advance of the submission of an application for regulatory approval to begin repository construction. At each of these milestone events, there will be a need to freeze site investigation database inputs and report the current status of data and interpretations.

A decision to complete site investigations, to narrow the range of potential sites during a screening stage in the siting process, may be entirely within the control of the RWMO. Alternatively, the decision making responsibility might rest with an outside agency or committee, typically nominated by national government. Decisions about siting should be made by the responsible entity within the framework of a defined and accepted decision making process established during the planning phase of the repository development programme. In advance of a decision to conclude investigation at a site during site screening, the basis for either discarding a site, collecting further data to support a decision or advancing a site to the next stage of the siting process should already have been set out, reflecting the social, environmental and technical attributes required of a site and the application of exclusionary, avoidance and preferential criteria. Deliberations and judgements would be based on a comparative assessment of each site, taking into account whether the data and understanding available is sufficient to meet specified requirements and of sufficient quality to adequately inform decision makers. More frequently, deliberations are being supported by the use of tools such as GIS and multiattribute decision analysis algorithms. Where site conditions indicate exclusionary criteria are not applicable, several sites may have to be evaluated against one another and it will be important to recognize that much uncertainty will likely remain concerning the characteristics of a site. However, during the screening process, the degree of uncertainty associated with a site should be balanced against a need to rationally assess preferences between sites on the basis of existing data and understanding. Also, depending on the openness of the process and national circumstances, the degree to which stakeholder participation is involved would also be an important factor during screening, such as community veto rights. Other aspects might also come into the decision making process including cost considerations and funding availability and other practical constraints such as the need to maintain momentum within a programme and limitations on access or planning rights. For those sites taken forward to the next stage of a siting process, investigations would resume at the earliest opportunity.

The basis for a decision to move from entirely surface based investigations at a site towards additional work carried out underground might be more difficult to assess objectively unless specific criteria have been defined in advance. Ultimately though, the decision would reflect the sufficiency of existing data and information in relation to the costs and anticipated benefits associated with investigating underground. Even in the face of a high degree of uncertainty, to merit going underground there would have to be a reasonably high level of confidence in the prospective performance of a site and its potential suitability to host a repository, given the significant direct and ongoing costs likely to be incurred in constructing and maintaining underground structures intended to further characterize the rock mass. Another consideration is that investigation underground would very likely require planning permission from the local authority and also environmental permits. Obtaining these would not be trivial. Whether or not the regulatory authority responsible for the authorization of activities related to radiological protection or nuclear installations would also become involved at this stage would likely depend on a number of factors including, for example, whether the underground investigation is to be carried out in a facility intended to be the first phase of a repository development (should the investigations demonstrate site suitability) or if it is to be a stand-alone facility independent of the repository, should one be constructed later. Also, the type of radiological/nuclear authority licensing permit being applied for would reflect whether in situ experiments involving the use of radionuclides in the subsurface environment are required to be carried out. Clearly, regulatory requirements will vary depending on national circumstances.

A decision to conclude the detailed site characterization phase, the last stage of a siting process, would be directly linked to an internal decision within the RWMO. This would reflect an overall assessment of site suitability and confidence in there being a good prospect for the regulatory authority granting a repository construction licence. The degree of political and public support could also be a contributing factor. The submission of a request for a construction licence would logically follow sometime after the final phase of site investigation inputs had been made available to the internal end users of site data and information. Consequently, in theory, there might be a significant lag between the conclusion of site investigation activities and the conclusion of safety assessment and design studies providing inputs into the safety case.

Note that it would be prudent for the RWMO to ensure contingency planning allows for the easy resumption of comprehensive site investigations, should a construction licence not be immediately forthcoming because of unresolved uncertainties, if the regulatory authorities were to approve this step, rather than rejecting a site entirely. Such conditional approval to resume investigations might be provided on the basis that a site continues to hold promise, but further data and understanding are required to strengthen arguments and build further confidence.

It is also noteworthy to mention it would be expected that repository development programmes would likely have certain site investigation activities continue beyond any data freeze established prior to a submission for a construction licence. The data and understanding derived from these activities would be required for additional licences to be applied for at a later date, such as for waste emplacement and closure.[40] As an example, continued environmental monitoring would be required to supplement a baseline description of the site and additional data would usefully be collected to further refine and validate models.

There are several criteria and approaches that could be used to inform the decision to conclude the main phase of site investigations and apply for a repository construction licence, as discussed in the sections that follow.

[40] It is important to recognize the significance of a graded approach. For some repository development programmes involving small volumes of radioactive waste posing a low hazard, it may be appropriate for a regulatory authority to issue a single licence to authorize repository construction, operations and closure. This might be the case in Member States requiring disposal of Category 4 and 5 DSRS, for example. In such cases there may be no justification for further environmental monitoring or site investigation data.

8.2. DEMONSTRATING THAT REQUIREMENTS HAVE BEEN MET

The simplest approach to support a decision to conclude the site investigations in support of a construction licence involves a simple audit, checking that sufficient data and information of appropriate quality in terms of accuracy, precision and representativity, have been collected to address the requirements defined at the start of the project or as amended during ongoing site investigations.

Based on good practices in Member States with advanced capabilities and significant experience in site investigations for disposal facilities, the requirements that drive the investigations can be specified in a number of ways and at various levels of detail. Section 3 sets out two examples as to how requirements might be organized at a high level. Firstly, on the basis of end user needs (e.g. for meeting international obligations, regulatory requirements or for use within the RWMO by addressing the needs of the internal customers involved in safety assessment, repository design and EIAs) and, secondly, on the basis of the work to be undertaken in specific disciplines. Such high level requirements categorizations can be split into lower level more detailed requirements resulting in a hierarchy of requirements until the stage when a site investigation project plan can be defined at an appropriate level of detail (e.g. as described in the UK Case Study in Annex I). The project plan would establish the link between requirements and site activities, defining what data and information are to be collected and why, what methods will be used to collect the data and information, and where and when will they be collected. The plan would also set out the basis for subsequent data processing, interpretation and integration. It would be monitored and updated as the site investigations progress.

As suggested in Section 4, the use of SIFDs to represent the movement of data and information from their collection through to end use would be a very useful tool to support a demonstration that requirements have been met.

Even though data and information might be provided to the internal end users in sufficient quantities to address their needs, without providing a robust demonstration that the data and information is of sufficient quality and that the people, processes and tools involved in the investigations have operated in line with the RWMO's management system, it cannot be demonstrated that the data and information are appropriate. Consequently, there should be continuous quality checking during the activities and adequate auditing after the activity and documentation to provide these assurances.

This approach to defining when a site investigation project can be concluded is essentially not much more than simply an audit addressing a checklist. However, the major benefit is that it is a relatively simple exercise and it is, in theory, easy to demonstrate compliance so long as the requirements have been predefined at a satisfactory level of detail at the outset of the project and as long as good records management has been maintained during the site investigations.

8.3. DEMONSTRATING SITE UNDERSTANDING HAS STABILIZED

At the start of a site investigation project, when there is a general lack of data and understanding, various hypotheses will exist to explain the nature of the observable features and the events and processes that may have occurred or are occurring at a site. These hypotheses can be captured in simple conceptual models of the site and of the wider region. The conceptual models might not be very detailed at this early stage, given the high degree of uncertainty and general lack of knowledge about a site. At the same time, many alternative conceptualizations might be possible for the same reasons.

As the site investigations proceed and more knowledge is gained about a site, the number of alternative conceptual models will likely reduce significantly until there is a much smaller number that are consistent with the data and information available at that stage. On the basis of feedback from internal end users, further investigations would be undertaken to increase the database of information and to address safety relevant or repository design relevant uncertainties and, in doing so, the conceptualization of the site would be further refined. At some stage, one can foresee that the collection and interpretation of further site data and information would have no discernible impact on the evolution of the conceptual models of a site. In other words, the reduction in uncertainty associated with the identification of additional individual features, more parameter values and improved process understanding and such would not fundamentally alter site understanding for the geoscientists involved.

As long as any equally valid and stabilized conceptual models are supported by an adequate database of data and information to address predefined requirements, as discussed in Section 6.16, this approach to support decision making is at face value appealing as it appears to establish the level of confidence in scientific understanding that is required about a site, as judged by the Earth scientists involved in data acquisition and interpretation.

8.4. DEMONSTRATING CALCULATED DOSE IS INSENSITIVE TO ADDITIONAL DATA AND UNDERSTANDING

The third approach that might be used to support a decision to conclude site investigations involves explicit input from the end users, particularly those responsible for safety assessment modelling. Figure 49 illustrates how overall uncertainty (and hence confidence associated with site understanding) will likely rise and fall over time, but in general will reduce as more data and information become available. As discussed in Sections 6.16 and 6.17, at some stage in the investigations the site database of property values and FEPs will contain sufficient data and information to satisfy a checklist based on predefined requirements, supplemented by one or a small number of equally valid conceptual models of the site. The safety assessment modellers will use this information, together with inputs on the waste characteristics, the repository design and any generic data and information as needed, to ultimately derive radiological dose and associated measures of risk for illustrative environmental conditions and representative groups and individuals that might be postulated to exist in the future. These dose and risk evaluations typically represent the central tendency of multiple realizations of radionuclide migration from the repository to the biosphere, based on probabilistic sampling of distributions of parameter values.

When the dose or risk limit or target is consistently being met within an acceptable margin, taking into account all of the known uncertainties associated with parameter values, modelling and future scenarios, it is reasonable to argue that incorporating further data and information into the safety assessment calculations will have no significant impact on the evaluation of the safety of the site and the overall disposal concept. The robustness of this conclusion could possibly be demonstrated by obtaining yet further new data and information but, at some point, a judgement would have to be made that the costs and effort associated with acquiring yet more data and information do not improve the confidence in the safety assessment of the site.

In the same manner, when feedback from the repository design engineers and those responsible for EIAs indicate they have sufficient suitable high quality data, then the site investigation phase leading to a regulatory submission for authorization to construct can justifiably be concluded.

This is a very powerful approach to use in decision making as it relates directly to the needs and judgement of the end users. However, while the use of dose and risk results might be considered a quantitative measure, there will always be a degree of subjectivity involved in the safety analysis. Therefore, it is to be emphasized that the results of a safety assessment, in terms of dose and risk, are not sufficient alone to make a case for the safety of a disposal facility. Consequently, the approach proposed here cannot be used in isolation; other approaches are necessary to provide additional confidence and argumentation about the suitability of a site.

8.5. SATISFYING ADDITIONAL STAKEHOLDER EXPECTATIONS

The fourth approach that might be adopted to indicate when site investigations can be concluded requires that the needs and expectations of other stakeholders are met (i.e. stakeholders other than safety

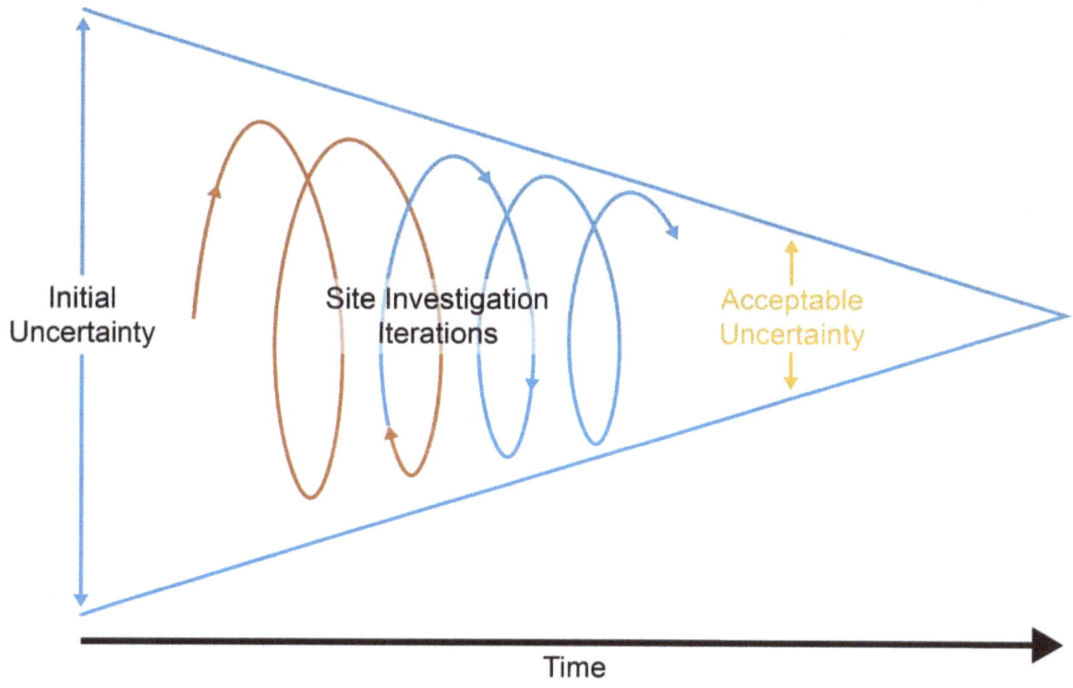

FIG. 49. Fluctuation in the overall degree of uncertainty associated with the site investigations over time (adapted, courtesy of NWS).

assessment, repository design and environmental impact personnel). These stakeholders would include the regulatory authorities and the general public.

It can be recalled from Section 3.5, that the various factors that might promote confidence in the site investigations and in the overall safety case were suggested to relate to internal factors within the control of an RWMO and external factors, reflecting the role of the regulator(s), the government, the overall decision making process and precedent set by good practices elsewhere. The internal factors influencing stakeholder confidence are dominated by an individual's evaluation of the safety of the concept, possibly guided by results from the safety assessment, and also trust in the RWMO based on its credibility and past behaviours, as well as the people, tools, processes, data and information used in support of the safety case (Fig. 10).

When there is: (a) a reasonable expectation that external stakeholders consider the RWMO to be trustworthy and credible, based on a culture of honesty, openness and transparency; and (b) when it is possible to demonstrate site understanding beyond that strictly necessary for the post-closure safety assessment, repository design requirements and EIA, in line with external stakeholder expectations and needs, then it is reasonable to believe stakeholders would have confidence in the technical work undertaken on the site investigations and the resulting conclusions. It can be readily appreciated that a judgement about trust in the RWMO and sufficiency of site understanding will be almost entirely subjective, but could be informed by carefully monitoring and recording interactions with all stakeholders, ensuring queries and concerns are addressed at an early stage and to the satisfaction of the stakeholders.

8.6. CONCLUDING COMMENTS

In addition to the discussion about when site investigations during screening should be suspended, or when the site investigations should advance to a stage of underground characterization, four approaches have been provided above, illustrating the different justifications that might be used to determine when site investigations in support of a repository construction licence application can be concluded. They are as follows:

- Demonstrating predefined requirements have been met (an audit approach);
- Demonstrating site understanding has stabilized (an approach largely based on the judgement of the data providers (i.e. the site investigation personnel));
- Demonstrating radiological dose and risk are insensitive to additional data and understanding (an approach that takes into account feedback from the safety assessment end users, but also from repository design engineers and EIA personnel);
- Satisfying additional stakeholder requirements that might be expected to be met by the RWMO (an approach that seeks to enhance the confidence that external stakeholders would have in the site investigations and the wider safety case).

In addition to these approaches, it is clearly the case that a site investigation project can only be considered to be concluded when all of the administrative responsibilities have been completed, such as formally compiling, reviewing and reporting the data and information obtained from the site investigation activities.

It is very important to be aware that the four approaches are not mutually exclusive. In fact, it is recommended that all four approaches be used in parallel to support a decision to conclude site investigations at the end of the detailed site characterization phase. Because regulator and other stakeholder behaviours cannot be predicted, and due to the degree of subjectivity involved in all four approaches, there will always be a risk that site investigations have been concluded too early. This should be recognized by the RWMO and dealt with as part of the project risk management approach. Consequently, opportunities to avoid this risk should be actively exploited and mitigating and contingency actions should be established so they can be enacted should the risk arise. This might mean being adaptable and able to re-establish the investigation project at short notice, should it prove necessary. Having highlighted this risk, it is considered appropriate to emphasize the need for the sensible application of a graded approach to the site investigation efforts. Ultimately, it is the judgement of the regulator, through granting an approval to construct a repository (or not), that demonstrates whether the effort expended on-site investigation is sufficient.

9. CONCLUSIONS

The discussion, guidance and examples provided in this publication are intended to illustrate many of the most common management considerations and challenges encountered when planning and implementing a site investigation project for a radioactive waste disposal facility. They are based on experiences derived from a number of site investigation projects undertaken since about 1990 and they represent input from a number of RWMOs around the world. The case studies in the Annexes provide good examples of more focused good practice in several areas relevant to a site investigation project.

Each Member State will have its own radioactive waste inventory for disposal and every repository development programme will be unique in terms of the geological settings available to host a repository, as well as legislation and regulatory requirements, stakeholder acceptance, the resources available and the range of disposal concept options available. Taking all of this into account, an RWMO and the relevant regulatory authorities responsible for licensing in a Member State are advised to very carefully consider the application of a graded approach to the site investigation effort and the nature of the demonstration of safety that will be required from a site [20]. Furthermore, because the disposal of radioactive waste is a national challenge, constructive and regular communications between the RWMO and the regulatory authorities is highly recommended to be established early on and maintained over the duration of the repository development programme.

The ultimate goal of site investigations during the siting process is to identify a suitable site for a disposal facility and to acquire the necessary authorizations for construction and disposal operations. The paramount concern during site investigations is to ensure the health and safety of workers and the public and to minimize adverse environmental impacts, but also to ensure the protection of people and the environment in the future, after repository closure. In order to achieve this, the RWMO will need to develop and implement an appropriate safety culture for all those involved in the site investigation project.

In planning a site investigation study, a very high priority should be assigned to defining data and information needs. Emphasis is to be placed on establishing clear relationships between:

— The information needed;
— The uses to be made of the information;
— The methods to be used to obtain the information;
— The timing of activities to address the information needs.

As a result, a site investigation project plan will be required that acknowledges a hierarchy of requirements established by different stakeholders and how those needs will be addressed. This is termed a requirements driven approach. Within a RWMO there are three key end users of the data and information to be obtained from a prospective repository site. These comprise safety assessment modellers, repository design engineers and those responsible for the EIA. The ultimate end users, though, are the regulatory authorities and wider stakeholder groups, including the public, who will judge the acceptability of any proposals. At the highest level, the requirements that are established will reflect international obligations and treaties, national policy and strategy, legislation and regulatory needs, and disposal system requirements. These will drive the types of data and information required from the site investigations, as well as the quality and confidence associated with them.

Comprehensive and detailed strategic planning in advance of the investigation to be undertaken at a site is essential so events do not drive the project and to ensure the effective use of resources. The planning of the site investigations should reflect a stepwise implementation approach, with clear decision points identified in advance, together with clear criteria for decision making. Experience indicates a comprehensive project risk management strategy should be established as a priority and the identified risks should be regularly reviewed to avoid, reduce or mitigate the various challenges that might arise on the basis of likelihood and potential impacts.

It can be appreciated that because the planning and implementation of a site investigation project requires a complicated mix of technical and traditional management experience and skills, it may be difficult to obtain all of the required services and personnel in a timely manner from a domestic market. Nevertheless, and regardless of the staffing model to be employed (e.g. full in-house capability or mainly outsourcing), the RWMO's task is to ensure there is a strong management team in place that possesses expertise and experience in all aspects of the investigations, such that it can effectively plan the work programme and manage the various specialist contractors that will be required.

Although not addressed in this report, the significance of stakeholder engagement and maintaining good relationships with all interested parties should not be forgotten, especially because siting is possibly the most contentious phase of a repository development programme.

The site investigation project should be designed as an integrated network of complementary activities that reinforce the overall understanding of a site, and work should be undertaken in parallel as far as is practicable. Critical activities that impact on other activities should be recognized and carefully managed.

Ensuring that a suitable and robust DMS is in place at the outset of a site investigation project is considered to be essential to accommodate the significant volumes of data and information that will rapidly be generated over the lifetime of a site investigation project. Experience indicates that this is something that is very difficult to adequately achieve as the effort and degree of complexity to be managed is typically underestimated.

There should be established channels of communication between all of the personnel involved in the site investigations and also between the site investigation specialists and those from safety assessment, repository engineering and those responsible for the EIA. These end user teams would provide feedback to the site investigation personnel, identifying remaining uncertainties of safety or design relevance and where additional data and information are required. A site investigation project would, therefore, be carried out as a series of iterative loops, with the site understanding advancing progressively as refined hypotheses (in the form of conceptual models) are developed. These conceptual models are to be tested by updated predictions cross-checked against further data collection and analysis.

Typically, an understanding of the natural environment increases with the effort expended over time. Initially, in the absence of significant data, understanding is low but tends to increase as more site specific data are collected. However, periodically it is typical that new data will be collected that are inconsistent with already established assumptions, hypotheses and conceptual models. Consequently, the understanding of one or more aspects of a site may fall and this can impact on the RWMO's confidence in understanding, especially where there may be significant consequences for safety evaluations. However, rigorous examination of the reasons for any anomalous but valid data generally leads to a greater understanding of the site; hence, confidence is ultimately enhanced by these challenges. Several of these dips in understanding and associated data confidence should be expected during the lifetime of a site investigation project.

Drilling is often the single most expensive activity undertaken during a site investigation project. This is a highly specialist technical area and when considering alternative borehole designs, drilling methods and drilling fluids, appropriate advice should be sought from well-qualified and experienced experts in the field. As a general principle, whichever drilling technique is selected, it has to ensure that the data being sought is collected with an acceptable degree of QA. Any technique employed should be safe, proven and designed to minimize environmental derogation. For all site investigation activities, the RWMO should ensure that it remains in overall control of planning and implementation. This will require that it acts as an educated customer for the services it contracts out and that it has its own staff involved in managing the data acquisition activities as much as is possible.

Delays can be expensive and can frustrate end users. Experience suggests they are particularly prone to occur during the process of obtaining permits and licences, when awaiting laboratory results and due to bottlenecks in formal reporting. Delays and overruns can come about during the investigations as a result of inadequate contracts and adverse field conditions or surprises, such as an unexpected discovery about the site.

A decision to complete the siting stage of site investigations prior to an application for a repository construction licence should be undertaken on the basis of logical reasoning. Section 8 suggests four complementary approaches. The final reporting associated with the completion of investigations prior to construction would include not just the complete set of written reports arising from the site investigations (designs, factual, interpretive and integrated reports), but also a specific set of deliverables passed on from the site investigation personnel to the safety assessment, repository design and environmental impact teams. These deliverables would essentially comprise:

(i) A comprehensive database of site wide and regional parameter values and observations together with their associated uncertainty limits, confidence assessments and contextual metadata.
(ii) A series of conceptual and descriptive models presented at various length scales to illustrate and document the characteristics of a site and its evolution to date, as well as the relationships between the key FEPs over the domain of interest. These would be underpinned by robust analysis, interpretation and the integration of data and information from all relevant disciplines.

The documentation associated with the reporting will be an important component of the safety case [20]. Thus, the QA procedures, DMSs and any other data and information that supports the safety case (e.g. that demonstrates wider site understanding and strengthens confidence) should also be reported.

The management of a site investigation project leading to the identification and confirmation of a site suitable for hosting a repository is essentially a research endeavour as it requires not only a site description, but also site understanding. Consequently, and depending on the scale and complexity of the investigations, there are bound to be unexpected discoveries and challenges involving the characteristics of the site, as well as the management of personnel, funding, equipment and timing. Many competing constraints and expectations will have to be managed and it will be imperative to ensure that the ongoing situation is constantly monitored in terms of the activities being carried out and in terms of the external environment and its potential to impact on the work. The project design and the people involved will require a robust and structured framework within which to operate, however. They will also have to be responsive to any changing conditions that might arise. To do so, they should employ good communication skills, flexibility and tact, as well as technical competence.

REFERENCES

[1] EUROPEAN ATOMIC ENERGY COMMUNITY, FOOD AND AGRICULTURE ORGANIZATION OF THE UNITED NATIONS, INTERNATIONAL ATOMIC ENERGY AGENCY, INTERNATIONAL LABOUR ORGANIZATION, INTERNATIONAL MARITIME ORGANIZATION, OECD NUCLEAR ENERGY AGENCY, PAN AMERICAN HEALTH ORGANIZATION, UNITED NATIONS ENVIRONMENT PROGRAMME, WORLD HEALTH ORGANIZATION, Fundamental Safety Principles, IAEA Safety Standards Series No. SF-1, IAEA, Vienna (2006), https://doi.org/10.61092/iaea.hmxn-vw0a

[2] INTERNATIONAL ATOMIC ENERGY AGENCY, IAEA Nuclear Safety and Security Glossary, IAEA, Vienna (2022), https://doi.org/10.61092/iaea.rrxi-t56z

[3] INTERNATIONAL ATOMIC ENERGY AGENCY, Site Investigations for Repositories for Solid Radioactive Wastes in Shallow Ground, Technical Reports Series No. 216, IAEA, Vienna (1982).

[4] INTERNATIONAL ATOMIC ENERGY AGENCY, Techniques for Site Investigations for Underground Disposal of Radioactive Wastes, Technical Reports Series No. 256, IAEA, Vienna (1985).

[5] INTERNATIONAL ATOMIC ENERGY AGENCY, Site Investigations for Repositories for Solid Radioactive Wastes in Deep Continental Geological Formations, Technical Reports Series No. 215, IAEA, Vienna (1982).

[6] INTERNATIONAL ATOMIC ENERGY AGENCY, Experience in Selection and Characterization of Sites for Geological Disposal of Radioactive Waste, IAEA-TECDOC-991, IAEA, Vienna (1997).

[7] INTERNATIONAL ATOMIC ENERGY AGENCY, Nuclear Geophysics and Its Applications, Technical Reports Series, No. 393, IAEA, Vienna, (1999).

[8] INTERNATIONAL ATOMIC ENERGY AGENCY, Hydrogeological Investigations of Sites for the Geological Disposal of Radioactive Waste, Technical Reports Series No. 391, IAEA, Vienna (1999).

[9] INTERNATIONAL ATOMIC ENERGY AGENCY, Characterization of Groundwater Flow for Near Surface Disposal Facilities, IAEA-TECDOC-1199, IAEA, Vienna (2001).

[10] INTERNATIONAL ATOMIC ENERGY AGENCY, Communication and Stakeholder Involvement in Radioactive Waste Disposal, IAEA Nuclear Energy Series No. NW-T-1.16, IAEA, Vienna (2022).

[11] INTERNATIONAL ATOMIC ENERGY AGENCY, Disposal of Radioactive Waste, IAEA Safety Standards Series No. SSR-5, IAEA, Vienna (2011).

[12] INTERNATIONAL ATOMIC ENERGY AGENCY, Geological Disposal Facilities for Radioactive Waste, IAEA Safety Standards Series No. SSG-14, IAEA, Vienna (2011).

[13] INTERNATIONAL ATOMIC ENERGY AGENCY, Borehole Disposal Facilities for Radioactive Waste, IAEA Safety Standards Series No. SSG-1, IAEA, Vienna (2009).

[14] INTERNATIONAL ATOMIC ENERGY AGENCY, Near Surface Disposal Facilities for Radioactive Waste, IAEA Safety Standards Series No. SSG-29, IAEA, Vienna (2014).

[15] INTERNATIONAL ATOMIC ENERGY AGENCY, Nuclear Energy Basic Principles, IAEA Nuclear Energy Series No. NE-BP, IAEA, Vienna (2008).

[16] INTERNATIONAL ATOMIC ENERGY AGENCY, Radioactive Waste Management Objectives, IAEA Nuclear Energy Series No. NW-O, IAEA, Vienna (2011).

[17] INTERNATIONAL ATOMIC ENERGY AGENCY, Classification of Radioactive Waste, IAEA Safety Standards Series No. GSG-1, IAEA, Vienna (2009).

[18] NUCLEAR DECOMMISSIONING AUTHORITY, Geological Disposal: Proposed Strategy for the Geo-Scientific Aspects of Site Characterisation, NDA, UK (2011).

[19] McEVOY, F.M., SCHOFIELD, D.I., SHAW, R.P., NORRIS, S., Tectonic and Climatic Considerations for Deep Geological Disposal of Radioactive Waste: A UK Perspective, Sci. Total Envi. **571** (2016) 507–521, https://doi.org/10.1016/j.scitotenv.2016.07.018

[20] INTERNATIONAL ATOMIC ENERGY AGENCY, The Safety Case and Safety Assessment for the Disposal of Radioactive Waste, IAEA Safety Standards Series No. SSG-23, IAEA, Vienna (2012).

[21] NUCLEAR DECOMMISSIONING AUTHORITY, RADIOACTIVE WASTE MANAGEMENT, Geological Disposal: Overview of International Siting Processes 2017, NDA, UK (2017).

[22] SVENSK KÄRNBRÄNSLEHANTERING, What requirements does the KBS-3 repository make on the host rock? Geoscientific suitability indicators and criteria for siting and site evaluation, SKB, Stockholm (2000).

[23] SVENSK KÄRNBRÄNSLEHANTERING, Geoscientific programme for investigation and evaluation of sites for the deep repository, SKB, Stockholm (2000).

[24] INTERA ENGINEERING, Geoscientific Site Characterization Plan DGR — Bruce Nuclear Site, OPG Report No. 00216-REP-03902-00002-R00, Ontario Power Generation, Toronto (2006).

[25] AGENCE NATIONALE POUR LA GESTION DES DÉCHETS RADIOACTIFS, Evaluation of the feasibility of a geological repository in an argillaceous formation. Dossier 2005 Argile Synthesis. Meuse/Haute Marne Site, Andra, France (2005).

[26] OTA, K., SAEGUSA, H., KONDO, H., GOTO, J., KUNIMARU, T., YAMADA, S., Site characterisation and synthesis into SDMs for NUMO safety case (Proc. ANS 17th Int. High-Level Radioactive Waste Mgmt Con. — IHLRWM 2019), Knoxville, TN (2019).

[27] NATIONALE GENOSSENSCHAFT FÜR DIE LAGERUNG RADIOAKTIVER ABFÄLLE, The Nagra Research, Development and Demonstration (RD&D) Plan for the Disposal of Radioactive Waste in Switzerland, Nagra, Switzerland (2016).

[28] INTERNATIONAL ATOMIC ENERGY AGENCY, Costing Methods and Funding Schemes for Radioactive Waste Disposal Programmes, IAEA Nuclear Energy Series No. NW-T-1.25, IAEA, Vienna (2020).

[29] TEXIER, D., DEGNAN, P., LOUTRE, M.F., PAILLARD, D., THORNE, M., Modelling BIOsphere systems under climate change for radioactive waste disposal (Proc. Int. High-Level Waste Mgt Con.), Las Vegas, NV (2003).

[30] MILLER, W., ALEXANDER, R., CHAPMAN, N., McKINLEY, I., SMELLIE, J.A.T. (Eds), Geological Disposal of Radioactive Wastes & Natural Analogues, Pergamon Press Waste Management Series Volume 2, (2000).

[31] OECD NUCLEAR ENERGY AGENCY, Confidence in the Long-term Safety of Deep Geological Repositories, OECD, France (1999).

[32] INTERNATIONAL ATOMIC ENERGY AGENCY, Leadership, Management and Culture for Safety in Radioactive Waste Management, IAEA Safety Standards Series No. GSG-16, IAEA, Vienna (2022).

[33] INTERNATIONAL ATOMIC ENERGY AGENCY, Factors Affecting Public and Political Acceptance for the Implementation of Geological Disposal, IAEA-TECDOC-1566, IAEA, Vienna (2007).

[34] NUCLEAR DECOMMISSIONING AUTHORITY, RADIOACTIVE WASTE MANAGEMENT, Geological Disposal: Generic Disposal System Specification, Part A High-Level Requirements, NDA, UK, (2016).

[35] NUCLEAR DECOMMISSIONING AUTHORITY, RADIOACTIVE WASTE MANAGEMENT, Geological Disposal: Design Information Requirements from the RWM Site Characterisation Programme, NDA, UK (2013).

[36] NUCLEAR DECOMMISSIONING AUTHORITY, RADIOACTIVE WASTE MANAGEMENT, Geological Disposal: Safety Case Information Requirements from the RWM Site Characterisation Programme, NDA, UK (2013).

[37] NUCLEAR DECOMMISSIONING AUTHORITY, RADIOACTIVE WASTE MANAGEMENT, Geological Disposal: EIA Information Requirements from the RWM Site Characterisation Programme, NDA, UK (2013).

[38] INTERNATIONAL ATOMIC ENERGY AGENCY, The Management System for the Development of Disposal Facilities for Radioactive Waste, IAEA Nuclear Energy Series No. NW-T-1.2, IAEA, Vienna (2011).

[39] INTERNATIONAL ATOMIC ENERGY AGENCY, Stakeholder Involvement Throughout the Life Cycle of Nuclear Facilities, IAEA Nuclear Energy Series No. NG-T-1.4, IAEA, Vienna, (2011).

[40] INTERNATIONAL ATOMIC ENERGY AGENCY, Leadership and Management for Safety, IAEA Safety Standards Series No. GSR Part 2, IAEA, Vienna (2017).

[41] INTERNATIONAL ORGANIZATION FOR STANDARDISATION, Quality Management Systems — Requirements (ISO Standard No. 9001:2015), ISO, Switzerland (2015).

[42] FARHOUD, R., TABANI, P., CHABIRON, A., GILARDI, N., MANGEOT, A., CRABOL, V., Andra's underground research laboratory as a tool to prepare monitoring system of Cigéo project —18070 (Proc. WM2018: 44 Ann. Waste Mgt Con.), Pheonix, AZ (2018).

[43] NATIONALE GENOSSENSCHAFT FÜR DIE LAGERUNG RADIOAKTIVER ABFÄLLE, Geosynthese Wellenberg 1996: Ergebnisse der Untersuchungsphasen 1 und 2: Textband. Nagra Technischer Bericht NTB 96-01. Nagra, Switzerland (1997).

[44] INTERNATIONAL ORGANIZATION FOR STANDARDISATION, Space Data and Information Transfer Systems — Open Archival Information System (OAIS) — Reference Model, ISO, Switzerland (2012).

[45] ANDERSSON, J., SKAGIUS, K., WINBERG, A., LINDBORG, T., STRÖM, A., Environmental Earth Sciences, Site-descriptive modelling for a final repository for spent nuclear fuel in Sweden, EES, Sweden (2013), https://doi.org/10.1007/s12665-013-2226-1

[46] INTERNATIONAL ATOMIC ENERGY AGENCY Sealing of Underground Repositories for Radioactive Wastes, Technical Reports Series No. 319, IAEA, Vienna (1990).

[47] SVENSK KÄRNBRÄNSLEHANTERING, Sealing of investigation boreholes, Phase 4 — Final report, SKB, Sweden (2011).

[48] NATIONALE GENOSSENSCHAFT FÜR DIE LAGERUNG RADIOAKTIVER ABFÄLLE, Long Term Stability of Potential System Components for Sealing Deep Investigation Boreholes — Experience and approaches at Nagra, Nagra, Switzerland (2012).

[49] JACKSON, C.P., et al., Sealing Deep Site Investigation Boreholes: Phase 1. Report to the Nuclear Decommissioning Authority RWMD/03/042, International Association for Measurement and Evaluation of Communication (AMEC) Nuclear Decommissioning Authority, UK, (2014).

[50] RIDER, M.H., KENNEDY, M., The Geological Interpretation of Well Logs, 3rd edn, Rider-French Consulting, Scotland (2011).

[51] STERRETT, R.J. (Ed.), Groundwater and Wells: A comprehensive guide for the design, installation, and maintenance of water well, 3rd edn, Johnson Screens, New Brighton, MN (2007).

[52] KRUSEMAN, G.P., DE RIDDER, N.A., Analysis and Evaluation of Pumping Test Data, 2nd edn, International Institute for Land Reclamation and Improvement, Netherlands (1991).

[53] FREEZE, R.A., CHERRY, J.A., Groundwater, Prentice Hall, Englewood Cliffs, NJ (1979).

[54] DE MARSILY, G., Quantitative Hydrogeology: Groundwater Hydrology for Engineers, Academic Press, San Diego, CA (1986).

[55] FETTER, C.W., Applied Hydrogeology, 3rd edn, Macmillan College Publishing, New York, NY (1994).

[56] UNITED STATES DEPARTMENT OF ENERGY, Yucca Mountain Science and Engineering Report: Technical Information Supporting Site Recommendation Consideration, Rep. DOE/RW-0539, USDOE, Washington, DC (2001).

[57] DEGNAN, P., et al., PADAMOT: Project Overview Report, Nirex, UK (2005) 85 pp.

[58] BARKER, J.A., A generalized radial flow model for hydraulic tests in fractured rock, Water Res. Res. **24** (1988) 1796–1804,
https://doi.org/10.1029/WR024i010p01796

[59] FETTER, C.W., Contaminant Hydrogeology, Macmillan College Publishing, USA (1993).

[60] ZHAO, Z., JING, L., NERETNICKS I., Evaluation of hydrodynamic dispersion parameters in fractured rocks, J. Rock Mech. Geotech. Eng. **2** (2010) 243–254,
https://doi.org/10.3724/SP.J.1235.2010.00243

[61] BLECHSCHMIDT, I., LANYON, G.W., Nationale Genossenschaft für die Lagerung Radioaktiver Abfälle: GTS Phase VI — CF1 Phase 1: Field Activities 2006–2008, Nagra Working Report NAB 08-33, Switzerland, (2008).

[62] NATIONALE GENOSSENSCHAFT FÜR DIE LAGERUNG RADIOAKTIVER ABFÄLLE, Colloid Formation and Migration Project: Modelling of Tracer, Colloid and Radionuclide/Homologue Transport for Dipole CFM 06.002 — Pinkel Surface Packer, Nagra Technical Report 16-06, Switzerland (2016).

[63] ÖHBERG, A., Posiva Groundwater Flow Measuring Techniques, Posiva Report 2000-12, Posiva, Finland (2000).

[64] FITTS, C.R., Groundwater Science, Elsevier Science: Academic Press, London (2002).

[65] LUSCZYNSKI, N.J., Head and flow of ground water of variable density, J. Geophys. Res. **66** 12, (1961) 4247–4256,
https://doi.org/10.1029/JZ066i012p04247

[66] MILODOWSKI, A.E., et al., The petrology and paragenesis of fracture mineralization in the Sellafield area, west Cumbria (Proc. Yorkshire Geo. Soc.) **52** (1998) 215–241,
https://doi.org/10.1144/pygs.52.2.215

[67] DEGNAN, P.J., LITTLEBOY, A.K., MICHIE, McL.U., JACKSON, C.P., WATSON, S.P., Fracture-dominated flow in the Borrowdale volcanic group at Sellafield, NW England: The identification of potential flowing features, development of conceptual models and derivation of effective parameters, Geo. Soc., London, Special Publications **214** 1 (2003) 197–219,
https://doi.org/10.1144/GSL.SP.2003.214.01.12

[68] FRANKLIN, J.A., Suggested methods for determining water content, porosity, density, absorption and related properties and swelling and slake-durability index properties. Int. J. Rock Mech. Min. Sci. Geomech. Abstr. **16** (1979) 141–156,
https://doi.org/10.1016/0148-9062(79)91453-0

[69] INTERNATIONAL ATOMIC ENERGY AGENCY, Isotope Methods for Dating Old Groundwater, Non-serial Publications, IAEA, Vienna (2013).

[70] CASCOYNE, M., Methods of Sampling and Analysis of Dissolved Gases in Deep Groundwaters, Posiva Working Report 2002-21, Posiva, Finland (2002).

[71] SVENSK KÄRNBRÄNSLEHANTERING, Development and Testing of Gas Samplers in Tunnel Environments, SKB Report R-16-16, SKB, Sweden, (2017).

[72] OECD NUCLEAR ENERGY AGENCY, The Role of Underground Laboratories in Nuclear Waste Disposal Programmes, OECD, France (2001).

[73] INTERNATIONAL ATOMIC ENERGY AGENCY, The Use of Scientific and Technical Results from Underground Research Laboratory Investigations for the Geological Disposal of Radioactive Waste, IAEA-TECDOC-1243, IAEA, Vienna (2001).

[74] MACKINNON, R.J., The use of underground research laboratories to support repository development programs: A roadmap for the underground research facilities network, Sandia Report SAND2015-0427, Sandia National Laboratories, USA (2015),
https://doi.org/10.2172/1225846

[75] BLECHSCHMIDT, I., VOMVORIS, S., "Underground research facilities and rock laboratories for the development of geological disposal concepts and repository systems", Geological Repository Systems for Safe Disposal of Spent Nuclear Fuels and Radioactive Waste (AHN, J., APTED, M.J., Eds), Woodhead Publishing, Boca Raton, FL (2010) 82–118,
https://doi.org/10.1533/9781845699789.1.82

[76] MATHERON, G., The Theory of Regionalised Variables and Its Applications, Volume 5 of Centre de Morphologie Mathematique Fontainebleau: Les cahiers du Centre de Morphologie Mathematique de Fontainebleau, Ecole national superieure des mines, France (1971).

[77] BONANO, E.J., DAVIS, P.A., BRAS, R.L., KITANIDIS, P.K., "Methodology for estimating groundwater travel times at a nuclear waste repository using a physically based geostatistical approach", Geostatistics (ARMSTRONG, M., Ed.). Quantitative Geology and Geostatistics **4**, Springer, Netherlands (1989),
https://doi.org/10.1007/978-94-015-6844-9_53

[78] SRIVASTAVA, R.M., DAVIS, B.M., Geostatistics for fracture characterization, high level radioactive waste management (Proc. 5th Ann. Int. Conf.), American Nuclear Society, Reno, NV (1994).

[79] SVENSK KÄRNBRÄNSLEHANTERING, Assigning probability distributions to input parameters of performance assessment models, SKB Report TR-02-11, SKB, Sweden (2002).

[80] GILLESPIE, M.R., BARNES, R.P., MILODOWSKI, A.E., British Geological Survey scheme for classifying discontinuities and fillings, BGS Research Report RR/10/05, Nottingham, UK (2011).

[81] ALEXANDER, W.R. (Ed.), A natural analogue study of the Maqarin, hyperalkaline groundwaters. Source term description and thermodynamic database testing. Nagra Tech. Rep. NTB 91–10, Switzerland (1992).

[82] DEGNAN, P., et al., "Tectonic setting, geology and hydrogeology of Maqarin", Maqarin Phase IV Report (PITTY, A., ALEXANDER, R., Eds), NDA Technical Report (2013).

[83] DERSHOWITZ, W., et al., FracMan Interactive discrete feature data analysis, geometric modeling, and exploration simulation: User Documentation Version 2.5. Golder Associates, Redmond, WA (1995).

[84] NIEMI, A., KONTIO, K., KUUSELA-LAHTINEN, A., POTERI, A., Hydraulic characterization and upscaling of fracture networks based on multiple-upscaling and well test data, Water Res. Res. **36** (2000) 3481–3497, https://doi.org/10.1029/2000WR900205

[85] SVENSK KÄRNBRÄNSLEHANTERING, SR97 — Alternative models project, SKB Report RR-99-43, SKB, Sweden (1999).

[86] SHAW, R.P., et al., Interpretation and Modelling: Geology, to support the development of an integrated site descriptive model, British Geological Survey Open Report, OR/11/057, Nottingham, UK (2011) 154 pp.

[87] STREETLY, M.J., HEATHCOTE, J.A., DEGNAN, P.J., Estimation of vertical diffusivity from seasonal fluctuations in groundwater pressures in deep boreholes near Sellafield, NW England. Geological Society, Special Publications **263** (2006) 155–167, https://doi.org/10.1144/GSL.SP.2006.263.01.08

[88] BARTON, N., LIEN R., LUNDE J., Engineering classification of rock masses for the design of tunnel support. Rock Mech. Rock Eng **6** (1974) 189–236, https://doi.org/10.1007/BF01239496

[89] NORWEGIAN GEOTECHNICAL INSTITUTE, Using the Q-system: Rock mass classification and support design. NGI Handbook, Norway (2015).

[90] BIENIAWSKI, Z.T., Engineering Rock Mass Classification, Wiley-Interscience, Hoboken (1989).

[91] HOEK, E., KAISER, P.K., BAWDEN, W.F., Support of underground excavations in hard rock, Balkema, The Netherlands (1995).

[92] LINDBORG, T., et al., Landscape development during a glacial cycle: Modeling ecosystems from the past into the future, Ambio. **42** 4 (2013) 402–413, https://doi.org/10.1007/s13280-013-0407-5

[93] INTERNATIONAL ATOMIC ENERGY AGENCY, The Use of Numerical Models in Support of Site Characterization and Performance Assessment Studies of Geological Repositories, IAEA-TECDOC-1717, IAEA, Vienna (2013).

[94] THORNE, M.C., DEGNAN, P., EWEN, J., PARKIN, G., Validation of a physically based catchment model for application in post-closure radiological safety assessments of deep geological repositories for solid radioactive wastes, J. Radiol. Prot. **20** (2000) 403–421, https://doi.org/10.1088/0952-4746/20/4/304

[95] NIREX, Sellafield Geological and Hydrogeological Investigations: An assessment of the influence of basinal processes on the hydrogeology of the Sellafield area, Nirex Report SA/97/055, UK (1997).

[96] NUCLEAR WASTE MANAGEMENT ORGANIZATION, OPG's Deep Geologic Repository for Low and Intermediate Level Waste: Geosynthesis, NWMO Report DGR-TR-2011-11, NWMO, Toronto (2011).

[97] HUDSON, J.A., Rock Engineering Systems: Theory and practice, Ellis Horwood, New York (1992).

[98] OECD NUCLEAR ENERGY AGENCY, Nirex methodology for scenario and conceptual model development, OECD/NEA, Paris (1999).

[99] US DEPARTMENT OF ENERGY, OST&I 2005 Annual Report, Office of Civilian Radioactive Waste Management, DOE/RW-0581, LBNL-59067, USDOE, USA (2005) 220 pp.

[100] DEGNAN, P., LITTLEBOY, A., Nirex 97: An assessment of the post-closure performance of a deep waste repository at Sellafield, Volume 1: hydrogeological conceptual model development conceptual basis and data, Nirex Science report S—97/012/1, Hartwell, UK (1997).

[101] ANDERSSON, J., Site descriptive modelling — strategy for integrated evaluation, SKB Report R-03-05, Svensk Kärnbränslehantering, Sweden (2003).

[102] MUNIER, R., STANFORS, R., MILNES, A.G., HERMANSON, J., TRIUMF, C.-A., Geological site descriptive model: A strategy for the model development during site investigations, SKB Report R-03-07, Svensk Kärnbränslehantering, Sweden (2003).

[103] NUCLEAR WASTE MANAGEMENT ORGANIZATION OF JAPAN, The NUMO Pre-siting SDM-based Safety Case Report, NUMO-TR-21-01, NUMO, Tokyo (2021).

Annex I

THE DEVELOPMENT OF INFORMATION REQUIREMENTS FOR A SITE INVESTIGATION PROJECT — A CASE STUDY FROM THE UK

I–1. CONTEXT

The UK has been accumulating higher activity radioactive wastes since the 1940s as a result of the civil nuclear, military and research programmes. This waste is held in safe and secure interim storage at a number of sites across the country.

In July 2006, the Committee on Radioactive Waste Management (known as CoRWM) recommended that geological disposal, coupled with safe and secure interim storage, was the best available approach for the long term management of the legacy of higher activity radioactive wastes in the UK. In October 2006, the UK government and the devolved administrations of Scotland, Wales and Northern Ireland published a response to the committee's recommendations, explaining that geological disposal, preceded by safe and secure interim storage, should be the means by which the higher activity radioactive waste is managed in the UK, long term.

The 2014 white paper "Implementing Geological Disposal" [I–1] updates an earlier 2008 white paper [I–2] and confirms that the UK government is committed to the policy of geological disposal. It continues to favour a siting process for a geological disposal facility (GDF) that is based on the willingness of local communities to participate.

Once a community comes forward with a potential area in which a facility could be hosted, this area will need to be investigated and characterized to such a degree that a decision can be made on its suitability. The ultimate objective of site characterization is to provide the information that is needed to assess site suitability and provide input into the facility design, safety case and environmental assessments. This assessment of suitability is to ensure all disposals are made in a way that protects the health of people and the integrity of the environment at the time of disposal and in the future.

The UK Nuclear Waste Services (NWS, formerly Radioactive Waste Management) is developing a needs-driven site characterization programme in preparation for work to be undertaken during the siting process for a GDF. Establishing geoscientific information requirements will ensure that only information required for the development of the safety case, engineering design or environmental assessment is collected (i.e. the information gathering is needs-driven and focused in relation to its end use). This will, in turn, ensure timely delivery and efficient use of funds.

The information requirements will form part of a hierarchy of requirements to link to and justify the parameters to be measured and data to be collected during the site characterization programme. Nuclear Waste Services has developed the disposal system specification that sets out the requirements on the disposal system as part of a requirements-led approach to the iterative development of the disposal system. The requirements on the disposal system are set out in two parts: Part A (high level requirements) [I–3] and Part B (technical requirements) [I–4]. They include the needs from the government's 2014 white paper [I–1] and the wider regulatory and policy framework. This needs-driven approach is illustrated in Fig. I–1.

I–2. MANAGEMENT CHALLENGES

One of the biggest challenges for the definition of information requirements is that no sites have yet been identified in the current generic stage of the siting programme. Due to the range of geological settings in the UK and the voluntary nature of the UK siting process, it is best to ensure the designs and

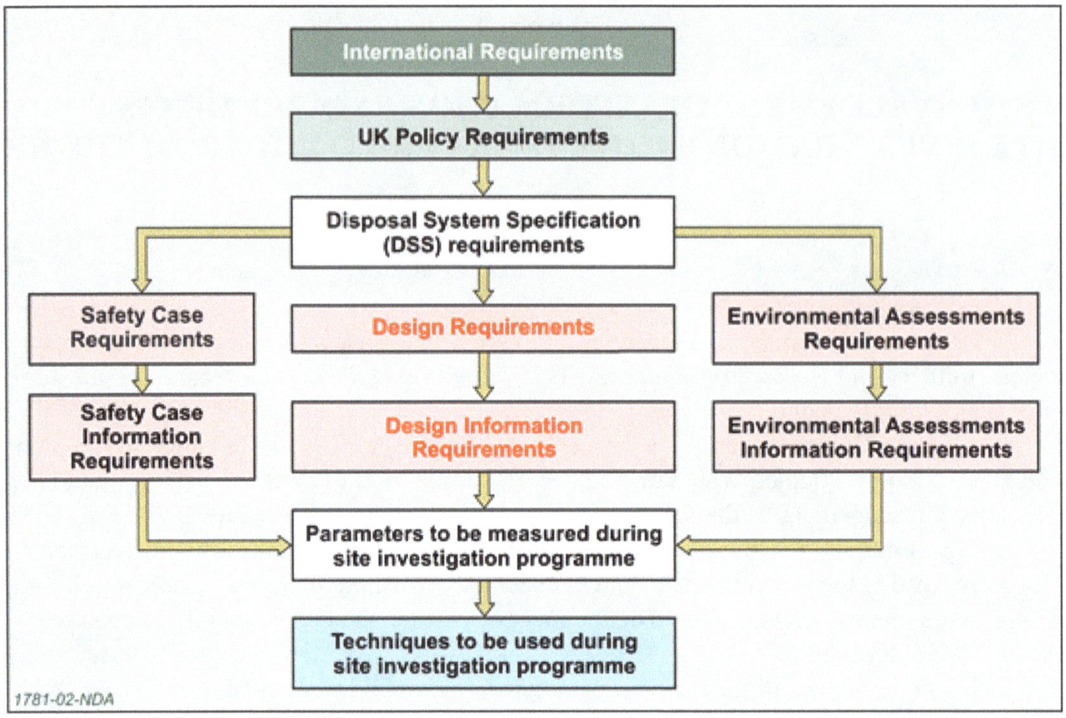

FIG. I–1. The needs driven approach to the NWS site characterization programme. The term policy here encompasses law and regulations (courtesy of NWS).

assessments cover a range of possible disposal environments and facility designs. Nuclear Waste Services has developed a generic disposal system specification, generic engineering designs, a generic disposal system safety case and generic environmental assessments.

The current illustrative GDF concept examples (Table I–1) that support NWS's disposal system safety case (DSSC) are based on concepts adopted by other waste management organizations (known as WMOs) that were supported by site characterization programmes.

A risk associated with this approach identified by some stakeholders is that NWS may use the full set of generic information requirements to plan the site investigations for specific sites resulting in an unnecessarily large investigation programme. Nuclear Waste Services recognizes this risk and has undertaken the development of generic information requirements as a learning exercise in advance of the identification of site specific information requirements. The generic information requirements also provide a comprehensive checklist to advise the site specific exercise in due course.

Therefore, as the siting process progresses, the site investigation information requirements will need to be progressively re-evaluated with careful consideration of the specific environment in the areas and then sites being studied, and with inputs from the communities affected by the project and from regulators through the permissioning process. At a particular site, the iterative process of site characterization, design and assessment will also progressively refine the information requirements — as illustrated in Fig. I–2.

It was agreed at the highest level in the organization that the information requirements are to be owned by the eventual end users of the information to demonstrate a truly needs driven approach to site investigations. Therefore, another key challenge in the development of the information requirements related to obtaining the time commitment of senior NWS staff from these end users at an early stage in the process. Without the alignment and buy-in of these senior staff, the value of the exercise would have been significantly reduced.

TABLE I–1. ILLUSTRATIVE GEOLOGICAL DISPOSAL CONCEPT EXAMPLES

Host rock	Illustrative Geological Disposal Concept Examples[d]	
	LHGW	HHGW
Higher strength rocks[a]	UK LHGW concept *(Radioactive Waste Management, UK)*	KBS-3 V concept *(SKB, Sweden)*
Lower strength sedimentary rock[b]	Opalinus clay concept *(Nagra, Switzerland)*	Opalinus clay concept *(Nagra, Switzerland)*
Evaporites[c]	WIPP bedded salt concept *(US-DOE, USA)*	Gorleben salt dome concept *(DBE Technology, Germany)*

Notes:

[a] Higher strength rocks — the UK LHGW concept and KBS-3V concept for spent nuclear fuel were selected due to availability of information on these concepts for the UK context.

[b] Lower strength sedimentary rocks — the Opalinus clay concept for disposal of long lived ILW, HLW and spent fuel was selected because an OECD Nuclear Energy Agency review regarded the Nagra (Switzerland) assessment of the concept as state of the art with respect to the level of knowledge available. However, it should be noted there is similarly extensive information available for a concept that has been accorded strong endorsement from international peer review. Although we will use the Opalinus clay concept as the basis of illustrative example, we will also draw on information from the Andra programme. In addition, we will also draw on information from the Belgian super container concept, based on the disposal of HHGW in Boom clay.

[c] Evaporites — the concept for the disposal of transuranic wastes (TRU) (long lived ILW) in a bedded salt host rock at the Waste Isolation Pilot Plant (WIPP) in New Mexico was selected because of the wealth of knowledge available from this facility. The concept for disposal of HHGW in a salt dome host rock developed by DBE Technology (Germany) was selected due to the level of concept information available.

[d] For planning purposes, the illustrative concept for depleted, natural and low enriched uranium is assumed to be the same as for ILW/LLW and for plutonium and highly enriched uranium is assumed to be the same as for HLW/SNF.

LHGW = low heat generating wastes (adapted from Ref. [I–5] courtesy of NWS).

I–3. STRATEGIC PLANNING

Development of the siting process in the UK is ongoing and no site has yet been identified for the GDF. Therefore, the disposal system development process is currently generic, in that it is not specific to any site. NWS's generic site investigation information requirements were first developed and published in 2013 based upon the requirements of the 2010 generic disposal system specification, design, safety case and environmental assessments. The current update was timed to coincide with the 2016 update to the generic disposal system specification, design, safety case and environmental assessments that takes into account feedback on the 2010 document suite, the increased knowledge developed since 2010 and the revised inventory published by the NDA in 2013.

Significant progress has been made by NWS in developing its disposal system specification in a form that captures the different levels of disposal system requirements. Figure I–3 illustrates the high level structure of these requirements.

Nuclear Waste Services intends to use a requirements management system (sometimes known as an RMS)[1] to manage how individual requirements link back to higher level requirements contained in NWS's disposal system specification. The requirements management system will eventually capture the verification

[1] Requirements management systems can be used to capture, trace, analyse and manage changes to information and to demonstrate compliance with legal and other requirements.

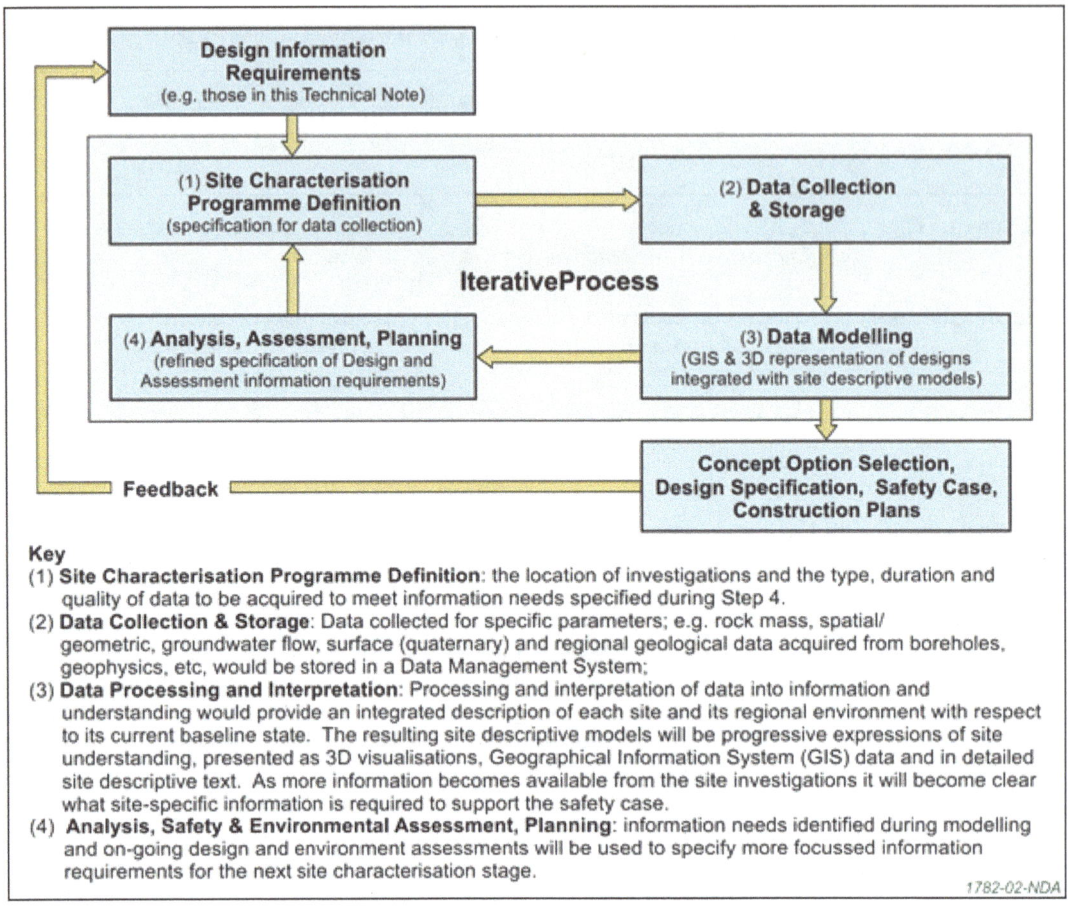

FIG. I–2. The iterative process between design information requirements and geoscientific data acquisition (courtesy of NWS).

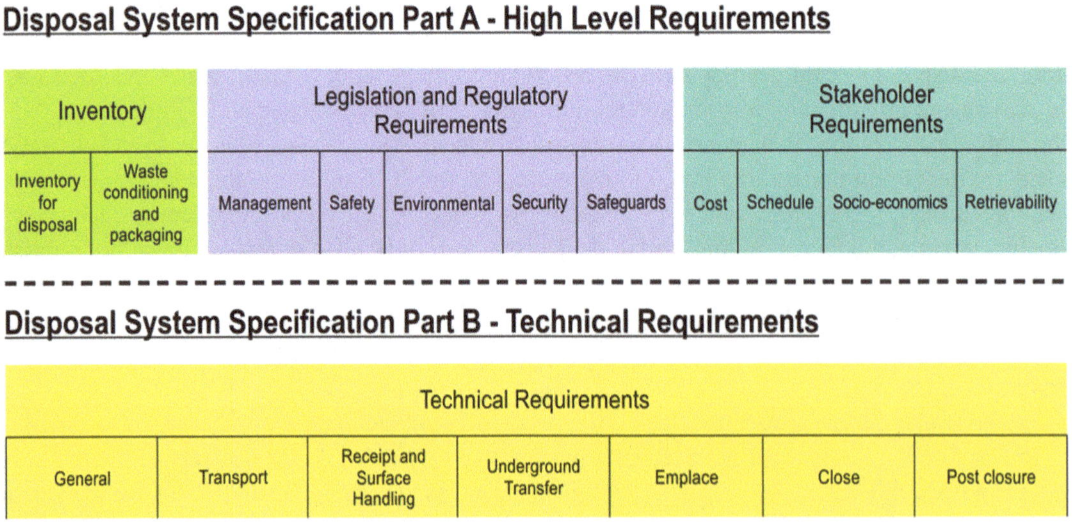

FIG. I–3. High level structure of the disposal system specification (reproduced from Ref. [I–3] courtesy of NWS).

that individual requirements have been met and ultimately demonstrate the validation that the disposal system provides a complete solution, as illustrated in Fig. I–4.

I–4. IMPLEMENTATION

I–4.1. Project team

In order to obtain the greatest value from the exercise, senior staff in the engineering design, safety case and environmental assessment teams were identified as team leads for the development of the generic site investigation information requirements. However, to deliver the process in a robust and consistent manner across the three areas, a geoscientist with site investigation experience in support of geological disposal projects was enlisted from the supply chain to support the NWS project team.

I–4.2. Methodology

The identification of the information requirements was largely undertaken through three interactive workshops facilitated by the external geoscientist, one for each of engineering design, safety case and environmental assessments. The requirements defined in the disposal specification were considered one by one and the geoscientific information and understanding needed to meet that requirement were identified. The workshops were attended by the team lead for the relevant area and selected team members, the disposal system specification manager and the head of site characterization.

I–4.3. Results

The generic information requirements were captured in the design, safety case and environmental assessment information requirements reports [I–6 to I–8]. Table I–2 summarizes the number of requirements and information requirements at different levels in the hierarchy for the safety case.

Once the information requirements of the three end users had been defined in the three reports referred to in Table I–2, a list of consolidated information requirements (CIRs) presented below was

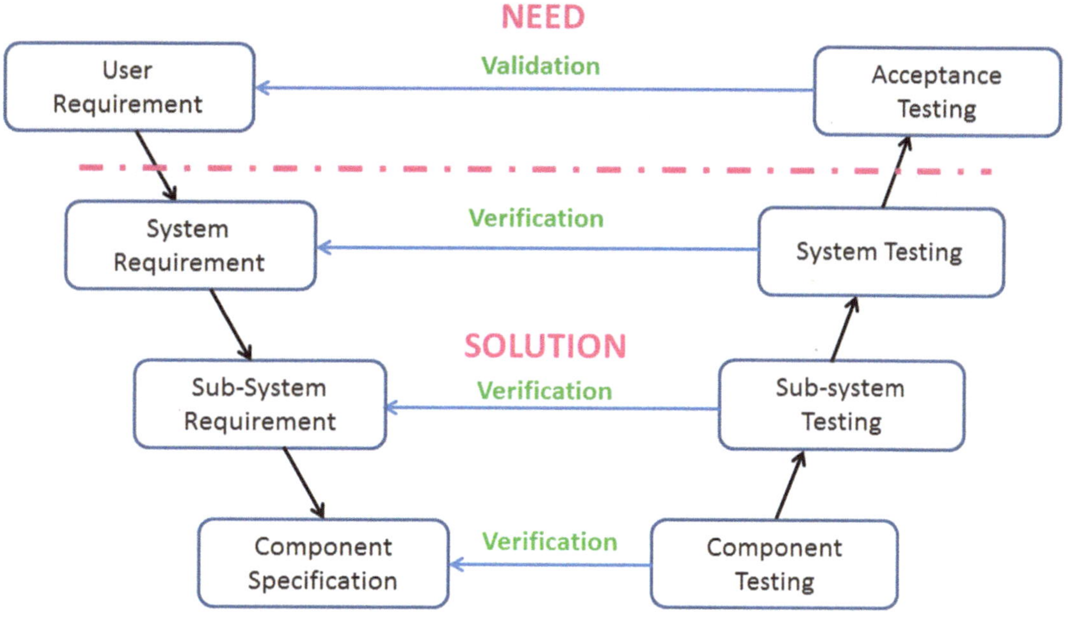

FIG. I–4. A requirements management system approach to verification and validation.

compiled by consolidating overlapping information requirements. There are 12 consolidated information requirement groups consisting of 63 specific information requirements. Although the information requirements can be consolidated at this relatively high level generic stage, it is recognized that the specifics of the data needed to meet those requirements may vary significantly between the different end users in terms of spatial distribution, resolution and accuracy.

TABLE I–2. THE NEEDS DRIVEN NWS SITE CHARACTERIZATION PROGRAMME HIERARCHY FOR AN ENVIRONMENTAL SAFETY CASE

Title	Contents
International legal requirements [IL]	3 legal requirements to address radioactive waste facility management
UK policy requirements [PR]	13 high level requirements contained within the UK government white paper and Environment Agency guidance on requirements for authorization of geological disposal facilities on land for solid radioactive wastes [I–9]
NWS DSS requirements [SR]	23 UK policy links to DSS requirements
ESC requirements [ESCR]	25 requirements applicable to the generic GDF disposal system derived from the DSS Part A (high level requirements) and B (technical specification)
ESC information groups [IG]	12 ESC information groups that are categories for more specific information requirements [IR]
ESC information requirements [IR]	60 ESC information requirements that identify specific site characterization information required to address ESC requirements
Parameters	To be specified by the site characterization team and approved by modellers to provide information needed by safety assessment, design and environmental assessment teams
Techniques	To be specified by the site characterization team to measure/derive parameters specified

Note: ESC = environmental safety case; DSS = disposal system specification. The abbreviations in this table apply to Annex I only.

Building a 3-D geological framework

CIR1 Assess the topography of the surface at specified site areas
CIR2 Assess the geomorphological evolution for specified site areas
CIR3 Assess the lithology of the superficial deposits
CIR4 Assess the 3-D geometry of the superficial deposits
CIR5 Assess the lithology of the bedrock
CIR6 Assess the mineralogy of the bedrock
CIR7 Assess the stratigraphy of the superficial deposits and the bedrock
CIR8 Assess the 3-D geometry of the bedrock
CIR9 Assess the 3-D geometry of the of discontinuities[2]
CIR10 Assess the geometric properties of discontinuities
CIR11 Assess the mineralogical properties of the discontinuities
CIR12 Assess the statistical properties of the discontinuities
CIR13 Assess the natural evolution of the geosphere

[2] CIR9 concerns the large scale 3-D geometry of discontinuities, such as the location of faults, folds and regions of heavy fracturing. CIR10 concerns the small scale features of these discontinuities.

Understand the surface hydrology

- CIR14 Assess the 3-D geometry of surface and near surface hydrological features, including flow in superficial deposits
- CIR15 Assess the hydraulic properties of surface and near surface hydrological features
- CIR16 Assess sea levels in the past
- CIR17 Assess the site specific meteorological conditions
- CIR18 Assess net site infiltration, recharge sources, recharge pathways and discharge

Understand the groundwater properties

- CIR19 Assess the distribution of hydrogeologically significant features in the bedrock
- CIR20 Assess the 3-D distribution of hydraulic head
- CIR21 Assess the hydraulic properties of the bedrock
- CIR22 Assess the hydraulic properties of the discontinuities in the bedrock
- CIR23 Assess the groundwater temperature
- CIR24 Assess the local and regional groundwater flow boundaries
- CIR25 Assess the distribution of groundwater bodies

Understand the hydrogeochemistry of the subsurface environment

- CIR26 Assess the nature and distribution of groundwater salinity
- CIR27 Assess the groundwater chemistry across the site
- CIR28 Assess the nature and distribution of natural colloids
- CIR29 Assess the nature of water–rock interactions, including the potential for leachate generation and the buffering capacity within the superficial deposits and geological formations
- CIR30 Assess the chemistry of the water discharge and recharge
- CIR31 Assess how groundwater chemistry has changed over time
- CIR32 Assess the radionuclide transport properties of the bedrock

Understand the stability of the geological environment

- CIR33 Assess the radionuclide transport properties of the discontinuities
- CIR34 Assess the orientation and magnitude of in situ stress fields in the potential host rock environment and its surroundings
- CIR35 Assess the seismicity (neotectonics) now and in the past
- CIR36 Assess the uplift / subsidence rate at the site and how it has changed over the past
- CIR37 Assess the erosion / deposition rate at the site now and in the past
- CIR38 Assess the extent of permafrost during previous glaciations
- CIR39 Assess past climate over the timescale relevant to post-closure

Understand the geotechnical properties of the bedrock and superficial deposits

- CIR40 Assess the physical and geotechnical properties of the superficial deposits (including rock strength and deformation properties)
- CIR41 Assess the physical and geotechnical properties of the bedrock (including rock strength and deformation properties)
- CIR42 Assess physical and geotechnical properties of discontinuities

Understand the thermal properties of the subsurface

- CIR43 Assess the in situ rock temperature in the potential host rock environment and its surroundings (including geothermal gradient)
- CIR44 Assess the thermal properties of the bedrock

Understand the potential for gas generation and migration

- CIR45 Assess the nature and distribution of natural gases present as mobile gas
- CIR46 Assess the nature and distribution of natural gases dissolved in the groundwater

CIR47 Assess the gas transport properties of the superficial deposits and bedrock
CIR48 Assess the gas transport properties of the discontinuities

Understand the natural radioactivity in the subsurface environment

CIR49 Assess the nature and distribution of natural radioactivity in the groundwater
CIR50 Assess the nature and distribution of natural radioactivity in the superficial deposits and bedrock
CIR51 Assess the nature and distribution of natural radioactivity in the minerals of superficial deposits and bedrock

Understand the ecological and biological processes

CIR52 Assess the nature and distribution of non-human biota in the environment at and near the surface
CIR53 Assess the nature and distribution of microbial populations in the potential host rock environment and its surroundings
CIR54 Assess the potentially exposed human groups
CIR55 Assess the habits of the potentially exposed human groups

Understand the potential for geohazards

CIR56 Assess 3-D geometry and condition of existing underground excavations
CIR57 Assess the frequency and severity of historical flooding
CIR58 Assess the distribution of gases that could present a hazard during excavation or operation including radioactive, flammable and noxious gases likely to be emitted
CIR59 Identify potential surface and underground geohazards resulting from superficial deposit or underground rock instability (e.g. rock falls, landslides, soil liquefaction, faults and fracture zones, etc.)

Understand the natural resources and human communities at the site

CIR60 Assess the nature and distribution of human communities, including historical development
CIR61 Assess the nature and distribution of natural resources
CIR62 Assess the history of exploration and exploitation natural resources
CIR63 Assess the current activities associated with exploration and exploitation of natural resources

I–4.4. Issues arising during development

The key issue arising during the development of the generic information requirements was the need for clarity in the terminology to describe the different elements of the requirement hierarchy. This was compounded by the need to group requirements at different levels in the hierarchy to make them more accessible to target audiences. As a result, clearly defined terms were developed and applied across the three information requirement reports, examples of which are provided in Table I–3.

Another issue that arose relates to how the information requirements and the links between them and the requirements they address are best communicated in an accessible manner. This is partly because individual information requirements often address more than one system requirement, a situation that will only become harder during the site specific stage when many more levels of requirements and information requirements are identified and managed in the requirements management system. Table I–4 provides an example of how the links are currently presented in a themed group approach in the current information requirement reports.

Key lessons

The most important lesson learned from the development of the generic information requirements was that the integration and knowledge gained by those carrying out the work was of similar value to the lists of information requirements produced.

A related lesson was, it is important that this development work is undertaken by the team that will go on to plan and implement the site investigation programme. The roles of the RWMO and the supply chain in such an exercise should, therefore, be carefully considered.

Although the development of generic information requirements is an extremely useful exercise, it is very important that its role in the siting process is clearly understood and communicated to relevant stakeholders. Without this there is a risk that key stakeholders perceive that the site specific information requirements have not been considered appropriately.

Finally, the succinct and clear definition of requirements in a way that can be agreed by all can be hard to achieve but is essential. Equally, the information requirements for site investigation can only be developed robustly if a good set of disposal system requirements is available. A good requirements set should include the following strengths:

— State the initial problem and the need that is to be satisfied;
— Provide a clear statement of objectives;
— Define the characteristics of the set of acceptable solutions;
— Provide guidance in the selection of the most appropriate solution.

Additional information

The development of a generic of information requirements for site investigations is not a particularly expensive task in the wider context of radioactive waste management. For NWS, the exercise took two years to complete involving two person years of effort and 15 staff from across the specification, engineering, safety case and site characterization teams. A geoscientific expert from the supply chain facilitated the exercise.

TABLE I–3. EXAMPLE DEFINITIONS OF TERMINOLOGY USED IN THE INFORMATION REQUIREMENTS REPORTS

Term	Definition	Example
Environmental safety case requirements	What needs to be assessed to comply with the regulatory and policy framework to meet stakeholder expectations, and to support the siting process	Determine the 3-D geometry of geological formations In order to make this assessment NWS needs to…
Information requirements groups	Broad categories of baseline information needed from the site characterization programme	Understand the 3-D geometry of superficial deposits and geological formations In order to develop this understanding NWS needs to…
Information requirements	Specific baseline information needed from the site characterization programme	…determine the spatial distribution of the host geological formation that is usable for the GDF (i.e. avoiding faults and other unsuitable ground within the host formation)
Data	Data are numbers, words or images that have yet to be organized or analysed to answer a specific question	Identification of geological units from borehole records Needed to understand…
Information	Information is produced through processing, manipulating and organizing data to answer questions, adding to the knowledge of the receiver	Engineering properties of identified geological units to support the construction of a GDF

TABLE I–4. INFORMATION REQUIREMENTS AND SUPPORTING ENVIRONMENTAL SAFETY CASE REQUIREMENTS FOR INFORMATION GROUP IG2

Environmental safety case requirements	
ESCR7	Assess the baseline groundwater flow system
ESCR8	Assess the potential for groundwater to move through discontinuities in the rocks
ESCR9	Assess the evolution of the groundwater flow system
ESCR10	Assess the potential for gas flow in the hydrogeological system
ESCR11	Assess the transport properties of the groundwater flow system
ESCR12	Assess the chemistry of the groundwater
ESCR16	Establish the baseline environmental radioactivity concentrations in the groundwater
ESCR20	Establish the baseline environmental chemotoxic concentrations in the groundwater

Information requirement group		Information requirements	
IG2	Understand groundwater flow and solute transport now and in the past	IR2.1	Understand the distribution of hydrogeologically significant features
		IR2.2	Understand the local and regional groundwater flow boundaries
		IR2.3	Understand the groundwater heads
		IR2.4	Understand the hydraulic properties of the rocks
		IR2.5	Understand the hydraulic properties of the discontinuities
		IR2.6	Understand the nature and distribution of natural colloids
		IR2.7	Understand the nature and distribution of the groundwater salinity
		IR2.8	Understand the groundwater temperature
		IR2.9	Understand the radionuclide transport properties of the rocks
		IR2.10	Understand the radionuclide transport properties of the discontinuities
		IR2.11	Understand the groundwater age

Note: ESCR = environmental safety case requirements; IG = information group; IR = information requirements.

The generic information requirements have also allowed a consideration of the specific parameters that may need to be measured and, therefore, the data acquisition and interpretation techniques that may need to be undertaken during site investigations (see Table I–5 for an example).

The generic information requirements have subsequently been used to underpin a suite of documents that justify the resource and budget requirements for needs driven site investigations (see Fig. I–5).

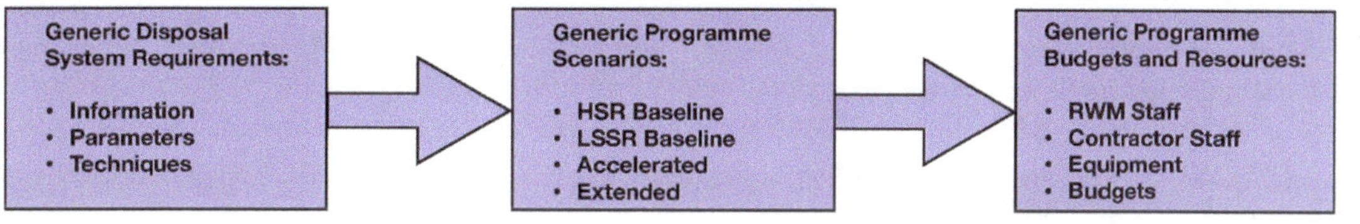

FIG. I–5. Underpinning of resource and budget requirements for needs driven investigations (courtesy of NWS).

TABLE I–5. CONSOLIDATED INFORMATION REQUIREMENTS, PARAMETERS AND DATA ACQUISITION TECHNIQUES FOR UNDERSTANDING THE HYDROCHEMICAL PROPERTIES OF THE SUBSURFACE

CIR No.	Information requirement	Parameter(s) measured	Principal data acquisition techniques
CIR1	Assess the nature and distribution of groundwater salinity	3-D data of saline groundwater bodies; salinity (TDS)	Groundwater, both near the surface and deep, and pore water salinity measured by TDS method (evaporation and gravimetric analysis), electrical conductivity
CIR2	Assess the groundwater chemistry across the site	Determination of alkalinity, chloride, TDS, Br, Na, Ca, Mg, K, SO_4, Si, Al, pH, Eh, tritium, nitrate, TOC, TIC, $\delta^{34}S$, $\delta^{13}C$, $\delta^{2}H$, $\delta^{18}O$, U-series elements, radiogenic Sr, inert gases	Interpretation and modelling of results from a suite of laboratory geochemical tests on groundwater across the site, including ICP-AES / ICP-MS, pH, Eh, radioisotope analysis, stable isotope analysis, TIC / TOC analysis, etc.
CIR3	Assess the nature and distribution of natural colloids	3-D data of the nature of natural colloids including size, composition, abundance	Filtration of groundwater using membrane filters or ultrasonic filtration, analysis using XRD, SEM-EDX
CIR4	Assess the nature of water/rock interactions, including the potential for leachate generation and the buffering capacity within the superficial deposits and geological formations	Determination of alkalinity, chloride, TDS, pH, Na, Ca, Mg, K, SO_4, Si, Al, Eh conditions, $\delta^{34}S$, $\delta^{13}C$, $\delta^{18}O$, U-series elements and radiogenic Sr in groundwater; mineralogical, elemental and isotopic ($\delta^{34}S$, $\delta^{13}C$, $\delta^{18}O$, U-series elements, and radiogenic Sr) compositions of geological units	Porewater analysis Interpretation and modelling of results from a suite of laboratory geochemical tests (ICP-AES / ICP-MS, pH, Eh, radioisotope analysis, stable isotope analysis, TIC / TOC analysis, etc.) on groundwater, superficial deposits and geological formations, and mineralogical analysis (CIR6 and CIR11)

Note: CIR = consolidated information requirements; TDS = total dissolved solids; TOC = total organic carbon; TIC = total inorganic carbon; ICP-AES = inductively coupled plasma atomic emission spectroscopy; ICP-MS = inductively coupled plasma mass spectroscopy; XRD = X ray diffraction; SEM-EDX = scanning electron microscopy with energy dispersive X ray spectroscopy.

REFERENCES TO ANNEX I

[I–1] DEPARTMENT OF ENERGY AND CLIMATE CHANGE, Implementing Geological Disposal: A Framework for the Long-term Management of Higher Activity Radioactive Waste, London (2014).

[I–2] DEFRA, BERR, WELSH ASSEMBLY GOVERNMENT, DEPARTMENT OF THE ENVIRONMENT NORTHERN IRELAND, Managing Radioactive Waste Safely: A Framework for Implementing geological disposal, Cm 7386, UK (2008).

[I–3] RADIOACTIVE WASTE MANAGEMENT, Geological Disposal — Generic Disposal System Specification Part A: High Level Requirements, NDA Report No. DSSC/401/01, Nuclear Decommissioning Authority, Didcot (2016).

[I–4] RADIOACTIVE WASTE MANAGEMENT, Geological Disposal — Generic Disposal System Specification Part B: Technical Specification, NDA Report No. DSSC/402/01, Nuclear Decommissioning Authority, Didcot (2016).

[I–5] RADIOACTIVE WASTE MANAGEMENT, Geological Disposal: Generic Disposal Facility Design, NDA Report No. DSSC/412/01, Nuclear Decommissioning Authority, Didcot (2016).

[I–6] RADIOACTIVE WASTE MANAGEMENT, Design Information Requirements from the RWM Site Characterisation Programme, NDA Technical Note No. 19656068, Nuclear Decommissioning Authority, Didcot (2013).

[I–7] RADIOACTIVE WASTE MANAGEMENT, Safety Case Information Requirements from the RWM Site Characterisation Programme, NDA Technical Note No. 19700021, Nuclear Decommissioning Authority, Didcot (2013).

[I–8] RADIOACTIVE WASTE MANAGEMENT, Environmental Assessment Information Requirements from the RWMD Site Characterisation Programme, NDA Technical Note No. 19657261, Nuclear Decommissioning Authority, Didcot (2013).

[I–9] NORTHERN IRELAND ENVIRONMENT AGENCY, Geological disposal facilities on land for solid radioactive wastes: Guidance on requirements for authorisation, EA, Belfast (2009).

Annex II

ESTABLISHING A STEPWISE APPROACH TO THE PLANNING AND IMPLEMENTATION OF A SITE INVESTIGATION PROJECT — A CASE STUDY FROM JAPAN

II–1. CONTEXT

Site investigations are generally implemented in a series of stages during the siting process in national programmes. In Japan, where siting has been initiated by a call for volunteers, the site investigation programme is planned to proceed in a stepwise manner to identify a suitable repository site. Three stages following the nationwide screening of scientifically favourable areas by the government are specified by law [II–1]: the initial literature survey, the following preliminary investigation and the final detailed investigation, as illustrated in Fig. II–1. NUMO's structured site investigation process is thus characterized by the iterative refinement of a site specific knowledge base, which is developed through quality assured geosynthesis and integrated within an SDM [II–2]. At project milestones at the end of each stage, the SDM supports decisions to be made on which (if any) sites to carry forward to the next stage.

The literature survey involves a desk based study using a site specific, state of the art geoscientific knowledge to exclude certain areas owing to the likelihood of potentially significant impacts of natural disruptive events and processes (e.g. volcanic activity, fault movement, occurrence of deep-seated fluids, significant uplift and erosion), the presence of thick sequences of unconsolidated Quaternary deposits, or the existence of significant mineral resources. The subsequent preliminary investigation and detailed investigation for non-excluded target area(s) involve field characterization at a progressively increasing level of detail: from early surface geological and geomorphological surveys through geophysical surveys and shallow to deep borehole investigations to the final more detailed subsurface characterization after

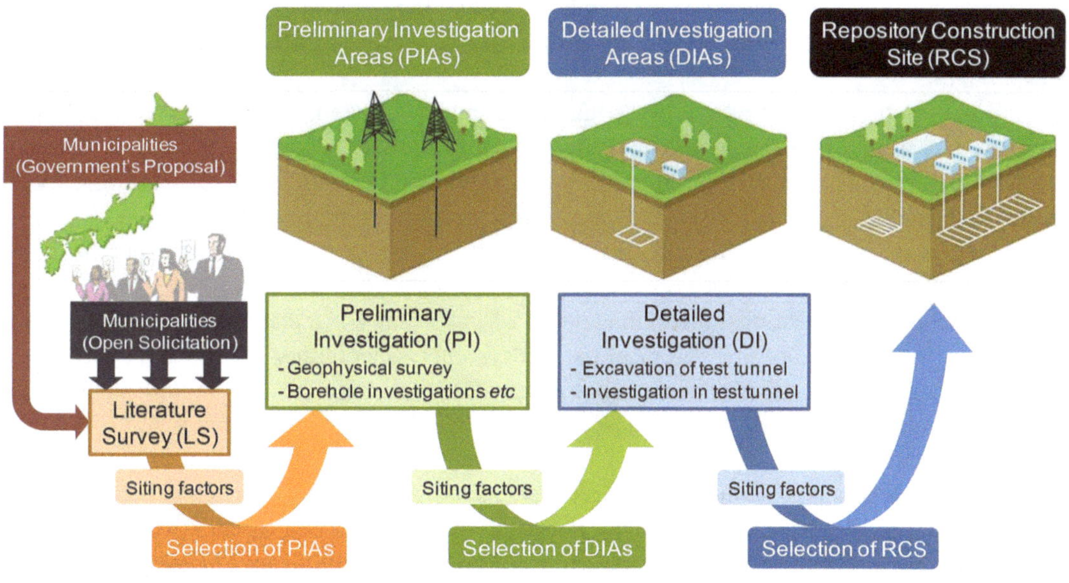

FIG. II–1. Staged selection of a suitable repository site in Japan (prepared based on Ref. [II–3], courtesy of NUMO).

the construction of an underground investigation facility [II–2]. The basic strategy for site investigations during the preliminary investigation and detailed investigation involves the following iterative process:

- On the basis of knowledge of the site geological environment from preceding investigations, key aspects are characterized and the type and degree of uncertainties are identified. These are particularly sensitive to the key safety functions of the specific host geological environment.
- The investigation targets and associated priorities are specified.
- The series of field investigations with a supporting laboratory programme are planned.
- The planned investigations are carried out.
- The knowledge gains are assessed to determine how well the site fulfils the requirements for repository design (RD) and safety assessment (SA). Such requirements are also developed in a stepwise manner to tailor the repository concept and the associated safety case to the specific siting environments under consideration.

To facilitate this in a logical and effective manner, which is one of the key challenges, site investigations within individual stages will be implemented in work steps (or milestones) based on a 'geosynthesis methodology.' This methodology clearly defines the goals of individual investigations and, finally, synthesizes a site specific knowledge base into a consistent SDM together with associated datasets (as required for RD and SA) ensuring the transparency, traceability, and quality of all output to be used to support decision making. The geosynthesis methodology specifies for each work step the requirements in terms of the following:

- Investigation and data acquisition;
- Data interpretation;
- Conceptualization;
- Numerical modelling and simulation;
- Derivation of output in a required form (e.g. SDM and associated data sets).

The rationale for the structured approach includes ensuring the impact of limitations in site specific knowledge and the significance of uncertainties can be explicitly assessed by the RD and SA. More importantly, information on the sensitivity of particular characteristics to the key safety functions of the host geological environment can be fed back at the end of each work step to guide focusing or optimizing investigations on reducing uncertainties that have the greatest impact in the subsequent work steps. Indeed, if appropriate, identifying 'hopeless' cases as early as possible avoids wasting limited resources on them.

During the implementation of site investigations at any work step, technical constraints associated with the characterization of the host geological environment should be recognized. It is thus of importance that site investigations have sufficient flexibility to respond to not only the expanding site specific knowledge base, but also the surprises that inevitably occur during the investigations. The stepwise approach provides the opportunity to enhance tailoring of investigations to the site specific conditions from this practical point of view.

In stepwise site investigations, the management of uncertainty is a critical issue, which is also one of the challenges, as uncertainty is inherent in natural systems. Uncertainty is inevitably associated with data acquisition and interpretation and the subsequent SDM development over a wide range of spatial scales and, particularly, very long temporal scales. In general, many uncertainties are expected to be reduced, in an iterative manner, to a level either deemed acceptable or unimportant in the context of the key safety functions of the host geological environment as the site specific knowledge base expands. This results in an enhancement of confidence in the decisions made at the end of each work step. However, there will always be a certain residual uncertainty that cannot be further reduced by any measures and, hence, should be captured in the geosynthesis process and its impact directly assessed by close interaction with the RD and SA.

To date, there has been no selection of either host rock or a repository site in Japan. NUMO does, however, benefit from a long history of geoscientific research and development (R&D) that has been undertaken by fundamental R&D organizations in Japan (e.g. JAEA) and also the ongoing repository site investigations in other countries, as extensive knowledge and hands-on experience have been gained and developed to date in such programmes. The JAEA's generic URL projects, which have been being implemented in a granite basement with sedimentary overburden at Mizunami in Gifu [II–4] and in a thick sedimentary formation at Horonobe in Hokkaido [II–5], are particularly appropriate. Although these URLs are NOT in selected repository sites, they have been developed in this kind of stepwise manner with a site specific knowledge base being integrated within SDMs. In the following case study, the stepwise approach to the planning and execution of surface based investigations at the Mizunami URL site is described.

II–2. MANAGEMENT CHALLENGES

The boundary conditions for the Mizunami URL project are rather unique. As the project has been implemented only for geoscientific R&D purposes, specific challenges have focused on the development of the technical basis for the characterization of the geological environment by applying various technologies relevant to site investigations. For the execution of the surface based investigations prior to URL construction, spatial and temporal constraints had to be considered. Since rough terrain prevails, the limited flat land available should primarily be used for URL construction at the Mizunami URL site (approximately 78 000 m^2) and, hence, only a few locations were available for borehole investigations. In addition, the surface based investigations were required to be completed within three years.

Despite such constraints, an approach involving an iterative loop of planning, execution, and geosynthesis of investigations was adopted to execute surface based investigations in an effective manner. This allowed assessing whether the outcomes of each investigation step actually meet the requirements (or particular aims), specified on the basis of earlier findings, leading to the identification and prioritization of key issues to be addressed in the subsequent investigation step. The details of the following investigations (e.g. where and how to investigate the key issues) were then drawn up. A significant management challenge, therefore, involved making decisions to proceed to the next investigation steps in as appropriate and prompt a manner as practicable, taking all the boundary conditions into consideration.

Although feedback from the RD and SA were not expected in such a generic URL project, the surface based investigations aimed at understanding the key characteristics of, and processes ongoing in, the geological environment in a comprehensive manner, which would be of great relevance to the RD and SA. In particular, the spatial variability of magnitude of hydraulic gradient and hydraulic properties of rock mass / major geological structures (e.g. faults of a larger scale) has a great impact on site suitability. The main focus was thus on geological / hydrogeological characterization and synthesis.

For making an appropriate and prompt decision in a practical manner during such iterative loops, geological / hydrogeological conceptualization was assessed to ensure it met the main aims and then particular investigation targets were specified and prioritized. This process based on conceptualization is different from that to be adopted in site investigations during the preliminary investigation and detailed investigation. Here, the management challenge also involved balancing the level of details of the next investigations with practical, temporal, and financial constraints, which eventually required trade-offs to be made in the optimization of the next investigation programme. It should be noted that trade-offs would reflect quite different boundary conditions in actual site investigations.

II–3. STRATEGIC PLANNING

The surface based investigations were planned to include, in an early phase, the characterization of sedimentary overburden (with thickness of up to 250 m) over a wider area around the Mizunami URL

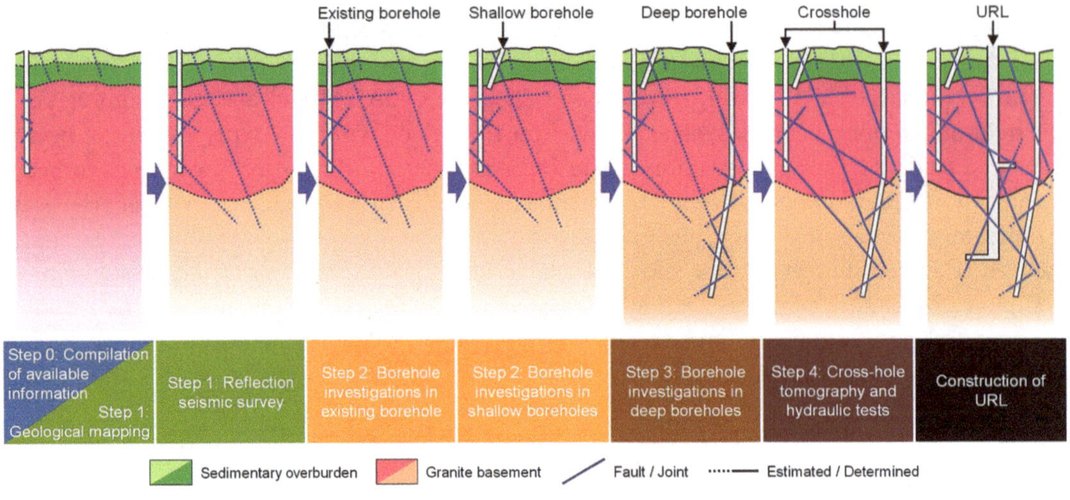

FIG. II–2. Stepwise surface based investigations at Mizunami URL site (modified from Ref. [II–6]).

site and, in a later phase, the granite basement by deep borehole investigations [II–6]. Attention was paid here to the timing and location of investigations, so any interactions between borehole drilling and other investigations could be precluded. Specifically, the planned investigations were categorized into five steps, as illustrated in Fig. II–2.

Although the particular investigation targets were specified on the basis of feedback from the previous step, the general aims of investigations in each step were defined in advance as follows:

— Step 0: Compilation of available information to understand the 3-D distribution of geological structures and groundwater flow characteristics on a regional scale;
— Step 1: Geological mapping and seismic reflection survey to estimate the 3-D distribution of faults and geological boundaries;
— Step 2: Borehole investigations in an existing 500 m deep borehole and four newly drilled 100 m to 200 m deep boreholes to characterize rock mass and faults;
— Step 3: Deep borehole investigations in two new boreholes (1300 m and 1000 m deep) to characterize rock mass and faults;
— Step 4: Cross-hole tomography and hydraulic tests in these deep boreholes to determine the locations, geometry, hydraulic properties and continuity of faults.

Because of temporal constraints, the borehole investigations in the existing deep borehole and the new shallow boreholes were planned to be executed in parallel in Step 2.

Taking into consideration the number and resolution of the surface based investigations, the reduction of the uncertainties associated with the output of geosynthesis was expected to be limited. The primary output from the stepwise surface based investigations was, therefore, defined to clarify through the 2-D conceptualization and 3-D SDM development processes particularly, the relationship between the type, quality and quantity of the investigations, the corresponding level of detail that was expected to be achieved and the type and degree of uncertainties with respect to the particular investigation target. For the testing of conceptualization and SDMs developed during the surface based investigations, performance measures were identified and included in the programme for investigations during URL construction.

II–4. IMPLEMENTATION

Following the review of relevant pre-existing information, field investigations commenced with geological mapping and a seismic reflection survey. These were followed by a large programme of investigations in boreholes (100 m to 1300 m deep), with comprehensive study of core and groundwater samples. This programme was extended by subsequent cross-hole tomography and hydraulic tests between the deep boreholes.

II–4.1. Step 0: Compilation of available information

Information from the literature, geological maps and geoscientific studies that had been conducted at the scale of regional groundwater flow from areas of recharge to discharge (approximately a few tens of square kilometres) prior to the initiation of the Mizunami URL project were compiled and reviewed. Based on the results of this review, the step 0 geological conceptualization and SDM were developed. The geological SDM shown in Fig. II–3(a) describes best estimates of the geometry of significant geological / stratigraphical units and a well-studied 8 km long major fault within the granite basement.

It was shown from groundwater flow simulations based on the step 0 hydrogeological SDM that a major fault zone acts as a hydraulic barrier, as can be seen in Fig. II–3(a). In addition, fluxes and flow velocities in the basement are considered to be related to the density of smaller water-conducting fractures that are either distributed throughout the rock mass or concentrated in damaged zones alongside the major fault. A key goal of subsequent geological / hydrogeological characterization and synthesis was, therefore, to better define the role of major faults on local groundwater flow.

II–4.2. Step 1: Geological mapping and seismic reflection survey

Detailed geological mapping and a seismic reflection survey provided information on the spatial extent and geometry of newly identified major faults and lithostratigraphical boundaries at the scale of the local groundwater flow from areas of recharge to discharge to a depth of 1000 m at the Mizunami URL site. Based on such information and complementary lineament interpretations, the initial geological conceptualization and SDM were refined. The step 1 geological SDM is shown in Fig. II–3(b). As the hydraulic properties of the newly identified major faults were not understood, these were grouped into five sets on the basis of their orientations and then represented as discrete hydraulic features in the updated hydrogeological conceptualization and SDM.

Groundwater flow simulations that focus on uncertainties in the hydraulic anisotropy of faults involved 32 sensitivity cases, in which the model variants assume that either all or particular sets of the faults act as hydraulic barriers or otherwise major conduits. Visualizations of the 3-D hydraulic head distribution demonstrate the uncertainty in the hydraulic anisotropy of the faults that trend NNW and NE has the greatest impact on modelled heads, with significantly reduced heads being presented when the faults trending NNW act as hydraulic barriers. As an example, the simulated hydraulic head distribution for the case that provides the best fit to the hydraulic measurements is illustrated in Fig. II–3(b). Geological / hydrogeological characterization of the faults trending NNW was, therefore, identified as a key issue for further steps.

II–4.3. Step 2: Borehole investigations in existing borehole and new shallow boreholes

A combination of techniques, including core logging, borehole wall imaging, geophysical and fluid logging, hydraulic tests and hydraulic pressure monitoring was applied in the existing 500 m deep borehole and the new shallow boreholes. The geological characteristics and hydraulic properties of sedimentary overburden and granite basement to a depth of 500 m were determined, with special emphasis on the geological / hydrogeological significance of the faults trending NNW and EW.

Such input allowed the interpretations of orientation and continuity of major faults to be refined and, hence, the number of fault groups to be reduced in the modified geological conceptualization and SDM, as can be seen in Fig. II–3(c). The hydrogeological conceptualization and SDM were also revised as the hydraulic conductivities of the faults were explicitly defined, which resulted in decreasing the number of variants to a total of 12 sensitivity cases. Groundwater flow simulations clearly show that the faults trending NNW, NW and EW have the significant impact on groundwater flow conditions at a depth of 1000 m in the granite basement. The simulated 3-D hydraulic head distribution for the 'best fit' case is illustrated in Fig. II–3(c). Geological / hydrogeological characterization of such faults was, therefore, identified as a key focus of surface based investigations in step 3.

II–4.4. Deep borehole investigations in new boreholes

Comprehensive field investigations were executed in newly drilled boreholes 1300 m deep deviated and 1000 m deep vertical, located at / near the Mizunami URL site, respectively. These investigations provided the detailed characterization of the faults trending NNW, NW and EW and also the background fractured granite basement. In particular, the hydrogeological characteristics of the major faults were estimated on the basis of the groundwater pressure responses monitored at the existing deep and shallow boreholes during the drilling of the deep boreholes and the subsequent investigations (e.g. hydraulic pumping tests) in these boreholes.

Through the geosynthesis of such information, the geological / hydrogeological conceptualization and SDMs developed in step 2 were further revised, with most of the faults trending NNW and EW being defined as either being hydraulic barriers or conduits. The step 3 geological SDM is shown in Fig. II–3(d). The refined hydrogeological conceptualization and SDM required eight variant cases for groundwater flow simulations, which resulted in decreasing the resultant range of the hydraulic head distributions, thereby significantly reducing remaining uncertainties in hydrogeological characteristics of key groups of faults. Fig. II–3(d) illustrates the 3-D hydraulic head distribution for the 'best fit' case.

II–4.5. Step 4: Cross-hole tomography and hydraulic tests in deep boreholes

The execution of a variety of cross-hole tests allowed more precise interpretation of the geological / hydrogeological connectivity of major faults and subhorizontal fractured zones developed in the upper domain of the granite basement. In particular, cross-hole seismic and resistivity tomography surveys, multioffset VSP surveys and additional seismic reflection surveys on two lines provided information on the distribution of the faults trending NNW, NW and EW and the fractured zones with more confidence. In addition, cross-hole hydraulic tests characterized all major faults between the boreholes as either being hydraulic barriers or conduits.

The hydrogeological SDM was then calibrated by focusing on geometry and hydrogeological characteristics of the major faults illustrated in the updated geological SDM. The step 4 geological SDM is shown in Fig. II–3(e). As the only remaining major uncertainty is associated with a minor fault that lies outside the Mizunami URL site, the number of model variants was greatly reduced and, hence, quite similar hydraulic head distributions derived from only two cases; the simulated hydraulic head distribution for the 'best fit' case is illustrated in Fig. II–3(e). These results are consistent with other site information, which indicates the geosynthesis process is leading to convergence on a consistent SDM.

II–5. ISSUES ARISING DURING IMPLEMENTATION

The most significant issue to be noted here is that an unavoidable delay in drilling the deep deviated borehole in step 3, which was mainly due to the inadequacies of systems and procedures for drilling, necessitated the revision and extension of the ongoing deep borehole investigation programme [II–7]. Specifically, as the deepest target fault was predicted to intersect the planned URL shafts subvertically, the direction and deviation of the borehole was optimized so the fault could be characterized in detail as early as practicable, which allowed providing relevant input to the revision of detailed design and construction plan of the Mizunami URL. The methodologies for deviated borehole drilling and subsequent investigations were, thus, refined and the amount of time available for these activities maximized. It should be noted here such revision did not cause a decrease in the quality and quantity of the remainder of the investigations carried out to fulfil the overall goals and specific aims of the deep borehole programme. This did, however, result in the postponement of the planned investigations in step 4.

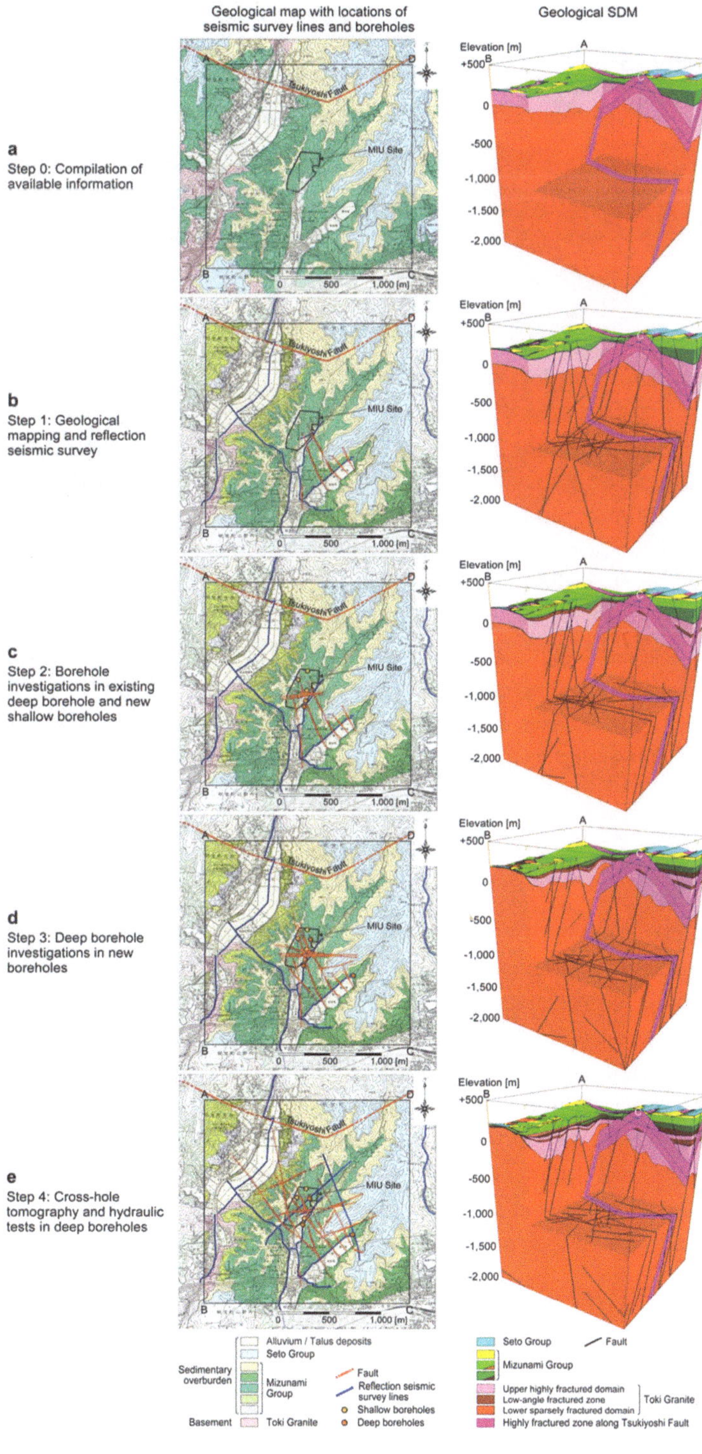

FIG. II–3. Improvement in understanding of geological/hydrogeological environments by stepwise surface based investigations at/around the Mizunami URL site (modified from Ref. [II–6]).

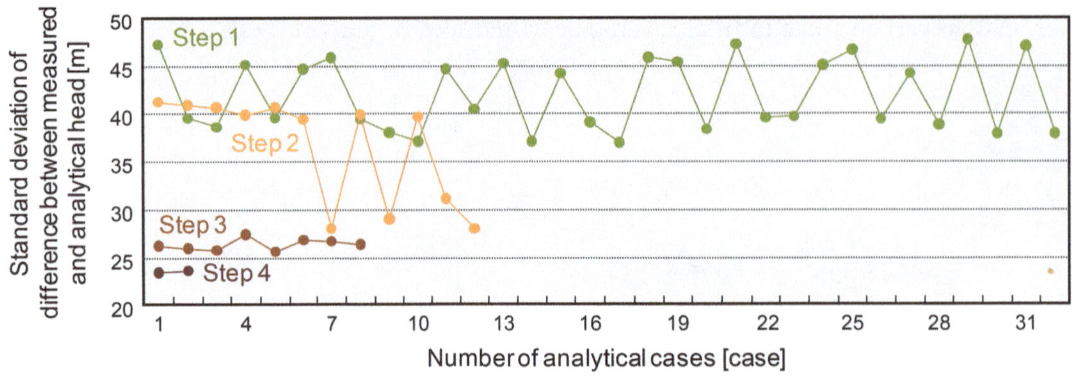

FIG. II–4. Reduction of uncertainties in groundwater flow simulation by stepwise surface based investigations at/around the Mizunami URL site (modified from Ref. [II–6]).

II–5.1. Key lessons

Feasibility of reducing key uncertainties in a stepwise manner

The most important lesson learned from the stepwise surface based investigations at or around the Mizunami URL site, in which the entire geosynthesis process was incorporated, is this approach is indeed effective in reducing key uncertainties that greatly impact safety relevant aspects captured in the geological / hydrogeological conceptualization and SDMs [II–6]. Site understanding has, therefore, been substantially improved as shown in Fig. II–3. This is clearly shown by the comparison of the calculated hydraulic heads of groundwater flow variants in each step with the best measured hydraulic heads from monitoring intervals in the existing deep and shallow boreholes at the Mizunami URL site. The standard deviation of differences between measured and calculated hydraulic heads, which is illustrated in Fig. II–4, clearly shows the number of model cases, the variation of results from sensitivity analyses, and the difference between measured and calculated hydraulic heads have been greatly reduced as the surface based investigations proceeded.

Need to optimize investigation steps

The number of investigation steps to be involved in individual stages should be carefully optimized when the stepwise site investigations are planned, taking temporal and financial constraints into consideration. Specifically, it would be appropriate to establish only a few investigation steps with an increasing level of detail (e.g. from general understanding over a wider area to detailed characterization to greater depth) and, in each step, the planned activities executed as much in parallel as practicable and, more importantly, in such a way that they cannot interfere with each other. As a number of steps require a long time to complete the planned investigations, any major unexpected setback encountered in an earlier step could produce significant temporal and financial impacts on remaining the investigation steps, thus reducing the flexibility in optimizing the investigations. In addition, the accumulation of small delays in individual steps could eventually result in unacceptable overall delay.

II–6. ADDITIONAL INFORMATION

To facilitate stepwise surface based investigations in which the geosynthesis methodology was effectively introduced at both URL sites, a 'Geosynthesis Data Flow Diagram' (GDFD) was formulated for each investigation step [II–6, II–8]. This provides a systematic and comprehensive framework for the geosynthesis, illustrating a sequence of processes, in a logical manner, that can guide which type and combination of investigation techniques are employed, which type of data are acquired, how to interpret the range of acquired data, how to derive a consistent data set, how to integrate relevant information that was obtained from different disciplines, etc. Application to coastal geological environment has also been illustrated within the framework of JAEA's Horonobe URL project [II–9]. The geosynthesis methodology and these GDFDs will be refined on the basis of accumulated knowledge and experience and further optimized for application to NUMO's site investigations.

As the fluxes of information from modern investigation techniques and the sophistication of preprocessing technology increase at an exponential rate, the capacity of geosynthesis is expanding rapidly. A smart system has thus been developed, which provides specific functionalities for a user-friendly interface to the wide range of personnel involved in site investigations. In particular, know-how or tacit knowledge developed mainly in JAEA's URL projects is managed in a knowledge base and can be transferred not only to make the geosynthesis process faster and more interactive but also to facilitate planning and tailoring the investigation programme in a more effective manner to improved site understanding and changes in site environmental conditions. This could also allow risks of costly mistakes to be minimized in such a process.

REFERENCES TO ANNEX II

[II–1] NUCLEAR WASTE MANAGEMENT ORGANIZATION OF JAPAN (NUMO), The Repository Site Selection Process: The geological disposal program, https://www.numo.or.jp/en/jigyou/new_eng_tab03.html

[II–2] OTA, K., SAEGUSA, H., KONDO, H., GOTO, J., KUNIMARU, T., YAMADA, S., Site Characterisation and Synthesis into SDMs for NUMO Safety Case. Proc. ANS 17th Int. HL Radioactive Waste Mgmt Conf. (IHLRWM 2019), Knoxville, TN (2019).

[II–3] NUCLEAR WASTE MANAGEMENT ORGANIZATION OF JAPAN, Requirements Management Systems (RMS): Status and Recent Developments, Information Exchange Meeting Report, RMS2010, NUMO-TR-10-07, NUMO (2011), https://www.numo.or.jp/en/reports/pdf/TR-10-07.pdf

[II–4] JAPAN ATOMIC ENERGY AGENCY (JAEA), Mizunami Underground Research Laboratory: Welcome to the MIU, https://www.jaea.go.jp/04/tono/miu_e/index.html

[II–5] JAPAN ATOMIC ENERGY AGENCY (JAEA), Horonobe Underground Research Center, https://www.jaea.go.jp/english/04/horonobe/index.html

[II–6] SAEGUSA, H., MATSUOKA, T., (Eds), Final report on the surface-based investigations (Phase I) at the Mizunami Underground Research Laboratory Project, JAEA-Research 2010-067, JAEA, Tokai, Japan (2011).

[II–7] TAKEUCHI, S., OTA, K., Working Programme for MIZ-1 Borehole Investigations: Revision of work procedures after Phase IV, JNC Technical Report, JNC TN7400 2005-010, JAEA, Tokai, Japan (2005), https://jopss.jaea.go.jp/pdfdata/JNC-TN7400-2005-010.pdf

[II–8] OTA, K., ABE, H., KUNIMARU, T., (Eds), Horonobe Underground Research Laboratory Project; Synthesis of Phase I Investigations 2001–2005, Volume 'Geoscientific Research,' JAEA-Research 2010-068, JAEA, Tokai, Japan (2011).

[II–9] OTA, K., AMANO, K., NIIZATO, T., ALEXANDER, W.R., YAMANAKA, Y., Development of comprehensive techniques for coastal site characterisation: (1) Strategic overview. Proc. ASME 2010 13th Int. Conf. Envi. Remed. Radioactive Waste Mgmt, 3–7 October 2010, Vol. 2, ASME, Tsukuba, Japan (2010) 123–132, https://doi.org/10.1115/ICEM2010-40056

Annex III

DATA COLLECTION AND SITE DESCRIPTIVE MODELLING TO SUPPORT REPOSITORY SITE CHARACTERIZATION FOR DESIGN, SAFETY AND ENVIRONMENTAL IMPACT ASSESSMENTS — A CASE STUDY FROM SWEDEN

III–1. STRATEGIC PLANNING

At the time when the preinvestigations for the Äspö HRL were carried out, there was no structured goal for how the data and information was to be used for engineering and safety assessment. With the outcome of preinvestigations and construction in hand, it was possible to define the contents of the site investigations to be carried out at the selected repository candidate sites [III–1].

Many different investigation techniques and methods were tested and evaluated as the construction of the tunnel gradually revealed the conditions in the rock mass. This was made systematically by comparing predictions and outcome. The overall experience from this work was the selected investigation methods could give a correct description of the rock and the relevant processes in the rock on the basis of the preinvestigations [III–2, III–3]. The onset of the site descriptive modelling was based on this outcome.

The siting work leading to the selection of Forsmark as the repository site in 2009 was initiated in 1992 by asking all municipalities in Sweden (284 municipalities) whether they would be interested in discussing how to host the repository for the SNF within their municipality. Discussions started in many municipalities and feasibility studies were carried out in eight municipalities. These studies were mainly focused on the future development options of the municipalities, with or without a repository. Efforts were also made to compile and assess the scientific data (especially the geoscientific data) available for the potential sites within the municipalities. By the year 2000, SKB selected Oskarshamn, Östhammar and Tierp as candidates for the site investigations. Oskarshamn and Östhammar both accepted, and the site investigations were started in both municipalities in 2002. Additional research and safety assessment studies were made in parallel during 1990 to 2000.

In 2002, a permanent organization was established in both municipalities with the mission to conduct the investigation programme. Due to a good detailed planning, it was mostly a question of executing the plans. The simple task was to conduct the investigations, the more demanding tasks were to create real multidisciplinary exchange of knowledge and to have fruitful feedback from the repository engineering and safety assessment work.

A necessity and, simultaneously, a challenge during the site investigation stage was the individual projects (investigations, SDM, repository design and safety assessment) were run in parallel. Investigations had to be done before the data could be evaluated and the integrated evaluation (SDM) had to be made before the repository engineering design and the safety assessment could make use of the outcome. The investigations, evaluation, engineering and safety assessment were looped twice while the individual projects had several loops. For the SDM, there were four loops where the second and the fourth fed into safety assessment; all of them fed into repository engineering. There was always feedback from repository design to the SDM and to investigations but, due to the time delay, there was little feedback from safety assessment to SDM and none to investigations.

III–2. SITE DESCRIPTIVE MODELS

Site descriptive modelling is a collective term for all evaluation and interpretation of the data collected by the investigations carried out. On the first hand, it was relevant to make the modelling work within the different disciplines of investigation/evaluation, which were geology, hydrogeology, hydrogeochemistry, geomechanical and thermal properties, and radionuclide transport properties. Within each of these several types of investigations were carried out. Together with the on-site activity leader of the investigations the interdisciplinary evaluation was made before the integrated modelling between the disciplines started.

The results of measurements were analysed and evaluated to provide a description of the site that can be used for design and safety assessment. Primary investigation data are collected and stored in a dedicated site characterization database (known as SICADA). For the most part, the data mainly comprise parameter values for single measurement points or limited measurement objects. Primary data are subjected to discipline specific and integrated analysis and interpretation. Based on this, the model is developed and subdivided into suitable geometric units and assigned discipline specific properties. As a result, a 3-D SDM of the rock volume and ground surface can be constructed. The SDM is presented through the use of GIS and, above all, a CAD based computer tool, called the rock visualization system. This system also used as an instrument used for the interpretation of information, especially to support a judgement concerning the relative locations of different deformation zones.

The ultimate goal of the SDM is to make a visualization/understanding of the site useful for repository design, long term safety assessment and EIA.

III–2.1. Modelling steps

The SDM provides a multidisciplinary interpretation of geology, rock mechanics, thermal properties, hydrogeology, hydrogeochemistry, transport properties and ecosystems using data and understanding derived from deep boreholes and surface based investigations. Model development comprises: an evaluation of primary data followed by descriptive and quantitative modelling in 3-D and an overall evaluation of confidence Refs [III–4, III–5].

The measured (primary) data comprise a wide range of measurement results that need to be checked for consistency and processed into a format more amenable for use in 3-D modelling. Examples of data evaluations reflect data sets used in the estimation of surface geology, lineament interpretation, geological single hole interpretation, assessment of reflection seismics, hydrogeological single hole interpretation, assessment of hydrogeochemical data and single hole assessment of rock stress measurements. The data are first evaluated within each discipline and then the evaluations are checked to ensure consistency between the disciplines (see Fig. III–1).

3-D modelling to provide estimates of the distribution of parameter values in space is made in a sequence, where the geometrical framework is first taken from the geological model and this in turn is used for the rock mechanics, thermal and hydrogeological modelling purposes. The hydrogeochemical description is to some extent developed independently of the geology, but its consistency with the hydrogeological description is closely checked. The description of transport properties is based on the hydrogeological and hydrogeochemical descriptions, although additional data are assessed as well. Lastly, a description of the near surface environment is made that is essentially independent of the

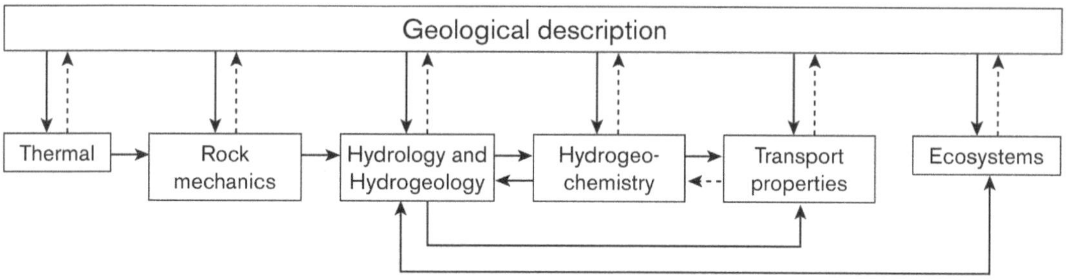

FIG. III–1. The different discipline descriptions are interrelated with several feedback loops and where geology provides the essential geometrical framework (courtesy of SKB, Sweden).

other descriptions, although the interface between the geosphere and biosphere is handled together with geosphere hydrogeology.

After the individual discipline specific modelling and uncertainty assessment is carried out, a phase of overall confidence evaluation follows. The various modelling teams assess the suggested uncertainties and evaluate dependencies and feedbacks between disciplines, in particular between the geology, hydrogeology and hydrogeochemistry. After revisions, the teams reassess the model description together with uncertainty and confidence statements. These discussions assess overall confidence by:

- — Checking that all relevant data are used (and appropriately);
- — Checking that information used in previous versions of the model is considered;
- — Checking that the different kinds of uncertainty are adequately addressed;
- — Checking whether suggested alternative descriptions make sense and whether there is the potential for additional alternatives;
- — Discussing, if appropriate, how additional measurements (i.e. more data) would affect confidence;
- — Making an assessment of uncertainty and confidence.

There are always uncertainties associated with measurements and rock parameters, which vary in space. A 3-D description should present the parameters based on their an evaluation of spatial variability over a relevant and specified length scale, with the uncertainty included in this description. Alternative descriptions may be required. However, instead of trying to only find alternative (hypothetical) explanations for data, the modelling work was focusing on describing the confidence in the model predictions and identifying weaknesses to be strengthened in the next batch of investigation and data evaluation.

There are several challenges that need to be addressed when dealing with alternative descriptions of a site and disciplines. In particular, it may be difficult for modelling teams to develop more than one alternative model in parallel. Furthermore, as models are used as input to other models, the total number of possible alternatives may become impractical to handle — a situation known as 'variant explosion.' Alternative model generation is typically seen as a component of model development and as a means of exploring the confidence. However, all alternatives need not be equally probable. Different alternatives can focus on clear differences in the geometrical framework and on clear differences in the description of rock and hydrogeological properties.

It is important to assess the potential benefit of undertaking additional measurements. Clearly, if new data compare well with the predictions made in advance of measurements, the need for yet additional data may diminish.

III–2.2. Documentation

Modelling results and findings are to be documented in a site description. This description encompasses the different databases and digital models developed, as well as a model report with supporting documentation. It is essential that the model report includes the following information:

- — Data sources and identification of previous model versions;
- — Methods of primary data evaluation for each discipline and for data previously evaluated, as well as the means of interdisciplinary comparisons;
- — The 3-D modelling approach used, including comparisons with approaches used to generate previous model versions, uncertainty estimates for individual data sets and joint uncertainty, and confidence evaluations;
- — The SDM (discipline by discipline) together with its uncertainties and alternatives;
- — Comparison of results with previous model versions, an assessment of overall confidence and discussion concerning potentially fruitful additions to the measurement programme;
- — Implementation details.

Site investigations consisted of different investigation activities: geological (surface mapping and drillcore logging), geophysical (survey and borehole logging), hydrogeological (recharge–discharge mapping, a whole set of borehole investigations), hydrogeochemical (fracture mineral and groundwater and surface water chemical investigations) and surface ecological (flora, fauna and marine) investigations. Drilling (percussion and core drilling), infrastructure (roads, electricity, drilling water handling, data communication and authorization/permissions) and landowner communication activities were supported at the same level of the other investigation activities.

The site investigation activities were managed at the site office where all activity leaders were gathered. Many of the activities required several activity leaders to manage the workload. The site investigation team consisted of 35 people.

To execute the investigations, the main stages of initial and complete site investigation were further broken down into smaller steps. Stepwise execution provides better opportunities for a site-adapted investigation methodology and more effective feedback from evaluation. In general, each new step consists of confirming or rejecting the main results of the preceding step, answering questions that have been raised and achieving the goals set for the particular stage. Each step builds further on the description that emerged from the preceding investigation step. In each step, all disciplines interact in the planning of the investigations and the evaluation of results.

Additional knowledge needed for the safety assessment had to be gathered through research carried out in parallel with the investigations, but was not considered part of the site investigations. Such knowledge was mostly taken directly into the safety assessment, outside of the SDM. This is particularly the case for the development of the engineered barriers (canister and clay) and for better understanding of the source term (i.e. the dissolution of the SNF).

The main purpose of interpreting measurement data is to obtain values for the different parameters in the descriptive model. At the same time, an evaluation of the uncertainties in individual parameter values and an assessment of whether the geometric subdivision is reasonable are made.

III–2.3. Related issues

Investigations were carried out according to a detailed planned generic programme [III–1]. The objective of the investigation(s) was to collect relevant data for further evaluation of site suitability as host rock for the repository. These data collection activities gave practical as well as conceptual ambiguities.

An example of a practical obstacle was insufficient rock stability which caused breakouts to fall into the borehole.. As there was a need to be able to pass up and down the hole and to pack it off and investigate the hydraulic properties, a method was needed to stabilize the borehole without sealing it with grout. A service team with good technical skill was always available to support such cases. In this case, a rather sophisticated method was developed for emplacing a perforated stainless steel plate in the borehole wall where there was a risk of rock pieces falling out. This method turned out to be reliable.

An example of a conceptual ambiguity was a few very high uranium concentrations in groundwater samples collected in Forsmark. Careful checking of the sampling and analytical procedures ruled out errors due to bad handling of samples. The highest observed concentration of dissolved uranium could only be explained by oxidized uranium (VI) being dissolved from the bedrock or from fracture infillings. It was concluded that the very high uranium concentration had no impact on the outcome of the safety assessment. However, for the confidence in the overall understanding of processes in the rock, it was important to get deeper in this assessment. In a PhD thesis [III–6], the most plausible explanation is presented to be that uranium (VI) mineral present ever since the formation of the rocks had been mobilized due to the drilling.

During the design assessment, the thermal properties of the rock were observed to vary more than expected with a large impact on the repository layout at both sites (Laxemar and Forsmark). This led to more sampling of rock cores for analyses of thermal properties. In that way, it was possible to obtain enough data to correlate the thermal properties to rock types and, thus, use the 3-D model of rock types to describe the thermal properties.

The intensive investigations brought a huge amount of data to be reported. This issue was handled by introducing a strict hierarchy for reporting. The first level of reporting was the primary data set together with relevant information about the quality of the data. These reports were to be published and accessible shortly after completing the data collection (days to week timescale). For the site description, there was a need to define the data sets to be included into the present loop of data evaluation. These so-called data freezes were very strict, as no data later than this point could be used for the ongoing SDM work.

III–2.4. Lessons learned

The modelling methodology was developed to focus on early data assessment, visualization and conceptual model building. It was necessary to take a freeze of data at four specific points, after which modelling and evaluation was conducted. The final SDM reporting placed more emphasis on the documents presenting the main disciplines; the evaluation and results were further summarized in the SDM report, the main document which focused on presenting a synthesis of all information in an integrated model and its associated descriptions.

Different conceptual approaches were intended from the onset of the site descriptive modelling. Uncertainty in interpretation in combination with little or non-existing data made it very difficult to come up with alternative conceptual suggestions. The integrated modelling, therefore, focused on managing the uncertainty in a strict, traceable manner. Uncertainty workshops held in the beginning of the SDM work became integrated confidence workshops by the end.

At the start of the investigations, it took a long time before SDM provided feedback to support the planning of the investigations. There was a strong need for such feedback as there were many different tracks to follow. The pragmatic approach from the investigation point of view was then to focus on the obvious most important object to investigate. This strategy applied to the drilling activities resulted in a target on the most relevant unknown potential fracture zone that could disqualify the intended area from being a repository site. This strategy was supported by the fact that, even though the target for the borehole was a specific potential fracture zone, the main part of the borehole passed through the potential intact rock mass giving ample data for representing the rock mass.

The SDM is used as input to repository design and safety assessment. However, an additional interpretation was needed for the information transfer. A separate product, the site engineering report, was compiled for design, construction and operation planning. For the long term safety assessment, all site data are assessed in a data report, using the SDM as input.

These and more lessons learned are presented in Ref. [III–7].

III–2.5. Note on resource use

The Äspö Hard Rock Laboratory was the dress rehearsal for the entire KBS-3 repository development. Site investigations prior to the construction of the Äspö facility were done to test investigation methods, techniques, equipment and to train people for the upcoming repository candidate site investigations. The experiences from Äspö were very useful for the planning of the site investigation stage. The programme for this stage was put on different levels depending on whom to approach: internally SKB, regulator, municipality, general public and experts. Regardless of which level, the programme was always driven by the requirements to be met.

The programming of the site investigation stage was done during the site selection phase, 1992–2000. The site selection phase ended in the selection of three candidate municipalities to be investigated. At this point, the programming could be made site specific. When this was done and the planning for investigation activities could start, it became obvious that running the site investigation would require roughly three to four times more money and people for each site compared to what was required for Äspö HRL. The planning of resources was later proven accurate once the site investigation stage had been completed.

REFERENCES TO ANNEX III

[III–1] SVENSK KÄRNBRÄNSLEHANTERING, Site Investigations. Investigation Methods and General Execution Programme, SKB Technical Report TR-01-29, SKB, Sweden (2001).

[III–2] STANFORS, R., OLSSON, P., STILLE, H., Äspö HRL — Geoscientific evaluation 1997/3. Results from pre-investigations and detailed site characterization: Comparison of Predictions and Observations. Geology and Mechanical Stability, SKB Technical Report TR-97-04, SKB, Sweden (1997).

[III–3] RHÉN, I., BÄCKBLOM, G., GUSTAFSON, G., STANFORS, R., WIKBERG, P., Äspö HRL — Geoscientific Evaluation 1997/2: Results from Preinvestigations and Detailed Site Characterization, Summary Report SKB Technical report TR-97-03, SKB, Sweden (1997).

[III–4] SVENSK KÄRNBRÄNSLEHANTERING, Site Description of Forsmark at Completion of the Site Investigation Phase. SDM-Site Forsmark, SKB Technical Report TR-08-05, SKB, Sweden (2008).

[III–5] SVENSK KÄRNBRÄNSLEHANTERING, Site description of Laxemar at completion of the site investigation phase: SDM-Site Laxemar, SKB Technical Report TR-09-01, SKB, Sweden (2009).

[III–6] KRALL, L., The Mobility of Natural Uranium at Forsmark, Sweden, Through Time, PhD Thesis, Department of Geological Sciences, Stockholm University (2017).

[III–7] ANDERSSON, J., SKAGIUS, K., WINBERG, A., LINDBORG, T., STRÖM, A., Site-descriptive modelling for a final repository for spent nuclear fuel in Sweden, Envi. Earth Sci. **69** 1045 (2013), https://doi.org/10.1007/s12665-013-2226-1

Annex IV

SITE INVESTIGATIONS AND PERFORMANCE ASSESSMENT MODELLING — A CASE STUDY FROM THE USA

IV–1. CONTEXT

Collection of site investigation data, for characterizing a radioactive waste disposal site, is fundamental to safety case development. Performance assessment modelling to quantify long term safety of the repository is also fundamental to the safety case. This report discusses the observed relation between these two components of the overall safety case. The AMIGO project (short for Approaches and Methods for Integrating Geological Information into the Safety Case) summary discussed the findings of three working group meetings on incorporation of geoscientific information into the safety case for radioactive waste disposal facilities [IV–1]. Here, the discussion on the relationship from the modelling point of view is drawn from the findings of the AMIGO project.

The safety case is a collection of evidence and technical arguments in support of the safety of a radioactive waste disposal facility. The evidence ranges from the suitability of the site to the design and operation of the facility, including the long term assessment of risk related to release of radionuclides. Site conceptual models are part of the safety case, integrating multiple lines of evidence to illustrate an understanding of past and current site conditions. Performance assessment modelling is also an integral part of the safety case, involving a quantification and forecast of the likely radiation dose and risk to the environment under a range of nominal and off-nominal future scenarios. The safety case, site conceptual models and performance assessment model evolve with the siting, characterization, development and construction of the disposal facility.

IV–2. MANAGEMENT CHALLENGES

It is important for data collected during the characterization phase of the repository development programme to be effectively incorporated into the modelling and prediction component of the repository development programme. This discussion defines the following forms of modelling, which range from informal to formal and high level to detailed. At the beginning of any study, conceptual models, sometimes called the *geosynthesis*, are used first. These models assemble geologic understanding, expert knowledge, field data and natural analogues into a consistent story that explains observed site conditions. Conceptual models are the framework within which site investigation experts assemble data they collect, and through which they motivate and justify further data collection. Data collection and analysis may lead to alternative conceptual models or interpretations for the same data. More formal models (i.e. mathematical and numerical) are used throughout any modern study and are sometimes involved in the geosynthesis stage. Mathematical and numerical models constrain conceptual models through governing equations or relationships between the inputs and outputs. Inputs to numerical models may be uncertain, but typically numerical and mathematical models have certain outputs, given a set of inputs. Numerical models should be consistently reproducible, even if they incorporate some aspect of randomness in their formulation. Different researchers using the same inputs should get the same outputs, as the process is formalized mathematically and expressed in a programme implementation.

Numerical modelling is here further subdivided into data analysis models, process models and performance assessment models. Figure IV–1 shows how these four types of modelling fit into the spectrum of abstraction and process integration (i.e. complexity). The arrows indicate the primary pathway during programme life, but a programme should re-evaluate and improve conceptual models given insights learned from process modelling and development of performance assessment models.

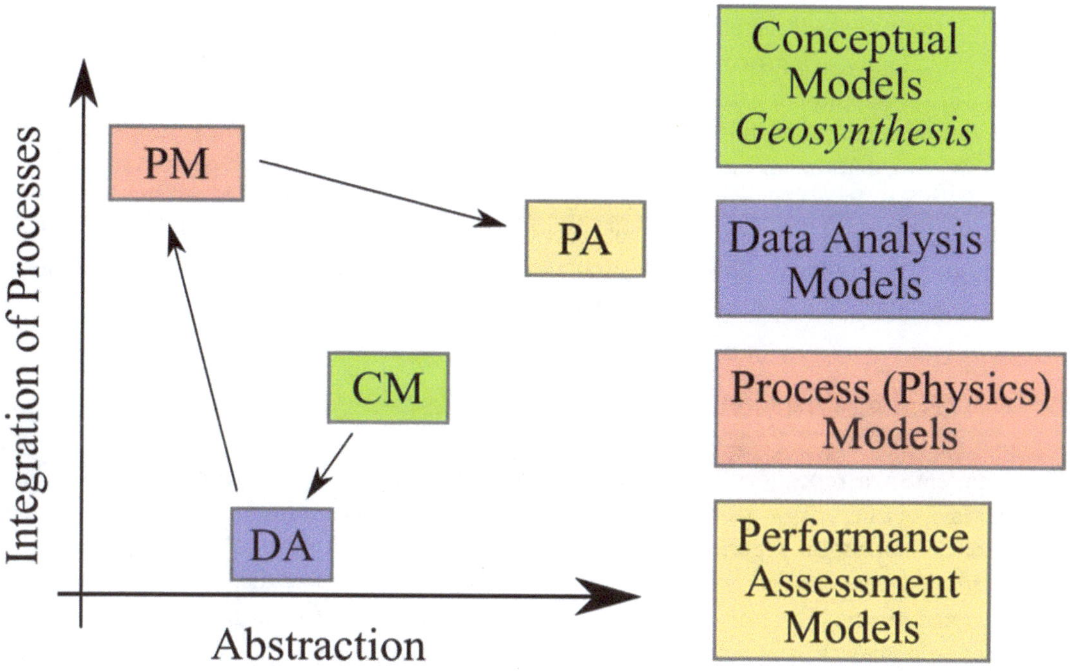

FIG. IV–1. Relationships among modelling types discussed in this annex.

Data analysis models are used to analyse field or laboratory data, are the least complex and typically involve little integration across multiple processes. They may be analytical solutions or approximations with simple problem geometry. The models are required to determine values of parameters (such as hydraulic conductivity or diffusion coefficient) not directly observable in the field or laboratory. They are often well established, rarely change and are sometimes handled directly by software built into instruments for field or laboratory data collection. Examples of field data interpretation models include aquifer test solutions for interpreting pumping tests and approximate methods for quickly interpreting geophysical data in the field. In the laboratory, a simple model may be used to infer geomechanical properties, such as bulk modulus from applied loads and observed strains on a rock core.

Near the other end of the complexity spectrum are what are referred to here as ‘process models’ or models based on physics. Process models typically include more relevant processes explicitly and often minimize abstraction. These gridded numerical models often can include multiple coupled physical processes, include complex non-linear constitutive models, represent physically realistic geometry and allow parameter variability in space. State of the art process models will evolve and change, based on advances in domain and computational sciences. Advances may allow incorporation of higher resolution discretizations, more physical processes, increasingly mechanistic representations of processes or non-linear coupling and constitutive laws. In terms of complexity, performance assessment models typically fall between the simplest data analysis models and the most complex or mechanistic process models. However, along with continued advances in domain and computational sciences, performance assessment models are approaching the complexity and degree of process integration of process models. Domain experts and process models are essential in addressing ‘what if’ questions that arise in repository safety case development. Process models sometimes point to the need for more experimentation or field data collection, either through sensitivity analysis and the desire to reduce parameter or conceptual model uncertainty, or because questions can’t be adequately answered with the current models and data.

A performance assessment model may be implemented using different software than the process models upon which it is based, or it may be a simpler form of a process model using (for example) a coarser mesh, simpler constitutive relations and the same software. One defining feature of performance assessment models is they explicitly consider model parameter uncertainty. Considering parameter uncertainty often means a large number of model runs, and the performance assessment model is typically abstracted away from the high level detail included in the process model to achieve practical ensemble runtimes. The uncertainty quantification may be handled by a separate driver programme that treats each individual forward simulation as a black box (e.g. using Monte Carlo sampling of parameter distributions) or they can consider uncertainty in their formulation directly (e.g. solve for mean and variance of the outputs given the distributions of the inputs). Caution should be exercised when performance assessment models are used to ask 'what if' questions during safety case development, because they are often built on more assumptions or simplifications than process models. When process models and performance assessment models are run with the same simulator (software), bridging the gap between the two is simpler than when different numerical models are used.

IV–3. STRATEGIC PLANNING

Development of a repository is a complex process and takes years to complete. The high level strategic plan laid out here is necessarily idealized. The path outlined in Fig. IV–1 shows a typical path from initial conceptualization to performance assessment required for final regulatory approval. In reality, this is not a simple linear path but will involve backtracking to re-evaluate the conceptual model, collect additional data and reconcile any discrepancies that may become apparent during the overall process.

All types of numerical and conceptual models are typically used in the development of the safety case for a repository. Field and laboratory data are initially interpreted and worked into the context of the site conceptual models. Model parameters will be interpreted from raw data and those parameters may be put into process models to investigate the relative importance of various physical processes. Process models may involve different spatial scales and time horizons different from those used in the performance assessment model. The performance assessment model will likely include a large physical domain and cover a long period of time due to regulatory requirements. Effective performance assessment models incorporate the important processes at the appropriate fidelity and address uncertainty in model parameters, uncertainty in underlying processes and multiple possible conceptualizations, and possible future scenarios. Finalizing what processes are important enough to include and what spatial and temporal fidelity to use are part of the balance that should be investigated during performance assessment model development. Geometric detail and mechanistic process representation should be balanced against model runtime and conceptual parsimony.

Performance assessment involves the development, parameterization, execution and peer review of an interconnected system of numerical models. Site investigations, data collection and performance assessment are necessarily performed by different groups of staff, with different areas of expertise. To ensure effective cooperation between these two groups, integration and collaboration are essential; one group's task cannot be performed in isolation from the other group.

IV–4. IMPLEMENTATION

Site investigation efforts are obviously required as part of the site selection process. The role of site investigations in establishing numerical model parameter ranges for performance assessment is also well-known. Beyond simply specifying model parameters and parameter ranges, site investigation efforts also help define location and nature of model features and boundaries and indicate which processes should be included in final performance assessment modelling. Site conceptual models, built directly from geologic understanding and site observations, provide evidence to illustrate the investigators' understanding of the geosphere, which is a prerequisite to communicating the results of performance assessment to stakeholders and during peer review. Often site data contributed to the geological conceptual model include qualitative, complementary or 'soft' data. Even if it is difficult to incorporate these types of observations directly into performance assessment models, confidence is enhanced by using multiple lines of reasoning to construct a waste isolation safety case. Performance assessment is an important aspect of the overall safety case and is likewise bolstered by a firm site specific foundation of defensible geologic, geohydrologic, geochemical and geomechanical understanding.

IV–4.1. Using numerical models to guide site characterization

Site investigation efforts typically begin early in the overall life of a repository project, and early stage numerical modelling is often so generic that it may not be very helpful in guiding the site investigation process. As the conceptual models of a site mature, they will motivate new data collection locations and data types. The uncertainty associated with the conceptual models will tend to decrease as additional data are collected (i.e. additional data are less likely to be surprising), while the complexity and level of process integration in the conceptual models will tend to increase (i.e. more processes and higher-order effects are incorporated). A natural way to further verify and mature conceptual models is to implement and validate aspects of them using numerical models. The numerical modelling effort will also become more site specific as the database of site specific information grows and the site conceptual models mature. Once numerical models are site specific enough, they may become useful to help plan and guide additional site investigation activities. Any field test of significant cost or complexity should be simulated before execution, to optimize sensor location or data collection frequency. Models should also be compared to data post-experiment to calibrate and improve their accuracy and assess uncertainty. Models being used for experiment design would likely require the level of complexity found in process models. When applicable, modifications made to process models to improve model–data agreement are also propagated to performance assessment models. As models gain site-specificity and represent observed processes more realistically, they contribute back to the site conceptual model and contribute to overall confidence in the safety case.

IV–4.2. Process models vs. performance assessment models

Ideally, performance assessment models would have all relevant physical processes represented mechanistically and include sufficiently fine temporal and spatial resolution to resolve transient processes, heterogeneity and geometrical complexities observed in reality. Likewise, process models used to interpret site observations and assess the importance of physical processes should explicitly address uncertainty. In reality, these two goals (physical realism and uncertainty quantification) are at odds with one another, because of the finite computational resources available. Increasing spatial and temporal resolution increases execution time and memory footprint for each model that is run, as does increasing additional physical coupling or non-linear constitutive laws. Assessing uncertainty in predictions requires many model evaluations with different combinations of parameters, which increases the overall execution time, typically demanding the analyst use lower-resolution or simplified models to achieve acceptable runtimes across an ensemble of simulations.

The distinction between process models and performance assessment models may seem artificial, but it often exists because domain experts and performance assessment experts may use different modelling tools and because of computational costs and constraints. From a practical point of view, it is beneficial to remove ancillary differences between different process models and between performance assessment and process models (i.e. finite volume vs. finite element, and differences in how boundary conditions are specified) and use the same numerical models for the performance assessment and process model stages. On the other hand, it is useful to validate numerical models with different formulations against one another, as this provides more confidence in the model formulation. It is useful to use the same numerical model for process modelling and performance assessment modelling and, if possible, this same model could be used to interpret field data. This uniformity would improve consistency, since putting parameters inferred using one set of assumptions into a different numerical model with another set of assumptions might lead to subtle inconsistencies. This step makes more explicit the upscaling of field and laboratory data into the framework of the process model, and necessarily the performance assessment model.

IV–4.3. Key lessons

Realistically, site investigation efforts will often only indirectly feed into performance assessment models through specifying parameter ranges and development of a robust site conceptual model from which the numerical models can be constructed. Collaboration is critical between groups running numerical models and groups conducting the geosynthesis of site information. Collaboration is also critical between domain experts (process modellers) and performance assessment modellers. Each of these efforts may proceed individually, but they should strive to integrate as much as possible, sharing data, lessons learned and site specific understanding. An exchange of information between site investigation, process modelling and performance assessment modelling will increase the consistency and completeness of site specific data used in long term performance assessment forecasts, increase confidence in the site conceptual model and lead to a more robust safety case.

REFERENCE TO ANNEX IV

[IV–1] OECD NUCLEAR ENERGY AGENCY, Geoscientific Information in the Radioactive Waste Management Safety Case: Main Messages from the AMIGO Project, NEA No. 6395, OECD/NEA, France (2010).

Annex V

THE DEVELOPMENT AND APPLICATION OF A MONITORING STRATEGY AS PART OF SITE INVESTIGATIONS FOR RADIOACTIVE WASTE REPOSITORIES — A CASE STUDY FROM FRANCE

V–1. INTRODUCTION

As stated in the IAEA Safety Standards [V–1], different kinds of monitoring activities are necessary during the lifetime of a radioactive waste disposal facility. Facility lifetime is divided into four periods as referenced in the standards: preoperational, operational, closure and post-closure.

The preoperational period includes concept definition, site evaluation (i.e. site investigation period comprising selection, verification and confirmation), safety assessment and design studies. The preoperational period also includes the development of programmes and procedures required in support of the licence application for construction and initial operation of a disposal facility.

The operational period will begin when waste is first received at the facility. This annex describes the monitoring strategy that has been developed during the preoperational period in Meuse/Haute-Marne districts within the framework of the Cigéo project, from the first phases of investigations up to the licence application, which was submitted in January 2023.

For the French case, besides the conventional surface investigations carried out during the early phases, three major scientific tools were used to support the monitoring strategies: (i) deep boreholes, (ii) the URL, and (iii) the perennial observatory of the environment (OPE).

However, these tools are not only used for monitoring, they are also used for site characterization and design of a disposal facility. The OPE is also a tool used for establishing initial reference state, as well as keeping memory of the environmental state prior the construction of the disposal facility. Thus, the overall monitoring strategy presented below has been developed during the siting phase which lasted 30 years. It should be noted that the monitoring strategy is fully integrated in the characterization and experimental programme. In addition, the monitoring programme aimed mainly to measure disturbances created by 'natural' factors such as seasons and rainfall but also due to changes in agricultural practices or climate. Monitoring was also developed in order to follow-up disturbances created by the URL (shaft and drift constructions), identifying intrinsic properties of the rocks as well as the impact on the environment of constructing and operating a large underground structure.

V–2. OVERVIEW OF THE CIGÉO PROJECT MONITORING PROGRAMMES

V–2.1. General framework

Since it was set up in 1991, Andra has been carrying out research linked to the management of HLW and intermediate level, long lived radioactive waste (IL-LLW) in France. The French Industrial Centre for Geological Disposal project is called 'Cigéo' (Centre Industriel de stockage GEOlogique).

Since 1991, the development of the Cigéo project has been supported by an R&D programme. Surveys of the clay formation included surface investigations (geological mapping, seismic survey, etc.) and boreholes. Actually, four borehole drillings and seismic survey campaigns were carried out throughout the preoperational phase. More than 50 deep geological boreholes have been drilled.

Prior to construction of and then during operation of the URL, geoscientific research programmes have been carried out to provide data for regional geological and hydrogeological modelling. The URL (also called a Bure URL) was licensed for construction in November 1999. Its main purpose was to give access to, in real repository-like conditions, the argillaceous rock formation identified in the region and expected to be suitable for deep geological disposal to guarantee the long term safety of high level and intermediate level, long lived radioactive waste.

From 2004 to 2019, 1700 m of scientific and technical drifts up to 9 m in diameter were excavated in the URL. Four technical excavation techniques have been used (drill and blast, pneumatic hammer, road header and tunnel boring machine with segment erectors). The name Cigéo was adopted in 2009.

In 2007, Andra set up OPE with an aim of establishing the initial state of the current environment around the future disposal facility for a 10 year period and then to track its evolution during Cigéo's construction and throughout its operating life. It should be noted that the OPE is currently not part of the regulatory environmental monitoring of the URL.

The knowledge gradually built up by Andra has been used to progressively develop the Cigéo project in line with the major milestones.

V–2.2. Monitoring in R&D programmes and link with safety assessment

Monitoring is an activity that may be required for a range of reasons at each of the phases of repository development and implementation. At those various phases, it may address requirements related to the need for baseline information, to contribute to operational safety, to ensure practical operability of the disposal facility or to confirm that conditions and evolutions remain consistent with the envelope considered for the assessment of long term safety.

The general objectives of the monitoring programme for the Cigéo project during the construction/operational phases are connected to:

— Compliance with operational safety and regulatory requirements;
— The acquisition of data to develop a deeper understanding of models and parameters underlying;
— The long term safety assessment;
— Reversibility, as well as demonstration or retrievability of waste canisters.

These objectives, however, have not resulted in three disconnected monitoring programmes. For example, drift deformation provided information pertaining to reversibility (adequate space to transfer canisters), operational safety (risk of drift collapse) and long term safety (excavation damaged/disturbed zone evolution).

Actually, the monitoring programme should be seen as a way to inform the safety case as it evolves during the various stages of the operations; in this sense, it can provide additional confidence in the assessments. Part of it is designed to assess regulatory compliance during operation and so is largely inspired by practices in other facilities (nuclear and others).

However, operational monitoring is not used to support the preoperational safety case. Its purpose is not to compensate for a lack of confidence in site characterization data, in URL experiments or in the safety case.

V–2.3. Monitoring programme during the preoperational phase

The French Agency for Nuclear Safety (known as ASN) published a 2008 'safety guide', in which it explicitly requests the implementation of a 'surveillance' programme (i.e. a monitoring programme) [V–2]. As per the requirements of that guide, monitoring is to also be carried out in the phase before operation.

The general objective of monitoring programmes during the preoperational period is to establish a baseline of pre-existing levels of contaminants, to enable an evaluation of the impact of the waste disposal system and to identify parameters that may be indicative of performance in the post-closure period.

As stated by the IAEA safety guide on monitoring and surveillance [V–1], site characterization programmes (as conducted in the preoperational period) typically establish natural characteristics of the FEPs. These FEPs occurring in the environment of the disposal facility (e.g. water table fluctuations) may significantly influence the design and subsequent short term performance (i.e. in the operational period) and long term performance (in the post-closure period) of the facility.

In this regard, the monitoring programme should be closely integrated with the safety case and supporting safety assessment and procedures for construction and operation. The baseline should be developed to allow for the identification of trends. The baseline should also enable the impact of the facility as it evolves over time to be determined; this information can then be used to update the safety case.

Therefore, for Andra, the monitoring programmes during the preoperational phase were intended to provide the following:

- Additional information on site characterization (ongoing site characterization, hydrogeology, spatial variability of rock properties, etc.), mainly through specific surface survey and setting up hydrogeological networks;
- Additional information on short term disturbance of the host formation, mainly by the monitoring of the URL construction and its impacts on the environment in the short term.

V–2.4. The French siting R&D approach

The basic strategy and road map for defining scientific and technological R&D programmes related to the URL, as part of Andra's overall R&D programmes, were driven by the legal framework adopted by the French parliament. The laws required submitting reports, which paved the way to Andra's technical and scientific approach for the development of the Cigéo project (Fig. V–1).

The activities related to the Cigéo project were broken down into successive research iterations in order to achieve continuous progress towards an industrial solution. The completion of each development phase of Cigéo has been confirmed, usually by a project review.

During these reviews, the performance of the overall system was assessed by compiling the information obtained. In addition, they assessed the ways in which the remaining uncertainties could jeopardize the functions of the disposal facility. The results were reported to the assessment bodies and

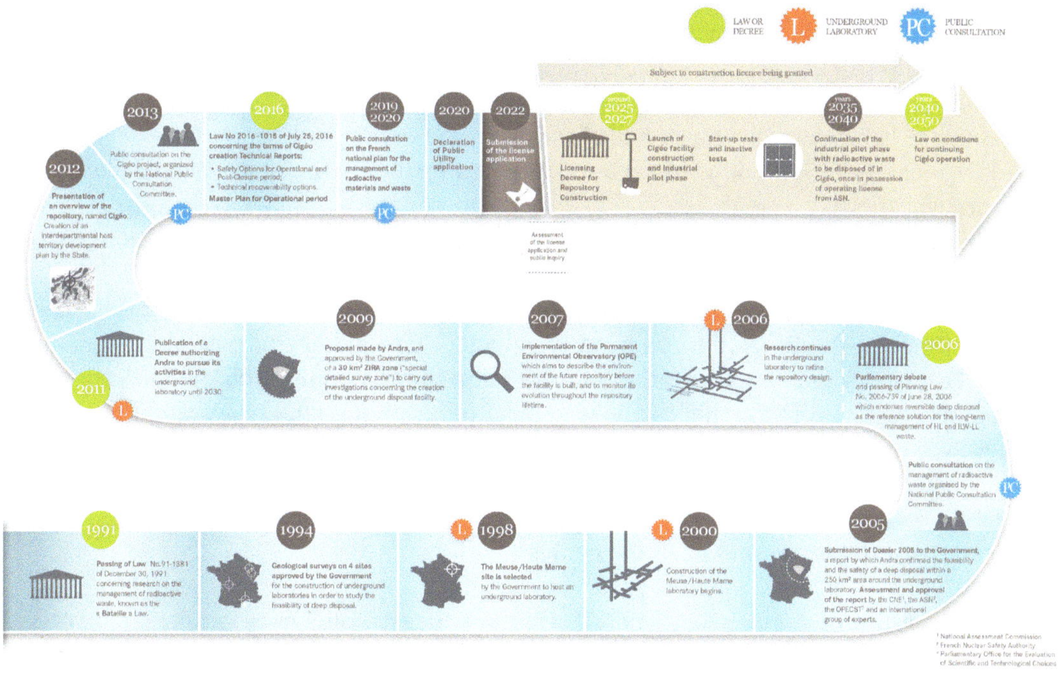

FIG. V–1. Global overview of the Cigéo development process (courtesy of Andra).

stakeholders. The objectives of the following iteration were then updated. To date, seven main research iterations of the Cigéo project have been carried out, as follows:

(1) Within the framework of the 1991 Act:
   - 1996: Licence application for three URLs;
   - 2001: Initial design option and international peer review (NEA);
   - 2005: Principle demonstration of Cigéo (long term safety and operation) and international peer review (NEA).

(2) Within the framework of the 2006 Act:
   - 2009: Operating safety options and choice of locations for an underground facility (ZIRA) and surface facilities;
   - 2011: Conceptual design;
   - 2015: Basic design and international peer review (IAEA).

(3) Within the framework of the 2016 Act:
   - 2016: Cigéo operating safety options report (called the DOS) in preparation for the licence application to be submitted in 2020.

In April 2016, Andra sent ASN the safety options report for the Cigéo project. Submission of the report meant that the project became part of a process governed by the regulations concerning basic nuclear installations (shortened to INB), more specifically by Article 6 of the Decree of 2 November 2007. Furthermore, ASN submitted the safety options report for review by experts from foreign safety regulators, coordinated by the IAEA.

The iteration currently in progress is the Cigéo project technical design, safety case and licence application, which has been submitted in January 2023.

V–3. DEVELOPING A MONITORING STRATEGY

Within the framework of the site characterization process, during the different phases of the site investigation, Andra worked on the development and implementation of monitoring programmes designed for the study of fundamental parameters (data acquired periodically and continuously) that were aimed at:

- Selecting the disposal facility site and its design;
- Assessing the behaviour of components of the disposal facility system;
- Assessing the impacts of the facility and its operation on the environment (need to define the initial reference state first);
- Supporting decision making during the disposal process to enhance confidence in geological disposal.

In addition, Andra's monitoring programme comprises environmental monitoring activities that will be undertaken after a disposal facility has been constructed (post-siting and post-closure environmental monitoring). The agency is currently working on planning future optimization options, observation and control of the disposal facility through the development of techniques and observation resources for the disposal facility (sensors, etc.) which will eventually demonstrate the ability to control operations in the context of stepwise management, reversibility and facility governance.

V–3.1. Surface investigations and deep boreholes

Following on from the 1991 Act, Andra's activities focused on surveying the Callovo-Oxfordian argillaceous rock in the sector south of the Meuse and north of the Haute-Marne departments, whose general characteristics were already known.

The following sections present the monitoring strategy developed using the boreholes during initial characterization of the site and up to the period preceding submission of the licence application.

Characterization and monitoring activities in the sector bordering the Meuse and Haute-Marne departments (1994 to 1996)

The preliminary investigation phase had as primary objective to verify the existence and characteristics of the selected host formation as well as the parameters of the surrounding rocks, especially the hydrogeological parameters of the Oxfordian and Dogger formations. Only limited knowledge was available at the time on the deep hydrogeology of the sector, therefore survey work focused on this aspect. An initial understanding of the sector was largely based on previous work performed for oil prospecting, as well as the use of relatively conventional site investigation tools such as surface mapping, seismic geophysics, borehole drilling and analyses of drill cores. From the outset, specific techniques were implemented that were designed for geochemical monitoring and hydrogeological testing in boreholes drilled in very low permeability formations.

Preliminary research in the sector involved:

— Detailed surface geological surveys;
— Deep borehole drillings;
— 2-D seismic profiles;
— Systematic and exhaustive search for data resulting from previous geological surveys and borehole drilling operations carried out in the sector.

Further to the systematic sampling of geologic formations, the scientific programme also included the following activities in specific formations.

In the limestone formations (Oxfordian, Dogger):

— Overall transmissivity was determined with hydrogeological testing with and without packers;
— Piezometric heads were measured;
— Geochemical characterization was obtained from water samples;
— Lithology and porosity of the Oxfordian carbonated formations were characterized with geophysical logging;
— Seismic profiles were adjusted and spatial variability of porosity was assessed based on VSP.

In the Callovo-Oxfordian formation:

— Lithological profile was acquired with geophysical logging;
— Sedimentation modes and associated paleo-geographic reconstruction were characterized based on core sampling results.

Sensors were installed in both boreholes and this marked the beginning of long term hydrogeological pressure monitoring.

V–3.2. Characterization and monitoring activities from late 1999 to mid-2002: implementing the site monitoring network

Shallow boreholes monitoring

Prior to a seismic data acquisition programme, 15 boreholes 41–62 m deep were drilled as part of the 3-D seismic campaign for surface correction purposes. These boreholes intersect all the Barrois

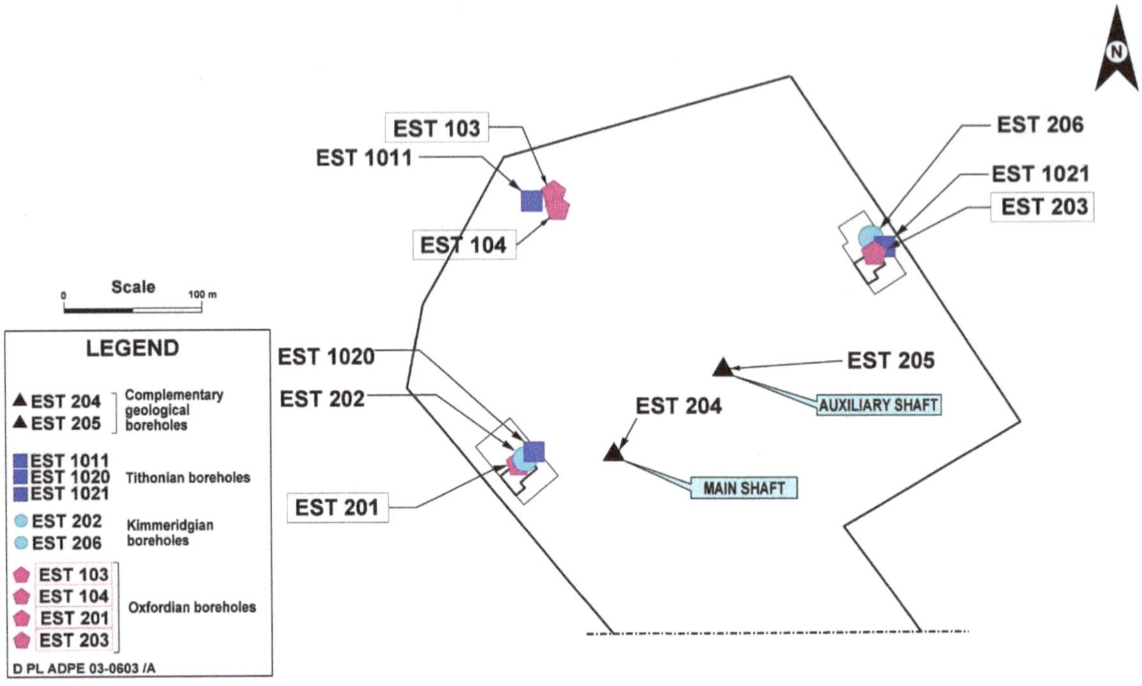

FIG. V–2. Location of shaft drawdown monitoring boreholes at the Meuse/Haute-Marne site studied for the Cigéo geological disposal facility (courtesy of Andra).

formations (Tithonian) and penetrate approximately 30 m into the Kimmeridgian age formation. The monitoring led to obtaining a piezometric map of the base of the Barrois limestone.

Deep boreholes at the URL location

Since the construction of access shafts to the URL were expected to affect all formations, the effects of drainage on the Kimmeridgian marls and Oxfordian limestone were monitored. In March 2000, boreholes for monitoring hydrogeological and geochemical disturbances were air drilled using an inverse circulation method that minimized contamination from drilling fluids (Fig. V–2).

Two 420 m deep boreholes (EST201 and EST203) were drilled and equipped with five measuring chambers to monitor the Oxfordian formation, thereby supplementing boreholes EST103 and EST104 located north of the laboratory site.

For the Kimmeridgian marls, a single borehole (EST202) equipped with three measuring chambers was initially planned. However, this borehole was completely dry and displayed a very low permeability. Since slight seepage was observed during the drilling of borehole EST203, a borehole into the Kimmeridgian marls was drilled from the same platform (borehole EST206).

V–3.3. Characterization and monitoring activities 2003–2005: Implementing a regional network

Hydrogeological measurements were made in boreholes during the 1996 campaign. This information, linked with studies on outcrops, allowed initial estimates of regional water flows to be made. In 2001, Andra developed a number of groundwater flow models that took into account all the hydrogeological and geological data then available for the Meuse/Haute-Marne sectors. These models integrated data obtained from 2-D and 3-D seismic profiles, geological and hydrogeological exploratory boreholes drilled from the MSE, HTM and EST platforms between 1994 and 1996 (Meuse/Haute-Marne

URL site) and boreholes drilled in the underground laboratory site in 2000. However, the underpinning hydrogeological and geological conceptual models included uncertainties associated with the following:

- Hydrodynamic properties of the Dogger and Oxfordian limestone formations (spatial distribution of permeability);
- Hydrodynamic properties of the tectonic discontinuities in the sector;
- Variation of rock lithology and their physical properties with depth.

As a result, the scientific objectives of the complementary borehole campaign involved consolidating the hydrogeological model, completing a lateral survey of the formations, performing additional permeability measurements and taking new samples in order to:

- Supplement the input data for hydrogeological modelling;
- Determine the geological model and the geometry of the transposition zone proposed in the geological baseline;
- Reduce uncertainties and better assess the variability of geochemical, hydrogeological, petrophysical and stratigraphic parameters.

Borehole locations were selected in accordance with the following criteria:

- Location within the sector of the URL covering the equivalent transposition zone or the local springs of the Oxfordian formation;
- Location within at least 5 km of each of the three existing boreholes;
- Location on or in the immediate vicinity of the seismic lines used during oil prospecting in order to improve the calibration parameters used for subsequent seismic data reprocessing and thereby obtain a better understanding of the geometry of the sedimentary structures (layer boundaries, faults, etc.);
- Location to collect adequate data for estimating groundwater residence durations using isotopic tracers.

During this campaign, a total of eight boreholes (Fig. V–3) were drilled from five platforms distributed throughout the sector.

The scientific programme included hydrogeological and geochemical monitoring in the carbonated formations (Oxfordian, Dogger) in order to collect uncontaminated water and gas samples.

V–3.4. Overall layout of the deep borehole monitoring network (1993–2019)

It is obvious that the monitoring programme should not compromise the passive safety of a GDF. Thus, it was decided to drastically limit the number of boreholes going through the host rock, meaning reaching the Dogger formation in order to mitigate any potential risk of a leak through an inadequate borehole seal.

Figure V–4 shows the location of the boreholes drilled during the main campaigns of the site characterization phase, including the URL construction and operation. It should be noted that the zone selected for the underground installations (Zira: zone d'investigation pour la reconnaissance approfondie [i.e. disposal footprint]) is free of any deep boreholes.

Due to the French legal framework, most of these boreholes were dismantled (filled with cement) one year after their construction to comply with administrative requirements. Thus, the monitoring period lasted (with the exception of regulatory monitoring of the URL) less than one year.

This also means that a new network will have to be built after the licensing of Cigéo.

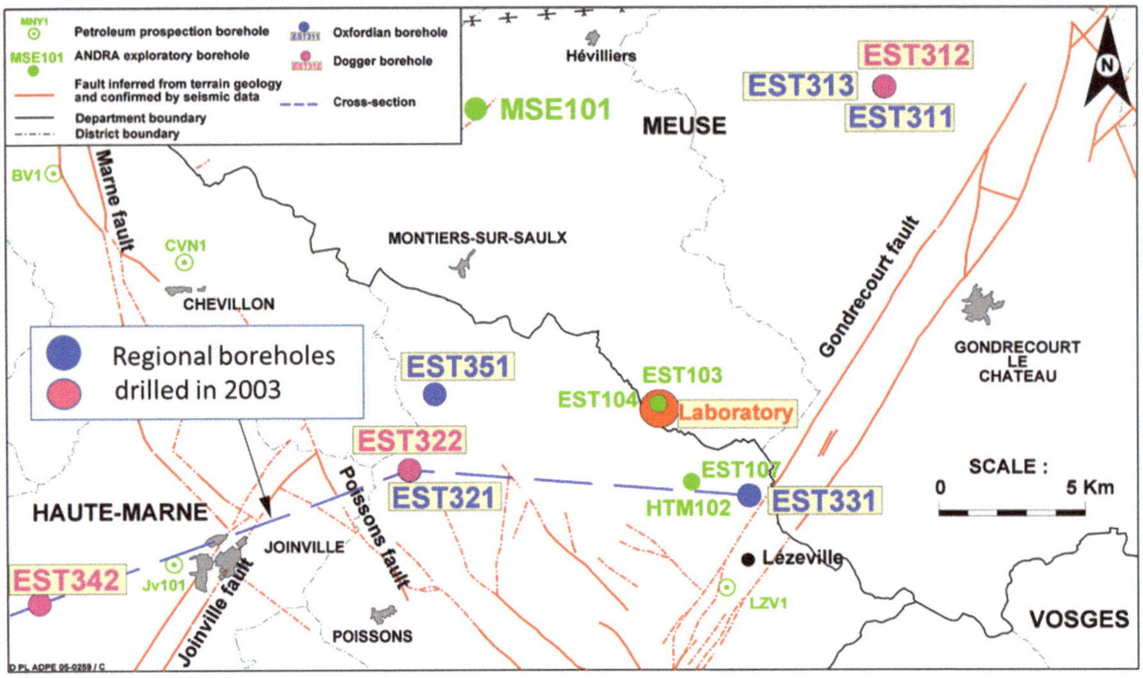

FIG. V–3. Location of sector survey boreholes at the Meuse/Haute-Marne URL (courtesy of Andra).

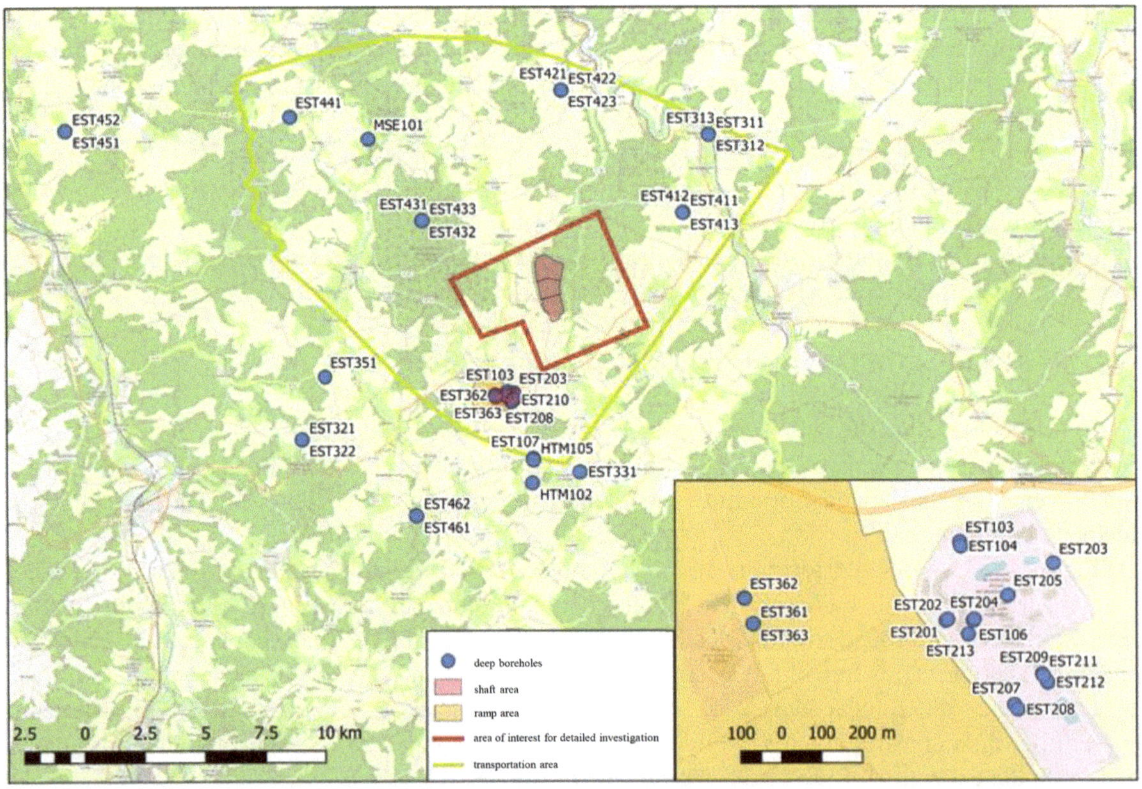

FIG. V–4. Deep boreholes drilled during Cigéo's preoperational phase (1993–2019) (courtesy of Andra).

V–4. THE UNDERGROUND RESEARCH LABORATORY

Currently, the most important technical information regarding monitoring is being compiled by the Bure URL because of follow-up on the construction and the need to conduct experiments lasting several years or decades. Bure URL is a site specific URL that could be seen as a precursor or the initial stage of Cigéo, the facility that will be implemented in the same geological formation. Consequently, the Bure URL provides data for scientific purposes (monitoring of experiments) and for characterizing the host formation (monitoring of the site). In other words, Bure URL is a subject of experiments, as well as a tool for experiments.

There are three main challenges specifically involved in monitoring in the URL. The first is related to construction of the URL itself, which may cause disturbances at a reduced or similar level as the industrial centre for geological disposal.

The second is an intrinsic factor in the experimental programme. It deals with the long term behaviour of the host rock and the long term behaviour of drifts, cells and vaults. One part of these studies deals with the durability of the materials used (concrete, bentonite plug structures, etc.) and the geochemical interaction with the pore water and the host rock. It involves understanding the processes occurring within the system, including potential coupling effects.

The third deals with strategic studies such as selecting the parameters to be monitored during disposal facility operating, long term data management and selecting representative parameters for long term facility surveillance. Facility operation monitoring comprises the objectives or retrievability.

These three challenges will be illustrated in the following sections by hydraulic and geomechanical disturbance monitoring.

The main level of the URL (490 m) is implemented in the middle of the Callovo-Oxfordian formation where the clay content is at its maximum (55%). This level is also the level defined for the waste disposal vaults. In the URL, the drift at 455 m is aimed at starting the experiment programme in the upper part of the layer.

V–4.1. Monitoring hydraulic disturbances during shaft excavation

As noted in Section V.2.3, the shafts leading to the underground facilities were expected to affect all formations, and the impact of drainage on the Kimmeridgian marls and Oxfordian limestone was monitored in boreholes drilled for hydrogeological and geochemical purposes (Fig. V–2).

Investigations on the laboratory site prior to the sinking of two shafts started in November 1999. The sinking of the shafts started in September 2000.

The excavation of the Oxfordian limestone started on 11 November 2001 in the main shaft and on 29 March 2002 in the auxiliary shaft. The excavation lasted 909 days in the main shaft and 748 days in the auxiliary shaft. Overall outflow measured just after the excavation of the Oxfordian limestone was 9.7 L/min in the main shaft and 8.4 L/min in the auxiliary shaft.

Pressure monitoring in borehole EST201 provides a good example of hydraulic interference between shafts and monitoring boreholes. Figure V–5 shows penetration versus time for the upper part of the main shaft. The hydraulic head versus time is shown in the lower part of the figure.

V–4.2. Monitoring the rock behaviour during drift excavation

Knowledge of the fracturing occurring during the early stage of the drift excavation in claystone is one of the major questions to be answered in order to estimate the geometry and extension of the damaged zone. Geological mapping of the front face of the drift from the beginning of excavation and systematic analysis of borehole cores provides a significant amount of data on the orientation of the fracturing planes induced by excavation processes.

It was thus possible to build a 3-D model of the geometry of the fracture network (Fig. V–6). The damage to the rock is characterized by the initiation of shear fractures from the excavation front face and

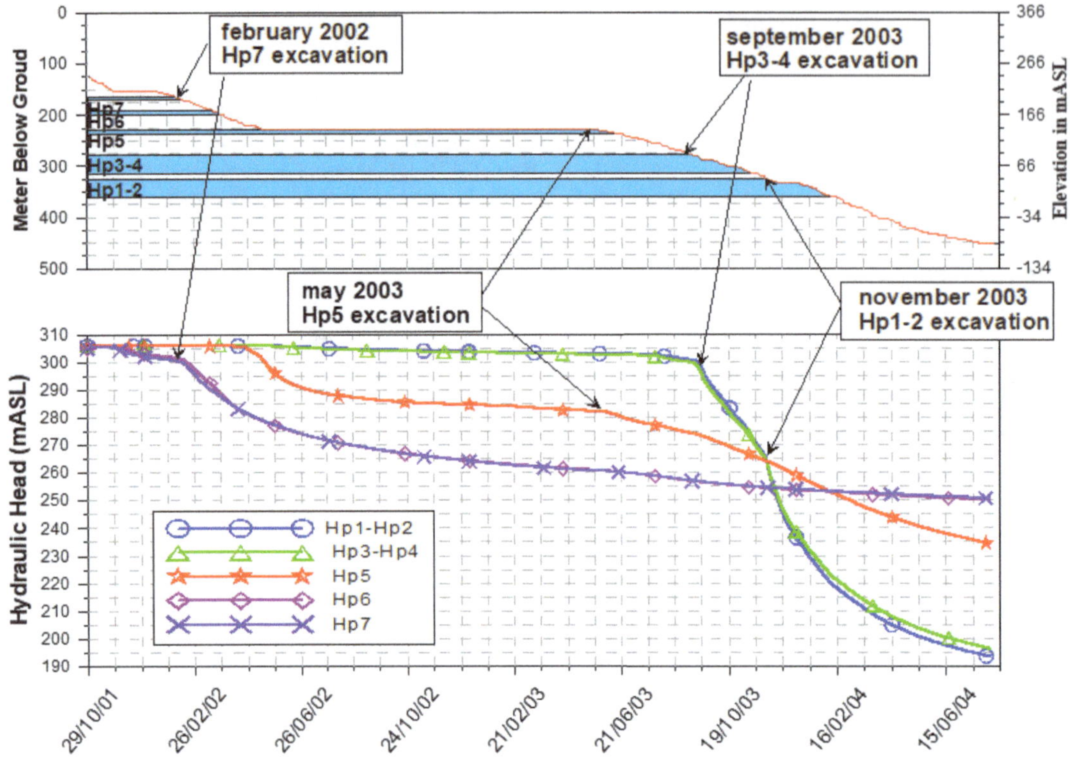

FIG. V–5. Shaft sinking and hydraulic head measurements in borehole EST201. Porous levels are represented by blue bars corresponding to the depth and time the boreholes were drilled during shaft sinking (courtesy of Andra).

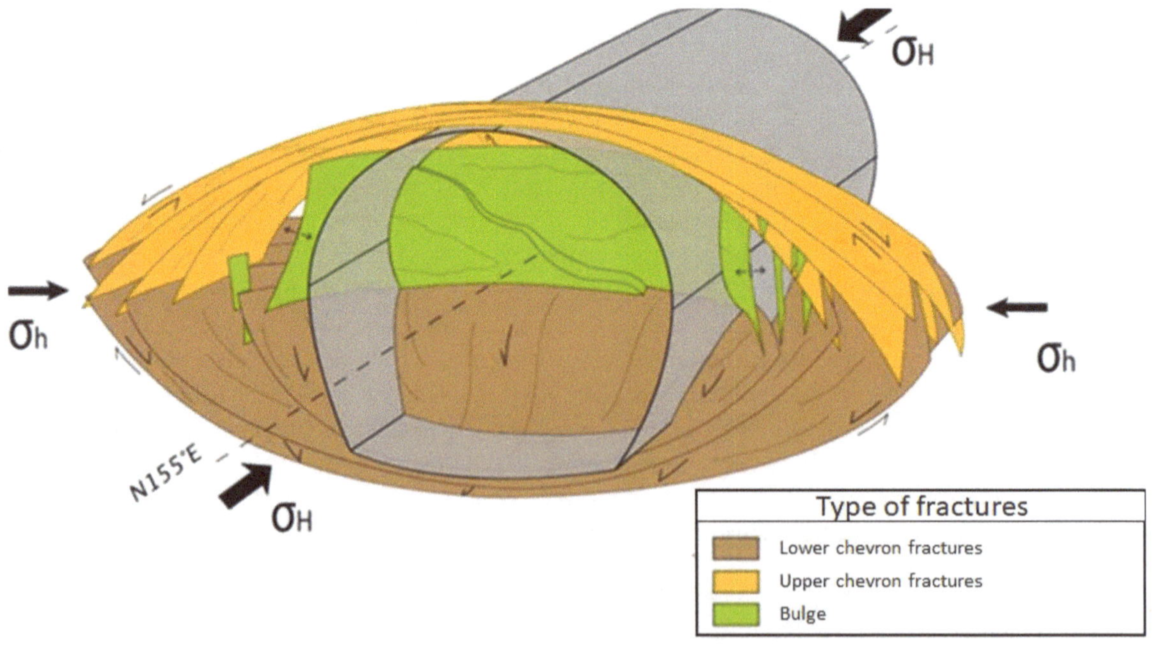

FIG. V–6. Fractures induced during excavation processes (courtesy of Andra).

extensional/unloading fractures after the excavation front has passed. If the nature of induced fracturing does not depend significantly on the direction of the excavation, its expression — and particularly the extension of the zone where the different kind of fractures coexist — depends on the orientation of the drift with respect to the major horizontal stress.

Systematic characterization of rock behaviour (the OHZ experiment) around drifts (deformation, interstitial pressure) and rock/support interaction (the ORS experiment) is carried out in the URL during the excavation of every new drift and is monitored over time. This experimental strategy allows for the comparison of the behaviour of drifts built with different excavation/support methods.

When possible, measuring equipment in the rock mass is installed before excavation (a mine-by test) in order to monitor the deformation and the interstitial pressure evolution throughout the entire construction activity period.

V–4.3. Planning monitoring tools and strategies during disposal facility operating: The modern project

Associated with almost all the experiments carried out in the URL, development of the monitoring tools and strategies is an integral part of the experimental R&D programmes carried out in Bure.

As already presented, the monitoring programmes to be implemented within Cigéo will have to fulfil three objectives: operating safety, long term safety and reversibility. However, the programmes are not unrelated. For instance, the drift deformation monitoring programmes may provide information pertaining to reversibility (adequate space to transfer canisters), operating safety (risk of rock fall or drift collapse) and long term safety (excavation damaged/disturbed zone evolution and fracture healing).

The monitoring devices are subjected to specific repository constraints:

- Devices are required not to interfere with operating safety (cables, data acquisition stations, operator traffic, etc.);
- Devices are required to not significantly reduce long term safety (mechanical, hydrological or chemical footprint, etc.);
- Devices need to operate under expected environmental conditions (heat, humidity, pressure, radiation, lack of access, etc.);
- The cost needs to be acceptable (choice of device, density and distribution of monitoring equipment, etc.).

They are specific requirements for the devices that could be used in a disposal facility, for example:

- Equipment should not jeopardize safety:
  - (No significant long term) disturbance of containment barriers;
  - (No significant long term) chemical disturbance;
  - Adapted to operating safety (ease of construction, no interference with construction and operating activities, etc.).
- Equipment should demonstrate sufficient robustness over time:
  - Accuracy (no drift):
    - Lifetime (50 years is common, 100 years appears reasonable), otherwise maintenance needs to be possible.
  - Resistance to disposal facility environment (temperature, humidity, radiation only inside waste emplacement tunnels, etc.):
    - The environment inside a waste emplacement tunnel is demanding.
- The equipment should be easy to integrate in an automated data acquisition network.

Considering this issue is of critical importance for the Cigéo project, Andra has coordinated two European projects: MoDeRn project [V–3] (FP7, 2009–2013) and MoDeRn 2020 [V–4] (Horizon 2020 and 2014–2019).

The MoDeRn project was designed to provide a reference framework for the development and possible implementation of monitoring activities and associated stakeholder engagement during the various phases of a radioactive waste disposal programme (i.e. during site characterization, repository construction, operations and during a staged closure of the repository, as well as for a post-closure institutional control phase).

The project was implemented through six work packages (known as WPs). The first four work packages were designed to (i) define key objectives and propose viable monitoring strategies, based on technical and stakeholder considerations; (ii) establish the state of the art and provide information concerning technical developments to reflect specific repository requirements; (iii) conduct in situ monitoring demonstration experiments using innovative techniques; and (iv) undertake a case study on monitoring and its integration within staged disposal operations, including specific scenario analysis aimed at providing guidance on how to handle and communicate monitoring results, in particular when these results provide 'unexpected' information.

A main objective was to provide a shared international understanding on how monitoring can be developed within a given national context. Work package 5 undertook dissemination activities and work package 6 provided a reference framework for integrating the results of the project and describing feasible monitoring activities. It also suggested relevant stakeholder engagement activities and illustrated how monitoring results could be used to support decision making.

The overall objective of the Modern2020 Project was based on the outcomes of the MoDeRn Project and aimed at providing the means for developing and implementing an effective and efficient repository operational monitoring programme, driven by safety case needs and taking into account the specific national context requirements — including waste inventory, host rock characteristics and alternative repository concepts and regulations, all of which differ between Member States, as well as public stakeholder expectations, particularly those of local public stakeholders at potential disposal sites.

Modern2020 focused on monitoring of the near field in a repository during the operational phase. The work addressed the following issues:

(i) Strategy: development of detailed methodologies for screening safety cases to identify repository monitoring methods driven by needs and to develop operational approaches for responding to monitoring information;
(ii) Technology: carry out R&D to solve outstanding technical issues associated with repository monitoring which are related to wireless data transmission technologies, alternative long term power supplies, new sensors, geophysics, reliability and qualification of components;
(iii) Demonstration and practical implementation: to enhance knowledge relating to operational implementation and demonstrate the performance of state of the art and innovative techniques by running full-scale and in situ experiments;
(iv) Societal concerns and stakeholder involvement: develop and evaluate methods for integrating public stakeholder concerns and societal expectations into repository monitoring programmes.

All the publications resulting from the two projects are available on their respective web sites.

V–5. THE PERENNIAL OBSERVATORY OF THE ENVIRONMENT

V–5.1. Objectives

The OPE founded in 2007 is designed to respond to the need for the Cigéo project to record the reference state of the surface environment prior to starting construction of Cigéo. It aims to assess potential environmental changes that could occur, to identify their origins and to discriminate between changes due to the construction and operation of Cigéo surface and underground facilities and changes due to other natural or human reasons [V–5].

The OPE will provide access to environmental dynamics from the local scale up to an area of 900 km^2, with the objective of recording integrated responses on the state of the environment at the scale of this area and beyond, thanks to multiple collaborations with national and international environmental observatories, this for around 100 years. Within this area, more detailed studies are being conducted on a reference sector of around 240 km^2, encompassing the zone proposed by Andra for siting the Cigéo industrial centre.

Andra also decided to open access to the OPE to the scientific community in order to develop innovative programmes addressing environmental questions (e.g. long term changes caused by global climate change, ecosystem responses to land use, critical zone dynamics, biogeochemical and biochemical cycles, as well as the socioeconomic impacts of industrial projects).

In addition, the OPE has a role to play in local economic development.

The OPE is a unique facility which reflects the exceptional nature and operating lifetime of the Cigéo project. The OPE, along with its measurement equipment and protocols, was defined in collaboration with expert scientific organizations recognized in their fields (INRA (National Institute for Agricultural Research), AirLorraine, etc.). Many of these organizations are tasked with performing measurements and analyses.

V–5.2. Monitoring tools and activities

The OPE is composed of several observation units and networks which are located within a 900 km^2 area (Fig. V–7).

The observatory units are located to enable the study of the main agricultural land types (forest, grassland and crops).

The OPE encompasses two ecosystem observatory units, a forest observatory unit (fully equipped biochemical stations, including an Integrated Carbon Observation System (shortened to ICOS) flux tower for biomass budget) and a grassland observatory unit (included in that ecosystem).

In addition, the OPE comprises:

— Four observation units and networks dedicated to specific environmental compartments (water, atmosphere, soil, biodiversity);
— A network for water quality observation (surface and groundwater);
— An atmospheric observatory unit (including an Integrated Carbon Observation System mast for monitoring greenhouse gases);
— A soil quality monitoring network (grid sampling of 1.5 km, integrated within national soil quality survey network (known as RMQS) in France managed by the National Institute for Agricultural Research);
— A biodiversity monitoring network including flora, fauna, and agricultural practices.

Satellite data and aerial photos are also acquired in order to track land cover changes. All data are collated within a single databank designed to facilitate cross-checking and interpreting data acquired by the different observation units.

V–5.3. The environmental specimen bank

In 2014, Andra opened the Environmental Specimen Bank located in Bure. This hi-tech building is designed and equipped for the preparation and long term conservation of samples taken as part of the work carried out at the OPE. The Environmental Specimen Bank will thus preserve the memory of the environment surrounding the Cigéo facility throughout its operating lifetime of around a hundred years.

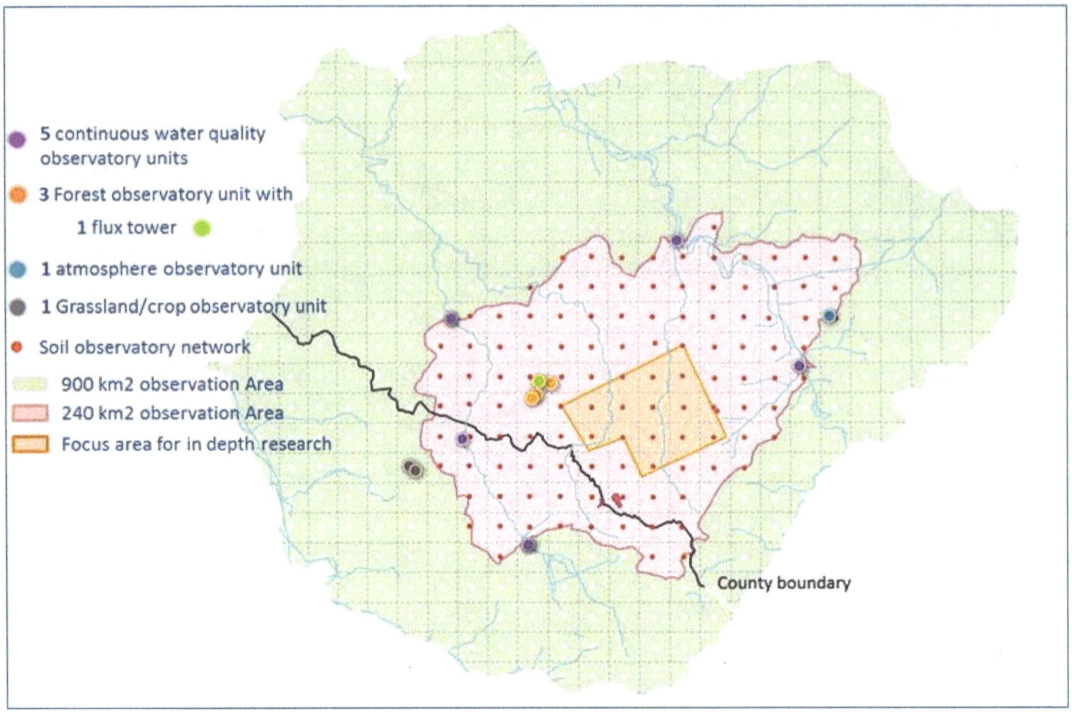

FIG. V–7. OPE monitoring network (courtesy of Andra).

Furthermore, the Environmental Specimen Bank will make it possible to track environmental changes by allowing analysis at a secular scale. Thus, Andra will be able to verify and repeat further analysis in the future, chronicling variations over the long term and correlations between parameters.

In the 1400 m² Environmental Specimen Bank, two new spaces have been opened: a 'clean room' for preparing specimens destined for cryogenic conservation (–180°C) and a public information centre that introduces people to the local environment and the work carried out at the OPE.

V–5.4. Example of Andra's overall environmental monitoring strategy

The OPE, which is scheduled to continuously monitor the environment for around a hundred years, is designed to facilitate the efficient integration of data at the scale of the observation area by providing detailed data from each of its specialized observatory units.

The OPE specifically encourages multidisciplinary research programmes that may benefit from a broad spectrum of data: physical and chemical parameters, biodiversity, land use and land cover, etc., all on a single and well defined area.

Since the outset, the OPE has developed strong links with other similar observatories through national and international environmental research collaborations. All the observations made by Andra and its associated research partners are used to understand, at different scales (local to global), the dynamics of individual environmental components (soil, hydrosphere, biodiversity, etc.) in response to natural changes or ones due to humans.

The data can be integrated and used to evaluate their impact on forest/grassland ecosystem health (the influence of climate change on forest/grassland ecosystem health, for instance). The OPE observations will also help to finesse environmental monitoring strategies for the future, taking into account spatial and temporal environmental variabilities, instrumentation limitations, human and financial costs.

The overall monitoring strategy is a long term iterative process that includes different steps presented in Fig. V–8.

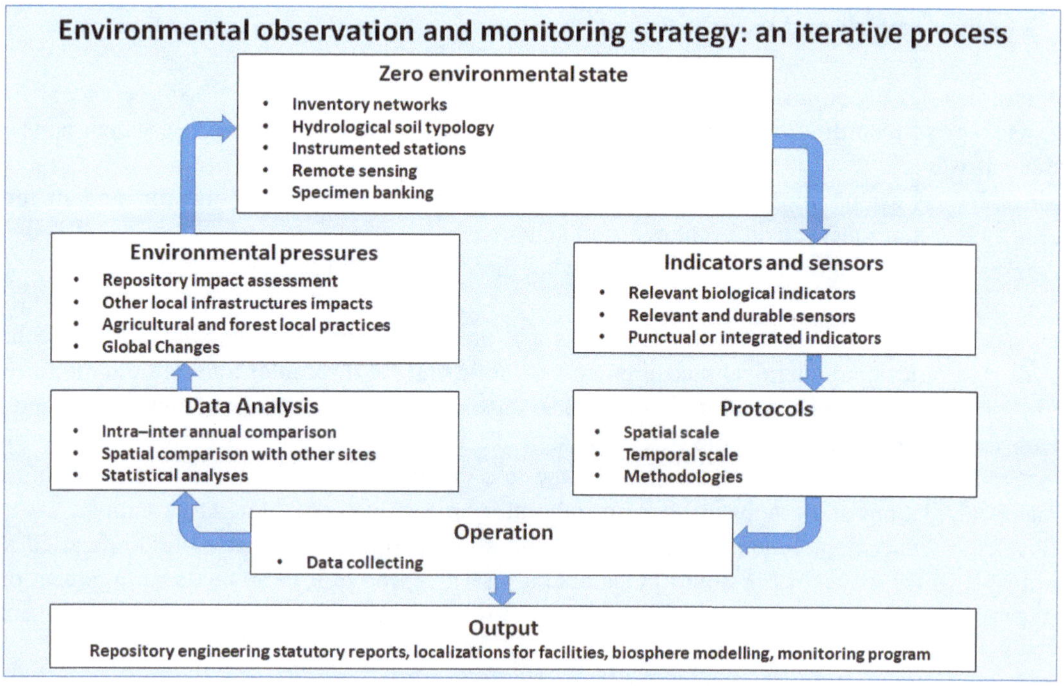

FIG. V–8. Environmental observation and monitoring strategy: an iterative process.

The outputs of this strategy aim to answer objectives already identified, and output deliverables include EIA reports, potential locations for facilities, hydrological and biosphere modelling, and requirements for disposal monitoring programmes.

V–6. DATA MANAGEMENT

For the scientific needs of the Cigéo project, Andra has developed its own scientific information system designed to gather all the acquired data in one place and ensure, over the long term, traceability of the information recorded.

The scientific investigation programme carried out from the surface in boreholes and in the URL makes it possible to provide the data necessary for qualification of the geological site, the disposal facility design and optimization, and the safety assessment for the deep GDF.

The data accumulated over 25 years is exceptional in terms of its quantity and diversity. Billions of values were acquired and millions of new values were recorded daily on hydrogeological parameters, logs of geochemical, environmental, geo-mechanical and geological data, seismic data, etc. All this data acquired over many years required the set-up of a georeferenced scientific information system capable of managing it all.

V–6.1. The geoscientific information system

The raw data collected manually (through boreholes and measurements on samples) are georeferenced and integrated into the information system using the Meuse/Haute-Marne Centre's geoscientific database called GEO. As for data acquired via sensors installed in boreholes and in the URL, this is georeferenced and integrated into the information system using the SAGD database (Système d'Acquisition et de Gestion des Données: a data acquisition and management system).

V–6.2. The geodatabase and its applications

Andra has developed an integrated methodology for the traceability of its samples: identification, coding, packaging, recording, localization and consultation, using the GEO database which is used with dedicated software.

Since it was set up in 1992, this database has constantly evolved technically and in terms of functionality to better meet the different needs of scientists.

At present, the GEO database has the following three main features:

- Data acquisition and management functions and data files relating to geological, geomechanical, hydrogeological and chemical measurements on solid and fluid samples and on field measurements (logging, measurements on samples, geological logs, etc.), as well as observations of flora and fauna;
- Consulting functions, accessible by all the staff on Andra's internal network, by selective displays that make it possible to search the data related to a drilling, a sample or a monitoring point, and to make calculations and graphs on series of laboratory measurements related to a sample;
- Physical management functions of the fluid and solid samples stored in a building called a 'core library', which makes it possible to locate a core sample, move it to an analysis organization (and return a sample not completely used after analysis) and vice versa, and to carry out regular inventories.

At of early 2022), the GEO database gathers data on the following:

- Over 3000 listed works (drifts, boreholes, instrumented zones, geological drift surveys);
- Over 70 000 solid samples;
- Over 15 000 fluid samples.

These objects include over 50 000 photos of samples and works, as well as over 20 000 measurement files acquired during geological survey and logging operations.

V–6.3. The SAGD data acquisition and management system

Andra is directly responsible for all scientific data acquired at the Meuse/Haute-Marne URL and has developed and installed the SAGD data acquisition and management system to fulfil the requirements for acquisition, storage and display of real time data. SAGD was initially deployed in November 2004 for the first experimental drift at –445 m. Its use was subsequently extended to the underground Mont Terri Rock Laboratory in Switzerland in 2005, to the entire surface logging network of the Meuse/Haute-Marne Centre in 2008 and to the environmental network in 2011.

SAGD fulfils its objectives by performing the following:

- Real time availability through a single system for all data acquired at the Meuse/Haute-Marne Centre and the Mont Terri Rock Laboratory;
- Recorded data display in temporal windows and for specific time steps;
- Allowing remote control of the experiments;
- Ensuring the traceability of all recorded information;
- Ensuring data storage in a database.

The sensors are connected to data loggers, connected to a computer network through a set of servers and computer communication devices (switches, optical fibre, copper cable, etc.). The SAGD's computer network is an autonomous network of fibre optic links that brings the data from experiments located in the URL or at the surface to a control room where servers and computers are located. Today, the SAGD allows scientists to visualize in real time from any computer connected to the Internet, the evolution of different experimental parameters thanks to a user-friendly graphical interface called Geoscope.

This software, which has been continuously developed for more than 15 years, manages and stores the following information:

— Over 200 data loggers;
— 20 000 measurement points (pressure sensors, deformation, displacement, convergence, etc.);
— 2 million values/day recorded on average in the SAGD database;
— 3 billion values recorded in the SAGD database;
— Over 5000 logbooks (data related events such as a loss of power, maintenance on a sensor, dismantling of an experiment, etc.).

V–6.4. Geographic information system and data analysis

A GIS server, in addition to GEO's and SAGD's software, has been set up to allow users to benefit from a graphical visualization of georeferenced data from a single interface and to check data consistency. However, there is no software that can integrate all available data formats, and conversions often lead to a loss of information. It became important to set up an efficient and scalable spatial data infrastructure to manage the geographic information available at Andra. This infrastructure serves as a basis for data consultation and provides data in standard formats. By centralizing data, it also ensures a unified vision for all the different users.

The application server is the central component of the spatial data infrastructure. It allows processing and publishing geographic data. It generates map representations in raster or vector formats using data from different sources and provides them to users using Open Geospatial Consortium (known as OGC) web services such as their Web Mapping Service, Web Feature Service and Web Coverage Service (known as WMS, WFS and WCS respectively). Other data (from sensors, observations, analysis, etc.) can also be published though web services using a well defined application programming interface (better known as an API). Furthermore, the server provides the following features:

— Filters: Entities can be requested with specific conditions (simple or spatial queries).
— Styling: Geographical data does not inherently have any visual component. It is therefore necessary to apply a style to them to visualize this data (i.e. define the colour, thickness and other visible attributes used to render data on a map).
— Processing: Operation on the topology, aggregations and statistics.
— Data caching: The map server generates all the images needed for a map immediately during a user query. This task requires server time and resources. In order to improve server performance, it is possible to set up a cache system that stores each image made for a user request and deliver that for future requests, making results faster.
— Security management: It is possible to define users and assign them read/write rules on the different data available on the server.

The OGC web services provide an interface for server features with quick data access for all users (Fig. V–9). Using the dashboard, spatial data can be combined with other data on a single screen which gives a quick overview of a monitoring station or an experiment with predefined indicators.

It is also possible to access server data using desktop GIS software such as MapInfo or BentleyMap, or 3-D viewers such as Google Earth or ArcGIS Explorer, and even non-spatial software like a spreadsheet or Grapher. These programmes enable advanced users to create and edit data and to perform specific data processing or analysis. Finally, data scientists have access to all the collected data though an API integrating many different programming languages, and are able to process data to discover new insight.

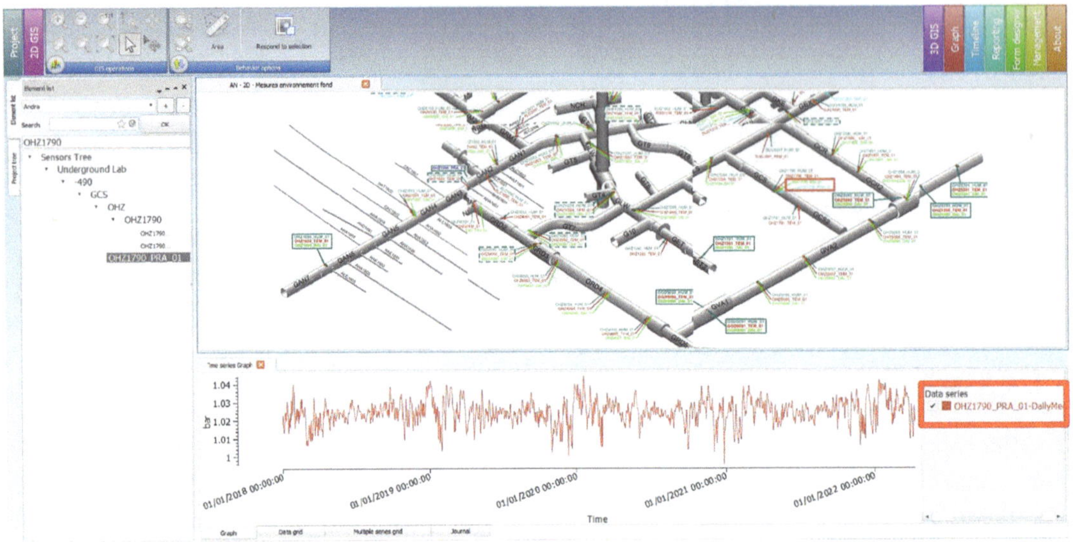

FIG. V–9. Sample visualization of data from drifts, boreholes and sensors produced using the web application (courtesy of Andra).

V–7. CONCLUSION

Most of the data obtained from monitoring will be used for the environmental assessment. In general, environmental and health impact studies contribute to the regulatory, scientific and technical studies designed to provide information and support decision making by assessing the impact a project could have on the environment and the health of local populations. Assessments are carried out throughout the life of a project, from construction to closure.

Andra has set up local consultation forums to involve civil society, particularly on subjects related to environmental and regional integration such as the water cycle, the energy supply to Cigéo, transport infrastructure, spatial planning and quality of life. These consultations will be fed into the EIA.

The EIA is part of an ongoing iterative process throughout the life of the project. It will be updated as the project progresses, particularly in liaison with subsequent administrative stages scheduled on the issue of its reversibility. Its content is proportionate to the environmental sensitivity of the area, the scale and nature of the work and any foreseeable impact of the work on the environment and human health.

REFERENCES TO ANNEX V

[V–1] INTERNATIONAL ATOMIC ENERGY AGENCY, Monitoring and Surveillance of Radioactive Waste Disposal Facilities, IAEA Safety Standards Series No. SSG-31, IAEA, Vienna (2014).

[V–2] NUCLEAR SAFETY AUTHORITY, Guide de sûreté relatif au stockage définitif des déchets radioactifs en formation géologique profonde, ASN, France (2008), https://www.asn.fr/Reglementer/Guides-de-l-ASN/Guide-de-l-ASN-n-1-Stockage-definitif-des-dechets-radioactifs-en-formation-geologique-profonde

[V–3] IMPLEMENTING GEOLOGICAL DISPOSAL OF RADIOACTIVE WASTE TECHNOLOGY PLATFORM, MoDeRn: Monitoring Developments for Safe Repository Operation and Staged Closure, IGD-TP (2015), https://igdtp.eu/activity/modern-monitoring-developments-for-safe-repository-operation-and-staged-closure/

[V–4] MODERN2020, About Modern2020, https://cordis.europa.eu/project/id/232598/de

[V–5] AGENCE NATIONALE POUR LA GESTION DES DECHETS RADIOACTIFS (ANDRA), Observatoire Pérenne de l'Environnement (OPE) Labellisé (2010), https://www.andra.fr/lobservatoire-perenne-de-environnement-ope-labellise

GLOSSARY

conceptual model. A set of qualitative assumptions used to describe a system (or part thereof). These assumptions would normally cover, as a minimum, the geometry and dimensionality of the system, initial and boundary conditions, time dependence, and the nature of the relevant physical, chemical and biological processes and phenomena.

containment. Methods or physical structures designed to prevent or control the release and the dispersion of radioactive substances.

data. Factual records of observations and measurements, but often without context. Raw (primary) data are differentiated from processed (secondary or derived) data and from metadata. Modelling and interpretations can result in the generation of further data. Data can generally be easily encoded for record keeping and further use.

data management system. A system usually comprising hardware and a structured software platform that interacts with data providers and users as well as other applications, to capture, store and present data. (may also be used implicitly for the management of information).

disposal. Emplacement of waste in an appropriate facility without the intention of retrieval.

disposal concept. A description of an idea or plan for disposal (i.e. its facility and its setting that has not yet been implemented or realized) implicitly containing many components, including: the type of waste to be disposed of and its form; engineered barrier materials and their functions; a description of the engineered system; the geological environment and natural barrier functions; and potential radionuclide pathways. May include provisions for the reversal of decisions and retrieval of the waste prior to final closure.

disposal facility. An engineered facility where waste is emplaced for disposal.

high level waste. The radioactive liquid containing most of the fission products and actinides present in spent fuel — which forms the residue from the first solvent extraction cycle in reprocessing — and some of the associated waste streams; this material following solidification; spent fuel (if it is declared as waste); or any other waste with similar radiological characteristics.

information. When data have been analysed, interpreted, structured and combined with other data, so their context and meaning are better understood, they become useful *information*. Information can be communicated as symbols (including numbers), words and figures. Information can require certain assumptions to be made and may include subjective interpretation. Therefore, unlike verified data, information may not be factually correct in all cases.

intermediate level waste. Radioactive waste that, because of its content, in particular its content of long lived radionuclides, requires a greater degree of containment and isolation than that provided by near surface disposal.

isolation. The physical separation and retention of radioactive waste away from people and from the environment.

knowledge. Data, information and ability gained by training or experience; a theoretical or practical understanding. Knowledge that resides in individuals may be difficult to encode and communicate, while corporate or institutional knowledge is the sum of knowledge in an organisation that should be effectively shared and stored as necessary, hence the need for knowledge management.

low level waste. Radioactive waste that is above clearance levels, but with limited amounts of long lived radionuclides.

metadata. Data that describe other data to provide information on their source and derivation, as well as any other contextual information, and to facilitate the proper use and interpretation of the underlying data.

model application. The specific or general area of use that is to be made of the model.

model calibration. The process whereby predictions by a model are compared with field observations and/or experimental measurements from the system being modelled, and the model is adjusted for bias if necessary to achieve a best fit to the measured and/or observed data.

model estimation. A stage of model development in which the generates early stage output based on a limited set of input parameters or an otherwise incomplete model set-up.

model validation. The process of determining whether a model is an adequate representation — within a predefined envelope of tolerance — of the real system being modelled, by comparing the predictions of the model with observations of the real system.

model verification. The process of determining whether a computational model correctly implements the intended conceptual model or mathematical model.

parameter uncertainty. The degree to which the value of a sampled parameter deviates from the true value it is intended to represent.

prediction. A prediction is defined as a statement about future events, quantities or circumstances that are expected to happen. Following this definition, the term prediction can only be applied when one is very confident about a result such that it is probable with a high degree of certainty. Furthermore, a prediction requires that it can later be checked for accuracy against the outcome, otherwise a prediction becomes no better than any other statement about the future. Predictions made as part of the numerical modelling of a system or system component at a site may be entirely appropriate, for example when evaluating anticipated groundwater inflow conditions into a tunnel still to be excavated.

projection. Outcome of a simulation which is considered possible, as opposed to probable, i.e. for which the confidence in the likelihood of occurrence is less than for a prediction and therefore the term projection is preferable. This is because the uncertainty associated with scenarios expands non-linearly into the future and any future conditions used to populate a model or generated during its simulations are unlikely to match the actual conditions that will exist at a site over the simulation timescale. In addition, such long term simulation results cannot be proven or verified during our lifetime. A projection is only one among many possible estimates of a plausible outcome rather than an exact prediction.

radioactive waste management organization. Radioactive waste management organization; the entity responsible for planning and implementing a radioactive waste disposal programme. It requires an appropriate mandate and sufficient resources to carry out its mission. The form of ownership and scope of responsibility varies depending on national circumstances.

repository. An engineered facility where waste is emplaced for disposal. Synonymous with disposal facility.

requirement. Any project constraint (i.e. a quantitative or qualitative boundary condition or goal to be satisfied by the site investigation project).

requirements management. The identification, structuring and response to the suite of requirements that the disposal system is expected to meet. It further provides for the management of evolving requirement specifications as the disposal programme progresses from initial studies to licensing and on to implementation, as well as a basis for clear and factual communication on how requirements are being met.

site characterization. Detailed surface and subsurface investigations and activities at a site to determine the radiological conditions at the site or to evaluate candidate disposal sites to obtain information to determine the suitability of the site for a disposal facility and to evaluate the long term performance of a disposal facility at the site.

site descriptive model. An integrated description of the site and its regional environment with respect to the current geometry and properties of the bedrock and water, as well as the naturally ongoing and interacting processes and mechanisms, to generate a better and more holistic understanding of a site and its evolution and to make results more widely available for use. Integration may be undertaken at several stages in a site investigation programme and can be generated for individual disciplines (e.g. geology or hydrogeology) or for all disciplines needed to describe the site as a whole, thus providing what is sometimes termed geosynthesis.

site investigation. A general term for an activity or series of activities undertaken to understand and describe the nature of a site for a prospective radioactive waste disposal facility.

site investigation flow diagram. A diagram that demonstrates the links between data collection methods and the ultimate use to be made of the data.

siting. The process of selecting a suitable site for a disposal facility, including appropriate assessment and definition of the related design bases. The siting process for a repository is particularly crucial to its long term safety; it may therefore be a particularly extensive process, and when using a screening approach is divided into stages of concept and planning, area survey, site investigation and detailed site characterization.

spent nuclear fuel. Nuclear fuel removed from a reactor following irradiation that is no longer usable in its present form because of depletion of fissile material, poison buildup or radiation damage.

storage. The holding of radioactive sources, radioactive material, spent fuel or radioactive waste in a facility that provides for their/its containment, with the intention of retrieval.

very low level waste. Radioactive waste that does not necessarily meet the criteria of exempt waste, but that does not need a high level of containment and isolation and, therefore, is suitable for disposal in landfill type near surface repositories with limited regulatory control.

very short lived waste. Radioactive waste that can be stored for decay over a limited period of up to a few years and subsequently cleared from regulatory control according to arrangements approved by the regulatory body, for uncontrolled disposal, use or discharge.

ABBREVIATIONS

DGR	deep geological repository
DMS	data management system
DSC	detailed site characterization
DSCS	detailed site characterization stage
DSS	disposal system specification
DSSC	disposal system safety case
EIA	environmental impact assessment
ESC	environmental safety case
FEP	features, events and processes
GDF	geological disposal facility
GIS	geographical information systems
HLW	high level waste
ILW	intermediate level waste
JAEA	Japan Atomic Energy Agency
LLW	low level waste
NDA	Nuclear Decommissioning Authority
NUMO	Nuclear Waste Management Organization of Japan
NWS	Nuclear Waste Services
OPE	perennial observatory of the environment
QA	quality assurance
QC	quality control
RWMO	radioactive waste management organization
SDM	site descriptive model
SIFD	site investigation flow diagram
SIS	site investigation stage
SKB	Swedish Nuclear Fuel and Waste Management Company
SNF	spent nuclear fuel
URL	underground research laboratory
VLLW	very low level waste
VSP	vertical seismic profiling

CONTRIBUTORS TO DRAFTING AND REVIEW

Aaltonen, I.	Posiva, Finland
Alexander, R.	Consultant, Switzerland
Bajorinas, V.	Radioactive Waste Management Agency, Lithuania
Bauer, J.	Federal Company for Nuclear Waste Management, Germany
Beilecke, T.	Federal Institute for Geosciences and Natural Resources, Germany
Blechschmidt, I.	Nagra, Switzerland
Chaplow, R.	Consultant, United Kingdom
Degnan, P.	Consultant, Australia
Delay, J.	Andra, France
Dia, A.F.	National Nuclear Energy Council, Brazil
Dima, M.A.	Nuclear and Radioactive Waste Agency, Romania
Do, I.W.	Korea Radioactive Waste Agency, Republic of Korea
Friedrich, C.	Federal Company for Nuclear Waste Management, Germany
Frieg, B.	Nagra, Switzerland
Gobel, A.	Federal Office for Radiation Protection, Germany
Gundelach, V.	Federal Institute for Geosciences and Natural Resources, Germany
Hammer, J.	Federal Institute for Geosciences and Natural Resources, Germany
Haverkamp, B.	Federal Company for Nuclear Waste Management, Germany
Huret, E.	Andra, France
Hyun, S.	Korea Institute of Nuclear Safety, Republic of Korea
Hyun, S.P.	Korea Institute of Geoscience and Mineral Resources, Republic of Korea
Ichiki, T.	National Institute of Advanced Industrial Science and Technology, Japan
Ito, K.	National Institute of Advanced Industrial Science and Technology, Japan
Jung, H.	International Atomic Energy Agency
Karastanev, D.	Bulgarian Academy of Sciences, Bulgaria
Kihm, Y.H.	Korea Institute of Geoscience and Mineral Resources, Republic of Korea
Kuhlman, K.	Sandia National Laboratory, USA
Lee, S.K.	Korea Institute of Geoscience and Mineral Resources, Republic of Korea
Liu, P.	Consultant, China

Makowski, K.	National Atomic Energy Agency, Poland
Manuilova, E.	Consultant, Russian Federation
Mayer, S.	International Atomic Energy Agency
Molnar, P.	Public Limited Company for Radioactive Waste Management, Hungary
Mrugalla, S.	Federal Institute for Geosciences and Natural Resources, Germany
Nasser, M.	Consultant, Pakistan
Nieder-Westermann, G.H.	International Atomic Energy Agency
Ota, K.	Nuclear Waste Management Organization of Japan, Japan
Park, E.S.	Korea Institute of Geoscience and Mineral Resources, Republic of Korea
Parkes, A.	Nuclear Waste Services, United Kingdom
Patasius, Z.	State Nuclear Power Safety Inspectorate, Lithuania
Rybalka, N.	Consultant, Ukraine
Saftiuc, M.	Consultant, Romania
Schubarth-Engelschall, N.	Federal Office for Radiation Protection, Germany
Schuster, K.	Federal Institute for Geosciences and Natural Resources, Germany
Shim, T.	Korea Institute of Nuclear Safety, Republic of Korea
Shybetskyi, I.	National Academy of Sciences of Ukraine, Ukraine
Sun, Q.	China Institute for Radiation Protection, China
Tetreault, T.	Consultant, United States of America
Tonkunas, A.	Consultant, Lithuania
Tretiak, O.	Consultant, Ukraine
Van Marke, P.	International Atomic Energy Agency
Veinovic, Z.	University of Zagreb, Croatia
Weitkamp, A.	Federal Institute for Geosciences and Natural Resources, Germany
Wikberg, P.	Swedish Nuclear Fuel and Waste Management Company, Sweden

Consultants Meetings

Vienna, Austria: 11–15 April 2016,

20–24 February 2017, 19–23 February 2018

Technical Meetings

Vienna, Austria: 7–11 November 2016,

Peine, Germany: 15–19 May 2017

Structure of the IAEA Nuclear Energy Series*

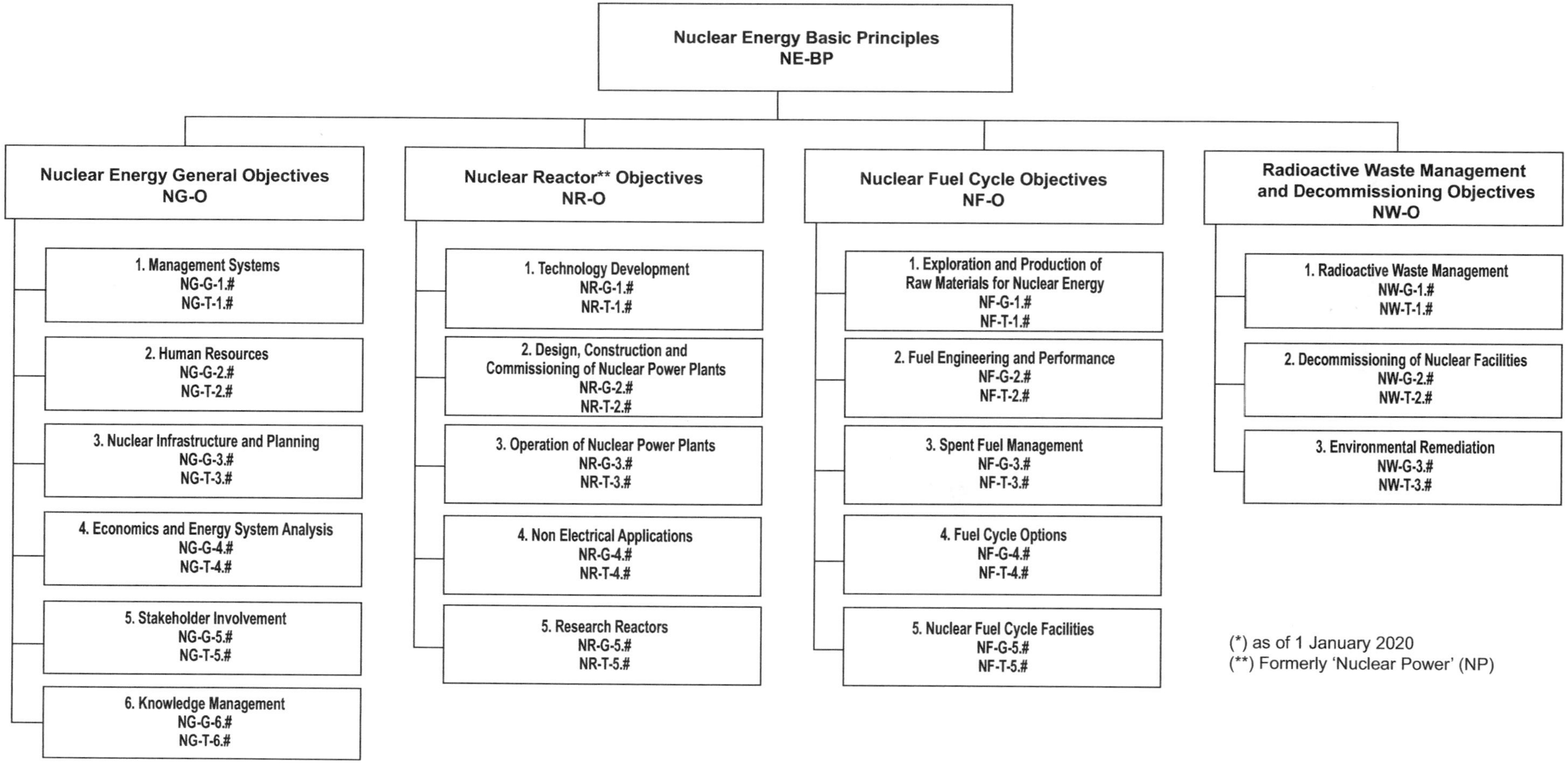

(*) as of 1 January 2020
(**) Formerly 'Nuclear Power' (NP)

Key

BP: Basic Principles
O: Objectives
G: Guides and Methodologies
T: Technical Reports
Nos 1–6: Topic designations
#: Guide or Report number

Examples

NG-G-3.1: Nuclear Energy General (**NG**), Guides and Methodologies (**G**), Nuclear Infrastructure and Planning (topic **3**), **#1**
NR-T-5.4: Nuclear Reactors (**NR**), Technical Report (**T**), Research Reactors (topic **5**), **#4**
NF-T-3.6: Nuclear Fuel (**NF**), Technical Report (**T**), Spent Fuel Management (topic **3**), **#6**
NW-G-1.1: Radioactive Waste Management and Decommissioning (**NW**), Guides and Methodologies (**G**), Radioactive Waste Management (topic **1**) **#1**

IAEA
International Atomic Energy Agency

No. 27

ORDERING LOCALLY

IAEA priced publications may be purchased from our lead distributor or from major local booksellers.

Orders for unpriced publications should be made directly to the IAEA.

Orders for priced publications

Please contact your preferred local supplier, or our lead distributor:

Eurospan
1 Bedford Row
London WC1R 4BU
United Kingdom

Trade orders and enquiries:
Tel: +44 (0)1235 465576
Email: trade.orders@marston.co.uk

Individual orders:
Tel: +44 (0)1235 465577
Email: direct.orders@marston.co.uk
www.eurospanbookstore.com/iaea

For further information:
Tel. +44 (0) 207 240 0856
Email: info@eurospan.co.uk
www.eurospan.co.uk

Orders for both priced and unpriced publications may be addressed directly to

Publishing Section
International Atomic Energy Agency
Vienna International Centre
PO Box 100
1400 Vienna, Austria
Telephone: +43 1 2600 22529 or 22530
Email: sales.publications@iaea.org
www.iaea.org/publications

24-01983E-T

FSC
www.fsc.org
MIX
Papier | Fördert gute Waldnutzung
FSC® C083411

Zeitfracht Medien GmbH
Ferdinand-Jühlke-Straße 7
99095 Erfurt, Deutschland
produktsicherheit@kolibri360.de

COATED PARTICLE FUELS FOR HIGH TEMPERATURE GAS COOLED SMALL MODULAR REACTORS

The following States are Members of the International Atomic Energy Agency:

AFGHANISTAN
ALBANIA
ALGERIA
ANGOLA
ANTIGUA AND BARBUDA
ARGENTINA
ARMENIA
AUSTRALIA
AUSTRIA
AZERBAIJAN
BAHAMAS, THE
BAHRAIN
BANGLADESH
BARBADOS
BELARUS
BELGIUM
BELIZE
BENIN
BOLIVIA, PLURINATIONAL STATE OF
BOSNIA AND HERZEGOVINA
BOTSWANA
BRAZIL
BRUNEI DARUSSALAM
BULGARIA
BURKINA FASO
BURUNDI
CABO VERDE
CAMBODIA
CAMEROON
CANADA
CENTRAL AFRICAN REPUBLIC
CHAD
CHILE
CHINA
COLOMBIA
COMOROS
CONGO
COOK ISLANDS
COSTA RICA
CÔTE D'IVOIRE
CROATIA
CUBA
CYPRUS
CZECH REPUBLIC
DEMOCRATIC REPUBLIC OF THE CONGO
DENMARK
DJIBOUTI
DOMINICA
DOMINICAN REPUBLIC
ECUADOR
EGYPT
EL SALVADOR
ERITREA
ESTONIA
ESWATINI
ETHIOPIA
FIJI
FINLAND
FRANCE
GABON
GAMBIA, THE
GEORGIA
GERMANY
GHANA
GREECE
GRENADA
GUATEMALA
GUINEA
GUYANA
HAITI
HOLY SEE
HONDURAS
HUNGARY
ICELAND
INDIA
INDONESIA
IRAN, ISLAMIC REPUBLIC OF
IRAQ
IRELAND
ISRAEL
ITALY
JAMAICA
JAPAN
JORDAN
KAZAKHSTAN
KENYA
KOREA, REPUBLIC OF
KUWAIT
KYRGYZSTAN
LAO PEOPLE'S DEMOCRATIC REPUBLIC
LATVIA
LEBANON
LESOTHO
LIBERIA
LIBYA
LIECHTENSTEIN
LITHUANIA
LUXEMBOURG
MADAGASCAR
MALAWI
MALAYSIA
MALI
MALTA
MARSHALL ISLANDS
MAURITANIA
MAURITIUS
MEXICO
MONACO
MONGOLIA
MONTENEGRO
MOROCCO
MOZAMBIQUE
MYANMAR
NAMIBIA
NEPAL
NETHERLANDS, KINGDOM OF THE
NEW ZEALAND
NICARAGUA
NIGER
NIGERIA
NORTH MACEDONIA
NORWAY
OMAN
PAKISTAN
PALAU
PANAMA
PAPUA NEW GUINEA
PARAGUAY
PERU
PHILIPPINES
POLAND
PORTUGAL
QATAR
REPUBLIC OF MOLDOVA
ROMANIA
RUSSIAN FEDERATION
RWANDA
SAINT KITTS AND NEVIS
SAINT LUCIA
SAINT VINCENT AND THE GRENADINES
SAMOA
SAN MARINO
SAUDI ARABIA
SENEGAL
SERBIA
SEYCHELLES
SIERRA LEONE
SINGAPORE
SLOVAKIA
SLOVENIA
SOMALIA
SOUTH AFRICA
SPAIN
SRI LANKA
SUDAN
SWEDEN
SWITZERLAND
SYRIAN ARAB REPUBLIC
TAJIKISTAN
THAILAND
TOGO
TONGA
TRINIDAD AND TOBAGO
TUNISIA
TÜRKİYE
TURKMENISTAN
UGANDA
UKRAINE
UNITED ARAB EMIRATES
UNITED KINGDOM OF GREAT BRITAIN AND NORTHERN IRELAND
UNITED REPUBLIC OF TANZANIA
UNITED STATES OF AMERICA
URUGUAY
UZBEKISTAN
VANUATU
VENEZUELA, BOLIVARIAN REPUBLIC OF
VIET NAM
YEMEN
ZAMBIA
ZIMBABWE

The Agency's Statute was approved on 23 October 1956 by the Conference on the Statute of the IAEA held at United Nations Headquarters, New York; it entered into force on 29 July 1957. The Headquarters of the Agency are situated in Vienna. Its principal objective is "to accelerate and enlarge the contribution of atomic energy to peace, health and prosperity throughout the world".

IAEA-TECDOC-2090

COATED PARTICLE FUELS FOR HIGH TEMPERATURE GAS COOLED SMALL MODULAR REACTORS

PROGRESS IN DESIGN, MANUFACTURING, EXPERIMENTATION, MODELLING AND ANALYSIS TECHNOLOGIES

INTERNATIONAL ATOMIC ENERGY AGENCY
VIENNA, 2025

COPYRIGHT NOTICE

Publishing Section
International Atomic Energy Agency
Vienna International Centre
PO Box 100
1400 Vienna, Austria
tel.: +43 1 2600 22529 or 22530
email: sales.publications@iaea.org
www.iaea.org/publications

For further information on this publication, please contact:

Nuclear Fuel Cycle and Materials Section
International Atomic Energy Agency
Vienna International Centre
PO Box 100
1400 Vienna, Austria
Email: Official.Mail@iaea.org

Printed by the IAEA in Austria
May 2025
https://doi.org/10.61092/iaea.zyya-k9sd

IAEA Library Cataloguing in Publication Data

Names: International Atomic Energy Agency.
Title: Coated particle fuels for high temperature gas cooled small modular reactors : progress in design, manufacturing, experimentation, modelling and analysis technologies / International Atomic Energy Agency.
Description: Vienna : International Atomic Energy Agency, 2025. | Series: IAEA TECDOC series, ISSN 1011-4289 ; no. 2090 | Includes bibliographical references.
Identifiers: IAEAL 25-01758 | ISBN 978-92-0-107625-0 (paperback : alk. paper) | ISBN 978-92-0-107525-3 (pdf)
Subjects: LCSH: Nuclear reactors — Design and construction. | Nuclear reactors — Technological innovations. | Nuclear fuels. | Gas cooled reactors.

FOREWORD

The high temperature gas cooled reactor (HTGR) concept is considered a promising nuclear reactor technology owing to its inherent safety and operational features. A typical example of a design feature unique to the HTGR core that imparts inherent safety to the system is the inert, single-phase coolant (helium gas) coupled with a high temperature, high heat capacity moderator (graphite), which can provide a longer time to respond to core heat-up accidents. With the potential of the HTGR to supply high temperature process heat, many other applications such as hydrogen production in addition to electricity generation are also viable.

Significant efforts have been made by several Member States to develop HTGRs for process heat, hydrogen production and electricity generation. With the feature of enhanced safety and modularity, HTGRs have recently been identified as a potentially viable small modular reactor (SMR) technology line. These reactors are primarily fuelled by tristructural isotropic (TRISO) coated particles. The main advantage of TRISO fuel is its capability to withstand extreme conditions like high temperature and prolonged irradiation.

Research programmes have been pursued in these Member States on the development of coated particle fuel technology including fuel design, manufacture, characterization, irradiation behaviour, accident tests and simulations of fuel behaviour.

The international experts who participated in the Consultancy Meeting on the Design, Fabrication and Operation of Small and Medium Sized or Modular Reactor Fuels, held in 2020, identified the need for a comprehensive review of HTGR particle fuel technologies (e.g. design, fabrication, characterization, irradiation performance, performance modelling) to support the development of high temperature gas cooled SMRs. This publication is intended to provide up-to-date information on coated particle fuel technologies to be used as a baseline reference to support the development and deployment of TRISO fuel technologies for HTGR type SMRs.

The IAEA is grateful to all those who actively contributed to the preparation of this publication. The IAEA officers responsible for this publication were K. Sim and A. Khaperskaia of the Division of Nuclear Fuel Cycle and Waste Technology.

CONTENTS

1. INTRODUCTION

1.1. BACKGROUND

High temperature gas cooled reactor technology has the following operational and safety distinctive characteristics compared to light water reactors (LWRs):

- Higher thermal efficiencies due to higher operating temperatures;
- Limited fuel temperature after a loss-of-coolant accident (LOCA) due to lower power-density and higher heat-capacity core;
- No release of explosive hydrogen gas under accident conditions due to the absence of Zr and water;
- Capability of on-power refuelling (depending on design).

The HTGR technology has evolved over several decades through the development of several designs. The historical path of the HTGR technology can be summarized as follows [1]:

- Deployment of early Gas Cooled Reactors (GCRs), which was characterized by graphite moderator and CO_2 gas coolant. Examples included Magnox reactors and Advanced Gas Cooled Reactors (AGRs) in UK, the Uranium Naturel Graphite Gaz reactor in France and GCRs in Italy, Japan and Spain;
- Deployment of HTGR prototype plants, which was demonstrated by UK's Dragon reactor, German Arbeitsgemeinschaft Versuchs Reaktor (AVR) and US Peach Bottom 1;
- Deployment of HTGR demonstration plants, which was demonstrated by Fort St. Vrain GCR in US, and Thorium High Temperature Reactor 300 (THTR-300) in Germany;
- Development of large HTGR steam cycle plants, which was demonstrated by German High Temperature Reactor HTR-500, Russian VG-400, and US HTGR steam cycle plants;
- Development of modular HTGR steam cycle plants, which was demonstrated by HTR-Module and HTR-100 in Germany, Russian VGR, Modular HTGR (MHTGR) in US, High Temperature Test Reactor (HTTR) in Japan, and HTR-10 and HTR-PM in China;
- Development of Gas Turbine-Modular Helium Reactor (GT-MHR). Examples included Pebble Bed Modular Reactor (PBMR) developed in South Africa, GTHTR-300 in Japan and ACACIA in Netherlands;

.

Two types of HTGRs exist: pebble bed and prismatic types. The pebble bed type contains pebble shape fuel elements in a cylindrical pressure vessel. The prismatic type contains cylindrical compacts as fuel elements inserted in the hexagonal graphite blocks in a cylindrical pressure vessel. Each spherical pebble or cylindrical compact contains thousands of coated particles.

Significant efforts have been made by several Member States to develop HTGRs for process heat, hydrogen production and electricity generation. Currently designed HTGRs have been considered as SMRs with significantly enhanced safety features compared to older large HTGR designs and modularity. (Annex I lists high temperature gas cooled SMRs and also advanced HTGR-type reactors.)

These high temperature gas cooled SMRs are characterized with the following features:

- Small to medium reactor power ranging from 100[1] to 300 MWe;

---

[1] Power of demonstration plants (HTR-10, HTTR-30) were lower than 100 MWe.

— Modularity[2] to target the economics of serial production with shorter construction time;
— Use of tri-structural isotropic (TRISO) particle fuel.

From fuel performance aspects, power density, burnup and fast neutron fluence are important operating parameters to be taken into account to ensure the safe and reliable operation of the fuel in a reactor. High power density can lead to high operating temperature and consequently can increase the probability to be at the risk of fuel failure and undesirable release of fission products at extended burnups during operation.

The main advantage of TRISO fuel is its capability to withstand extreme conditions like high temperature and prolonged irradiation. The components of TRISO fuel such as multiple protective layers, a fissile kernel, a buffer layer and a silicon carbide layer have been refined significantly for showing structural integrity, fission product retention and mechanical strength under irradiation. The fuel kernel of TRISO particles consists of uranium dioxide (UO_2), uranium carbide (UC), or uranium oxycarbide is the primary source of fission energy. The buffer carbon layer absorbs fission gases and mitigates the mechanical stresses caused by kernel swelling. By enhancing the porosity and thickness, fission gas retention capacity has been improved without deteriorating the structural integrity. The inner and outer pyrolytic carbon layers (IPyC & OPyC) provide structural support, fission gas barrier, and protect the silicon carbide the layer, the main barrier for metallic fission products. Its excellent characteristics showing improved strength, thermal conductivity, and radiation resistance make it an essential element of TRISO fuel.

Research programmes on TRISO particle fuel development have been pursued in several Member States focusing on design, fabrication, characterization, in-reactor behaviour and its prediction, accident tests and simulation under normal operating and accident conditions.

Earlier IAEA publications [2–5] documented available results of such research programmes on TRISO particle fuel up to 2010.

1.2. OBJECTIVE

The main objective of the publication is to provide an up-to-date compilation of information on technological progress in design, manufacturing, experimentation, modelling, and analysis technologies for HTGR coated particle fuels, and to identify the major challenges and prospects for use of these coated particle fuels in high temperature gas cooled SMRs.

1.3. SCOPE

This publication addresses:

— General description of major design features along with design bases of coated particle fuels;
— Overall information on fuel manufacturing technology;
— Overall information on fuel particle characterization methods, quality assurance and quality control;
— Overview on irradiation tests and annealing tests;
— Irradiation performance, fission product release under accident conditions, and failure mechanisms;
— Capabilities of fuel performance analysis codes.

[2] The term 'modularity' refers to the ability to fabricate major components of the nuclear steam supply system in a factory environment and ship to the point of use.

1.4. STRUCTURE

The rationales and purpose of developing this publication are described above in this section.

Major design characteristics of TRISO particle fuel are described in Section 2 along with design bases.

R&D activities are underway in some Member States to set up manufacturing process for TRISO particle fuel and to develop advanced coated particle fuel (e.g. fully ceramic microencapsulated fuel). The status of such R&Ds is described in Section 3.

The manufacturing processes used to produce TRISO particle fuel and related production quality control practices are described in Section 4 and Section 5, respectively.

There have been a number of irradiation tests and post-irradiation heating tests conducted on coated particle fuels since 1970s. Important tests are summarized in Section 6 with a description of highlighted post-irradiation examination (PIE) results. Key phenomena affecting the in-reactor performance of coated particle fuel, observed from these irradiation and heating tests, are discussed in Section 7, with respect to fuel failure mechanisms.

Fuel modelling and computer codes for predicting the thermal-mechanical behaviour of TRISO particle fuel in the reactor are described in Section 8. International benchmarking exercises of fuel performance analysis codes are briefly described.

Annexes contain useful supplemental information:

- Annex I provides a list of high temperature gas cooled SMRs and also advanced HTGR-type reactors;
- Annex II provides the design specifications of TRISO particle fuels developed by various fuel vendors or developers.

2. DESIGN DESCRIPTION AND DESIGN REQUIREMENTS

2.1. GENERAL DESCRIPTION OF COATED PARTICLES AND FUEL ELEMENTS

The essential component of HTGR fuel is the TRISO particle. A TRISO particle contains a ceramic kernel of fuel materials, surrounded by four coating layers that function collectively as an effective fission product barrier. These TRISO coatings begin with a low density, porous, non-load bearing carbon layer (i.e. buffer layer) that surrounds the fuel kernel. The buffer layer is surrounded by three load-bearing composite layers that consist of two high density, isotropic pyrocarbon (PyC) layers (i.e. the inner PyC (IPyC) and outer PyC (OPyC)), and a very dense silicon carbide (SiC) layer in between. These load-bearing composite layers are intended to withstand the pressure of the generated fission products and to form an essentially leak-tight barrier to fission product release.

The TRISO particles are embedded in a matrix made of special grade graphite in the form of spheres for a pebble bed HTGR or in the form of cylindrical compacts to be loaded into the hexagonal blocks in a prismatic HTGR (see Fig. 1).

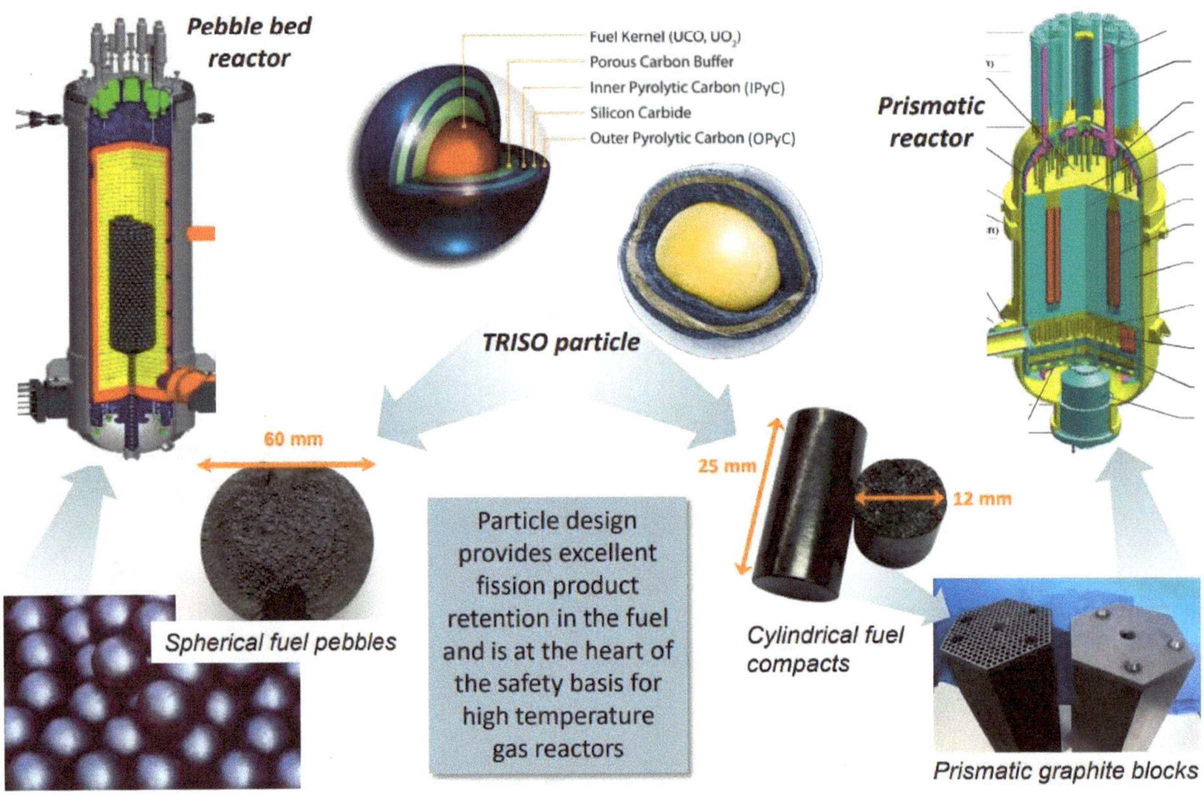

FIG. 1. TRISO particle in a pebble and in a compact made from special grade graphite matrix (reproduced from Ref. [6], courtesy of Idaho National Laboratory (INL)).

Figure 2 shows the cross-sectional view of a TRISO particle with the description of the intended role of each layer. In the figure, some typical values are shown, though values may vary from one design to another. For example, while the kernel diameter varies between 350 to 500 µm in the US, German and Chinese designs, it is 600 µm in the Japanese HTTR design where an annular fuel compact design is adopted.

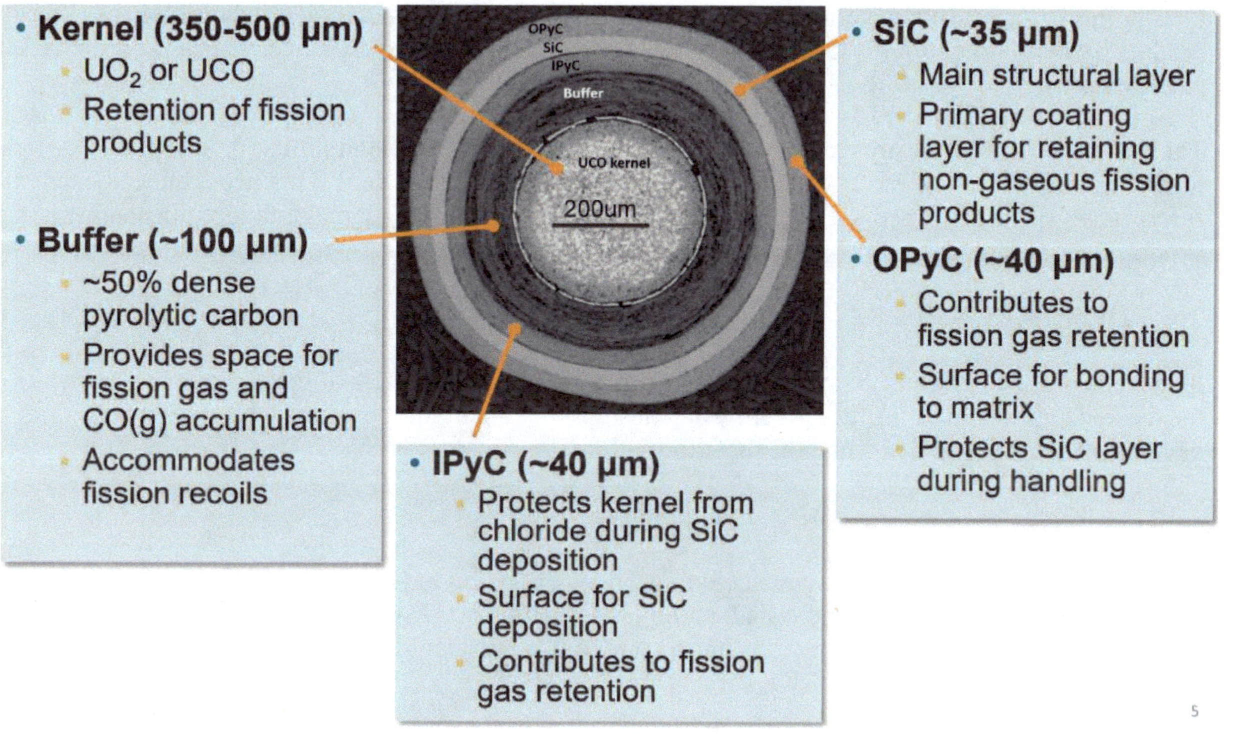

FIG. 2. Cross-sectional view of a TRISO particle fuel with typical dimensions (reproduced from Ref. [6], courtesy of INL).

2.2. DESIGN BASES

According to the IAEA Safety Standard, Specific Design Requirements SSR-2/1 (Rev. 1) [7], the fuel and reactor core are required to be designed based on the concept of defence in depth, to fulfil the fundamental reactor safety functions. The fundamental safety functions as they apply specifically to the design of the reactor core are:

(a) Control of reactivity;
(b) Removal of heat from the reactor core;
(c) Confinement of radioactive material.

In order to achieve the fundamental safety function (c) above, the fuel is designed and manufactured to minimize the release of fission products from the fuel, and together with other means, such as purification systems, the radionuclide activity in the coolant is maintained 'as low as reasonably achievable (ALARA)'.

Defective TRISO particles may be produced during the fabrication process or in the reactor during normal operation (e.g. a missing layer due to the flaw in the quality assurance of the manufacturing process can produce a defect in the particle during normal operation). A rigorous quality assurance and quality control approach is applied to the manufacturing process to ensure that the manufactured fuel quality specifications are met.

To ensure good performance of TRISO particle fuel, all mechanisms that result in the degradation of the collective fission product barrier (that is, the fuel kernel, the TRISO coating of particles and the matrix surrounding these particles in a spherical fuel pebble or a cylindrical fuel compact) need to be examined.

Design limits[3] or acceptance criteria[4] are established to prevent significant degradation of these fission product barriers. (See Ref. [9] for some examples of design limits or acceptance criteria used for TRISO particle fuel.)

The design of the fuel ensures that a sufficient margin exists to the design limits (or acceptance criteria) to prevent any failure of the collective fission product barrier of the fuel. This necessitates design and safety analysis tools based on the fundamental understanding of damage/failure mechanisms and supporting experiments.

2.3. DESIGN CONSIDERATIONS

In this section, design considerations for determining critical parameters of the different components of the coated particle fuel are described. Fuel design specifications developed by various vendors or developers are described in Annex II for information.

2.3.1. Kernel

A variety of fissile and fertile kernels including ThC_2, ThO_2, PuO_2, $(Th,U)O_2$, UC_2, UO_2, UCO [10] and UN [11] have been fabricated and tested; among these, UO_2 and UCO (a two-phase mixture of UO_2 and UC_2) are the fissile materials that are most commonly considered for recent reactor designs. Much experience has been gained with UO_2 kernels.

UCO kernels were originally proposed to better control CO gas production and thereby mitigate its negative impacts on fuel performance [12]. A significant amount of performance data has been collected in the US in the last two decades [13]. This fuel type is considered, in particular, for applications where fuel will experience relatively high burnup (10–20% fissions per initial metal atoms (FIMA)[5]) and high temperatures (i.e. ~1200°C).

The two phases, UC_2 and UO_2, interact in the following way [14]:

- As oxygen is liberated from fissioning, it first oxidizes the UC_2 and rare earth elements (e.g. Sm, La, Nd, Pr, Ce, Y) because they have the greatest affinity for oxygen;
- Once UC_2 and rare earth elements are oxidized, oxygen is available for some of the elements with less affinity, such as Sr, Eu, Zr, and Ba, which formed originally carbides from the liberated carbon from UC_2;
- Finally, only at the end of life, there is enough oxygen for CO production;

The goal is to balance this final CO production point (e.g. by defining the burnup accordingly or at the end of life).

---

[3] The term 'design limits' is used in SSR-2/1 (Rev.1) [7]: In Requirement 15, it is stated that "a set of design limits consistent with the key physical parameters for each item important to annealing for the nuclear power plant shall be specified for all operational states and for accident conditions".

[4] The term 'acceptance criteria' is defined in the IAEA Safety Glossary [8] as "Specified bounds on the value of a functional indicator or condition indicator used to assess the ability of a structure, system or component to perform its design function".

[5] FIMA stands for fissions per initial metal atom. Depending on the neutron energy spectrum and the proportions of fast fissions in ^{238}U and in $^{235}U/^{239}Pu/^{241}U$, the conversion factor ranges from 9.38 to 9.75 GWd/tU per 1% FIMA, and a mid-range value of 9.5 GWd/tU/FIMA is often used.

For the fuel utilizing TRISO particle with High Assay Low Enriched Uranium (HALEU), we may assume that the conversion factor from FIMA to GWd/tU established for LWR fuel with up to 5 wt.% enrichment for the thermal energy spectrum corresponding to LWR type of reactors are applicable to thermal gas cooled reactor fuel.

At low to moderate burnups (e.g. 10% FIMA) under normal operating conditions, the kernel can retain ~50% of total fission products generated. Gas retention of the kernel diminishes with burnup and is often assumed to be nil for accidents [14].

Several issues have been identified with the production of CO, which is associated with an increase in the oxygen potential in the fuel kernel (by fissioning the U in UO_2) and subsequent reaction with the carbon buffer layer:

- Increase of pressure in the particle due to CO (see Section 2.3.1.1);
- Corrosion of the SiC layer by CO;
- 'Amoeba effect' or the kernel migration (see section 2.3.1.2).

2.3.1.1. CO pressure

The fission products, carbon and uranium, all compete for the liberated oxygen from UO_2 (due to fissioning) in this closed system, and the system oxygen potential determines which elements are oxidized for a limited amount of oxygen. The carbon buffer layer can react with the liberated oxygen from UO_2 to form CO. CO production can increase the particle pressure and hence its failure likelihood. Furthermore, in the presence of a thermal gradient, CO can cause kernel migration.

Another reason for the control of the oxygen potential is to ensure that the rare earth elements are oxidized and thus immobilized in the kernel, preventing migration to the SiC layer and reaction with it.

Consequently, the oxygen potential is controlled either by limiting burnup or by tailoring the kernel composition. For the latter, UCO kernels has been developed and employed since the 1970s [14].

In pure UO_2 fuel the oxygen potential increases as a function of burnup and results in the production of CO, which in turn increases the pressure in the particle. The effect of burnup and temperature on the CO pressure is described in [14] and is represented in Table 1.

TABLE 1. EFFECT OF BURNUP AND TEMPERATURE ON CO PRESSURE

UO_2 burnup	**Oxygen potential (kcal/mol)**	**T(K)/(°C)**	**P_{CO} (atm)**
Low (< 10% FIMA)	-100	1300 / 1027	1
		1600 / 1327	5
High (> 10% FIMA)	-75	1300 / 1027	580
		1600 / 1327	1300

To prevent overpressure failure of the SiC, a simple calculation was carried out using the stress equation [14] to define an allowable particle internal pressure as:

$$\sigma = \frac{rP}{2t} \rightarrow P_{max} = \frac{UTS_{SiC}2t}{r} \approx 80 \text{ MPa} \quad (1)$$

where P_{max} includes all the pressure that is applied by fission gases Xe and Kr, and CO; UTS_{SiC} is the ultimate tensile strength of SiC and it is 350 MPa; t is the SiC thickness and it is 35 μm; r is the SiC layer radius and it is 310 μm.

It is noted that the krypton and xenon pressures depend on burnup. This equation simply sets a limit (for designers) for internal pressure to avoid overpressurization of the SiC layer, and the result will be dependent on particle dimensions and burnup, which has an impact on the fission gas inventory.

To avoid the problems associated with CO production, several approaches are considered:

- Making a sub-stoichiometric kernel and thus limiting the amount of oxygen available for CO production. This is possible with plutonium, but not with uranium [14];
- Including a 'getter' material (e.g. ZrC) in or near the kernel to absorb the released oxygen and make it unavailable for CO production, though it involves greater complexity in particle production [14,15];
- Making a two-phase kernel consisting of both carbide and oxide phases, which is known as UCO [14,15];
- Reducing the kernel size; reducing kernel size as enrichment/burnup increases will reduce CO and fission gas pressures;
- Replacing the SiC layer with ZrC. ZrC was indicated [15] to have great potential/performance capability as a coating for particle fuel, based on some testing.

2.3.1.2. Amoeba effect – Carbon transportation from the hot side to the cool side

In the presence of a temperature gradient, carbon transfers from one side of the kernel to another, known as 'Amoeba' effect (see Fig. 3). This effect can also be observed in UC_2, though it is particularly important for UO_2. The net effect of this carbon transport is the gradual movement of the kernel in the direction of increasing temperature so that the kernel moves toward the SiC layer and it may damage the layer. Particle failure is assumed to occur when the kernel touches the SiC layer.

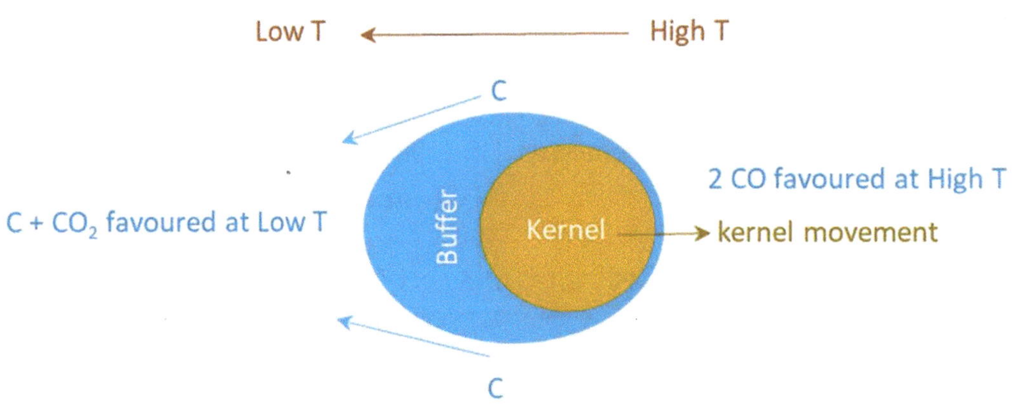

FIG. 3. Illustration of the amoeba effect.

The kernel migration/movement, Δx in meters, can be calculated [14] as:

$$\Delta x = KMC \frac{1}{T^2} \frac{dT}{dx} t \qquad (2)$$

where t is time in seconds, T is the temperature in Kelvin, and KMC is the kernel migration coefficient in m^2K/s. Typical KMC values at 1300 K are [14]:

- LEU (low enriched uranium) UCO: 9×10^{-13} m^2 K/s;
- LEU UO_2: 6×10^{-11} m^2 K/s;
- UC_2: 9×10^{-13} m^2 K/s.

The amoeba effect has been identified to be a concern for the fuels using UC_2 and UO_2 kernels in large (1000 MWe) HTGR designs [14]. For the reactor systems with low power and low burnup below ~10% FIMA, the amoeba effect may not be of significant concern for UO_2 fuel kernel [14]. Since CO production is minimized in the UCO kernel and solid-state carbon transport through the UCO is very low, it is expected that kernels with UCO will significantly reduce the amoeba effect.

2.3.1.3. Asphericity

In a study to investigate the impact of various particle properties on calculated failure probability [16], it is stated that "aspherical particles can generate tensile stresses in the SiC layer and cause failure". The failure probability of SiC layer for deviations from particle sphericity (indicated as 'aspect ratio') is shown in Fig. 4 [16]. The probability of failure increases with increasing aspect ratio (or asphericity), where 1 indicates perfect spherical shape. Then, the cumulative failure fraction was calculated by numerical integration of the SiC failure (due to asphericity) probability curve shown in Fig. 4.

On the cumulative failure fraction curve, the point at which the particle becomes increasingly susceptible to failure is determined as the critical limit [16]. For the aspect ratios varying between 1.0 and 1.24, of the value 1.06 was determined as the critical limit for SiC failure due to particle asphericity [16], above which the failure probability increases significantly.

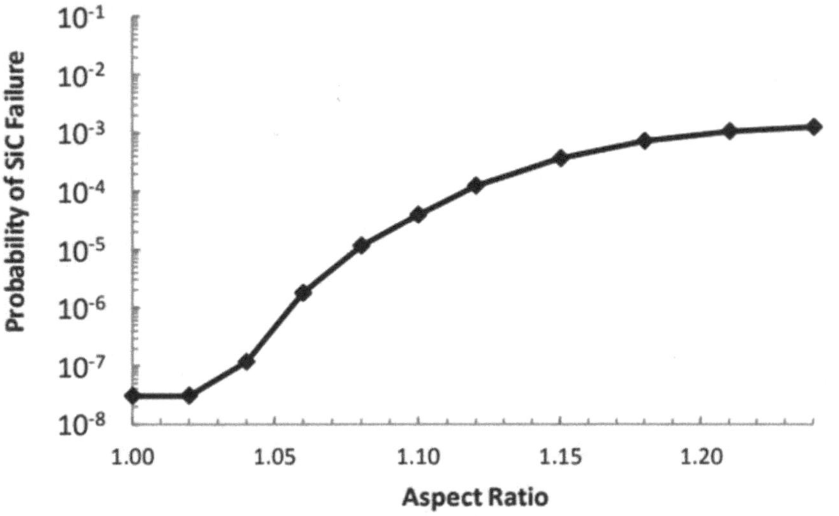

FIG. 4. Probability of SiC failure due to particle asphericity (reproduced from Ref. [16], courtesy of INL).

2.3.1.4. Kernel design parameters

Specifications for kernels are somewhat different from LWR fuel pellets. For example, for TRISO particle kernels there is less concern with grain structure or size and pores. Important kernel design parameters and their rationales are described in Ref. [14] and are represented in Table 2 (based on the information in Ref. [14]).

TABLE 2. KERNEL DESIGN PARAMETERS

Kernel design parameters	Rationale
Diameter	Diameter affects the power generation and hence generation of fission products.
Density	Density affects the amount of fissile material available in the particle as well as the degree to which the kernel can retain fission products (at least at lower burnup). In addition, it affects the kernel reactivity with chlorine during the SiC coating process.
Sphericity	Sphericity affects the coating uniformity (stress distribution in coating layers). Significant deviations from sphericity can result in high local stress in the coating layers. It is desirable to have a round kernel or a kernel with little deviations from sphericity.
Stoichiometry for UO_2: uranium to oxygen ratio	Stoichiometry in UO_2 kernel affects the CO production, resulting in higher pressures in the particle and kernel migration. Hypo-stoichiometry (i.e., oxygen to metal ratio $O/M < 2$) may be preferable to decrease the C-O reaction, but it is difficult to achieve hypo-stoichiometry in UO_2.
Stoichiometry for UCO: uranium to oxygen ratio and uranium to carbon ratio	Stoichiometry in UCO kernel affects CO production and oxidation of rare earths (e.g. Sm, La, Nd, Pr, Ce, Y), which is directly related to the retention of rare earth elements in the kernel.
Purity	Impurities may affect the chemical (i.e., reactions) and nuclear (i.e., neutron absorber, poison) behaviour of the fuel
Enrichment	Enrichment affects the power density, burnup, and radiotoxicity in the spent fuel.

2.3.2. Buffer layer

As the first layer surrounding the kernel, the thickness and the density of the buffer layer plays an important role in the attenuation of fission product recoil and in the accommodation of gases. In the majority of current TRISO particle designs, the thickness of the buffer layer is 100 μm, and the density is approximately 50% theoretical density (TD). In case of LWR fuel, high energetic fission products are slowed down and stopped by dense UO_2 in the range of 10 μm. However, in low-density materials, such as carbon, the range of the recoils can be longer. The buffer layer captures fission-produced recoils originating on the surface of the kernel and shields the IPyC from recoil damage. This indicates that it is important to have a thicker buffer layer. While a denser carbon layer may be preferable to slow down and stop the fission-produced recoils in a shorter distance, low-density carbon is desirable for the buffer layer to accommodate gases, serving as a gas plenum or void volume. This discussion supports thicker and low-density buffer layer to have a better control on particle pressure. On the other hand, a thinner buffer layer is also taken into account from heat transfer aspect: the thermal conductivity of the buffer is not as high as the other coatings and there is a more substantial temperature drop across this layer. Therefore, the buffer layer thickness is optimized for gas pressure and heat transfer. Finally, the buffer layer can serve as a sacrificial layer, as it can distort to accommodate kernel swelling. Important buffer layer design parameters and their rationales are described in [14] and summarized in Table 3 (based on the information in Ref. [14]).

TABLE 3. BUFFER LAYER DESIGN PARAMETERS

Buffer layer design parameters	Rationale
Thickness	Both thickness and density affect: 1) Void volume available for gases – low density and thicker layer may be preferable to accommodate gases; 2) Fission product recoil attenuation – thicker and high-density layer may be preferable to reduce the fission product attenuation range; 3) Flexibility for distortion to accommodate kernel swelling – thicker and low-density layer may be preferable to accommodate kernel swelling.
Density	Heat transfer is also affected by the density and the thickness of the buffer layer. Denser and thinner layer may be preferable for a better heat transfer. Overall, density and thickness need to be optimized for better gas retention and heat transfer.

With the increase of the buffer thickness, the free volume for storing fission gas increases. As a result, pressure becomes lower and stress in the SiC layer becomes lower. Probability of SiC layer failure decreases with increasing layer thickness (as shown in Fig. 5).

A lower critical limit of 50 μm for the buffer layer thickness was determined for SiC failure, below which the failure probability increases significantly. Note that the cited study [16] was based on a nominal particle design with a 425 μm diameter UCO kernel with 15.5% ^{235}U enrichment and a peak burnup of 16% FIMA.

Different particle designs and operating conditions will produce unique results with regard to coating layer optimization. The intent is to demonstrate that such analysis can be applied to optimize the parameters of interest (e.g. layer thickness) for a given layer for TRISO particles that would be employed under different conditions.

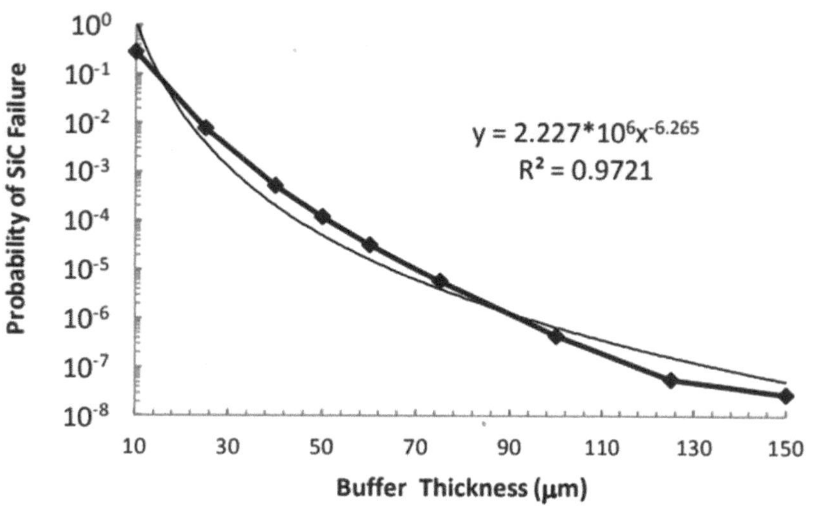

FIG. 5. Probability of SiC failure due to buffer layer thickness (reproduced from Ref. [16], courtesy of INL).

2.3.3. Inner pyrocarbon layer

The high density inner pyrocarbon layer (IPyC) is deposited on the buffer layer and provides a smooth surface for SiC deposition. It is the layer that protects both the kernel and SiC layer. In the absence of the inner pyrocarbon layer, during the SiC deposition process, liberated chlorine (in the form of hydrochloric acid, HCl) from the process would easily migrate to the kernel through the buffer layer and react with uranium, producing volatile chlorides. In this case, it would be highly likely that these chlorides would transport the uranium out of the kernel and contaminate the coatings [14]. Consequently, uranium fissioning would occur outside the kernel in the contaminated layers resulting in increase of fission product transport and releases.

The IPyC is an effective layer for gas retention and delays the transport of fission products to the SiC layer. The IPyC thickness influences stress in the IPyC layer and thus its propensity to fail: the failure probability increases as the IPyC layer thickness increases [16]. This is due to the increase in pyrocarbon material causing an increase in shrinkage early in irradiation, increasing the tensile stress in the IPyC layer and resulting in a higher IPyC cracking probability, which in turn can cause localized stress concentrations in the SiC layer [16].

An analysis of the effect of IPyC thickness on particle failure probability [16] was performed to determine the critical value for the IPyC layer thickness. As the IPyC layer thickness increases, the probability of SiC layer failure increases as shown in Fig. 6. An upper critical limit of 52 μm for the IPyC layer thickness was determined for SiC failure, above which the failure probability increases significantly. It is noted that the study was performed under a specific condition, but similar analysis can be applied to optimize the parameters of interest (e.g. layer thickness).

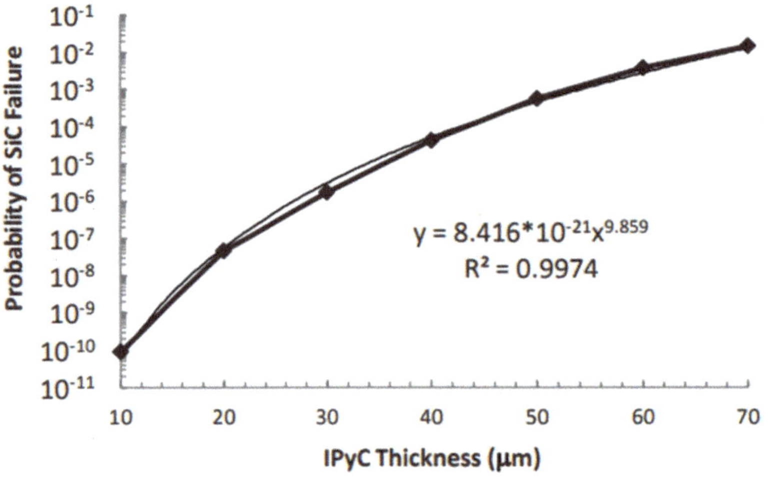

FIG. 6. Probability of SiC failure due to IPyC layer thickness (reproduced from Ref. [16], courtesy of INL).

IPyC density determines dimensional change and thus determines IPyC failure probability [16]. The failure probability decreases with increasing IPyC density, as shown in Fig. 7. An upper[6] critical limit of 2.0 g/cm^3 for the IPyC layer density was determined for SiC failure.

---

[6] It was not clarified in Ref. [16] why 2.0 g/cm^3 was chosen to be the upper limit and there is no information on failure probabilities beyond 2.0 g/cm^3; however, the attainable density of 2.0 g/cm^3 for pyrocarbon can be interpreted as both the upper and lower limit because, practically, higher densities cannot be achieved and the probability of failure is the lowest for 2.0 g/cm^3.

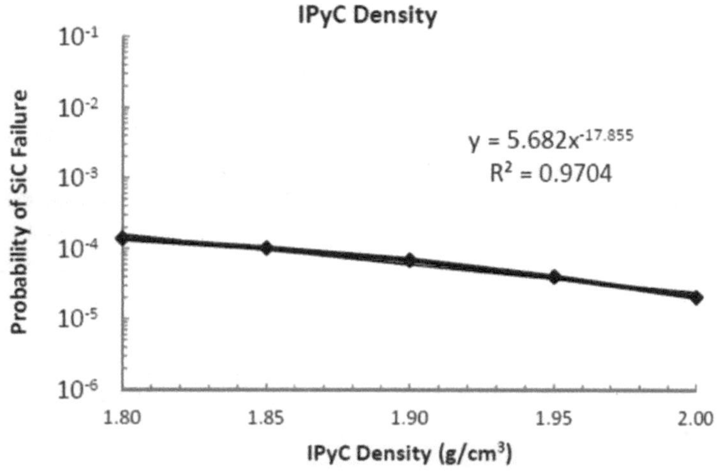

FIG. 7. Probability of SiC failure due to IPyC layer density (reproduced from Ref. [16], courtesy of INL).

High density pyrolytic carbon behaves similarly with a polycrystalline material. The properties of each crystallite are inherently anisotropic, as represented by, for example, thermal expansion and fast neutron induced shrinkage. For macroscopic graphite samples, macroscopic material anisotropy is defined by Bacon Anisotropy Factor (BAF) that can be measured by X-ray diffraction [4]. (Note that because of the very small dimensions of the coated particle layers, normal X-ray diffraction is not effective.) It is noted from [4] that "The intensity of reflected and polarized light differs depending on the orientation of the polarization direction relative to the crystallographic axes of the graphite crystal. Measurement of the ratio of the reflected intensities of a light beam, polarized first along one direction and then perpendicular to that direction, will yield an Optical Anisotropy Factor (OAF). It can be shown that this OAF can be related in a consistent way to the BAF, which in turn relates to actual expected anisotropy and fuel performance". In addition, an advanced ellipsometry technique, developed by Oak Ridge National Laboratory (ORNL) and called the Two-Modulator Generalized Ellipsometry Microscope, was applied to the measurement of pyrocarbon anisotropy in TRISO particle fuels, determining the polarization effect on light reflected out of a polished pyrocarbon cross-section [4].

BAF was also described using the coefficient of thermal expansion (CTE) [17]:

$$BAF = \frac{\alpha_{AG}}{\alpha_{WG}} \tag{3}$$

or [18]

$$BAF = \frac{\alpha_X}{\alpha_Z} \tag{4}$$

where α is CTE; subscripts, AG, WG indicate against grain, with grain; X and Z are the X axis and Z axis (of two orthogonal sample axes), respectively. It is indicated in Ref. [17] that "BAF is usually quoted at ~400°C". In Ref. [18], the BAF obtained using CTE data was correlated with BAF estimated by measuring the graphite texture using transmission X-ray diffraction (XRD), as

$$BAF_{CTE} = BAF_{XRD} = \frac{\bar{\alpha}_X}{\bar{\alpha}_Z} = \frac{1}{2}\frac{\int_0^{\frac{\pi}{2}} I(f)\cdot\sin^3 f\cdot df}{\int_0^{\frac{\pi}{2}} I(f)\cdot\cos^2 f\cdot\sin f\cdot df} \tag{5}$$

where α_Z is the constant stress estimate for CTE of the polycrystalline graphite at 400°C in the Z direction, where Z is the symmetry axis; α_X is the constant stress estimate for CTE of the polycrystalline graphite at 400°C in the X direction, where X is one of the other two orthogonal axes; $I(\phi)$ is the density of [002] plane normal per unit solid angle about the angle ϕ; ϕ is the angle between the [002] normal and the sample horizontal during transmission XRD measurement, in this case the WG direction. $I(\phi)$ describes the proportion of crystallites oriented at a given angle to Z, the sample symmetry axis.

It is indicated in Ref. [14]: "Good irradiation behaviour requires that the pyrocarbon layer exhibit similar dimensional changes in the longitudinal and latitudinal directions for the fast fluence of interest. That is, it is desired that the carbon layer material and physical properties be anisotropic. This can be achieved by ensuring that the deposited carbon has a random rather than a preferred macroscopic crystal orientation. A measurement of anisotropy is known as the BAF. A BAF of 1 is completely isotropic, greater than 1 implies increasing crystal orientation. Some argue that a BAF of less than about 1.05 is necessary for good irradiation performance, with a BAF in the range of 1.02–1.05 as the desired target."

The shrinkage and swelling behaviour of the PyC layers is a function of BAF and determined by the fabrication process of the layers. The dimensional change in the radial and tangential directions are equal for an isotropic material [19]. When the PyC is considered slightly anisotropic (BAF > 1.0), less shrinkage of the PyC in the radial direction is expected in the early stages of irradiation [20]. It is also indicated in Ref. [19] that, at some point during the irradiation, the shrinkage in the radial direction changes to swelling. Therefore, the dimensional change behaviour in the tangential direction becomes more important than that in the radial direction.

Based on BAF measurements on TRISO particles based on ThO_2 and fabricated by General Atomics in the US, the degree of crystallite anisotropy in IPyC layers (1.065 < BAF < 1.098) was observed to be greater than anisotropy in OPyC layers (1.018 < BAF < 1.036) [21]. The relatively large IPyC BAF values were thought to be caused by the exposure of IPyC layer to coating temperatures of 1550 to 1650°C during SiC deposition [21]. This hypothesis was tested by measuring BAF at the bi-structure isotropic (or BISO) stage (i.e., kernel + buffer layer + pyrocarbon) for specimens heat-treated at different temperatures, and was shown that BAF increases with increasing temperature, but when the sample was held at a constant temperature for a long period of time, a decrease in BAF was observed. The results indicate that higher temperatures cause larger deviations from isotropy, and longer hold times result in smaller crystallite anisotropy values (Table 4, based on data from Ref. [21]).

TABLE 4. HEAT TREATMENT OF BISO FOR THE EFFECT OF TEMPERATURE ON BAF

Thermal Treatment	BAF
As-coated through IPyC stage (1300–1400°C)	1.051
1 h at 1550°C	1.088
2 h at 1550°C	1.086
3 h at 1550°C	1.082
3 h at 1550°C + ½ h at 1650°C	1.087
½ h at 1850°C	1.112

In a computational analysis of TRISO fuel behaviour, it was demonstrated in Ref. [16] that anisotropy is critical to determining the magnitude of the dimensional change and the SiC failure probability due to stress concentration in the layer caused by an irradiation-induced shrinkage crack in the IPyC layer. However, based on the analysis results (see Fig. 7 for the probability of the SiC layer failure), there was little change in the failure probability when BAF was varied between 1.0 and 1.16, though an upper critical limit of 1.09 was determined. The lack of response was attributed to the uncertainties in the material properties, especially irradiation-induced creep.

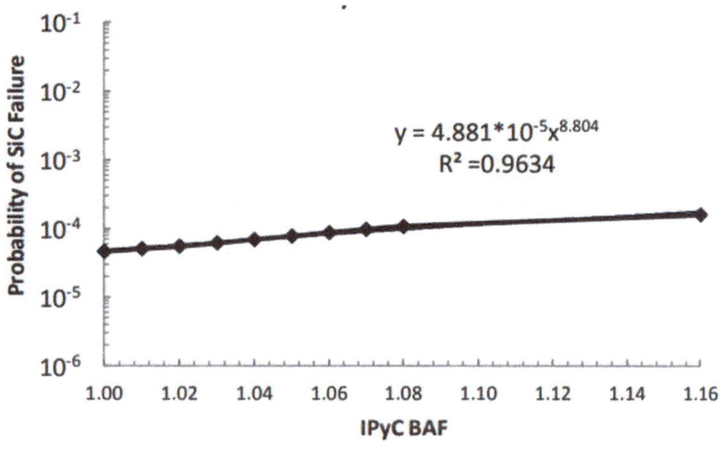

FIG. 8. Probability of SiC failure due to IPyC BAF (reproduced from Ref. [16], courtesy of INL).

Another sensitivity analysis was performed by using three different BAF values (i.e., 1.0, 1.03 and 1.06), resulting in three different PyC irradiation-induced dimensional change rates for the radial and tangential directions [19]. When it is considered as isotropic (BAF=1), the dimensional changes in the radial and tangential directions are the same as shown in Fig. 9. The effect of BAF on PyC dimensional changes is the largest at the largest BAF value considered and it increases with increasing fast fluence (or burnup for a constant power). Figure 9 shows that the difference in dimensional change values in the radial and tangential directions increases with increasing BAF value and with increasing fast fluence, though for BAF above 1.03 in the radial direction, the PyC dimensional change makes a minimum with fast fluence [19].

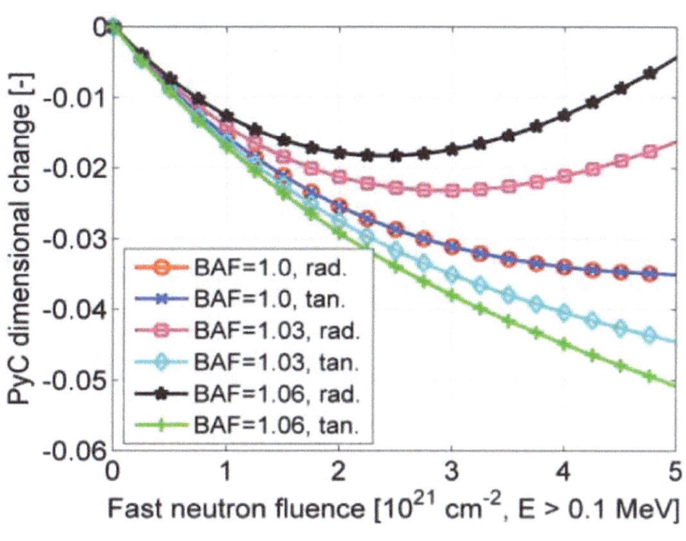

FIG. 9. Irradiation-induced dimensional change as a function of the fast fluence (E > 0.1 MeV) for three different BAF values (reproduced from Ref. [19], courtesy of INL).

Figure 10 shows the stress levels in IPyC, SiC and OPyC coating layers for three different BAF values (i.e., 1.0, 1.03 and 1.06). The stress in the SiC layer is especially sensitive to the dimensional change of the PyC, in particular at high fast fluences (or high burnups for a constant power), while the maximum stress of the IPyC or OPyC is relatively insensitive.

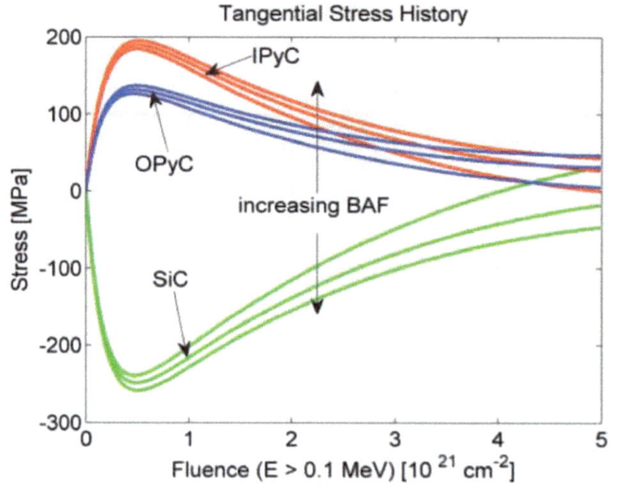

FIG. 10. Stresses in the IPyC, SiC, OPyC coating layers as a function of the fast fluence (E > 0.1 MeV) for three different BAF values (1.00, 1.03, 1.06) (reproduced from Ref. [19], courtesy of INL).

Figure 11 shows the effect of BAF on elastic modulus as a function of irradiation, or fast neutron fluence, f ($\times 10^{25}$ n/m^2, E > 0.18 MeV): the elastic modulus in the radial direction is the highest for isotropic (BAF=1.0) PyC (for the BAF values ranging between 1.0 and 1.2), and it decreases with increasing BAF value, while in the tangential direction, it is the lowest for BAF=1.0 and it increases with increasing BAF [22].

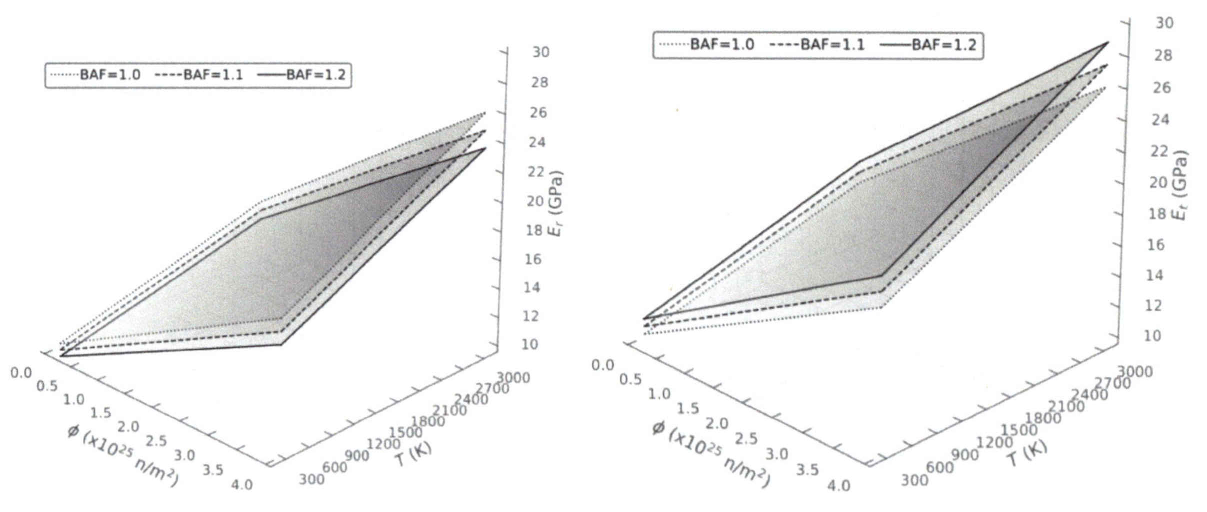

(a) (b)

FIG. 11. Effect of BAF on PyC elastic modulus: (a) Radial, (b) Tangential (reproduced from Ref. [22], courtesy of INL).

Figure 12 shows the effect of BAF on CTE α: α increases in the radial direction with increasing BAF, whereas it decreases in the tangential direction with increasing BAF [22].

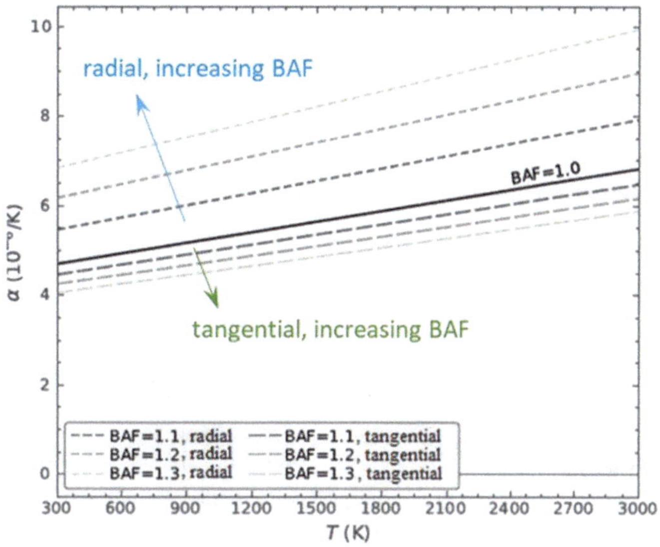

FIG. 12. Effect of BAF on CTE α (reproduced from Ref. [22], courtesy of INL).

Figure 13 shows the effect of BAF on irradiation strain as a function of irradiation, or fast neutron fluence, f ($\times 10^{25}$ n/m^2, E > 0.18 MeV) [22]: The eigenstrain increases with increasing BAF in the radial direction, while it is decreasing in the tangential direction. The effect of irradiation on eigenstrain in both directions (i.e., radial and tangential) is more pronounced at the highest temperature (i.e., 1350°C) and at the highest BAF (i.e., 1.25).

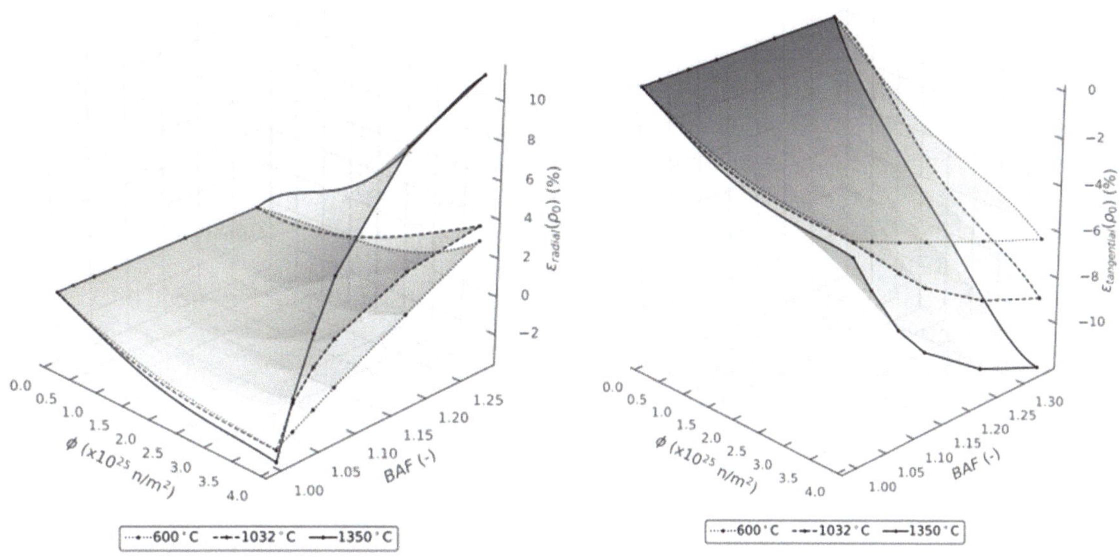

(a) (b)

FIG. 13. Effect of BAF on PyC irradiation strain at ρ_0 = 1:96g/cm^3: (a) Radial, (b) Tangential (reproduced from Ref. [22], courtesy of INL).

BAF is presumed to accelerate radial swelling and tangential shrinkage with fast neutron fluence in a fairly but not completely constant volume, as shown in Fig. 14, where 'Iso Vol' indicates isotropic volume changes and 'Vol' indicates anisotropic volume changes. At BAF=1.16, dimensional change in the radial direction switched from shrinkage to swelling at the fast neutron fluence of 5x10^{25} n/m^2, whereas the same behaviour for BAF=1.036 was observed at a higher fast neutron fluence (i.e., 10x10^{25} n/m^2).

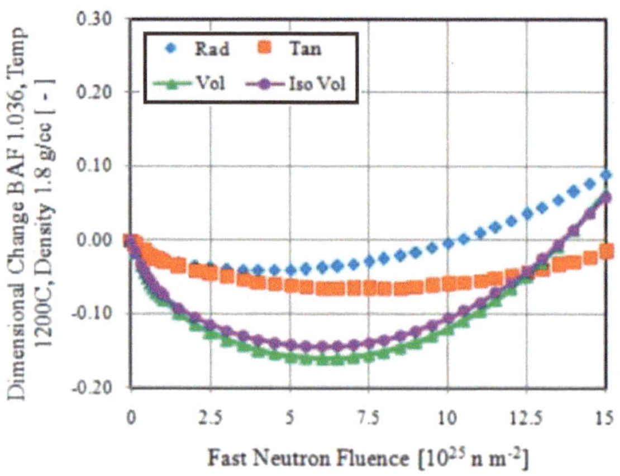

(a)

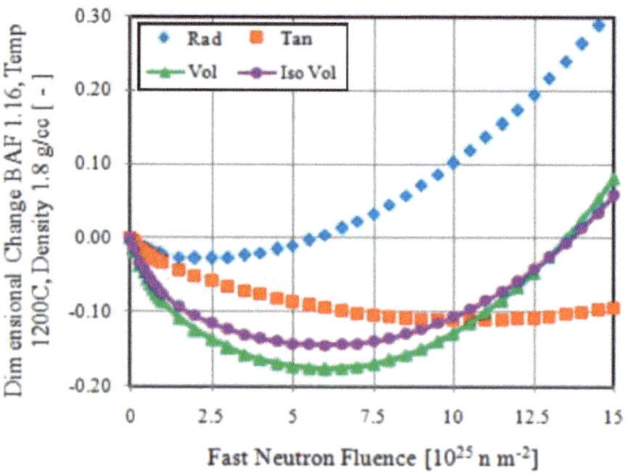

(b)

FIG. 14. Effect of BAF on dimensional and volume changes at 1200℃: (a) BAF=1.036, (b) BAF=1.16 (reproduced from ref. [23], courtesy of INL).

The following material and physical properties were identified for IPyC design, though the requirements for some of them may conflict with the requirements of the others:

— Density;
— Permeability and anisotropy:
  - Permeability (i.e., chlorine transport to the kernel during SiC layer coating) tends to be less with a higher BAF, but BAF with a value ranging between 1.02 and 1.05 is desirable for good irradiation performance [14]. Therefore, these permeability and anisotropy need to be optimized for irradiation stability and controlling coating contamination by chlorine reactivity in the kernel;
  - Increased IPyC thickness would not resolve the permeability problem since thicker IPyC can cause higher irradiation induced stresses, leading to higher failure probabilities of the particle.

— Creep and shrinkage:
  - The IPyC layer (influenced also by BAF) shrinks due to the fast flux, but it is also relieved by IPyC creep [14]. The tangential stresses in the IPyC and OPyC layers provide radial stress components on the SiC, thereby affecting the tangential stress in that layer [24]. Cracks (or debonding) in the IPyC can cause locally high tensile stresses in the SiC layer if the between layer bond is strong [14]. Such cracks can also provide a passage for CO to oxidize the SiC at high temperatures. Therefore, dimensional stability is important as cracking can lead to particle failure [14]. During irradiation both PyC layers shrink, compressing the SiC coating (see Fig. 15) and counteracting the hoop stresses created in it by the internal gas pressure [25]. These compressive stresses have to not be so large that the tangential stresses in the PyC layers exceed the fracture stress. In addition, the radial stress between the IPyC and SiC layers have to not be so large that debonding between the two occurs. This requires that these two layers are to be approximately isotropic [25].

— Strength.

The IPyC design parameters of interest to the fuel design to avoid failures are described in Ref. [14], and they are shown in Table 5 (based on information from Ref. [14]).

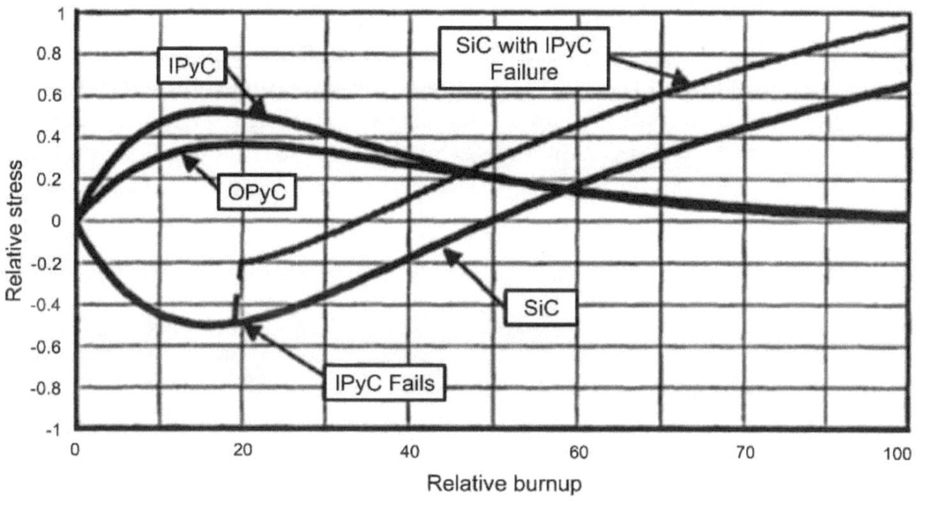

FIG. 15 Qualitative stresses in coated particle layers (reproduced from Ref. [25], courtesy of IAEA).

TABLE 5. IPYC DESIGN PARAMETERS

IPyC design parameters	**Rationale**
Thickness	Thickness affects structural properties of the IPyC, gas retention, and control of possible hydrochloric acid attack of kernel during fabrication.
Density	Density indirectly determines material properties.
Anisotropy	The irradiation stability of the layer is determined by the crystalline structure.

2.3.4. SiC layer

SiC layer is considered as the main barrier against release of fission products, including metallic fission products - serving as an impermeable layer, and gases - serving as a pressure vessel. It is dimensionally stable under irradiation and shows significant tensile strength.

Strains between the pyrocarbon layers and the SiC layer can result in SiC layer failure. Stresses are caused in the PyC layers by gas pressure and irradiation induced shrinkage, and in the SiC layer by gas pressure and the relative stress distributions between the layers [14]. It is desirable for pyrocarbon layers to keep the SiC in compression as long as possible. However, the stress distributions will change upon the failure of pyrocarbon layers, and the stress in the SiC layer will change from compression to tension even at low burnups.

Under normal operating conditions, it is expected that SiC layer retains all fission products, though silver is an exception, having a high release rate above 1100°C. At higher temperatures (>1250°C) palladium is of concern to diffuse and attack the SiC layer as it is not retained in the kernel like lanthanide elements, which form oxides that are well retained in the kernel. Therefore, this behaviour sets a limit on the normal operating temperatures (< 1300°C) [14]. At accident temperatures, above 1800°C, fission products release quickly. Above ~2000°C, thermal decomposition of SiC is a dominant failure mechanism. However, even at 1600°C, formation of porosities was observed, indicating the decomposition of SiC [14]. Therefore, the conditions are set such that the SiC layer is kept in compression during most of the irradiation and the operating (and accident) temperatures are limited to control SiC layer corrosion and decomposition.

A computational analysis was performed to determine the critical value for the SiC thickness. It was determined that the failure probability increases as the thickness decreases (see Fig. 16 for the probability of the SiC layer failure), which is attributed to less structural material to retain the fission gas pressure and subsequent increase in tangential stress in the layer. The analysis indicated a lower critical limit of 20 μm for the SiC layer thickness, for the thickness values varying between 10 and 60 μm.

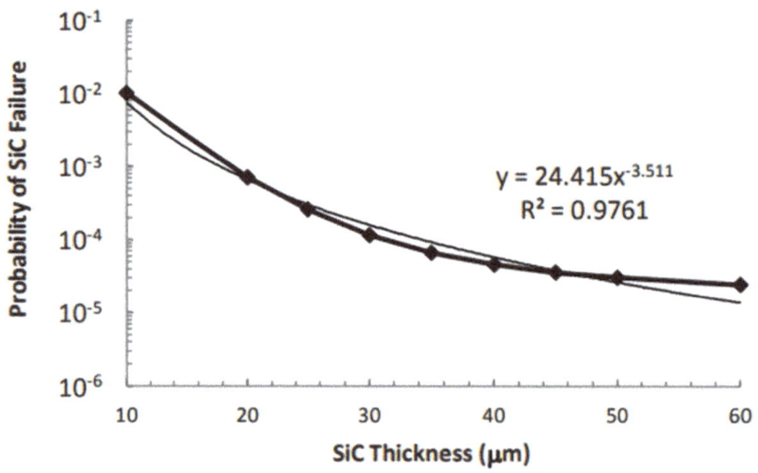

FIG. 16. Probability of SiC failure due to SiC layer thickness (reproduced from Ref. [16], courtesy of INL).

The SiC parameters of particular interest to the fuel design to avoid failures are described in Ref. [14], and they are represented in Table 6 (based on information from Ref. [14]).

TABLE 6. SIC DESIGN PARAMETERS

SiC design parameters	**Rationale**
Thickness	Thickness affects the strength of the layer. Uniform distribution of the thickness is desirable.
Density	Density determines material properties. It is desirable to have a high density SiC with small grains.
Fraction with defective SiC layers	Initially defective particles affect in-reactor performance and fabrication cost.

2.3.5. Outer pyrocarbon layer

The OPyC is the last layer, serving as a cushion to protect the SiC layer from mechanical damage during handling, and binding the particle to the matrix. It has similar requirements as discussed for IPyC layer.

Permeability of the OPyC has been identified to be important due to the potential intrusion of matrix material into the pores of the coating [14]. An OPyC layer that is too permeable would result in coating failure due to the strong bond formed between the OPyC and the fuel element matrix material, and differential shrinkage of these materials from irradiation exposure.

The computational analysis performed in Ref. [16] determined that calculated particle failure rates were insensitive to the OPyC BAF or density. Therefore, no critical limits for these properties were determined in that study.

Similar to the IPyC design parameters, the OPyC design parameters are listed in Table 7, based on information described in Ref. [14]. Originally, density and anisotropy were thought to be important, though the importance was deemed not significant in a separate study [16]; these parameters are kept in Table 7 because a separate experimental study can be performed to confirm either of the statements.

TABLE 7. OPYC DESIGN PARAMETERS

OPyC design parameters	**Rationale**
Thickness	Thickness affects the strength of the OPyC and gas retention in the case where SiC layer fails.
Density	Density indirectly determines material properties.
Anisotropy	The irradiation stability of the layer is determined by the crystalline structure.

2.3.6. Fuel element (spherical pebble or cylindrical compact)

While fabricating fuel elements, pressure is applied to compact the fuel. Consequently, cracks can form in the layers of the TRISO particle due to direct particle-to-particle contact. In order to avoid particle-to-particle interaction, particles are overcoated with matrix material before they are pressed to fabricate fuel elements. The thickness of the overcoating will vary considerably depending on the desired particle volumetric packing fraction in the final fuel element, though a typical value would be in the range of 200 μm.

The compacted fuel elements, either in the form of spherical pebbles or cylindrical compacts stabilizing the particles in the matrix (so that particles cannot move), allow the fuel to be handled without damage to the particles and provide a medium for heat transfer from particles to the coolant.

Traditionally, graphite matrix-based spherical fuel pebbles are used in a pebble bed reactor, and graphite matrix-based cylindrical fuel compacts are used in hexagonal blocks in a prismatic reactor. However SiC-based matrix fuel elements, called fully ceramic microencapsulated (FCM), have been proposed to be used in SMRs and some micro-modular reactors (MMRs). The FCM fuel consists of TRISO particles in a SiC matrix, offering the following potential advantages over a graphite matrix, which were taken from Ref. [26]:

- “Improved irradiation stability;
- Incorporation of yet another effective barrier to fission product release;
- Environmental stability under operating (steady state) and transient conditions as well as long-term storage;
- Proliferation resistance.”

The high thermal conductivity of SiC may allow thermal conductivity of the FCM fuel to be comparable to that of traditional graphite-matrix fuel.

Power of a fuel element depends on the number of coated particles in the fuel element (i.e. more particles are desirable for higher power); however, too many particles (or heavy fuel particle loading) in the fuel matrix will result in damaging the particles during the fabrication of the element. In addition, inhomogeneous distribution of particles will result in peculiar power distribution in the element, creating hot spots in the element, which may eventually lead to the fuel element failure. Design parameters related to the fuel element are described in [14] and represented in Table 8, based on the information from Ref. [14].

TABLE 8. FUEL ELEMENT DESIGN PARAMETERS

Fuel element design parameters	Rationale
Matrix material (i.e. graphite, SiC)	Matrix material contributes to the fuel element properties, such as thermal conductivity.
Particle packing fraction (particle load)	Particle packing fraction determines not only the power that can be extracted from the fuel element, but also impacts the likelihood of particle damage during fabrication; the higher the particle packing fraction, the higher the power and higher the possibility of damaged particles.
Particle distribution in the fuel element	Inhomogeneous particle distribution within fuel elements can result in hot spots.
Particle overcoating	Another layer on TRISO particles is applied in particular for pebble bed spherical fuels. This layer provides a particle-to-particle spacing function.
Fuel free zone	A fuel-free zone, surrounding the fuelled core that contains the TRISO particles, provides a protective shell at the periphery of the fuelled core in an element. This is typically used in spherical fuel pebbles, which will experience continual pebble-to-pebble contact and abrasion during their operational lifetime in a pebble-bed reactor core; however, it is also considered for cylindrical fuel elements.
Unconfined heavy metal outside SiC layer	It is undesirable to have the fuel elements contaminated with heavy metals (outside the particle), which will increase the release of fission products into the coolant.

2.4. SUMMARY AND DISCUSSION

Design bases for determining the dimensions and specifications of coated particle fuel components (i.e. kernel and coating layers) are well understood. Taking into account the current practice that fuel design limits or acceptance criteria form the basis for margin assessment to ensure the good performance and safe operation of fuels in LWRs, a similar approach needs to be taken for coated particle fuels for HTGRs. Hence, a complete list of design limits or criteria needs to be developed.

3. UPDATES IN MEMBER STATES

A survey during the Technical Meeting on the Design, Fabrication and Irradiation Behaviour of Small Modular Reactor Fuels, held virtually from 18 to 22 October 2021 indicated that more than 10 countries consider high temperature gas cooled SMRs for prospects in their energy planning. Some countries are also interested in developing FCM[7] fuel with intention to use in LWRs[7] or MMRs.

In this section, R&D activities in some Member States are described to set up a domestic manufacturing capability for TRISO particle fuel and to develop advanced coated particle fuel, i.e. FCM fuel.

3.1. CANADA

Four vendors have expressed their interest in building HTGRs in collaboration/partnership with the Canadian nuclear industry or the government. These include:

- Ultra Safe Nuclear Corporation (USNC) – MMR-5/MMR-10: 5-10 MWe;
- U-Battery Canada Ltd. – U Battery HTGR: 4 MWe (project started pending as of 2022);
- X-Energy LLC – Xe-100 HTGR: 80 MWe;
- StarCore Nuclear – StarCore Module HTGR: 10 MWe.

In addition, other activities related to coated particle fuel have also been pursued. One example is the joint research between Canadian Nuclear Laboratories (CNL) and USNC on FCM fuel in support of MMRs. The research project has explored manufacturing process for FCM fuel.

The first concept of FCM fuel developed by USNC is based on layered structure (see Fig. 17). One layer consists of SiC crust, which is pressed SiC powder, a TRISO plane, and a SiC lid. Once one layer is complete, the layer is pressed again. At CNL, the first fabrication trials were made using a surrogate material of ZrO_2 in the kernel instead of UO_2 or UCO. The matrix material was nano-sized SiC powder.

Figure 18 shows the sintered FCM fuel with surrogate kernels. Several analyses for characterization of the fuel were performed, using:

- X-Ray Computed Tomography with which it is possible to identify defective particles as a result of manufacturing, how densely packed the particles are inside the fuel form, and how close the particles are to the edges;
- Particle Size Analyser that is used to measure the average TRISO particle diameter;
- Optical Microscopy/Analytical Software that is used to determine the size, shape and distribution of TRISO particles in the FCM;
- Dilatometer that is used to determine the SiC thermal expansion;
- Differential Scanning Calorimeter that is used to determine the SiC matrix melting point, heat capacity and phase diagrams;
- Laser Flash Apparatus that is used to determine the SiC matrix specific heat capacity, thermal diffusivity – currently not on irradiated fuel;
- Pycnometer to determine the density of TRISO particles.

Compared to the conventional TRISO particle fuel, the FCM fuel is characterized by the use of a large diameter kernel containing an increased density of uranium, SiC matrix and the cladding made of advanced steel, SiC clad or refractory alloys. When coupled with SiC cladding, FCM fuel eliminates zirconium altogether and is expected to drastically reduce hydrogen generation during a beyond-design basis accident.

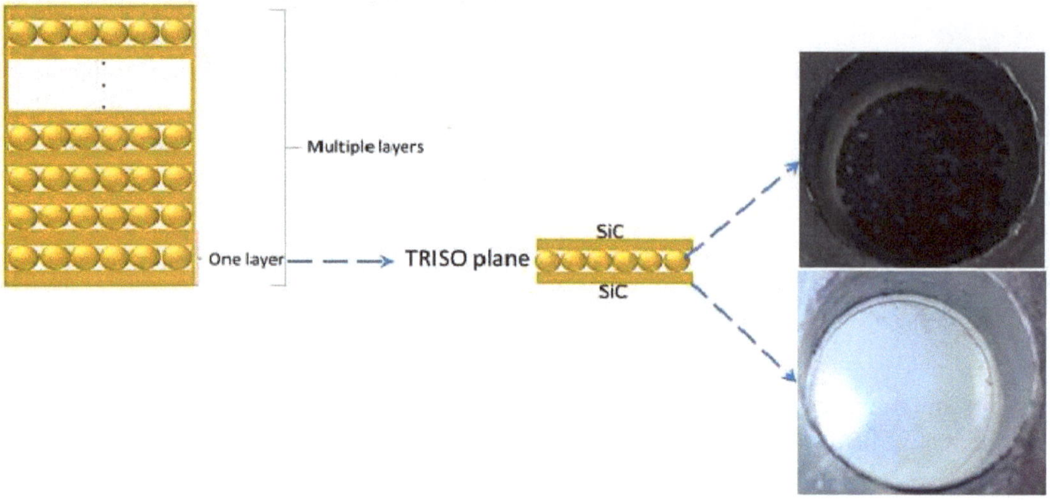

FIG. 17. Layered structured of fully ceramic microencapsulated fuel concept and one layer of the layered structure (courtesy of CNL).

FIG. 18. Sintered FCM fuel with surrogate kernels (courtesy of CNL).

3.2. CHINA

Coated particle fuel has been developed in China since the 1970s, in several stages (see Fig. 19).

In the first stage, from the 1970s to 1986, the principles of coated particle fuel fabrication were developed, including the sol-gel method for the kernel, the fluidized bed chemical vapour deposition (CVD) for coated particles, and the module press for the fuel sphere. The principal research work was accelerated by the international cooperation with Jülich Research Centre (FZJ), Germany.

In the second stage, from 1986 to 2003, coated particle fuel was developed for the HTR-10. Based on the research work, the HTR-10 reactor project was approved by the government and constructed in the Institute of Nuclear and New Energy Technology (INET) located in Tsinghua University. Comprehensive coated particle fuel fabrication technology was developed, including examination methods, quality control, and quality assurance. A lab scale manufacturing line was established in INET, which manufactured 23 000 spheres for the first loading of HTR-10 including the irradiation test.

In the third stage (2003-2012), large scale manufacture technology of coated particle fuel was developed for HTR-PM. From 2006 to 2016, based on the development of HTR-10, the HTR-PM project was approved, and the coated particle fuel was developed. A prototype manufacturing line with the capability of producing 100 000 spheres in one year was established in INET based on the lab scale process.

Meanwhile, the fuel fabrication technology was transferred to China Nuclear North Fuel Company and an industrial manufacturing line was designed and constructed based on the prototype manufacturing line in INET, with a capability of producing 300 000 spheres per year. This manufacturing line produced the first loading for the HTR-PM. By the end of 2019, a total of 860 000 spheres were fabricated for HTR-PM. The first batch of fuel was loaded in the HTR-PM in 2021, which allowed the HTR-PM to reach the criticality that year.

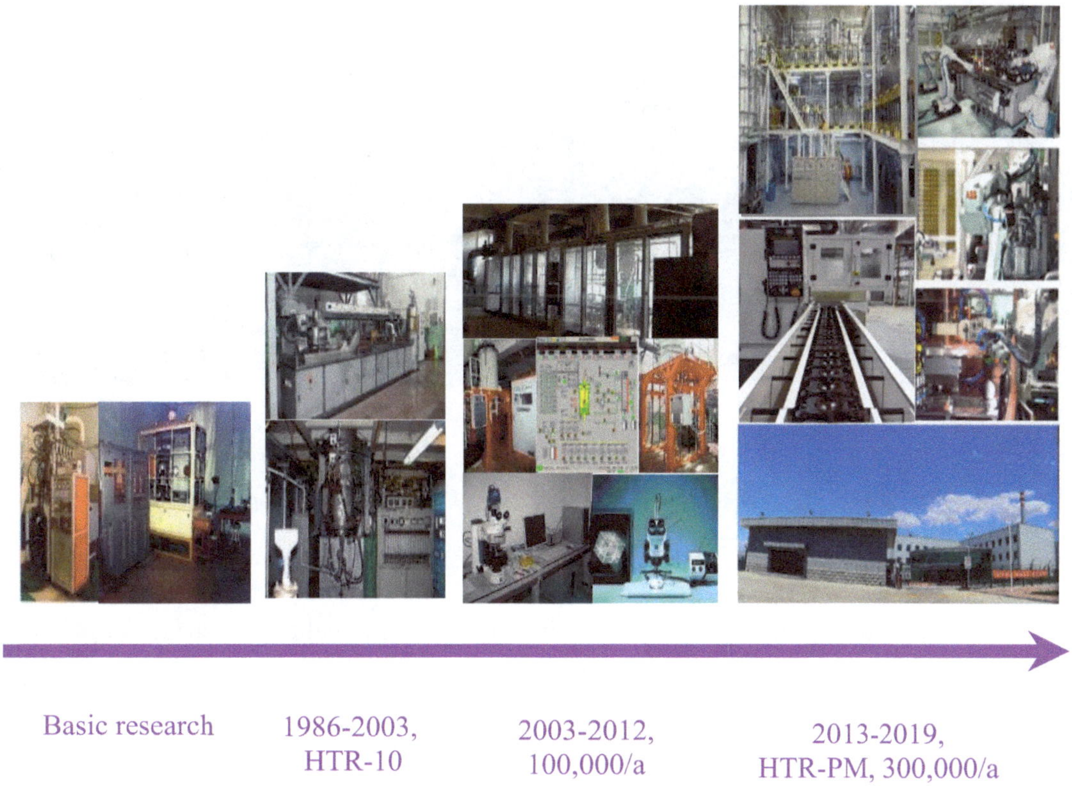

FIG. 19. The development map of coated particle fuel in China (courtesy of Tsinghua University).

Now the coated particle fuel technology development is in the next stage, which includes the design for a large-scale industrial manufacturing line with the capability of producing more than 3 000 000 spheres per year. This fuel manufacturing capability is intended to support the deployment of the HTR-PM600 reactor.

In addition, a new type of coated particle with new fuel kernels and new coating layers is being developed for the application of coated particle fuel in SMRs. The new coated particle fuel includes new kernels such as UCO, UN, and a new coating layer of ZrC.

3.3. REPUBLIC OF KOREA

In 2019, the Korean government announced its plan to foster a hydrogen related industry such as hydrogen-powered vehicles and hydrogen fuel cells. Hydrogen demand by 2040 is expected to reach 5.26 million tonnes per year. Under this circumstance, VHTR technology has been considered as one potential option for 'green' technologies for the production of hydrogen gas.

An intensive R&D programme to develop important technologies required for VHTR has been launched, which focuses on design and analysis technologies, thermal fluid experiments, TRISO fuel, high temperature materials database, and high temperature heat application.

The Korea Atomic Energy Research Institute (KAERI) has successfully established a manufacturing process for TRISO particle fuel at the laboratory scale (20~30 g UO_2/batch). During this development, KAERI made progress in kernel fabrication, TRISO coating, overcoating, and compaction of the coated fuel particles using graphite powder. Figure 20 shows kernel fabrication equipment installed at KAERI.

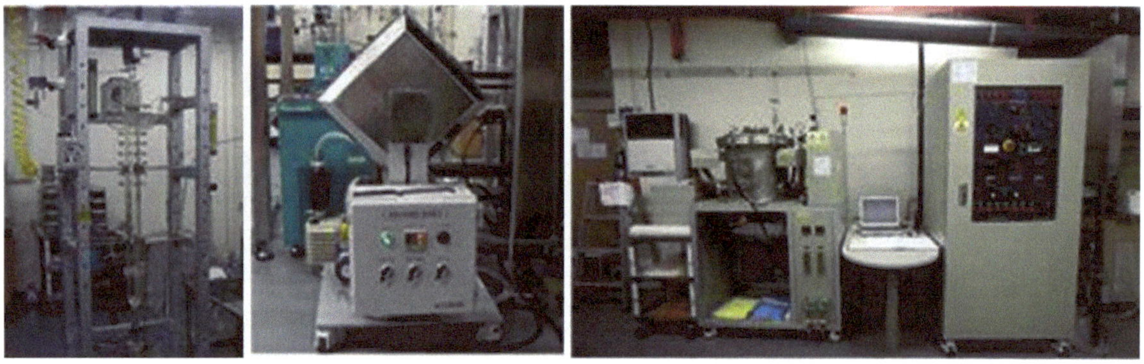

FIG. 20. Kernel fabrication system at KAERI (courtesy of KAERI).

KAERI uses gel supported precipitation for the fabrication of spherical UO_2 kernels. Associated process parameters have been determined as appropriate.

KAERI uses fluidized bed-chemical vapour deposition (FB-CVD) technology for TRISO coating (Fig. 21) with a capacity of 20~30 g/batch scale.

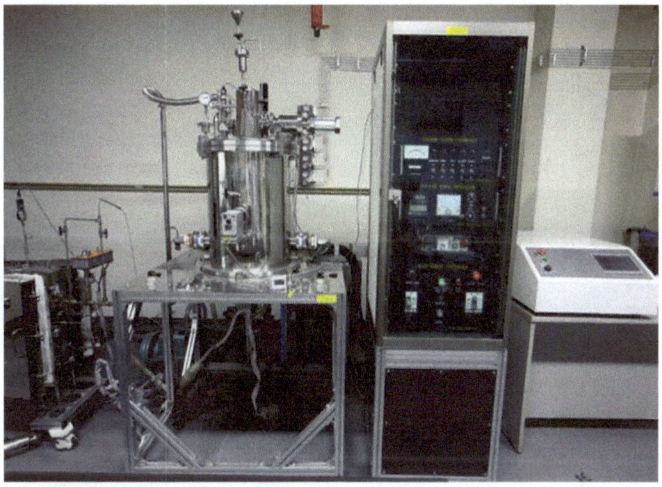

FIG. 21. TRISO coating system (courtesy of KAERI).

KAERI's TRISO fuel has been irradiated at the HANARO material test reactor, followed by PIE of the irradiated fuel.

Research and development for advanced coated particle fuel technology has been undertaken at KAERI, which includes UCO kernel fabrication and ZrC/SiC double layer coatings. As regards UCO kernel fabrication, a process to dissolve the carbon powder in the viscous solution using an ultrasonic and mechanical blending method has been established.

As regards ZrC coating, $ZrCl_4$ powder as a Zr source was vaporized in the sublimation system equipped with a screw feeding system. A lab-scale ZrC coating process capable of controlling the stoichiometry of ZrC was established using FB-CVD. The high temperature stability of ZrC/SiC double layers was confirmed by measuring microstructural changes (composition and grain size) and mechanical properties (elastic modulus and hardness) after heat treatment.

TRISO-SiC-composite fuel, widely called FCM fuel, was conceptualized as an accident tolerant fuel for LWRs. Through the Korea-US International Nuclear Energy Research Initiative, the feasibility of LEU FCM replacement fuel for current LWRs was investigated. Using enriched UN kernels with a higher TRISO particle volume fraction was suggested to meet uranium loading and fuel cycle length of LWRs.

As a promising processing method other than hot-pressing and spark-plasma sintering, non-pressurized sintering was applied to fabricate FCM fuels. A new quaternary additive system of AlN-Y_2O_3-Sc_2O_3-MgO was chosen to make SiC matrix densified at as low as 1850°C without applied pressure. FCM fuels with > 35 vol.% TRISO particles were fabricated. Thermal conductivity of the FCM pellets was measured as a function of TRISO particle content (see Fig. 22).

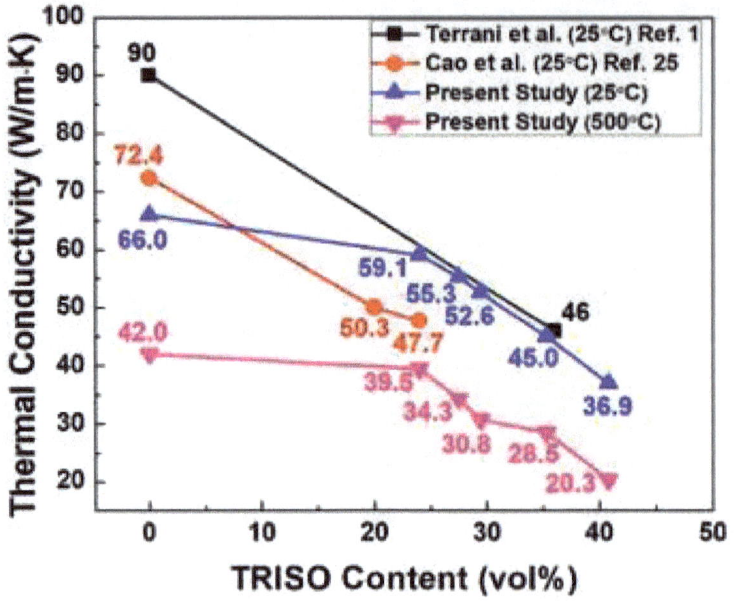

FIG. 22. Thermal conductivity of non-pressurized sintered FCM pellet characterization (courtesy of KAERI).

4. COATED PARTICLE FUEL MANUFACTURING

Coated particle fuel manufacturing technology has been developed for over 60 years. The technology includes the fuel kernel, coated particle and the pebble or compact manufacture, and the associated examination methods. This section provides an overview of coated particle fuel fabrication technology together with advancements made.

4.1. FUEL KERNEL MANUFACTURING

Fuel kernel manufacture has two processes: the powder metallurgical process and the sol-gel process.

4.1.1. Powder metallurgical process

The powder metallurgical process, known as the 'dry route process' is widely used in LWR fuel manufacturing. The first fuel charges of the AVR, Peach-Bottom and Dragon reactors were made from carbide kernels produced by such a powder metallurgical process. UO_2 or U_3O_8, alone or mixed with ThO_2, is granulated to form green particles (carbon powder and a binder are added during this process). The green particles embedded in graphite powder form kernels of UC_2 or $(Th,U)C_2$ at a temperature above 2500°C [27,28]. After sieving, these kernels become suitable for coating. Oxides in the form of UO_2 or $(Th,U)O_2$ are formed by burning off the carbon in air. Following sintering at 1600°C in hydrogen environment, porous kernels (e.g. with 90% theoretical density) are produced. Note that liquid processes (refer to [29–37]) are necessary for producing dense kernels. Some of aqueous processes are described in the subsequent sections.

4.1.2. Sol-gel process

The sol-gel process known as the wet process is widely used in kernel manufacturing for coated particles. There are two primary ways of making spherical kernels by wet route procedures: external gelation (or gel precipitation) - where the gelation proceeds from the outside, and internal gelation - where the gelation proceeds from the inside.

Figure 23 shows the external gelation sol-gel process, represented based on information from Ref. [38]. A sol is prepared using $UO_2(NO_3)_2$, polyvinyl alcohol, and other additives. Then, the sol is dropped into a liquid from small nozzles to form microspheres. The sol contains small crystallites (~5 nm) of an oxide of the fuel; it is possible for oxides to obtain high densities after the sintering at 1600°C in a reducing atmosphere with no applying pressure.

External gelation is done by extracting water from the droplet, while internal gelation is achieved by a chemical reaction inside the droplets. In the chemical reaction ammonia is formed by thermal decomposition of hexamethylenetetramine to initiate gelation. Internal gelation is fast and uniform.

As a result of the sol-gel process, spherical and high density $(Th,U)O_2$ or UO_2 fuel kernels are produced with diameters of more than 500 μm. Carbide microspheres are formed from the mixtures of UO_2 sol and carbon black through the pre-sintering process at 1000°C in an argon atmosphere. These UO_2+C microspheres are sintered in the graphite powder at 1900°C to obtain porous UC_2 kernels.

The microspheres are dried and calcined at temperatures higher than 700°C. Highly spherical and high-density kernels of UO_2 are obtained by sintering in a reducing atmosphere at 1600°C. Of the techniques used to synthesize UN from UO_2, carbothermic reduction is commonly used in industrial applications.

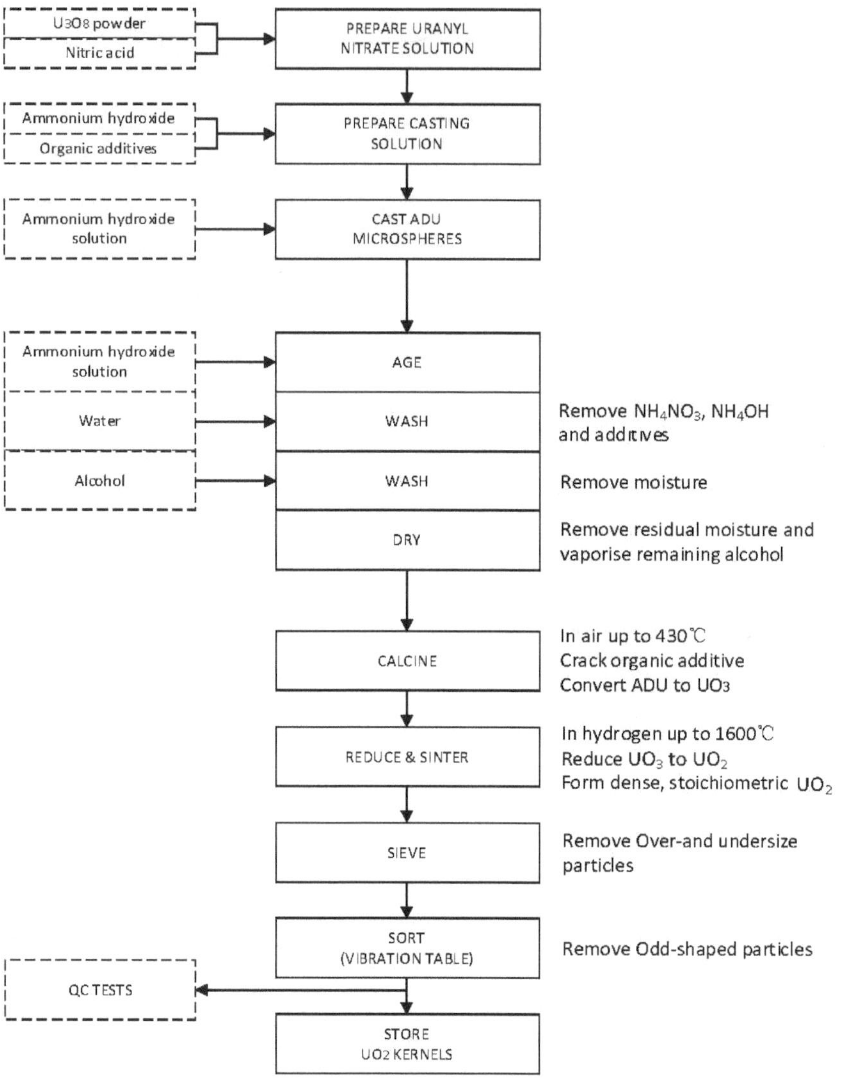

FIG. 23. Illustrative flow diagram of the external sol-gel process for kernel fabrication.

4.1.3. Gel precipitation process for UO_2 kernels

A modified sol-gel external gelation process using a drip casting method to produce UO_2 microspheres was developed by NUKEM [38,39], which is also known as 'gel precipitation'. The production of UO_2, ThO_2 or mixed oxide kernels by this external gelation process is relatively simple due to a few process steps involved and a small volume of effluent and waste generated and is prone to automated manufacturing. The overview of this process is represented in Fig. 23. Important steps are described below:

— Prepare feed solution. U_3O_8 powder dissolved in nitric acid to form a uranyl nitrate solution is fed:

$$3U_3O_8\ (s) + 20HNO_3\ (aq.) \rightarrow 9UO_2(NO_3)_2\ (aq.) + 10H_2O + 2NO\ (g)$$

The uranyl nitrate solution is pre-neutralized with ammonium hydroxide just prior to precipitation:

$$2UO_2(NO_3)_2\ (aq.) + NH_4OH\ (aq.) \rightarrow 2UO_2(NO_3)1.5(OH)_{0.5}\ (aq.) + NH_4NO_3\ (aq.)$$

— Prepare casting solution. Small amounts of polyvinyl alcohol and tetrahydrofurfuryl alcohol are added to the pre-neutralized uranyl nitrate solution and a casting solution is prepared. By adjusting the surface tension and viscosity in this process, the formation of proper droplets is facilitated. Figure 24 illustrates the casting equipment of the gel-sol.

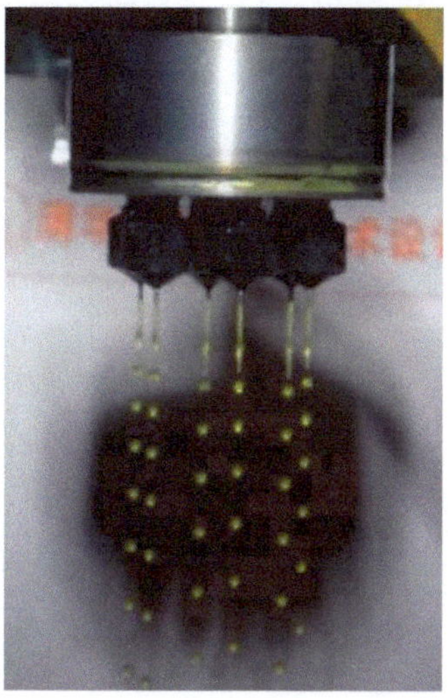

FIG. 24. Casting equipment of the sol-gel in INET (courtesy of Tsinghua University).

— Cast acid-deficient uranyl nitrate (ADUN) microspheres in aqueous solution. This step is performed in a glass column that is filled with the ammonium hydroxide precipitation solution. Figure 25 shows a schematic diagram of the casting process: (a) The casting solution is pneumatically supplied to the nozzles (up to five) through a silicone tube at the top of the column. A vibrator is used to shake off droplets at a rate of 100 droplets per second from the feed stream; (b) The droplets travel through air to form a spherical shape owing to surface tension; (c) Then, the microspheres pass through an ammonia atmosphere, causing it to precipitate as ammonium diuranate (ADU) in the outer layer of the microspheres, through a chemical reaction with the uranyl nitrate on the surface of the microspheres. The diameter of a cast gel sphere is about 1.8 mm.

— Age. The microspheres along with the accompanying precipitation solution are transferred from the casting column to a jacketed rotary flat tank for aging. Steam is used to heat the vessel to 80°C during ageing. The gel spheres are fully converted to solid ADU kernels, and a complete crystal growth takes place.

— Wash. The solution after ageing remains drained from the vessel. Water is used to wash the ADUN kernels in the vessel to remove the ammonium nitrate, ammonium hydroxide and tetrahydrofurfuryl alcohol. The kernels are washed again with isopropyl alcohol to remove moisture, and any remains from ammonium nitrate, ammonium hydroxide and tetrahydrofurfuryl alcohol.

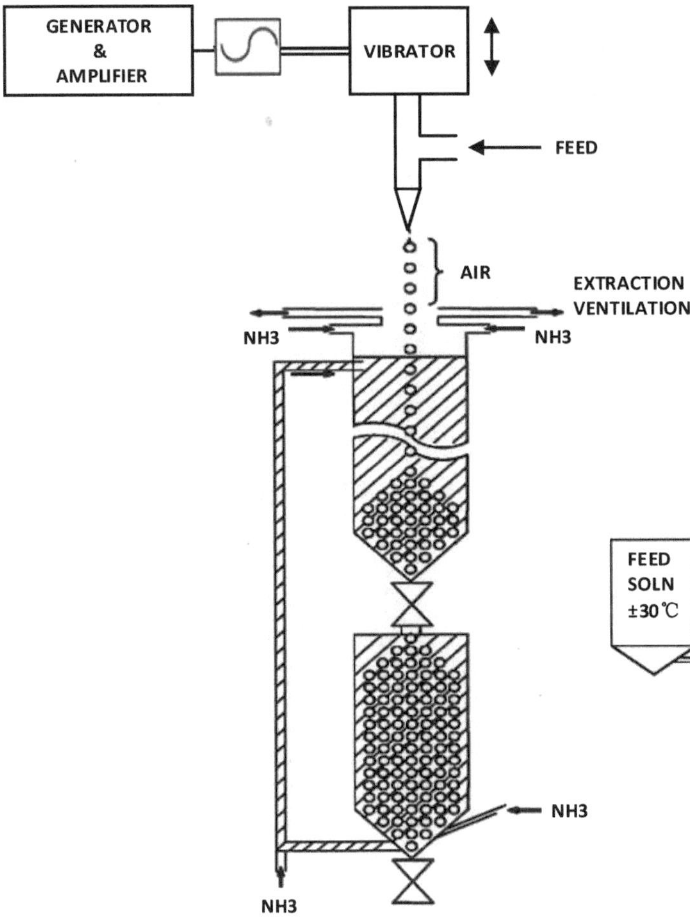

FIG. 25. Illustrative casting process for nuclear fuel kernel production (courtesy of Tsinghua University).

- Dry. The kernels are dried at 80°C under vacuum. The diameter of a dried ADU kernel becomes about 1 mm and the bulk density 1 g/cm^3.
- Calcine. After drying, the ADUN kernels are calcined in air at up to 430°C. Above 400°C the ADU is converted to UO_3:

$$(NH_4)U_2O_7\ (s) + O_2\ (g) \rightarrow 2UO_3\ (s) + 2H_2O\ (g) + NO\ (g)$$

The diameter of a calcined kernel becomes ~0.8 mm and the bulk density ~ 2 g/cm^3.

- Reduce and sinter. After calcining, reduction and sintering processes at high temperature are followed. As a result, the remaining impurities are removed, and the kernels are densified. The process is conducted under hydrogen atmosphere and UO_3 is converted to UO_2:

$$UO_3\ (s) + H_2\ (g) \rightarrow UO_2\ (s) + H_2O\ (g)$$

The temperature is maintained up to 1600°C so that dense and stoichiometric UO_2 kernels are formed with a diameter of 500 μm and a density close to the theoretical value of 10.96 g/cm3.

Figure 26 shows a composite picture of mixed oxide $(Th,U)O_2$ fuel gel spheres and kernels after various stages of processing.

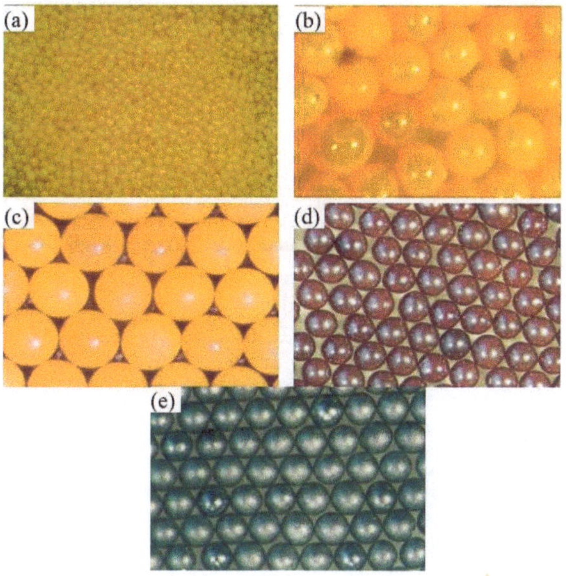

FIG. 26. Kernels in different steps: (a) casted gel particle, (b) aged particle, (c) dried particle, (d) calcined particle, and (e) reduced and sintered particle (courtesy of Tsinghua University).

— Sieve. The final production steps are sieving and sorting. Sieving is used to remove kernels of undersized and oversized.
— Sort. Kernels with odd shapes are then removed through sieving. On a vibrating sorting table that is slightly inclined, spherical kernels roll down-hill and odd-shaped particles are transported along a perpendicular direction by means of vibration and are collected for recycling.

In addition to UO_2 kernels used widely in HTGR fuel, UCO, UC and UN kernels are considered for use in HTGRs and other SMRs. These kernel compositions have higher thermal conductivity and higher uranium loading fraction compared with UO_2. UCO kernels have been used in coated particles in the US for several decades, including in the Next Generation Nuclear Plant (NGNP) project. Based on the UO_2 kernel manufacturing process, UCO, UC and UN kernel manufacturing technologies are being developed. Figures 27 and 28 show flowsheets of the internal gelation and kernel conversion processes for UC and the UN kernels. Now the new process is under testing to some manufacture scale.

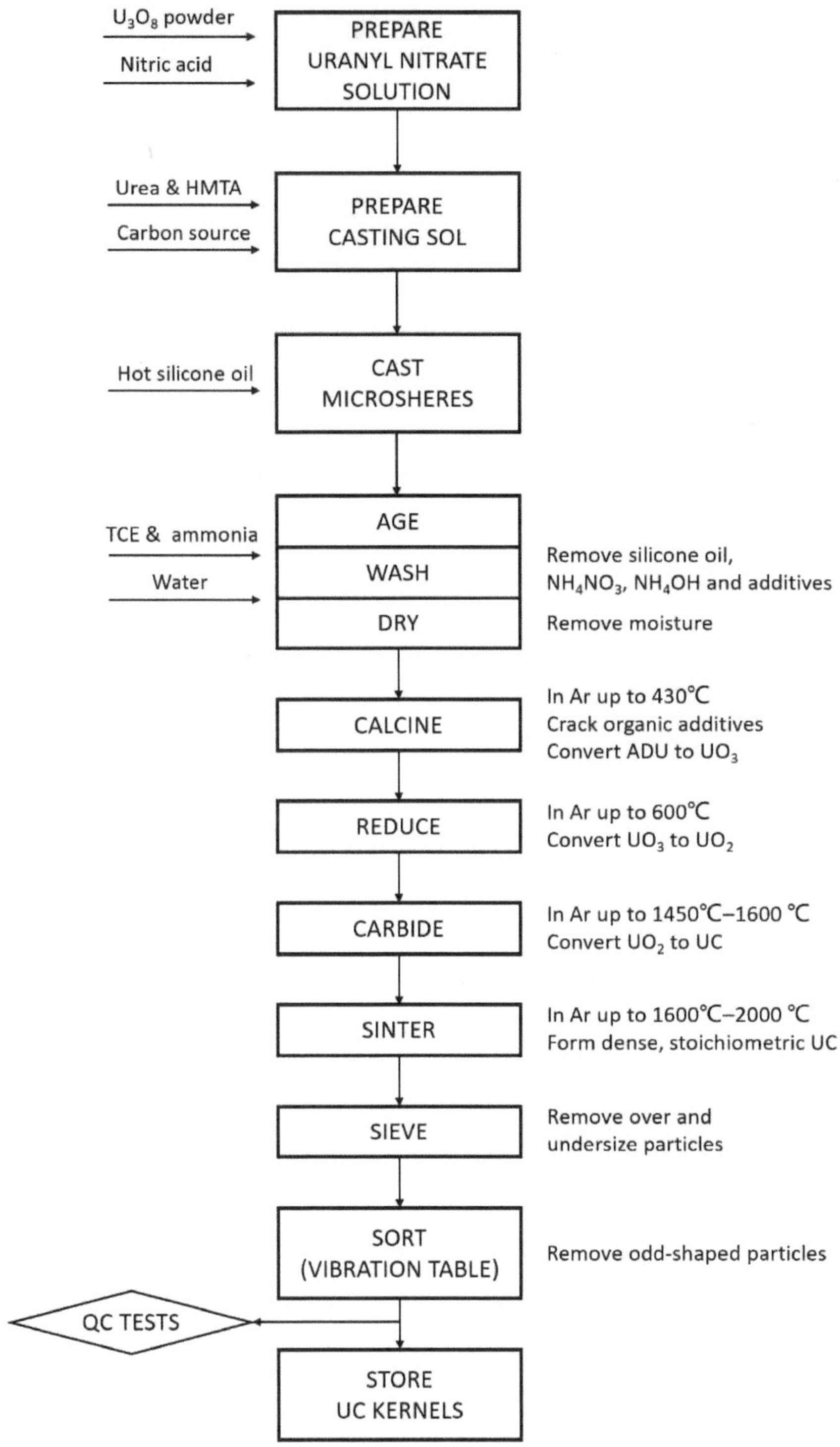

FIG. 27. Illustrative flowsheet for UC kernel fabrication by the sol-gel process (courtesy of Tsinghua University).

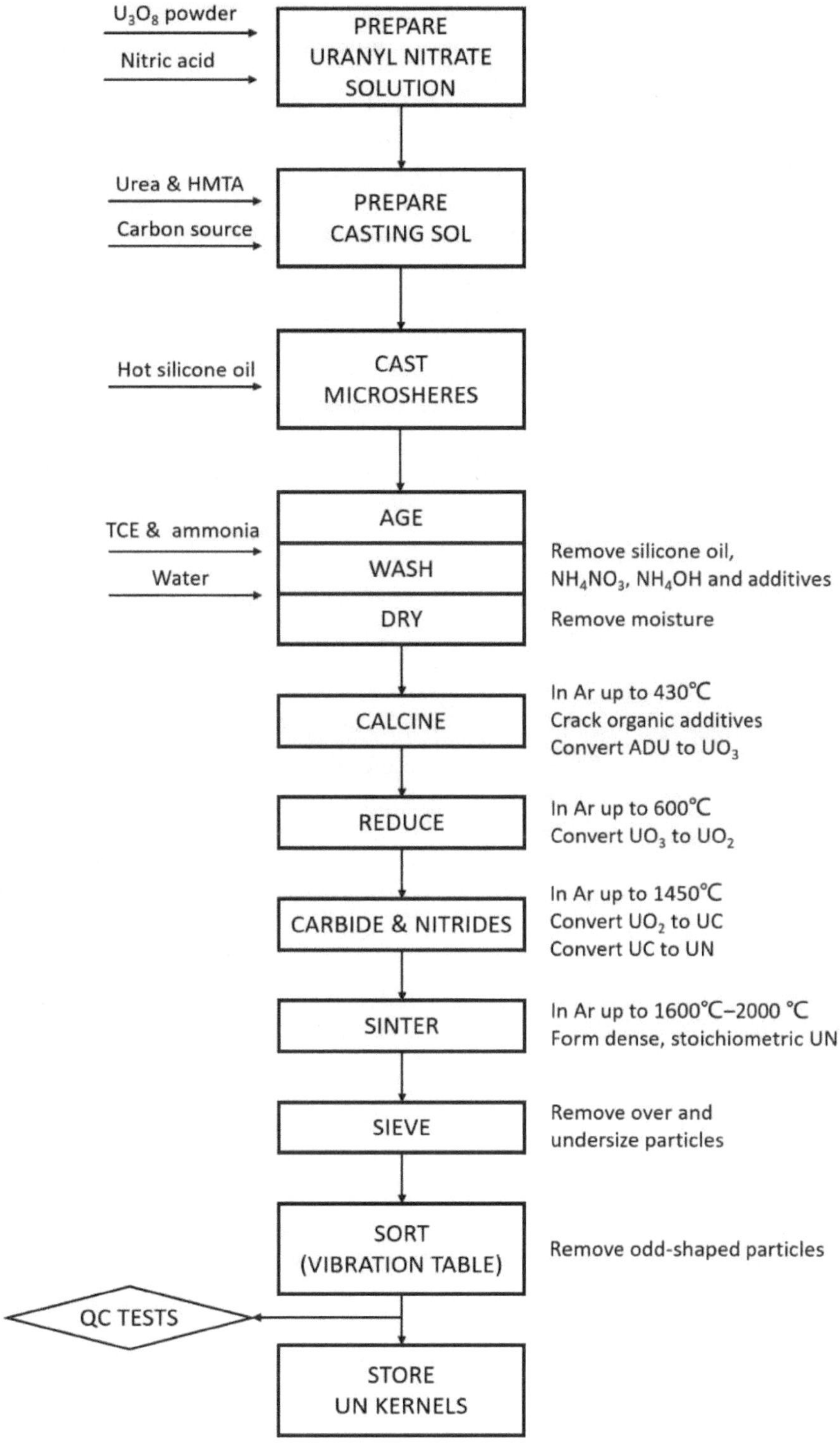

FIG. 28. Illustrative flowsheet for UN kernel fabrication by the sol-gel process (courtesy of Tsinghua University).

4.2. CERAMIC COATINGS BY FLUIDIZED BED CHEMICAL VAPOUR DEPOSITION

The sequence of ceramic coating layers is specific to coated particle fuel: since it is primarily responsible for guaranteeing intact fuel, the design, material selection and processing methods of the layers is immensely important. But this also makes the fuel largely independent of fuel chemistry in the kernel and provides a wide range of burnup and temperature with intact fuel.

In the manufacturing of TRISO particles, spherical fuel kernels are coated with multiple layers of PyC and an additional layer of SiC. Each layer is deposited by the FB-CVD method. Flowing gases are fed into a bed of kernels in a reaction chamber in a high temperature furnace to suspend the kernels so that they form a fluidized bed. The coating gases thermally decompose at temperatures from 1200°C to 1600°C and deposit their constituent materials onto the surface of the particles.

The low-density buffer layer of pyrolytic carbon is obtained by the decomposition of acetylene (C_2H_2) at temperatures from 1200 to 1400°C, whilst the dense pyrolytic layers are obtained by the decomposition of a mixture of ethyne and propene (C_3H_6) at temperatures from 1300 to 1400°C. The deposition rate of the dense pyrolytic layers is lower than that of the buffer layer.

The SiC layer is also obtained by a CVD process that includes the decomposition of methyl trichlorosilane (CH_3SiCl_3) with H_2 at 1500–1650°C. The deposition rate is correlated with the SiC density, which is very important for the permeability of fissions products.

In the fabrication of the TRISO coating layers, it was noticed that coating temperature was the main parameter affecting the content of free silicon, density, crystallite size, and microstructure [39–44].

The temperature range in which stoichiometric SiC is produced depends on gas flow rates, coater characteristics and particle loading. Deposition at lower temperatures results in the co-deposition of excess silicon and the formation of some porosity, and hence is avoided. Note that the grain size grows considerably with thickness in the coating production, and also affects the retention of fission products as well as the mechanical properties of ceramic materials. Some new technologies have been developed to make fine SiC grain sizes. The crushing strength of SiC coated particles decreases with increasing coating temperature and the strength is correlated with grain size. Xu and co-workers at the Tsinghua University studied coating temperature effect on various parameters of SiC layer such [45] as the microstructure, density, strength and elastic modulus, and concluded that strength and elastic modulus reached the highest values at 1500–1550°C. This is in full agreement with the original DRAGON studies of the 1960s and early 1970s [46,47]. Ref. [43] shows the lower SiC deposition temperatures may be used.

Fuel particle concepts using ZrC as a substitute or in combination with SiC have been studied to a considerable degree since the 1970s. The major advantage of ZrC over SiC is its ability to withstand temperatures in the 2000–2400°C range. However, in modern modular HTGRs the peak fuel temperature is limited by design to ≤1600°C and ZrC may be no longer interesting as an alternative particle coating, particularly in view of the substantial development effort that would be needed to achieve similar quality as SiC particles.

A potential benefit of ZrC is an increased resistance to chemical corrosion caused by fission products, particularly by Pd. Some coated particles with ZrC layers have been shown to survive irradiation at elevated temperatures higher than those with SiC. Some results reveal that particles with a thin ZrC coating on the kernel serve as an oxygen absorber, thereby preventing kernel migration through coatings. ZrC may retain caesium and possibly silver better than SiC. However, the retention of certain fission products by SiC is better than by ZrC. In particular, ^{106}Ru has been found to have higher diffusion rates in ZrC than in SiC, and because Pd has not been found concentrated at the inner surface of ZrC layers, there is concern that it could migrate through the coating.

The correlation of coating structure with its irradiation stability has been studied over decades. Different microstructure and density of PyC or SiC coatings can be obtained depending on the concentration of the hydrocarbon or silane coating gases and the temperature of their decomposition (pyrolysis). Figure 29 illustrates a schematic of the CVD process for depositing PyC and SiC coatings.

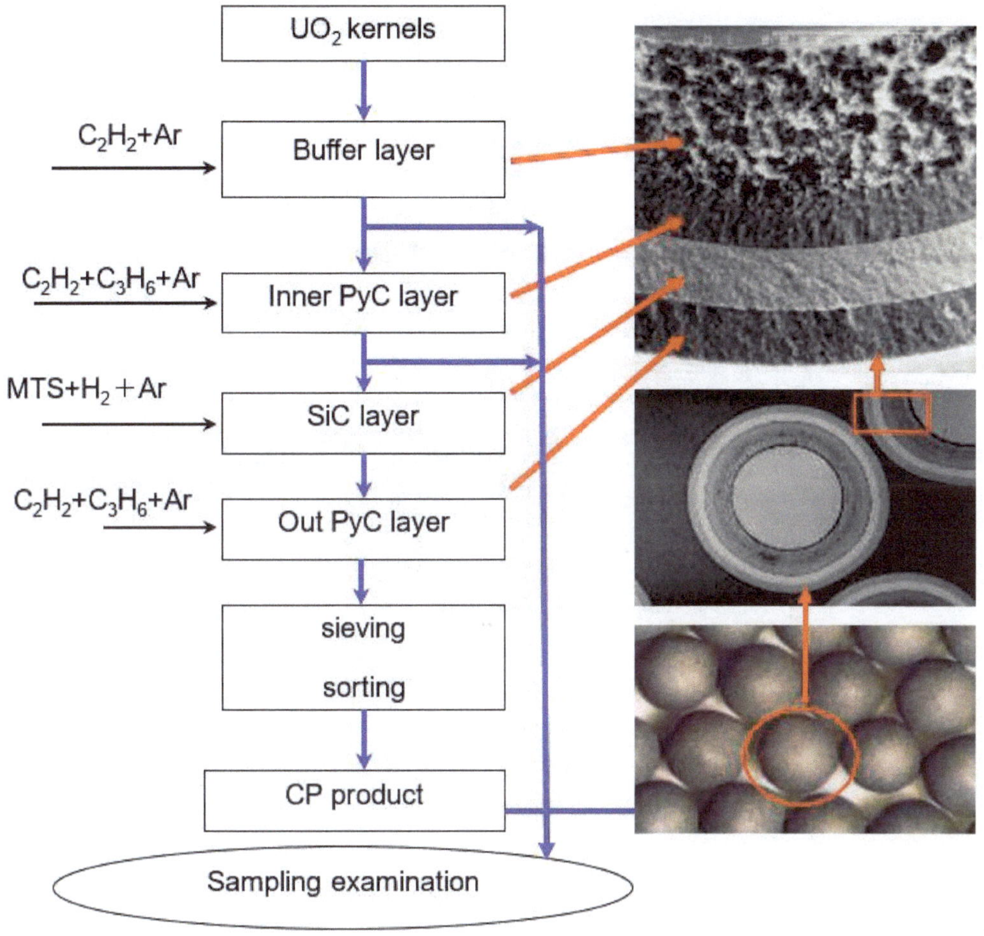

FIG. 29. Illustrative schematic of the deposition of PyC and SiC coatings on dense spherical UO_2 fuel kernels by the CVD process;MTS = methyltrichlorosilane, CP = coated particle (courtesy of Tsinghua University).

A schematic of a typical fluidized bed coating furnace with a so-called 'spouting nozzle' [48] is shown in Fig. 30 and the spouted bed principle [48] in Fig. 31.

The fluidized bed coating furnace is a vertically installed, cylindrical stainless-steel vessel and is double-walled with insulation and water cooled. In the centre of this vessel, a graphite resistance heating furnace is mounted which contains a graphite particle-bed tube in which the fuel kernels are fluidized and coated. Various hydrocarbons or silane-coating gases are added to the carrier gas (usually argon gas), responsible for fluidizing the particle-bed, depending on the targeted structure and density of the coating. The temperature for coating remains in the range 1200–1600°C, depending on the nature of the coatings. PyC-coated particles were produced at Battelle with a fluidized bed as early as 1959 [49,50].

The process for ZrC layer fabrication is very like the process of a SiC coating layer, but using $ZrCl_4$ or $ZrBr_4$ as the precursor for the source of Zr, and ethyne or propene as the source of C.

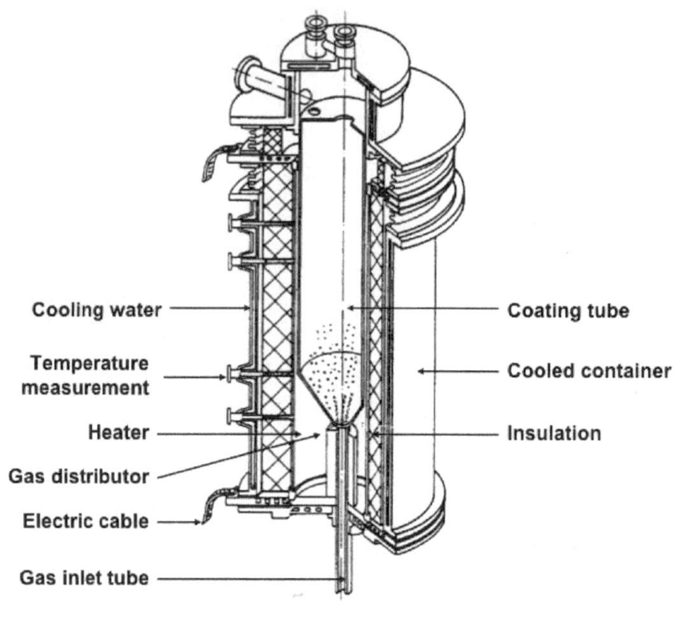

FIG. 30. Typical conical coater with spouted bed (courtesy of IAEA).

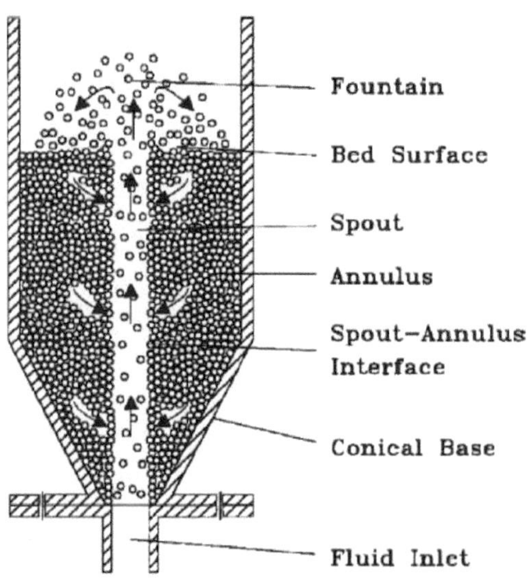

FIG. 31. Principle of the spouted bed characterized by low pressure drop, excellent heat transfer, and good mass transfer (courtesy of IAEA).

4.3. FABRICATION OF SPHERICAL FUEL ELEMENTS

The fabrication process for HTGR spherical fuel elements (pebbles) includes mainly the following steps as described in Ref. [51,52]:

— “Resonated graphitic matrix powder preparation;

— Overcoating of particles;
— Pre-moulding of fuel zone;
— High-pressure isostatic pressing of the complete fuel element;
— Machining;
— Carbonization at 800°C;
— Final heat treatment at 1900–1950°C."

These processes are illustrated in Fig. 32.

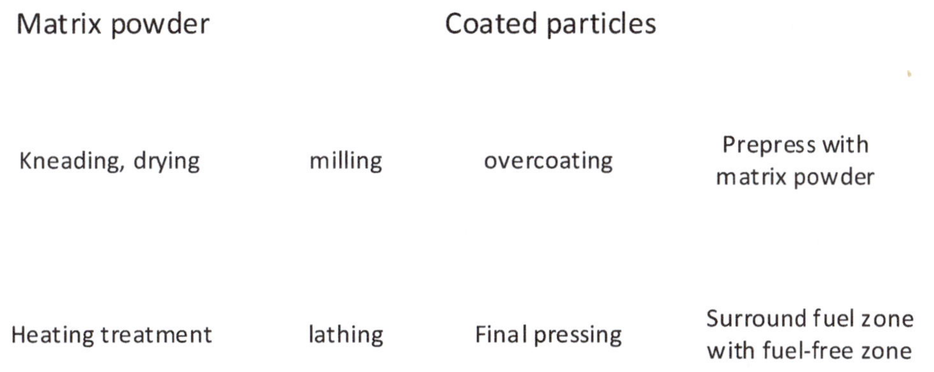

FIG. 32. Illustrative flow chart of processes involved in spherical fuel element fabrication (courtesy of Tsinghua University).

4.3.1. Preparation of resonated graphitic matrix powder

Graphitic matrix is important for the properties of spherical fuel, because more than 90% wt. of the spherical fuel is composed of the graphitic matrix, as shown in Fig. 33, which is based on HTR-10 fuel with 5 g U per pebble loading.

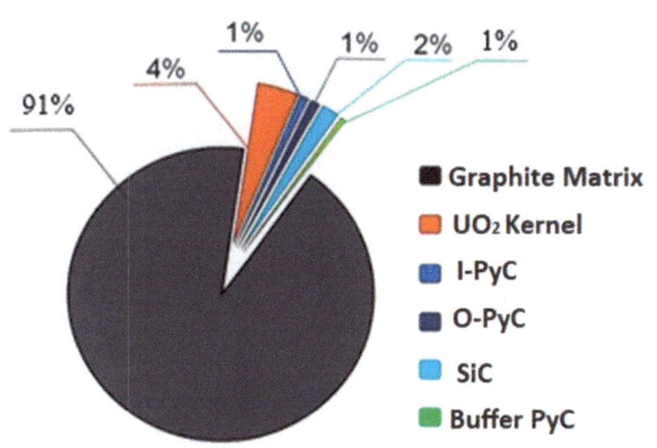

FIG. 33. The weight composition of the spherical fuel (with 5 g U loading of HTR-10 fuel) (courtesy of Tsinghua University).

Table 9 represents differences in composition and fabrication conditions employed by two types of matrix graphite, A3-3 and A3-27, developed for spherical fuel element production by NUKEM [53]. Both types of matrix graphite comprise ~64% natural graphite, ~16% electro-graphite powders, and ~20% phenol resin binder. The primary difference between both types is related to binder.

TABLE 9. COMPOSITION AND FABRICATION PARAMETERS OF GRAPHITE MATRIX MATERIALS USED IN SPHERICAL FUEL ELEMENT FABRICATION

Materials and fabrication	Standard matrix, A3-3	Matrix with synthesized resin, A3-27
Composition of raw materials:		
— Natural graphite	64 wt.%	62.4 wt.%
— Petroleum coke graphite	16 wt.%	15.6 wt.%
— Resin binder	20 wt.%	22.0 wt.%
Binder	Phenolic resin Prefabricated from phenol and formaldehyde	Synthesized resin 2 Synthesized from phenol and hexamethylenetetramine formation
Moulding method	Quasi-isostatic cold moulding	
High temperature treatment:		
— Fuel elements	1800 or 1950°C	1950°C
— Fuel-free matrix spheres	1800°C	1950°C
• For AVR fuel elements: A3-3 and A3-27 • For THTR fuel element production: A3-3.		

The manufacturing process for the graphitic matrix powder is as follows, as described in Ref. [53]:

— "Natural graphite and electro-graphite powders are mixed in a four-to-one ratio in a conical mixer. Depending on the type of matrix required, either A3-3 or A3-27, the binder materials are added in a different manner and the binder is synthesized differently:
  - Standard A3-3 matrix: a phenolic resin is dissolved in alcohol (methanol) to form the binder in a separate process step, added to the natural- and electro-graphite powders and then the mixture homogenized. This mixture is then fed into a kneading machine;
  - A3-27 matrix: all of the raw materials, including the natural- and electro-graphite powders, are warm-mixed together with the binder components—phenol and hexa methylene tetramine—at a temperature of ~130°C where the binder synthesized. This process eliminates the need for kneading and steps 3 and 4 below.

— The paste-like mixture is extruded through a punched screen creating strings that are cut into small pieces.
— These small pieces are placed in drying trays which are heated to approximately 100°C.
— The graphitic mass is then transferred into a hopper that feeds a hammer mill used to grind the material into powder of the desired grain size.
— The dried graphitic mass is transferred into a hopper that feeds a hammer mill used to grind the material into powder of the desired grain size.
— The milled powder is homogenized and ready for pressing."

The two resin binders show differences in terms of the binder type and cross-linking. As described in Ref. [53], "the phenolic resin binder used for the standard A3-3 matrix graphite is thermoplastic and the polymers are cross-linked primarily two dimensionally. In the A3-27 matrix material, the binder synthesized from phenol and hexamethylenetetramine is duopolistic and the polymers are cross-linked primarily three dimensionally. Thus, the binder cokes formed from the resin binders during the carbonization and heat treatment processes are of different structure."

The material properties of the two fuel matrix types are described in Ref. [53] and are represented in Table 10. As described in Ref. [53], "a comparison of the test data on the standard A3-3 matrix shows an improvement in the corrosion rate with the higher 1950°C heat-treatment compared to the lower 1800°C heat-treatment data. The falling strength in the drop and tumble test and the corrosion rate for the A3-27 matrix material are significantly better than those of the standard A3-3 at the higher heat-treatment temperature of 1950°C."

TABLE 10. MATERIAL PROPERTIES OF GRAPHITE MATRIX USED IN SPHERE FABRICATION

Material property	**Standard matrix A3-3**		**Matrix w. synth. Resin A3-27**
	1800°C	1950°C	1950°C
	Older AVR fuel elements	AVR fuel elements THTR prod.	AVR fuel elements
Young's modulus (kN/cm^2)			
— Parallel*	1020	1000	1070
— Perpendicular*	991	970	1020
Geometrical density (g/cm^3)	1.70	1.73	1.74
Coefficient of linear thermal expansion (10^{-6}/K), 20-500°C			
— Parallel	2.80	2.89	2.43
— Perpendicular	2.92	3.45	2.69
Quotient of coefficient of thermal expansion (ration of 'perpendicular' over 'parallel')	1.07	1.19	1.11
Thermal conductivity @ room temperature (W/cm/K)			
— Parallel	0.59	0.70	0.69
— Perpendicular	0.63	0.63	0.64
Thermal conductivity @ 1000°C (W/cm/K)			
— Parallel	0.38	0.41	0.44
— Perpendicular	0.38	0.37	0.39
Spec. electrical resistance (10-3 Ω.cm)			
— Parallel	1.56	1.46	1.43
— Perpendicular	1.60	1.48	1.48
Falling strength (number of falls to fracture), fall of a tests sphere from a height of 4 m onto A3-3 spheres	521	437	652
Corrosion rate (mg/cm^2/h), @ 1000°C in He of 1 bar with 1 vol.% H_2O (10 h)	1.19	0.97	0.73

* Parallel and perpendicular to the equatorial plane of the matrix sphere.

4.3.2. Overcoating of the TRISO particles

The overcoating of the TRISO particles takes place in a rotating drum as illustrated in Fig. 34. The overcoating is intended to prevent direct contact between particles that may cause the cracking of coating layers during sphere formation. The overcoating is of ~200 µm thickness on the rigid TRISO coated particles. It has the same composition with the graphite matrix. As described in Ref. [53], "the dry graphitic matrix material and a solvent are added simultaneously into the rotating drum in order to maximize adherence and obtain a uniform thickness. The moist overcoated particles are then dried at 80–90°C to remove any of the remaining solvent. The dried overcoated particles are sieved to select the proper sized particles within the range of 1.1 mm and 1.5 mm and are once again sorted on an inclined vibrating table to remove odd shaped, twin, or non-spherical overcoated particles."

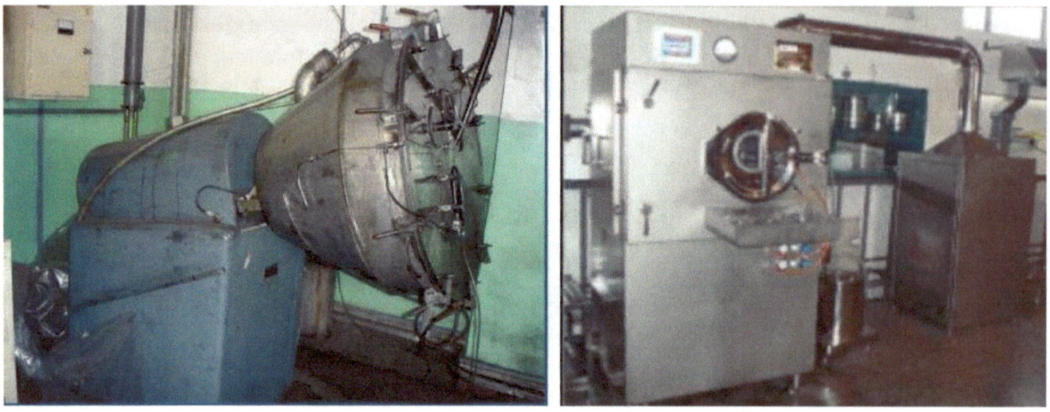

(a) (b)

FIG. 34. (a) An old, manual overcoating drum and (b) a new, automated overcoating facility located at the INET, Tsinghua university (courtesy of Tsinghua University).

4.3.3. Moulding and pressing of fuel spheres with TRISO particle

Silicon rubber moulds are used to manufacture the fuel spheres by means of quasi-isostatic pressing at room temperature. The pressing operation first includes pre-pressing operation such as taking overcoated TRISO particles together with graphite matrix powder and moulding them to form the internal fuelled spherical zone, ~50 mm diameter. Then the fuel-free shell of ~5 mm thickness is formed by adding additional matrix material around the fuelled core using a final high pressure moulding process. Figure 35 shows the sphere pressing line of the HTR-PM fuel manufacture, Baotou, China.

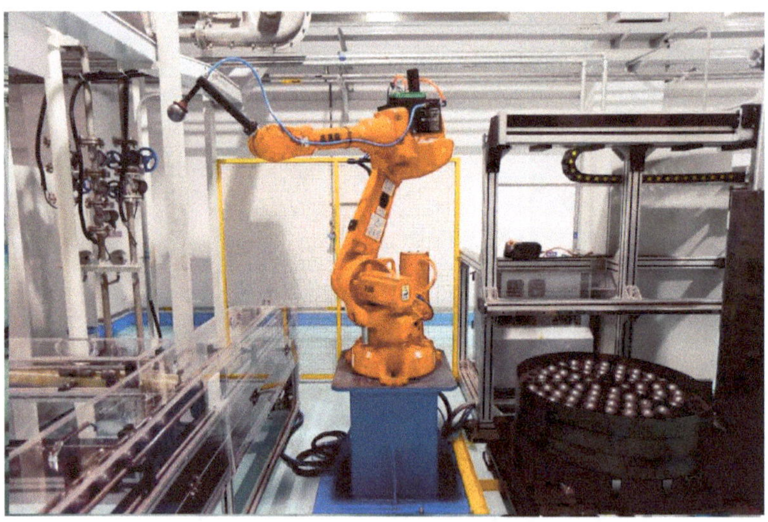

FIG. 35. Moulding and pressing lines for green fuel sphere at the HTR-PM fuel manufacture line, Baotou, China (courtesy of Tsinghua University).

The sphere moulding and pressing process consists of the following steps:

— Firstly, combine overcoated particles with matrix graphite powder and form the fuel zone. As described in Ref. [53],
  - "The matrix graphite powder volume is carefully controlled along with the overcoated particle volume and the mixture is homogenized. The homogenized mixture is injected into the pre-pressing mould and pressed at 5 MPa pressure;
  - The pressed fuel zone spheres are then transferred into the final mould. The lower half of the final mould contains matrix graphite powder. The fuel zone sphere is placed into the centre of the bottom mould and the second half of the mould is placed on top. More matrix material is added through a feeder tube to completely fill the internal annulus between fuel zone sphere and top final mould. Final pressing process is performed at 300 MPa pressure."

— Secondly, carbonization.
  - After pressing, the green fuel spheres are transported to the carbonization equipment and are heated to 800°C in an inert argon atmosphere furnace to carbonize the phenolic resin binder to provide strength. During this process, the phenolic resin pyrolyzes to carbon materials, which form the matrix strength. Volatile gases are evolved from the sphere at this stage.

— Thirdly, lathing the elements. As described in Ref. [53],
  - "After carbonization, the fuel spheres are transported to the lathing equipment, where they are machined in a two-step process to obtain uniform spheres with specified dimensions."

— At last, high temperature annealing.
  - The annealing process is conducted in a vacuum at a temperature between 1800–1950°C for one hour with the intention of eliminating residual impurities in the matrix graphite. A higher temperature (1950ºC) is preferred to improve corrosion resistance and mechanical strength. Following a cool down phase, the spherical elements are inspected.

4.4. FABRICATION OF COMPACT FUEL ELEMENTS

Fuel compacts are fabricated in cylindrical forms and loaded into the graphite blocks of prismatic HTGR. Solid fuel compacts are mostly used except for Japanese design where annular fuel compact is adopted. Fuel compacts are fabricated in a way generally similar to spherical fuel elements.

Resinated graphite matrix powder is prepared by mixing natural graphite powder, synthetic graphite powder, and resin binder. TRISO particles are overcoated with resinated graphite powder. The overcoating prevents particle-to-particle contact. The coated particles are warm pressed to form cylindrical compacts. Sometimes unfuelled end caps are introduced at both ends of a fuel compact to reduce the likelihood of damage to TRISO particles during compaction and handling. The green compacts are subjected to heat treatments. The resin is carbonized in the range of 800–950°C in an inert nitrogen atmosphere. A subsequent sintering process is made to form compacts and eliminate gaseous impurities by baking at 1650–1950°C under vacuum condition. Cylindrical compacts are typically fabricated to net shape meeting the product specification during this process and are not machined.

4.5. SUMMARY AND DISCUSSION

Technologies for fabricating coated particle fuels have been developed for around half century continued by several countries, which includes the sol-gel technology for the kernel manufacture of UO_2, UCO and other kinds of Uranium compounds, the fluidized bed chemical vapour deposition technology for the coated fuel particles, and the press technology for both sphere fuel and compact.

Adaption of new technologies, such as additive manufacturing technology, may facilitate a large-scale industrial manufacturing line.

5. QUALITY CONTROL FOR FUEL PARTICLE CHARACTERIZATION

5.1. QUALITY CONTROL METHODS

In the fuel qualification, quality characteristics are specified and then proven by examinations. The quality characteristics examined for coated particle fuel and associated testing methods are represented in Table 11, based on data from Ref. [54].

TABLE 11. CHARACTERIZATION METHODS FOR HTGR FUEL INCLUDING KERNEL, COATED PARTICLE AND THE SPHERICAL FUEL

Inspection item	Method	**Sampling rate**
Starting and raw materials		
Compression density of graphite powder	Density measurement under defined load	Powder in forging die
Rebound of graphite powder	Measurement of height difference of powder column during and after load	Powder in forging die
Spec. electric resistance of graphite powder	Measurement of voltage drop along powder column	Powder in forging die
Impurities in graphite powder	Chemical analysis after incineration, emission and absorption spectrometry, photometry, fluorimetry	Representative quantity
Impurities in uranyl nitrate solution	Chemical analysis after incineration, emission and absorption spectrometry, photometry, fluorimetry	Representative quantity
Isotope composition	Mass spectrometry with regards to ^{234}U, ^{235}U, ^{238}U	Representative quantity
Fuel kernel		
Heavy metal loading	Transfer of kernels into a stoichiometrically well-defined state and chemical analysis	
Oxygen to metal ratio (O/M)	Potential controlled coulometry	
Isotope composition	Mass spectrometry with regards to ^{234}U, ^{235}U, ^{238}U	
Carbon content	Oxidation of kernels and chemical analysis of CO_2	
Oxygen content (UCO)	Hot extraction of oxygen, transfer into CO, chemical analysis of CO, infrared spectrometry	
Dopant material content	Spectral photometry, atom absorption spectrometry	
Diameter	Optical imaging with particle size analyser, X ray micro-radiography	
Sphericity	Counting of fraction of odd-shaped particles, multiple measurement of maximum and minimum diameter, micro-radiography, stereomicroscopy	
Density	Optical particle size analyser or V-slot to measure mean diameter. Mercury pycnometer or air pycnometer to measure volume	Larger number of kernels
Structure	Measurement of reflection on defined lattice planes, X ray with Debye-Scherrer goniometer	Ground kernels
Sieve fraction	100% sieving with DIN sieves	
Weight	Weight of counted number of kernel and determine mean weight	
Impurities	Spectral photometry, atom absorption spectrometry	
Coated fuel particle		
Diameter	Optical particle size analyser	

TABLE 11. CHARACTERIZATION METHODS FOR HTGR FUEL INCLUDING KERNEL, COATED PARTICLE AND THE SPHERICAL FUEL

Layer thickness	X ray projection micro-radiography (only OPyC and SiC), X ray contact microradiography, microscopy analysis of ceramographic sections, optical particle size analyser, fluid pycnometer (in case of discontinuous coating of single layers at a time)	
Density	Weight of counted number of particles and determine mean weight	
Density of highly dense layers	Liquid density gradient column with calibration bodies; Gas pycnometer	Volume of ~50 g
OAF, BAF	Ceramographic sections exposed to polarized light. OAF is ratio of reflected light intensity vertically to deposition direction over reflected light intensity in deposition direction. Correlation between OAF and BAF.	
Growth features, size, and distribution	Etching of ceramographic sections by wet oxidation, plasma oxidation, or ion bombardment, scanning electron microscopy (SEM), transmission electron microscopy (TEM)	SEM on fractured coating, TEM on thinned coating specimens
Polygonity of layers	X ray diffraction, stereomicroscope, measurement of layer thickness in 300–900 position	
Heavy metal content	Grinding of particles and transfer into distinct compounds of U by oxidation, quantitative chemical analysis of U	
Surface contamination	Leaching of particles with HNO_3, quantitative chemical analysis of U	
Defective SiC layers	Burn-leach method	
Heavy metal migration	Micro-radiography, visual inspection of buffer layer	
Tightness of IPyC	Micro-radiography, visual inspection of buffer and IPyC layers after leaching with HNO_3 compared to before	Samples taken after IPyC coating process
Micro-porosity	Determination of fractions of layer, fibre, mosaic components in PyC by X ray small-angle diffraction	Fragments of PyC layers
Pore structure	Quantitative image analysis and determination of pore size distribution	
Ultimate tensile strength of SiC and PyC	Determination of fracture load by crushing between sapphire plates, hemispherical bursting, ring compression test	Single SiC or PyC rings prepared from layers, single SiC half shells
Micro-hardness	Vickers or Brinell hardness	
Young's modulus: PyC, SiC	Crushing between sapphire plates and recording stress-strain curve	PyC or SiC specimens
Fuel sphere		
Matrix density	Dimension and weight measurements	Matrix specimens 5 mm × 5 mm × 35 mm
Thermal expansion coefficient and anisotropy	Measurement of temperature and dimensional change with dilatometer; Anisotropy is ratio of coefficient (parallel) to coefficient (vertical)	Matrix specimens 5 mm × 5 mm × 35 mm
Dynamic elasticity modulus	Elastomat or frequency generator $E = 4 \times f^2 \times l^2 \times \rho$ where f =resonance frequency, l =length, ρ =density.	Matrix specimens 5 mm^3 × 5 mm^3 × 35 mm^3
Bending strength	3-point test on bending device $\sigma = (Fm \times l)/W$ where Fm = fracture strength, l = support span, W = resistance momentum	Matrix specimens 5 mm × 5 mm × 35 mm

TABLE 11. CHARACTERIZATION METHODS FOR HTGR FUEL INCLUDING KERNEL, COATED PARTICLE AND THE SPHERICAL FUEL

Compressive strength	σ = Fc/Q where Fc = crushing strength, Q = cross section of specimen.	Matrix specimens 5 mm × 5 mm × 35 mm
Tensile strength	σ = Ft/Q where Ft =breaking force, Q = cross section of specimen.	Matrix specimens 8 mm diameter × 30 mm.
Specific electrical resistance	R = (U×Q) / (I×l) where U = voltage drop, Q = cross section, I = electric current, l = length.	Matrix specimens 5 mm × 5 mm × 35 mm
Thermal conductivity @ room temperature	Direct measurement	Matrix specimens 5 mm × 5 mm × 35 mm
Thermal conductivity @ 40°C	Thermo-conductometer after Schröder setting a stationary temperature difference by means of the boiling temperatures of two liquids and measuring the time required for the vaporization of a certain quantity of liquid, comparison with calibration standard	Matrix specimens 5 mm × 5 mm × 35 mm
Thermal conductivity @ 1000°C	Radial flux method λ = (Q×ln(r2/r1))/(2×l×ΔT) where Q = power of central heater, r1, r2 = distances of thermocouple (TC) from specimen axis, l = active length of specimen, ΔT = temperature difference between TC Modified Kohlrausch procedure by setting an almost parabolic axial temperature profile with maximum in specimen centre and small drop to the sides (< 10°)	Matrix specimens 40 mm diameter × 25 mm Matrix specimens 6 mm diameter × 32 mm with axial bore hole of 1 mm diameter
Impurities, ash contents, B equivalent	Spectral photometer, atomic absorption spectrometry	20–50 g of matrix material
Number of falls	Falls onto pebble bed of the same spheres until fracture	Sphere
Fracture load	Direct measurement	Sphere
Corrosion velocity	Determination of mass loss after 10 h heating @ 900°C or 1000°C in flowing helium at 0.1 MPa and 1% steam: Corrosion velocity K = Δm/(F×t) where Δm = mass loss, F = sphere surface, t = time	Sphere
Fuel-free zone	Examination of particle-free shell by X ray and visual inspection	Sphere
Abrasion	Determination of mass loss in mg/h in abrasion drum	Sphere
Surface appearance	Visual inspection	Sphere
Released heavy metal (matrix contamination)	Electrolytic deconsolidation of matrix material with HNO_3 and quantitative chemical analysis of U in electrolyte and leach solution	Sphere
Defective SiC layers	Burning of spheres in muffle furnace and leaching of uranium	Sphere
Heavy metal content	Burning of spheres in muffle furnace, destruction of SiC layers	
Pore size distribution	Quantitative image analysis or determination of pore radius by mercury porosimeter: P×r = 2×s×cos q, where P = pressure, r = pore radius, s = surface tension, q = border angle between specimen and mercury	

5.1.1. Particle size measurement

The size (diameter) of the kernel and the particle is measured by using the automated optical particle size analyser (PSA). It can measure at a rate of about 50 particles per second maximum, depending on the feed rate [55]. The sphericity of the kernel and the coated particle is determined by passing through one particle in the light beam many times at randomly changed positions. The PSA can also be used in the optical image analysis of a photograph of many particles on a tray.

5.1.2. Kernel, buffer and coating layer density and thickness measurement

Density significantly affects the retention of fission gases, thermomechanical properties and irradiation behaviour in reactor. Densities of SiC and pyrolytic carbon can be measured with the sink-float method [56] by gradient density columns. In a column, two miscible liquids are mixed and a uniform density gradient is developed. Representative test samples are extracted from the coater after the layer is completed. Pieces of that layer are broken, sink and settled in the column. The column is calibrated by using standards of known density.

This method is not applicable to UO_2 kernels whose density is high and not applicable to the porous buffer layer because of the possibility of liquid infiltration. For these components, volume can be measured using the PSA or a mercury porosimeter.

The sample's mass is determined prior to its passage through the PSA. Subsequently, the cumulative volume of all kernels within the sample is divided by the sample mass to derive the mean 'apparent' density of the kernels. Following a similar methodology, the density of the buffer layer is determined, taking into account the subtraction of the mean kernel volume from the mean total volume and employing the corresponding mass values. To ascertain the real or 'theoretical' density, assuming a specimen devoid of internal voids, the specimen is crushed into powder using particles of approximately 1 µm in size. A pycnometer is then utilized to measure the volume of the powder. The total porosity is subsequently calculated as 1 minus the ratio of apparent density to theoretical density.

Mercury porosimetry involves surrounding each kernel or buffer-coated particle in a sample with mercury within a container of known volume such that the weight of mercury displaced by the sample defines the envelope volume of the sample. This provides a more accurate measurement of volume compared with the PSA method, especially in the case of non-spherical particles.

A technique for measuring layer thickness involves X-ray micro-radiography, providing robust statistical data on the inherent variation in layer thickness across a substantial number of particles (100–200). In this method, a monolayer of particles is placed directly on a high-resolution photographic film and subjected to X-ray exposure. The analysis of layer thickness on the film is performed using a transmission light microscope and standard image processing software. Alternatively, another approach involves optical image analysis of a monolayer of particles that has been polished to the midplane.

5.1.3. Anisotropy measurement

High density pyrocarbon is a polycrystalline graphitic material and therefore has anisotropic material properties, such as thermal expansion and fast neutron induced shrinkage. The material should be as isotropic as possible to allow uniform heat conduction and minimize dimensional changes under irradiation.

The BAF, outlined in Equations (3) and (4) in section 2.3.3, serves as a direct measure of the anisotropy in macroscopic material. This factor is primarily defined by the coefficients of thermal expansion at 400°C in parallel and perpendicular directions to the preferred orientation. However, obtaining BAF measurements on coated fuel particles is challenging. To overcome this, it becomes necessary to deposit pyrolytic carbon on graphite, producing small disk-type specimens.

Given the non-uniform structure of pyrolytic carbon on spherical particles, an optical method leveraging the bi-reflection of crystalline graphite can be employed. This involves directing a beam of linear polarized light through the equatorial area of the coated particle onto a ceramographic cut.

The procedures for the fabrication and quality control of spherical fuel elements based on the experience of the Institute of Nuclear and New Energy Technologies, Tsinghua University, Beijing, China are shown in Fig. 36 [57].

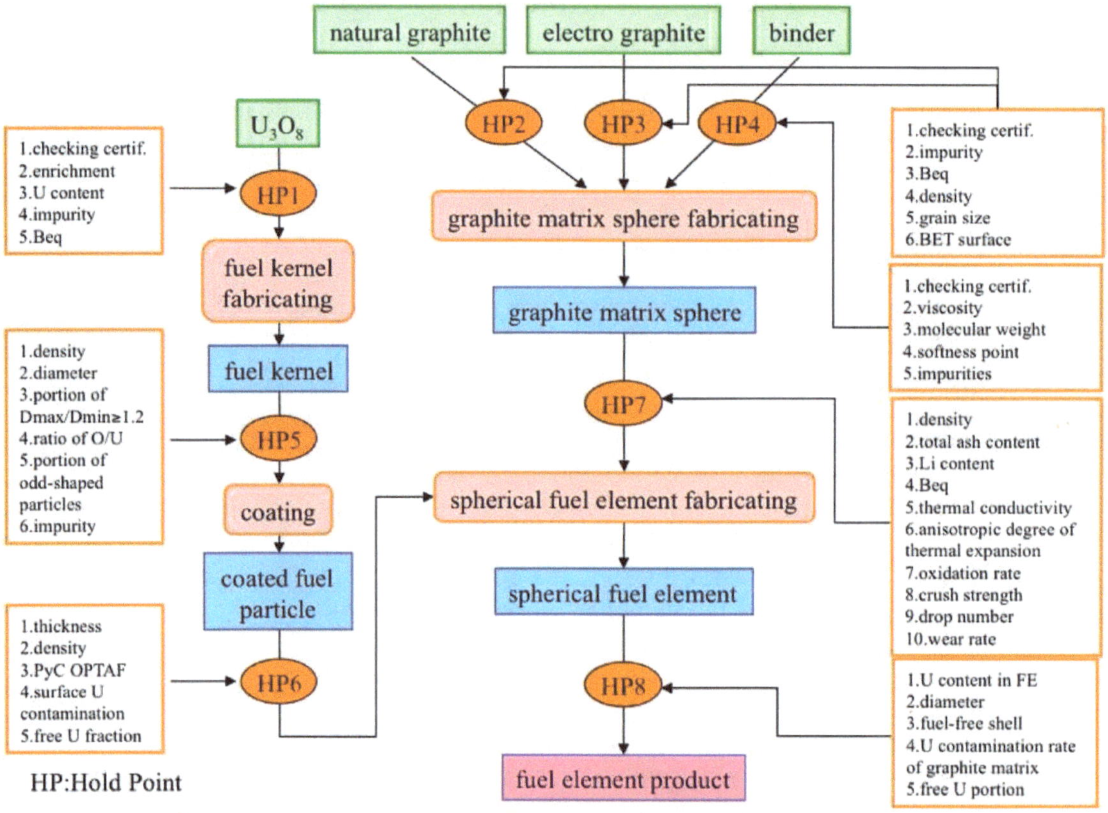

FIG. 36. Fabrication process and quality control diagram for modern TRISO fuel; FE = fuel element, Beq = activity, OPTAF = Optical Anisotropy Factor (courtesy of Tsinghua University).

5.1.4. Phase contrast radiography

Phase contrast radiography is an emerging non-destructive testing method where X-ray radiation from a monochromatic synchrotron or a microfocus source is passed through the TRISO coated microsphere for obtaining the image of various coating layers. This technique is best suited for the detection of boundaries between various coating layers. The boundaries between buffer layer, IPyC layer, SiC layer and the OPyC layer can be clearly distinguished by employing this technique. Using appropriate image analysis software like contrast enhancement or edge enhancement, the thickness of the various coatings can also be measured with good accuracy. Additionally, the uniformity of coating and the shape of the coated layers can also be evaluated.

Phase contrast radiography has been developed at KAERI where they have acquired images with very good contrast using ZrO_2 simulated TRISO particles with a microfocus X-ray source and digital imaging system [58]. From the acquired images the coating thickness was non-destructively measured using proper calibration. AREVA has developed an X-ray technique that enables the non-destructive control of both the geometry and density of coating layers, replacing the previously employed destructive methods [59,60]. This technique has been explored in the Bhabha Atomic Research Centre (BARC), India to study the pyrocarbon coating on surrogate ZrO_2 kernels as part of the Compact High Temperature Reactor programme in India [61]. This information thus obtained may be used for optimization of the various parameters of the deposition process during TRISO coating development. The phase contract radiography technique using thermal neutrons instead of X-rays is also being explored at BARC. The interference between neutrons that have been scattered and those transmitted without scatter has been employed to get a contrasting image on neutron detectors. At thermal energies, phase shifts of 2π can occur after passing through only a few tenths of μm for many materials. This work is in the preliminary stage at present.

Such new promising techniques for the time being developed only at laboratory level and is required to be industrialized. Though this technique is non-destructive, non-contaminating and yielded comparable results to those obtained with other methods like ceramography, it may find difficult to integrate into a production line. However, this technique may be useful for benchmarking the coating thickness measurement during the coating process development stage.

5.1.5. Burn-leach testing

A crucial technique for quality assurance is burn-leach testing, particularly in the examination of coated particle fuel for HTGRs. In a 'burn-and-leach' test, the graphite within the sample (which may be loose coated particles, fuel spheres, fuel compacts, or coupons) undergoes combustion in a chamber at 800–850°C in air until reaching the SiC layer, maintaining a constant weight (typically around 90 hours for a fuel sphere). The remaining ash and particles are then treated with a nitric acid solution at ~90°C, and the dissolved uranium content is analysed.

Due to the corrosion-resistant nature of the SiC layer, the uranium detected in the solution comprises both the natural uranium content of the matrix material and the uranium content of particles with a defective SiC layer. Additionally, particles with incomplete coating can be identified. Test results are presented as the ratio of measured free uranium to the inserted uranium, denoted as U_{free}/U_{total}. The detection limit typically reaches a level of $(1–3)\times10^{-6}$, depending on the uranium content of the sample. This is significantly lower than the uranium content of a single defective coated particle. In a fuel element containing approximately 1 g of ^{235}U per sphere and 0.07 mg per particle, the content of one defective particle corresponds to a U_{free}/U_{total} value of 7×10^{-5}.

5.1.6. Fuel elements

Sphere fuel (pebble) will be recycling in the pebble-bed reactor, which has some additional requirements in some properties, such as the fall test, the abrasion test and the free fuel zone examination.

The free-fall drop test is a special requirement for the pebble fuel. During the fall test, the sphere fuel usually represented by graphite ball which was fabricated used the same materials and the same process, the graphite ball will be dropped from 4 m height onto the pebble bed composed of the same spheres. Usually, the test cycle will continue more than 50 times for each graphite ball, or until the graphite ball fractures. After fall test, every graphite ball should be checked, and also the graphite balls in the pebble bed.

Abrasion is another special requirement for the sphere fuel, due to the sphere fuel movement in the reactor. The abrasion test is conducted in the special designed abrasion drum, usually the fuel is represented by the graphite ball. The abrasion drum is designed to simulate the sphere fuel move path, both abrasion between fuels in the reactor, and the abrasion between the fuels and the metal pipes, which used to upload and download the fuels out of the reactor. The graphite balls were put into the drum, and the drum rotate at a designed speed. After 100 h, the graphite balls will be checked. And the total ash will be collected and weighted, the mass loss is the determination of the abrasion properties. The detection mass loss limit is typically at a level of 6.0 mg/h per ball, but usually the test results is less than 2.0 mg/h per ball.

As mentioned above, the sphere fuel moves in the reactor from the top to the bottom and recycle in the handing pipe out of the reactor. The fuel free zone of the sphere fuel is designed for the protection of the coated particles in the fuel zone when the sphere fuel cycles. The free fuel zone is examined by X-ray method while the sphere fuel makes a complete rotation to find the coated particles out of the fuel zone. The free fuel zone is designed with a 5 mm thickness. All the sphere fuels should be examined after manufactured.

5.2. QUALITY CONTROL RESULTS

There exists a well-defined set of criteria for contemporary high-quality nuclear fuel, encompassing stringent standards for minimal levels of heavy metal contamination, prevention of manufacture-induced particle defects during the fuel body and fuel element production, mitigation of irradiation/accident-induced particle failures, and restrictions on fission product release from intact particles [62,63].

While the design of HTGRs remains flexible, incorporating configurations for prismatic and spherical fuel elements, there is a global consensus on the characteristics of high-quality fuel. This typically involves a 500 µm diameter UO_2 kernel enveloped by a substantial sacrificial buffer layer, succeeded by a dense inner pyrocarbon layer, a high-quality silicon carbide layer with close to theoretical density, and another dense outer pyrocarbon layer.

The development and manufacturing details of UO_2 TRISO fuel elements in Germany are described in [64] and are represented in Tables 12 and 13. Good performance has been demonstrated both under operational conditions to 12% FIMA and more and under accident conditions to a maximum 1600°C.

TABLE 12. DEVELOPMENT OF THE LOW ENRICHE UO_2 TRISO FUEL ELEMENTS IN GERMANY

Characteristic	**Before 1985 production**			**After 1985 production**	
Year of manufacture	1981-83	1981	1983	1985–87	1988
Designation	AVR 19	LEU Phase 1	AVR 21-1	AVR 21-2	Proof Test Phase 2
Matrix material	A3-27	A3-27	A3-27	A3-3	A3-3
Irradiation tests	GLE 3	HFR-K3 FRJ2-K13 HFR-P4 SL-P1 FRJ2-P27 FRJ2-P28	GLE 4 FRJ2-K15	GLE 4/2 HFR-EU1 HFR-EU1bis	HFR-K5 HFR-K6
No. of spheres manufactured	24 600	100	20 500	14 000	200
Number of fuel element lots (1)	14	-	11	8	-
Number of coating batches	65	1	54	29	8
Coating batch size (kg)	5	5	3	3	5
Coated particle batch designation	HT232-245	EUO 2308	HT354-383	HT385-393 HT395-404 HT406-423	EUO 2358-2365
Number of kernel sets	5	1	2	5	1

TABLE 13. MANUFACTURING DETAILS OF LEU UO_2 TRISO FUEL IN GERMANY

Characteristic	**Before 1985 production**			**After 1985 production**		**Typical modern specs**
Designation	**AVR 19**	**LEU Phase 1**	**AVR 21**	**AVR 21-2**	**Proof test Phase 2**	
Kernel diameter (μm)	500	497	501	502	508	500
Kernel density (g/cm³)	10.80	10.81	10.85	10.86	10.72	> 10.4
Coating thickness (μm)						
- Buffer	93	94	92	94	102	95
- IPyC	38	41	38	41	39	40
- SiC	35	36	33	36	36	35
- OPyC	40	40	41	40	38	40
Coating density (g/cm³)						
- Buffer	1.01	1.00	1.01	1.0	1.02	< 1.05
- IPyC	1.86	~1.9	1.9	1.87	1.92	1.9
- SiC	3.19	3.20	3.20	3.20	3.20	≥ 3.18
- OPyC	1.89	1.88	1.88	1.87	1.92	1.9
Loading						
Heavy metal (g/fuel element)	10	10	6	6	9.4	9
^{235}U (g/fuel element)	1	1	1	1	1	0.86
Enrichment (%^{235}U)	9.82	9.82	16.76	16.76/ 16.67	10.6	9.6
Number of particles/fuel element	16 400	16 400	9560	9560	14 580	14 440
Free uranium fraction (10^{-6})	50.7	35	43.2	7.8	13.5	< 60

5.2.1. UO_2 kernel results

The fuel kernel, in a spherical form, comprises stoichiometric uranium dioxide (UO_2). This kernel functions as a partial barrier, effectively immobilizing a significant portion of fission products and slowing down the diffusive release of others. Consequently, this mechanism significantly diminishes the release of radionuclides from the particle, as a considerable portion undergoes radioactive decay before being released from the kernel.

Table 14 represents the quality control results from the latest large scale AVR fuel manufacture type GLE-4/2 (reload 21-2) from December 1985 to August 1987 [65,66].

TABLE 14. SIMPLIFIED QUALITY CONTROL DATA FROM KERNEL MANUFACTURE OF THE AVR 21-2 FUEL WITH ~ 17% ENRICHMENT

Property	Test frequency	Specification	Lot number				
			#4	#5	#6	#7	#8
Sieve fraction (μm)	100%	450–560					
Kernel diameter (μm)	1000 kernels per lot	(480 ≤ d ≤ 520) 90	502 ± 1	500 ± 1	501 ± 1	501 ± 1	507 ± 1
		(d > 600) 95/95	$T_U^{(1)}$ = 520	T_U = 519	T_U = 520	T_U = 521	T_U = 522
Sphericity ≤ 1.2	800 kernels	≤ 11	0	1	0	0	0
Sphericity ≤ 1.5	800 kernels	≤ 3	0	0	0	0	1
Other shapes	800 kernels	≤ 3	0	0	1	1	1
Density (g/cm^3)	1 determination per lot	mean ≥ 10.4	10.81	10.85	10.90	10.89	10.89
Stoichiometry	1 determination per lot	O/U ≤ 2.015	2.00	2.00	2.01	2.00	2.01
Equivalent boron content (ppm)	1 sphere per lot	≤ 2	0.64	0.54	0.60	0.54	0.59

(1) T_U = one-sided 95% confidence/ 95% coverage upper limit.

In manufacturing processes, there are two types of tests and specifications:

- Comprehensive non-destructive inspection tests are conducted on 100% of the fuel kernels. Techniques such as sieving and other methods are employed to ensure that diameters fall within the range of 450 to 560 μm;
- Sampling tests are also implemented, where a portion of the product undergoes testing. All destructive tests fall under the category of sampling, representing a subset of the overall production.

For sampling tests, sampling statistics has been employed to deal with variations of properties. For the kernel diameter, the one-sided 95% confidence/ 95% coverage upper limit T_U is correlated with mean value m and standard deviation , as it follows:

$$T_U = m + K_U \sigma \qquad (6)$$

where K_U is displayed as function of sample size n (see Fig. 37); K_U serves as the multiplying factor applied to the standard deviation to define an upper tolerance limit in production; the 'K_U 95/95' values are calculated using the non-central t-distribution, as detailed in Ref. [67].

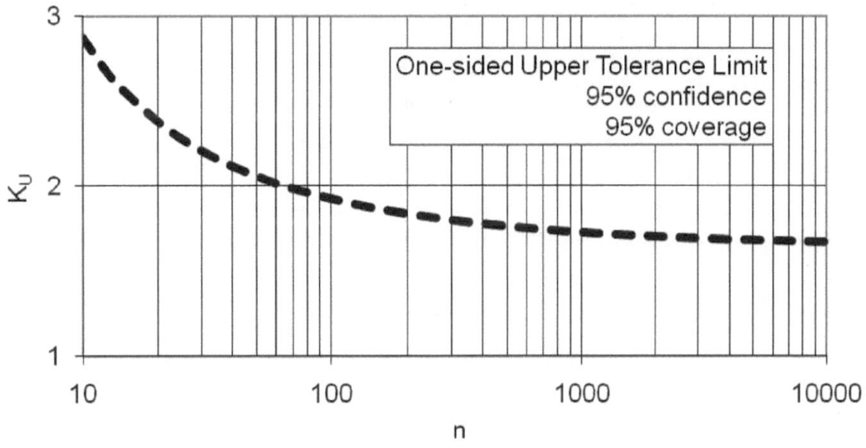

FIG. 37. Factor K_U as a function of sample size n for covering a one-sided upper 95% of the cases with 95% confidence (e.g. K_U = 1.727 for n=1000).

This correlation allows for the derivation of standard deviations for kernel diameters, a crucial parameter in model calculations predicting the irradiation performance of coated particles. In the case of the AVR 21-2 fuel, the model would incorporate kernels with a diameter of 502.2 ± 10.6 µm, taking into account the specified standard deviation. This information is vital for accurate and comprehensive assessments of irradiation behaviour based on the specified kernel size and its associated variability.

5.2.2. Coated particle results

The first layer in contact with the kernel is the buffer layer deposited from C_2H_2. Temperature and other conditions in the fluidized bed are adjusted to produce a porous layer with approximately 50% of theoretical density 2.2 g/cm^3.

The buffer layer serves a dual purpose in coated particles. Firstly, it provides void volume for gaseous fission products, preventing undue pressure build-up within the particle. Additionally, it mechanically decouples the kernel from the inner pyrocarbon layer, accommodating kernel swelling and minimizing stress accumulation in the outer coating layers during irradiation. The buffer layer also acts as a shield, absorbing energetic fission products recoiling from the kernel surface and protecting the inner pyrocarbon layers.

The inner layer of high-density, isotropic IPyC is deposited from a mixture of C_3H_6 and C_2H_2, boasting an average density of around 1.9 g/cm^3. As the first load-bearing barrier, the IPyC layer mitigates pressure exerted by fission products within the fuel kernel and buffer layer, thus reducing the stress on SiC. During irradiation, the IPyC and OPyC layers initially shrink, potentially expanding again with sufficiently high fast neutron dose levels. The interaction between the high-density pyrocarbon layers and the SiC layer is crucial for maintaining compression on the SiC layer during irradiation.

While an intact IPyC layer forms an effective barrier against fission gases and iodine, its permeability increases at higher temperatures, allowing passage of palladium, caesium, silver, and strontium. SiC, deposited from methyl trichlorosilane $Si(CH_3)Cl_3$ under specific conditions, attains a density of 3.20 g/cm^3, slightly below the theoretical density of β SiC with a 3C structure (3.216 g/cm^3).

At elevated temperatures, the ability of the IPyC and OPyC layers to contain caesium, silver, and strontium diminishes. The SiC layer plays a critical role in preventing the release of these metallic fission products into the graphite matrix and subsequently into the reactor helium stream. Thus, the SiC layer acts as the primary retention barrier for metallic fission products in the coated particle.

With a numerical property like the coating layer thickness with mean m and the standard deviation σ, the upper tolerance limit T_u is given by:

$$T_u = m + k\sigma \quad (7)$$

The lower tolerance limit T_l is given by:

$$T_l = m - k\sigma \quad (8)$$

where:

$$k = \sqrt{\frac{(n-1)\left(1+\frac{1}{n}\right)z^2_{(1-p)/2}}{\chi^2_{\gamma,n-1}}} \quad (9)$$

$\chi^2_{\gamma,n-1}$ represents the critical value of the χ^2 distribution with $(n-1)$ degrees of freedom, indicating the threshold beyond which this distribution is exceeded with a probability of γ; and $z_{(1-p)/2}$ represents the critical value of the standard normal distribution that is exceeded with a probability of $(1-p)/2$. This value is often utilized in statistical analyses to establish confidence intervals or significance levels. The k values data as a function of sample size are shown in Fig. 38.

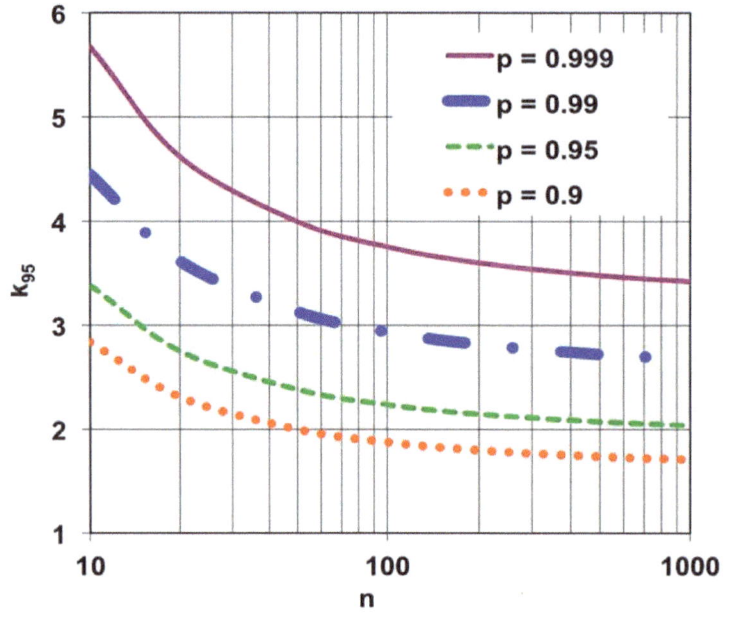

FIG. 38. K-value for two-sided tolerance limits in the case of a two-sided 95% confidence for 99.9%, 99%, 95% and 90% coverage (or proportion) p as a function of sample size n (courtesy of Tsinghua University).

From this, the necessary particle data with standard deviation as required for modelling purposes are derived. A summary is given in Ref. [68]. These data are used in mechanical performance benchmark calculation of the IAEA Coordinated Research Programme #6 on Advances in HTGR Fuel Technology [5]. Table 15 represents these data.

TABLE 15. GERMAN LEU PARTICLE DATA DERIVED FROM MANUFACTURER'S QUALITY CONTROL DATA

Test	HFR-K3 with fuel from AVR GLE3	FRJ2-K15 with fuel from AVR GLE4	HFR-EU1 with fuel from AVR GLE4/2
Coating batch number	EUO 2308	HT 354-383	HT 385-393 HT 395-404 HT 406-423
Kernel composition	LEU UO_2	LEU UO_2	LEU UO_2
Enrichment (U-235 wt.%)	9.82	16.76	16.76/16.67
Kernel diameter (μm)	497 ± 14.1	501 ± 10.8	502.2 ± 10.6
Buffer layer thickness (μm)	94 ±10.3	92 ± 14.3	94.3 ± 13.0
IPyC layer thickness (μm)	41 ± 4.0	38 ± 3.4	40.6 ± 3.7
SiC layer thickness (μm)	36 ± 1.7	33 ± 1.9	35.9 ± 2.2
OPyC layer thickness (μm)	40 ± 2.2	41 ± 3.8	39.8 ± 3.3
Kernel density (g/cm^3)	10.81	10.85	10.86
Buffer density (g/cm^3)	1.00	1.013	1.012
IPyC density (g/cm^3)	1.9	1.9	1.87
SiC density (g/cm^3)	3.20	3.20	3.20
OPyC density (g/cm^3)	1.88	1.88	1.87
IPyC anisotropy BAF	1.053	1.029	1.02
OPyC anisotropy BAF	1.019	1.020	1.02

5.2.3. Fuel elements

In the pre-1985 category, particle overcoating was carried out manually, while in the post-1985 category, an automated process was implemented using a specifically designed mixer operated by a robot. This transition to automation significantly enhanced fuel quality, as illustrated in Table 16 (from [69]).

To create a fuel sphere, overcoated particles, along with matrix material, are pressed into a 50 mm diameter inner sphere, accommodating between 9000 and 17,000 coated particles. Subsequently, this inner fuel sphere is encased within a protective layer of A3 matrix material, measuring 5 mm in thickness. The outer layer is formed through an isostatic pressing process and precision machining to achieve the final dimensions.

TABLE 16. GERMAN AVR 21-2 (GLE-4/2) BURN-LEACH DATA

Number of particle defects in sphere (see Note)	Lot number							
	#1	#2	#3	#4	#5	#6	#7	#8
#1	0	0	0	0	0	0	0	0
#2	0	0	0	1	0	0	0	0
#3	0	0	0	0	0	0	0	0
#4	0	0	0	0	0	2	0	0
#5	0	0	0	0	0	0	0	0
Total number of defective particles in 5 spheres	0	0	0	1	0	2	0	0
Defect fraction (expected level) = $\frac{1\times1+1\times2+0\times3+\cdots}{5\times8\times9560}=\frac{3}{382400}=7.8\times10^{-6}$								
Defect fraction (upper 95% conf. limit) = 2.0×10^{-5}								

Note: Evaluation of defect fraction (mean= expected and one-sided upper 95% confidence limit).

Only 8,740 fuel elements from the 14,000 spheres of the GLE-4/2 production were inserted as reload AVR 21-2 in October 1987, because no further fuel was needed due to AVR shutdown on 31 December 1988. Some of the remaining elements underwent further investigations including the insertion of five spheres into the HFR-EU1bis irradiation test and three spheres into the HFR-EU1 irradiation test [69].

Results shown in Table 17 represent a record low level in terms of particle defects during HTGR fuel manufacture. This is due to improved tabling of kernels, coated particles and overcoated particles in the selection of spherical shapes and also due to advances in the design and operation of the overcoating drum in producing a perfected product for the cold isostatic pressing of spheres. Selected data from overcoating runs are shown in Table 17.

TABLE 17. SELECTED RESULTS FROM AVR 21-2 OVERCOATING RUNS

Overcoating charge	**TU208**	**TU209**	**TU210**	**TU212**	**TU213**	**TU214**	**Mean**
Diameter of overcoated particles (μm)	1327	1306	1279	1319	1308	1332	1312
Diameter of particles (μm) after removal of overcoating	917	915	913	915	917	916	916
Overcoating density (g/cm³)	1.60	1.59	1.61	1.57	1.59	Not measured	
Overcoating thickness (μm)	205	195.5	183	202	195.5	208	198
Observations in 1500 overcoated particles:							
Missing overcoating	0	0	0	0	0	0	
Partial overcoating	0	0	0	0	0	0	
Overcoating < 50 μm	0	0	0	0	0	0	
Twins	0	0	0	0	0	0	
Overcoated graphite lumps	0	1	0	0	0	1	
Odd shapes	0	0	0	0	0	0	

The compact fuel is designed and used in the HTTR in Japan, and in the NGNP in US. In the production of HTTR fuel, a total of 4770 fuel rods were manufactured using the fuel kernel, coated fuel particle, and fuel compaction processes. Subsequently, in December 1997, these fuel rods were assembled into 150 fuel elements within the reactor building. NFI developed the measurement equipment for the coated layer thickness of a coated fuel particle and the uranium content of a fuel compact. The uranium content examination method is leach-burn-leach, which is similar for both sphere fuel and compact fuel. The leach-burn-leach process is described in Section 5.1.5. The initial loading fuel met the design specifications completely. The fuel compacts exhibited an average bare uranium fraction of 2×10^{-6} and a SiC defective fraction of 8×10^{-5}, both in accordance with the specified criteria [67].

5.3. SUMMARY AND DISCUSSION

The quality assurance/quality control system for the TRISO coated particle fuel was also established based on the examination methods. The industrial scale manufacture technology established in some countries can provide the high quality TRISO coated fuel particle needed for SMRs deployment using this type of fuel.

6. IRRADIATION TESTS, HEATING TESTS AND POST IRRADIATION EXAMINATIONS

A number of irradiation tests, post-irradiation annealing tests at elevated temperatures and follow-up PIEs have been conducted on coated particle fuels in several countries. These irradiation tests were done in operating HTGRs or in various material test reactors where gas loop techniques enable to detect failed coated particles. Typical irradiation conditions were of normal operation in a wide range of burnup, temperature and fast fluence.

6.1. TESTS CONDUCTED BEFORE THE EARLY 2000S

6.1.1. Spherical pebbles

6.1.1.1. Irradiation tests

Table 18 shows irradiation tests performed with spherical pebbles up to the early 2000s. These irradiations were found from various sources such as [3–5,70–74]. The information includes fuel types, irradiation time in full power days (FPDs), fast fluence, burnup, operating temperature, and the fractional release of ^{85m}Kr (i.e., the ratio of release (R) to birth (B) of ^{85m}Kr). All test elements in Table 18 remained intact as described in Ref. [3], which reveals that spherical pebbles can withstand normal operating conditions in the reactor that covers burnup to 14% FIMA and higher operating temperatures than 1000°C.

6.1.1.2. Heating tests

Heating tests on irradiated coated particle fuel at elevated temperature are performed to simulate accident conditions of HTGRs. For a small HTGR, the maximum temperature during a LOCA with complete depressurization is estimated to slightly exceed 1600°C for around 30 h, as shown in Fig. 39 [3].

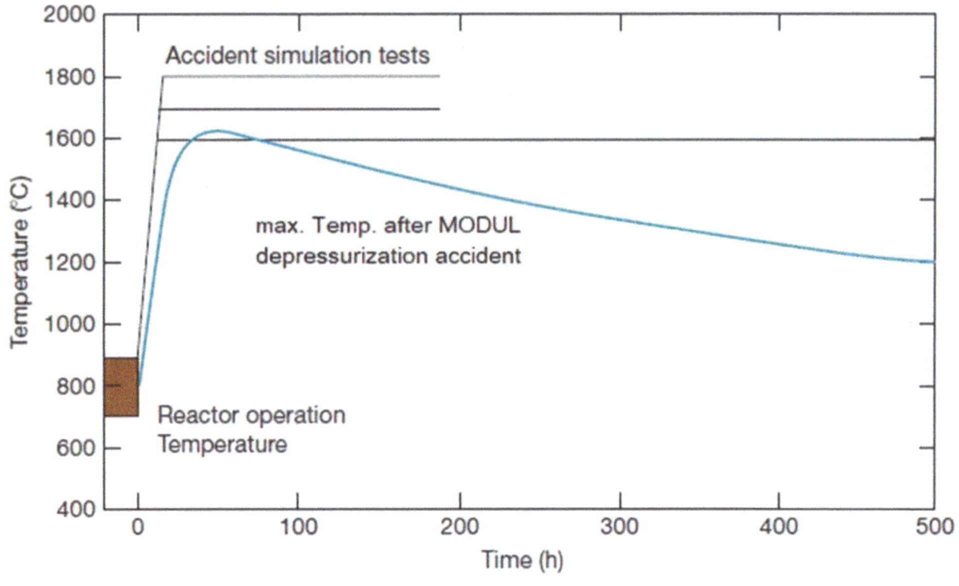

FIG. 39. Temperature evolution during a loss-of-coolant accident in a small HTR, and in heating tests to simulate the accident (taken from Ref. [3], courtesy of IAEA).

TABLE 18. IRRADIATIONS OF SPHERICAL PEBBLES CONDUCTED BEFORE THE EARLY 2000s

Test reactor	Experiment ID	No. of specimen and fuel type (mm)	Country	Irradiation time (FPDs)	Fast fluence (10^{21}n/cm^2, E>0.10 MeV)	Burnup (% FIMA)	Operating temp. (°C)	Defective SiC layer fraction (U/U-total)	BOL ^{85m}Kr *R/B* EOL ^{85m}Kr *R/B*	Reference
R2 (Studsvik, Sweden)	R2-K12	4 spheres (dia. 59.9, fuel zone 47) HEU (Th,U)O_2 TRISO	Germany	28 Nov. 1978 – 12 Feb. 1980 (308)	5.6, 6.9	11.1, 12.4	Centre; 1100, 1280 Surface; 950, 1120	$< 1\times10^{-5}$	3.9×10^{-9}, 4.6×10^{-9} 3.0×10^{-7}, 2.0×10^{-7}	[70]
R2 (Studsvik, Sweden)	R2-K13	2 spheres (dia. 59.77, fuel zone 47) HEU (Th,U)O_2 TRISO	Germany (12 compacts, USA)	22 Apr. 1980 – 19 Sep. 1982 (517)	8.5, 6.8	10.2, 9.8	Centre; 1170, 980 Surface; 960, 750	$< 5\times10^{-6}$	2.2×10^{-9}, 1.5×10^{-9} 7.0×10^{-8}, 5.0×10^{-8}	[3,70]
BR2 (Mol, Belgium)	BR2-P25	12 small spheres (spherical fuel zone 20 mm in graphite compact; dia. 26.58-27.74, height 29.87-30.03) HEU (Th,U)O_2 TRISO	Germany	30 Oct. 1978 – 19 Sep. 1981 (350)	62-8.1	13.9-15.6	Centre; 1070 Surface; 1010	$< 1\times10^{-5}$	2×10^{-7} 1×10^{-6}	[70]
HFR (Petten, Netherlands)	HFR-P4	12 small spheres (spherical fuel zone 20 mm in graphite compact; dia. 23-29, height 32) LEU UO_2 TRISO	Germany	10 Jun. 1982 – 28 Nov. 1983 (350)	8.0 (peak)	14.7, 14.9, 14.0 (max.)	940, 945, 1075 (max) 915, 920, 1050 (min.)	$< 1\times10^{-6}$	3.5×10^{-9},-, 3.6×10^{-8} 8×10^{-9}, 8×10^{-9}, 8×10^{-9}	[3,5,70,74]
Siloë (Grenoble, France)	SL-P1	12 small spheres (spherical fuel zone 20 mm in graphite compact; dia. 30.1, height 30.8) LEU UO_2 TRISO	Germany	24 Jun. 1982 – 23 Dec. 1983 (330)	5.0-6.8	8.6-11.3	743-794 (compact mean temp.)	$< 1\times10^{-6}$	5.8×10^{-7} 1.2×10^{-6}	[3,5,70]

TABLE 18. IRRADIATIONS OF SPHERICAL PEBBLES CONDUCTED BEFORE THE EARLY 2000s (CONT)

Test reactor	Experiment ID	No. of specimen and fuel type (mm)	Country	Irradiation time (FPDs)	Fast fluence (10^{21}n/cm^2, E>0.10 MeV)	Burnup (% FIMA)	Operating temp. (°C)	Defective SiC layer fraction (U/U-total)	BOL ^{85m}Kr *R/B* EOL ^{85m}Kr *R/B*	Reference
HFR (Petten, Netherlands)	HFR-K3	4 spheres (dia. 59.89, fuel zone 47) LEU UO_2 TRISO	Germany	15 Apr. 1982 – 5 Dec. 1983 (359)	4.0, 5.8, 5.9, 4.9	7.5, 10.0, 10.6, 9.0	Centre; 1200, 920, 920, 1220 Surface; 1020, 700, 700, 1020	< 4×10^{-5}	1×10^{-9}, 9×10^{-10}, 9×10^{-10}, 2×10^{-9} 2×10^{-7}, 1×10^{-7}, 1×10^{-7}, 3×10^{-7}	[3,5,70,71]
HFR (Petten, Netherlands)	HFR-K4	2 spheres LEU UO_2 TRISO	Germany	- (667)	10.0	13	Centre; 1250	-	-	[70,71]
DIDO (Jülich, German)	FRJ2-K13	4 spheres (dia. 59.89, fuel zone 47) LEU UO_2 TRISO	Germany	24 Jun. 1982 – 12 Dec. 1984 (396)	0.2, 0.2, 0.2, 0.2	7.5, 8.0, 7.9, 7.6	Centre; 1125, 1150, 1150, 1120 Surface; 985, 990, 990, 980	4×10^{-5}	2×10^{-9}, 2×10^{-10}, 8×10^{-10}, 8×10^{-10} 2×10^{-8}, 2×10^{-8}, 7×10^{-9}, 7×10^{-9}	[3,5,70]
DIDO (Jülich, German)	FRJ2-K15	3 spheres (dia. 60.04, fuel zone 47) LEU UO_2 TRISO	Germany	4 Sep. 1986 – 20 May 1990 (590)	0.2, 0.2, 0.1	14.1, 15.3, 14.7	Centre; 970, 1150, 990 Surface; 800, 980, 800	< 5×10^{-5}	2×10^{-10},2.47×10^{-10}, 2.0×10^{-10} 1×10^{-8}, 5×10^{-9}, 3.0×10^{-9}	[3,5,70]
HFR (Petten, Netherlands)	HFR-K6	4 spheres (dia. 60, fuel zone 47) LEU UO_2 TRISO	Germany	21 Jun. 1990 – 8 April 1993 (634)	3.2, <4.8, 4.8, <4.8	7.2, 9.3, 9.7, 9.2	Centre; 800 (1/3 of a cycle), 1000 (2/3 of a cycle)	-	EOL $\leq 3\times10^{-7}$	[5,70]
HFR (Petten, Netherlands)	HFR-K5	4 spheres (dia. 60, fuel zone 47) LEU UO_2 TRISO	Germany	8 Jan. 1991 – 16 May 1994 (564)	2.9, <4.3, 4.3, <4.3	6.7, 8.8, 9.1, 8.7	Centre; 800 (1/3 of a cycle), 1000 (2/3 of a cycle)	-	EOL $\leq 3\times10^{-7}$	[5,70]
IVV-2M (Zarechny, Russia)	Vostok-1~6 (6; long term irrad.)	4 spheres/each LEU UO_2 TRISO	Russia	1982-1989 (208-283)	0.4-2.2 (E>0.1 MeV)	8.8-21.5	900-1400; irradiation temp. (1600 shortly)	-	at 1000°C $\leq 1\times10^{-6}$ at 1200°C 1×10^{-5} at 1400°C $\leq 1\times10^{-5}$	[3,5]
MR (Moscow, Russia)	PG-100 (KVG-1~-5)	Spheres	Russia	56-637	0.075-2.3 (E>0.2 MeV)	1.5-18	440-1350; irradiation temp.	-	-	[3]

TABLE 18. IRRADIATIONS OF SPHERICAL PEBBLES CONDUCTED BEFORE THE EARLY 2000s (CONT)

Test reactor	Experiment ID	No. of specimen and fuel type (mm)	Country	Irradiation time (FPDs)	Fast fluence (10^{21}n/cm^2, E>0.10 MeV)	Burnup (% FIMA)	Operating temp. (°C)	Defective SiC layer fraction (U/U-total)	BOL ^{85m}Kr *R/B* EOL ^{85m}Kr *R/B*	Reference
VVR-C (Obnisk, Russia)	Cashtan-1~ - 4	Spheres 0.5-1.5% LEU UO_2 TRISO	Russia	393-1370	0.22-0.57 (E>0.2 MeV)	6.8-41	1000-1700; irradiation temp.	-	-	[3]
RBT-6 (Dimitrovgrad, Russia)	Udar	Spheres 21-45% LEU UO_2 TRISO	Russia	-	0.1 (E>0.18 MeV)	1.0	300 (up to 2000 cycles) (1200 max.); irradiation temp.	-	-	[3,4]
RBT-6 (Dimitrovgrad, Russia)	BKS	Sphere prototype (10 mm dia.)	Russia	-	2.0-5.1	4.5-11	1250-1400; irradiation temp.	-	-	[5]
RBT-6 (Dimitrovgrad, Russia)	RBT	Spheres	Russia	-	1.6-2.1	10-12	1250-1400; irradiation temp.	-	-	[5]

Heating tests at 1600–1800°C were performed with spherical pebbles containing 16,400 particles. Details of some of these tests are described in Ref. [3] and are summarized in Table 19.

TABLE 19. RESULTS OF POST-IRRADIATION HEATING TESTS WITH PEBBLES

Fuel element UO_2 TRISO	Burnup [% FIMA]	Fast fluence [$10^{25}/m^2$]	Heating test		No. of failed particles (2)		Fractional release				
			Temp. [°C]	Time [h]	Manuf.	Heating	Kr-85	Sr-90	Ag-110m	Cs-134	Cs-137
AVR 71/22	3.5	0.9	1600	500	-	-	$4.0\cdot10^{-7}$	$5.3\cdot10^{-6}$	$9.0\cdot10^{-4}$	$6.9\cdot10^{-5}$	$2.0\cdot10^{-5}$
HFR-K3/1	7.7	3.9	1600	500	-	-	$1.8\cdot10^{-6}$	$1.8\cdot10^{-7}$	$2.7\cdot10^{-2}$	$1.3\cdot10^{-4}$	$1.1\cdot10^{-4}$
FRJ2-K13/2	8.0	0.1	1600	138 (160)	-	-	$6.4\cdot10^{-7}$	$3.3\cdot10^{-7}$	$2.8\cdot10^{-3}$	$1.0\cdot10^{-4}$	$3.9\cdot10^{-5}$
AVR 82/20	8.6	2.4	1600	100	-	-	$1.5\cdot10^{-7}$	$3.8\cdot10^{-6}$	$4.4\cdot10^{-3}$	$1.2\cdot10^{-4}$	$6.2\cdot10^{-5}$
AVR 82/9	8.9	2.5	1600	500	-	-	$5.3\cdot10^{-7}$	$8.3\cdot10^{-5}$	$1.9\cdot10^{-2}$	$5.9\cdot10^{-4}$	$7.6\cdot10^{-4}$
AVR 89/13	9.1	2.6	1620(1) 1620(1)	≈ 10 ≈ 10	-	-	$2.0\cdot10^{-7}$ $1.3\cdot10^{-9}$	(3) (3)	$8.3\cdot10^{-4}$ $1.5\cdot10^{-2}$	$1.3\cdot10^{-5}$ $1.6\cdot10^{-6}$	$1.1\cdot10^{-5}$ $1.4\cdot10^{-6}$
AVR 85/18	9.2	2.6	1620(1)	≈ 10	-	-	$1.4\cdot10^{-7}$	(3)	$6.5\cdot10^{-3}$	$1.0\cdot10^{-5}$	$1.3\cdot10^{-5}$
AVR 90/5	9.2	2.7	1620(1) 1620(1)	≈ 10 ≈ 10	-	-	$1.9\cdot10^{-7}$ $6.6\cdot10^{-9}$	(3) (3)	$1.1\cdot10^{-3}$ $9.0\cdot10^{-4}$	$7.7\cdot10^{-6}$ $3.5\cdot10^{-6}$	$9.0\cdot10^{-6}$ $3.3\cdot10^{-6}$
AVR 90/2	9.3	2.7	1620(1)	≈ 10	1	2	$1.0\cdot10^{-4}$	(3)	$3.7\cdot10^{-2}$	$5.0\cdot10^{-5}$	$4.6\cdot10^{-5}$
AVR 90/20	9.8	2.9	1620(1)	≈ 10	2	3	$2.4\cdot10^{-4}$	(3)	$7.6\cdot10^{-2}$	$5.6\cdot10^{-6}$	$6.5\cdot10^{-6}$
AVR 91/31	9.0	2.6	1700(1)	≈ 10	2	18	$1.2\cdot10^{-3}$	(3)	$6.2\cdot10^{-1}$	$3.7\cdot10^{-3}$	$2.4\cdot10^{-3}$
AVR 74/11	6.2	1.6	1700	185	1	-	$3.0\cdot10^{-5}$	$7.2\cdot10^{-6}$	$4.8\cdot10^{-2}$	$8.4\cdot10^{-5}$	$7.6\cdot10^{-5}$
FRJ2-K13/4	7.6	0.1	1600 1800	138 100	-	- 2	$3.0\cdot10^{-7}$ $7.2\cdot10^{-5}$	$2.0\cdot10^{-8}$ $1.4\cdot10^{-3}$	$4.5\cdot10^{-4}$ $5.3\cdot10^{-1}$	$5.7\cdot10^{-6}$ $9.7\cdot10^{-3}$	$2.5\cdot10^{-6}$ $9.9\cdot10^{-3}$
AVR 88/33	8.5	2.3	1600 1800	50 20	-	-	$1.0\cdot10^{-7}$ $1.8\cdot10^{-4}$	$8.4\cdot10^{-6}$ $2.3\cdot10^{-4}$	$1.2\cdot10^{-3}$ $2.1\cdot10^{-1}$	$1.1\cdot10^{-4}$ $4.4\cdot10^{-4}$	$1.2\cdot10^{-4}$ $4.6\cdot10^{-4}$
AVR 88/15	8.7	2.4	1600 1800	50 20	-	- ≈ 4	$6.3\cdot10^{-8}$ $2.9\cdot10^{-4}$	(3) $1.1\cdot10^{-2}$	$9.1\cdot10^{-3}$ $8.1\cdot10^{-1}$	$8.8\cdot10^{-6}$ $1.3\cdot10^{-2}$	$1.2\cdot10^{-5}$ $1.4\cdot10^{-2}$
AVR 76/18	7.1	1.9	1800	200	-	≈ 3	$1.2\cdot10^{-4}$	$6.6\cdot10^{-2}$	$6.2\cdot10^{-1}$	$5.3\cdot10^{-2}$	$4.5\cdot10^{-2}$
AVR 88/41	7.6	2.0	1800	24	-	-	$2.4\cdot10^{-7}$	$1.2\cdot10^{-4}$	$7.7\cdot10^{-2}$	$1.4\cdot10^{-4}$	$1.5\cdot10^{-4}$
HFR-K3/3	10.2	6.0	1800	100	-	≈ 12	$6.5\cdot10^{-4}$	$1.5\cdot10^{-3}$	$6.7\cdot10^{-1}$	$6.4\cdot10^{-2}$	$5.9\cdot10^{-2}$

(1) Simulation calculated core heat up curve.
(2) Out of 16,400 particles.
(3) Not measured.

6.1.2. Cylindrical compacts

6.1.2.1. Irradiation tests

Table 20 shows irradiation tests performed with cylindrical compacts up to the early 2000s. These irradiations were collected from various references such as Refs. [3, 71-82]. The information includes fuel types, irradiation time in FPDs, fast fluence, burnup, operating temperature, and the fraction releases (i.e., the ratio of release (R) to birth (B) of ^{85m}Kr). The irradiation results showed the good performance of cylindrical compacts under normal operating conditions with a peak burnup to about 27% FIMA and to high operating temperature, e.g. 1000°C.

6.1.2.2. Heating tests

Heating tests at 1600–1800°C were performed with cylindrical compacts containing 1600 particles. Details of these tests are described in Refs. [3,70] and are summarized in Table 21.

Two experiments, HRB-17 and HRB-18 were also conducted with Highly Enriched Uranium (HEU) TRISO UCO unbonded, irradiated particles at the ORNL [83]. Eighty particles each were heated at 1400 and 1600°C for 300 and 100 h, respectively. Four experiments with HRB-22 fuel were also reported [3,70,80].

Post-irradiation heating tests at temperatures up to 1500°C for long durations were performed to accelerate the diffusional release of fission gas atoms from various fuel types including UO_2, UC_2, UCO, UO_2*(1), and UO_2*(2) fuel particles irradiated at the HRB-15B capsule installed in the High Flux Isotope Reactor (HFIR) [70,84]. Note that in the UO_2*(1) fuel the kernel was enveloped by a ZrC layer, which replaced the graphite buffer. In contrast, the UO_2*(2) fuel featured a buffer layer surrounding the kernel that included dispersed ZrC.

The distribution of fission product releases among particles within the fuel batches indicated that the release of caesium from the UO_2 fuel particles was from two of the ten UO_2 particles and from only one of the ten UC_2 particle [70] (see Table 22).

6.1.3. Loose TRISO particle fuels

Irradiation tests of loose coated particle fuels were also carried out in various material test reactors to support the design of coated particle fuels. The historical irradiation testing, conducted in variety of material test reactors in Germany, USA, Japan, and Russia, of loose coated particle fuels is summarized in Table 23 which contains information on fuel types, irradiation time (in FPDs), fast fluence, burnup, irradiation temperature, and ^{85m}Kr R/B.

6.2. PROGRESS ON TESTS SINCE THE EARLY 2000S

Since the early 2000s, VHTR, one of the advanced reactor systems, has been paid attention, and irradiation tests have been conducted to qualify TRISO coated particle fuels under such conditions. In this section, progress of such irradiation tests since the early 2000s is described.

6.2.1. Spherical pebbles

Table 24 shows irradiations of spherical pebbles performed since the early 2000s.

6.2.1.1. HFR-EU1bis irradiation experiment

The irradiation experiment HFR-EU1bis [5,69,86] was intended to demonstrate a low failure fraction of coated particles under normal operating condition, with specific objectives of (1) successfully operating at increased central temperature of 1250°C (in comparison with 1000–1200°C in earlier irradiation tests), (2) successfully operating at burnups nearing 16%[8], and (3) affirming low failure fraction at demanding conditions such as increased temperature, burnup and neutron fluence conditions.

[8] The target value was double the license limit of the HTR-Module. Due to a neutronics data processing error, the experiment could not continue beyond 11.07% FIMA.

TABLE 20. IRRADIATIONS OF COMPACTS CONDUCTED BEFORE THE EARLY 2000s

Reactor	Experiment ID	No. of specimen and fuel type (mm)	Country	Irradiation time (FPDs)	Fast fluence (10^{21}n/cm^2, E>0.18 MeV)	Burnup (%FIMA)	Operating temp. (°C)	Defective SiC layer fraction (U/U-total)	BOL ^{85m}Kr *R/B* EOL ^{85m}Kr *R/B*	Reference
GETR (GE(Pleasanton), USA)	F-30	13 compacts (dia. 12.45, length 18.54 & 49.28); Cell 1,3,4,5 HEU (Th,U)C_2 TRISO and ThC_2 TRISO	USA	15 May 1972 – 5 Apr 1994 (269)	8, 11.5, 9.5, 12	15, 20, 18, 12 (fissile) 3, 5, 4, 1.5 (fertile)	1100, 1120, 1100, 1200 (time ave. peak temp. (TAPT))	$< 5\times10^{-4}$ – 1.5×10^{-3} (fissile particles) 3×10^{-4} – 1.0×10^{-3} (fertile particles)	2×10^{-6}, 8×10^{-7}, 7×10^{-7}, 2×10^{-6} 8×10^{-6}, 1×10^{-5}, 2×10^{-5}, 2×10^{-5}	[70]
HFIR (ORNL, USA)	HRB-4	6 compacts (dia. 12.4, length 25.4) LEU WAR (weak acid resin) UC_2 TRISO and ThO_2 BISO	USA	8 Oct. 1972 – 26 Jun. 1972 (244)	10.5 (peak)	27.7 (peak fissile) 13.4 (peak fertile)	1250 (peak)	-	1.4×10^{-5} 3.2×10^{-4}	[70]
HFIR (ORNL, USA)	HRB-5	6 compacts (dia. 12.4, length 25.4) LEU WAR UC_2 TRISO and ThO_2 BISO	USA	8 Oct. 1972 – 3 Feb. 1973 (107)	4.7 (peak)	15.7 (peak fissile) 4.3 (peak fertile)	1250 (peak)	-	3×10^{-6} 1×10^{-4}	[70]
HFIR (ORNL, USA)	HRB-6	6 compacts (dia. 12.4, length 25.4) HEU (Th,U)C_2 TRISO and ThO_2 BISO	USA	27 Feb. 1973 – 8 Sep. 1973 (183)	7.9 (peak)	26.6 (peak fissile) 9.3 (peak fertile)	1100 (peak)	$< 5\times10^{-4}$	5.0×10^{-7} (min.) 2.7×10^{-4} (max.)	[70]

TABLE 20. IRRADIATIONS OF COMPACTS CONDCUTED BEFORE THE EARLY 2000s (CONT)

Reactor	Experiment ID	No. of specimen and fuel type (mm)	Country	Irradiation time (FPDs)	Fast fluence (10^{21}n/cm^2, E>0.18 MeV)	Burnup (%FIMA)	Operating temp. (°C)	Defective SiC layer fraction (U/U-total)	BOL ^{85m}Kr *R/B* EOL ^{85m}Kr *R/B*	Reference
ORR (ORNL, USA)	OF-2	16 compacts (dia. 15.75, height 25.4) 48 compacts (dia. 15.75 OD, 3.30 ID, height 12.70) 24 compacts (dia. 15.75, height 50.8) WAR UCO UC_2 (Th,U)O_2 TRISO and ThO_2 BISO	USA	21 Jun. 1975 – 1 Aug. 1976 (352)	5.86-8.91 1.94-8.36	75.9-79.6 50.0-795.	1350 (max.)	-	2×10^{-5}, 1×10^{-4} 1×10^{-4}, 5×10^{-6}	[70]
HFIR (ORNL, USA)	HRB-14	20 compacts (dia. 12.5, height 9.52) LEU UCO UO_2 (Th,U)O_2 TRISO and ThO_2 BISO	USA	20 May 1978 – 4 Jan. 1979 (214)	8.3 (peak)	28.6 (peak fissile) 8.5 (peak fertile)	1190 (max.), 895 (min.)	7.0×10^{-7} – 1.3×10^{-4} (fertile particles)	3.8×10^{-7} (min.) 3.0×10^{-4} (max.)	[3],[70]
R2 (Studvik, Sweden)	R2-K13	12 compacts (dia. 12.52, height 25.4) LEU UCO TRISO and ThO_2 TRISO	USA (2 spheres, Germany)	22 Apr. 1980 – 19 Sep. 1982 (517)	7.8, 7.4 (peak)	22.5, 22.1 (peak fissile) 7.8, 7.4 (peak fertile)	1190, 985 (ave.)	1.9×10^{-4} – 4.4×10^{-4} (fissile particles) $< 2\times10^{-6}$ – 1.6×10^{-5} (fertile particles)	1×10^{-5}, 2×10^{-7} 8×10^{-5}, 8×10^{-6}	[3],[70]
HFIR (ORNL, USA)	HRB-15A	20 compacts (dia. 12.52, 3 compacts; height 9.53, 17 compacts; 19.05) LEU UCO UC_2 UO_2 TRISO and ZrC-TRISO and ThO_2 TRISO and Si-BISO	USA	26 Jul. 1980 – 29 Jan. 1981 (174)	6.5 (peak)	29.0 (peak fissile) 6.4 (peak fertile)	1150 (ave.)	1.4×10^{-5} – 7.4×10^{-2} (fissile particles) 6.7×10^{-5} – 1.4×10^{-3} (fertile particles)	6.96×10^{-6} 3.76×10^{-4}	[3],[70]
	HRB-15B	LEU UCO UC_2 UO_2 TRISO			6.6	27	815–1000			[3]

TABLE 20. IRRADIATIONS OF COMPACTS CONDCUTED BEFORE THE EARLY 2000s (CONT)

Reactor	Experiment ID	No. of specimen and fuel type (mm)	Country	Irradiation time (FPDs)	Fast fluence (10^{21}n/cm^2, E>0.18 MeV)	Burnup (%FIMA)	Operating temp. (°C)	Defective SiC layer fraction (U/U-total)	BOL ^{85m}Kr *R/B* EOL ^{85m}Kr *R/B*	Reference
HFIR (ORNL, USA)	HRB-16	18 compacts (dia. 12.54, height 18.70) LEU UCO UC_2 UO_2 (Th,U)O_2 TRISO and ZrC-TRISO and ThC_2 ThO_2 TRISO and BISO	USA	21 Jun. 1981 – 23 Dec. 1981 (170)	6.3 (peak)	28.7 (peak fissile) 6.1 (peak fertile)	1150 (ave.)	4.6×10^{-7} – 4.4×10^{-4} (fissile particles) 1.6×10^{-5}, 5.0×10^{-4} (fertile particles)	2.44×10^{-5} 2.08×10^{-4}	[3],[70]
	HRB-17, -18	18 compacts (dia. 12.54, height 18.70) LEU UCO UC_2 TRISO			4.3-4.6	25	700–1000			[3]
DIDO (Jülich, German)	FRJ2-P27	9 compacts (dia. 27.9–28.03, height 29) LEU UO_2 TRISO	Germany	17 Feb. 1984 – 10 Feb. 1985 (232)	1.4, 1.7, 1.3 1.5–2 (E>0.10 MeV)	7.6, 8.0, 7.6	Centre; 1080, 1320, 1130 Surface; 880, 1220, 1080	< 3×10^{-6}	1.0×10^{-6}, 8.6×10^{-7}, 2.0×10^{-8} 1.6×10^{-6}, 1.0×10^{-5}, 1.2×10^{-7}	[3],[70]
	FRJ2-P28				1.5–2	11–13	950–1150			[3]
HFR (Petten, Netherlands)	HFR-B1	36 fuel rods LEU UCO and ThO_2 TRISO	Germany USA	Apr. 1987 – Jul. 1989 (445)	6.5 (E>0.10 MeV)	19	1470	-	EOL 4.1×10^{-3} (10% failed particles)	[3],[71]
HFIR (ORNL, USA)	HRB-21	24 compacts (dia. 12.27–12.51, height 49.13–49.35) LEU UCO and ThO_2 TRISO-P	USA	20 Jun. 1991 – 21 Nov. 1991 (105)	3.5 (peak)	22 (peak)	950 (ave.), 1300 (peak)	$\leq5.4\times10^{-6}$ (fissile particles) 1.7×10^{-5} (fertile particles)	1×10^{-8} 2×10^{-4}	[70]
HFIR (ORNL, USA)	NPR-1	16 compacts (dia. 12.43, height 49.42)	USA	25 Jul. 1991 – 29 May 1992 (170)	3.75 (peak)	79 (peak)	974 (ave.), 1240 (peak)	3×10^{-6}	1×10^{-8} 3×10^{-4}	[5],[70],[72]

TABLE 20. IRRADIATIONS OF COMPACTS CONDCUTED BEFORE THE EARLY 2000s (CONT)

Reactor	Experiment ID	No. of specimen and fuel type (mm)	Country	Irradiation time (FPDs)	Fast fluence (10^{21}n/cm^2, E>0.18 MeV)	Burnup (%FIMA)	Operating temp. (°C)	Defective SiC layer fraction (U/U-total)	BOL ^{85m}Kr *R/B* EOL ^{85m}Kr *R/B*	Reference
		HEU UCO TRISO-P								
HFIR (ORNL, USA)	NPR-2	16 compacts (dia. 12.43, height 49.42) HEU UCO TRISO-P	USA	28 Aug. 1991 – 29 May 1991 (172)	3.75 (peak)	79 (peak)	753 (ave.), 1030 (peak)	3×10^{-6}	5×10^{-9} 6×10^{-5}	[5],[70],[72]
ATR (INL, USA)	NPR-1A	20 compacts (dia. 12.37-12.50, height 49.33) HEU UCO TRISO-P	USA	2 Oct. 1991 – 3 Jan. 1992 (64)	2.1 (peak)	64 (peak)	977 (ave.), 1220 (peak)	3×10^{-6}	4×10^{-9} 1.8×10^{-5}	[5],[70],[72]
JMTR (Oarai, Japan)	OGL-1 1st (TRISO)	54 compacts (OD 24, ID 8, height 40) LEU UO_2 TRISO	Japan	JMTR cycle 39–40 Feb. 1977 – May 1977 (39)	0.12 (max.) (E>0.1 MeV)	0.08 (min.) 0.62 (max.)	1380 (time ave. max. temp.) (calculated by STPDSP-2) (max.-min. 1356–897)	-	ave. 3.7×10^{-7}	[5],[73–76]
JMTR (Oarai, Japan)	OGL-1 2nd (TRISO)	60 compacts (OD 24, ID 8, height 36) LEU UO_2 TRISO	Japan	JMTR cycle 42–45 Dec. 1977 – Feb 1979 (71.6)	0.20 (max.) (E>0.1 MeV)	0.12 (min.) 1.0 (max.)	1430 (time ave. max. temp.) (calculated by STPDSP-2) (max.-min. 1425–935)	-	ave. 7.3×10^{-7}	[5],[73–76]
JMTR (Oarai, Japan)	OGL-1 3rd (TRISO)	20 compacts (OD 36, ID 18, height 36) LEU UO_2 TRISO	Japan	JMTR cycle 46–47 Mar. 1979 – Jul. 1979 (40.7)	0.09 (max.) (E>0.1 MeV)	0.08 (min.) 0.45 (max.)	1320 (time ave. max. temp.) (calculated by STPDSP-2) (max.-min. 1355–965)	-	ave. 3.1×10^{-5}	[5],[73–76]
JMTR (Oarai, Japan)	OGL-1 4th (TRISO)	60 compacts (OD 24, ID 8, height	Japan	JMTR cycle 48–51	0.23 (max.) (E>0.1 MeV)	0.22 (min.) 1.96(max.)	1340 (time ave. max. temp.)	-	ave. 5.9×10^{-7}	[5],[73–76]

TABLE 20. IRRADIATIONS OF COMPACTS CONDCUTED BEFORE THE EARLY 2000s (CONT)

Reactor	Experiment ID	No. of specimen and fuel type (mm)	Country	Irradiation time (FPDs)	Fast fluence (10^{21}n/cm^2, E>0.18 MeV)	Burnup (%FIMA)	Operating temp. (°C)	Defective SiC layer fraction (U/U-total)	BOL ^{85m}Kr *R/B* EOL ^{85m}Kr *R/B*	Reference
		36) LEU UO_2 TRISO		Nov. 1979 – Jun. 1980 (78.0)			(calculated by STPDSP-2) (max.-min. 1470–873)			
JMTR (Oarai, Japan)	OGL-1 5th (TRISO)	60 compacts LEU UO_2 TRISO	Japan	JMTR cycle 52–58 Dec. 1980–Apr. 1983 (142.3)	0.38 (max.) (E>0.1 MeV)	3.3 (max.)	1350 (time ave. max. temp.) (calculated by STPDSP-2)	-	-	[5],[73],[76]
JMTR (Oarai, Japan)	OGL-1 6th (TRISO)	20 compacts LEU UO_2 TRISO	Japan	JMTR cycle 59 (21.9)	0.04 (max.) (E>0.1 MeV)	0.4 (max.)	-/1480×280 (°C×h; transient, (calculated by STPDSP-2))	1×10^{-3}	- 1.5×10^{-6}	[3],[5],[73],[77],[78]
JMTR (Oarai, Japan)	OGL-1 7th (TRISO)	60 compacts LEU UO_2 TRISO	Japan	JMTR cycle 61,61,63 (58.0)	0.16 (max.) (E>0.1 MeV)	1.4 (max.)	1380 (time ave. max. temp.) (calculated by STPDSP-2)	3.0×10^{-3} 5.8×10^{-3}	- 4.0×10^{-7}	[3],[5],[73],[77],[78]
JMTR (Oarai, Japan)	OGL-1 8th (TRISO)	20 compacts LEU UO_2 TRISO	Japan	JMTR cycle 64–66 (53.8)	0.12 (max.) (E>0.1 MeV)	1.0 (max.)	1390 (time ave. max. temp.) (calculated by STPDSP-2)	4.6×10^{-4} 4.8×10^{-4}	- 1.1×10^{-7}	[3],[5],[73],[77],[78]
JMTR (Oarai, Japan)	OGL-1 9th (TRISO)	20 compacts LEU UO_2 TRISO	Japan	JMTR cycle 67–73 (145.5)	0.28 (max.) (E>0.1 MeV)	2.7 (max.)	1340 (time ave. max. temp.) (calculated by STPDSP-2)	3.9×10^{-3}	-	[5],[73],[77],[78]
JMTR (Oarai, Japan)	OGL-1 10th (TRISO)	20 compacts LEU UO_2 TRISO	Japan	JMTR cycle 74–79 May 1986 – Jun. 1987 (130.2)	0.24 (max.) (E>0.1 MeV)	2.8 (max.)	1330 (time ave. max. temp.)/ 1500×95 (°C×h; transient) (calculated by STPDSP-2)	9.5×10^{-4} 5.4×10^{-4}	- 3.2×10^{-6}	[3],[5],[73],[77],[78]

TABLE 20. IRRADIATIONS OF COMPACTS CONDCUTED BEFORE THE EARLY 2000s (CONT)

Reactor	Experiment ID	No. of specimen and fuel type (mm)	Country	Irradiation time (FPDs)	Fast fluence (10^{21}n/cm^2, E>0.18 MeV)	Burnup (%FIMA)	Operating temp. (°C)	Defective SiC layer fraction (U/U-total)	BOL ^{85m}Kr *R/B* EOL ^{85m}Kr *R/B*	Reference
JMTR (Oarai, Japan)	OGL-1 11th (TRISO)	54 compacts LEU UO_2 TRISO	Japan	JMTR cycle 80–82 Feb. 1988 – Jul. 1988 (62.3)	0.17 (max.) (E>0.1 MeV)	1.6 (max.) (calculated by STPDSP-2)	1350 (time ave. max. temp.)	1.1×10^{-3} 7.5×10^{-4}	-	[5],[73],[77],[78]
JMTR (Oarai, Japan)	OGL-1 12th (TRISO)	19 compacts (OD 26, ID 10, height 39) LEU UO_2 TRISO	Japan	JMTR cycle 83–91 Oct. 1988 – Jun. 1990 (195.0)	0.31 (max.) (E>0.1 MeV)	3.9 (max.) (calculated by STPDSP-2)	1340 (time ave. max. temp.)	3.6×10^{-4} 5.4×10^{-4}	- 3.2×10^{-6}	[3],[5],[73],[77],[78]
JMTR (Oarai, Japan)	OGL-1 13th (TRISO)	19 compacts (OD 26, ID 10, height 39) LEU UO_2 TRISO	Japan	JMTR cycle 92–102 (243.0)	0.49 (max.) (E>0.1 MeV)	3.7 (max.)	1340 (time ave. max. temp.)/ 1500×30.5 (°C×h; transient)	3.9×10^{-6}	- 6.8×10^{-8}	[3],[5],[73]
JMTR (Oarai, Japan)	OGL-1 14th (TRISO)	19 compacts (OD 26, ID 10, height 39) LEU UO_2 TRISO	Japan	JMTR cycle 103–106 (73.0)	0.07-0.12 0.14 (max.) (E>0.1 MeV)	0.73-1.17 1.2 (max.)	1160, 1339 (max.) 1122, 1305 (ave.) 1350 (time ave. max. temp.)/1500×47 (°C×h; transient)	6.1×10^{-5}	BOL ave. 4.1×10^{-5} EOL ave. 1.0×10^{-4}	[3],[5],[70],[73]
JMTR (Oarai, Japan)	OGL-1 15th (TRISO)	19 compacts (OD 26, ID 10, height 39) LEU UO_2 TRISO	Japan	JMTR cycle 107–115 (217.8)	0.44 (max.) (E>0.1 MeV)	3.1 (max.)	1345 (time ave. max. temp.)	5.1×10^{-6}	- 3.8×10^{-7}	[3],[5],[73]
JMTR (Oarai, Japan)	79F-1A	5 compacts (smaller than those for HTTR)	Japan		0.027 (E>0.1 MeV)	0.5	normal 1197 transient 1510			[79]
JMTR (Oarai, Japan)	80F-5A	5 compacts (smaller than those for HTTR)	Japan		0.097 (E>0.1 MeV)	1.0	normal 1236 transient 1520			[79]

TABLE 20. IRRADIATIONS OF COMPACTS CONDCUTED BEFORE THE EARLY 2000s (CONT)

Reactor	Experiment ID	No. of specimen and fuel type (mm)	Country	Irradiation time (FPDs)	Fast fluence (10^{21}n/cm^2, E>0.18 MeV)	Burnup (%FIMA)	Operating temp. (°C)	Defective SiC layer fraction (U/U-total)	BOL ^{85m}Kr *R/B* EOL ^{85m}Kr *R/B*	Reference
JMTR (Oarai, Japan)	83F-2A	5 compacts (smaller than those for HTTR)	Japan		0.012 (E>0.1 MeV)	1.2	normal 1373 transient 2000			[79]
JMTR (Oarai, Japan)	91F-1A	2 compacts (OD 26, ID 10, height 39) LEU UO_2 TRISO	Japan	(~360)	6.0, 4.6 (E>0.1 MeV)	9.6, 7.9	1120, 1340			[70]
JMTR (Oarai, Japan)	94F-9A (TRISO)	2 compacts (OD 26, ID 10, height 39) LEU UO_2 TRISO	Japan	(~360)	3.1, 2.3 (E>0.1 MeV)	7.0, 5.5	1120, 1190 (max.)	-	-	[5],[70]
HFIR (ORNL, USA)	HRB-22	12 compacts (OD 26.03-26.07, ID 10.00-10.02, height 38.95-39.09) HEU UO_2 TRISO	Japan	10 Apr. 1994 – 22 Aug. 1994 (88)	1.3–2.8 (E>0.1 MeV)	4.1-6.7	821–1071 (ave.), 898–1156 (max.)	-	1.0×10^{-6} 3.0×10^{-6}	[5],[70],[74],[80],[81]
JMTR (Oarai, Japan)	78F-4A (ZrC layer)	Compacts (OD 18, ID 8, height 18) LEU UO_2 TRISO	Japan	81.2	2.2 (E>0.1 MeV)	4.0	1100	-	-	[82]
JMTR (Oarai, Japan)	80F-4A (ZrC layer)	Compact (OD 18, ID 8, height 18) LEU UO_2 TRISO	Japan	79.9	1.2 (E>0.1 MeV)	1.5	900	-	^{88}Kr *R/B* at ~900°C 2×10^{-7} – 4×10^{-7}	[82]
JMTR (Oarai, Japan)	88-3A (ZrC layer)	Compact (OD 24, ID 8, height 36) LEU UO_2 TRISO	Japan	99.9	0.9 (E>0.1 MeV)	4.5	1400–1650	-	-	[82]
JRR-2 (Oarai, Japan)	ICF-26H (ZrC layer)	Compact (OD 18, ID 8, height 18) LEU UO_2 TRISO	Japan	134.7	1.0 (E>0.1 MeV)	1.8	1400–1500			[74],[82]

TABLE 20. IRRADIATIONS OF COMPACTS CONDCUTED BEFORE THE EARLY 2000s (CONT)

Reactor	Experiment ID	No. of specimen and fuel type (mm)	Country	Irradiation time (FPDs)	Fast fluence (10^{21}n/cm^2, E>0.18 MeV)	Burnup (%FIMA)	Operating temp. (°C)	Defective SiC layer fraction (U/U-total)	BOL ^{85m}Kr *R/B* EOL ^{85m}Kr *R/B*	Reference
IVV-2M (Zarechny, Russia)	MT-2	6–8 pellets	Russia	1982–1989 (63, 167, 333)	0.21–2.7 (E>0.1 MeV)	1.4–14	1100–1950; irradiation temp.	-	-	[3],[5]
IVV-2M (Zarechny, Russia)	MTI	6–8 pellets	Russia	1982–1989 (~221)	0.6–2.0 (E>0.1 MeV)	4–12	1400; irradiation temp.	-	-	[3],[5]
IVV-2M (Zarechny, Russia)	MT-3	6–8 compacts	Russia	1982–1989 (129)	1.5–2.3 (E>0.1 MeV)	11–17	750–1400; irradiation temp.	-	-	[3],[5]
MR (Moscow, Russia)	ASU-8	1 compact	Russia	-	0.7–2.7 (E>0.1 MeV)	5–15.3	1000–1300; irradiation temp.	-	-	[5]
SM-2 (Dimitorvgrad, Russia)	Carat-2~6	Pellets 0.7–45% LEU UO_2 TRISO	Russia	-	0.75–2.2 (E>0.1 MeV)	3–24	800–1700; irradiation temp.	-	-	[3],[5]

TABLE 21. RESULTS OF POST-IRRADIATION HEATING TESTS WITH FUEL COMPACTS AT 1600-1800°C

Fuel compact	Irradiation conditions			Heating test		Fractional release			
	Burnup (% FIMA)	Fast fluence (10^{25} n m^{-2})	Temperature (°C)	Temperature (°C)	Duration (h)	^{85}Kr	^{90}Sr	^{134}Cs	^{137}Cs
SL-P1/6	10.7	6.7	790–800	1600	304	7.3×10^{-7}	4.3×10^{-5}	7.5×10^{-4}	3.9×10^{-4}
HFR-P4/1/12	11.1	5.5	900–940	1600	304	5.4×10^{-7}	6.0×10^{-6}	3.0×10^{-4}	2.6×10^{-4}
HFR-P4/1/8	13.8	7.2	900–940	1600	304	5.4×10^{-5}	1.5×10^{-4}	1.5×10^{-3}	2.0×10^{-3}
HFR-P4/2/8/	13.8	7.2	900–945	1600	304	8.1×10^{-5}	1.1×10^{-4}	1.5×10^{-3}	1.4×10^{-3}
HFR-P4/3/7	13.9	7.5	1050–1075	1600	304	9.9×10^{-4}	2.4×10^{-4}	3.5×10^{-3}	3.9×10^{-3}
SL-P1/10	10.3	6.0	790–800	1700	304	9.1×10^{-5}	2.1×10^{-2}	9.3×10^{-2}	1.0×10^{-1}
SL-P1/9	10.7	6.3	790–800	1700	304	3.7×10^{-5}	5.8×10^{-2}	1.3×10^{-1}	9.8×10^{-2}
HFR-P4/3/12	9.9/12.0	5.5	1050–1075	1800	279	1.0×10^{-3}	Not measured	5.2×10^{-1}	5.2×10^{-1}

TABLE 22. DISTRIBUTION OF FISSION PRODUCT RELEASE WITHIN PARTICLE BATCHES DURING POST IRRADIATION ANNEALING

Annealing temperature (°C)	TRISO particle type[a]	Release breakdown from the ten particles within a test batch for:			
		^{134}Cs ^{137}Cs	^{110m}Ag	^{154}Eu	^{144}Ce
1500	UC_2 with laminar SiC	9 : 0% 1 : 99% 10 : 12%	1 : 0% 1 : 50% 85%<2<95% 6 : 100% 10 : 82%	15%<5<25% 45%<3<55% 2 : 100% 10 : 46%	12%<3<18% 18%<3<25% 70%<3<80% 1 : 99% 10 : 45%
1500	UO_2 with columnar SiC	8 : 0% 2 : 99% 10 : 24%	10 : 100%	Uniform release of 16%	10 : 0%
1500	$UC_{0.4}O_{1.6}$ with laminar SiC	10 : 0%	7 : 0% 10%<3<20% 10 : 3%	Uniform release of 37%	10 : 1%[b]
1500	$UC_2^*(2)$ with laminar SiC	10 : 0%	7 : 0% 70%<3<80% 10 : 27%	2 : 0% 0%<5<10% 15%<3<25%	10 : 0%
1350	$UC_{0.4}O_{1.6}$ with laminar SiC	10 : 0%	10 : 0%	Uniform release of 23%	10 : 0%
1350	$UC_2^*(2)$ with laminar SiC	10 : 0%	7 : 0% 45%<3<55% 10 : 19%	10 : 4%[b]	10 : 0%
1200	$UC_{0.4}O_{1.6}$ with laminar SiC	10 : 0%	10 : 0%	Uniform release of 6%	10 : 0%
1200	$UC_2^*(2)$ with laminar SiC	10 : 0%	10 : 2%[b]	10 : 0%	10 : 0%

According to Ref. [70],

[a] "There was zero release within about ±5% as determined from individual particle counting before and after annealing for all isotopes from each of the ten particles in all test combinations not listed, that is, $UC_2^*(1)$ at all temperatures, and UC_2, UO_2, and $UC_{0.4}O_{1.6}$ at 1350 and 1200°C. As no release on collectors was detected at the 0.01% level from the combined ten particles within each of these test batches, it can be assumed that release from any one of these particles was certainly <0.01% and was probably not more than 0.001%.

[b] These total releases from ten particles were too small and too uniformly distributed among particles to allow the determinations of individual release values."

TABLE 23. IRRADIATIONS OF LOOSE COATED PARTICLE FUELS BEFORE THE EARLY 2000s

Reactor	Experiment ID	No. of specimen and fuel type (mm)	Country	Irradiation time (FPD)	Fast fluence ($10^{21} n\cdot cm^{-2}$, E>0.18 MeV)	Burnup (%FIMA)	Operating temp. (°C)	Defective SiC layer fraction (U/U-total)	BOL ^{85m}Kr *R/B* EOL ^{85m}Kr *R/B*	Ref.
HFR (Petten, Netherlands)	HFR-P5	116 coupons LEU UO_2 TRISO	Germany	(142)	5.5 (E>0.10 MeV)	7.2	Centre; 1730	-	EOL 1.1×10^{-4} (amoeba effect)	[70],[71]
GETR (GE(Pleasanton), USA)	F-30	Loose particles; Cell 2 & 5 HEU $(Th,U)C_2$ TRISO and ThC_2 TRISO	USA	15 May 1972 – 5 Apr 1994 (269)	10.5, 12	19, 12 (fissile) 4.5, 1.5 (fertile)	1100, 1200 (time ave. peak temp.; TAPT)	$< 5\times10^{-4}$ – 1.5×10^{-3} (fissile particles) 3×10^{-4} – 1.0×10^{-3} (fertile particles)	7×10^{-7}, 2×10^{-6} 1×10^{-4}, 2×10^{-5}	[70]
HFIR (ORNL, USA)	HRB-15B	184 particle trays (OD 22.3-23.6, ID 11.1) 116 loose particles/tray LEU UCO UC_2 UO_2 TRISO	USA	6 Jul. 1978 – 4 Jan. 1979 (169)	6.6 (E>0.10 MeV)	26.7 (peak fissile) 6.0 (peak fertile)	Centre; 815–915 (time ave. temp.)	-	2.9×10^{-8} 5.1×10^{-6}	[70]
DIDO (Jülich, Germany)	FRJ2-P27	2 coupons (dia. 29, height 2.2, fuel annulus dia. 23, 34 particles/coupon) LEU UO_2 TRISO	Germany	17 Feb. 1984 – 10 Feb. 1985 (232)	1.4, 1.7, 1.3 (E>0.10 MeV)	7.6, 8.0, 7.6	Centre; 1080, 1320, 1130 Surface; 880, 1220, 1080	$< 3\times10^{-6}$	1.0×10^{-6}, 8.6×10^{-7}, 2.0×10^{-8} 1.6×10^{-6}, 1.0×10^{-5}, 1.2×10^{-7}	[5],[70]
HFIR (ORNL, USA)	HRB-16 HRB-17, -18	27 loose particle trays (110 particles/tray, 2 particles /holes) LEU UCO UC_2 UO_2 $(Th,U)O_2$ TRISO and ZrC-TRISO and ThC_2 ThO_2 TRISO and BISO	USA	21 Jun. 1981 – 23 Dec. 1981 (170)	6.3 (peak)	28.7 (peak fissile) 6.1 (peak fertile)	1150 (peak)	4.6×10^{-7} – 4.4×10^{-4} (fissile particles) 1.6×10^{-5}, 5.0×10^{-4} (fertile particles)	2.44×10^{-5} 2.08×10^{-4}	[3],[70]

TABLE 23. IRRADIATIONS OF LOOSE COATED PARTICLE FUELS BEFORE THE EARLY 2000s (CONT)

Reactor	Experiment ID	No. of specimen and fuel type (mm)	Country	Irradiation time (FPD)	Fast fluence (10^{21}n·cm^{-2}, E>0.18 MeV)	Burnup (%FIMA)	Operating temp. (ºC)	Defective SiC layer fraction (U/U-total)	BOL ^{85m}Kr *R/B* EOL ^{85m}Kr *R/B*	Ref.
HFIR (ORNL, USA)	HRB-21	24 encapsulates piggyback specimens (absorptivity specimens or loose fuel particles) LEU UCO and ThO_2 TRISO-P	USA	20 Jun. 1991 – 21 Nov. 1991 (105)	3.5 (peak)	22 (peak)	950 (ave.), 1300 (peak)	$\leq 5.4\times10^{-6}$ (fissile particles) 1.7×10^{-5} (fertile particles)	1×10^{-8} 2×10^{-4}	[70]
HFIR (ORNL, USA)	NPR-1	12 encapsulates piggyback specimens (loose fuel particles) HEU UCO TRISO-P	USA	25 Jul. 1991 – 29 May 1992 (170)	3.75 (peak)	79 (peak)	974 (ave.), 1240 (peak)	3×10^{-6}	1×10^{-8} 5×10^{-9}	[70],[72]
HFIR (ORNL, USA)	NPR-2	16 encapsulates piggyback specimens (loose fuel particles) HEU UCO TRISO-P	USA	28 Aug. 1991 – 29 May 1991 (172)	3.75 (peak)	79 (peak)	753 (ave.), 1030 (peak)	3×10^{-6}	3×10^{-4} 6×10^{-5}	[70],[72]
JRR-2 (Oarai, Japan)	VOF-8H (ZrC layer)	Loose particles in graphite disk	Japan	156.4	<0.1	1.6	1370 (15ºC/min; temp. gradient)			[85]
JRR-2 (Oarai, Japan)	VOF-14H (ZrC layer)	Loose particles in graphite disk	Japan	138.7	<0.1	1.6	1600 (15ºC/min; temp. gradient))			[85]
JRR-2 (Oarai, Japan)	ICF-51H	Loose particles in graphite disk	Japan	142.6	-	2.1–2.3	1318-1363(ave.) (1173-1471; min. –max.)			[3],[82]
MR (Moscow, Russia)	ASU-8	Loose coated particles Fuel coupons (5250–7500)	Russia	-	0.7-2.2 (E>0.1 MeV)	5–15.3	1000–1300; irradiation temp	-	-	[5]
RBT-6 (Dimitrovgrad, Russia)	BKS	Loose coated particles Fuel coupons	Russia	-	2.0-5.1 (E>0.1 MeV)	4.5–11	1250–1400; irradiation temp	-	-	[5]

TABLE 24. IRRADIATIONS OF SPHERICAL PEBBLES PERFORMED SINCE THE EARLY 2000s

Reactor	Experiment ID	No. of specimen and fuel type (mm)	Country	Irradiation time (FPDs)	Fast fluence (10^{21}n·cm^{-2}, E>0.18 MeV)	Burnup (%FIMA)	Operating temp. (°C)	Defective SiC layer fraction (U/U-total)	BOL ^{85m}Kr *R/B* EOL ^{85m}Kr *R/B*	Ref.
IVV-2M (Russia)	(HTR-10 fuel)	4 spheres (dia. 60.2, fuel zone dia. 50) LEU UO_2 TRISO	China	Jul. 2000 – Feb. 2003	1.10, 1.31, 1.30, 1.06 (E>0.10 MeV)	11.9, 13.1, 11.6 (10.4, 11.4, 10.9, 10.1)	1000±50 (centre temp.) 1200 (max.) at 4.7% FIMA 1250 (max.) at 12% FIMA	-	EOL 10^{-6}–8×10^{-5}	[5],[70],[87–89]
HFR (Petten, Netherlands)	HFR-EU1bis	5 spheres (dia. 60.2, fuel zone dia. 50, AVR) LEU UO_2 TRISO	Germany	9 Sep. 2004 – 18 Oct. 2005 (249.9)	3.0–4.0 (peak)	9–11 (max.)	1200 (max.), 900 (min.)	-	1.0×10^{-6} 4.0×10^{-6}	[5],[69],[70]
HFR (Petten, Netherlands)	HFR-EU1 (AVR & HTR-10 fuel)	3 spheres (dia. 60.2, fuel zone dia. 50, AVR) 2 spheres (dia. 60, fuel zone dia. 50, INET) LEU UO_2 TRISO	Germany China	29 Sep. 2006 – 24 Feb. 2008 (332.8)	3.7 (peak)	9–11 (max.)	1000 (max.), 800 (min.)	-	1.0×10^{-8}, 1.0×10^{-8} 4.0×10^{-8}, 1.4×10^{-7}	[4],[70],[88]
HFR (Petten, Netherlands)	(HTR-PM fuel)	5 spheres for HTR-PM LEU UO_2 TRISO	China	8 Sep. 2012 – 30 Dec. 2014 (355)	3.79–4.95	11.1–13.7 (estimated) >12.3	1050±50 (centre temp.) 1023, 1040, 1020, 1017, 1067 (ave. central temp.)	-	2.4-3.3×10^{-9} Fig. 10 in Ref. 13	[5],[12],[90–92]

Table 25 shows specifications of the spherical pebbles irradiated for HFR-EU1bis experiment [88]. Burnup measurements on the HFR-EU1bis spheres were measured and described in Refs. [69,70], which is shown in Table 26 [88].

TABLE 25. NOMINAL CHARACTERISTIC DATA OF AVR GLE-4/2 PEBBLES

Coated particle	**AVR**
Particle batch	HT 384–393
Kernel composition	UO_2
Kernel diameter (μm)	502
Enrichment (^{235}U wt.%)	16.76
Coat thicknesses (μm)	
Buffer	92
Inner PyC	40
SiC	35
Outer PyC	40
Particle diameter (μm)	916
Pebble	
Heavy metal loading (g/pebble)	6.0
^{235}U content (g/pebble)	1.00 ± 1%
Number of coated particles per pebble	9560
Volume packing fraction (%)	6.2
Free U fraction (U/U_{tot})	$7.8 \cdot 10^{-6}$
Matrix graphite grade	A3-3
Matrix density (kg/m^3)	1750
Temperature at final heat treatment (°C)	1900

TABLE 26. MEASURED CS-137 ACTIVITIES, CORRECTED FOR DECAY TO END OF IRRADIATION, RESULTING BURNUP AND NEUTRON FLUENCE

Pebble	**^{137}Cs (Bq)**	**^{137}Cs (Bq) at EOI**	**Burnup (% FIMA)**	**Thermal fluence (n/m^2, E<0.638 eV)**	**Fast fluence (n/m^2, E>0.1 MeV)**
HFR-EU1bis/1 (top)	6.19×10^{10}	6.48×10^{10}	9.34%	1.34×10^{25}	2.41×10^{25}
HFR-EU1bis/2	N/A due to destructive PIE				
HFR-EU1bis/3	7.34×10^{10}	7.68×10^{10}	11.07%	1.59×10^{25}	2.86×10^{25}
HFR-EU1bis/4	7.34×10^{10}	7.69×10^{10}	11.07%	1.59×10^{25}	2.86×10^{25}
HFR-EU1bis/5 (bottom)	6.43×10^{10}	6.73×10^{10}	9.70%	1.40×10^{25}	2.51×10^{25}

Some observations of interest from post-irradiation examinations are noted below:

— Gamma scanning of the pebbles showed inhomogeneities in the distribution of coated particles within the pebbles.
— Visual inspection of the pebbles confirmed the absence of corrosion phenomena through their shiny surface.
— High porosity and a significant number of metallic inclusions were generated due to irradiation in the kernel [93,94]. These porosities in the UO_2 kernel were more biased at the colder side of the particle from outer region of the pebble. Other particles showed no such bias and evenly distributed pores and metallic inclusions.

— It is noted from Ref. [94] that "larger metallic islands and larger porosity are observed within kernels from particles at the centre of the fuel pebble compared to kernels from particles from the outer region of the fuel pebble". As also described in Ref. [95], "the presence of the larger metallic islands in the centre of the pebbles is in agreement with the observation on other fuels (e.g. LWR fuel) where it is also observed that the higher temperature (in transient conditions, for example) and thus the larger thermal diffusion coefficient increases the size of these precipitates".

— As described in Ref. [96], "The network of intergranular porosities created in outer particles is to be only partially interconnected. In the fuel particles located in the centre of the pebble, larger porosities are observed compared to outer fuel particles. The pores on the cold side of the particles at the outer radius of the pebble fuel zone have a more irregular shape than the pores in the kernel bulk. Additionally, the UO_2 structure at the cold side of these kernels is different: there are hardly any grain boundaries visible, and metallic islands do not occur. This could indicate displacement of UO_2 from the kernels to the cold side, leaving irregular pores as a result. However, such UO_2 migration should be associated with the amoeba-effect (carbon migration), but for the particles considered it was clear that the kernels remained in the centre of the coating, and the buffer layer remained of uniform thickness". From this point of view, the particles irradiated in this test showed a very different aspect from that shown in Ref. [96] that is described in such a manner that "an example of very extensive amoeba attack was identified on a UO_2 TRISO particle after irradiation conditions rather similar to an irradiated German TRISO fuel (9% FIMA at 1373 K, 207 days). The difference in behaviour could be assumed to be linked to the difference of enrichment (9% versus 17% ^{235}U here). In less enriched fuel, a significant fraction of the fissions occurs via plutonium leading to release of a larger amount of oxygen and greater $CO + CO_2$ internal pressure."

— As additional information obtained from electron probe microanalyses (EPMA) [94]:
  - At room temperature, xenon either adhered to the buffer layer or presented within enclosed pores beneath the pyrolytic carbon. This supported a significant release of xenon from the kernel, and also was consistent with the report in Ref. [96].
  - While caesium was primarily detected in the buffer layer, its presence in the kernel was minimal, potentially originating from xenon decay.
  - Ruthenium was very largely contained within the metallic islands in the kernel, but some ruthenium was also found in the coatings. Palladium was distinguished as a small trace near the IPyC/SiC interface, but palladium was not present in the metallic inclusions.
  - Important observations were (1) palladium had accumulated near the IPyC/SiC, but not at the SiC layer (no Pd attack was observed), and (2) most caesium was found in the buffer layer (not in the other coating layers). The latter observation coincides with the presence of xenon in the buffer [5]. More evaluation and interpretation can be referred to Refs. [86,94].

Palladium is characterized with its high vapour pressure and affinity with variety of fission products to form inter-metallic compounds or mixed phases with volatile fission products [97]. For the outer-pebble zone samples, palladium is more clearly distinguished, albeit on the cold side of the sample, confirming similar observation presented in Ref. [96].

Comparison of destructive PIEs on samples taken from the centre and outside of the fuel zone of the pebble revealed a more deterioration of the SiC coatings in the central zone. The central zone sample showed apparent break-out of the SiC coating and the outer zone sample no cracking of the SiC coating [5].

Four spherical pebbles (HFR-EU1bis/1, HFR-EU1bis/3, HFR-EU1bis/4, and HFR-EU1bis/5) irradiated at the High Flux Reactor (HFR) in Petten for the HFR-EU1bis irradiation campaign were transported to JRC-ITU for PIEs and heating tests at the KÜFA facility.

Fractional release of fission products during the heating tests on these pebbles are described in Ref. [98] and are represented in Figs. 40 and 41. From these figures, the following observations were made [99]:

— During reactor normal operation, simulation of both ^{134}Cs and ^{137}Cs fractional release remained below 10^{-4};

— HFR-EU1bis/5 showed the highest release of Cs among the entire series due to high temperature (1800°C) and rapid cycling of temperature. (Note that HFR-EU1bis/5 was exposed to extended heating phases at temperatures of 1800°C as well as rapid temperature transients to 1800°C);

— HFR-EU1bis/4 yielded a release around 10^{-2} after extensive accident testing at 1700°C. The release was similar to the release from HFR-EU1bis/5 which was subjected to more extreme heating conditions, which might be attributed to an unfavourable operation of the test equipment for HFR-EU1bis/4.

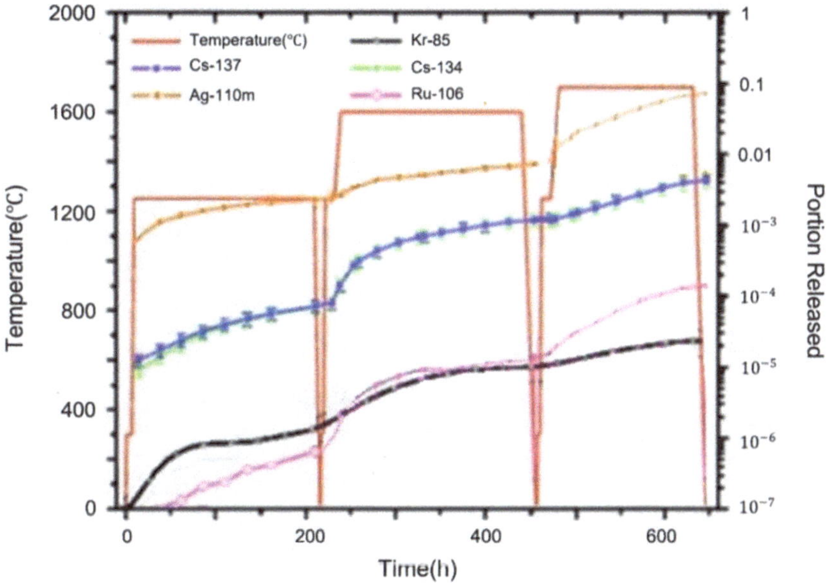

FIG. 40. Fractional release of various fission products during KÜFA test on HFR-EU1bis/1 through 15.

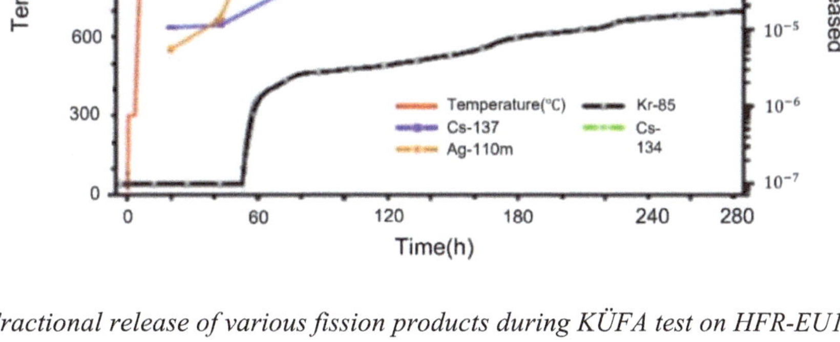

FIG. 41. Fractional release of various fission products during KÜFA test on HFR-EU1bis/4 and 5.

6.2.1.2. HFR-EU1 irradiation experiment

The irradiation experiment HFR-EU1 [5,88] was conducted with three spherical pebbles, each containing 9500 particles in the Petten HFR reactor. It aimed at exploring the possibility for high power and high burnup operation of the German AVR pebbles and of newly developed fuel for VHTR.

Figure 42 shows the schematic configuration of the test rig.

Table 27 shows specifications of test pebbles. Three AVR pebbles were produced by Hochtemperaturreaktor-Brennelement GmbH (HOBEG), Germany and two by INET, China.

Conditions for the irradiation test are described in Ref. [100] and are represented in Table 28 [100]. During the irradiation, fission gas releases were measured through sweep loop facility. Detailed information on irradiation conditions, facilities and test pebbles can be found from [5,88,100].

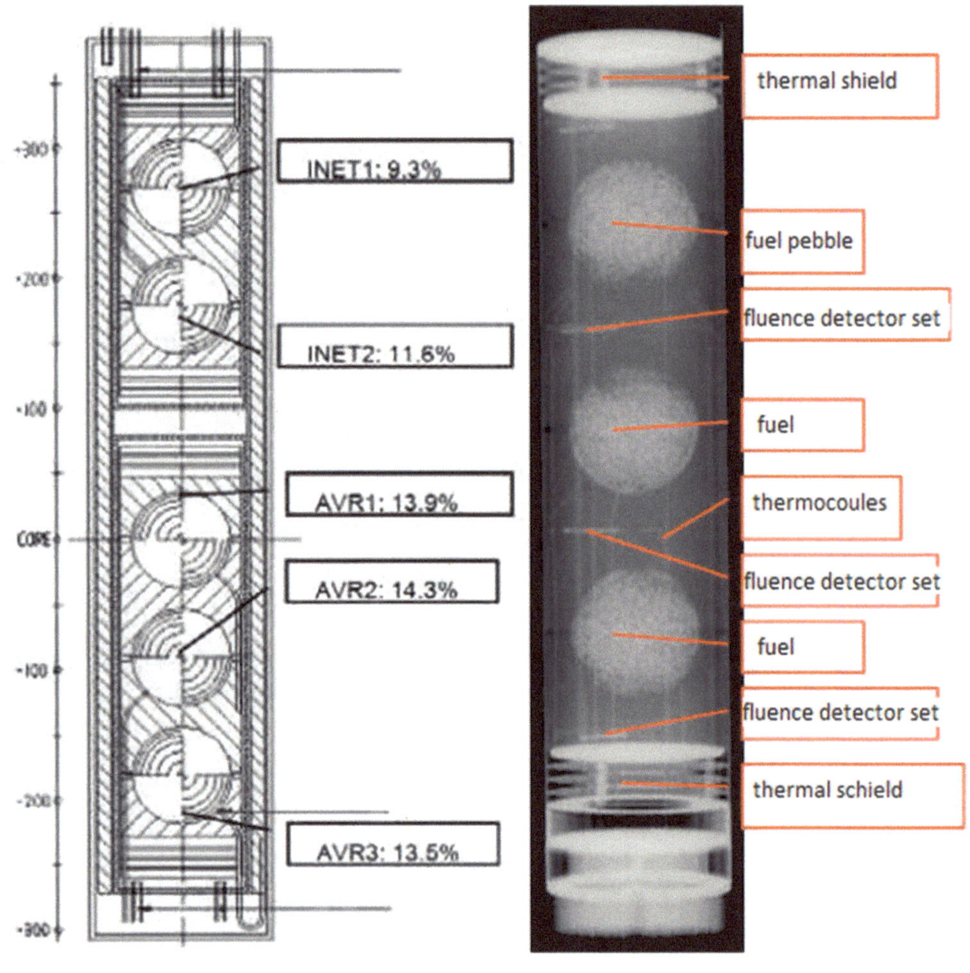

(a) (b)

FIG. 42 (a) Cross-section of the HFR-EU1 sample holder with upper Capsule 1 containing two Chinese spheres and lower Capsule 2 containing three German spheres; (b) pre-irradiation X ray of German capsule, Capsule 2.

TABLE 27. NOMINAL CHARACTERISTICS OF INET AND AVR GLE-4 PEBBLES OF HFR-EU1 IRRADIATION

	INET	AVR
Coated particle		
Particle batch	V000802	HT384-393
Kernel composition	UO_2	UO_2
Kernel diameter (µm)	490.3	502
Enrichment (^{235}U wt.%)	17.08	16.76
Coat thicknesses (µm)		
Buffer	97	92
Inner PyC	7 42	40
SiC	37.8	35
Outer PyC	40.8	40
Particle diameter (µm)	926.9	916
Pebble		
Heavy metal loading (g/pebble)	5.02	6.0
^{235}U content (g/pebble)	0.858	1.00 ± 1%
Number of coated particles per pebble	8500	9560
Volume packing fraction (%)	5.0	6.2
Free U fraction (U/U_{tot})	2.3×10^{-7}	7.8×10^{-6}
Matrix graphite grade	A3-3	A3-3
Matrix density (kg/m^3)	1760	1750
Temperature at final heat treatment (°C)	1900	1900

TABLE 28 HFR-EU1 IRRADIATION EXPERIMENT

	HFR-EU1
Pebble number and type	2 INET + 3 AVR
Start	30.09.2006
End	19.02.2010
Duration (EFPD)	445
Burnup (% FIMA)	11–14
Temperature (°C)	900 (INET) – 950 (AVR)
R/B at end of irradiation (based on ^{85m}Kr release)	Approx. 5.5×10^{-8} (INET) Approx. 1.6×10^{-7} (AVR)

Figure 43 shows the fractional release of fission products at end of irradiation [5]. The fractional release in Fig. 43 was approximately 2.5×10^{-7} based on ^{85m}Kr for AVR fuel [5,88] and was at least two orders lower than the fractional release that could come out in theoretical maximum from a single particle in the capsule (approximately 2.8×10^{-4} for the AVR capsule). Also note that HFR-EU1bis experiment showed a higher fractional release of approximately 4×10^{-6} [69,101].

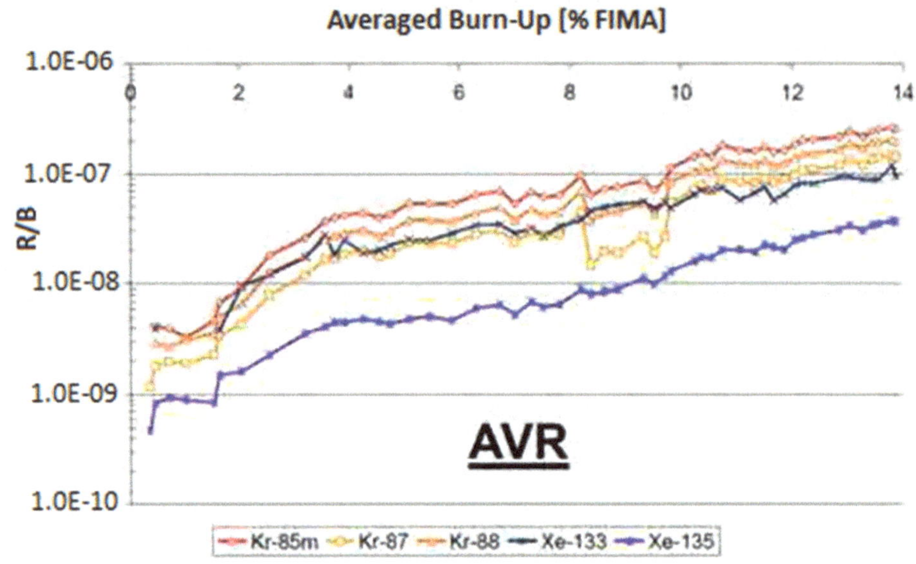

FIG. 43. R/B vs. burnup for AVR pebbles (reproduced from Ref. [5], courtesy of IAEA).

Figure 44 shows the fractional release of fission products from INET pebbles as a function of burnup [5]. The fractional release of ^{85m}Kr at the end of irradiation was approximately 8×10^{-8}, which was lower than those from the earlier experiments HFR-K5 and HFR-K6 (5×10^{-7}) and HFR-EU1bis experiment (approximately 4×10^{-6}) [69,101].

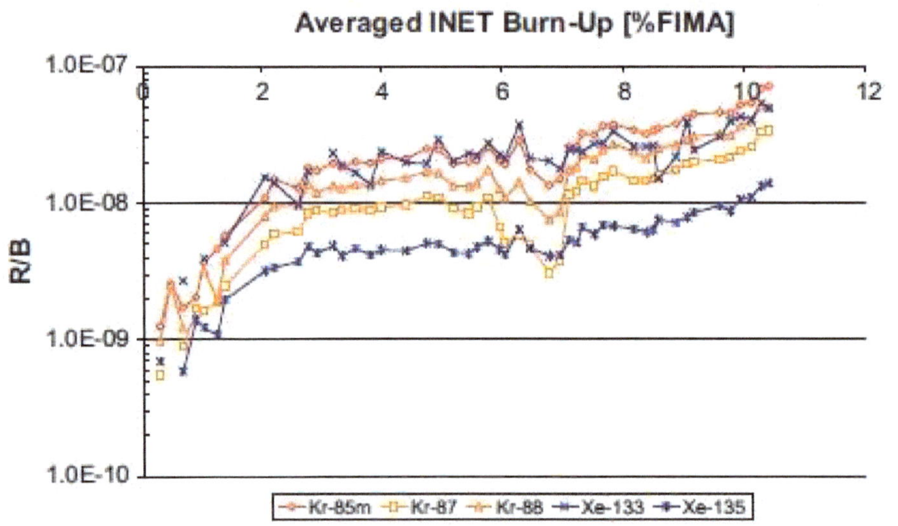

FIG. 44. R/B vs. burnup for INET pebbles (reproduced from Ref. [5], courtesy of IAEA).

6.2.1.3. HTR-10 fuel irradiation at HFR

The HTR-10 fuel was irradiated at the IVV-2M reactor during 2000–2003 [87]. Five capsules were contained in the irradiation rig. These capsules were numbered 1 through 5 from top to bottom. Each capsule contained:

— Capsule 1: 13,500 loose particles plus 60 samples of matrix graphite;
— Capsule 2: a spherical pebble (identification no: SFE5);
— Capsule 3: a spherical pebble (identification no: SFE12);
— Capsule 4: a spherical pebble (identification no: SFE8);
— Capsule 5: a spherical pebble (identification no: SFE7).

Table 29 [102] represents specifications of test pebbles. Note that test pebbles SFE5 and SFE7 were taken from batch 1 and test pebbles SFE8 and SFE12 from batch 2. Irradiation conditions are described in Refs [87,102] and are summarized in Table 30.

TABLE 29. MAIN CHARACTERISTICS OF THE HTR-10 PEBBLES

Batch no.	**1**[c]	**2**[d]
UO_2 kernel		
Enrichment (%)	17	17
Diameter (μm)	500.9	497.0
Density (g/cm^3)	10.81	10.90
Sphericity (D_{max}/D_{min})	1.04	1.06
O/U ratio	2.00	2.00
B equivalent (μg/g)	0.49	0.90
Coated fuel particle		
Buffer PyC layer thickness (μm)	91.0	96.0
IPyC layer thickness (μm)	48.1	44.7
SiC layer thickness (μm)	35.0	36.6
OPyC layer thickness (μm)	41.6	42.1
Buffer PyC layer density (g/cm^3)	1.02	1.05
IPyC layer density (g/cm^3)	1.84	1.84
SiC layer density (g/cm^3)	3.20	3.20
OPyC layer density (g/cm^3)	1.90	1.84
IPyC layer OPTAF[a]	1.03	1.02
OPyC layer OPTAF	1.03	1.03
Fuel ball		
Density (g/cm^3)	1.73	1.72
Total ash (μg/g)	91.0	94.5
Li content (μg/g)	<0.01	<0.02
B equivalent (μg/g)	1.14	0.87
Corrosion rate (He with 1% H_2O, 1000°C, 10 h) (mg/cm^{2}h)	1.0	1.2
Erosion rate (μg/g)	3.0	3.0
Number of drops (4 m in height)	375	>200
Breaking loading (KN)	23.9	22.0
Thermal conductivity (1000°C) [W/cm·K]	29.5	30.8
CTE[b] anisotropy ($\alpha_{\perp}/\alpha_{//}$)	1.1	1.1
U loading (g)	5.00	4.98
U contamination	1.1×10^{-6}	7.6×10^{-7}
U_{free}/U_{total}	1.4×10^{-4}	2.3×10^{-4}

[a] Optical anisotropy factor.
[b] Coefficient of thermal expansion.
[C] Test pebbles SFE5 and SFE7 were taken from batch 1.
[d] Test pebbles SFE8 and SFE12 were taken from batch 2.
* Information taken from Ref. [102].

TABLE 30. HTR-10 TEST FUEL IRRADIATION DATA

Start date	July 2000			
End date	February 2003			
Irradiation temperature (°C)	1000±50			
Capsule number	2	3	4	5
Fuel batch number/sphere ID	1/SFE5	2/SFE12	2/SFE8	1/SFE7
Burnup (% FIMA)	10.4	11.4	10.9	10.1
Fast neutron fluence (10^{25} n/m^2)[a]	1.10	1.31	1.30	1.06
EOL ^{85m}Kr R/B[b]	1.7×10^{-5}	6.7×10^{-7}	Not measured	7.8×10^{-6}
Capsule notes		In-pile heat-up test to 1200 and 1250°C for 200 h each	Capsule failed after 3.9% FIMA	In-pile heat-up test above 1600°C at end of irradiation

[a] Cut-off energy for fast neutron fluence not specified.

[b] The tabulated end of life (EOL) R/B for capsule 5 was measured at the end of the irradiation test phase and prior to the start of the high-temperature heat-up test.

Figure 45 shows the levels of fractional release (R/B ratio) of ^{135}Xe from four irradiated pebbles as a function of the burnup [5]. From Fig. 45, it can be noted [87]:

- Overall, the R/B values did not show significant increases and varied with only small fluctuations at temperatures of ~1000°C during irradiation. This implies that most likely irradiation-induced failures of the coated particles can be eliminated during normal operation.
- The high R/B levels were mainly attributed to the high failure fractions of fuel particles easily encountered in the early stage of fuel particles fabrication. The measured amount of the average tramp uranium from the four pebbles indicated the failures of some particles during fabrication.

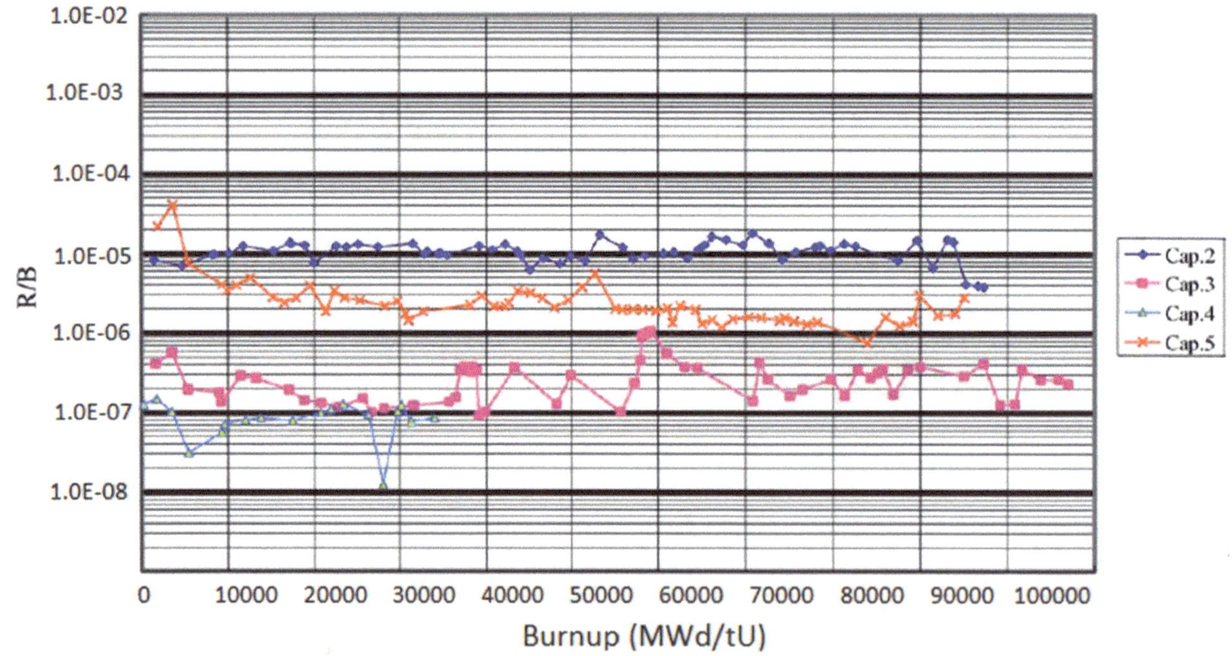

FIG. 45. R/B of ^{135}Xe of four irradiated fuel elements as a function of the burnup (reproduced from Ref. [5], courtesy of IAEA).

Table 31 demonstrates irradiation induced deformation of pebbles [89]; see dimensional changes of test pebbles (SFE5, SFE12 and SFE7) after the irradiation [89].

TABLE 31. NEUTRON-INDUCED DIAMETER CHANGE OF THE FUEL ELEMENTS AT 1000°C

Fuel element		Fast neutron fluence (E>0.1 MeV)/(10^{21} cm^{-2})	Before irradiation (mm)	After irradiation (mm)	Relative change (%)
Parallel to equatorial plane of fuel sphere	SFE5	1.10	60.03	59.62	-0.68
	SFE12	1.31	60.04	59.67	-0.62
	SFE7	1.06	60.06	59.66	-0.67
Perpendicular to equatorial of fuel sphere	SFE5	1.10	59.90	59.48	-0.70
	SFE12	1.31	60.06	59.62	-0.73
	SFE7	1.06	60.10	59.67	-0.72

During the irradiation, Capsule 4 failed due to a chemical reaction of the matrix graphite contained in test pebble SFE8 with impurities ingressed through the helium loop, e.g. oxygen of ingressed air. Due to the heat released from this chemical reaction, thermocouple was damaged, and the integrity of the graphite matrix was lost. This reaction caused the matrix graphite of the test pebble SFE8 to form a graphite powder agglomerate which was unbreakable during PIE. The loss of integrity of Capsule 4 resulted in the failure of all particles embedded in the pebble SFE8.

Figure 46 shows the fission gas release of SFE8 in Capsule 4, indicating a massive failure of coated particles [5].

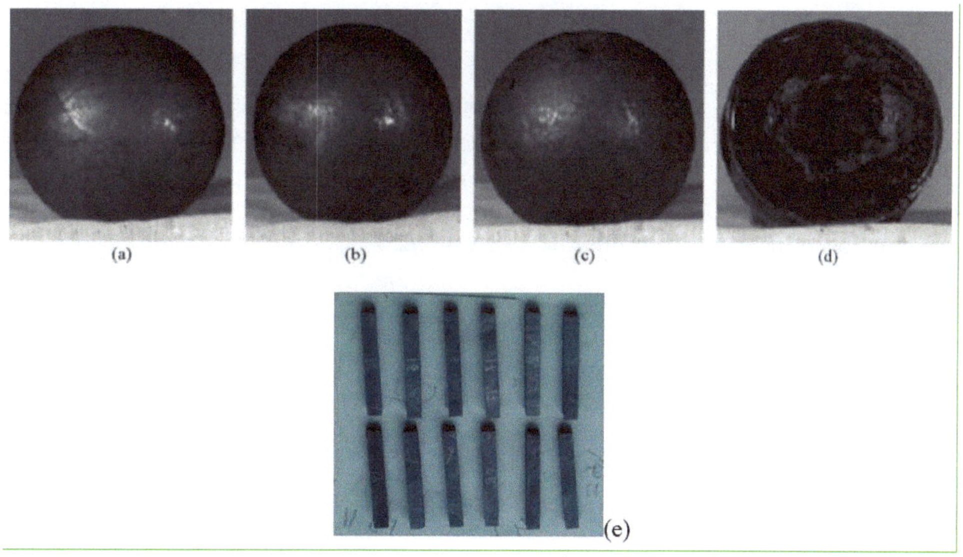

FIG. 46. Appearance of irradiated fuel elements in (a) Capsule 2; (b) Capsule 3; (c) Capsule 5; (d) Capsule 4;and in (e) matrix graphite specimens (courtesy of IAEA).

The microstructure of the well-coated particles was examined both before and after irradiation, extracted from Capsule 2. Remarkably, the PyC and SiC layers in these particles remained fully intact. No observable cracks or other radiation-induced defects were noted in these layers. The buffer layers retained an adequate thickness, and there was no evidence of damage in the IPyC layers, which could have potentially resulted from fission fragments.

The temperature of test pebble SFE12 in Capsule 3 was increased from 1050 to 1200°C and then up to 1250°C, which resulted in an increase of fission gas release. When the temperature returned to 1050°C, the R/B ratio was restored to its initial level.

Failures during irradiation were determined by using the criterion of activity ratio between caesium (^{134}Cs or ^{137}Cs) and ^{144}Ce. The activity ratio indicates the ratio of activities between a volatile fission product and a non-volatile fission product. The boiling point of Cs is 678°C, and Ce 3470°C. During the high temperature irradiations, Cs escaped much more readily from a defective coating than Ce. If the activity ratio between Cs and Ce from a particle is less than A_{mean} -3σ (where A_{mean} is the mean average of the activity ratios and σ standard deviation), this particle is considered failed.

A total of 2,014 particles from the fragmented pebble SFE12 were scrutinized using the specified criteria, resulting in the identification of one failed particle among the 2,014 examined. Ceramographic sections, ranging from 200 to 400 coated particles, were meticulously prepared within a hot cell. A detailed examination of these sections using an optical microscope revealed that the PyC and SiC layers in these particles remained unaltered. No cracks, exfoliation, or other irradiation-induced defects were evident in these layers. There were no signs of corrosive impact from fission products on the SiC layer. Notably, the buffer layer maintained an ample thickness, and no damages were observed in the subsequent IPyC layer, as well as in the SiC and OPyC layers.

As stated earlier, test pebble SFE7 contained in Capsule 5 showed a significant number of failures of particles during the in-pile heating test. Test pebble SFE7 was disintegrated by electrolysis to obtain loose coated particles for the examination of failed particles. By examining the activity ratio of ^{134}Cs to ^{144}Ce, 127 coated fuel particles were determined to have failed, amounting to a failure fraction of approximately 6%. Such massive failures were believed to have caused by the operation at higher temperatures than 1600°C, from PIE results. Cracked layers that were most likely caused by the very high heating temperature and the resultant high stress.

6.2.1.4. HTR-PM fuel irradiation at HFR

To validate the manufactured fuel for the HTR-PM, an in-pile irradiation test was conducted. INET produced ten HTR fuel pebbles, which were subsequently delivered to NRG for irradiation at HFR in 2012. These HTR fuel pebbles are composed of a moulded graphite matrix with a diameter of 60 mm, encompassing coated particles within an inner zone of 52 mm diameter. The fuel-free shell surrounding this inner zone is more than 4 mm in thickness. Table 32 shows the specification of HTR-PM pebbles [91].

To create a cylindrical structure housing approximately 60,000 TRISO particles, a stack of five fuel pebbles was enveloped in graphite half-shells. This assembly was then enclosed within a stainless-steel containment, serving as the initial barrier for the fuel pebbles. Within this containment, 24 thermocouples were placed near the pebbles and in the graphite half-shells for temperature measurements. Additionally, continuous gas flow was maintained within the containment to enable online measurements of fission gas release.

The irradiation plan aimed to achieve burnup levels exceeding 12.3% FIMA with central temperatures of 1050±50°C for the upper four pebbles. This extended over a duration of 355 full power irradiation days. Neutronics calculations indicated total fast fluence levels ranging between 3.79 and 4.95×10^{25} n/m^2 (E > 0.1 MeV). The estimated burnup was 11.1% FIMA for Pebble 5 and between 12.6 and 13.7% FIMA for the remaining pebbles. Central temperatures consistently hovered around the specified target boundaries of 1050±50°C.

TABLE 32. NOMINAL CHARACTERISTIC DATA OF INET PEBBLES

Pebble properties	
Particle batch	1
Kernel composition	UO_2
Kernel diameter (μm)	500
Enrichment (U-235 wt.%)	17
Thickness of coatings (μm):	
Low density graphite (buffer)	95
Inner high-density Pyrolytic	40
Carbon or "PyC"	35
High density SiC	40
Outer high density PyC	
Particle diameter (μm)	920
Pebble	
Heavy metal loading (g/pebble)	7
U-235 content (g/pebble)	1.19
Number of coated particles (-)	
per pebble	12,000
Matrix graphite grade (-)	A3-3
Matrix density (g/cm^3)	1.75
Temperature at final heat treatment (°C)	1900

All five spheres underwent irradiation in a single test rig, with online fission gas measurements capturing cumulative releases from the entire set. Remarkably, R/B ratios remained very low ($(2.4–3.3)\times10^{-9}$ at the end of irradiation), consistently below the calculated value (1.1×10^{-7}) for ^{85m}Kr from a single failed particle. This outcome strongly suggested the absence of manufacturing defects and the absence of in-pile particle failures.

From PIEs, the dimensional shrinkage in all 5 pebbles was found in the range between 0.88% and 1.25% [92].

Burnups were determined based on measured activities of ^{137}Cs [92] and are represented in Table 33.

TABLE 33. CS-137 ACTIVITY AT EOI AND CALCULATED BURNUP OF FUEL ELEMENTS FOR HTR-PM

Fuel element	**Cs-137 activity at EOI (Bq)**	**Calculated burnup (%FIMA)**
HTR-PM 1	9.27×10^{10}	11.64
HTR-PM 2	9.97×10^{10}	12.53
HTR-PM 3	1.01×10^{11}	12.66
HTR-PM 4	9.47×10^{10}	11.89
HTR-PM 5	8.05×10^{10}	10.11

Heating tests were carried out with three irradiated pebbles (HTR-PM 1, 2 and 4) in the KüFA facility. Figure 47 shows the temperature histories and measured fractional releases during the heating test with HTR-PM 1 [102].

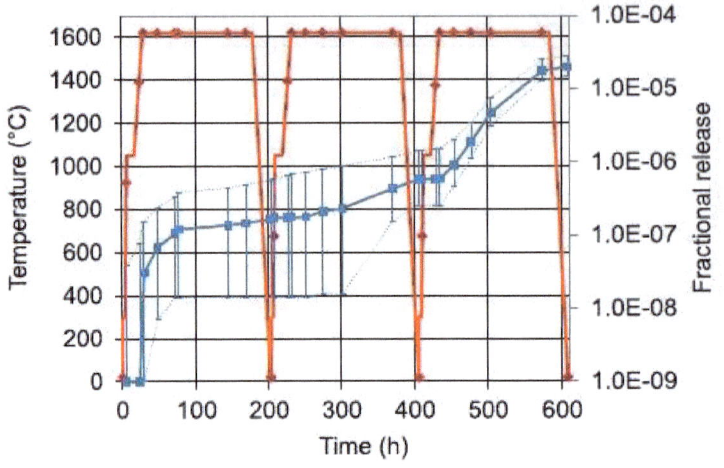

FIG. 47. HTR-PM 1 heating programme (red) and fractional release of Cs-137 (blue) (courtesy of INL).

Figure 48 shows the temperature histories and measured fractional releases during the heating test with HTR-PM 4 [102].

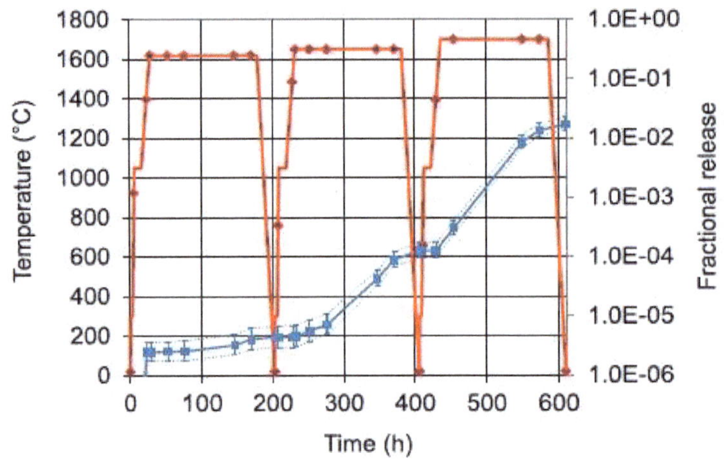

FIG. 48. HTR-PM 4 heating programme (red) and fractional release of Cs-137 (blue) (courtesy of INL).

Figures 49 and 50 show the temperature histories and measured fractional releases during the heating test with HTR-PM 2 [102].

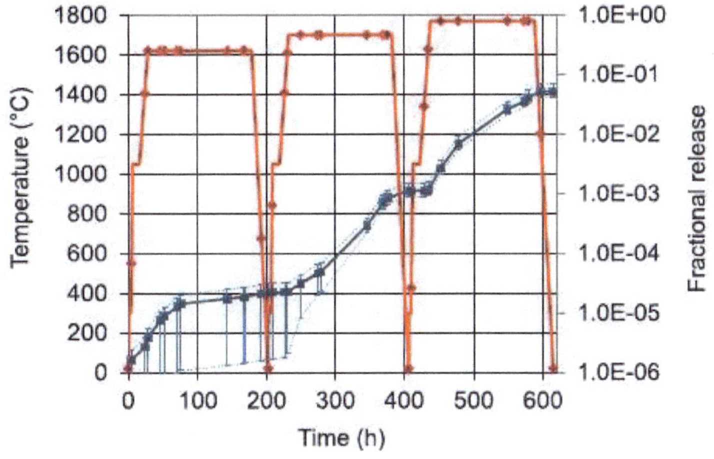

FIG. 49. HTR-PM 2 heating programme (red) and fractional release of Cs-134 (blue) and Cs-137 (grey) (courtesy of INL).

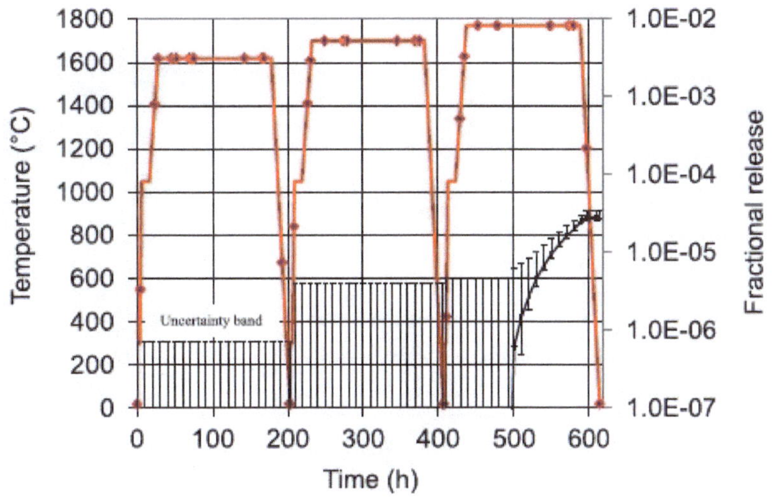

FIG. 50. HTR-PM 2 Kr-85 fractional release plus uncertainty (courtesy of INL).

6.2.2. Cylindrical compacts

Irradiations of cylindrical compacts have been pursued since the early 2000s mainly for the qualification of coated particle fuels for use in VHTRs. Table 34 represents relevant information.

In the United States, the Advanced Gas Reactor (AGR) Fuel Development and Qualification Program was initiated to qualify coated particle fuels within an operational range that encompasses both pebble bed and prismatic reactors. The objective was to develop a fuel form capable of withstanding peak fuel temperatures averaging 1250°C and achieving high burnups ranging from 150 to 200 GWd/tHM or 16.4 to 21.8% FIMA. UCO was chosen as the fissile kernel for this program due to its ability to restrict carbon monoxide (CO) production and prevent kernel migration under irradiation. These factors were deemed potential life-limiting issues for traditional UO_2 TRISO fuels when operating at elevated temperatures (~1250°C) and extended burnup, characteristic of a prismatic HTGR.

Four irradiation tests have been performed at the Advanced Test Reactor (ATR) at INL (AGR-1, AGR-2, AGR-3/4, and AGR-5/6/7; see Table 34 for irradiation experiment details). PIE including high-temperature annealing tests of compacts from the AGR-1 experiment followed. PIE including annealing tests of AGR-2, AGR-3/4 and AGR-5/6/7 are planned.

The AGR program has already been described in the literature [4,5,12,70,103–106], and is not described in detail here. The progress and results of irradiation testing, PIE, and annealing testing for each AGR experiment are briefly described in the following sections.

6.2.2.1. AGR-1 irradiation experiment

Figure 51 [110] shows a schematic diagram of the AGR-1 irradiation rig.

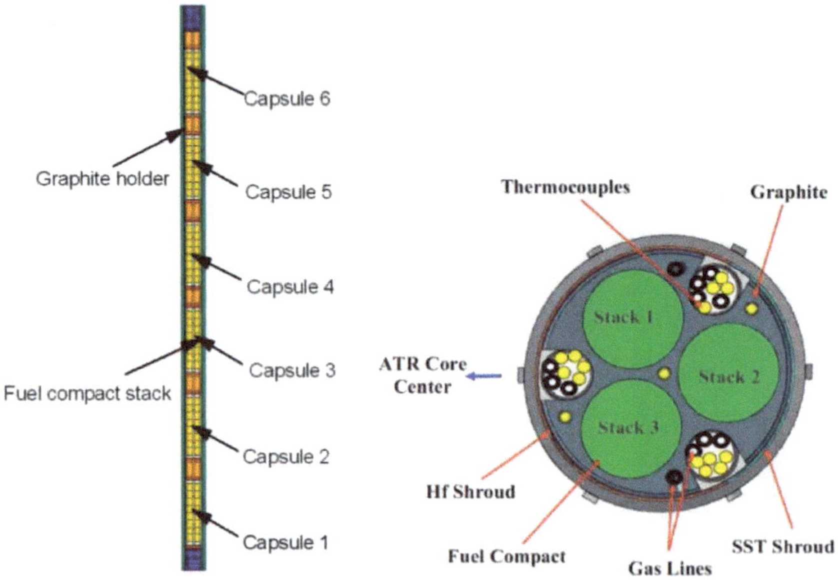

FIG. 51. Axial and radial cut schematics of the AGR-1 capsule (courtesy of INL).

The experiment used coated particle and fuel compacts fabricated at the laboratory scale at ORNL. One of the goals of the experiment was to explore the impact of coating microstructure and properties on fuel performance by including several different types of TRISO-coated particles. There were six separate capsules, each containing 12 fuel compacts. The baseline fuel underwent irradiation in Capsules 6 and 3, Variant 1 in Capsule 5, Variant 2 in Capsule 2, and Variant 3 in Capsules 1 and 4.

TABLE 34. IRRADIATIONS OF CYLINDRICAL COMPACTS SINCE THE EARLY 2000s

Reactor	Experiment ID	No. of specimen and fuel type (mm)	Country	Irradiation time (FPDs)	Fast fluence (10^{21}n·cm^{-2}, E>0.18 MeV)	Burnup (%FIMA)	Operating temp. (°C)	Defective SiC layer fraction (U/U-total)	BOL ^{85m}Kr *R/B* EOL ^{85m}Kr *R/B*	Reference
ATR (INL, USA)	AGR-1	UCO; 72 compacts (dia. 12.34–12.36, height 25.0-25.3) LEU UCO TRISO	USA	24 Dec. 2006 – 30 Oct. 2009 (621)	3.2, 3.8, 4.1, 4.0, 3.7, 3.0 (peak)	17.5, 18.9, 19.5, 19.3, 18.3, 16.2 (max.)	1136, 1118, 1118, 1180, 1152, 1117 (time-ave. peak. Temp.) 1029, 991, 980, 1041, 1005, 980 (time-ave. vol.-ave. temp.)	< 4×10^{-5}	8×10^{-8}, 1×10^{-8}, 6×10^{-9}, 9.0×10^{-9}, 1×10^{-8}, 1×10^{-8} 9×10^{-8}, 4×10^{-8}, 1×10^{-8}, 5×10^{-8}, 2×10^{-7}, 1×10^{-7}	[5],[12],[70], [107–108]
ATR (INL, USA)	AGR-2 included French compacts and South Africa compacts	UCO; 36 compacts UO_2; 12 compacts LEU UCO and UO_2 TRISO	USA France South Africa	Jun. 2010–Nov. 2013 (560)	3.47 (UCO) 3.05–3.53 (UO_2)	7.3–13.2 (UCO) 9.0–10.7 (UO_2)	1080–1360 (UCO) 1072–1105 (UO_2) (time-ave. peak temp.)		~10^{-6} (UCO) ~10^{-7} (UO_2) (3 cycle irradiation)	[12],[82], [103],[108]
ATR (INL, USA)	AGR-3/4 (designed to fail (DTF) fraction; 1×10^{-2}	UCO; 48 compacts 20 DTF particles/ compacts LEU UCO TRISO	USA	Dec. 2011 – Apr. 2014 (370)	1.19–532	4.9–15.3	845–1276 (time-ave. peak temp.)		10^{-4} ~ 10^{-3}	[12],[103]
ATR (INL, USA)	AGR-5/6/7	194 compacts LEU UCO TRISO	USA	Feb. 2018 – Jul. 2020 (~500)	1.5–4.0	7.4–18.6 (ave. ≥6)	AGR-5/6; 750-1350 (time-ave. peak temp.) AGR-7; 1350-1550 (time-ave. peak temp.)			[12],[103]
HANARO (KAERI, Republic of Korea)	12F-01K	14 compacts (dia. 8.0, height 10.0) LEU UO_2 TRISO	Republic of Korea	Aug. 2013–Mar. 2014 (132)	0.3	3.81 (max.)	1083 (calculated max.)	-	-	[5],[109]

A brief description of the coating property variations in these fuel types is given below [56]:

— Baseline fuel — Due to its outstanding irradiation performance, the coating process conditions employed in the fabrication of the historical German fuel were selected as the initial basis for the baseline fuel. Subsequent parametric studies fine-tuned these conditions, specifically tailored to the coater utilized for coating AGR-1 fuel. The goal was to generate coatings deemed most likely to exhibit successful performance during the AGR-1 irradiation, as well as in PIE and annealing testing. These adjustments were informed by assessments of the relative performance of previously irradiated TRISO fuels [10].
— Variant 1 — The temperature for coating the IPyC was elevated compared to the baseline process. This adjustment was anticipated to potentially improve the irradiation dimensional stability of the pyrocarbon, albeit with a potential trade-off of increased uranium dispersion due to higher permeability. The IPyC layer exhibited a slightly lower density than the baseline, and the anisotropy of the IPyC layer was also marginally reduced.
— Variant 2 — The IPyC coating gas fraction was increased relative to the baseline process for this variant. This change was expected to possibly reduce permeability, which could decrease uranium dispersion, but possibly at the expense of higher anisotropy reducing the irradiation dimensional stability of the pyrocarbon. The IPyC layer density was slightly higher than the baseline density and the IPyC anisotropy was also slightly higher after 1800°C heat treatment of the compacts.
— Variant 3 — The carrier gas composition for the SiC layer deposition was changed from hydrogen to an argon-hydrogen mixture. The introduction of the argon diluent resulted in more re-nucleation during SiC deposition and the optimal deposition temperature was lowered. This lower temperature was expected to possibly reduce SiC defects resulting from uranium dispersion, and the SiC microstructure had significantly finer and more equiaxed grains than in the other fuel types above [111], which could impact irradiation performance.

Tables 35, 36 and 37 represent properties of kernel, particles and compacts, respectively, taken from Refs. [110, 113].

TABLE 35. SELECTED PROPERTIES FOR AGR-1 KERNEL LOT G73D-20-69302

Kernel property	**Specified range for mean value**	**Actual mean value ± population standard deviation**
Diameter (μm)	350±10	349.7 ± 9.0
Density (kg/m^3)	≥ 10.4	10.924 ± 0.0015
U-235 enrichment (wt %)	19.80±0.10	19.736 ± 0.047
Carbon/uranium (atomic ratio)	0.50±0.20	0.3253 ± 0.0028
Oxygen/uranium (atomic ratio)	1.50±0.20	1.3613 ± 0.0064
(Carbon + oxygen)/uranium (atomic ratio)	Not specified	1.6850 ± 0.0093
Total uranium (wt. %)	Not specified	90.059 ± 0.086
Sulfur impurity (ppm – wt.)	Not specified	608
All other impurities	Various	Below minimum detection limits and within specification

TABLE 36. SELECTED PROPERTIES FOR AGR-1 COATED PARTICLE COMPOSITES

Property	Specified range for mean	Actual mean value ± Population standard deviation			
		Baseline	Variant 1	Variant 2	Variant 3
Buffer thickness (μm)	100±15	103.5±8.2	102.5±7.1	102.9±7.3	104.2±7.8
IPyC thickness (μm)	40±4	39.4±2.3	40.5±2.4	40.1±2.8	38.8±2.1
SiC thickness (μm)	35±3	35.3±1.3	35.7±1.2	35.0±1.0	35.9±2.1
OPyC thickness (μm)	40±4	41.0±2.1	41.1±2.4	39.8±2.1	39.3±2.1
Buffer density (Mg/m^3)	0.95±0.15	1.10±0.04	1.10±0.04	1.10±0.04	1.10±0.04
IPyC density (Mg/m^3)	1.90±0.05	1.904±0.014	1.853±0.012	1.912±0.015	1.904±0.013
SiC density (Mg/m^3)	Not specified	3.208±0.003	3.206±0.002	3.207±0.002	3.205±0.001
OPyC density (Mg/m^3)	1.90±0.05	1.907±0.008	1.898±0.009	1.901±0.008	1.911±0.008
IPyC anisotropy[(a)] (BAF_o) and diattenuation[(b,c)]	Not specified	1.022±0.002 (0.0074±0.0007)	1.014±0.001 (0.0047±0.0005)	1.023±0.002 (0.0075±0.0006)	1.029±0.002 (0.0098±0.0007)
OPyC anisotropy (BAF_o) and diattenuation[(b,c)]	Not specified	1.019±0.003 (0.0063±0.0009)	1.013±0.002 (0.0044±0.0006)	1.018±0.001 (0.0059±0.0004)	1.021±0.003 (0.0071±0.0008)
IPyC anisotropy (BAF_o) and diattenuation post compact anneal[(b,c)]	Not specified	1.033±0.004 (0.0109±0.0013)	1.021±0.003 (0.0071±0.0006)	1.036±0.001 (0.0121±0.0004)	1.034±0.003 (0.0114±0.0009)
OPyC anisotropy (BAF_o) and diattenuation post compact anneal[(b,c)]	Not specified	1.033±0.003 (0.0111±0.0009)	1.030±0.003 (0.0100±0.0011)	1.029±0.004 (0.0097±0.0013)	1.036±0.002 (0.0120±0.0007)
Sphericity (aspect ratio)	Mean not specified[(d)]	1.054±0.019	1.056±0.019	1.053±0.019	1.055±0.018
Particle diameter[(e)] (μm)	Mean not specified	800±14	804±13	798±14	795±15
Particle mass (g)	Mean not specified	$7.27 \cdot 10^{-4}$	$7.33 \cdot 10^{-4}$	$7.24 \cdot 10^{-4}$	$7.26 \cdot 10^{-4}$

(a) Specification does not apply to Variants 1 and 2.

(b) The first number is the equivalent-BAF_o, defined as 1+3N, where N is the diattenuation. This value was discarded in favour of the optical anisotropy factor (OAF = [1+N]/[1-N]) in later AGR programme irradiation experiments but is listed here because the AGR-1 fuel specification was defined in terms of equivalent-BAF_o.

(c) Values in parentheses are the diattenuation (N) that was measured with the Two-Modulator Generalized Ellipsometry Microscope (2-MGEM) developed at ORNL [112].

(d) Critical region is specified such that ≤ 1% of the particles shall have an aspect ratio ≥1.14.

(e) Based upon mean average particle measurements, not sums of mean layer thickness.

TABLE 37. SELECTED PROPERTIES FOR AGR-1 COMPACTS

Property	Specified range for mean value	Actual mean value ± Population standard deviation			
		Baseline	**Variant 1**	**Variant 2**	**Variant 3**
Compact mass (g)	Not specified	5.4789	5.3371	5.3736	
Mean uranium loading (g U/compact)	0.905±0.04	0.917	0.915	0.904	0.912
Diameter[b] (mm)	12.22–12.46	12.36±0.01	12.36±0.01	12.36±0.01	12.34±0.01
Length[b] (mm)	25.02–25.40	25.066±0.080	25.066±0.030	25.077±0.065	25.227±0.037
Number of particles per compact[a]	Not specified	4154	4145	4095	4132
Particle volume packing fraction[a] (%)	Not specified	36.99	37.42	36.26	36.04
Effective overall compact density[a] (Mg/m^3)	Not specified	1.822	1.771	1.786	1.854
Compact matrix density[a] (Mg/m^3)	Not specified	1.297	1.219	1.256	1.344
U contamination fraction[c] (g exposed U/g U in compact)	Not specified	0	0	0	0
Defective SiC coating fraction[I]	Not specified	4.0×10^{-5}	0	2.0×10^{-5}	0
Defective IPyC coating fraction©	Not specified	0	0	0	0
Defective OPyC coating fract©[c]	Not specified	0	9.6×10^{-4}	0	0

(a) Value derived from other characterized properties.

(b) Allowable range corresponding to upper and lower critical limits specified with no compacts exceeding the limits, which require 100% inspection of all compacts.

(c) Values are measured fraction in each analysed sample, and are an estimate of an attribute property, not the mean of a variable property.

Figure 52 shows the diametric changes of AGR-1 compacts. The data indicate the increase of compact diameter shrinkage up to a fast fluence of approximately 3×10^{25} n/m^2. Above 3×10^{25} n/m^2, the observed trend shows minimal dependence on fast fluence, indicating either negligible dependence or a slight decrease in shrinkage as fast fluence increases. It is important to note that other factors, such as irradiation temperature, could potentially exert a significant influence on the overall dimensional change.

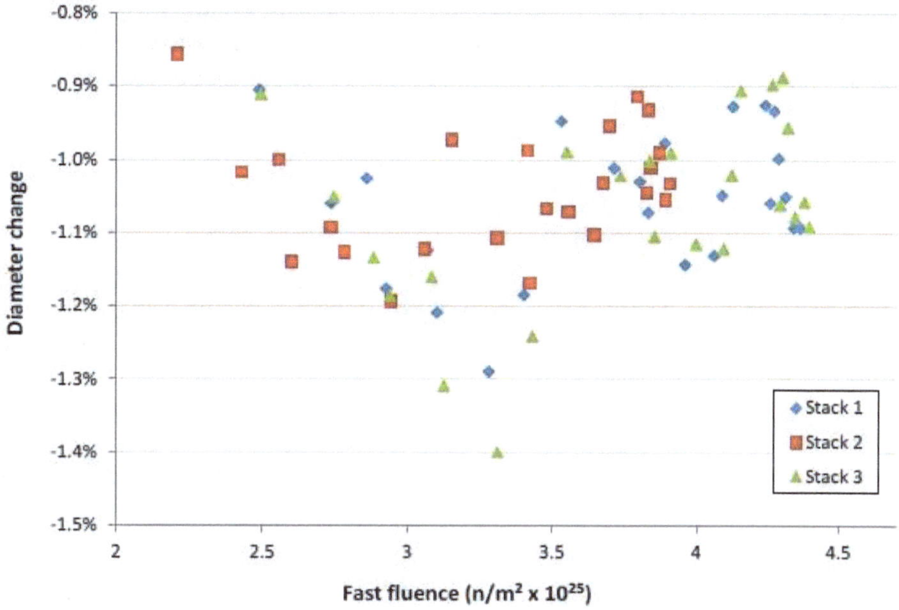

FIG. 52. Average diameter change for all AGR-1 compacts as a function of calculated fast neutron fluence (E>0.18 MeV) (courtesy of INL).

The prevalent characteristic observed was a radial gap between the buffer and IPyC layers, where the buffer underwent inward densification during irradiation, as illustrated in Fig. 53. AGR-1 buffers were initially deposited at a nominal density of 1.1 g/cm^3 [110], representing 49% of the theoretical density [114]. However, Fig. 53 suggests that substantial porosity persisted post-irradiation, indicating incomplete densification. As depicted by the curved red arrow in Fig. 53 [107], the buffer and its encapsulated kernel were not securely retained within the outer TRISO shells after the buffer and IPyC layers had completely debonded. To stabilize loose buffers and kernels, fluid epoxy needed to be vacuum-impregnated (back-potted) into these gaps during grinding through the saw-damaged thickness on each mount. Failure to perform frequent back-potting posed the risk of dislodged kernels breaking into coarse abrasive fragments, potentially causing damage to the mount by notching the surface with deep scratches.

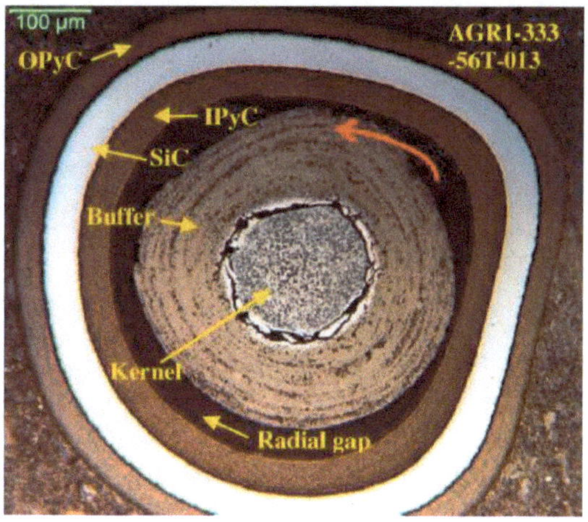

FIG. 53. Buffer and kernel that rotated before the radial gap was filled by back-potted epoxy (courtesy of INL).

The results of AGR-1 PIE and annealing testing are described in detail in Refs. [107,115–116]. The annealing test was undertaken in the FACS furnace at the INL and the Core Conduction Cooldown Test Facility at ORNL. Both systems can heat fuel specimens at temperatures up to 1800°C in inert atmospheres while monitoring the release of volatile fission products and gases.

A notable discovery was the remarkably low levels of caesium released through intact coatings, observed both during irradiation and in post-irradiation heating tests conducted at temperatures of 1800°C. It was determined that the release of caesium from a particle was predominantly linked to discrete failures in the SiC layer. In instances where the SiC layer experienced failure, a particle could release the majority of its caesium inventory. Conversely, in the absence of such failures, the release was exceptionally low, with capsule-average release from compacts (in the absence of SiC failures during irradiation) measuring less than 3×10^{-6}.

The primary cause of SiC failure was identified as a two-part mechanism involving (1) fracture of the IPyC layer due to densification of an adherent buffer layer, and (2) localized accumulation of palladium at the IPyC-SiC interface in the region of the IPyC fracture, leading to subsequent degradation of the SiC layer through chemical reaction with palladium. Instances of widespread chemical attack on the SiC layer by palladium were not observed in the absence of IPyC layer fracture. Four particles in the AGR-1 experiment experienced SiC layer failure during irradiation, resulting in an in-pile SiC failure fraction of $\leq 3.1\times10^{-5}$ at a 95% confidence level. Importantly, no particles experienced TRISO failure (failure of all three dense coating layers) during the irradiation, yielding a particle failure fraction of $\leq 1.1\times10^{-5}$ at 95% confidence.

Post-irradiation annealing test fuel performance was generally excellent. For instance, the maximum fractional releases from fuel compacts heated for 300 h at 1600°C were approximately 2×10^{-4} (^{134}Cs), 3×10^{-3} (^{154}Eu), 2×10^{-3} (^{90}Sr), and 4×10^{-6} (^{85}Kr). At 1800°C, the maximum releases for ^{154}Eu, ^{134}Cs, and ^{90}Sr were roughly an order of magnitude higher. In the case of ^{134}Cs, release was linked to an increase in the average number of SiC failures as the test temperature rose; in the absence of SiC failures, the maximum release at 1600°C was approximately 5×10^{-6}. The observed ^{85}Kr release at 1800°C increased due to elevated coating failure, with the maximum ^{85}Kr release after 300 h at 1800°C reaching approximately 5×10^{-4} and being associated with TRISO failures at this extremely high test temperature. Subsequent analysis of the fuel post-annealingtests revealed that SiC failure typically followed the same mechanism observed in in-pile failures, and the localized palladium attack on the SiC layer was accelerated at higher temperatures.

The release of silver from the particles and compacts during irradiation was often notably high and primarily influenced by variations in fuel temperature, aligning with findings from previous TRISO fuel irradiation experiments. Although silver release from fuel compacts during high-temperature post-irradiation annealing tests could be substantial (reaching as high as ~30%), it typically concluded within a few hours of attaining the target test temperature. The amount of released silver was determined by the inventory available in the outer OPyC layers or compact matrix that remained at the end of the irradiation. Minimal additional silver release from particles was measured during isothermal testing lasting up to 300 h at 1600 and 1700°C. Only at a test temperature of 1800°C was clear evidence of additional silver release through intact coatings observed. Additionally, a single test indicated higher silver release at intermediate temperatures (peaking at 1150°C), warranting further investigation in future tests.

For example, as illustrated in Fig. 54 [107], silver release patterns also showed a difference between the Variant 3 compacts and the compacts with coarser-grained SiC. Transport of silver from the fine-grained SiC (Variant 3) was apparently higher at 1800°C than from the coarse-grained SiC, as additional silver release was observed after ~100 h at this temperature only for the Variant 3 specimens. Since no additional silver release was observed in the later stages of the tests from any of the fuel specimens (including Variant 3), this indicates that diffusive behaviour at 1800°C is significantly higher than at 1600°C.

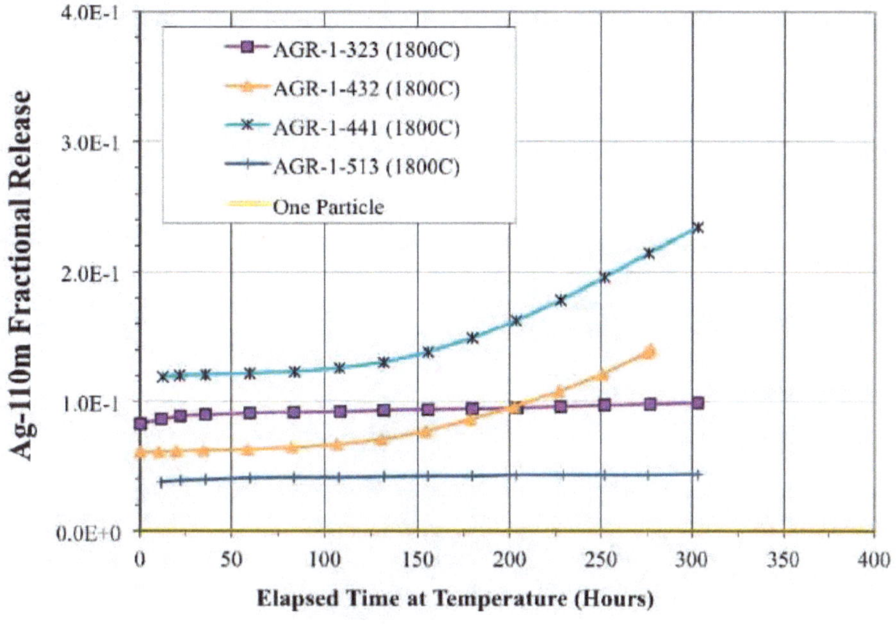

FIG. 54. Differences in silver behaviour observed between Variant 3 (compacts 4-3-2 and 4-4-1) and Baseline (Compact 3-2-3) and Variant 1 (Compact 5-1-3). (The results are qualitatively similar to the outcomes for Eu and Sr results) (courtesy of INL).

A comprehensive microanalysis of fuel particles was conducted to enhance the understanding of fission product migration under various conditions. This analysis involved conventional techniques such as optical microscopy and scanning electron microscopy, as well as more advanced methods like transmission electron microscopy and scanning transmission electron microscopy, coupled with several techniques for elemental and crystallographic analysis. In some instances, these analytical techniques were applied to irradiated TRISO fuel for the first time, allowing the analysis of fission products within the coating layer microstructures down to nanometre-length scales. For example, the particles subjected to 1600°C annealing tests and analysed by SEM (Scanning Electron Microscope) included particles from two Baseline compacts (6-4-3 and 6-2-1), two Variant 3 compacts (4-1-2 and 4-3-3), and one Variant 1 compact (5-3-3) [107].

Figure 55 [107] provides a comparison between a particle with very low silver retention (^{110m}Ag M/C[9] ≤0.16) and a particle with high silver retention (^{110m}Ag M/C = 1.12) from Compact 4-1-2. The exposure to 1600°C did not alter the characteristics of the inner IPyC-SiC interface, as evidenced by the intimately stitched IPyC-SiC interface observed in the secondary electron imaging (SEI) micrographs in Fig. 55. Backscattered Electron Composition (BEC) imaging of the TRISO layers reveals the accumulation of fission products at the IPyC-SiC interface and embedded fission product clusters in the SiC layers.

The distribution and frequency of embedded fission products in the SiC layer showed a dependency on silver retention behaviour. Particles with low silver retention exhibited a greater frequency of fission product clusters distributed deeper into the SiC layer relative to high silver-retention particles. A correlation with palladium-uranium (Pd-U) and palladium (Pd) features was observed, with Pd-U features being predominant in the first 10 µm of the SiC layer, and Pd clusters with no detectable uranium being more frequent in the outer half of the SiC layer. These observations align with the

[9] In some cases, individual particles taken from irradiated and heated fuel specimens were analyzed with gamma spectrometry to determine the inventory of fission products. These measured inventories were decay-corrected and compared to the calculated inventory from neutronics analysis to determine the ratio of measured to calculated (M/C) inventories. For ^{110m}Ag, this provides an estimate of the level of retention in the particles. See [107] for more details.

distributions of fission product clusters in the SiC layer of as-irradiated compacts discussed in Section 3.4.2.1 of Ref. [107].

In addition to palladium and uranium, other fission products, similar to those observed in the as-irradiated particles [107], were detected by energy-dispersive X-ray spectroscopy (EDS) and were primarily located near the IPyC-SiC interface.

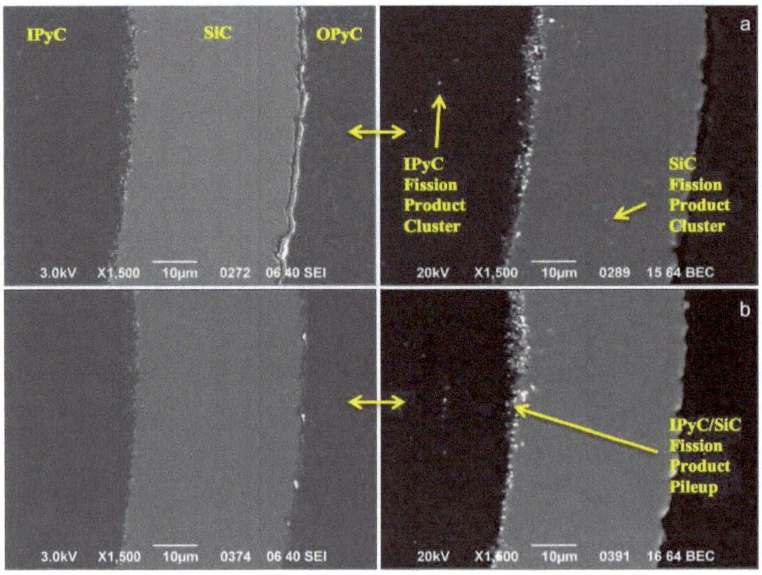

FIG. 55. SEI/BEC-paired images of Compact 4-1-2, annealed at 1600°C: (a) very low Ag retention particle (M/C ≤0.16) and (b) high Ag retention particle (M/C = 1.12). Areas of bright contrast in BEC micrograph indicate isolated fission product features (courtesy of INL).

6.2.2.2. AGR-2 irradiation experiment

The AGR-2 experiment was the second in the series of planned experiments designed to test TRISO-coated, LEU UCO fuel. The main objective of the AGR-2 test was to irradiate TRISO particles produced from pilot-scale coating equipment, both UCO and UO_2, to gather data on fuel performance under normal operation and accident conditions. The UCO compacts underwent a range of burnups and temperatures distributed across three capsules, while compacts containing UO_2 particles from the United States, South Africa, and France were separately irradiated in three capsules [117].

The burnups and temperatures applied to the UCO compacts were representative of expected service conditions in pebble bed reactors. It is important to note that the irradiation performance of the UO_2 fuel from France and South Africa has not been reported. Therefore, the following discussion focuses solely on the UCO fuel produced in the United States and irradiated in capsules 2, 5, and 6, as well as the UO_2 fuel irradiated in capsule 3.

Tables 38 and 39 show properties of AGR-2 coated particles and compacts, respectively, taken from Refs. [117,118]. Table 40 describes requirements for irradiations [119].

TABLE 38. SELECTED PROPERTIES OF THE AGR-2 UCO AND UO_2 FUEL PARTICLES

Property	Actual mean value ± Population standard deviation	
Kernel type	UCO	UO_2
Kernel diameter (μm)	426.7±8.8	507.7±11.9
U enrichment (wt.%)	14.029±0.026	9.600±0.010
Buffer thickness (μm)	98.9±8.4	97.7±9.9
IPyC thickness (μm)	40.4±2.5	41.9±3.2
SiC thickness (μm)	35.2±1.2	37.5±1.2
OPyC thickness (μm)	43.4±2.9	45.6±2.4
IPyC anisotropy (OAF) and diattenuation[a,b]	1.0236±0.0008 (0.0116±0.0004)	1.0225±0.0019 (0.0111±0.0009)
OPyC anisotropy (OAF) and diattenuation[a,b]	1.0177±0.0008 (0.0088±0.0004)	1.0147±0.0008 (0.0073±0.0004)
IPyC anisotropy (OAF) and diattenuation post-compact anneal[a,b]	1.0315±0.0033 (0.0155±0.0016)	1.0319±0.0025 (0.0157±0.0012)
OPyC anisotropy (OAF) and diattenuation post-compact anneal[a,b]	1.0290±0.0013 (0.0143±0.0006)	1.0246±0.0011 (0.0122±0.0005)
Particle diameter (μm)	873±23	953±28
Particle mass (mg)	1.032±0.003	1.462±0.005

a. The optical anisotropy factor (OAF) is equivalent to (1+N)/(1-N), where N is the diattenuation.
b. Values in parentheses are the diattenuation (N) that was measured with the Two-Modulator Generalized Ellipsometry Microscope (2-MGEM) developed at ORNL [112].

TABLE 39. SELECTED PROPERTIES OF AGR-2 COMPACTS

Property	Actual mean value ± Population standard deviation	
	UCO	**UO_2**
Compact mass (g)	6.293±0.011	6.103±0.014
Mean U loading (g U/compact)	1.257±0.03	0.993±0.006
Diameter (mm)	12.286±0.005	12.269±0.007
Length (mm)	25.141±0.017	25.134±0.018
Number of particles per compact[(a)]	3176	1543
Particle volume packing fraction (%)	37	23
Effective overall compact density[(a)] (Mg/m^3)	2.11	2.05
Compact matrix density (Mg/m^3)	1.589±0.005	1.677±0.006
U contamination fraction w/o exposed kernels (g leached U/g U in compact)	3.94×10^{-6}	9.66×10^{-7}

(a) Approximate calculated value derived from other characterized properties.

TABLE 40. AGR-2 FUEL IRRADIATION TEST CONDITION REQUIREMENTS

Parameter	Specification
Peak temperature for each capsule (°C)	≤1250 for the remaining capsules containing UCO fuel ≤1150 for each capsule
Time average, volume average temperature goal for each capsule (°C)	≥ime for the highest temperature UCO capsule ≥1000 for the remaining capsules containing UCO fuel ≥900 for each capsule containing UO_2 fuel
Minimum compact average burnup (% fissions per initial heavy metal atom)	>7 for UCO, U.S. UO_2 and South African UO_2 fuel >11 for French UO_2 fuel
Compact average burnup goal for majority of fuel compacts (% fissions per initial heavy metal atom)	>10 for UCO, U.S. UO_2 and South African UO_2 fuel >13 for French UO_2 fuel
Peak fast neutron fluence (n/m², E>0.18 MeV)	$<5\times10^{25}$
Minimum peak fast neutron fluence (n/m², E>0.18 MeV)	$>1.5\times10^{25}$
Instantaneous peak power per particle (MW/particle)	≤/pa

Actual irradiation data are shown in Tables 41 (for burnup), 42 (for fluence) and 43 (for temperature), taken from Ref. [120].

TABLE 41. MINIMUM, CAPSULE AVERAGE, AND PEAK COMPACT BURNUP (%FIMA) AT THE END OF IRRADIATION

Capsule	Minimum Compact	Capsule Average	Peak Compact	Goals
6	7.26	9.30	10.81	Burnup >7% for all compacts
5	10.07	11.68	12.88	
3	9.01	10.10	10.69	Burnup >10%[(a)] for the majority of compacts
2	10.80	12.17	13.15	
Thirty-six of 48 compacts (28/36 UCO and 8/12 UO_2) achieved a compact burnup of at least 10%.				

TABLE 42. MINIMUM, CAPSULE AVERAGE, AND PEAK COMPACT FAST FLUENCE (10^{25} N/M², E > 0.18 MeV) AT THE END OF IRRADIATION

Capsule	Minimum Compact	Capsule Average	Peak Compact	Goals
6	1.94	2.39	2.73	
5	2.77	3.18	3.42	Fluence >1.5 and <5.0 for all compacts
3	3.05	3.35	3.53	
2	2.88	3.25	3.47	

TABLE 43. TIME-AVERAGE (TA) MINIMUM, TIME-AVERAGE, VOLUME-AVERAGE (TAVA), AND TA PEAK TEMPERATURES (°C) AT THE END OF IRRADIATION

Capsule	TA Minimum	Capsule Average	TA Peak	Goals
6	868	1074	1183	TA peak < 1400 °C for one UCO capsule
5	923	1101	1210	TA peak < 1250 °C for other UCO capsules
3	889	1032	1105	TA peak < 1150 °C for UO_2 capsule
2	1034	1252	1360	TAVA > 1150 °C for one UCO capsule
				TAVA > 1000 °C for other UCO capsules
				TAVA > 900 °C for UO_2 capsule

In the initial three cycles of the AGR-2 irradiation (from 0 to 149 EFPD), the fission gas R/B (Release-to-Birth) ratios remained relatively low. As depicted in Fig. 56 [121] for isotopes such as ^{85m}Kr, ^{88}Kr, and ^{138}Xe, the R/B values in the UCO capsules were predominantly below 10^{-6}. The exception was observed in the hottest Capsule 2, where the ^{85m}Kr R/B reached 2×10^{-6}. The average ^{85m}Kr R/B values in Capsules 2, 5, and 6 were recorded as 1.3×10^{-6}, 5×10^{-7} and 6×10^{-7}, respectively [120]. In the UO_2 capsule (Capsule 3), the R/B values during the initial three cycles consistently remained below 10^{-7}. These values, while higher than the beginning-of-life (BOL) R/B for the AGR-1 irradiation capsules (with an average R/B of ~10^{-8}) [122], were influenced in part by increased uranium contamination in the AGR-2 compacts compared to AGR-1 (uranium contamination in the AGR-2 compacts was ~4×10^{-6}, compared to an average value of 3×10^{-7} for the AGR-1 compacts).

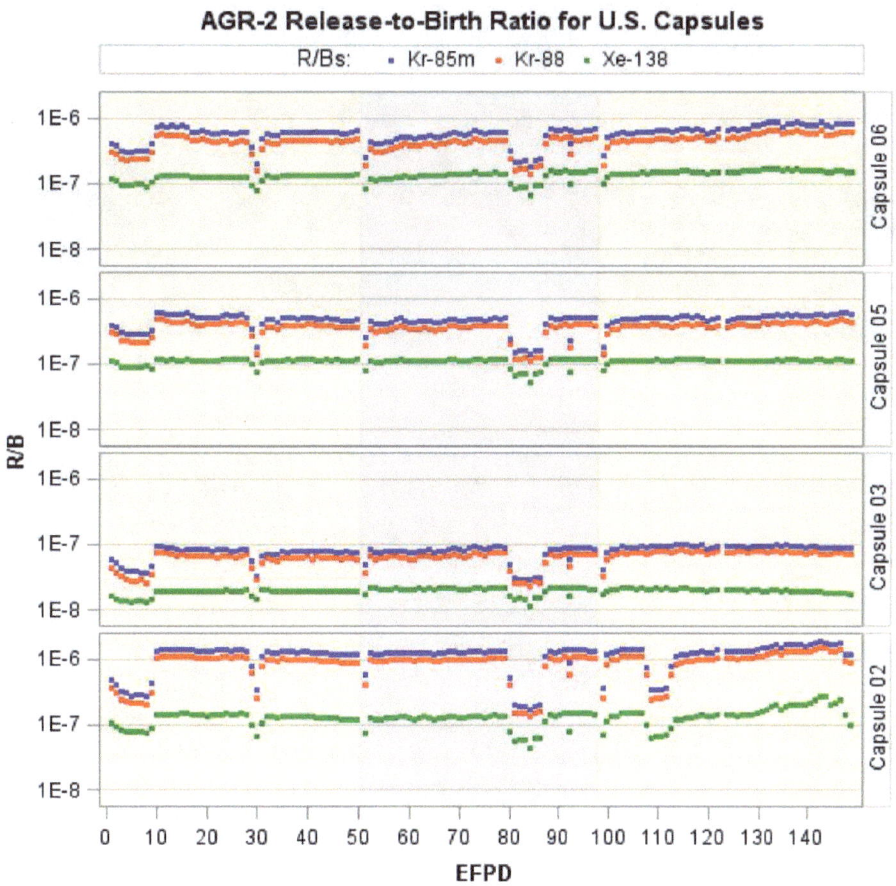

FIG. 56. R/B ratios from daily birth rates for Kr-85m, Kr-88, and Xe-138 versus time in EFPD during the initial three irradiation cycles of AGR-2 (courtesy of INL).

After the third irradiation cycle, problems with the gas lines in the capsules resulted in a loss of reliable fission gas R/B data, and therefore the EOL R/B values are not known [120]. Detailed PIE of the capsules and fuel specimens was performed to assist in determining if there were any in-pile particle failures. The results indicated that there were ≤4 UCO particle failures during the irradiation, with the uncertainty in the number arising from challenges in differentiating particles with (a) in-pile TRISO failures, (b) particles with in-pile SiC layer failures, and (c) particles with as-fabricated coating defects during the PIE process [121]. With a total of 114 000 UCO particles in the experiment, four in-pile particle failures would result in a calculated failure fraction of $\leq 8.1 \times 10^{-5}$ at 95% confidence, while the value would be $\leq 2.7 \times 10^{-5}$ if there were zero particle failures. No evidence of in-pile particle failure was observed for the UO_2 fuel in Capsule 3. With a total of 18,500 UO_2 TRISO particles in this capsule, this results in a calculated failure fraction of $\leq 1.7 \times 10^{-5}$ at 95% confidence.

The major components of the AGR-2 PIE and annealing testing have been completed, and results can be found in a number of publications [108, 121, 123–134]. The in-pile behaviour and the annealing test results for the AGR-2 UCO fuel were generally consistent with previous observations from the AGR-1 fuel. Kernel and coating microstructures appeared similar to the AGR-1 fuel, with characteristic kernel porosity and a gap between the buffer and IPyC layers (Fig. 57 [135, 136]).

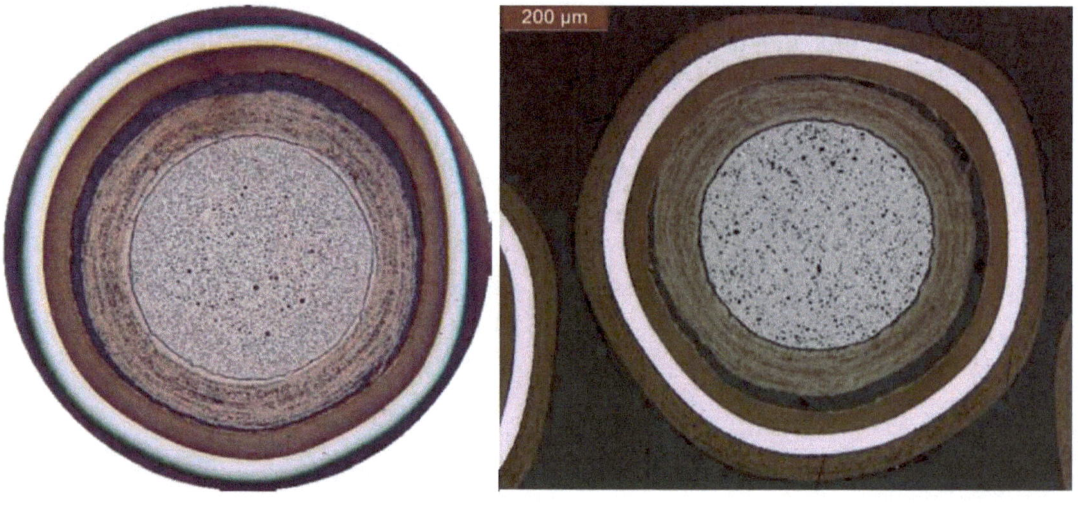

(a) (b)

FIG. 57. Cross-section micrographs for (a) an AGR-2 UCO TRISO particle from fuel compact 5-1-3 with time-average peak temperature 1177°C and burnup 11.1% FIMA and (b) an AGR-2 UO_2 TRISO particle from compact 3-1-2 with time-average peak temperature 1084C and burnup 10.7% FIMA (courtesy of INL).

Fuel compacts were annealing tested in a manner similar to the AGR-1 fuel, including a hold at the target test temperature for nominally 300 h while measuring fission product release. The AGR-2 UCO TRISO fuel exhibited performance similar to that demonstrated by the AGR-1 fuel tested previously. No particle failures were observed during 1600°C tests, and some failures occurred during 1800°C tests. The ^{85}Kr release from all fuel specimens was $\leq 7.7 \times 10^{-6}$ at 1600°C and approximately a factor of 10 higher at 1800°C (excluding a result from a compact where a particle failed due to degradation of the SiC layer from Mo that was introduced during the fabrication process).

A notable result of the AGR-2 annealing testing was the comparison of caesium release fractions for UCO and UO_2 fuel. The ^{134}Cs release from the UCO fuel for all tests was $\leq 6.2 \times 10^{-5}$ at 1600°C and $\leq 6.7 \times 10^{-4}$ at 1800°C, indicating good retention at these extreme temperatures and very limited degradation of the SiC layers. In contrast, ^{134}Cs release from the AGR-2 UO_2 fuel was in the range of approximately (0.2–1) × 10^{-2} after 300 h at 1600°C and 8.7 × 10^{-2} after 150 h at 1700°C. This behaviour is demonstrated in Fig. 58 [108], which shows ^{134}Cs release fractions for several of the AGR-2 annealing tests (note that not all AGR-2 tests are included in this figure). The caesium release from UO_2 is significantly higher than UCO at 1600°C for similar test durations, and significantly higher after 150 h at 1700°C than UCO release after 300 h at 1800°C. Note that the UCO data in Fig. 58 also included fuel from Capsule 2, which had a time-average peak irradiation temperature ~250°C higher than the UO_2 fuel. Extensive detailed post-test examination of the UO_2 particles indicated that the increase in caesium release throughout these tests was due to degradation of the SiC layers by reaction with CO(g) generated in the particle. The result of this SiC degradation is a gradual decrease in caesium retention by the particles, with the effect accelerating with increasing test temperature [121].

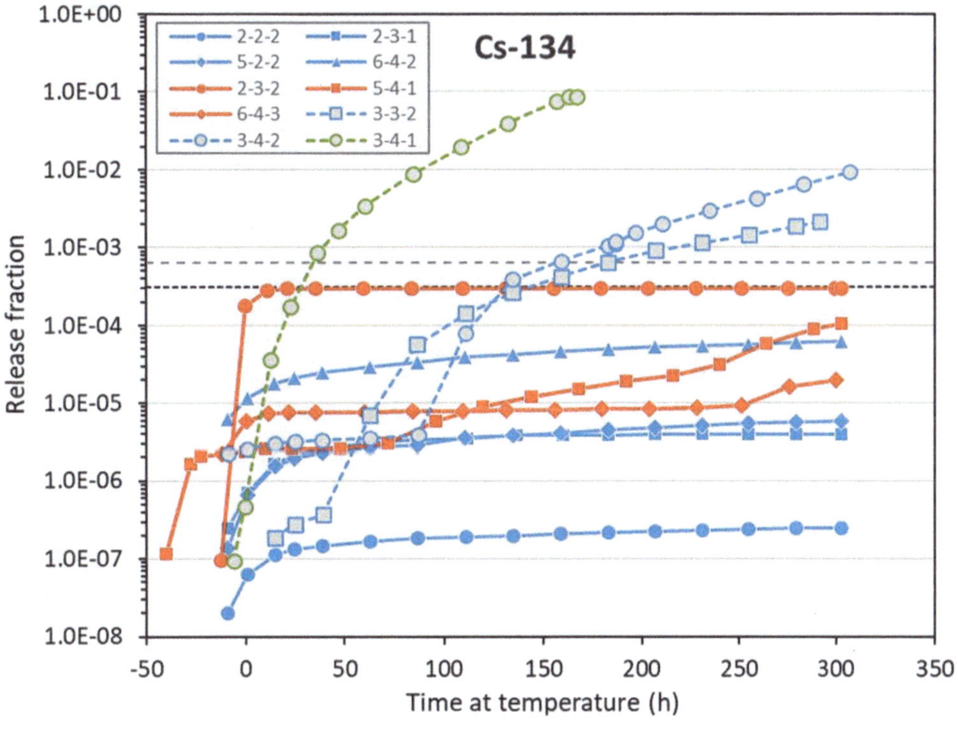

FIG. 58. ^{134}Cs release from AGR-2 fuel compacts during isothermal annealing tests, plotted as a function of hold time at the target test temperature. Plot colour indicates test temperature: 1600°C (blue), 1700°C (green), or 1800°C (red). UO_2 tests are indicated by dotted lines with grey-filled symbols (courtesy of INL).

Another notable result from the AGR-2 PIE and annealing tests was the impact of relatively high irradiation temperature on strontium and europium release from the fuel. The ^{90}Sr release fractions from the fuel compacts in Capsules 3, 5, and 6 (time-average peak temperatures <1210°C) were approximately 10^{-4}. In contrast, Capsule 2 contained UCO fuel intentionally operated at high temperature (time-average peak fuel temperature of 1360°C; see Table 40) and exhibited ^{90}Sr release of ~10^{-2}, two orders of magnitude higher than the other capsules. Primarily due to this higher in-pile release (where most of the inventory released from the particles in-pile remains in the OPyC layer or matrix), the release of ^{90}Sr from the Capsule 2 compacts after 300 h at 1600°C in post-irradiation annealing tests was appreciably higher than for the fuel in the other capsules [121]. This is demonstrated in Fig. 59, which shows ^{90}Sr release fractions for several of the AGR-2 annealing tests (note that not all AGR-2 tests are included in this figure), where the higher-temperature Capsule 2 fuel is highlighted. Europium exhibited generally similar trends for in-pile and annealing test releases. Thus, an important result from the AGR-2 experiment is that while higher irradiation temperatures (up to a time-average peak of

1360°C) do not significantly impact the level of coating layer failures in UCO fuel, there is a notable increase in release of certain fission products from intact fuel particles. These impacts should be considered during HTGR core design and safety analyses.

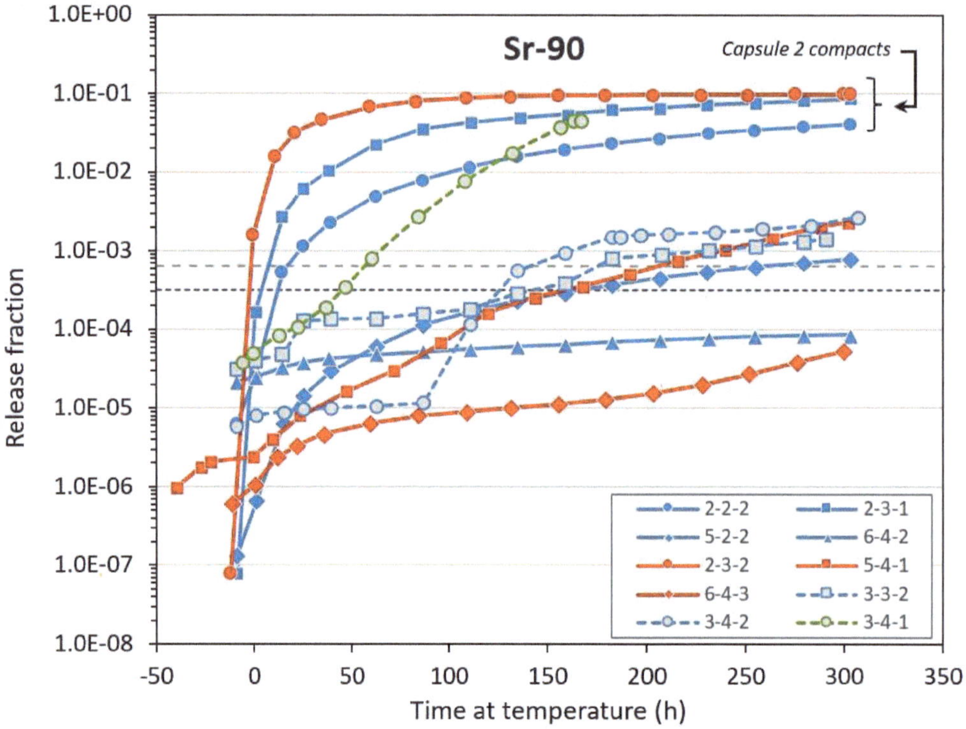

FIG. 59. 90Sr release from AGR-2 fuel compacts during isothermal annealing tests, plotted as a function of hold time at the target test temperature. Plot colour indicates test temperature: 1600°C (blue), 1700°C (green), or 1800°C (red). UO_2 tests are indicated by dotted lines with grey-filled symbols (courtesy of EPRI).

6.2.2.3. AGR-3/4 irradiation experiment

The AGR-3/4 irradiation experiment aimed to enhance fission product transport models and evaluate the impact of sweep gas impurities on fuel performance. This was achieved by measuring the spatial distributions of fission products within the fuel-compact matrix material and nuclear-grade graphite [137]. In addition to high-quality TRISO-coated fuel particles, each fuel compact in the experiment also contained DTF particles that would release fission products, with the objective of assessing fission product transport behaviour in core graphite and resinated graphite fuel matrix materials.

The test train comprised 12 distinct capsules, each equipped with independent control and monitoring systems (refer to Fig. 60 [137]). Within each capsule were four compacts, each measuring 12.51 mm in length and containing a mix of UCO TRISO-coated 'driver' fuel particles and UCO DTF fuel particles. The specified DTF fraction was set at 1×10^{-2} [137, 138].

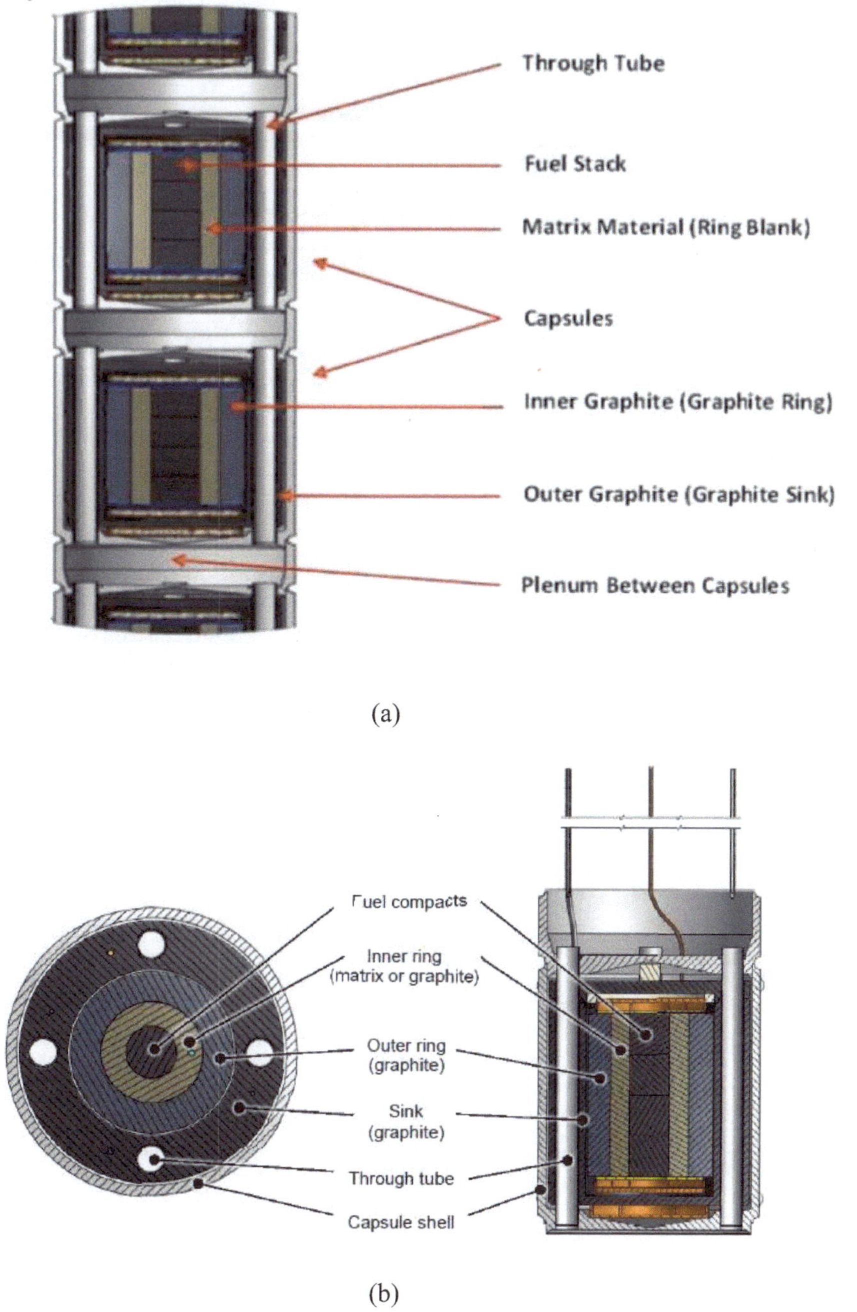

FIG. 60. (a) Axial and (b) radial cross-sectional views of AGR-3/4 capsules (courtesy of INL).

In Fig. 61, an X-radiograph reveals a cross-sectional view along the axial centreline of an AGR-3/4 compact, with the kernels of the DTF particles highlighted by red dots [139]. The 20 DTF particles were positioned along the central axis of the compact, with some apparent clustering observed after the removal of the delivery tube. However, this minor clustering toward the bottom of the compact was not anticipated to impact the outcome of the irradiation test.

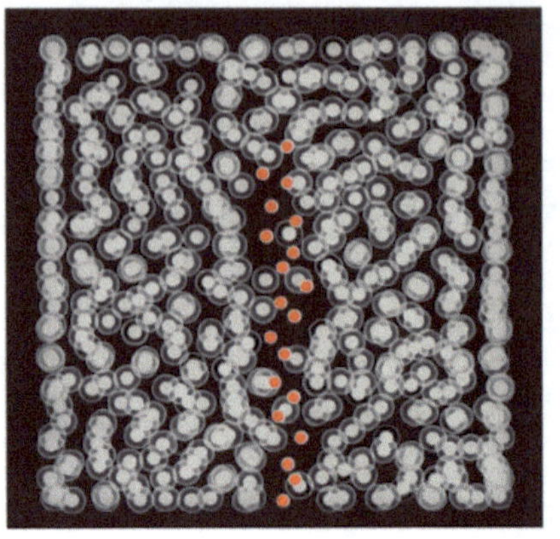

FIG. 61. X-ray radiograph of a 2.5-mm-thick section taken from the centre of an AGR-3/4 compact showing the location of 20 DTF particles (DTF kernels marked in red) (courtesy of INL).

Table 44 shows conditions for AGR-3/4 fuel test, taken from Ref. [140]. Note that with 20 DTF particles placed in each fuel compact, there were a total of 80 DTF particles in each capsule. The values in the final column of Table 44 are the best estimates for the number of DTF that failed based only on the results of fission gas measurements, but the uncertainty in this number was substantial. Thus, the deviation of these values from 80 are believed to be due to uncertainties in quantifying the particle failures based on fission gas measurements during irradiation and are not an accurate indication that some DTF particles did not fail during the irradiation. This conclusion is qualitatively supported by cross-section analysis of several irradiated AGR-3/4 compacts that included observations of the DTF particles [141].

TABLE 44. IRRADIATION DATA FOR EACH AGR-3/4 CAPSULE

Capsule	Burnup [(1)] (% FIMA)	Fast fluence [(1)] ($\times10^{25}$ n/m^2)	Time-averaged peak temperature (°C)			End-of-life numbers of failed particles
			Fuel	Inner ring[(2)]	Outer ring[(2)]	
12	5.4	1.50	888	802	748	40
11	9.1	2.87	1280	1166	975	69
10	11.8	3.94	1249	1055	986	47
9	13.7	4.65	1083	884	721	90
8	14.5	5.08	1257	1048	945	78
7	15.0	5.27	1418	1203	1045	52
6	15.2	5.31	1133	912	728	47
5	14.9	5.19	1102	858	706	54
4	14.2	4.85	1084	882	727	76
3	12.6	4.22	1242	1050	976	96
2	10.1	3.21	1113	977	875	91
1	6.1	1.76	978	889	785	41

(1) Burnup and fast fluence are capsule-average values.
(2) Ring temperatures are the volume average for a 1-in. tall axial section centred on the fuel stack.

The R/B values for fission gases reached levels in the range of 10^{-5} to 10^{-3} early in the AGR-3/4 irradiation as DTF particles began to experience failure during the initial cycle. The hotter Capsule 7 exhibited a higher R/B value, reaching approximately 3×10^{-3} [142, 143]. To assess the impact of uncertainty in failure estimates on the release relationship with fuel temperature and decay constants, regressions were conducted for three sets of data for the R/B per failed particle: best-estimate, maximum, and minimum failure estimates.

In Fig. 62, the R/B per failed particle for AGR-3/4 and its fitted function of reciprocal peak fuel temperature for ^{85m}Kr are depicted using best-estimated (blue), maximum (red), and minimum (green) failure counts [144].

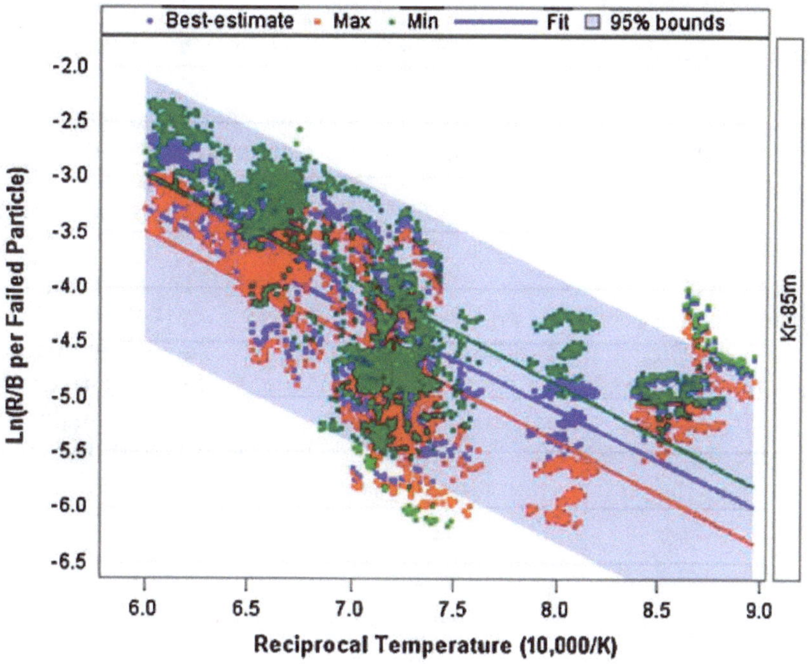

FIG. 62. AGR-3/4 R/B per failed particle and their fitted function of reciprocal peak fuel temperature for ^{85m}Kr using best-estimated (blue), maximum (red) and minimum (green) failure counts (courtesy of INL).

Dimensional changes after the irradiation were also reported as summarized below (details are described in Ref. [145]):

— The diameter of all compacts was measured to decrease during irradiation, with relative changes ranging from approximately -0.5 to -2.0%;
— Most compacts showed the decrease in their length, while several compacts in Capsules 4, 5, and 7 showed a net increase in length. Relative changes ranged from approximately +0.5 to -0.9%.

PIEs on the irradiated compacts are currently ongoing, with specific objectives outlined as follows [145]:

— Assessing the migration of fission product in the matrix and graphite rings during irradiation;
— Determining the fractional inventory of fission products in fuel kernels and compact matrix at the end of irradiation;
— Investigating the transport of condensable and gaseous fission product from fuel kernels and compact matrix during heating in pure helium.

6.2.2.4. AGR-5/6/7 irradiation experiment

The fourth and final irradiation in the AGR program, designated AGR-5/6/7, began in February 2018 and ended in July 2020. The AGR-5/6/7 irradiation test comprised both a fuel qualification test and a performance margin test for industrially produced UCO TRISO coated particle fuel developed under the AGR program. The primary aim of the AGR-5/6/7 irradiation experiment was to qualify UCO TRISO fuel fabricated using engineering-scale production equipment under various operating conditions, including conditions beyond normal operating parameters. Specific goals of the AGR-5/6/7 irradiation experiment were [105, 146, 147]:

— Irradiating reference design fuel containing low-enriched UCO TRISO fuel particles to support fuel qualification;
— Establishing operating margins for the fuel beyond normal operating conditions;
— Providing irradiated fuel performance data and irradiated fuel samples for PIE and safety assessments.

The AGR-5/6/7 experiment took place at the ATR at INL [146-148]. The test train included five separate capsules, each independently controlled and monitored (Fig. 63 [146]). The performance margin test, designated AGR-7, was located in the middle capsule, while the fuel qualification test, designated AGR-5/6, comprised the two bottom and two top capsules. The capsules were filled with fuel compacts containing UCO TRISO particles, using two different packing fractions (nominally 25 and 40%). Tables 45 – 47 provide properties of AGR-5/6/7 kernels (taken from Ref. [146]), particles (taken from Ref. [146, 149]) and compacts (taken from Ref. [146]), respectively.

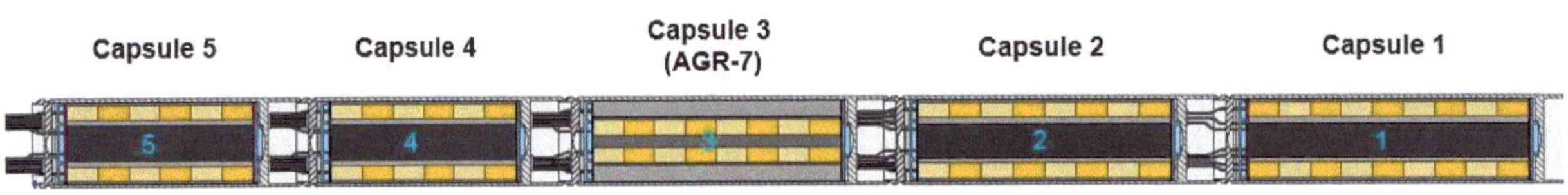

FIG. 63. Axial cross-sectional schematic of the AGR-5/6/7 test train (courtesy of INL).

TABLE 45. PROPERTIES FOR AGR-5/6/7 KERNELS

Property	Specified range for mean value	Actual mean value ± population standard deviation
Diameter (µm)	425±10	425.78±10.42
Density (g/cm^3)	Not specified	11.048±0.044
U-235 enrichment (wt.%)	15.5±0.1	15.477±0.013
Carbon/uranium (atomic ratio)	0.40±0.10	0.370±0.000
Oxygen/uranium (atomic ratio)	1.50±0.20	1.441±0.004
(Carbon+oxygen)/uranium (atomic ratio)	Not specified	1.811±0.004
Total uranium (wt.%)	Not specified	89.684±0.040
Sulfur impurity (ppm-wt.)	Not specified	247±14
Phosphorus impurity (ppm-wt.)	Not specified	25±8

(a) Upper limit of the 95% confidence interval

TABLE 46. PROPERTIES FOR AGR-5/6/7 COATED PARTICLES

Property	Specified range for mean value	Actual mean value ± Population standard deviation
Buffer thickness (μm)	100±15	100.37±5.55
IPyC thickness (μm)	40±4	39.24±1.26
SiC thickness (μm)	35±3	36.15±0.65
OPyC thickness (μm)	40±4	35.03±1.99
Buffer density (g/cm³)	1.05±0.10	1.031±0.022
IPyC density (g/cm³)	1.90±0.05	1.897±0.010
SiC density (g/cm³)	Not specified	3.195±0.002
OPyC density [g/cm³]	1.90±0.05	1.897±0.004
IPyC optical anisotropy factor and diattenuation(a,b)	Not specified	1.031±0.002 (0.0153±0.0010)
OPyC optical anisotropy factor and diattenuation(a,b)	Not specified	1.021±0.001 (0.0102±0.0006)
IPyC optical anisotropy factor and diattenuation post compact anneal(a,b)	Not specified	1.0388 (0.0191±0.0017)
OPyC optical anisotropy factor and diattenuation post compact anneal(a,b)	Not specified	1.0296 (0.0146±0.0009)
SiC aspect ratio(c)	Not specified(d)	1.053±0.009
Particle mass [mg]	Not specified	1.02

(a) The optical anisotropy factor (OAF) is defined as (1+N)/(1-N).
(b) Values in parentheses are the diattenuation (N) that was measured with the Two-Modulator Generalized Ellipsometry Microscope (2-MGEM) developed at ORNL [112].
(c) Aspect ratio is defined as the ratio of maximum to minimum diameter of the coated particle and is measured on SiC-coated particles following burn back of the OPyC layer.
(d) Critical region is specified such that ≤1% of the particles must have an aspect ratio ≥1.14.

TABLE 47. PROPERTIES FOR AGR-5/6/7 COMPACTS

Property	Specified range for mean value	Actual mean value ± Population standard deviation
	Compact mass (g)	
Capsule 1	Not specified	6.676±0.065
Capsule 2		6.182±0.026
Capsule 3		6.187±0.021
Capsule 4		6.100±0.034
Capsule 5		6.603±0.021
	Mean uranium loading (g U/compact)	
Capsule 1	1.36±0.10	1.362±0.014
Capsule 2	0.90±0.08	0.898±0.004
Capsule 3	0.90±0.08	0.898±0.003
Capsule 4	0.90±0.08	0.871±0.005
Capsule 5	1.36±0.10	1.346±0.004
	Diameter(a) (mm)	
Capsule 1	12.20 – 12.44	12.293±0.007
Capsule 2		12.241±0.007
Capsule 3		12.245±0.006
Capsule 4		12.248±0.006
Capsule 5		12.296±0.006
	Length(a) (mm)	
Capsule 1	24.40 – 25.30	24.947±0.219
Capsule 2		24.991±0.098
Capsule 3		25.000±0.078
Capsule 4		24.770±0.119

TABLE 47. PROPERTIES FOR AGR-5/6/7 COMPACTS

Property	Specified range for mean value	Actual mean value ± Population standard deviation
Capsule 5		24.675±0.059
	Estimated mean number of particles per compact(b)	
Capsule 1	Not specified	3434
Capsule 2		2265
Capsule 3		2265
Capsule 4		2197
Capsule 5		3392
	Particle volume packing fraction (%)	
Capsule 1	40	38.4
Capsule 2	25	25.5
Capsule 3	25	25.5
Capsule 4	25	24.9
Capsule 5	40	38.4
	Effective overall compact density(b) (g/cm^3)	
Capsule 1	Not specified	2.26
Capsule 2		2.10
Capsule 3		2.10
Capsule 4		2.09
Capsule 5		2.25
	Compact matrix density (g/cm^3)	
Capsule 1	≥ 1.65	1.748±0.007
Capsule 2		1.772±0.005
Capsule 3		1.771±0.005
Capsule 4		1.766±0.006
Capsule 5		1.747±0.007
	Dispersed uranium fraction(c) (g leached U/g U in compact)	
Nominal 25% packing fraction	$\leq 1.0\times10^{-5}$	$\leq 2.95\times10^{-5}$ (d)
Nominal 40% packing fraction	$\leq 1.0\times10^{-5}$	$\leq 3.80\times10^{-5}$ (d)
	Exposed kernel fraction(c) (kernel equivalent/particle count)	
Nominal 25% packing fraction	$\leq 5.0\times10^{-5}$	$\leq 1.48\times10^{-5}$
Nominal 40% packing fraction	$\leq 5.0\times10^{-5}$	$\leq 9.28\times10^{-5}$ (d)
	Defective SiC coating fraction(c)	
Nominal 25% packing fraction	$\leq 1.0\times10^{-4}$	$\leq 1.22\times10^{-4}$ (d)
Nominal 40% packing fraction	$\leq 1.0\times10^{-4}$	$\leq 9.66\times10^{-5}$

(a) Allowable range corresponding to upper and lower critical limits specified with no compacts exceeding the limits, which require 100% inspection of all compacts.
(b) Calculated value derived from other characterized properties.
(c) 95% confidence fraction.
(d) The 95% confidence fraction exceeds the specification; despite the non-conformance, it was decided to accept the fuel as it is.

The irradiation period for the AGR-5/6/7 experiment spanned 361 EFPD, with time-average fuel temperatures calculated to range between 467 and 1231°C in the AGR-5/6 capsules and between 989 and 1432°C in the AGR-7 capsule. The AGR-5/6 irradiation temperatures aimed to cover the operational temperature range in a HTGR core, encompassing fuel temperatures substantially lower than those observed in previous AGR experiments. In contrast, the AGR-7 capsule experienced higher temperatures as part of the performance margin test, emphasizing the importance of time at temperature in TRISO fuel performance. The AGR-7 experiment was intentionally designed to induce some level of measurable fuel failure and/or fission product release, enabling the assessment of fuel performance beyond normal operating conditions and demonstrating the safety margin between typical operational conditions and temperatures where the fuel particle failure rate becomes unacceptable [146].

The burnup achieved in the experiment ranged from 5.7 to 15.3% FIMA on a compact-average basis. The range for fast neutron fluence was 1.62 to 5.55×10^{25} n/m^2 (E>0.18 MeV).

^{85m}Kr R/B in the five AGR-5/6/7 capsules averaged 3.5×10^{-7} for the first five irradiation cycles (approximately half of the total experiment duration), with values slightly higher in Capsules 1 and 3 compared to the other capsules because of significantly higher fuel temperatures (Capsule 3) or relatively high fuel temperatures combined with a relatively high exposed kernel fraction in the 40% packing fraction fuel (Capsule 1) [148]. The R/B values were determined to be consistent with models based on the level of exposed kernel defects and dispersed uranium contamination in the fuel. The R/B values were similar to those of AGR-2 (which had similar exposed kernel defect and uranium contamination fractions) and appreciably higher than the values for AGR-1 (which had significantly lower exposed kernel defect and uranium contamination fractions).

After approximately 230 EFPD, there was a sharp increase in the fission gas release from Capsule 1, as well as smaller—but still quite significant—increases in fission gas release from the other capsules. The increase was apparently due to significant particle failure in Capsule 1, with the fission gas released in that capsule able to gradually diffuse into the other capsules because of the nature of the capsule and test train design. This effectively eliminated the ability to monitor fission gas release from Capsule 1 for the remainder of the experiment. In addition, the fission gas contamination that was able to diffuse from Capsule 1 severely complicated the measurement of fission gas release in the other capsules. As a result, the R/B data for capsules 2–5 was impacted and it was difficult to reliably estimate if particles failed in the other capsules during the remainder of the irradiation. It was determined that no particle failures occurred in capsules 4 and 5 (the furthest removed from Capsule 1 and therefore the least impacted by fission gas contamination), while the possibility of a small number of failures in capsules 2 and 3 could not be eliminated. The PIE of the capsules will help to clarify if any particle failures occurred.

The cause of particle failures in Capsule 1 remained unknown during the irradiation. Analysis of the internal components of this capsule during the initial stages of PIE indicated that the failures were caused by transition metal contamination (primarily Ni) from type N thermocouples embedded in the graphite holder adjacent to the fuel compacts [150]. The transport of Ni to the fuel compacts was caused by significant overheating of a region in the capsule, in which the fuel compacts and surrounding materials reached temperatures substantially above the intended range. This was determined to be due to a design flaw in the capsule not identified until after the irradiation was complete.

The PIE and annealing testing campaigns began in 2021 and 2022, respectively.

The primary objectives of the PIE and annealing testing for the AGR-5/6/7 experiment include:

- Evaluate and characterize the unexpected behaviour observed in Capsule 1;
- Determine the acceptability of fuel performance and behaviour under normal irradiation conditions, focusing on Capsules 2, 4, and 5, and integrate these findings with PIE results from earlier AGR irradiation;
- Assess the performance and characterize the behaviour of fuel under high irradiation temperatures in Capsule 3;
- Conduct post-irradiation high-temperature tests in helium to verify satisfactory fuel performance in accidents with conductive cooling;
- Conduct oxidation tests to characterize the behaviour of the fuel when exposed to air or moisture at normal and emergency temperatures [147].

6.2.2.5. Korean irradiation programme

The first irradiation test for TRISO-coated particle fuel to validate its application in a VHTR was conducted at the High-Flux Advanced Neutron Application Reactor (HANARO) of KAERI in 2014. Subsequently, a PIE of the irradiated TRISO-coated particle fuel was carried out. The irradiation apparatus comprised two test rods: one accommodating nine fuel compacts, and the other featuring five compacts and eight graphite specimens positioned between the fuel compacts (Fig. 64 [151]). Table 48 presents the characteristics of kernels, particles, and compacts irradiated at HANARO, sourced from [151].

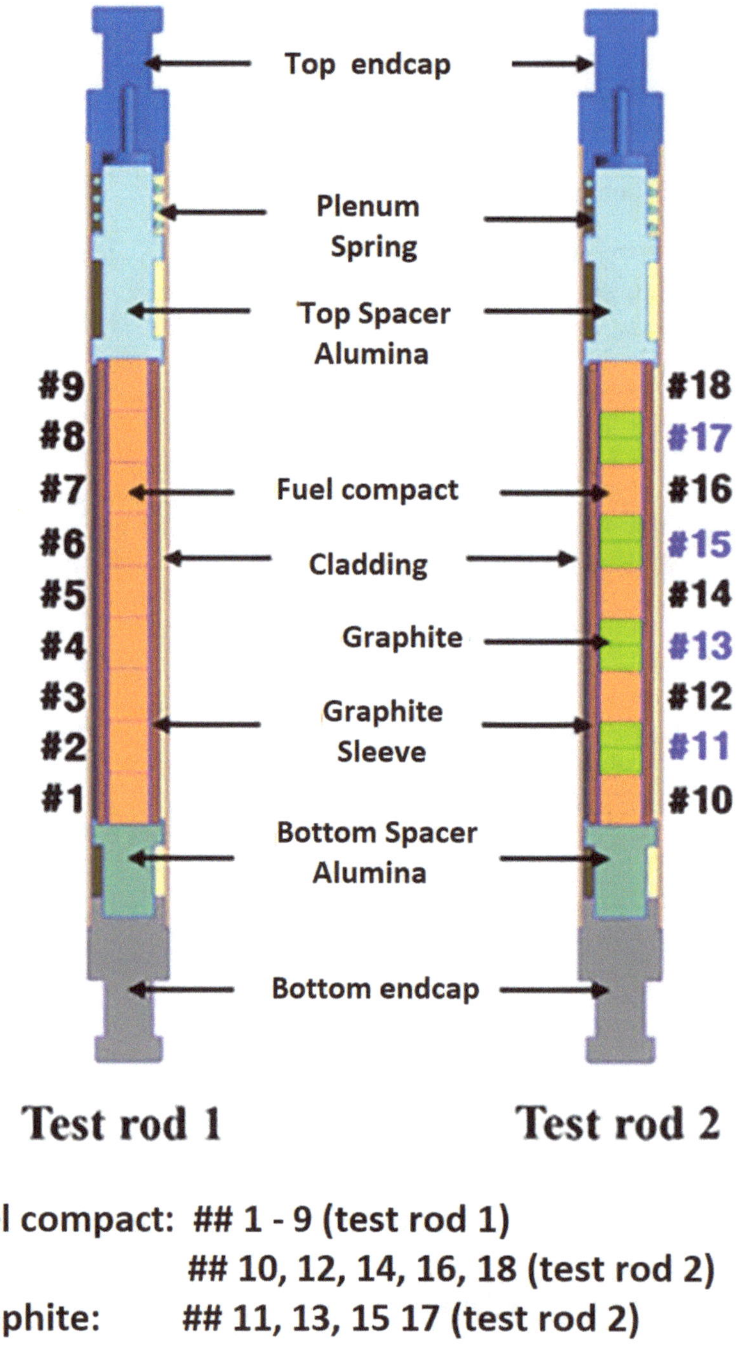

FIG. 64. Schematic layout of two kinds of test rods and position of compacts and graphite specimens in two test rods (courtesy of KAERI).

TABLE 48. CHARACTERISTICS OF TRISO-COATED PARTICLE FUEL AND GRAPHITE

Properties	Design value	Measured value	Remarks
Kernel			
Diameter (μm)	480±30	483.50±15	300-350 particles/batch (PSA*)
Density (g/cm^3)	10.65±0.25	10.68±0.023	6 g of sample weight (Pycnometer)
U-235 enrich. (wt.%)	4.5±0.10	4.504±0.004	Chemical analysis (TIMS)
O/U ratio	2.00±0.01	2.003±0.002	30 mg of sample weight (TG/DTA)
Total uranium (wt.%)	≥ 87.0	88.13	calculated value
Sphericity (aspect ratio)	< 1.2	≤ 1.04	Average
Coated fuel particle			
Buffer thickness (μm)	95±45	102.91±30	CG* (10 particles/batch)
Buffer density (g/cm^3)	1.00±0.10	1.052±0.002	X-ray (10 particles/batch)
IPyC thickness (μm)	40±20	40.55±0.53	CG* (10 particles/batch)
IPyC density (g/cm^3)	1.85±0.20	1.91±0.01	DC* (10 particles/batch)
Anisotropic index of IPyC	≤Ciso	1.0165±0.003	OM* (10 particles/batch)
SiC thickness (μm)	35±10	36.08±0.15	CG* (10 particles/batch)
SiC density (g/cm^3)	≥iC d	3.182±0.004	DC* (10 particles/batch)
OPyC thickness (μm)	40±20	46.3±3.05	CG* (10 particles/batch)
OPyC density (g/cm^3)	1.85±0.20	1.88±0.07	DC* (10 particles/batch)
Anisotropic index of OPyC	≤Ciso	1.019±0.004	OM* (10 particles/batch)
Particle dia. (mm)	0.90±0.10	0.95±0.011	PSA* (10 particles/batch)
Average wt. (g)	0.001392	0.001392	
Fuel compact			
Compact mass (g)	1.050	1.028±0.024	Average pf 14 compacts
Mean U loading (gU)	0.14	0.135	Average/compact
Diameter (mm)	8.0±0.2	7.981±0.006	Average of 14 compacts
Length (mm)	10.0±0.5	9.995±0.270	Average of 14 compacts
No. of compact	9(rod1)/5(rod2)	9(rod1)/5(rod2)	
Packing (volume) fraction (%)	20	19.703±0.44	263 particles/compact
Matrix graphite (M) & structural graphite (G) specimen			
Diameter (mm)	8.0±0.2	7.976±0.004	Average of 8 specimens
Length (mm)	5±0.2	5.028±0.008	Average of specimens
No. of specimens	8(rod2)	8(rod2)	
Density (g/cm^3)	1.7±0.1	1.770±0.004	Average of 8 specimens

*CG; Ceramography, DC; Density column, PSA; Particle size analysis; OM; Optical microscopy.

The irradiation test was conducted under an inert gas atmosphere without sweep gas. Post-irradiation, the maximum power of the fuel compact was estimated to be approximately 56 W at 25 EFPD, and the maximum power of the particle reached 215.4 MW. The highest discharged burnup was evaluated to be around 37,344 MWd/tU (3.99% FIMA). The maximum fast fluence of the graphite specimen was estimated as 2.99×10^{20} n/cm^2 (E > 0.18 MeV).

Figure 65 shows dimensional changes of fuel compacts measured after the irradiation [151]. Figure 66 shows ceramographic, cross-sectional view of test compacts [151]. Figure 67 shows EPMS images of kernels [151].

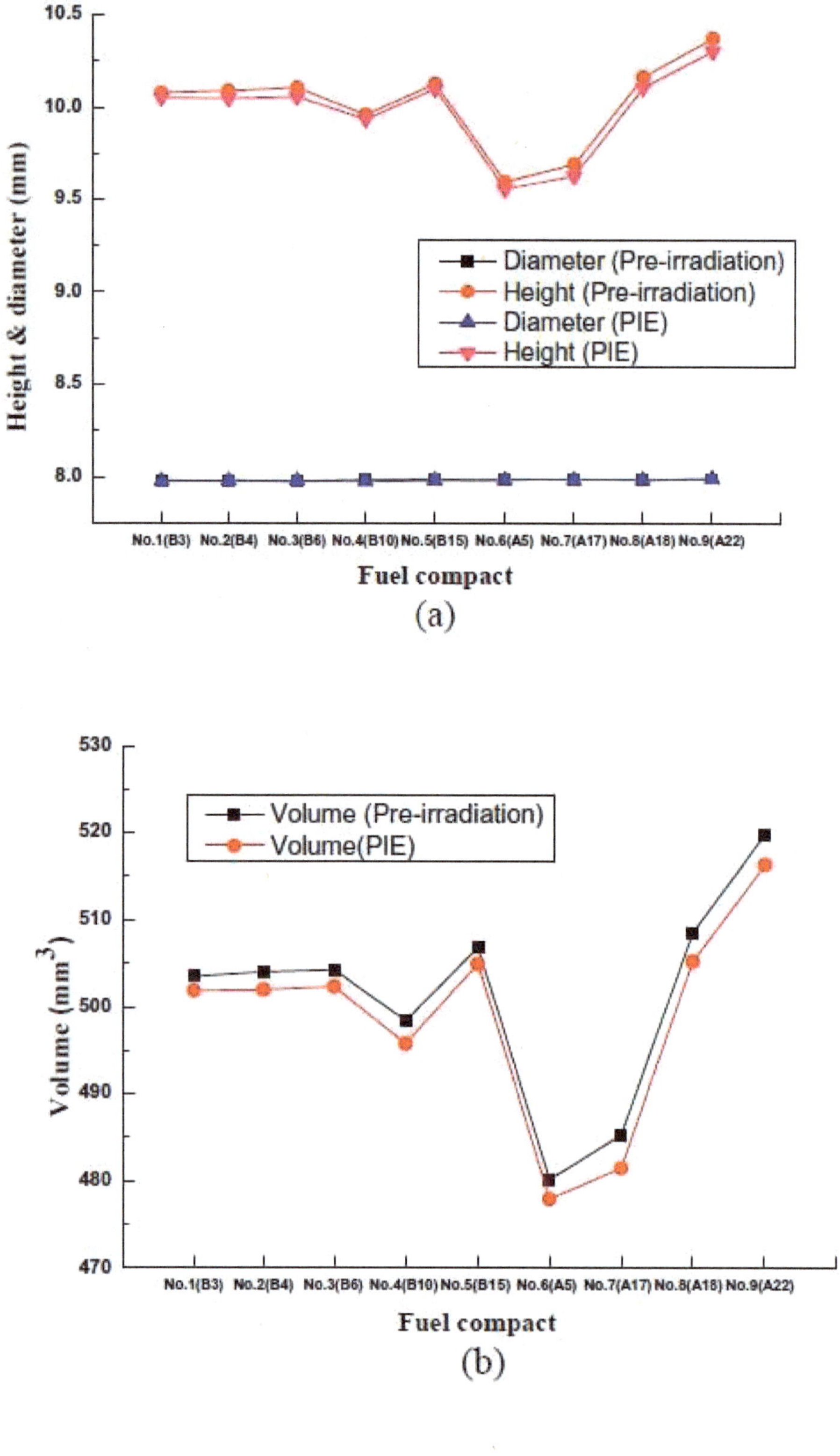

FIG. 65. Dimensional change of fuel compacts in test rod 1: (a) height and diameter and (b) volume (courtesy of KAERI).

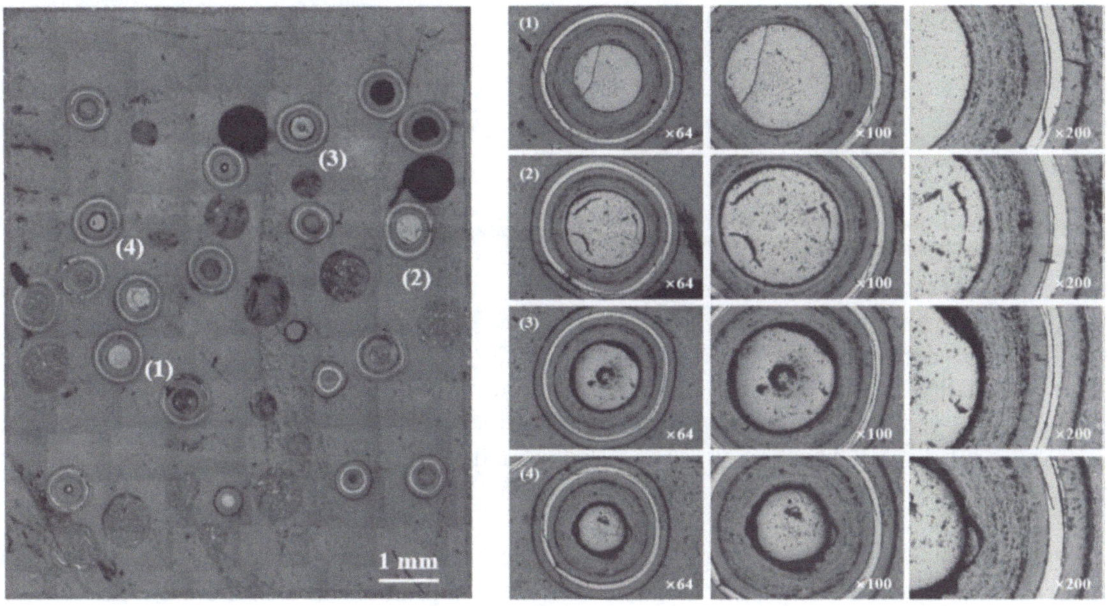

(a) (b)

FIG. 66. (a) Cross-section of one of fuel compacts in test rod 1, and (b) enlarged views of (1), (2), (3), and (4) in (a) (courtesy of KAERI).

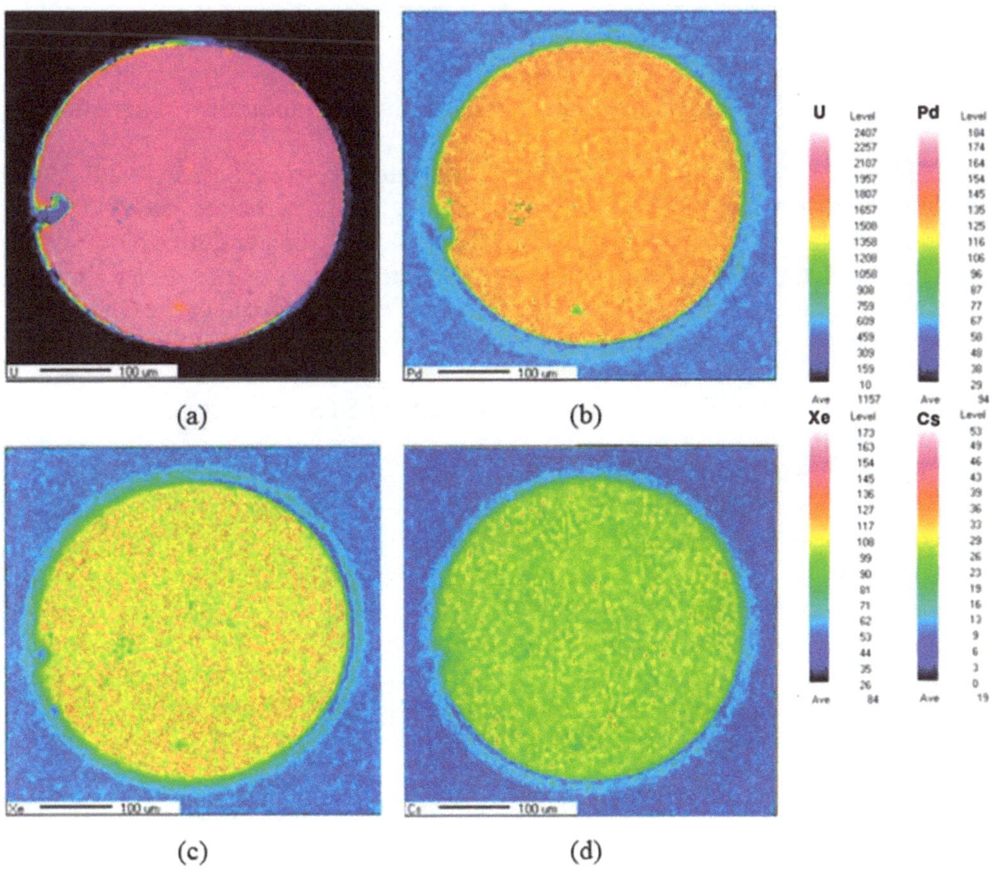

FIG.67. EPMA images from kernel: (a) U, (b) Pd, (c) Xe, and (d) Cs (courtesy of KAERI).

6.2.3. Loose TRISO particle fuels

Table 49 shows the status of irradiation test for loose TRISO fuel particles since 2000. There are three irradiation tests of loose coated particles; HTR-10 fuel particles in IVV-2M, PYCASSO-I and -II with ZrO_2 and Al_2O_3 surrogate kernel TRISO particles of CEA, JAEA, and KAERI in HFR.

6.2.4. Overall comparison of available irradiations in terms of fission gas release

Figure 69 [103] shows a comparison of the fractional release of ^{85m}Kr between German spherical pebbles and US historic cylindrical compacts, where US cylindrical compacts show more release than German spherical pebbles. This difference was attributed in [103] mainly to as-manufactured fuel quality. The variation was attributed to differences in the microstructures of the PyC and SiC layers within the TRISO coating and the bonding of these layers. These distinctions were found to be associated with the utilization of different fabrication processes in Germany and the United States. The manufacturing process for coating layers in the US has been improved.

Figure 69 also shows additional mapping of measured fractional releases from AGR-1 and AGR-2 compacts to those from US historic compacts and German pebbles. Measurements from new US fuel compacts are comparable with those from German pebbles, which revealed the significance of improved coating quality.

In Fig. 69, the release from AGR-1 is quite low, indicating no particle failures during the irradiation. Taking into account high burnup operation of AGR-1 compacts, AGR-1 result of fractional release supported the good performance of UCO even at high burnup, with observations of no kernel migration, no evidence of CO attack of SiC, and no indication of severe SiC attack by noble metal or lanthanide fission products.

The measured releases from AGR-2 were higher than those from AGR-1, reflecting a higher number of defective particles and uranium contamination, and indicating the possibility of a particle failure.

Figure 69 [103] shows fractional releases per failed particle irradiated in AGR-3/4 and AGR-2 experiments and their best fit. The fractional releases per failed particle of both krypton and xenon isotopes were less than 1% and seemed not sensitive to fuel temperature below 1050°C. At fuel temperature greater than 1050°C, however, the fractional releases per failed particle increased exponentially with increasing fuel temperature. The clear downward trend of the fitted lines for AGR-2 and AGR-3/4 R/B data support the above observations.

6.3. REQUIREMENTS FOR FUTURE IRRADIATION TESTS

There exist a few datasets obtained from in-pile annealing tests with TRISO fuel. These tests include reactivity-initiated accident (RIA) tests at the Nuclear Safety Research Reactor (NSRR), Japan and control rod withdrawal tests at the operating HTGRs including HTR-10 [156]. Even though a small number of datasets are available as shown in Table 50 [3, 157, 158], these annealing tests may not be particularly relevant to the advanced TRISO fuel (e.g. FCM fuel) proposed for fluoride salt-cooled high-temperature reactors (FHRs) or LWRs, due to different fissile materials (e.g. UCO or UN kernels) contained in the FCM fuel and longer or shorter pulse widths in FHRs or LWRs. Combined irradiation using the HFIR MiniFuel and power pulse at the TREAT reactor could narrow these data gaps by clarifying the appropriate energy deposition limits and any concomitant energy deposition rate effects.

Irradiation tests with instrumentations are also recommended to reduce the time required for PIEs at hot cell facilities.

TABLE 49. IRRADIATIONS OF LOOSE COATED PARTICLE FUELS SINCE THE EARLY 2000s

Reactor	Experiment ID	Fuel type	Country	Irradiation time (FPDs)	Fast fluence (10^{21}n/cm^2, E>0.18 MeV)	Burnup (%FIMA)	Irradiating temp. (°C)	Reference
IVV-2M (Russia)	(HTR-10 fuel particles)	Loose particle (capsule 1) (LEU UO_2 TRISO)	China	Jul. 2000 – Feb. 2003	-	-	1000±50	[70],[87],[89]
HFR (Petten, Netherlands)	PYCASSO-I	TRISO particles (surrogate kernel ZrO_2, Al_2O_3)	France, Japan, Republic of Korea	18 Apr. 2008 – 26 Apr. 2009 (144)	1.93, 2.02, 1.87, 1.49	-	923 (ave. JAEA, KAERI) 988 (ave. CEA) 1058 (ave. JAEA, KAERI) 988 (ave. CEA)	[5],[152],[153]
HFR (Petten, Netherlands)	PYCASSO-II	TRISO particles (surrogate kernel ZrO_2, Al_2O_3)	France, Japan, Republic of Korea	Jun. 2009 – Apr. 2010 (252)	3.66	-	855-1070 (ave.)	[5],[153–155]

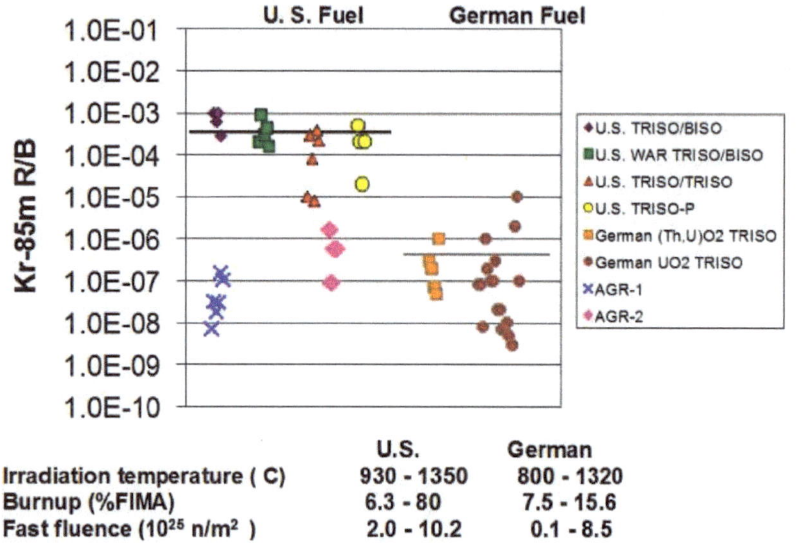

FIG. 68. End-of-life ^{85m}Kr fission gas release for AGR-1 and AGR-2 compared to historic performance in US and German TRISO fuel irradiations (courtesy of INL).

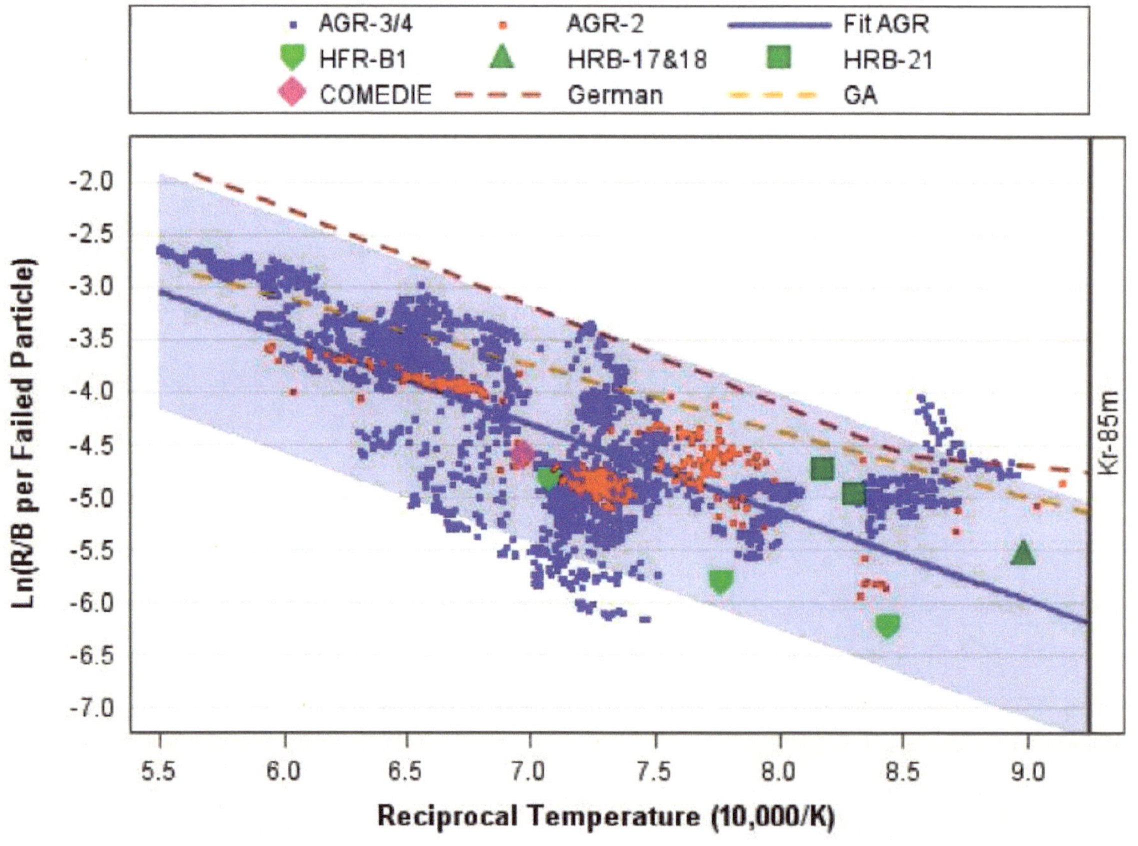

FIG. 69. Combined AGR fitted line and R/B per failed particle data for AGR irradiations, historical irradiations, and models (the blue shaded area is 95% bounds of the fitted line) (courtesy of INL).

TABLE 50. HISTORICAL TRANSIENT TESTS OF TRISO PARTICLE-BASED FUELS

Reference	Reactor	Kernel	Type of test	Energy deposition (J/g-fuel)	Pulse width (ms)	Failure
[157]	NSRR	UO_2	Element and loose particle	500 – 2300	~5	> 1400 J/g-UO_2 (fuel) > 2300 J/g-UO_2 (matrix)
[158]	NSRR	UO_2	Loose particle	500 – 1700	~5	> 1400 J/g-UO_2
[3]	HYDRA	UO_2	Element and loose particle	100 – 1700	1 – 2	> 1300 J/g-UO_2
[3]	IGR	UO_2	Element	> 10 000	7 – 30 000	Matrix

6.4. SUMMARY AND DISCUSSION

A number of datasets that enable to support the good performance of TRISO particle fuel in either spherical pebbles or cylindrical compacts are available from the irradiation tests under normal operating conditions and from out-of-pile heating tests to simulate a LOCA.

The followings are expert observations for further study:

— Irradiation tests under normal operating conditions and post-irradiation annealing tests at elevated temperatures are needed for a better understanding of the fission product diffusion/transport in a UCO kernel, due to the presence of different phases (UO_2 and UC);
— Irradiation tests under normal operating conditions and post-irradiation annealing tests at elevated temperatures are necessary to understand the effect of SiC used for the matrix (replacing graphite) on the integrity of fuel elements and on the nuclear physics characteristics (e.g. neutron absorption/scattering effect);
— The effect of the use of uranium mononitride (UN) in the kernel on the fuel performance needs to be understood as the oxygen potential will change; in particular, an impact on the oxidation (and hence immobilization) of rare earth elements can be expected;
— Irradiation tests under normal operating conditions and accident conditions are necessary to support a licensing of advanced coated particle fuel for use in LWRs, MMRs, nuclear thermal propulsion, and molten salt-cooled reactors, since with the advanced coated particle fuel in non-HTGRs the temperature transients may occur much faster than in HTGR depressurized loss-of-forced cooling (DLOFC) accidents.

7. FUEL PERFORMANCE ASSESSMENTS

7.1. FISSION PRODUCTS

7.1.1. Fission product yields and inventory

Figure 70 shows the inventory of some important elements contained in the TRISO particle fuel that achieved a burnup of 5.5% FIMA [4]. Zr and Mo constitute the larger inventory among metallic fission products, whereas Xe constitutes the larger inventory among gaseous fission products.

As burnup progresses, the concentration of Pu, in particular, ^{239}Pu, and minor actinides increases. In nuclear data libraries, fission yields are given per nuclide (and per incident neutron energy). Therefore, the effective fission yield of a mixture of actinides (e.g. U, Pu, Am) depends on burnup.

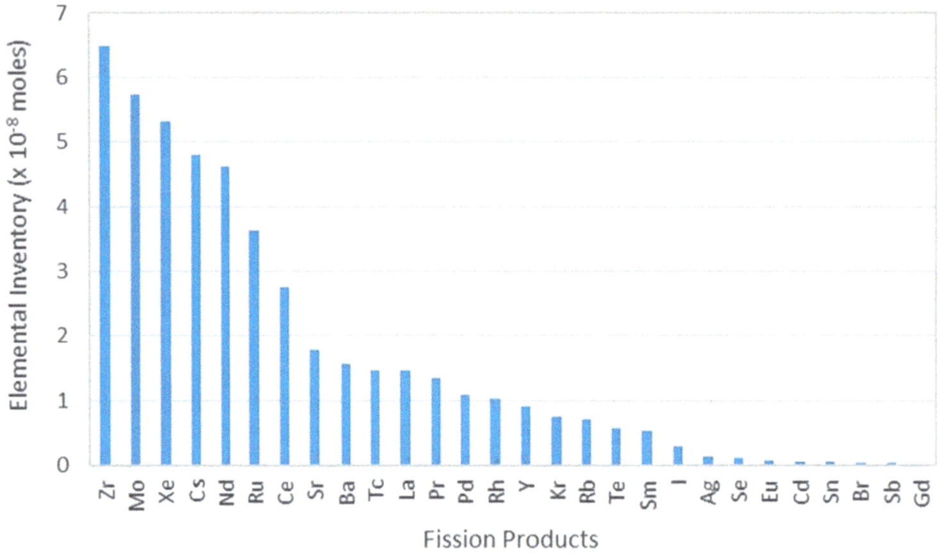

FIG. 70. Elemental inventory in a TRISO particle fuel achieving a burnup of 5.5% FIMA (courtesy of IAEA).

7.1.2. Fuel chemistry

The chemical state of fission product elements plays a crucial role in determining the physical properties of the fuel. This includes factors such as thermal conductivity, swelling, creep, and the onset of the melting point. Additionally, the chemical state influences the release behaviour of fission products from the fuel matrix. For instance, the relative volatility and mobility of these elements can be affected, leading to potential immobilization if rare earth elements oxidize and precipitate in the kernel as oxides. This influence is observed both under normal operating conditions and during transient events. Gaseous fission products (xenon and krypton) influence the thermal conductivity of a layer or the matrix (if it is released to the matrix in large amounts) and the gas pressure in the system.

Oxygen, liberated from uranium fissioning in UO_2, can combine with either the fission products or carbon from the PyC coatings to form CO and CO_2 gases. It is important to identify the oxide compounds formed from the fission products in the fuel. The oxygen potential, which is dependent on the pressure of oxygen in the gas phase within the fuel, is the critical parameter determining which elements (i.e., fission products) form oxides.

The oxygen potential is determined by the amount of oxygen in the system and affected by the affinity of particular elements for it. In this closed system, fission products, carbon, and uranium engage in competition for oxygen. The oxygen potential of the system ultimately dictates which elements undergo oxidation and which remain unoxidized.

The oxygen potential, μ_{O_2} is defined as:

$$\mu_{O_2} = RT \ln[P_{O_2}] \quad (10)$$

where

R is gas constant,
T is the absolute temperature,
P_{O_2} is oxygen partial pressure, atm.

In the reference case of pure UO_2 fuel, the oxygen potential rises with increasing burnup, leading to the generation of carbon monoxide (CO). Unirradiated stoichiometric UO_2 has an oxygen potential of roughly -400 kJ·mol^{-1} at nominal HTGR operating temperatures [4], see also Fig. 71 [159] for the oxygen potential of U^{4+} (i.e., stoichiometric fuel).

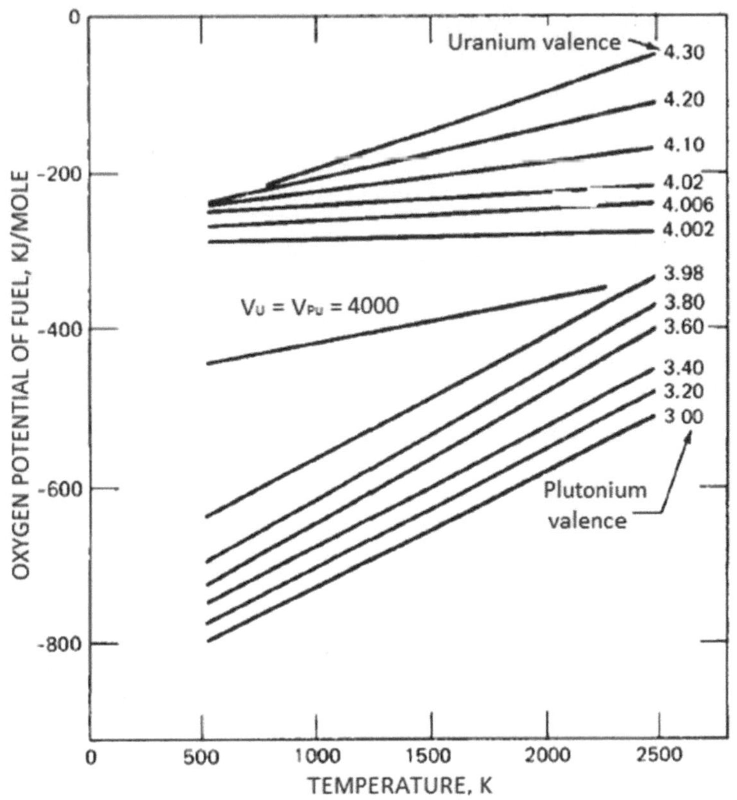

FIG. 71. Oxygen potentials of uranium and plutonium-based oxide fuels.

Depending on O/U ratio and the availability of excess C, two domains (see Fig. 72 [4]) were identified based on U-O-C phase diagram at 1573 K (1300°C):

— Triphasic domain with UC_2+UO_2+C when O/U ratio is < 2.0;
— Diphasic domain with UO_{2+x}+C when O/U ratio is > 2.0 [4].

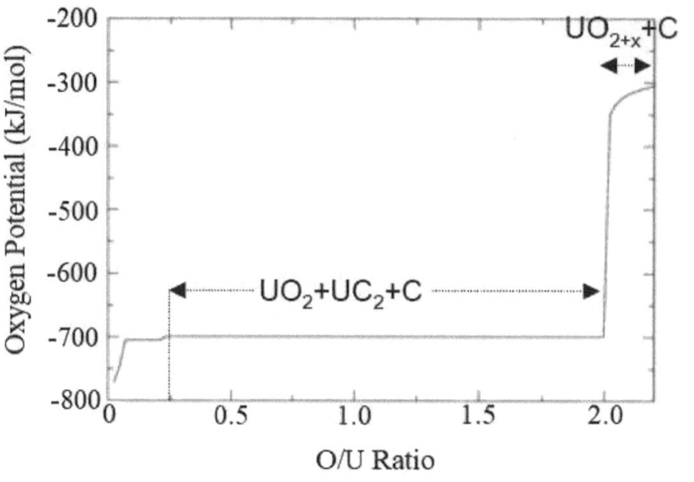

FIG. 72. Calculated oxygen potential as a function of O/U at 1300°C (courtesy of IAEA).

In an unirradiated fuel (or without taking into consideration of fission product affinity to oxygen), the oxygen potential, μ_{O_2}, calculated at 1573 K (1300°C) is effectively reduced to –700 kJ/mol in the triphasic domain UC_2+UO_2+C in comparison to the biphasic domain UO_{2+x}+C for which the calculated oxygen potential varied between -350 and -300 kJ/mol, as shown in Fig. 72. This shows that the oxygen affinity is increased in the UCO fuel in the triphasic domain (i.e., O/U ratio < 2).

In general, the system is assumed to be in thermodynamic equilibrium. However, during and after irradiation, the chemistry of the fuel–fission product system changes due to:

— Fission product accumulation: The concentration of fission products progressively rises during irradiation.
— Chemical transformation: Post-irradiation, the chemical properties of fission products undergo modifications due to radioactive β-decay.
— Oxygen potential shift: The oxygen chemical potential of the fuel changes with burnup, influencing the production of fission products that, in turn, can alter the oxygen-to-metal ratio of the fuel.
— Temperature-induced element distribution: Axial and radial temperature gradients within the fuel impact the distribution of elements across phases, leading to compositional gradients through material transport facilitated by thermal diffusion processes.

Some elements, especially volatile fission products, may undergo significant migration, leading to a gradient in the oxygen chemical potential within the fuel. As a result, equilibrium calculations can only serve as a foundational reference for estimating the chemical state.

In an irradiated fuel, fission products start to accumulate and compete with U and C for oxygen. The free energy of formation of the fission product oxide per mole of oxygen, ΔG_{FF}, is given by

$$\Delta G_{FF}[\text{kJ/mol of oxygen}] = RT\ln[P_{O_2}(\text{atm})] \quad (11)$$

Lewis illustrated the variation of ΔG_{FF} graphically in Ref. [160].

At a temperature of 1500 K, PdO, with a standard Gibbs energy of formation of +50 kJ/mol of O_2, results in a P_{O_2} (the partial pressure at which pure oxide and pure metal can coexist) of 50 atm. In contrast, for

La_2O_3, which has a standard Gibbs energy of formation of -960 kJ/mol of O_2, the equilibrium P_{O_2} is approximately ~10^{-34} atm [159].

If the free energy of formation of the fission product oxide is smaller than the oxygen potential of the fuel, the element will be capable of removing oxygen from the fuel matrix and forming a stable oxide; otherwise, the fission product will exist as an element in the fuel under all reactor conditions. Accordingly, it is expected that palladium should always be found as a metal and lanthanum as an oxide, in a fuel with an oxygen potential of -400 kJ/mol. Based on the fuel oxygen potential and on the basis of numerous elemental analyses of irradiated LWR and fast breeder reactor fuels, the chemical state of the fission products in the LWR and fast breeder reactor fuel were identified [160] as:

— Fission gases and other volatile fission products: Kr, Xe, Br, and I;
— Fission products forming metallic precipitates: Mo, Tc, Ru, Rh, Pd, Ag, Cd, In, Sn, Sb, and Te;
— Fission products forming oxide precipitates: Rb, Cs, Ba, Zr, Nb, Mo, and Te;
— Fission products dissolved in the fuel matrix (i.e., solid solution): Sr, Zr, Nb, and the rare earths Y, La, Ce, Pr, Nd, Pm, and Sm.

The states of fission products are as follows [4]:

— Noble gases demonstrate almost insolubility in the fuel. Iodine, bromine (Br), and tellurium (Te) also exhibit high insolubility, while caesium (Cs) and barium (Ba) show only low solubility in hyper-stoichiometric fuel conditions. Solubility increases for zirconium (Zr), depending on temperature, and strontium (Sr). The rare earth (RE) elements, such as cerium (Ce), attain very high solubility.
— Fission-gases (Xe, Kr) have very low solubility in the fuel. The gases precipitate primarily as intragranular bubbles. These bubbles can grow by gas atom and vacancy absorption, and their mobility is limited.
— For the UO_2 kernel in TRISO fuel particles, both the fission rate and temperature play crucial roles in limiting bubble size and maintaining a substantial population of fission gas atoms in the fuel lattice through the resolution of gas from bubbles. Fission gases can migrate to grain boundaries through atomic or bubble diffusion, forming intergranular bubbles (porosities) where they can grow by the addition of gas and vacancies. Similar to intragranular bubbles, high levels of irradiation-induced resolution significantly restrict this growth (and the bubble-induced swelling of the kernel), while stresses in the particle may also limit the equilibrium bubble size. Once intergranular bubble growth becomes significant, causing interconnection of porosity, a path from within the kernel to the buffer is created, leading to fission gas release through a percolation mechanism.
— Regarding the behaviour of solid fission products, considerations include solubility, possible oxidation, and diffusion. In HTGR fuel particles with expected oxygen potentials, REs are highly soluble and are incorporated into the fuel lattice as stable oxides (REO_2 or RE_2O_3). In UCO fuel, as the UC_2/UO_2 ratio increases, the mobility of the REs also increases since their trapping in the kernel in oxide form is reduced. Zirconium, Nb, and Sr exhibit significant solubility in UO_2, with Zr showing more solubility at higher temperatures, and these elements are easily oxidized (ZrO_2, Nb_2O_3, and SrO). Barium exists in the form of low-solubility BaO and can migrate to grain boundaries, forming separate complex ternary compounds with Sr, Zr, REs (and small amounts of Cs and Mo) of perovskite structure. Tellurium has low solubility and is challenging to oxidize; at grain boundaries, it forms intermetallic compounds (e.g. with Pd) and Cs_2Te, and is significantly released from the kernel. Many other metallic elements also have relatively low solubility and are difficult to oxidize, except at high oxygen potentials. After segregation at grain boundaries, they form metallic inclusions, with noble metals (Mo, Ru, Rh, Tc, and Pd) constituting the main type of inclusion. Other inclusions like Te, Sn, or Ag, Cd are also observed. The size of inclusions is temperature-dependent. Pd and Ag, being highly volatile, are significantly released from the fuel kernel. Molybdenum exhibits complex behaviour due to its oxidation potential being close to the value corresponding to stoichiometric

fuel; a significant amount can be oxidized to MoO_2 (limiting the increase of fuel stoichiometry), a low-solubility oxide. At high temperatures, Mo can form molybdate compounds with Cs or Ba. Caesium also has low solubility and might be oxidized to Cs_2O depending on the fuel temperature and oxygen potential, with this oxide having low solubility. Complex ternary phases can be formed at high temperatures and high burnups.

— In UCO-type fuel kernels, oxygen released during the fission process of UO_2 initially oxidizes the carbide UC_2 phase and then the RE fission products due to their high affinity for oxygen. When the UC_2 phase is present in the kernel, the oxygen potential is maintained at a fixed value, approximately -700 kJ/mol, at 1573 K (1300°C). Fission product elements with lower affinity for oxygen (Sr, Eu, Zr, Ba) exist as carbide compounds. As burnup increases, the UC_2 phase in the kernel decreases, leading to an increase in the oxygen potential. This makes oxygen available to form.

The oxygen potential in a TRISO particle with UO_2 kernel, calculated at 1200 K (927°C) and 2000 K (1727°C) for burnups up to 8% FIMA is shown in Ref. [161], and is illustrated in Fig. 73. The oxygen potential increases with burnup and temperature.

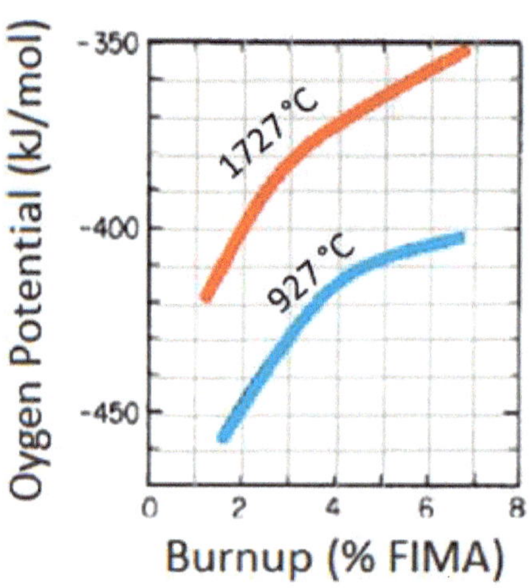

FIG. 73. Oxygen potential in the coated UO_2 fuel particle calculated at 1200 K (927°C) and 2000 K (1727°C).

7.1.3. Chemical state and location of fission products in a TRISO particle fuel

7.1.3.1. TRISO fuel with UO_2 kernel

Fission products in irradiated TRISO particle fuels with UO_2 kernel were investigated using shielded EPMA, which is equipped with both wavelength dispersive X-ray spectrometer (WDX) and energy dispersive X-ray spectrometer (EDX), and by performing thermodynamic analysis [161]. The diameter of the fabricated kernels was ~ 600 μm; the ^{235}U enrichment was either 4% or 8%; the thickness of buffer, IPyC, SiC and OPyC layers was on average 60 μm, 30 μm, 26 μm, 45 μm, respectively; the density of the kernels, buffer, IPyC, SiC and OPyC layers was 10.5 g/cm^3, 1.17 g/cm^3, 1.84 g/cm^3, 3.21 g/cm^3, 1.85 g/cm^3, respectively.

The particles were irradiated in the form of fuel compacts, achieving burnups up to 5.16% FIMA at time averaged irradiation temperatures of 1613 K (1340°C) – 1783 K (1510°C), though several samples achieved lower burnups or were exposed to lower time average irradiation temperatures [161]. The observations associated to the fission product composition, chemical state of the fission products and their locations in the fuel particle can be summarized as:

— In the UO_2 kernel: vastly, metallic precipitates consisting of molybdenum (Mo), technetium (Tc), ruthenium (Ru), rhodium (Rh) and palladium (Pd), were observed, though in some particles, the precipitates consisted of:
  - Mainly palladium and tellurium (Te);
  - Tin (Sn);
  - Predominantly molybdenum;
  - The Mo-Tc-Ru-Rh-Pd type reacting with silicon (Si).

— In the Pd-Te type precipitates, a minor presence of rhodium was noted. The Pd-Te precipitate exhibited a significantly higher concentration of palladium compared to the Mo-Tc-Ru-Rh-Pd precipitate, despite the Pd-Te precipitates being fewer in number than the Mo-Tc-Ru-Rh-Pd precipitates.

— The metallic precipitate containing Sn also included Mo, Pd and small amounts of Tc, Ru and Rh. (Mo, Tc, Ru, Rh, Pd) silicide precipitates were also observed when the precipitate predominantly consisted of Mo, though it was indicated [161] that the precipitates predominantly consisting of molybdenum were rarely observed in the experiment. The reason to observe (Mo, Tc, Ru, Rh, Pd) silicides was explained as the transfer of some silicon to the UO_2 kernel when the SiC layer of the particle is corroded by carbon monoxide due to the failure of the protective layer of the IPyC. It was also indicated [161] that oxide precipitates were not observed in the kernel.

— Within the coating layers, palladium tended to accumulate on the inner surface of the SiC layer, occasionally undergoing reactions with the SiC layer, particularly on the cold side of the particle. Unlike the fuel kernel, the other metal elements (i.e., Mo, Tc, Ru, Rh) accompanying Pd in the kernel were rarely observed in the coating, which is probably due to the high vapour pressure of Pd, compared to others.

— Tellurium was detected within the buffer PyC layer on the cold side of the particle, but unlike palladium, it did not permeate through the IPyC layer. Te did not form oxide but exists as elemental Te. Since the boiling point of Te is 1266 K (993°C), Te is present in the gas form at temperatures of interest to HTGR. Te was transferred into the buffer layer through the gas phase at irradiation temperatures (1340°C – 1510°C) and was condensed there during cooling. Although cerium and barium were often observed in the coating layers of the samples, they were detected more frequently in samples experienced higher irradiation temperatures (i.e., 1470–1510°C). Ce and Ba were observed in the buffer layer and at the interface of the IPyC-SiC layers in particles with both intact and breached IPyC layers. Ce and Ba were found in the coating layers in an oxide form.

— The coating layers did not show the presence of REs other than Ce, such as lanthanum and neodymium, as well as strontium, which belongs to the alkali earth metals along with barium. Based on the thermodynamic analysis, the vapour pressure of CeO_2 is the highest of the species containing the REs. Similarly, the vapour pressure of SrO is much lower than that of Ba-containing species, which may explain the absence of strontium.

— Caesium was observed in the buffer PyC and the IPyC layers, though the concentrations were low.

— Xenon was retained in the buffer PyC layer.

Another study was performed to investigate the TRISO particles for their potential for very high temperature performance and burnup. Five spherical pebbles of former German production with TRISO particles were irradiated, achieving a maximum burnup of 11% FIMA [93]. The central temperature of all pebbles was kept close to 1250°C and held constant during the entire irradiation. One of them was used for the analysis.

High porosity and a significant amount of metallic inclusions in the kernel were observed from particles at the centre of the fuel pebble (see Fig. 74 [93]), while particles at the periphery of the pebble fuel zone have porosities at the cold side of the kernel with hardly any visible grain boundaries and metallic islands, though a number of particles at the periphery of the pebble fuel zone have kernels with evenly distributed pores and metallic islands (see Fig. 74 [93]).

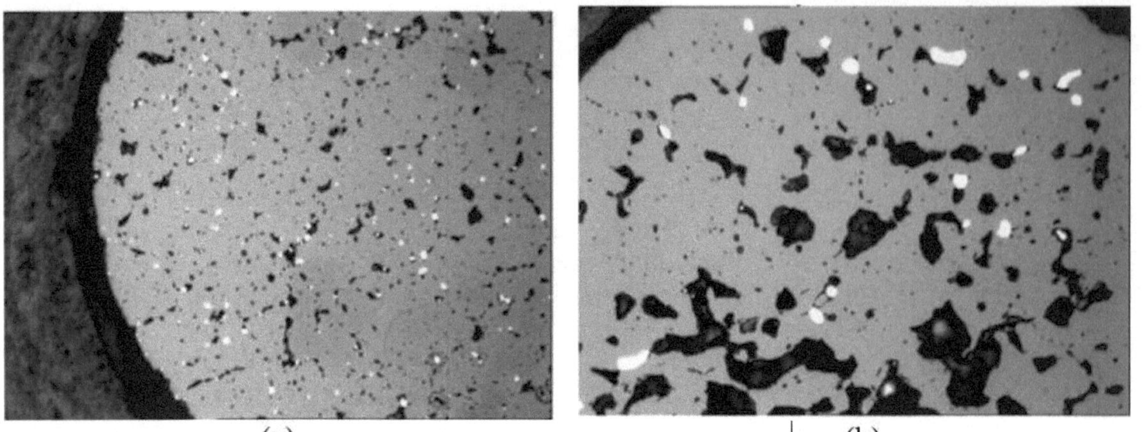

(a) (b)

FIG. 74. Metallic islands and pores in the kernel in a particle at the: (a) periphery of the pebble, and (b) centre of the pebble (courtesy of INL).

Additionally, the followings were also reported in Ref. [93]:

— In the UO_2 kernel:
  - Zirconium was retained in the kernel, which is expected, because Zr is a rather immobile fission product;
  - Ru and Mo are mainly contained in the metallic islands in the UO_2 kernel;
  - Pd was observed both in the metallic islands in the kernel and as metallic element distributed in the kernel.

— In the coating layers:
  - Pd was observed at the interface of the IPyC-SiC layers and the inside of the SiC layer;
  - Cs and Xe were found accumulated in the buffer layer.

Table 51 shows the location of fission products of a TRISO particle with UO_2 kernel that achieved burnups up to 11% FIMA.

TABLE 51. SUMMARY OF LOCATIONS OF FISSION PRODUCTS IN A TRISO PARTICLE WITH UO_2 KERNEL

Location	Fission products and their chemical state
Kernel	• Metallic precipitates consisting of Mo, Tc, Ru, Rh and Pd – Metallic • Pd-Te – Metallic • Metallic precipitate containing Sn, with Mo, Pd and small amounts of Tc, Ru and Rh – Metallic • (Mo, Tc, Ru, Rh, Pd) silicide precipitates, with high concentrations of Mo
Coating Layers	• Pd and Pd-Si at the inner surface of the SiC layer – Metallic and Silicide • Te in the buffer layer – Metallic • Ce and Ba in the buffer layer and at the interface of the IPyC-SiC layers (in particles with both intact and breached IPyC layers) – Oxide • Cs in the buffer and IPyC layers • Xe in the buffer layer

7.1.3.2. TRISO fuel with UCO kernel

Fission product distribution in UCO kernel

Fuel kernels from an unirradiated coated particle and irradiated AGR fuel compacts from the first and second irradiation test campaigns, achieving burnups of 17.4% FIMA (AGR-1; UCO kernel with an enrichment of 19.74%) and to 12.68% FIMA (AGR-2; UCO kernel with an enrichment of 14.03%), respectively, were examined (using scanning transmission electron microscopy (STEM), and energy dispersive spectroscopy (EDS) elemental mapping) to determine the microstructural and micro-chemical evolution of irradiated UCO fuel kernels, kernel-buffer interfaces, and recoil zones. In addition, the AGR-1 fuel was heated at 1600°C for 300 h, and a chemical analysis for fission products was performed to determine three-dimensional microstructure and chemical mapping (using atom probe tomography(APT) technique) at nanometric length scale [162]. The enrichment of the AGR-1 kernel (for the annealingtest) was 19.74%, and the burnup achieved before the annealing test was 18.6% FIMA. Results can be summarized [162] as follows:

— The unirradiated fuel kernel exhibited a microstructure composed of several phases: UO_2 and UC_2 with UC platelets, as shown in Fig. 75 [162]. Higher-magnification transmission electron microscopy (TEM) images (see Fig. 76 [162]) of the structure of the fuel taken from the edge of the fuel kernel indicated some strain build-up, presumably due to a lattice mismatch between the UC_2 and UC phases.
— The resultant microstructure after irradiation consisted of two primary phases, namely a 'high-Z (atomic mass)' UC(O) phase (rock salt; a=0.496 nm) and a 'low-Z' UO_2 I phase (fluorite structure; a = 0.547 nm).
— The UO_2 phase contained multiple precipitates of the UC phase, whereas no precipitates of the UO_2 phase were identified in the UC phase.
— The UO_2 phase contained trace amounts of I, Nd, Pr, Cs, Ce, and Er, whereas the UC phase contained Zr, Mo, and/or Ru; the elements Zr, Mo, and Ru were often observed as segregated into separate UC grains. Figure 77 shows a higher-magnification STEM image and elemental spectral images from the fuel kernel centre, exhibiting the segregation of Mo and Zr (in the high-Z material) into distinct grains; arrows in the Xe spectral image in Fig. 77 indicates Xe bubbles.
— Xe bubbles were located exclusively in the Zr containing part of the UC phase (see also Fig. 77 [162]).
— APT analysis on a annealing tested AGR-1 particle also showed Xe preference for the UC phase, although Xe was also identified in the UO_2 phase.
— Interaction between the fuel kernel and buffer layer was observed with sub-micron-sized carbon inclusions in the kernel side, and isolated UCO particles were located in the buffer layer.
— APT data revealed segregation of Te and Ag at the UO-UC interface.
— The APT results confirmed the selective association of specific elements for the two different U-phases. The UO phase showed presence of fission products such as Pd, Zr, Ce, Nd, Er, Dy, Gd, La, Y, Xe, and Sm, some of which are shown in Fig. 78, whereas the UC phase showed higher concentrations or segregation of Rh, Ru, Zr, Xe, Cd, In, Sn, Tc, and Nb, some of which are shown in Fig. 78 [162], Cs was not detected.
— The APT results confirmed the selective association of specific elements for the two different U-phases. The UO phase showed presence of fission products such as Pd, Zr, Ce, Nd, Er, Dy, Gd, La, Y, Xe, and Sm, some of which are shown in Fig. 78, whereas the UC phase showed higher concentrations or segregation of Rh, Ru, Zr, Xe, Cd, In, Sn, Tc, and Nb, some of which are shown in Fig. 78, Cs was not detected.

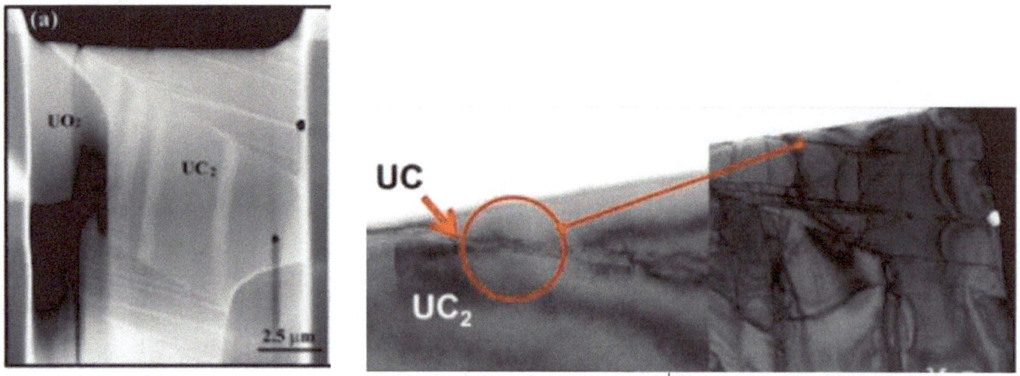

(a) (b)

FIG. 75. Unirradiated UCO kernel: (a) STEM Z-contrast image, (b) TEM image for the identification of the UC phase (courtesy of INL).

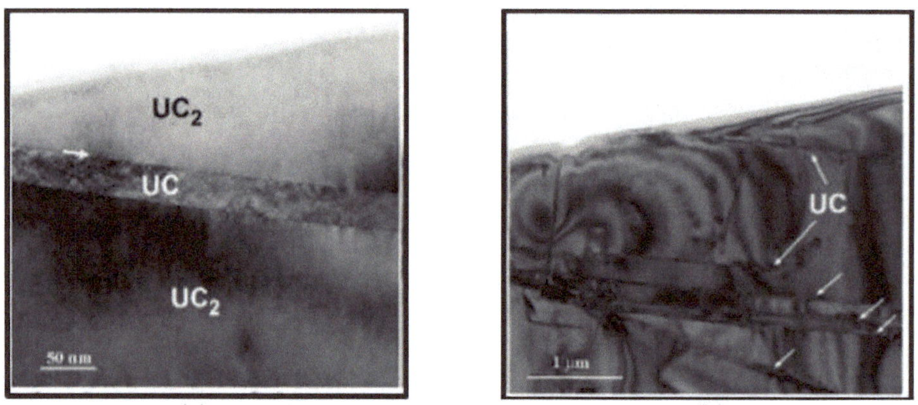

FIG. 76. Unirradiated UCO kernel TEM images of the structural arrangement of the UC platelets within the UC_2 host phase (a) at a higher magnification, and (b) at a lower magnification (courtesy of INL).

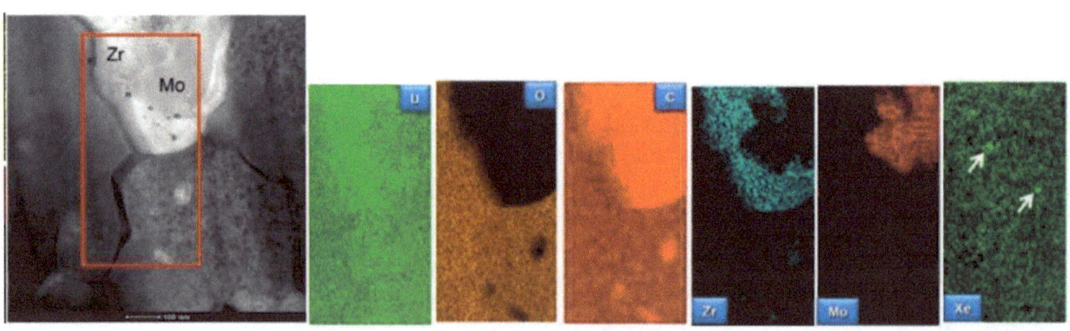

FIG. 77. Irradiated AGR-1 fuel: STEM Z-contrast image and corresponding spectral images for U, O, C, Zr, Mo and Xe from the indicated scan area of the fuel kernel (courtesy of INL).

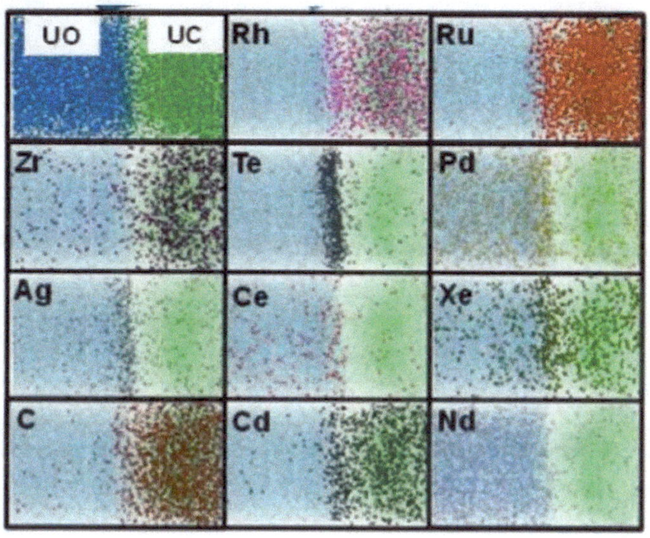

FIG. 78. 3-D APT data obtained at the centre of UCO Kernel; interfacial distribution of fission products Rh, Ru, Zr, Te, Pd, Ag, Ce, Xe, Cd, and Nd across the UO and UC phase (courtesy of INL).

Table 52 summarizes the location of the fission products in different phases of the UCO kernel.

TABLE 52. LOCATIONS OF FISSION PRODUCTS IN THE UCO KERNEL

Conditions	Fission products
Irradiated Fuel: AGR-1 and AGR-2	• In UO_2 phase: trace amounts of I, Nd, Pr, Cs Ce, and Er • In UC phase: Zr, Mo, and/or Ru; Xe in the Zr-rich areas (in the UC phase)
High temperature annealing tested (AGR-1) fuel after irradiation	• UO phase: Pd, Zr, Ce, Nd, Er, Dy, Gd, La, Y, Xe, and Sm • UC phase: Rh, Ru, Zr, Xe, Cd, In, Sn, Tc, and Nb • Te and Ag at the UO-UC interface

Another study showed that I, Xe, and Cs mainly resided in the recoil zone (i.e., outer perimeter of the kernel, where the kernel interfaces with the buffer layer, and the thickness of the recoil zone is about 25 µm), buffer layer, and IPyC layer [163]. Elements of Zr, Ce, La, Sm, Pr, and Nd nearly all remained in the fuel kernel [163].

Fission product distribution in the coating layers

Post-irradiation examination of fuel compacts irradiated under the AGR-1 test campaign focused on the fission product distribution in the coating layers, in particular in the SiC layer and at the IPyC-SiC interface, and revealed numerous clusters of fission product precipitates in the coating layers, as shown in Fig. 79 [107].

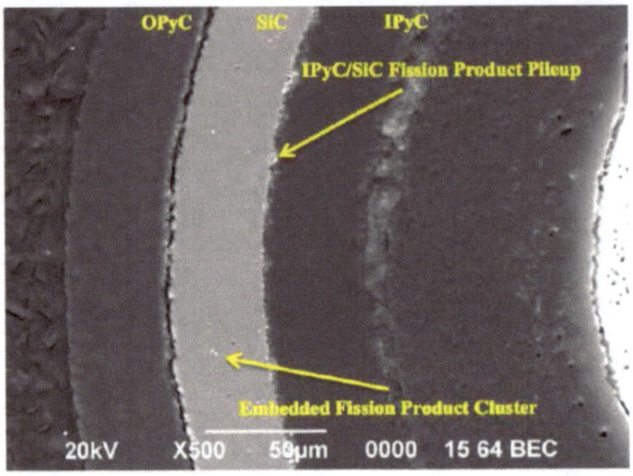

FIG. 79. Backscattered electron composition (BEC) micrograph of the TRISO layers (courtesy of INL).

Figure 80 shows the accumulation of fission products at the IPyC-SiC interface and fission product clusters embedded in the SiC layer [107]. The general observations can be summarized:

- — Presence of Pd in SiC layer;
- — Accumulation of Cs, Pd and U at the IPyC-SiC interface (see Fig. 80 [107]);
- — Presence of Ba, Ce, Cs, Nd, O and Zr and also Ag, Cd, Pu, and Sr at the IPyC-SiC interface.

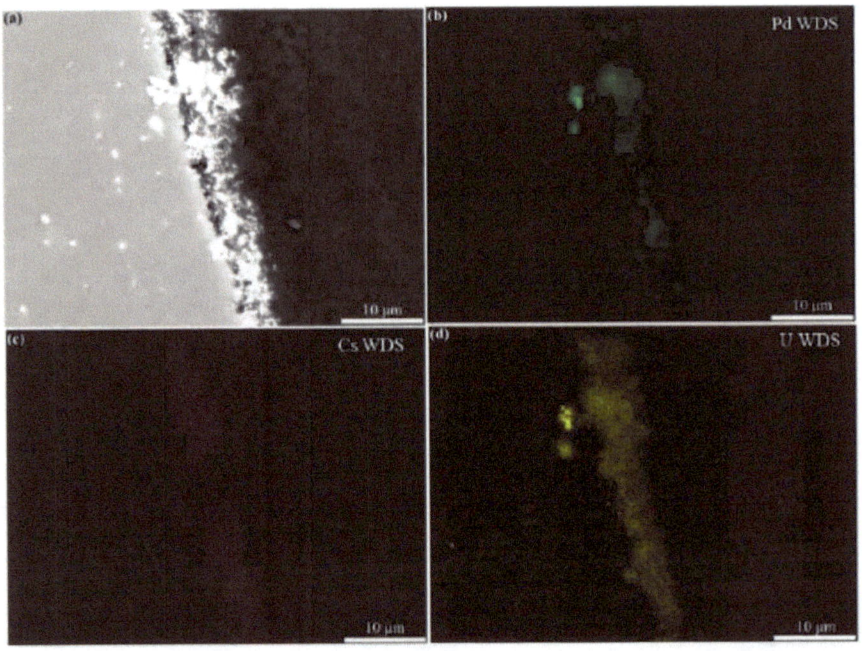

FIG. 80. Scanning electron microscopy (SEM) micrograph showing Pd/Cs/U rich fission product segment at the IPyC-SiC interface (courtesy of INL).

In the analysed particle, which represents the baseline fuel fabricated with coating conditions closely resembling the German reference fuel for the AGR-1 test campaign, U was consistently found in association with Pd in precipitates within the coating layers. Conversely, in another particle, fabricated under conditions similar to the baseline but with variations in the silicon carbide coating process (lower deposition temperature of 1425°C using argon and hydrogen gas for fluidization), uranium was identified separately in the SiC layer, and nearly all precipitates at the inner pyrolytic carbon-SiC interface contained uranium, as illustrated in Fig. 81 [107].

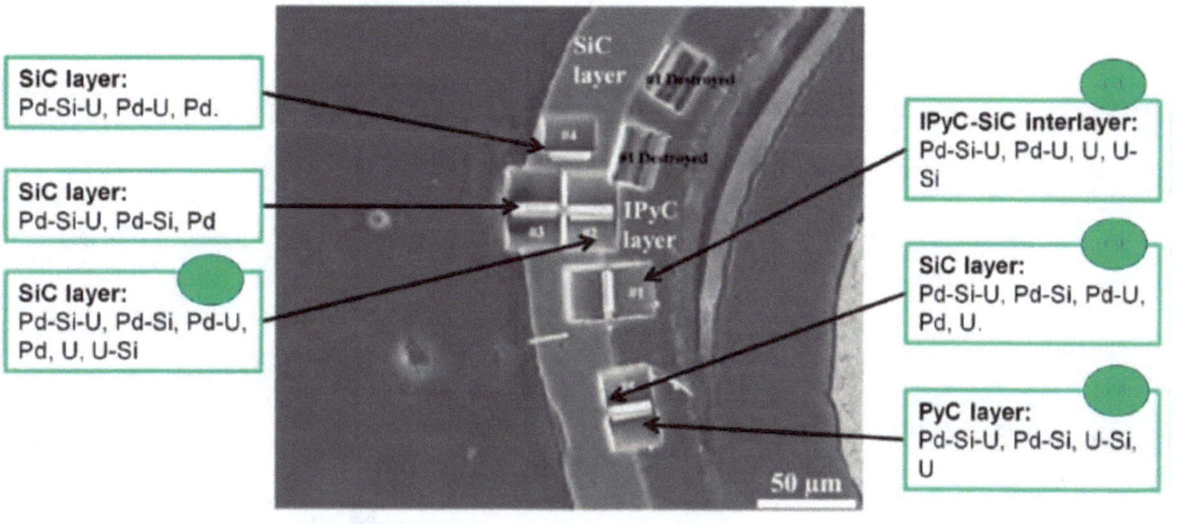

FIG.81. Fission product distribution in a particle with altered SiC coating conditions (courtesy of INL).

Ag was always found with Cd and Pd at grain boundaries and triple points (see Fig. 82 [164]).

Pd precipitates were found at stacking faults (see Fig. 83 [164]) in SiC layer; some black atoms (dark spots representing fission product atoms) are also observable in the SiC grain in Fig. 83 [164]. Phosphorus was also detected inside the SiC grain [165]; this may be due to Si (n, γ) reaction, which results in an unstable ^{31}Si with a half-life time of 157 minutes, followed by stable phosphorus by β decay.

Figure 84 [107] shows the high-resolution STEM image of an area in the IPyC close to the IPyC-SiC interface. Single and multiphase precipitates can be identified. In addition, single phase precipitates in Fig. 85 [166] are indicated by red arrows. It looks like the selected area is carbon rich and Si is spotted at the low concentration of C. In the multiphase region, Si, Pd, U, Pu, Zr were identified [107]. Despite that the pattern for U and Zr in Fig. 84 looks like U and Zr formed a two-phase, it is indicated by arrows that those are single phases.

A TRISO particle from the compact irradiated in the ATR at INL to an average burnup of 17.42% FIMA, with a time-averaged, volume-averaged temperature of 1059°C and an average fast fluence of 3.77×10^{21}n/cm^2 was used to investigate fission product behaviour in a TRISO particle [166]. X-ray tomography conducted on this particle revealed the presence of a crack extending through both the buffer layer and the IPyC layer. Additionally, localized degradation in the SiC layer was observed, specifically in a region connected to the tip of the IPyC crack. Notably, no crack was identified in the SiC layer. The initiation of the crack in the IPyC layer seems to be associated with the buffer layer pulling away from the IPyC layer as the buffer layer experienced shrinkage under irradiation. This crack is situated along a separation between the detached and attached portions of the buffer layer.

Close to the crack tip (see Fig. 85 for the SEM image), distribution and composition of fission products, and fission product transport mechanisms in the SiC layer of a TRISO particle exhibiting localized corrosion were investigated [166].

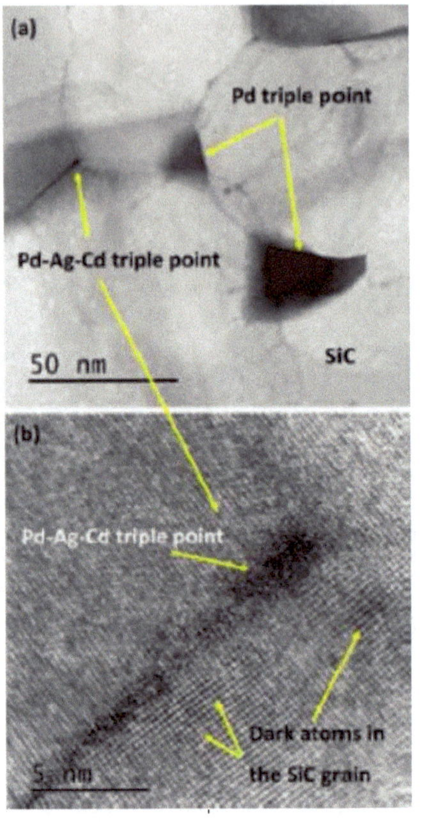

FIG. 82. Bright field STEM image (a) of a triple point containing Pd, Ag, and Cd, (b) high-Resolution bright field STEM image of the triple point containing Pd, Ag, and Cd with fission product atoms (dark spots) in the SiC grain (courtesy of INL).

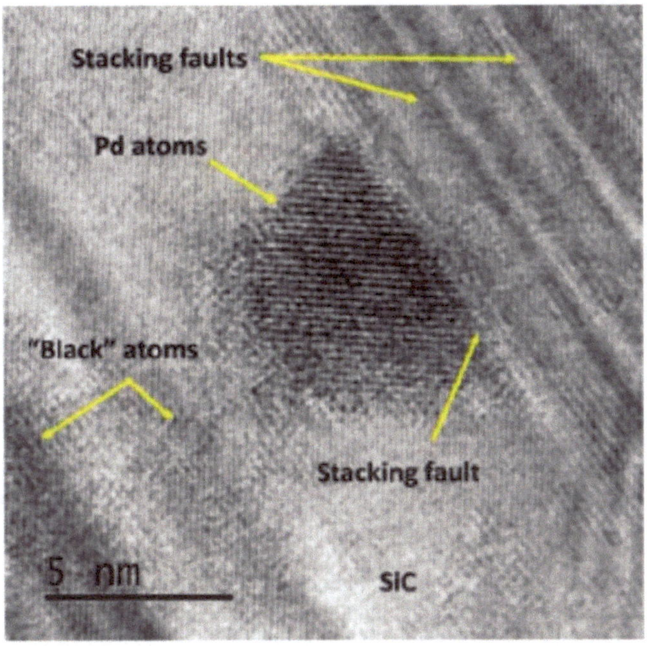

FIG. 83. High-resolution bright field STEM image of a Pd precipitate (courtesy of INL).

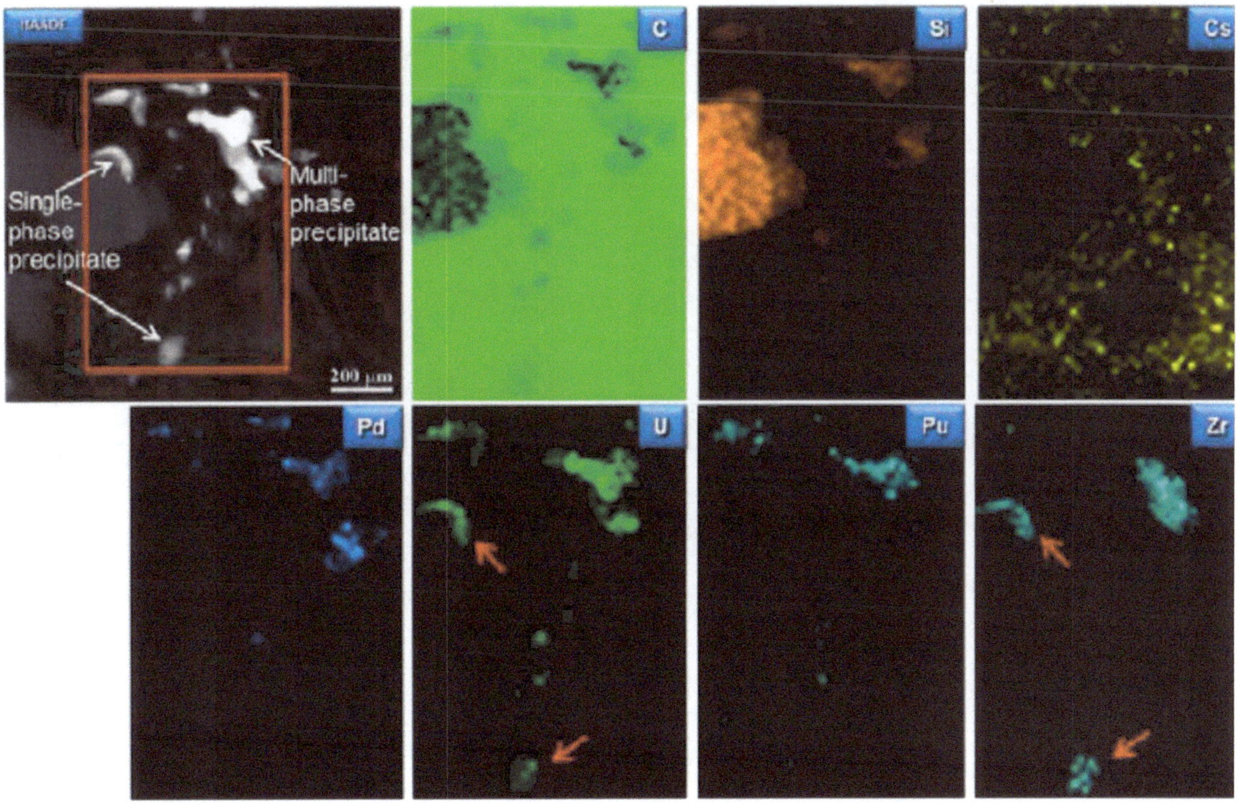

FIG. 84. High resolution STEM image and energy-dispersive X-ray spectroscopy (EDS) mapping in an area in the IPyC layer close to the IpyC-SiC interface (courtesy of INL).

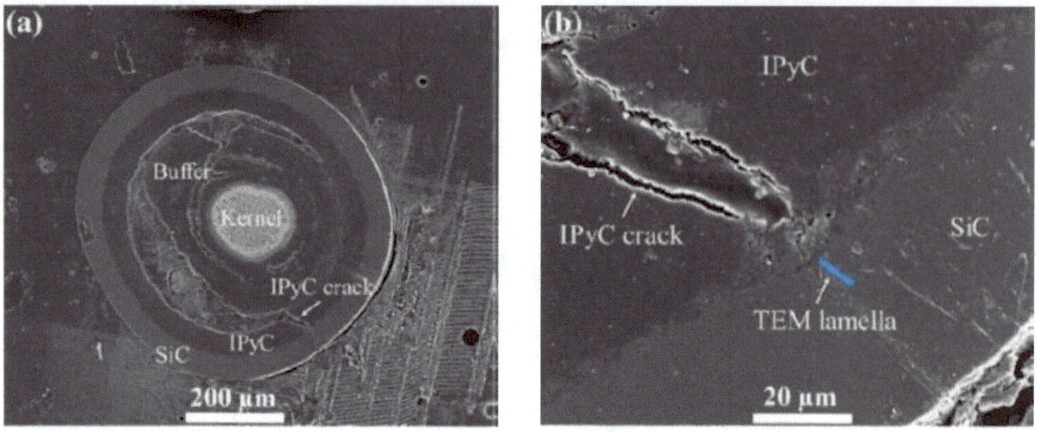

(a) (b)

FIG. 85. SEM images showing the IPyC crack: (a) Lower magnification, (b) higher magnification (courtesy of INL).

SEM back-scattered electron and secondary electron images at the crack tip are represented in Fig. 86: locations 5 and 6 – no evident accumulation of fission products; location 7 – accumulated fission products in the IPyC layer right along an edge of the crack; location 8 – fission products accumulation at the IPyC-SiC interface right at the crack tip; location 9 - a slight accumulation of fission products in the region where delamination between IPyC and SiC occurred due to the IpyC crack; location 10 – locally corroded area within the SiC layer at the IpyC crack tip.

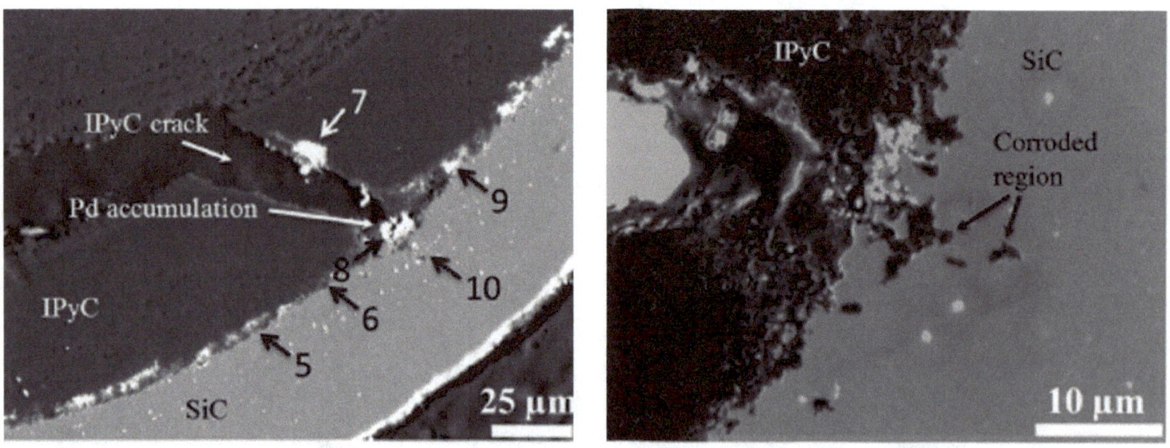

(a) (b)

FIG. 86. SEM images of the crack tip: (a) Back scattered electron image, (b) secondary electron image (courtesy of INL).

Fission products in the uncorroded areas

Figure 87 [166] shows the precipitates at the SiC grain boundaries in the uncorroded SiC areas. Precipitates were mostly identified as Pd-Si-U, Pd-Si, Pd-U, and Pd. Ag and/or Cd were observed in some Pd and Pd-Si, though sometimes by themselves. Caesium (Cs) was not detected in any precipitates and no detectable amount of cerium (Ce) was observed. The analysis showed the following results [166]:

— Pd: Precipitates containing Pd-U, Pd-Si, and Pd-Si-U were all situated at the grain boundaries of SiC. Additionally, pure Pd precipitates were identified both at the grain boundaries and within the interiors of SiC grains. At ~1000°C, both lattice diffusion and grain boundary diffusion became significant for Pd transport in SiC. Neutron irradiation also significantly enhanced the diffusion due to high concentrations of defects generated in excess of thermal equilibrium values. Pd precipitates at SiC grain boundaries were larger than those inside SiC grains, possibly due to the faster grain boundary diffusion compared with lattice diffusion. At the IPyC-SiC interface, to some extent, Pd accumulated and reacted with SiC, creating Pd-Si. The Pd-Si created at the IPyC-SiC interface might have migrated into the SiC layer, explaining the presence of Pd-Si in the SiC layer. Otherwise, there were no notable chemical reactions observed between Pd and SiC in regions of the SiC layer that were distant from the crack in the IPyC layer of this specific TRISO particle. Additionally, no such reactions were noted in the SiC layers of TRISO particles originating from other compacts.

— Ag: The mechanisms for Ag migration were anticipated to be grain boundary diffusion, neutron enhanced diffusion, Pd-assisted transport, and vapour-phase migration. Cd was found with Ag. Since Cd is also a direct fission product, it can also be from Ag decay. Of the Ag/Cd containing precipitates, > 90% contained Pd, suggesting that Pd assisted Ag/Cd transport. Ag containing precipitates were predominantly observed at grain boundaries, indicating that grain boundary diffusion is one of the mechanisms for Ag transport. Another mechanism can be the intragranular transport of Ag enabled by Pd, based on the observation of an Ag-Pd-Cd precipitate at a stacking fault inside a SiC grain. The presence of intragranular Ag-Pd-Cd precipitate at a stacking fault indicates that the intragranular transport of Ag occurs via relatively high-energy pathways that have lower activation energy for diffusion because stacking faults have higher energies than lattice sites. With grain boundary diffusion being the predominant mechanism, the lattice diffusivity of Ag in SiC is expected to be low, though lattice diffusivity might be enhanced due to high concentrations of defects in the lattice created by neutron irradiations.

— U: U-containing precipitates, including Pd-Si-U and Pd-U, were identified in the uncorroded SiC areas. For the majority of Pd-Si-U and Pd-U precipitates (see Fig. 87), the U concentrations are much lower than the Pd concentrations. A significant density of uranium-rich precipitates was identified in the buffer layer, subsequently migrating deeper into the IPyC layer. Results from scanning transmission electron microscopy with energy-dispersive X-ray spectroscopy (STEM-EDS) on transmission electron microscope (TEM) lamellae extracted from the IPyC layer revealed that precipitates located several microns away from the IPyC-SiC interface were predominantly rich in uranium, while those in close proximity to the interface were mostly composed of Pd-Si or Pd-Si-U, with minor concentrations of uranium. At the IPyC-SiC interface, an increased concentration of U was observed due to the accumulation of U as its diffusion slows down while approaching to SiC layer, which is a much stronger barrier to U diffusion compared with carbon. At the IPyC-SiC interface, Pd and U react with SiC, creating Pd-Si-U precipitates that migrate toward the SiC layer or toward the IPyC layer, though some U is available to directly migrate into the SiC layer. Interestingly enough, Pd-U was not observed in the IPyC layer, possibly because U may not need assistance from Pd, if their diffusivities in carbon are at the same order.

Owing to constrained diffusivity, the count of Pd-Si-U or Pd-U precipitates diminishes as the distance from the IPyC-SiC interface increases. It was noted that all Pd-Si-U and Pd-U precipitates in the SiC layer were situated at grain boundaries, indicating that grain boundary diffusion serves as the primary mechanism. It is expected that the lattice diffusivity of U in SiC is significantly lower than the grain boundary diffusivity, and even with the assistance of Pd. Intermittently, Pd-U or even U in isolation was identified throughout the majority of the SiC layer, hinting at the possibility that the diffusivity of uranium in the SiC might be augmented by the particular irradiation conditions or certain microstructural features of the SiC.

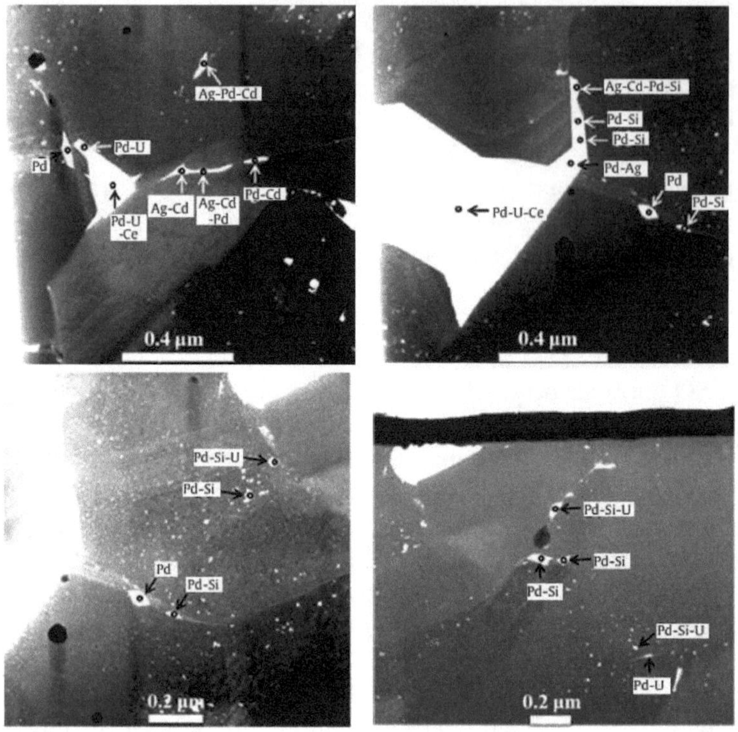

FIG. 87. Precipitates at SiC grain boundaries in unreacted SiC areas (courtesy of INL).

Fission products in the corroded areas

Apart from the primary crack in the IPyC layer and in the proximity of the IPyC-SiC interface, precipitates were typically small with a relatively sparse distribution, while a dense concentration of sizable precipitates was noted at or near the IPyC-SiC interface, particularly in the vicinity of the IPyC crack [167]. The crack across the buffer layer and the IPyC layer provided a pathway for fission products to migrate across the buffer and the IPyC layers and fission products distinctly accumulated at the IPyC-SiC interface where the crack terminated.

At the locations where fission products accumulated (locations 7 and 8 in Fig. 86), precipitates constituted of Pd-Si, Pd-Si-U, Pd-Si-Ce, Pd-Si-Ce-U. Only at one spot, the precipitate was composed of Pd-Ag. At the other locations, excluding corroded (location 10 in Fig. 86) and fission product accumulated locations (locations 7 and 8 in Fig. 86), precipitates constituted of Pd, Pd-Si, Pd-U, or Pd-Si-U at IPyC-SiC interface, whereas in the IPyC layer near the IPyC-SiC interface, precipitates comprised Pd-Si or Pd-Si-U, with no observations of Ag or Cd-containing precipitates.

In the corroded area (location 10 in Fig. 86) in the SiC layer close to the crack, C areas were detected with Pd silicides inside these areas (see Fig. 88(a)), indicating corrosion of SiC by Pd. For Pd chemical reaction with SiC, high Pd concentration is required. The elevated Pd concentrations in the SiC layer near the IPyC crack tip resulted from the localized accumulation of Pd, instigating the reaction with SiC. Silicides formed through this reaction migrated from the reaction site, leaving behind areas rich in carbon, and these silicides were observed in the IPyC layer. In C areas, the following precipitates were observed [167]:

— Pd
— Pd-Si; Pd-Ag; Pd-Cd
— Pd-Si-Ag; Pd-Si-U; Pd-Si-Cd; Pd-Si-Ce; Pd-Cd-Ag
— Pd-Si-Cd-U; Pd-Si-U-Ce; Pd-Si-Cd-Ag

— Pd-Si-Cd-U-Ce; Pd-Si-Ag-U-Ce; Pd-Si-Cd-Ag-U.

Xe was also observed (see Fig. 88(b)) in nano-cracks in the C areas in the SiC layer, indicating that nano-cracks provide short-range pathways for Xe transport. Cs was not detected in the corroded area, but it was hypothesized that the C areas in the SiC layer might have provided pathways for Cs transport, due to the nano-cracks, as indicated for Xe [167].

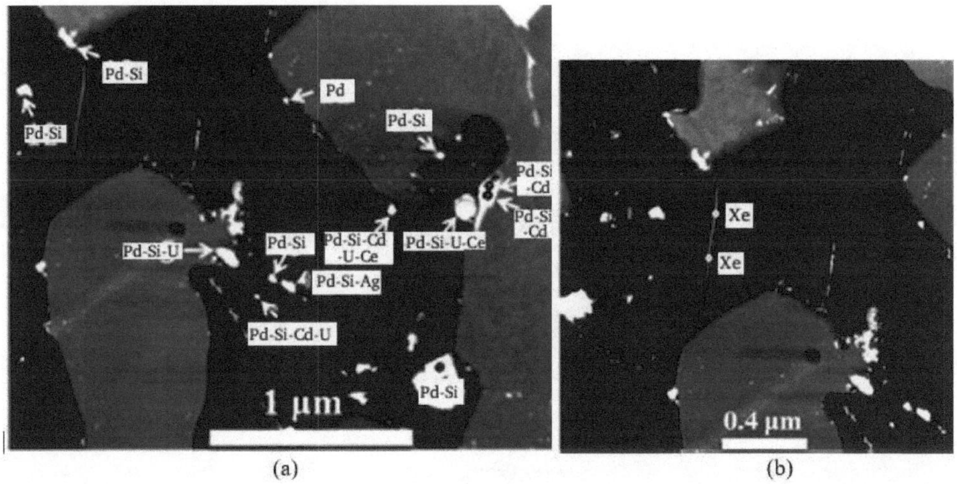

FIG. 88. Precipitates in the carbon areas in the SiC layer (a) Pd-Silicides, (b) Xe (courtesy of INL).

Fission product distribution in the coating layers after high temperature annealing tests

High temperature annealing tests were performed at 1600°C, 1700°C and 1800°C on some of the compacts after having been irradiated in the ATR reactor (i.e., after the first irradiation campaign, AGR-1) [107].

Figure 89 [107] shows the backscattered electron composition (BEC) images of the TRISO layers in particles with low and high Ag retention after the annealing tests performed at 1600°C, 1700°C and 1800°C. The images clearly demonstrate the accumulation of fission products at the IPyC-SiC interface and fission product clusters distributed in the SiC layers. Low Ag retaining particles had a larger number density of fission product clusters distributed deeper into the SiC layer relative to high Ag retaining particles. It was noted that Pd-U based precipitates were predominant in the first 10 µm of the SiC layer and in the outer half of the SiC layer, Pd clusters with no detectable U were more frequently observed. In addition to Pd and U, other fission products, similar to those in the as-irradiated particles were observed. Unlike to 'as-irradiated' fuels (i.e., fuels that were irradiated but not exposed to high temperatures), fission products including Pd-U precipitates were U-rich following the annealing tests (i.e., irradiated fuels exposed to high temperatures). Particles exhibiting low Ag retention displayed an augmented presence of U in substantial clusters within the SiC layer, accompanied by a relative reduction in Pd at the IPyC-SiC interface.

Under the high temperature conditions corresponding to those at the high temperature annealing tests, U is liberated from the kernel, and migrates into the IPyC layer, and segregates at the IPyC-SiC interface. An increase in temperature results in higher number density of fission products in the SiC layer, as noted by large bright spots in the inner half of the SiC layer in Fig. 89 [107]. In the case of annealing test at 1800°C, a large number density of fission products in SiC layer (in the half of the layer close to the IPyC-SiC interface) was observed though this seemed not to lead to significant degradation of the SiC layer, as the IPyC-SiC interface remained intact [8]. The large U-rich fission product precipitates in the half (close to the IPyC-SiC interface) of the SiC layer suggests that U transport in SiC is active at 1800°C. In addition, 'dark' spots in the half of the SiC layer close to the SiC-OPyC layer were detected. These areas were associated with 'pits' or possibly C-rich areas; similar observations, though with less number density of dark spots, were made for the particles tested at 1700°C [107]. It was argued [107] that voids or C-rich areas were left behind while Pd was leaving the SiC layer. In summary, U exhibits mobility within the graphite layers at temperatures of 1600°C and above, while both U and Pd demonstrate mobility within SiC at 1800°C and potentially at 1700°C [107].

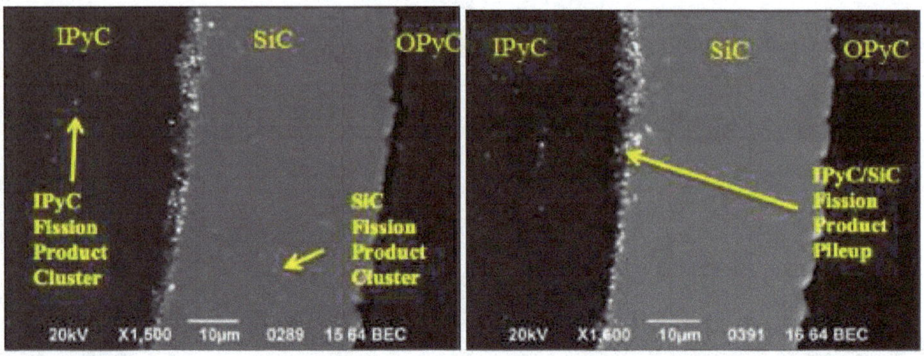

(1) Low Ag retaining particle (2) High Ag retaining particle
(a) annealing tested at 1600°C

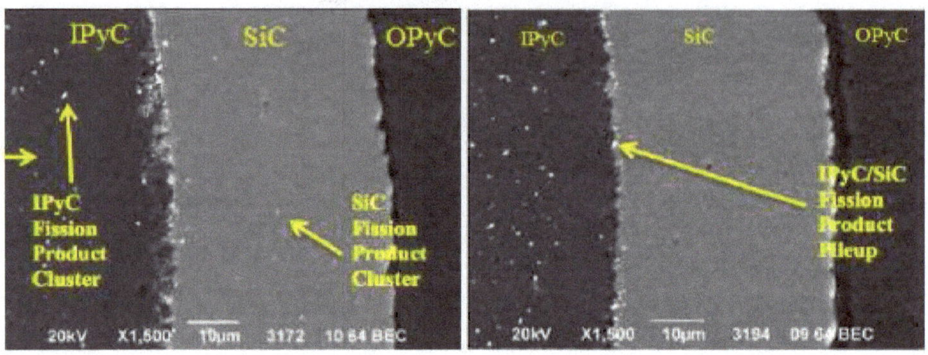

(1) Low Ag retaining particle (2) High Ag retaining particle
(b) annealing tested at 1700°C

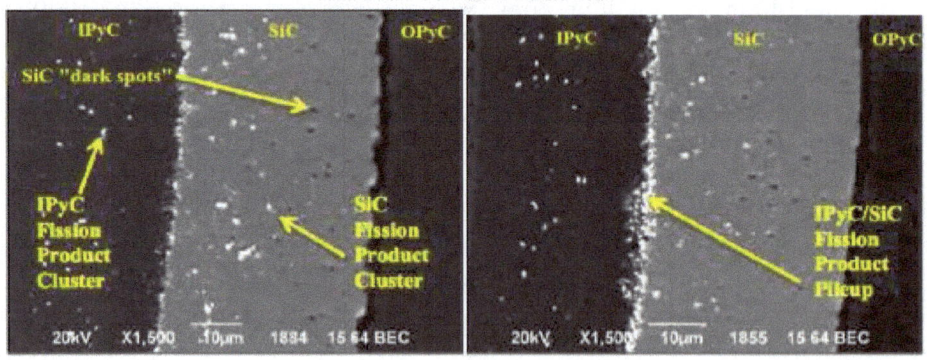

(1) Low Ag retaining particle (2) High Ag retaining particle
(c) annealing tested at 1800°C

FIG. 89. BEC images of annealing tested low and high Ag retaining particles at (a) 1600°C, (b) 1700°C and (c) 1800°C (courtesy of INL).

7.1.4. Fission product transport

Fission product transport is the main component of the source term calculation in the afety analysis. The source term is defined as the timing, fraction, and speciation of fission products released to the containment during an accident. Consequently, it is important to calculate the inventory of fission products and release during normal and abnormal operating conditions. During normal operating conditions, any defected elements (e.g. fabrication defects) can release small amounts of fission products into the coolant (i.e., helium gas). At very high temperatures and power surges, TRISO particles may start to fail, releasing fission products. Therefore, the release of fission product can be calculated with the knowledge of fission product transport in the coated particle. Calculations become complicated due to the chemical state of the fission products, the different mechanisms responsible for gaseous and metallic fission product transport in each layer, high burnups and fast neutron fluences.

Fission product transport includes:

- Release from failed or defective particles;
- Release from uranium contamination in the compact;
- Diffusive release through the intact coatings.

In the safety analysis, instead of evaluating all fission and activation products produced and transported in the fuel and released into the containment and environment, only the most significant radiological nuclides are analysed. The radiologically important fission products released from fuel elements are divided into two groups:

- Gaseous fission products;
- Metallic and activation products; activation products are formed when a fission product captures a neutron to form a new nuclide.

Key gaseous/volatile fission products that are selected based on the combination of their radiological hazard level, fission yield, and their transport and release properties, include ^{131}I, ^{133}I, ^{88}Kr, ^{138}Xe, ^{133}Xe, ^{90}Kr and ^{137}Xe and ^{132}Te [4]. The long-lived metallic fission and activation products include ^{137}Cs, ^{111}Ag, ^{134}Cs, ^{136}Cs, ^{110m}Ag (m indicates metastable) and ^{90}Sr. For normal operating conditions, the inventories of the fourteen nuclides listed above are calculated, though for abnormal conditions, ^{90}Sr, ^{131}I, ^{133}I, ^{137}Cs and ^{111}Ag are of significance [4].

To model the release of long-lived fission gases from the kernel, the diffusion-controlled release from a sphere is employed. As for short-lived gaseous fission product species, where production and release from the kernel rapidly reach a steady state, the Booth model [168] is commonly utilized to determine the ratio of the release rate to the birth rate (R/B). This ratio, R/B, is influenced by the diffusion coefficient of the radionuclide, decay constants of the fission product isotopes, and the surface-to-volume ratio ($S/V=3/a$) of a sphere equivalent to the representative grain of fuel with radius a. The Booth model is applicable only to the diffusive release of fission products from kernels and from matrix graphite grains (which contain the heavy metal contamination). In the most recent models, besides the Booth model, the effects of temperature, burnup, the mechanisms of athermal and vacancy migration diffusion, both irradiation- induced, intrinsic diffusion and a buffer-recoil contribution to release were incorporated [3].

Diffusion model can be applied to determine the transient release behaviour of metallic fission products. Excluding silver, under normal operation conditions, an intact TRISO particle coating prevents effectively metallic fission product release. Hence, estimating the release involves understanding the proportion of defective or failed particles and heavy metal contamination in the fuel element graphite, along with comprehending fission product transport in both the kernel material and the fuel element material. Beyond diffusion, other release mechanisms during normal operation include the recoil effect and the knockout effect at low temperatures, with diffusion becoming dominant at elevated temperatures.

Fickian equation is applied to solve the diffusion of fission product species in different kinds of reactor materials using effective diffusion coefficients. Effective implies that all possible transport mechanisms are considered in a simplified single transport process. The diffusion coefficients are usually given as an Arrhenius type equation as a function of temperature. Separate effects, such as fast neutron fluence or fission product concentration or burnup, or by combining diffusion processes with different activation energies in different temperature ranges, are also considered.

Fission product transport mechanisms in the kernel and layers of a TRISO particle are described in Ref. [14] and summarized in Table 53.

TABLE 53. MECHANISMS FOR FISSION PRODUCT TRANSPORT IN TRISO PARTICLE FUELS

Layer	Mechanisms
Kernel	• Recoil • Fuel restructuring due to kernel-buffer pyrocarbon chemical reaction • Diffusion of fission products to grain boundaries and to the surface of the kernel through the interconnected porosity of the kernel • Transport through vapour phases
Buffer	• Fission product transport in porous media: fission product transport in pores and through cracks created due to stresses in the buffer zone: binary diffusion (interdiffusion), pressure driven diffusion – Knudsen diffusion, the mass transport behaviour by: 1) free molecular flow, 2) viscous flow, and 3) diffusive flow
IPyC and OPyC	• Gas phase fission product transport: slow diffusion • Metallic fission product transport: diffusion and trapping
SiC	• Bulk diffusion – less likely • Grain boundary diffusion • Fission product transport along pre-existing nano-cracks • Pd assisted SiC corrosion

Parameters that influence fission product transport in a coated particle fuel include [14]:

— Macroscopic parameters, such as particle burnup, fast fluence (linked to radiation damage), layer temperature, and gas or vapour partial pressure;
— Microscopic parameters related to material structure, such as porosity, tortuosity of the porous medium, and grain boundary microstructure;
— Parameters linked to the chemical speciation of fission products, encompassing fuel stoichiometry and its variations during normal and accident conditions, thermochemical data like free energies of formation, vapour pressures, adsorption isotherms, and transport properties like binary gas phase diffusivities and heat of transport;
— Physical parameters leading to multidimensional and multicomponent effects, including segregation and concentration of fission products due to cracking and azimuthal temperature gradients.

Fission product transport phenomena that are applicable to kernel and the layers are described in Ref. [16] and summarized in Table 54. Some of the factors may be unique to more than one layer.

TABLE 54. FISSION PRODUCT TRANSPORT PHENOMENA

Factor	Definition
Condensed phase diffusion	Transport of condensable fission products by intergranular diffusion and/or intragranular solid-state diffusion (grain boundary and/or bulk diffusion)
Gas phase diffusion	Diffusion of gaseous fission products through layer (Knudsen and bulk diffusion through pore structure, and pressure driven permeation with factors including cracking, adsorption, permeability, sintering and
Thermodynamics of fission product-SiC system	Chemical form of fission products including the effects of solubility, inter-metallics and chemical activity
Intercalation	Trapping of species between sheets of the graphite structure
Trapping	Adsorption of fission products on defects
Fission product release through failures, e.g. cracking	Fission product migration from the buffer layer through failed SiC

7.1.4.1. Fuel kernel

Fission products can be transported in the kernel as a function of burnup (i.e., microstructure changes) and temperature (over the residing period in the reactor) due to:

- Recoil;
- Fuel restructuring due to UO_2 reaction with pyrocarbon buffer layer;
- Diffusion of fission products to grain boundaries and to the surface of the kernel through interconnected porosity of the kernel;
- Vaporization.

Recoil

Some of the fission products formed near the surface of a fuel kernel are released directly into the buffer region because of recoil, which is defined as the fission product movements in the surrounding material from their birth to their coming to a stop. The recoil range that a fission product travels in the material before coming to a stop is dependent on its mass, its kinetic energy and the material through which it penetrates.

The fractional release, F_{recoil}, of fission products from spherical kernel into the neighbouring buffer layer is independent from temperature and is calculated [14] as:

$$F_{\text{recoil}} = \frac{1}{4}\frac{r^3-(r-a)^3}{r^3} \tag{12}$$

where r is the radius of the fuel kernel and a is the average fission fragment[10] range. Using fission energies of 107 MeV for krypton and 72 MeV for xenon, the average range is 5.8 μm for krypton and 4.1 μm for xenon in UO_2 with a density of 10.5 g/cm^3 [14]. For a 500 μm kernel, the recoil release fraction is approximately 1.5%. For a 350 μm kernel, the recoil release fraction is about 2%.

Fuel restructuring

Some fractions of the impurities including fission products can be expelled from the kernel, during restructuring of the fuel as a result of fuel kernel's chemical reaction with the surrounding pyrocarbon, producing surface layers of uranium carbide and oxy-carbide, and carbon monoxide (CO) [14]. These fission products are considered to be released from the kernel, as during the recrystallization process,

[10] Note that fission fragments are distinguished from fission products by their kinetic energy. Fission fragments possess all or part of the energy received from the fission event, whereas fission products are stationary.

they move to the exposed surfaces of the fuel kernel. Fission gas release based on kernel reaction with the surrounding carbon becomes important as the temperature increases due to the temperature dependence of the reaction of graphite with the kernel. It should also be noted that the irradiated carbon is more reactive toward the fuel kernel than the unirradiated carbon.

Diffusion

One of the most important fission product transport mechanisms is the conventional release process of diffusion through grains to the grain boundaries and subsequent transport through the interconnected porosity. The Booth diffusion model has been used to describe fission product release from the kernel [14]. The fractional release due to diffusion, originally developed for a grain [168] is given by:

$$F_{\text{diffusion}} = 1 - \frac{6}{\pi^2} \sum_{n=1}^{\infty} \frac{1}{n^2} \exp\left(\frac{-n^2 \pi^2 D t}{a^2}\right) \quad (13)$$

where D is the diffusion coefficient of fission products in the grain and a is the grain radius, and t is the irradiation time. For the fuel kernel, the reduced diffusion coefficient, D', can be calculated by equating the kernel radius, r, to the equivalent grain radius [4] as

$$D' = \left(\frac{D}{a^2}\right)_{\text{grain}} = \frac{D_k}{r^2} \quad (14)$$

where D_k is the diffusion coefficient of the fission products in the fuel kernel.

When $\pi^2 D' t \geq 1$ in Equation (13), the infinite series converges rapidly and the first term becomes dominant, resulting in Equation (15) [168]:

$$F_{\text{diffusion}} = 1 - \frac{6}{\pi^2} \exp(-\pi^2 D' t) \quad (15)$$

When $\pi^2 D' t \leq 1$, the fractional release can be estimated as [168]:

$$F_{\text{diffusion}} = 6\sqrt{\frac{D' t}{\pi}} - 3D' t \quad (16)$$

When $\pi^2 D' t = 1$, either of the equations can be used [168].

Further enhancements to the model to take into account effects, such as temperature were suggested and discussed in Refs. [3,10].

Transport through vapour phases

Gas phases of fission products need to be understood for the estimation of fission product transport through the layers of coated particle fuel. The chemical environment of the fuel particle is reducing and the fission products are sensitive to the reducing conditions. Fission products can also be affected by the existence of carbon monoxide produced by the reaction of pyrocarbon with oxygen liberated by the fission process. Elemental vapour species and oxide vapour species need to be determined. In addition, impurities left from the manufacturing process such as HCl can affect fission products by the formation of chlorides. Chlorides are typically volatile, in the form of monochloride and dichloride vapour or oxychloride vapour [14]. Vapour phases of BaC, SrC, ZrC and RuC were also identified to form [14].

Vapour phase metal carbonyls, in particular vapour, mono-carbonyls (MCO) and di-carbonyls ($M(CO)_2$) may also be of interest for the analysis of fission product transport via vapour through the coated particle layers [14].

7.1.4.2. Buffer layer

Since the buffer layer is porous (~50% theoretical density in the as-fabricated layer) and can crack due to tangential stresses developed under irradiation, it has the lowest thermal conductivity of any layer in the particle, and hence it has the largest temperature drop. Consequently, the temperature gradient in the buffer may cause thermal (Soret effect) fission product diffusion in the layer [14]. Depending on the specific irradiation conditions, the nature of the shrinkage and densification of the buffer establishes the initial condition for fission product transport.

Figure 90 [14] shows the temperature drop across the TRISO particle layers for an average particle that generates ~ 62 mW of power, which is about the average power per particle in a pebble bed reactor core, with the thermal conductivities of UO_2 kernel, and buffer, IPyC, SiC and OPyC layers being 2.5, 0.5, 4.0, 13.9, 4.0 W/m·K, respectively. The highest value of 10.88 K for the temperature drop was calculated for the buffer layer, while the lowest value of 0.07 was calculated for the SiC layer.

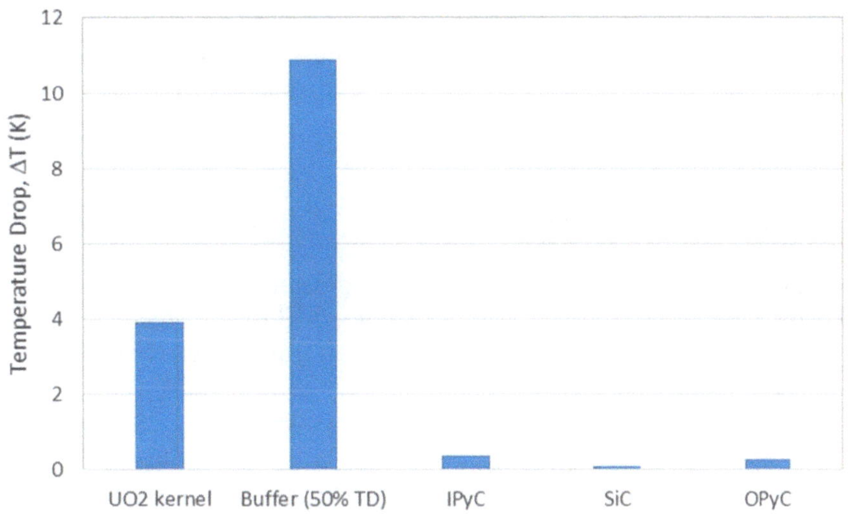

FIG. 90. Temperature drop across the TRISO particle layers for particle that generates a power of 62 mW (courtesy of US DOE).

As power levels increase, notable thermal gradients may arise, resulting in elevated thermal stresses within the layer. These stresses, particularly in the buffer, induced by thermal gradients and densification, have the potential to cause buffer cracking. Additionally, substantial thermal gradients exceeding approximately 1000 K/mm across the buffer can induce the thermal diffusion (Soret effect) of fission products within the layer [14]. Therefore, fission products can transport through cracks and by thermal diffusion. Table 55 shows the effects of particle power on the buffer layer. At relatively low powers, corresponding to the powers that a particle experiences in a pebble bed reactor, the buffer experiences uniform shrinkage. However, as the power increases and hence the thermal gradient, the particle experiences moderate to high tensile stresses and cracks, and at very high powers, excessive shrinkage of the buffer layer occurs, and the buffer is repositioned side-by-side with the kernel, as shown in Table 55 (from [14]).

TABLE 55. EFFECTS OF PARTICLE POWER ON THE BUFFER IN COATED PARTICLE FUEL

				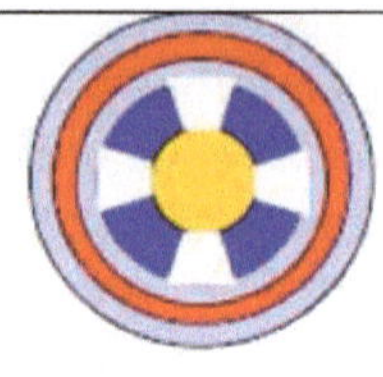	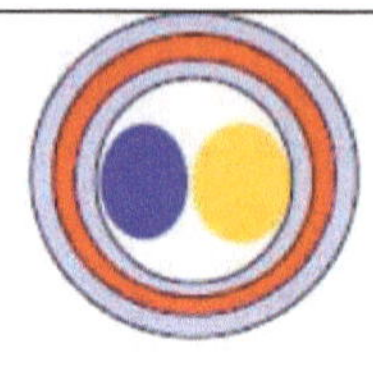
Description of location	PBMR low flux region; GT-MHR average	PBMR average	PBMR pebble and prismatic compact peak	Current prismatic irradiation limit	Very accelerated
Particle power	25 – 40 mW	60 mW	100 mW	400 mW	500–5000 mW
Thermal gradient across the buffer	< 50 K/mm	~ 100 K/mm	~ 250 K/mm	~ 750 K/mm	~ 9009000 K/mm
Condition of the buffer	Uniform shrinkage		Moderate tensile stress – some cracking	High tensile stress – many cracks	Excess shrinkage; buffer and fuel side by side

Fission product transport in a porous medium

Multicomponent gas phase mass transport can be used to determine the behaviour of fission gases and vapour in porous media. On the contrary, the multicomponent diffusion equation, as formulated in the Chapman-Enskog approximation [16], postulates that mass transfer ensues due to:

- — Temperature gradients (thermal diffusion or the Soret effect);
- — Chemical composition gradients;
- — Pressure gradients.

However, most investigators have used the Grad (the so-called '13-moment' method) version of mass transport through porous media because it allows the explicit consideration of the porous medium in the so-called 'dusty' gas' approximation [16].

In the model, the porosity of the material and tortuosity of the pore network were taken into account. In certain instances, a slip correction was incorporated for the pressure gradient term coefficient. This correction accommodates the relative velocity between the molecule and the surface, particularly when the characteristic size in the medium is comparable to the mean free path of the gas. In addition, for detailed analysis of fission product transport by gas phase mass transport, the vapour species transport properties such as binary diffusion coefficient, viscosity and thermal conductivity need to be known.

The molar flux of the substance through the porous medium is contingent on the pressure gradient across the material. Traditionally, three distinct regimes are recognized, contingent on the mean free path of the gas in relation to the characteristic size[11] of the medium, expressed as the Knudsen number ($Kn = \lambda/d_{pore}$, where λ denotes the mean free path and d_{pore} represents the average pore size in the medium) [20]:

- — For Kn >1, the mass transport behaviour can be described using the free molecular flow and the molar flux;
- — In the transition region, 0.01 < Kn < 1, both viscous flow and diffusive flow are considered;

[11] Characteristic sizes could range from nanopores, in a material like an as fabricated buffer, to microcracks as might be typical of a cracked buffer.

— In the continuum region, characterized by Kn < 0.01, the overall molar flux is determined by combining contributions from viscous flow and diffusive flow. It is important to note that molecular flow effects in this region are minimal.

When the pore diameter is smaller than the mean free path of diffusing gas molecules and the gas density is low, the molecules collide more frequently with the pore walls than with each other. A Knudsen number much greater than one indicates the Knudsen diffusion is important.

A sensitivity analysis was performed for pressure driven diffusion to determine effective diffusivities as a function of pore or crack size [167]. At conditions corresponding to normal operation and accidents, effective diffusivity of Kr gas at 1000° and 1600°C and pressures in the range of 0.5 MPa to 25 MPa, was analysed.

Figure 91 shows the calculated effective diffusivities at 1000°C and 1600°C, respectively [93]. Gas pressure becomes important for characteristic sizes greater than ~ 0.02 mm. In addition, the effect of temperature on effective diffusivity is moderate, based on comparison of Fig. 92(a) with (b). For nanopores (where Knudsen diffusion becomes important), effective diffusivities are on the order of 3 to 53×10^{-7} m^2/s. By contrast, transport through micropores or micron sized cracks (where viscous diffusion becomes important) is much faster, with effective diffusivities ranging between 10^{-4} and 10^{-2} m^2/s depending on the pressure of the gas involved [93]. Note that the German scientists assumed the diffusivity of all species in the buffer to be 10^{-8} m^2/s and the US used a value of 10^{-10} m^2/s in their evaluations [14]. Depending on the pore or crack sizes, rapid transport of fission gases and fission product vapours could be expected through the buffer layer in a coated particle. From mass transport point of view, it is also important to identify connected porosities.

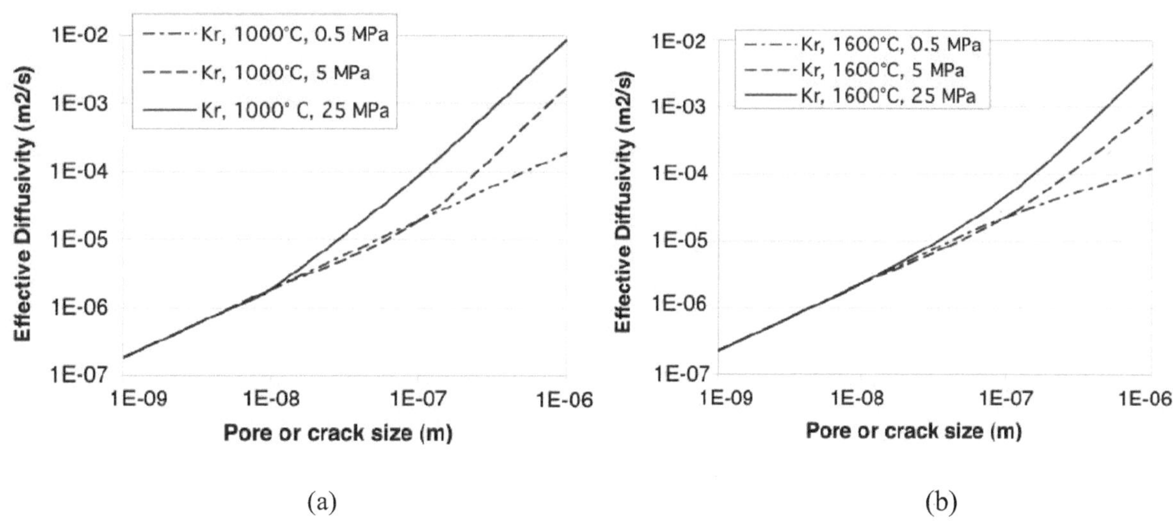

FIG. 91. Effective diffusivities for Knudsen and viscous diffusion at: a) 1000°C and (b) 1600°C (courtesy of INL).

Thermal diffusion

Besides the traditional concentration gradient driven Fickian diffusion across the layer, large thermal gradients in the buffer layer can lead to thermal diffusion (the Soret effect). Consequently, the combined diffusive flux for one species can be given by [14]:

$$J = -D\left(\underbrace{\nabla C}_{\substack{\text{Fickian}\\ \text{diffusion}}} + \underbrace{\frac{CQ}{RT^2}\nabla T}_{\substack{\text{Thermal diffusion}\\ \text{(Soret effect)}}}\right) \tag{17}$$

where D is the diffusion coefficient, Q is the heat of transport, T is the temperature, J diffusive flux, C is the concentration, and R is the ideal gas constant [14].

For low power/thermal gradients in pebbles in pebble bed reactors, thermal diffusion is expected to be much less important in fission product release, whereas at high powers, thermal diffusion becomes important.

7.1.4.3. Inner pyrocarbon and outer pyrocarbon layers

For gas phase fission product transport, the measured diffusion coefficients suggest very slow transport through the PyC layer.

Beyond 1900°C, it is advisable to consider rapid diffusive transport (similar to the buffer layer) for metallic fission products within the IPyC. In the case of high-temperature isotropic PyC, this critical temperature has been presumed to be as low as 1200°C, with the exception of caesium, for which experimental evidence indicates much better retention than strontium or silver [3].

The proposed mechanism for the transport of Cs, Sr, and even iodine and CO in the PyC involves intercalation, which is the insertion of guest atoms into a host structure. A conventional diffusion and trapping model has been suggested for modelling the transport, with trapping taking place at the carbon crystallite edges and defects in the graphitic material [14].

7.1.4.4. Silicon carbide layer

The mechanisms for fission product transport through SiC include [169]:

- Bulk diffusion through grains;
- Grain boundary diffusion;
- Diffusion along interconnected micro/nano cracks;
- Pd-assisted fission products transport.

Bulk diffusion

For the bulk diffusion, the first mechanism is fission product diffusion through the bulk of the SiC (single) crystal (i.e., grain).

While the SiC shell has the cubic (3C) crystal structure (with an ABC close-packed stacking of Si layers and C layers), SiC can form a wide variety of polymorphs, each characterized by a different stacking sequence of close packed planes, e.g. i) cubic SiC having zincblende structure, and ii) hexagonally packed (2H) SiC having a wurtzite structure [169]. Fission products can occupy two distinct types of interstitial sites (labelled as A and B) in both these crystal structures. In wurtzite, A is 6-fold coordinated by Si and by C, while the B site is 4-fold coordinated by Si and C. In cubic SiC, the A site is 6-fold coordinated by Si and 4-fold coordinated by C, while the B site is 4-fold coordinated by Si and 6-fold coordinated by C [169]. For example, for Ag, the site occupation energies were calculated using density functional theory, and then normalized to the cubic structure: for the interstitial site, the site occupation energy was the lowest, and hence Ag is expected to occupy an interstitial site in the cubic structure. Ag thermodynamically has a higher solubility in cubic SiC as compared to wurtzite.

The interstitial site energies of other fission products in cubic SiC, including I, Cs and Sr were calculated for two different interstitial sites: A and B [169]. Similar to Ag, I prefers A-site. Caesium prefers the B-site over the A-site, though the energy difference is small, in particular at high temperatures. Sr, while preferring the A site, only has a small energy penalty to overcome to occupy the B site. Consequently, I will have a similar interstitial diffusion coefficient as Ag, while Cs and Sr would have significantly higher interstitial diffusion coefficients.

Energies required to move fission product (atom) from their pure crystalline state into an isolated interstitial site of the SiC crystal were also calculated. The corresponding energies for Ag, I, Cs, Sr were 9.897, 21.654, 24.450 and 15.628 eV, respectively [169]. The required energies are quite high for I, Cs and Sr, indicating that these elements have low solubilities in SiC.

Since the interest is the calculation of the release rate of fission products through the SiC layer, the solubility of the fission products at the inner surface of the SiC layer when in contact with PyC and the diffusion coefficient for fission product transport through cubic SiC need to be taken into account. The integrated fission product flux, F, indicating fission products emerging at the outer surface of the SiC, can be calculated for a steady state concentration profile as [169]:

$$F = \frac{4\pi \quad C_0}{\frac{1}{R_1} - \frac{1}{R_2}} \tag{18}$$

where D is the fission product diffusion coefficient, C_0 is the fission product concentration on the inner surface of the SiC layer (number of atoms per unit volume of SiC), R_1 and R_2 are the inner and outer radii of the spherical SiC layer, respectively. The value of D is very sensitive to the diffusion mechanism, depending on whether fission product diffusion occurs through the bulk via an interstitial or a substitutional mechanism, or whether it occurs along grain boundaries or internal cracks.

Assuming interstitial solubility in the bulk and interstitial bulk diffusion, Ag release rate, F, was calculated using Equation (18), and plotted in Fig. 92. It is noted in Ref. [169] that bulk diffusion is highly unlikely as the transport mechanism for Ag passing through the SiC layer: the energy to take Ag from graphite to an interstitial SiC site is of the order of 6–7 eV, resulting in very low equilibrium Ag concentrations on the inner surface of the SiC layer, whereas the solubility of Ag along grain boundaries and internal surfaces was identified significantly high.

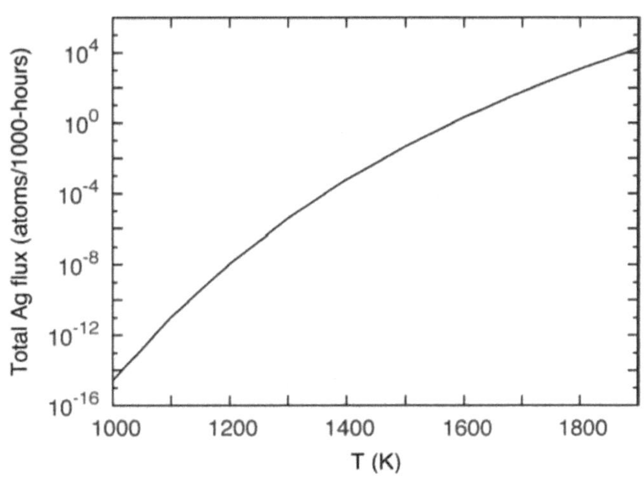

FIG. 92. Calculated Ag release rate from a TRISO particle.

Grain boundary diffusion

Despite that fission product escape through bulk diffusion is not likely, it may occur along short-circuit pathways such as grain boundaries. These boundaries, therefore, are expected to be a less constrained environment for diffusing fission products than the bulk, but more constrained than microcracks. More open structure of grain boundaries can raise the local thermodynamic solubility of fission products through segregation as well as their mobility [169].

Based on density functional theory (DFT) calculations, the formation energy for Ag in the grain boundaries was larger than for the C-rich surfaces, but much less than that for an interstitial in bulk SiC [169]. The energy required to move an Ag atom from pure Ag (precipitated metallic state) or from graphite to a site at the grain boundaries is still positive, indicating that there is no energetic driving force for Ag to segregate to SiC grain boundaries if it is already in the metallic state or inserted within graphite.

Fission product transport along pre-existing nano-cracks

In the presence of nano or micro cracks, very fast short-circuit diffusion pathways are provided to fission products. Transport along interconnected crack networks can occur either through surface diffusion or in gaseous form. SiC has both Si-rich and C-rich surfaces. Based on DFT calculations, adsorbed Ag has a formation energy near +2.0 eV for different sites on the carbon-rich surface, and at -1.246 eV for one site on the silicon-rich surface, which indicates that there is a thermodynamic driving force for Ag to segregate to Si-rich surfaces [169]. Based on first-principles calculations (i.e., DFT), when Ag is present in graphite (IPyC), it precipitates as metallic Ag particles in SiC or along grain boundaries in SiC. Only Si-rich surfaces exhibit an energetic attraction for Ag in the metallic state. Similarly, formation energies of other fission products (i.e., I, Cs, and Sr) were calculated to be lower at the Si-rich surface compared to C-rich, and even these formation energies are lower than the one for Ag, as shown in Table 56 (from [169]). The values for C were presented in a range, as calculations were performed for three sites for C. Caesium and strontium bind favourably to all of the surfaces and sites, while Ag binds favourably only to the Si-rich surface. Iodine is favourable in all locations except for one of the sites on the C-rich surface (with a positive value).

TABLE 56. FORMATION ENERGIES (IN EV) FOR AG, CS, I, AND SR ON SURFACES OF SIC-3C

Surface	Ag	Cs	I	Sr
C	1.974 – 2.050	(-1.436) – (-1.637)	(-0.242) – (+0.498)	(-1.868) – (-0.756)
Si	-1.246	-2.331	-2.280	-3.205

Palladium assisted fission product transport

There are two main factors that need to be understood under Pd assisted fission product transport: i) decohesion in SiC, and ii) the embrittlement effect of Pd.

Pd has a corrosive effect on SiC and it is believed to facilitate fission product transport in SiC. Pd makes SiC susceptible to crack opening in regions of high residual tensile stresses. The tensile stresses can emerge during growth or as a result of large temperature gradients [169]. An analysis was performed using the cohesive zone model[12] [168] to determine the response of the decohering[13] region to externally imposed loads or displacements: in the model, the energy of the decohering zone on a metric of

[12] The cohesive zone model in fracture mechanics treats fracture formation as a gradual process, where separation of the crack surfaces occurs along an extended crack tip, or cohesive zone, and is opposed by cohesive tractions.

[13] Decohesion is described as the separation of parallel crystallographic planes [169].

separation δ was described; the derivative of this energy with respect to separation yielded a traction-separation curve.

SiC is brittle and has very little susceptibility to plastic deformation through dislocation glide [169], and in the presence of a crack, intergranular and transgranular crack propagation will occur. Although the atomic planes ahead of the crack tip share crystallographic identity with the atomic planes in the crystal adjacent to the crack path, they experience higher normal stress owing to the broken symmetry caused by the presence of the crack. There is little inclination between decohering atomic planes for the slow variation in the crack opening behind and ahead of the tip [169].

When subjected to a constant Pd chemical potential and stress, the equilibrium state of the cohesive zone in SiC was determined [169]. The maximum stress for decohesion of the Pd saturated cohesive zone is substantially lower than that of pure SiC: Pd acts as an embrittlement agent of SiC, dramatically lowering the maximum stress of decohesion. If the residual stresses at the crystallographic level are greater than the maximum stress of decohesion when Pd segregates to the cohesive zone, cracks will propagate. These cracks can serve as short-circuit diffusion paths for fission products escape. Pd segregation and precipitation along grain boundaries may also facilitate fission product diffusion along these grain boundaries, as the solubility of fission products such as Ag is higher in Pd than in SiC. Ag, for example, has an energetic driving force to adsorb to the Si-rich surfaces that form upon decohesion. Furthermore, Ag readily alloys with Pd, indicating an energetic attraction to Pd atoms [169].

7.1.5. Fission product release

Compacts containing TRISO particles were irradiated in the ATR under the AGR program; AGR-1 refers to the first irradiation test campaign and AGR-2 refers to the second irradiation test campaign. Radioactive fission products were measured in both compacts and capsules (see Fig. 93 [107]). The UCO fuel kernels fabricated for AGR-1 irradiation had a nominal enrichment of 19.7% ^{235}U and a nominal diameter of 350 μm. For the AGR-2 irradiation, the UCO kernels had a nominal enrichment of 14% and a nominal diameter of 425 μm, whereas the UO_2 kernels had a nominal enrichment of 9.6% and a nominal diameter of 500 μm to be comparable with historic German fuel particles [108]. The coatings on both types of kernels featured a nominal thickness of 100 μm (buffer), 40 μm (inner and outer pyrolytic carbon), and 35 μm (SiC). These TRISO particles were encased in graphitic matrix material and shaped into cylindrical compacts with a diameter of about 12.3 mm and a length of 25.1 mm. The packing fractions for the UCO and UO_2 compacts were 37% and 23%, respectively [108]. The irradiated fuel compacts achieved burnups of 10% – 13%.

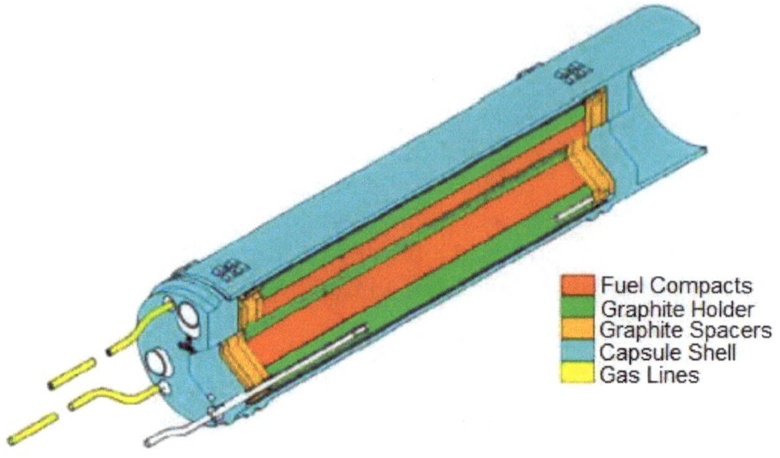

FIG. 93. Components of a Capsule used for AGR Irradiations (courtesy of INL).

Figure 94 [6] shows the fission product release pathway: 1) from kernel to coating layers, 2) from coating layers to compact matrix, 3) from compact matrix to structural graphite, 4) from structural graphite to capsule shell. Overall, the fission product release has been evaluated [6]:

— In fuel compacts;
— On capsule components;
— In compact matrix;
— In individual particles.

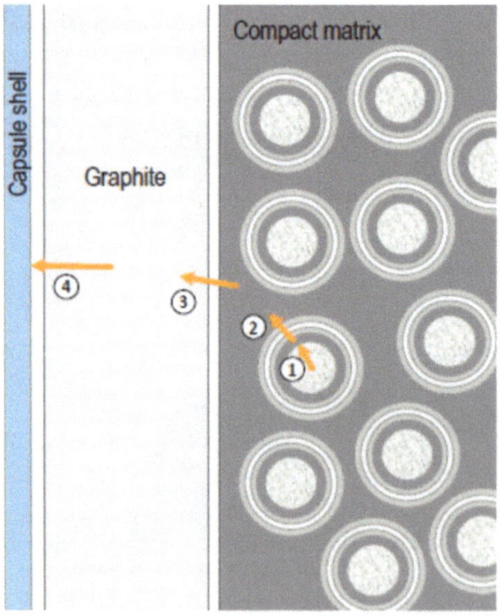

FIG. 94. In-pile fission product release evaluation (courtesy of INL).

Table 57 shows the fission product retained or released in the TRISO particles, taken from Ref. [6].

TABLE 57. FISSION PRODUCT RELEASE/RETENTION

Element	Retention/release in layers
Kr, Xe, I	• Retained by intact PyC or SiC layers • Released from uranium contamination and exposed kernels • Kr and Xe are key indicators of failed TRISO layers
Cs	• Retained by SiC but released through intact PyC • Key indicator of failed SiC layer
Sr	• Moderate retention in the fuel kernel • Modest release through intact coatings (T > 1100°C); significantly higher release for very high irradiation temperatures • Some retention in the compact matrix
Eu	• Similar to Sr, but slightly higher releases
Ag	• Significant release through intact SiC (T > 1100°C) • Relatively low retention in compact matrix

Similarly, Fig. 95 [6] shows the release of fission products from fuel compacts irradiated in the ATR (corresponding to the first and second irradiation test campaigns, AGR-1 and AGR-2):

— Cs release is minimal when SiC remains intact; increased releases are linked to a restricted number of particles with compromised SiC;
— Sr and Eu may display moderate releases, with significantly higher release observed at elevated in-pile temperatures (AGR-2 Capsule 2 time-average peak temperatures of 1360°C);
— Ag release is notably high.

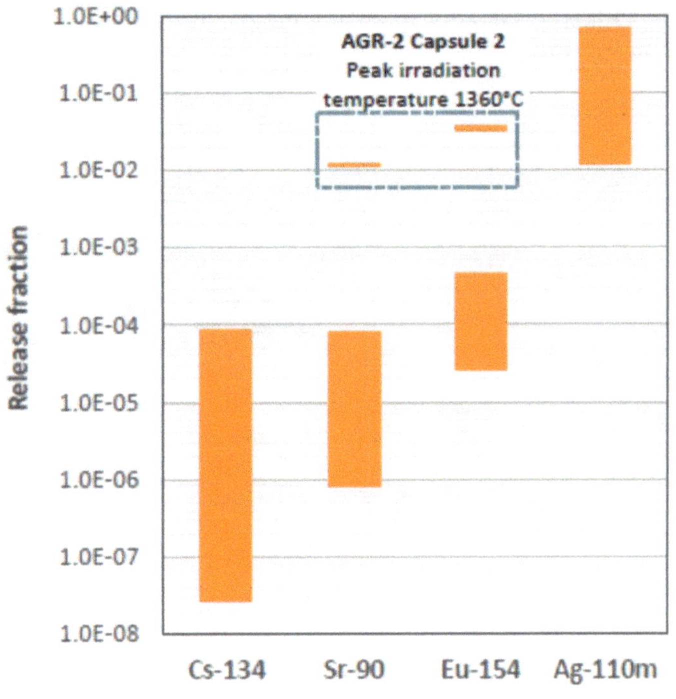

FIG. 95. Fission product release from AGR-1 and AGR-2 fuel compacts (courtesy of INL).

Figure 96 [108] shows the comparison of ^{85m}Kr (m indicates metastable isotope) release from AGR-1 and AGR-2 fuel compacts with historic performance in US and German TRISO fuel irradiations. The observed values are elevated in AGR-2 irradiations compared to AGR-1, attributed partly to increased uranium contamination in AGR-2 compacts relative to AGR-1 (AGR-2 compacts exhibited uranium contamination around 4×10^{-6}, whereas the AGR-1 average was 3×10^{-7}). It is worth noting that no particle failures were noted during the initial three cycles of the AGR-2 irradiation.

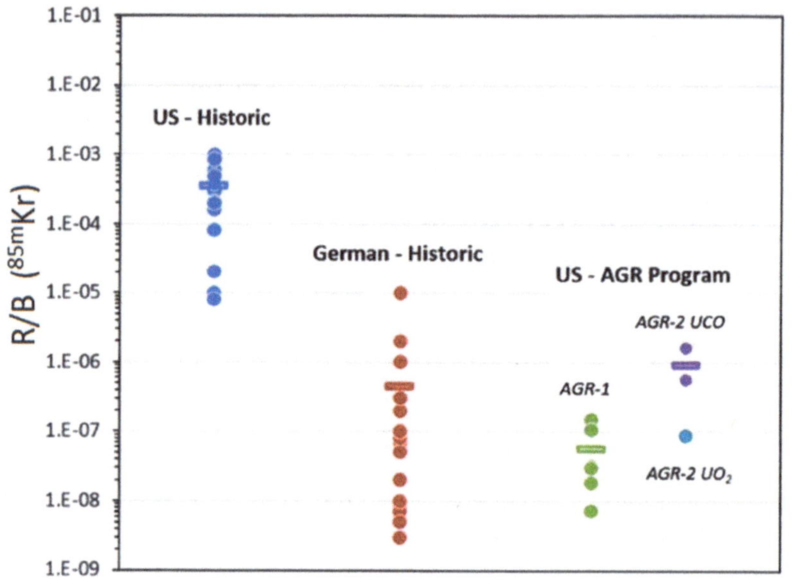

FIG. 96. ^{85m}Kr fission gas release for AGR-1 (end of life) and AGR-2 (after the first three irradiation cycles) Compared to historic performance in US and German TRISO fuel irradiations (courtesy of EPRI).

It needs to be noted that UO_2 kernels irradiated during the AGR-2 test campaign achieved relatively low burnups, as shown in Fig. 97 [108], and slightly lower irradiation temperatures, compared to UCO kernels.

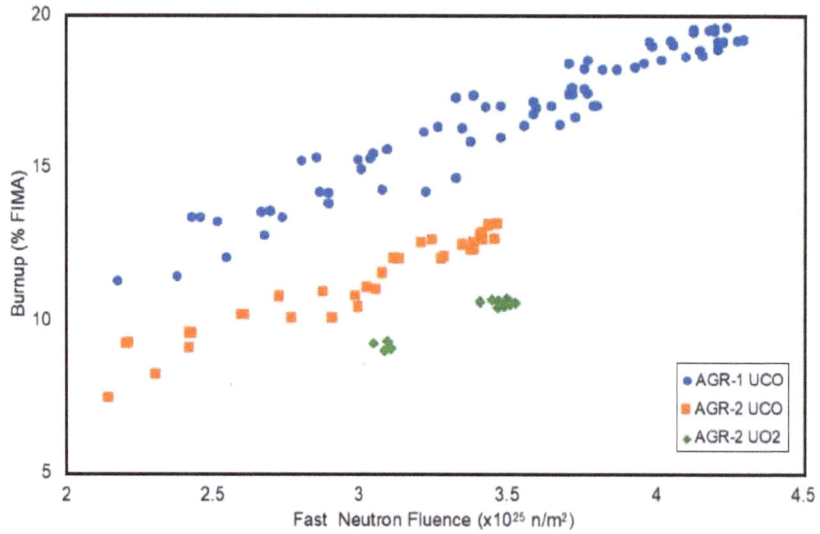

FIG. 97. Burnups achieved by AGR-1 and AGR-2 fuel compacts with UCO and UO_2 fuel kernels (courtesy of EPRI).

In addition, annealing tests, isothermal tests at a temperature of 1600°C, 1700°C, or 1800°C for a nominal duration of 300 h, were performed on irradiated (i.e., AGR-1 and AGR-2) fuel compacts. As shown in Fig. 98 [108], 1600°C test results in an AGR-2 UCO compact exhibited very low fractional release of Cs isotopes for the duration of the test, modest release of Eu and Sr isotopes and fairly high release of ^{110m}Ag.

^{134}Cs, ^{110m}Ag, ^{154}Eu and ^{90}Sr release were measured from heating of AGR-2 UCO and UO_2 compacts (different symbols represent different compacts, stacks and compacts[14]) at temperatures of 1600°C, 1700°C and 1800°C. The releases for ^{134}Cs and ^{90}Sr are presented in Figs. 58 and 59 (in Section 6.2.2.2), respectively. Figures 99 and 100 [108] show releases of ^{110m}Ag and ^{154}Eu, respectively. ^{134}Cs release was high for UO_2 kernels, even at the temperature of 1600°C, compared to UCO kernels. Ag release was high for UO_2 kernel at the temperature of 1700°C, though at the temperature of 1600°C, compacts with UO_2 kernels exhibited similar releases to the UCO kernels. ^{154}Eu and ^{90}Sr releases from Capsule 2 with a UCO kernel that was exposed to high irradiation temperatures[15] were followed by the release from the compact with UO_2 kernel exposed to a temperature of 1700°C during a post-irradiation annealing test.

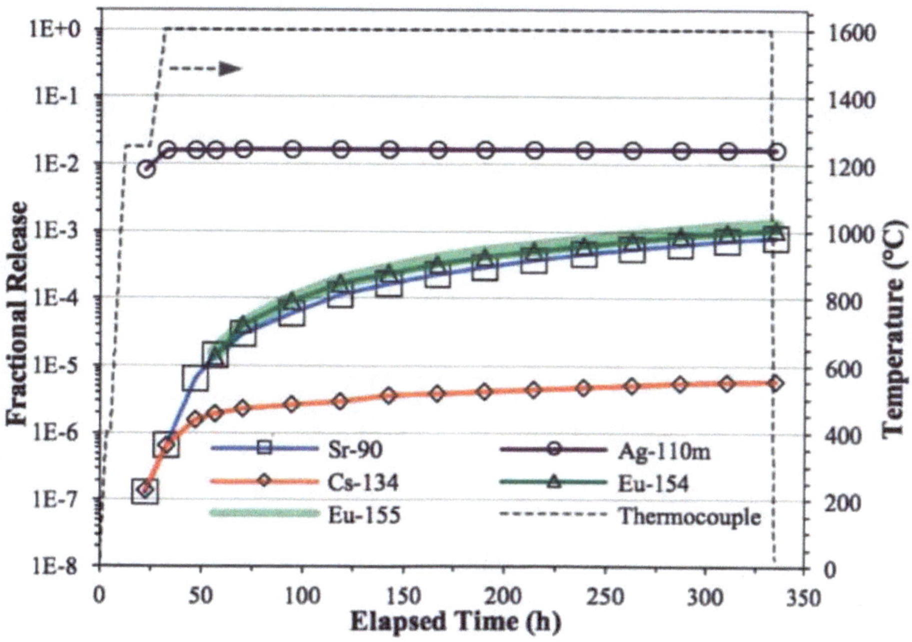

FIG. 98. Fission product release from heating of AGR-2 UCO compact at 1600°C (courtesy of EPRI).

---

[14] The AGR irradiation tests were multi-capsule, instrumented experiments. For each irradiation, six capsules, each independently controlled for temperature and independently monitored for fission product gas release, were used. Each capsule was 152.4 mm long and contained 12 fuel compacts arranged in three vertical stacks, with each stack containing four compacts.
[15] During irradiation in the reactor, while Capsule 2 was exposed to a time-averaged volume-averaged temperature of 1252°C, the other capsules including UCO kernels were exposed to 1101°C and 1074°C, and the capsule including UO_2 kernels was exposed to 1032°C, due to the location of the capsules in the reactor, affecting the neutron fluxes.

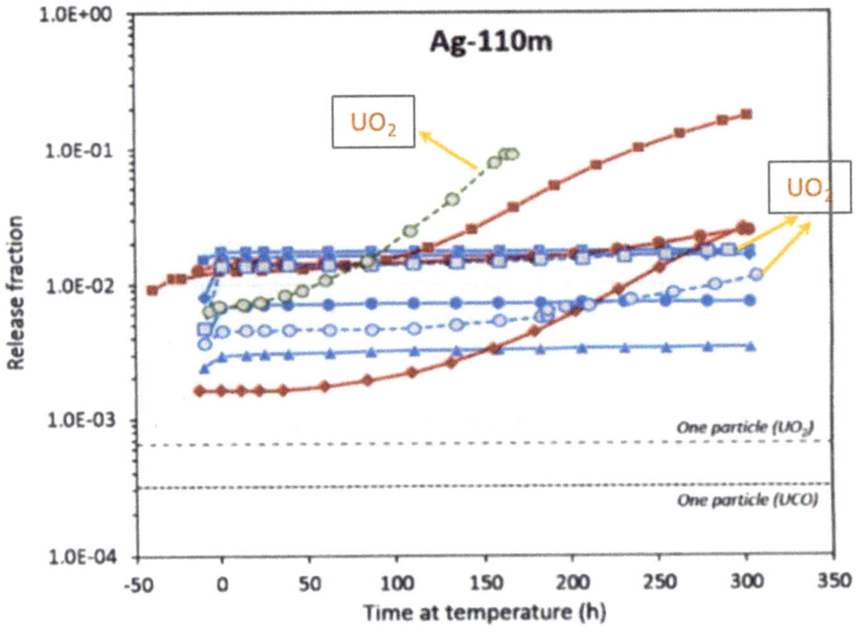

FIG. 99. ^{110m}Ag release from heating of AGR-2 UCO and UO_2 compacts at different temperatures. The plot colour indicates test temperature: 1600° (blue), 1700°C (green), or 1800°C (red) (courtesy of EPRI).

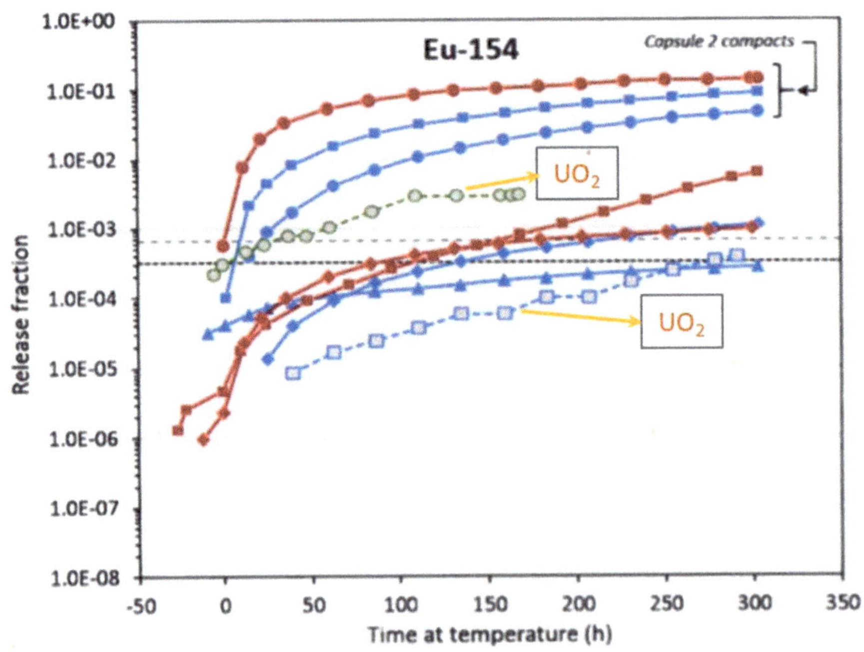

FIG. 100. ^{154}Eu release from heating of AGR-2 UCO and UO_2 compacts at different temperatures. The plot colour indicates test temperature: 1600° (blue), 1700°C (green), or 1800°C (red) (courtesy of EPRI).

7.2. EFFECT OF FABRICATION DIFFERENCES ON FUEL PERFORMANCE

Due to differences in the fabrication process, three specific differences in the coating layers, affecting the fuel performance, have been identified [10]:

- Pyrocarbon anisotropy and density;
- SiC microstructure;
- IPyC-SiC interface structure.

7.2.1. Pyrocarbon anisotropy and density

The density and anisotropy of PyC are dictated by the deposition conditions within the coater. Early US fuels produced under the New Production Reactor (NPR) program utilized low coating gas concentrations, leading to extremely low deposition rates for pyrocarbons, ranging from 1 to 4 μm/min. These conditions resulted in obtaining high-density but anisotropic pyrocarbons with a laminar-like structure, as shown in Fig. 101(a) [10]. Note that high density IPyC is desirable to prevent chlorine attack of the kernel during deposition of the SiC layer. In contrast, the German fuels were manufactured using higher coating gas concentrations and correspondingly high coating rates resulting in more isotropic pyrocarbons, as shown in Fig. 101(b) [10]. Different irradiation behaviours were observed from these different microstructures: the highly dense, more anisotropic US PyC was more susceptible to cracking under irradiation.

Based on PIE of many of the NPR capsules, it was concluded that large shrinkage cracks in the inner pyrocarbon layer led to stress concentrations in the SiC layer and subsequent failure of the layer [10]. Such irradiation-induced shrinkage crack is shown in Fig. 102 [10].

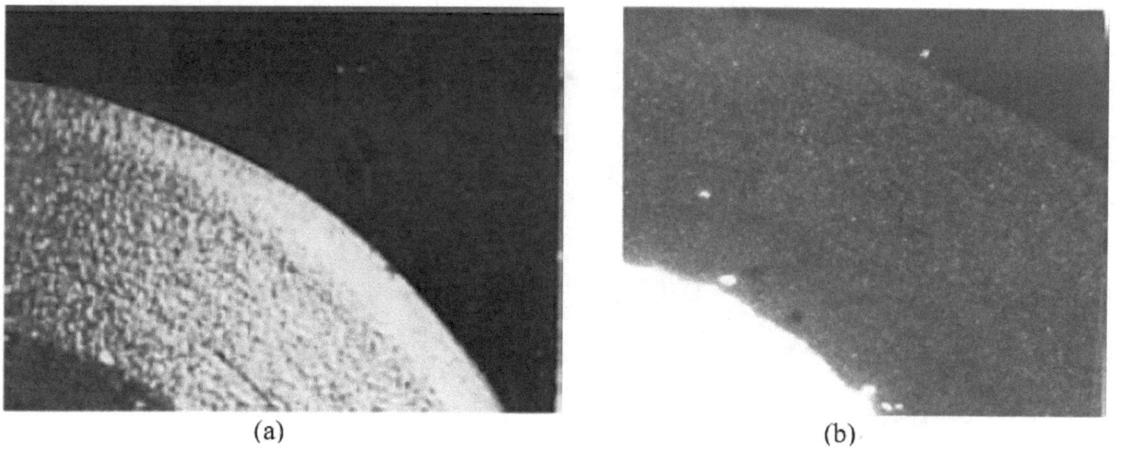

(a) (b)

FIG. 101. Effect of the coating rate on the PyC microstructure: (a) US laminar – high density/more anisotropic PyC: low coating gas concentrations; (b) German isotropic PyC: higher coating gas concentrations (courtesy of INL).

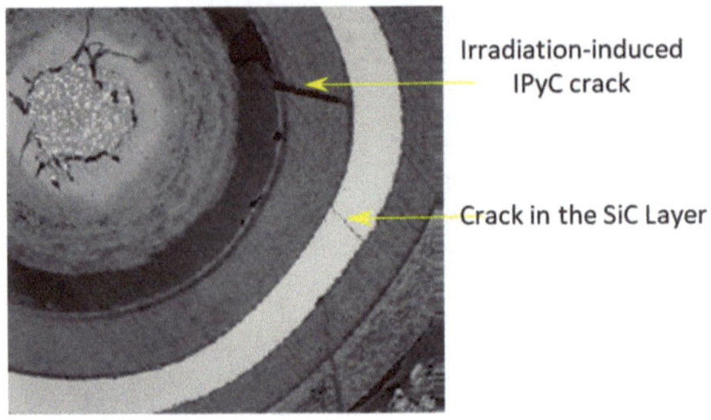

FIG. 102. Irradiation induced cracking of IPyC (courtesy of INL).

7.2.2. Nature of the IPyC-SiC interface

Due to differences in fabrication process (e.g. coating rates, temperatures) of the IPyC and SiC coating layers, differences in the surface porosity and microstructure between the German and US (NPR) IPyC were observed. The interface in German fuel was more tightly bonded because of the deposition of SiC into a PyC having more surface porosity (see Fig. 103(a) [10]). In the case of US NPR fuel for which SiC deposited on the reduced surface porosity in the IPyC layer, less strong bond between IPyC and SiC layer was observed (see Fig. 103(b) [10]). Under irradiation, debonding of the TRISO coating of German fuel was never observed but quite frequently observed for that of US NPR fuel [10]. The strength of the IPyC-SiC interface affects the debonding. Partial debonding can lead to stress intensification in the SiC layer that may cause failure.

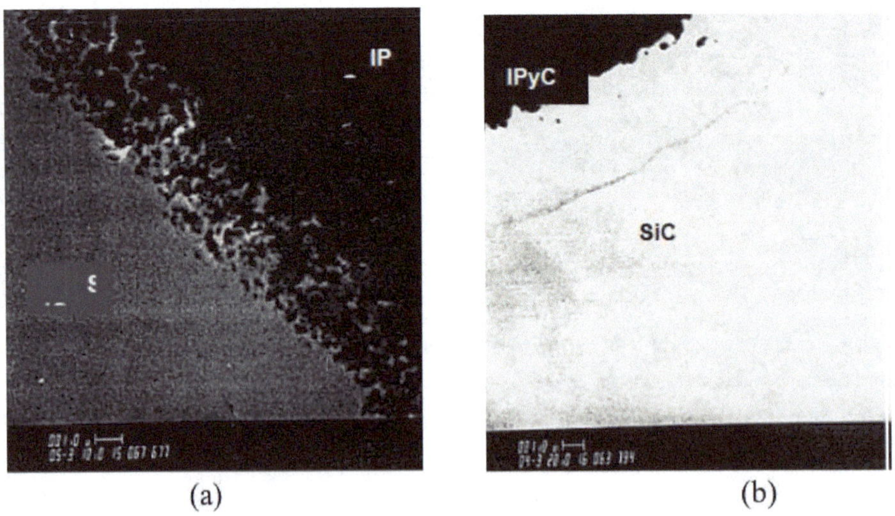

FIG. 103. SiC/IPyC Interface in (a) German and (b) US NPR Fuel (courtesy of INL).

7.2.3. SiC microstructure

Different temperatures used in the SiC coating process of US NPR and German TRISO fuels led to the different SiC microstructures: SiC layer was deposited at a temperature of 1500°C in the German process, whereas of 1650°C in the US NPR process. It was argued in Ref. [4]: "At higher temperatures the grains can be as large as the SiC layer thickness[16], which may result in poor retention of the fission products under high temperature conditions typical of accident situations. The smaller grained German SiC, with its longer grain pathway to the surface, should in principle retain metallic fission products better than the large columnar US SiC. Approximately 90% of the Ag was released from the large columnar grained SiC, whereas only ~ 30% was released in the smaller grained SiC microstructure". Figure 104 shows the different SiC microstructures of US coated particles with different kernels heated at 1500°C following irradiation [10]:

— The laminar SiC microstructures associated with UCO (see Fig. 104(a) [10]) showed very little release of Ag and none for Cs. The diffusivity of Cs through columnar SiC was given as an order of magnitude greater than through laminar SiC;
— Release of Ag was 100% from the UO_2 particles (see Fig. 104(b) [10]) with large columnar grained SiC, and 24% for Cs;
— The weaker laminar SiC structure observed for the UC_2 kernel (see Fig. 104(c) [10]) also showed very high Ag (82%) and Cs (12%) releases.

It is difficult to conclude the role of SiC microstructure in fission product release based on this US study using different kernels, because these data were obtained from different fuel kernel types whose ability to retain metallic fission products may be different.

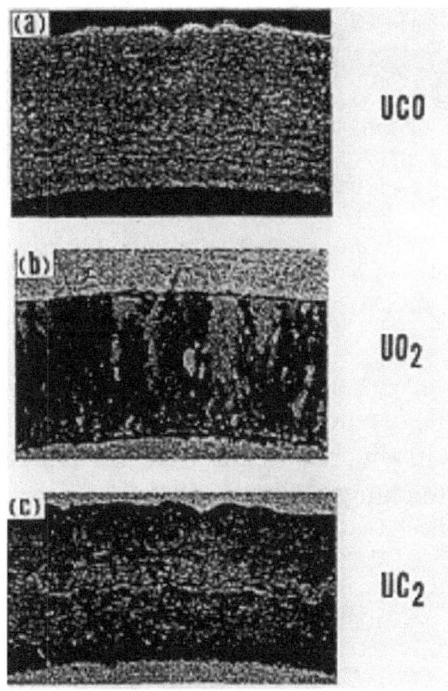

FIG. 104. Microstructures of different SiC layers on coated particles (courtesy of INL).

7.3. TENSILE STRESSES IN SIC LAYER AND FUEL FAILURES

During an irradiation, a coated particle fuel is subjected to several forces that put stress on the TRISO coatings. A particle failure is expected to take place, if the tangential stress reaches a tensile value that exceeds the strength of the SiC for that particle. Therefore, it is important to analyse the stresses in the SiC layer (see Fig. 105 [108]):

---

16 Columnar SiC grains as long as 30 μm (comparable with the thickness of the SiC layer) in the US fuel, while only smaller grains (about 2 μm) in the German fuel [10].

— Fission gas pressure builds up in the kernel and buffer regions, while the IPyC, SiC, and OPyC act to retain this pressure. As irradiation progresses, a portion of the gas pressure is transferred through the IPyC layer to the SiC, leading to an increase in pressure and generating tensile hoop stress in the SiC layer.
— Early during irradiation, pyrolytic carbon layers shrink, inducing an increasing compressive stress in the SiC layer.
— As irradiation progresses, the creep of the pyrolytic carbon layers relieves some of the compressive stress, reducing the beneficial effect of pyrocarbon shrinkage, as compressive stress in SiC is desirable.
— During irradiation of the particle, the SiC exhibits only elastic response, while the IPyC and OPyC layers both shrink and creep.

In summary, the pressure load is pushing outward on the SiC while the shrinkage of the IPyC and OPyC causes them to push or pull inward on the SiC (see Fig. 105). Due to anisotropy, the PyC shrinkage behaviour differs in the radial and tangential directions (see Section 2.3.3 in general). The shrinkage in the radial direction reverses to swelling at moderate fluence levels, whereas shrinkage in the tangential direction continues to high fluence levels.

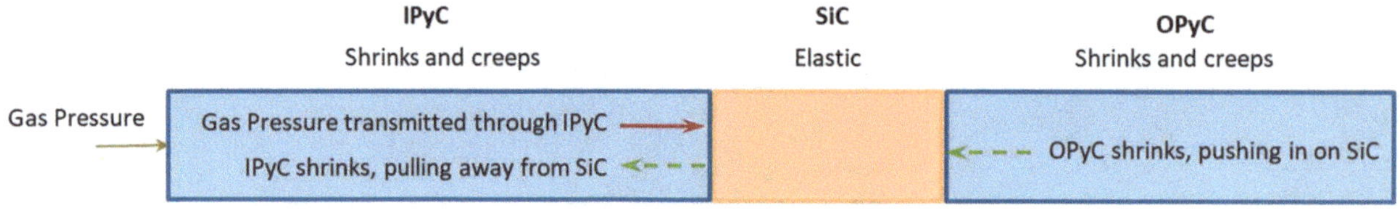

FIG. 105. Behaviour of coating layers in fuel particle (courtesy of EPRI).

Figure 106 shows tangential stress with burnup (fast fluence) at the inner surface of the SiC layer at three different temperatures (details can be found in Ref. [22]). The tangential stress in the SiC reaches a minimum value, and then steadily increases through the remainder of irradiation due to the PyC shrinkage and creep. In addition, the tangential stresses increase with increasing temperature.

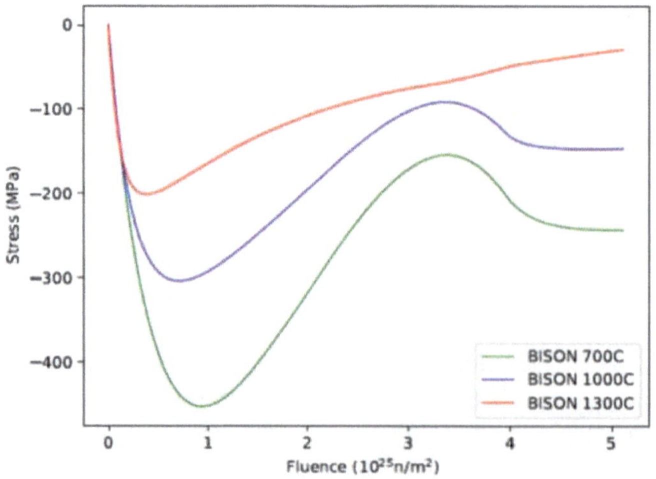

FIG. 106. Evolution of tangential stress with burnup (fast fluence) at the inner surface of the SiC layer (courtesy of INL).

The tensile stress in the SiC layer is contingent upon the pressure of fission gas in the buffer. Assuming complete fission gas release and maximum burnup, the stress in the SiC layer is directly proportional to the following parameters [170]:

$$\sigma \propto \frac{B \times V_k}{V_b} \frac{r_{SiC}}{t_{SiC}} \tag{19}$$

where σ is the tensile stress, B is the maximum burnup, V_k is the kernel volume, V_b is the buffer volume, r_{SiC} is the radius of SiC layer, and t_{SiC} is the thickness of SiC layer.

Based on Equation (19), an analysis was performed to compare tensile stress metrics of the fuels fabricated using different specifications and in different countries [108]. The parameters that are used in the analysis and the results (calculated values) normalized to the German tensile stress metric are listed in Table 58. Tensile stress metrics normalized to the German values are within 20% of the German value indicating the tensile stress component should be similar in all these particle designs [108]. It was also indicated in the analysis that the irradiation-induced PyC shrinkage resulting in a strong compressive component to the stress in the SiC layer was not credited [108].

TABLE 58. PARAMETERS CONTRIBUTING TO TENSILE STRESSES IN THE SIC LAYER

Parameter	**German**	**JAEA HTRR**	**JAERI Advanced**	**U.S. LEU**	**U.S. LEU Fertile**	**U.S. NPR**	**U.S. AGR**
			Particle design parameters				
Kernel composition	UO_2	UO_2	UO_2	UCO	UCO	UCO	UCO
Kernel diameter (μm)	500	600	550	350	500	200	425
Buffer thickness (μm)	95	60	100	100	65	100	100
IPyC thickness (μm)	40	30	35	35	40	50	40
SiC thickness (μm)	35	30	35	35	35	35	35
OPyC thickness (μm)	40	45	40	40	40	40	40
Enrichment (%)	10.6	6	10	19.9	0.7	93	14.0
Burnup (% FIMA)	10	2.6	10	26	6	80	17
			Calculated values				
Simple tensile stress metric	0.676	0.643	0.763	0.799	0.608	0.816	0.785
Normalized to German value	1.00	0.95	1.13	1.18	0.90	1.21	1.16

7.4. PARAMETERS AFFECTING FUEL PERFORMANCE

A radar chart of the five most important parameters affecting fuel performance is shown in Fig. 107 [108]; NGNP project was established by US Department of Energy (DOE) in 2005:

- Fuel temperature is the key parameter for potential failure mechanisms and fission product transport mechanisms during normal operating conditions and under postulated accident condition;
- Fuel burnup is the parameter that determines the concentration of fission products in the kernel and thus the gas pressure and fission product concentration in the particles;
- Fuel fast fluence is the factor governing the extent of radiation damage in the particles and the potential alterations in properties and dimensions in the layers;
- Power density, along with thermal conductivity and the geometry of the fuel (such as compact or pebble), determines the temperature gradient across the fuel specimen, as some potential failure mechanisms are contingent on this temperature gradient;
- Particle packing fraction is the parameter employed with the global power density and it can be used to establish the power per particle, which establishes the temperature inside the particle.

Figure 107 shows the higher envelope for the five key parameters for the US program, NGNP: higher packing fraction (40% vs up to 30% for German and Japanese fuel), higher power density (10 W/cm^3 vs up to 4 W/cm^3 for German and Japanese fuel), higher burnups (20% FIMA vs up to 10% FIMA for German and Japanese fuel), higher fast fluence (5x10^{25} n/m^2 vs 4x10^{25} n/m^2 for German and Japanese fuel), and higher time-average fuel temperatures (1250°C vs up to 1200°C for German and Japanese fuel) based on information from Ref. [108].

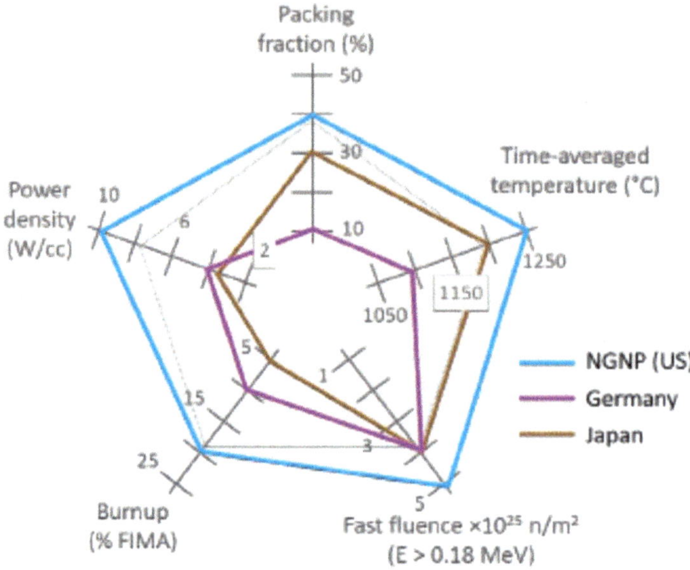

FIG. 107. Radar chart of key parameters for TRISO fuel performance (courtesy of EPRI).

7.5. FUEL FAILURE MECHANISMS

During normal and abnormal conditions, coating integrity is of paramount importance: coating layers should remain intact to retain fission products. Fission products can be retained in the kernel, in the intact layers (i.e., the coating integrity is maintained) and in the matrix. The main mechanism for the fission product transport is diffusion. In the case of SiC layer failure with at least one pyrocarbon layer intact, most condensable fission products will be released but fission gases will be retained. In the case of failure of all three dense coating layers of TRISO particle, both fission gas and condensable fission products will be released.

For the coating layers, a number of potential failure mechanisms have been identified [4,10,108], which are functions of temperature, burnup, fluence, and temperature gradient across the particle, as:

— Overpressure – SiC failure caused by internal gas pressure.
— IPyC related issues, leading to SiC failure:
  - Irradiation induced IPyC shrinkage and IPyC cracking;
  - Debonding between IPyC and SiC;
  - Thermal creep – thinning of IPyC;
  - Thermal shock – thermal stresses induced by the discontinuous fabrication process were well below the failure strength of IPyC.
— OPyC related issues, leading to SiC failure:

- Matrix-OPyC interaction – OPyC cracking;
- Matrix-OPyC interaction – OPyC debonding from the SiC layer;
- Thermal creep.

— Factors directly leading to SiC failures:

- Kernel migration (amoeba effect);
- Fission product attack;
- Non-retentive SiC: diffusive release through intact layers; SiC degradation resulting in permeability to fission products
-
- SiC thermal decomposition.

— Kernel-coating mechanical interaction (KCMI).

— As-manufactured defects produced during fabrication of fuel particles or during pressing of fuel compacts/spheres.

Table 59 summarizes the fuel service conditions and the parameters that influence the fuel failure mechanisms.

TABLE 59. FUEL FAILURE MECHANISMS AND PARAMETERS INFLUENCING FAILURE MECHANISMS

Failure mechanism	Service conditions	Parameters influencing failure mechanism
Overpressure	Temperature Burnup Fast fluence	Strength of SiC Buffer density (void volume) Fission gas release Kernel type Layer thicknesses IPyC and OPyC performance
Irradiation induced PyC failure	Fast fluence Temperature	Dimensional change of PyC Irradiation induced creep of PyC Anisotropy of PyC Strength of PyC PyC thickness PyC density Tensile stress in SiC at IPyC crack tip SiC strength
PyC thermal creep	Time at temperature	Thickness of PyC Stress state of PyC *Only, at very high temperatures (> 2000°C) and very long durations, where thermal creep can operate.*
Kernel migration	Temperature Burnup Temperature gradient	Kernel type (UO_2, UCO, etc.) Buffer and IPyC thickness
Fission product attack	Temperature Burnup Temperature gradient Time at temperature	Chemical state/transport behaviour of fission products Kernel type (UO_2, UCO, etc.) Microstructure of IPyC and SiC
Non-retentive SiC: Diffusive release through intact layers	Temperature Burnup Temperature gradient Time at temperature	Chemical state/transport behaviour of fission products Microstructure of SiC SiC thickness
Non-retentive SiC: corrosion of SiC by CO	Temperature Burnup Time at temperature	Kernel type (UO_2, UCO, etc.) IPyC integrity

TABLE 59. FUEL FAILURE MECHANISMS AND PARAMETERS INFLUENCING FAILURE MECHANISMS

Failure mechanism	**Service conditions**	**Parameters influencing failure mechanism**
Non-retentive SiC: SiC degradation (by Cs) resulting in permeability to fission products	Burnup Temperature Fast fluence	Microstructure of SiC SiC thickness Permeability of SiC SiC impurities from fabrication process
SiC thermal decomposition	Temperature Time at temperature	Microstructure of SiC SiC thickness
Kernel coating mechanical interaction	Burnup Fast fluence Temperature	Initial kernel-coating gas gap Buffer properties IPyC properties Kernel swelling rate

7.5.1. Overpressure

Fission gases that are released from the kernel to the porous buffer layer and CO gas[17], that is formed as a result of the reaction of the oxygen released during fission with the carbon buffer layer, exert tensile forces on the IPyC and SiC layer of the particle, as discussed in Section 7.2. The production of O, CO and fission gases depends primarily on burnup and temperature. To avoid the particle failure by overpressure, a large buffer layer is included. The particle failure is postulated to occur if particles are coated with an insufficient or missing buffer layer [4]. Fabrication specifications were imposed to limit the number of particles produced with thin or missing buffer layers and limits were imposed on the statistical variation in kernel diameter and buffer thickness. The pressure required to fail the SiC layer (or the particle) is given in Equation (1). Photomicrographs displaying overpressure failures for different fuel particles are presented in Fig. 108 [10].

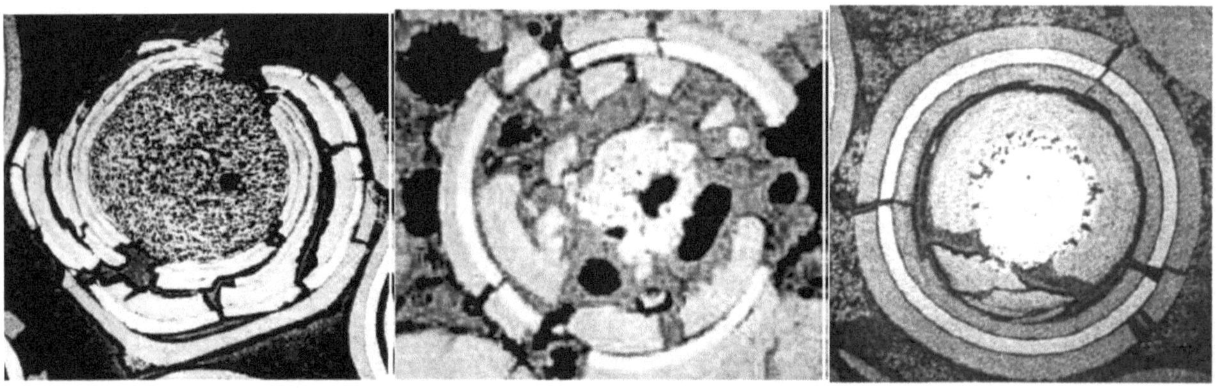

FIG. 108. Overpressure failures from different fuel particles (courtesy of INL).

7.5.2. Inner pyrocarbon layer related issues leading to SiC failure

IPyC related issues leading to SiC failure include:

- Irradiation induced IPyC shrinkage which ultimately leads to the failure of the SiC layer;
- Debonding between IPyC and SiC;

[17] "CO is not expected to be produced in a (UCO) kernel, provided enough uranium carbide is added to the kernel to ensure that there is no excess oxygen available from fission to react with the buffer layer over the burnup life of the fuel." [4]

— Thermal creep – thinning of IPyC layer;
— Thermal shock.

7.5.2.1. Irradiation-induced IPyC cracking

Early on the irradiation, PyC shrinks in both radial and tangential directions. At modest fluences (~ 2 × 10^{25} n/m^2), it begins to swell in the radial direction and continues to shrink in the tangential direction, resulting in PyC layers to be in tension in the tangential direction. Later-on in the irradiation, irradiation induced creep works to relieve the tensile stress in the PyC layer. If there exists a robust bond between PyC and the SiC layer, the shrinkage of PyC imparts a robust compressive stress in the SiC layer, counteracting the tensile stresses induced by gas production. In essence, the particles are engineered so that in undamaged particles, the SiC layer sustains compressive stress throughout the irradiation [10].

Based on many US irradiation tests[18], it was found that the shrinkage exceeded expectations, resulting in significant tangential stresses in the PyC layer, leading to its cracking. These cracks induced concentrated tensile stresses in the SiC layer, ultimately causing the failure of the SiC layer. The crack crosses the buffer layer and the IPyC layer and terminates at the IPyC-SiC interface.

High anisotropy in the PyC layer that was produced in the US fuels due to coating process with too low coating gas concentrations (corresponding to a very low coating rate) was found to be the reason for this failure mechanism, because this failure was not observed (i.e., excessive irradiation induced shrinkage was prevented) in the German fuels that achieved a better isotropy by coating at higher coating rates. In addition, the thicker the PyC layer, the larger are the stresses that may lead to failures [4].

Figure 109 shows the tangential stresses in the SiC layer of a TRISO particle with an initially cracked IPyC [22] (also see [171]). The stress in the SiC at the IPyC crack tip increases with irradiation time (fast fluence) as the IPyC shrinks, though it relaxes later due to irradiation-induced creep of the IPyC. When the IPyC is intact, SiC is under compression stresses. The higher the temperatures, the lower the compression stresses. Cracks in the IPyC causes the stresses in SiC layer changing from compression to tensile. The higher the temperatures, the lower the tensile stresses.

[18] "This is by far the most common fuel failure mechanism observed in the US GA NPR fuel…Thus, the experimental evidence to date suggests that this mechanism is most likely not important for very isotropic PyC – German fuel" [10].

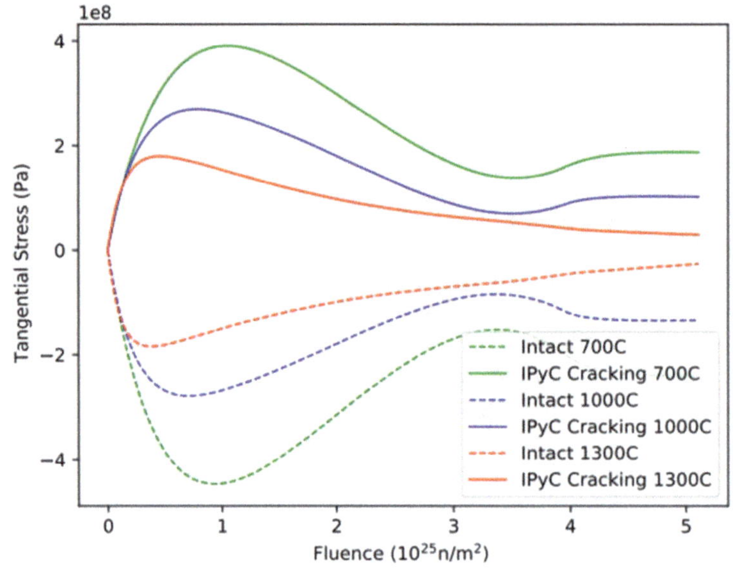

FIG. 109. Finite element model for TRISO-particle with cracked IPyC (courtesy of INL).

7.5.2.2. Debonding between IPyC and SiC

In numerous US irradiations, separation at the IPyC-SiC interface has been noted, primarily ascribed to the characteristics of the IPyC-SiC interface. Coating layers with weak bonds, as seen in US fuel, may undergo partial detachment due to the radial tensile stresses induced by PyC shrinkage during irradiation. Large tensile stresses in the SiC layer can develop in a particle with partially debonded IPyC from the SiC.

Figure 110 shows the tangential stresses in the SiC layer of a TRISO particle with partially debonded IPyC from the SiC layer [10]. Tensile stresses occur at the point of IPyC-SiC contact as the IPyC shrinks under irradiation. Later, irradiation induced creep relieves the stress. With the Weibull statistics, these stresses were used to calculate the SiC failure probability, for which the analysis concluded that the SiC fails at a low rate.

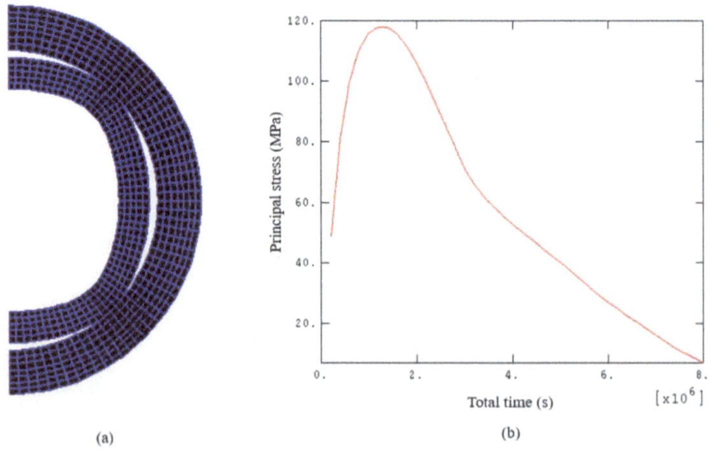

FIG. 110. (a) Finite element model for TRISO particle with partially debonded IPyC from SiC layer and (b) tangential stresses in the SiC layer (courtesy of INL).

7.5.2.3. *Thermal creep*

Some post-irradiation heating tests at very high temperatures (> 2000°C) and very long durations, where thermal creep can operate, revealed a thinning and failure of the IPyC, though such failures were not observed to lead to failure of the SiC layer [4].

7.5.2.4. *Thermal shock*

Unlike the German fuel, the US fabrication of TRISO particles includes a discontinuous coating process (i.e., after each coating layer, the process stops and is reinitiated for the other layer). Hence, the hypothesis posited that subjecting the particles to loading and unloading after depositing each layer in US fuel could induce thermal shock, potentially resulting in the formation of microcracks in the IPyC layer. Over time, these microcracks might propagate and contribute to fuel failure during irradiation [10]. An analysis was performed to quantify the effect of thermal stresses on the failure of IPyC with the conclusion that the thermal stresses induced by this discontinuous fabrication process were well below the failure strength of IPyC [10].

7.5.3. Outer pyrocarbon layer related issues leading to SiC failure

OPyC related issues leading to SiC failure include:

— Matrix-OPyC interaction – OPyC cracking;
— Matrix-OPyC interaction – OPyC debonding from the SiC layer;
— Thermal creep.

7.5.3.1. *Matrix-OPyC interaction – OPyC cracking*

Matrix-particle interaction involves in a strong bond between the outer coating and the matrix material. In the early US coated particle fuel development, OPyC layer failed by cracking (as shown in Fig. 111 [172]) because during the compact fabrication, the liquid carbonaceous matrix material intruded in the OPyC and both materials densified under irradiation, resulting in tensile forces that destroys coating integrity if their bond is not broken first. The German fuel did not exhibit such a behaviour because of the powdered matrix material, which was difficult to infiltrate into OPyC, and more isotropic OPyC [10].

FIG. 111. Particles broken during irradiation due to matrix-particle interaction (courtesy of ORNL).

7.5.3.2. Matrix-OPyC interaction – OPyC debonding from SiC layer

In the early US coated particle fuel development, OPyC layer failed by debonding from SiC layer because during the compact fabrication, the liquid carbonaceous matrix material intruded in the OPyC and shrunk under irradiation. The German fuel did not exhibit such a behaviour because of the powdered matrix material, which was difficult to infiltrate into OPyC, and more isotropic OPyC [10].

7.5.3.3. Thermal creep

Some post-irradiation heating tests at very high temperatures (> 2000°C) and very long durations, where thermal creep can operate, revealed a thinning and failure of the OPyC, though such failures were not observed to lead to failure of the SiC layer [4].

7.5.4. Factors directly leading to SiC failures

The factors directly leading to SiC failures include:

- Kernel migration (amoeba effect);
- Fission product attack;
- Non-retentive SiC: diffusive release through intact layers; SiC degradation resulting in permeability to fission products;
- SiC thermal decomposition.

7.5.4.1. Kernel migration (Amoeba effect)

As discussed in section 2.3.1, kernel migration, also known as the amoeba effect, is associated with carbon transport in the particle when there is a temperature gradient. In the absence of temperature gradient, there is an equilibrium between C, UO_2 and CO [10]. In the presence of a thermal gradient across the particle, different equilibrium conditions are reached on each side of the particle, resulting in the mass transport of carbon down the temperature gradient, which is visually evident in photomicrographs of the fuel as a displacement of the kernel up the temperature gradient. This phenomenon is referred to as "kernel migration," as illustrated in Fig. 112 [172], for UO_2 that achieved 79.8% FIMA at a fast fluence of 6×10^{21} n/cm^2, time-average temperature of 1200–1220°C, and a thermal gradient of 1000–1030°C/cm (based on fuel thermal conductivity).

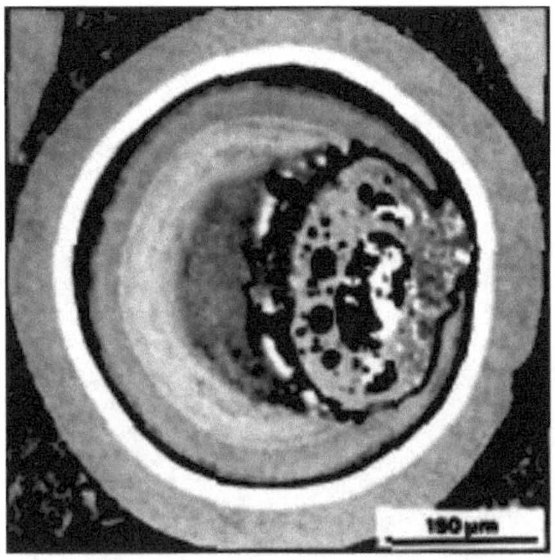

FIG. 112. UO_2 exhibiting the amoeba effect (courtesy of ORNL).

Figure 113 [172] shows the average kernel migration as a function of temperature, which can also be interpreted as the average thermal stability of kernels. The shaded area and the region above it indicate a performance deemed unacceptable. This determination involved calculating kernel migration coefficients associated with the movement of the kernel through the buffer layer during regular operation in various zones of a large HTGR core. Figure 113 illustrates that the UO_2 curve intersects the shaded region within the temperature range crucial for the large HTGR. In contrast, the curves for UC_2 and Th-based carbide and oxide consistently remain below the shaded area across all temperatures. Therefore, the favourable performance of UC_2 and Th-based fuel relative to UO_2 suggested dilution of UO_2 either with enough ThO_2 in Ref. [172] or with enough UC_2 (as adapted in the UCO kernel) for stabilization.

The temperature and temperature gradient in the fuel are the main factors, followed by burnup, causing the kernel migration. In pebble bed cores, the occurrence of the phenomenon was not observed in German irradiation experiments or during AVR and THTR operation. This absence is attributed to the lower power densities and, consequently, smaller thermal gradients in pebble bed cores, preventing the manifestation of the phenomenon. However, in prismatic cores with UO_2 fuel, where power densities in the particles are greater, kernel migration was observed in a number of US irradiations. Consequently, US decided to change the kernel design from UO_2 to UCO with the anticipation of no CO production, and elimination of re-equilibration and carbon transport phenomena. With a proper C/O ratio, kernel migration in UCO fuel was not observed in US irradiation experiments [10].

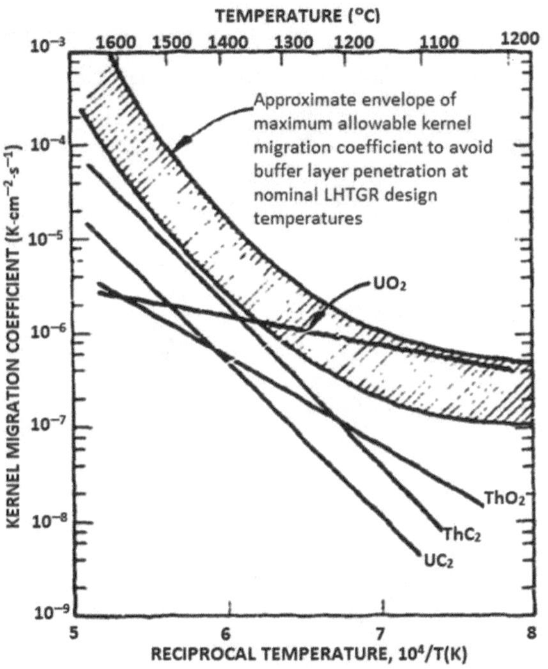

FIG. 113. Average fuel kernel migration coefficient with temperature (courtesy of ORNL).

7.5.4.2. Kernel extrusion

HRB-14 experiments were performed in the HFIR at ORNL) using fertile and fissile TRISO particle fuels: ThO_2, UC_xO_y, $(Th,U)O_2$, UO_2 [10]. Upon metallographic examination of fissile particles, findings revealed that 3% of the IPyC layers exhibited failure (cracking), and debonding from the SiC layer occurred in 7.7% of the particles. In UO_2 or $(Th,U)O_2$ fuel, the buffer layers remained crack-free; however, in UCO fuel, 71% of the particles experienced cracking in the buffer layers. Notably, kernel extrusion, as depicted in Fig. 114 [10], was exclusively reported in UCO fuel.

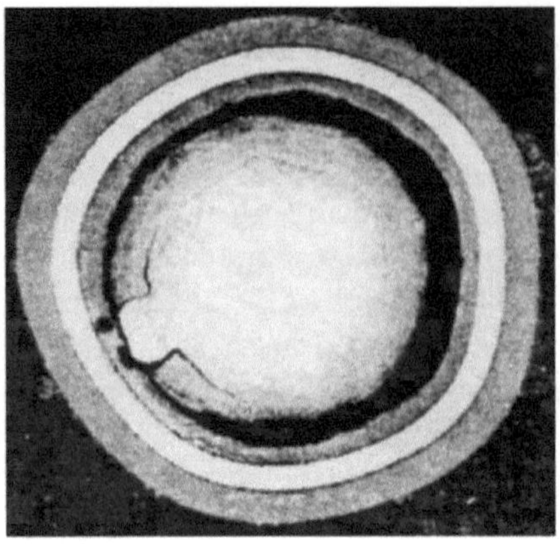

FIG. 114. Kernel extrusion (courtesy of INL).

7.5.4.3. Fission product attack

Fission products, including the REs (cerium, neodymium, lanthanum, samarium, praseodymium, and europium), palladium, ruthenium, and rhodium of the platinum family, and strontium, were observed to be transported from the kernel to the inner surface of the SiC layer [173]. The concern is that fission products interact and can damage the SiC layer (see Fig. 115 [4]). SiC-fission product reactions occur in two modes:

- Based on temperature gradient; REs-SiC reactions on the inner surface of the cooler side of SiC coatings;
- Independent of temperature gradient at the localized reaction sites.

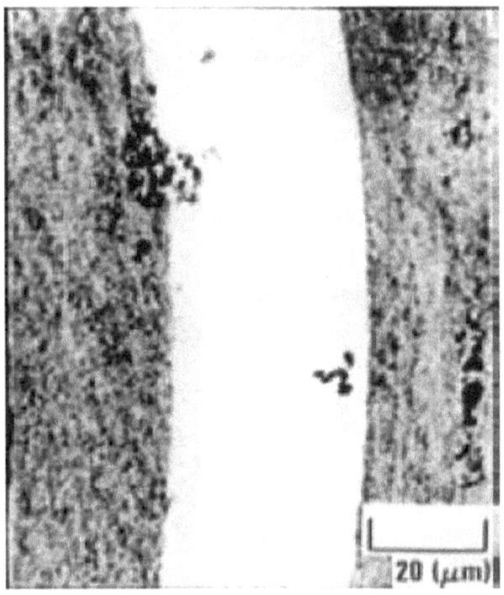

FIG. 115. Fission product attack of the SiC layer (courtesy of IAEA).

Although the second mode of attack is generally attributed to palladium, reactions of ruthenium, rhodium, strontium and silver with SiC reactions were also observed [173]. SiC corrosion by Pd was postulated to be a failure mechanism in TRISO fuels, however, the occurrence of a locally elevated Pd concentration, induced by a crack in the IPyC layer, is a crucial prerequisite for the chemical reaction between Pd and SiC. Without this condition, Pd undergoes migration in the SiC layer through diffusion without engaging in a chemical reaction with SiC [166]. Another noteworthy fission product is Cs, given its relatively high fission yield of radioactive isotopes ^{134}Cs and ^{137}Cs. Despite this, Cs is generally well retained in TRISO coated particles. Earlier studies indicated that even when half the SiC layer thickness was corroded by Pd, the integrity of the layer and its ability to retain volatile fission products, such as Cs, remained uncompromised [167].

Drawing from HRB-14 experiments, significant concentrations of fission products were identified in small, localized regions at the interface between the SiC and IPyC layers. Alongside fission product accumulation, instances of localized chemical attack were observed in the SiC layers of various $(Th,U)O_2$ and UO_2 fuel particles [10]. This localized attack, penetrating approximately 2 μm into the SiC, was attributed to palladium and was present in 8% of the particles. In UCO fuel particles without localized chemical attack, uniform attack along the inner SiC layer was observed. This uniform attack was attributed to REs. With optimized UCO stoichiometry, the kernel retained RE fission products, preventing kernel migration observed in non-optimized UCO kernels containing excess UC_2, leading to RE migration [10].

Palladium is not significantly retained in the kernel of any U-C-O composition. In irradiated kernels of dense LEU UO_2, PuO_{2-x}, $ThO_2 \cdot PuO_{2-x}$, a portion of Pd was found to be incorporated into noble metal inclusions, while the majority of Pd had been released from the kernel and was located within the SiC coating. Therefore, Pd is of concern for both UO_2 and UCO kernels. In addition, some other noble fission products are of concern for the UO_2 kernels, whereas in the UCO kernel, metals (Ag, Cs) are of concern. Other fission products are tied up in the UCO kernel either in carbide or oxide form, and hence their mobility is limited in the particle. Though it is not a failure mechanism, in both UO_2 and UCO, the migration of silver was observed and there are concerns related to maintenance (worker dose) implications: Ag can migrate through intact particles and be released into the reactor coolant system and deposit on cold surfaces, which may happen to be in the turbine for direct cycle gas reactors [10].

Separate effects

Enrichment, O/U ratio, density of the kernel and power density are some of the parameters affecting the concentration of the fission products or the permeability to fission product migration:

- Enrichment: The concentration of certain fission products that impact SiC integrity is influenced by the isotopic composition of HTGR fuel. This variation arises due to the differing ratios of fissioning isotopes (^{23}5U, ^{239}Pu, ^{241}Pu) with changes in uranium enrichment, and because the fission product yields from each isotope are not the same. Ag and Pd can be 25 to 50 times higher depending on the U or Pu concentrations in the fuel [173]. Therefore, the change in isotopic composition (and hence fission product concentrations) from HEU (93% enriched ^{235}U) to LEU (19.5% enriched ^{235}U) was investigated [173].
- In LEU fuels, where a substantial portion of fission arises from Pu at the end of life, the concentrations of Ag and Pd can be considerably higher than in HEU fuel with similar burnups [10]. At peak burnup (~25% FIMA), the fission products with great inventories in typical LEU fuel particles include Mo, Ru, Rh, Pd, Ag and Cd. Ag is 11 times, and Pd and Cd are 4.5 times higher in a 300 mm LEU kernel than in a 200 mm HEU kernel [173].
- O/U ratio: Based on the study performed at ORNL, RE fission products migrate down the temperature gradient to the cool side of the particle when O/U $\leq$ 1.1; however, at higher O/U ratios, the REs are retained in the kernel [173].
- Density: Kernels with lower densities do not retain any amount of Pd; palladium and strontium attack on SiC observed in LEU DRAGON TRISO UO_2 which had an 80% theoretical density;
- Power density (prismatic vs pebble bed): While temperature gradient is the most important parameter for the fission product migration, temperature and burnup also play a role in their migration. Fission product attack is expected to be more pronounced in particles in prismatic reactors due to larger power densities, compared to those in pebble bed reactors (see Section 7.4).

7.5.4.1. Non-retentive SiC: diffusive release through intact layers

Based on the heating test data, some of the fission products (e.g. caesium) that are usually retained by the TRISO coating can diffuse out of the particle during normal operation, if the fuel temperatures during normal operation approach 1300°C [4].

7.5.4.2. Non-retentive SiC: SiC degradation resulting in permeability to fission products

The permeability of the SiC layer to fission products can be compromised under irradiation and high temperatures. When the fuel was exposed to high fluences ($4 \times 10^{25} - 6 \times 10^{25}$ n/m^2 vs. 2–4 × 10^{25} n/m^2) and high burnups (14% FIMA vs 8–9% FIMA), greater release of fission products (e.g. caesium) in heating tests was observed [2]. Two different mechanisms were suggested for the degradation of SiC layer: caesium attack of the SiC and/or CO corrosion of the SiC [4]:

— Caesium vapour can attack SiC at temperatures in excess of 1500°C. Pitting of the SiC layer was observed when SiC samples were exposed to caesium vapour.
— CO generated during irradiation of UO_2 kernels can attack the SiC layer (see Fig. 116 [108]) if the IPyC layer is either permeable or cracked.

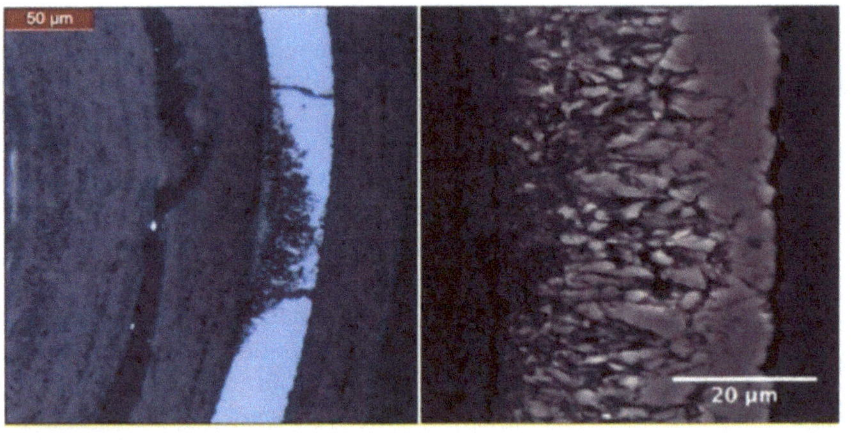

(a) (b)

FIG. 116. A region of the SiC layer corroded by CO in an irradiated UO_2 particle heated to 1600°C: (a) optical image, (b) magnified electron image (courtesy of EPRI).

7.5.4.3. SiC thermal decomposition

At temperatures above 2000°C, thermal decomposition of SiC occurs rapidly: SiC decomposes into its constituent elements, and silicon vaporizes, leaving a porous carbon structure. This failure mechanism is not expected to be an important contributor to fuel failure at normal operating or postulated accident conditions, based on calculations performed for previous core designs [108].

7.5.5. Kernel-coating mechanical interaction

The KCMI is expected to occur at sufficiently high burnups when all gas gaps between the kernel and coatings close due to the swelling of kernel during the irradiation. Based on modelling, SiC layer was predicted to fail shortly after the onset of KCMI [4].

7.5.6. As-manufactured defects

Heavy metal contamination outside of the SiC layer and initially defective particles (i.e., as fabricated fuel defects) can contribute to the releases of fission products. Initially defective particles can be due to undetected as-fabricated defective particles, attack of the particles during fabrication or irradiation (experiments) by impurity metals (e.g. Fe), or particles that have failed during compacting the particles into a matrix (i.e., pebble or fuel compact). To minimize heavy metal contamination and as-fabricated defects, stringent control of key parameters during the fabrication was implemented in US during the NGR program.

7.5.7. Failure mechanisms during accidents

During accident conditions, when the TRISO particle is subjected to air and/or water exposure, carbon materials and fuel kernels undergo reactions with air and water [14]. Water ingress predominantly impacts exposed kernels, leading to the release of a significant portion of their stored inventory, while reactions with carbon materials are relatively moderate, resulting in less aggressive fuel damage. Air reacts not only with the exposed kernels, but also reacts with the carbon materials at a greater rate than water. In both cases, reactions with the carbon materials will release absorbed fission products.

In a reactivity insertion accident, the abrupt release of high energies within the coated particle fuel can induce overheating, overpressure, and ultimately rupture, leading to the release of fission products. Based on analysis, coated particle fuel is expected to fail with energy depositions in the range of 1000–2000 J/g [14]. The duration of the reactivity insertion accident was estimated to be a few seconds with energies much larger than that required to damage fuel particles.

7.6. PARAMETERS AFFECTING PARTICLE PERFORMANCE DURING NORMAL OPERATING CONDITIONS

During normal operation, it is expected that the fuel failure rate is very low. Fuel failures occur during operation due to design, manufacture and operating environment. The sources of fission product releases during normal operation include [14]:

- As –fabricated particle defects (e.g. flawed SiC particles, particles broken during fuel element fabrication) and the presence of uranium impurities in the fuel element fabrication will result in uranium outside of the SiC coating, the pyrocarbon, or both;
- Releases from particles experiencing failure (attributed to missing or thin buffer layers or poor-quality pyrocarbon prone to cracking) during routine operation will lead to elevated internal pressure, culminating in eventual rupture at some point during the core's operational life;
- Hot spots in the core increases the probability of failure from other sources such as amoeba (for UO_2 in particular) or fission product corrosion of the SiC layer;
- Silver (^{110m}Ag) released from particles at temperatures above 1100°C (as SiC does not retain silver well above approximately 1100°C) is viewed as an occupational dose matter rather than a public health and safety issue in current designs.

Fission gases will transfer to the coolant and can be eliminated through the coolant gas clean up system, though metallic fission products will remain in the reactor internal components, because it appears to be no practical way to remove metallic releases from an operating reactor [14].

7.6.1. Kernel

Besides its functionality for power generation, the kernel is expected to control the oxygen potential to limit kernel migration and particle CO pressure, and to retain the REs [14]. The kernel contains gases to some extent, though at high burnups (and at high temperatures), the kernel may not be able to retain fission gases due to changes in the microstructure (i.e., void formation with burnup) and the change in lattice structure with burnup, increasing the diffusion of fission products.

Table 60 shows the operating parameters affecting the kernel performance and the performance of the particle via kernel during normal operating conditions.

TABLE 60. KERNEL OPERATING PARAMETERS

Kernel operating parameter	Rationale
CO production	CO production influences particle pressure and kernel migration.
Burnup	The level of burnup determines the chemical state in the kernel and its ability to retain fission products.
Swelling	The kernel swells because of fission products and the buffer layer is designed to accommodate the kernel distortion.
Microstructure changes	The crystal structure of the kernel can influence the retention of fission products.
Fission product chemical form	The chemical form of the fission products determines their mobility with the kernel and affects the CO pressure.
Buffer interaction	The periphery of the kernel may react with the buffer layer carbon.
Kernel migration (*fuel dependent*)	Temperature gradients can drive carbon transfer and result in the movement of the kernel towards the coatings.
Fission product generation	Different fissile isotopes produce different mixes of fission products. Isotopes that generate more noble metals will cause degradation of performance due to the noble metal attacks to the SiC.
Temperature gradient	The temperature gradient across the kernel drives the amoeba effect and REs fission product migration.
Isotope half-life	Isotope half-life determines which fission products will survive the diffusion to and through the coatings.

7.6.2. Buffer layer

The buffer layer is to provide a free space for the released fission gases and any generated CO. The particle can fail by overpressure if sufficient volume to accommodate the volume of generated gas is not provided by the buffer layer. An excessively thick (out of specification) buffer could increase thermal impedance, though this extra impedance does not appear to adversely affect particle performance. Nevertheless, based on some calculations, thermal gradients can drive fission product diffusion [14].

Table 61 shows the operating parameters affecting the buffer performance and the performance of the particle via the buffer during normal operating conditions.

TABLE 61. BUFFER OPERATING PARAMETERS

Buffer operating parameter	Rationale
Pressure	The buffer function is to provide void volume to control particle pressure.
Shrinkage and cracking	It is desirable to control buffer shrinkage in order to control cracks and to minimize kernel extrusion. Cracking may also allow recoil path to IPyC.
Temperature gradient	An excessive gradient can lead to higher kernel temperatures.
Condensed/gas phase diffusion	Buffer is the first layer that encounters the fission product releases.
Recoil effects	The buffer layer protects the IPyC layer by attenuating the fission product recoils.

7.6.3. Inner pyrocarbon layer

The IPyC is engineered to maintain dimensional stability and integrity throughout the fuel's lifespan. The occurrence of cracks or debonding in the IPyC may generate localized tensile stresses in the SiC, leading to potential cracking if the bond between the two layers is robust. Cracks can expose the SiC to CO, causing corrosion at elevated temperatures. Additionally, the IPyC serves as a barrier to the transport of fission products to the SiC; it exhibits high permeability to fission gases while allowing the diffusion of metallic fission products.

Table 62 shows the operating parameters affecting the IPyC performance and the performance of the particle via IPyC during normal operating conditions.

TABLE 62. IPYC OPERATING PARAMETERS

IPyC operating parameter	Rationale
Irradiation-induced creep	Irradiation-induced IPyC creep relieves some of the stress in the IPyC caused by irradiation-induced shrinkage and pressure.
Fast fluence	The fluence dose has an impact on the shrinkage, creep and layer stresses.
Dimensional change	The stresses caused by irradiation can result distortion and debonding of the IPyC from the SiC.
Anisotropy	The anisotropy plays an important role for the IPyC dimensional stability.

7.6.4. Silicon carbide layer

With the SiC layer, being the main barrier to fission product release, releases to the coolant are expected to be dominated by the initial defective fuels and failures caused by other phenomena during normal operation. The exception is silver. Above 1100°C, silver will diffuse through the SiC into the coolant. Furthermore, prolonged operation of the fuel at elevated temperatures exceeding 1300°C may result in SiC corrosion by noble metals and CO, particularly in cases where the IPyC is cracked. Palladium is the primary noble metal of concern due to its facile diffusion into the SiC and its propensity to react with SiC, leading to the formation of silicide. The type of kernel is important for noble metal inventory: depending on the fissile material, the kernel may have higher or lower yields of noble metals. If the kernel does not retain the REs as oxides, they can attack the SiC. In the event of heavy metal contamination in the SiC layer during the fabrication process, the fissioning of this material can cause SiC damage, ultimately resulting in premature failure.

Table 63 shows the operating parameters affecting the SiC performance and the performance of the particle via SiC during normal operating conditions.

TABLE 63. SIC OPERTING PARAMETERS

SiC operating parameter	Rationale
Kernel interaction with SiC layer	Temperature gradients can drive carbon transfer and result in the movement of the kernel towards the coatings.
Fission product corrosion	Noble metals present a corrosion hazard at high temperatures.
CO attack	CO can attack SiC with a cracked IPyC at high temperatures
Heavy metal attack	If fissile material is present in the SiC layer from fabrication, fission in the SiC layer will damage it.

TABLE 63. SIC OPERTING PARAMETERS

SiC operating parameter	Rationale
Cracking	Cracking will cause the particle to release metallic fission products, but not gases, if one of the PyCs remain intact.
Condensed/gas phase diffusion	SiC is the primary diffusion barrier and it retains both metallic and gaseous fission products, with the exception of silver.

7.6.5. Outer pyrocarbon layer

The OPyC layer is designed to remain dimensionally stable and intact over the fuel lifetime. Failure of the OPyC could increase the failure probability of the SiC. The shrinkage of the OPyC is limited by OPyC creep, keeping the stresses in the layer within acceptable limits. The OPyC limits the transport of fission gases, but not the metallic fission products. The control of OPyC-matrix interactions is desirable so that the OPyC is not damaged by matrix shrinkage or micro cracks that occur in the matrix material.

Table 64 shows the operating parameters affecting the OPyC performance and the performance of the particle via OPyC during normal operating conditions.

TABLE 64. OPYC OPERATING PARAMETERS

OPyC operating parameter	Rationale
Irradiation-induced creep	Irradiation-induced creep relieves some of the stresses caused by shrinkage and pressure.
Dimensional change	The stresses caused by irradiation can result distortion and possible breakage of the OPyC layer.
Anisotropy	Anisotropy is the factor affecting the OPyC dimensional stability.
Condensed/gas phase diffusion	OPyC does not retain well metallic fission products, though serves as a barrier for fission gases in the case of SiC failure.
Cracking	Cracking causes easy transport of fission products, in the case of SiC failure to the fuel matrix.

7.6.6. Fuel element

The main functions of the fuel element include [14]:

- Maintain dimensional stability;
- Not transmit undue stresses to the fuel particles;
- Withstand external damage mechanisms, such as the fall of pebbles into the reactor core from the entrance;
- Maintain good thermal properties;
- Act as a sink for any released metallic fission products.

Elements made with a high filler concentration tend to be more stable than those with low filler concentration. Microcracks and voids may form in the fuel element due to shrinkage of the material, though minor cracking is tolerable for intact particles. Corrosion by coolant impurities can affect fuel elements and result in "peeling" or "spalling" of small regions (<200 mm deep) on the surface of pebbles [14].

Due to irradiation damage, matrix material's mechanical and thermophysical properties change. For the fuel pebbles, matrix irradiation performance was very good for the operating temperatures (<1400°C) and fast fluences of interest ($<9\times10^{25}$ n/m^2) [14]. Irradiation temperature and accumulated fast fluence strongly influence property changes, but the pebbles maintain integrity under the irradiation conditions of interest. Under irradiation [14]:

- Pebbles exhibited a diameter shrinkage of less than 2%, and this shrinkage was approximately equal in both radial and tangential directions up to a fast fluence of 5×10^{25} n/m^2. Beyond this value, it deviated by about 30%.
- Strength increased by about 10% at moderate fast fluences, and the elastic modulus showed an increase with irradiation, reaching up to 70% at 700°C. Subsequently, it only experienced a slight decrease at higher fast fluences. Crush strength almost doubled.
- Thermal conductivity decreased about 60% and the coefficient of thermal expansion increased by about 40% with irradiation.

Matrix material either sorbs metallic fission products or delays their migration into the coolant at normal operating temperatures, though gases readily migrate through the matrix material.

Table 65 shows the operating parameters affecting the fuel element performance and the performance of the particle via fuel element during normal operating conditions [93].

TABLE 65. FUEL ELEMENT OPERATING PARAMETERS

Fuel element operating parameter	**Rationale**
Temperature	Temperature and fast fluence affect the matrix behaviour (e.g. shrinkage, cracks) and stability.
Fast fluence	Fast fluence affects the matrix behaviour (e.g. shrinkage, cracks).
Power density	Power affects the operating temperature and gradients.
Temperature difference	The behaviour of fuel particles can be affected by micro temperature gradients in the fuel matrix.
Temperature histories	Time at temperature is a factor determining corrosion and diffusion.
Condensed phase diffusion	The matrix material can sorb/trap significant amounts of less volatile fission products.
Gas phase diffusion	The element does not retain gases significantly.
Corrosion by coolant impurities	Coolant impurities can corrode/damage the fuel element matrix.

7.6.7. Summary

A total of thirteen parameters, characteristics and phenomena were identified to be of high importance during normal operating conditions for kernel, coatings and the fuel element, as listed in Table 66.

TABLE 66. OPERATING PARAMETERS OF HIGH IMPORTANCE

Layer	Parameter
Kernel	Carbon monoxide production
Buffer	Pressure Cracking Temperature gradient
IPyC	Cracking Condensed phase diffusion Gas phase diffusion
SiC	Fission product corrosion Cracking Condensed phase diffusion Gas phase diffusion
OPyC	Gas phase diffusion
Fuel element	Condensed phase diffusion

7.7. FUEL FAILURES DURING NORMAL OPERATIONS

During normal operating conditions, it is expected that the integrity of the coating layers is maintained.

7.7.1. German fuel

For the German fuel, among approximately 380 000 UO_2 and 80 000 $(Th,U)O_2$ particles subjected to testing, there were no in-pile failures, and only a few particles were considered 'damaged' due to experimental anomalies [10]. Gas release was attributed solely to as-manufactured defects and heavy metal contamination.

Similarly, accelerated irradiation tests conducted for German LEU UO_2 TRISO particles showed no particle failures for 754 000 particles; however, at high burnups (14% FIMA) and temperatures (1700°C) some releases were observed [108].

Based on numerous studies, it has been concluded [108] that the German fuels had low as-manufactured particle defect fractions, low particle failure fractions during irradiation and during post-irradiation heating tests at postulated accident temperatures.

7.7.2. Chinese fuel

During irradiation tests in Russia, it was observed that two of the irradiated spheres had one or two exposed kernels from the initial stages of irradiation, aligning with the free-uranium measurements obtained during the manufacturing of early production batches. Similarly, performed on two Chinese pebbles irradiated in the HFR Petten reactor, the ^{85m}Kr R/B ratio for the capsule was approximately 8×10^{-8} [108]. The calculated release fraction of ^{85m}Kr from a single coated particle is 3.26×10^{-3}. In a capsule containing two pebbles (~16,600 particles), the release fraction from a single failed particle is 1.96×10^{-7} [108]. This suggests that there was no complete particle failure during the irradiation, and the observed fission gas release is attributed to uranium impurities in the carbonaceous matrix of the pebbles and the graphite cups used for pebble retention. The early batches had a free-uranium content (measured by the burn-leach procedure) of typically 10^{-4}, whereas the last 15 batches were normally 10^{-5} and lower [108].

For the development of HTR-PM (with a capacity of 210 MWe) in China, irradiation testing of five HTR-PM fuel spheres was performed in the Petten High Flux Reactor (HFR). The free uranium fraction in the pebbles decreased from 5×10^{-4} for the HTR-10 (existing prototype reactor) to 6×10^{-5} for the HTR-PM. According to neutronics calculations, the total fast fluence levels were estimated between

3.79 and 4.95 × 10^{25} n/m^2 (E >0.1 MeV), and burnup was between 11.1 % FIMA and 13.7% FIMA. Measured ^{85m}Kr R/B values during the last cycle were between 2.4 and 3.3 × 10^{-9}, indicating no particle failure [108]. Based on the preliminary post-irradiation experiments, the dimensional shrinkage in all five pebbles was between 0.88% and 1.25% [108].

7.7.3. US fuel

In most experiments of US fuel performed prior to the year 2000, a significant fraction of the TRISO-coated particles experienced failure of one or more layers. Failures of the coating layers were ascribed to various factors [10], including: (a) SiC failure, (b) kernel migration (amoeba effect), (c) fission product attack on the SiC layer, (d) irradiation-induced IPyC cracking and/or debonding leading to cracking in the SiC layer, and (e) carbon matrix-OPyC interaction and irradiation-induced OPyC failure. The PyC-related mechanisms are strongly influenced by the anisotropy and porosity in the coatings. Anisotropy has a significant impact on the shrinkage and swelling behaviour of the PyC layers under irradiation.

Figure 117 shows the frequency of layer failures observed during PIE of early US coated particle fuel from different irradiations [10].

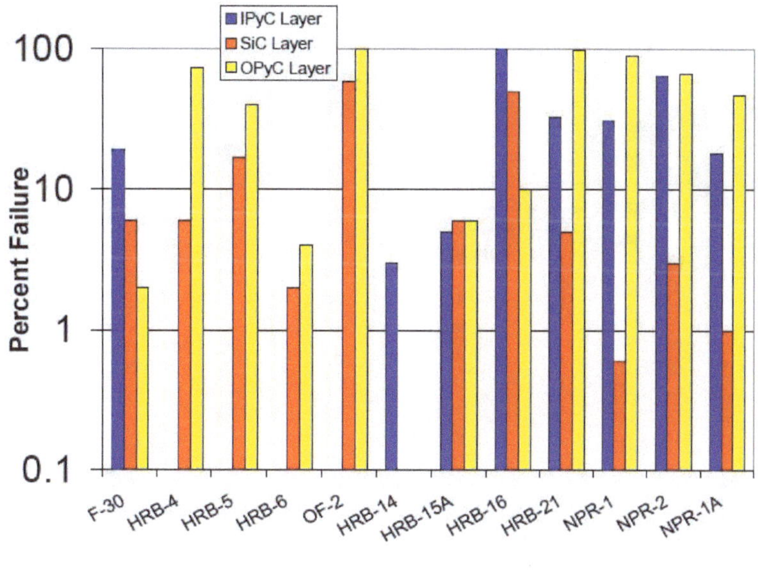

FIG. 117. Failures observed during PIE of US coated particle fuel from different irradiations (courtesy of INL).

Later under the AGR Program (with the objective to advance fabrication and characterization technologies and generate irradiation and annealing performance data necessary for licensing TRISO particle fuel for the NGNP and upcoming HTGRs) in the US, irradiation experiments were performed, and low-frequency SiC layer failure (i.e., failure of the SiC layer with at least one PyC layer remaining intact) was identified [108]. SiC failure was due to the buffer shrinkage contributed to IPyC fracture: IPyC fracture was due to the stress from the buffer pulling away from the IPyC and the IPyC fracture exposed the SiC layer to concentrated chemical attack of fission products (notably Pd), causing degradation through the entire layer. Very low fractions of TRISO failure (i.e., failure of all three dense coating layers such that fission gas is released) were observed [108].

Table 67 shows the specifications for particle defects and failure fractions for NGNP at the core outlet temperature of 750°C [6]. The design value of TRISO failure fraction during normal operating conditions at the core outlet temperature of 750°C is $\leq 2.0 \times 10^{-4}$.

TABLE 67. SPECIFICATIONS FOR PARTICLE DEFECTS AND FAILURE FRACTIONS FOR NGNP

Parameter	Maximum expected	Design
As manufactured fuel quality		
Heavy metal contamination	$\leq 1.0 \times 10^{-5}$	$\leq 2.0 \times 10^{-5}$
Defective SiC	$\leq 5.0 \times 10^{-5}$	$\leq 1.0 \times 10^{-4}$
In-service TRISO failure		
Normal operation	$\leq 5.0 \times 10^{-5}$	$\leq 2.0 \times 10^{-4}$
Accidents	$\leq 1.5 \times 10^{-4}$	$\leq 6.0 \times 10^{-4}$

The TRISO failure fraction in a fuel with UCO kernel after the AGR-1 and the AGR-2 capsule irradiation tests was $\leq 2.3 \times 10^{-5}$ [108]. This represents an approximately nine-fold reduction compared to the standard reactor design specifications for acceptable in-service TRISO failures during normal operational conditions (2×10^{-4}).

Figure 118 [108] shows the SiC layer and TRISO particle failure fractions for combined AGR-1 and AGR-2 UCO results during irradiation and during annealing tests. The combined (AGR-1 + AGR-2) SiC failure fractions in a fuel with UCO kernel were $\leq 3.6 \times 10^{-5}$ during irradiation [108]. Note that historic reactor performance requirements have not included specifications for SiC layer failure.

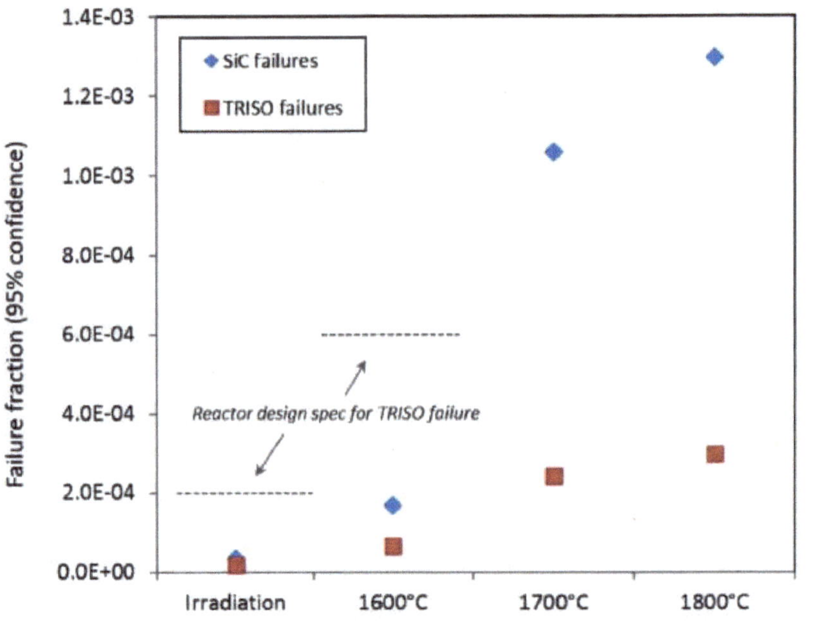

FIG. 118. SiC layer and TRISO particle failure fractions (upper limit at 95% confidence) for combined AGR-1 and AGR-2 UCO results during irradiation and during annealing tests (courtesy of EPRI).

There were no observed TRISO or SiC failures in the AGR-2 UO_2 fuel compacts following the irradiation. Despite a higher occurrence of SiC failures in annealing tests compared to UCO, there were no instances of TRISO failures in the three annealing tests conducted at 1600°C, 1700°C, and 1800°C [108].

Densification of the buffer layer and swelling of the kernel with related formation of gas filled bubbles were commonly observed [108]. The buffer layer experienced partial or complete debonding from the IPyC layer in most particles. This debonding was induced by buffer densification and volume shrinkage, resulting in the formation of a void between the buffer and IPyC layer. In the particles in which the buffer and IPyC layers remained completely bonded, the buffer densification resulted in the inner diameter increase while the kernel swelled to fill the increasing volume. Buffer fracture was common, and it was often accompanied by expansion of the kernel (i.e., kernel extrusion) into the gap formed at the point of fracture. Buffer failure fraction of 86% was observed for the compacts irradiated at a temperature of ~1100°C, whereas that of 1–2% was observed for those irradiated at a temperature above 1200°C, which is believed to be due to greater magnitude of thermal creep occurring at the higher temperatures, which relaxes stresses developed due to buffer densification and shrinkage [108]. However, buffer fracture seems not to pose a threat to the particle integrity, as there were very low SiC and TRISO coating failure fractions.

In the case where there was a buffer and IPyC layer separation and buffer fracture, no outer dense coating layer failure was observed, even in cases where the kernel was in direct contact with the IPyC layer [108]. When the buffer-IPyC interface remained intact, the fracture of the buffer layer coincided with the fracture of the IPyC layer. This often led to debonding of the IPyC layer from the SiC layer [108]. Without a buffer fracture, the partial debonding of the buffer-IPyC layer led to development of sufficient stress in the IPyC layer to cause fracture in some particles.

Caesium fractional release from compacts containing only particles with intact SiC was very low. The total Cs release from the fuel compacts will increase depending on the extent of SiC layer failure. The releases of Eu and Sr were modest through intact coatings, and they were significantly retained in the fuel matrix, though at higher irradiation temperatures (up to a time-average maximum of 1360°C), the releases of Eu and Sr became higher. Silver release was notably high, aligning with historical patterns. Interestingly, there was no widespread evidence of Pd attack or corrosion of SiC, even though measurable amounts of Pd were detected outside of the SiC layer [108].

7.7.4. Japanese fuel

Based on the first core irradiation in the HTTR, the measured R/B ratio of ^{88}Kr at full power and at the coolant outlet temperature of 950°C was 1.0×10^{-8}, corresponding to gaseous diffusion from heavy metal contamination and no significant in-reactor fuel particle failures [108].

From the first batch core loading of HTTR, the TRISO particle design was slightly modified (i.e., larger buffer layer: 90 vs 60 μm in the original core loading, and slightly larger SiC layer: 35 vs. 25–30 μm) to operate at higher burnups (5–10% FIMA vs 2.4% FIMA which was the end-of-life core average burnup in the original core loading and 3.6% FIMA was the design limit peak burnup) [174]. The fuel compacts utilizing UO_2 fuel kernels, featuring a slightly modified particle design, underwent irradiation in the 91F-1A capsule at the Japan Materials Testing Reactor (JMTR) operated by JAERI and in the HRB-22 capsule at the HFIR operated by ORNL [175]. The maximum burnup and the fast neutron fluence were about 10% FIMA and 3×10^{25} n/m^2, respectively [175]. Two through-coatings failed particles were detected at the beginning of irradiation for the HRB-22 capsules [175]. Post-irradiation examinations revealed the absence of kernel migration and SiC corrosion [174]. Within the deconsolidated coated fuel particles irradiated in the 91F 1A capsule, only one failed particle was identified, indicating that the OPyC layer experienced cracking due to tensile stress [174]. In addition, in the 91F-1A capsule irradiation test, it was estimated one fuel particle failure based on comparison of measured and calculated R/B ratios of ^{88}Kr. In the HRB-22 capsule irradiation test, four fuel particle failures were observed during irradiation. Calculation results showed [174] that no tensile stresses acted on the SiC layers even at the end of irradiation, and no SiC layer failure occurred in the intact particles even in a particle with thin buffer layer with failed OPyC layer. In conclusion, the presumed failure mechanisms were identified to be additional through-coatings failure of as-fabricated SiC failed particles, and an excessive increase of internal pressure by the accelerated irradiation [174].

The preliminary calculation findings indicate [175] that solely relying on mean fracture stress values for PyC and SiC does not allow for predicting particle failures, defined as the failure of all three load-bearing layers. In contrast, incorporating statistical variations in fracture stresses and particle specifications, as accomplished in the STAPLE code, enables the prediction of failures.

When high fracture stresses were employed for each of the three load bearing layers (IPyC, SiC and OPyC), namely 400 MPa, up to 8% FIMA, stresses in the SiC were always compressive, and neither of the PyC layers had failed [175]. However, experimentally, failures were observed. Therefore, for the failure of all three load bearing layers, one or both PyC layers must have fractured first. Setting the fracture stress of a PyC layer to 50 MPa resulted in early failure [175]. If the IPyC layer failed prematurely, the stress in the SiC layer at 8% FIMA became tensile, reaching 25 MPa. Premature OPyC layer failure increased the stress slightly to 36 MPa, while simultaneous IPyC and OPyC failure raised the tangential stress in the SiC layer at 8% FIMA to 125 MPa [175]. Clearly, failure of all three load-bearing layers is most likely if both the IPyC and OPyC layers fail first.

The compact contained 4400 particles, and in the calculation for better statistics, ten times that number of particles were used. Figure 119 [175] illustrates the number of calculated failed particles as a function of burnup, showing that 1 and 2 failures occurred at around 7.5 and 8% FIMA respectively, consistent with observations of 1–2 particle failures towards the end of irradiation.

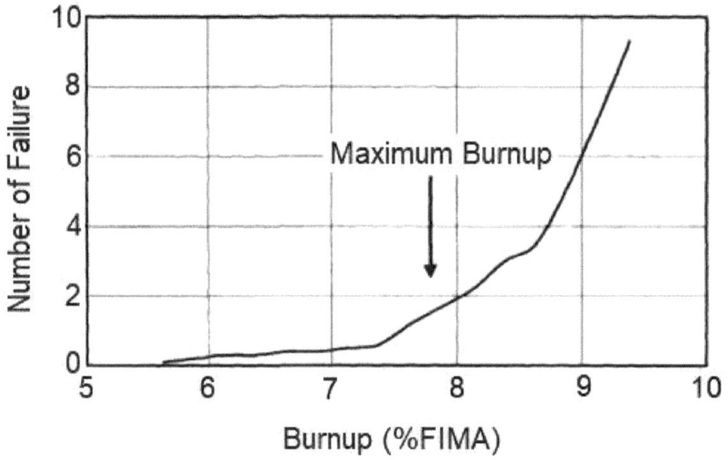

FIG. 119. Calculated number of fuel failures for the fuels irradiated in JMTR and HFIR (courtesy of JAEA).

Similarly, Fig. 120 shows the predicted failure fractions as a function of fuel burnup level for thorium and depleted uranium, based on Ref. [176]. Predicted fuel failure fraction exceeds the predefined safety limit of 10^{-4} when the fuel achieves a burnup of 8% FIMA. In the analysis, the dominant failure mode for TRISO particles was expected to be failure of the SiC layer, and Weibull theory was used for predicting failure probabilities. Weibull statistical strength theory represents a distribution of SiC strengths within a population of TRISO particles and allows failure probability predictions for this population:

$$P_f = 1 - \exp\left[-\ln 2\left(\frac{\sigma_{t,SiC\_cal}}{\sigma_{t,SiC\_ms}}\right)^m\right] \quad (20)$$

where P_f is the failure probability of the SiC layer, σ_{t,SiC_calc} is the maximum calculated tangential stress in SiC (MPa), which can be assumed to be equal to the tangential stress at the inner surface of the SiC, σ_{t,SiC_ms} is the SiC Weibull mean strength (MPa), and m is the SiC Weibull modulus [176].

Assuming that the manufacturing results for each parameter conform to a truncated Gaussian distribution with user-specified values for the standard deviation and upper and lower truncation bounds, the TRIUNE code was utilized for calculating failure probabilities and simulating the lifetime history of each particle. At each time step, following the completion of stress-strain-displacement calculations, Weibull failure probabilities were computed for each particle. The effective predicted batch failure fraction (Batch FF) for the entire particle population is then determined as [176]:

$$\text{Batch FF} = \frac{\sum_{n=1}^{N} P_f^n}{N} \tag{21}$$

where Batch FF represents the overall predicted failure fraction for a batch comprising N particles (specified by the user), P_f^n stands for the Weibull failure probability of the particle with the number n.

A failure fraction limit of 10^{-4} was adopted to ensure that radiation source terms, resulting from radiation escaping from the particles and passing through the pebbles into the coolant, do not surpass a predefined threshold [176].

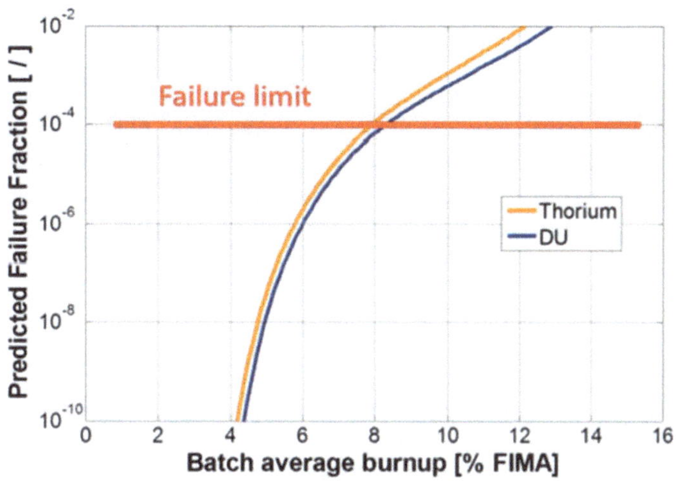

FIG. 120. Calculated failure fraction as a function of burnup for Th-based and depleted uranium fuels.

Furthermore, additional modifications in the HTTR particle design were made (e.g. increase in the buffer thickness: 95 μm), such that it closely resembled the standard TRISO particle design from the German programme [108]. Based on the irradiation tests of this fuel to a burnup of >9% FIMA in the WWR-K reactor (Kazakhstan), no in-pile particle failures occurred, and R/B data indicated a few exposed kernels, which were postulated to be related to as-fabricated SiC defects [108].

7.7.5. Summary

As discussed in Section 7.4, the operating parameters would affect the fuel performance: high temperature, large temperature gradient, high burnup, high power density and high particle packing fraction would adversely affect the fuel performance. Such differences have been identified for fuel compacts to be used in prismatic block HTGRs and spherical elements to be used in pebble bed HTGRs. Compared to the US fuel compacts, the German pebbles were being operated at lower burnups (10% FIMA for German pebbles vs 20% FIMA for US compacts), fast fluence (4 n/m^2 vs 5x10^{25} n/m^2), power density (4 W/cm^3 vs 10 W/cm^3), packing fraction (10% vs 40%) and fuel time average temperatures (1100°C vs 1250°C). Therefore, the US fuel compacts have been exposed to harsher conditions.

The German pebbles exhibited minimal as-manufactured particle defects and low particle failure fractions both during irradiation and post-irradiation heating tests at anticipated accident temperatures. Conversely, in the case of Chinese pebbles, the releases were attributed to as-manufactured defects, specifically uranium impurities present in the carbonaceous matrix of the pebbles.

During the initial stages of US fuel compact development, a significant number of fuel failures occurred, attributed to various factors, including: (a) SiC failure, (b) kernel migration (amoeba effect), (c) fission product attack on the SiC layer, (d) irradiation-induced cracking and/or debonding of the IPyC layer resulting in SiC layer fractures, and (e) graphite matrix-OPyC interaction leading to irradiation-induced OPyC failure. Under the most recent AGR fuel development program, SiC failure was observed due to the buffer shrinkage contributed to IPyC fracture, and the fracture in the IPyC layer was ascribed to stress induced by the buffer pulling away from the IPyC due to buffer shrinkage. This IPyC fracture exposed the SiC layer to concentrated chemical attack from fission products, particularly Pd, resulting in degradation throughout the entire layer. Cs was significantly retained in the kernel and the release was low in a particle with intact SiC layer. The Eu and Sr releases through intact SiC layer were modest, though at higher irradiation temperatures (up to a time-average maximum of 1360°C), the release of Eu and Sr became higher. High silver release was consistent with historical patterns, yet there was no pervasive Pd attack or corrosion of SiC observed, despite the presence of substantial amounts of Pd outside the SiC layer.

Release from the early design of the Japanese fuel were attributed to the heavy metal contamination of the matrix and no significant in-reactor fuel particle failures were observed. Later, the design was slightly modified (e.g. thicker buffer layer) for fuels to operate at higher burnups. OPyC layer crack by tensile stress was observed for one particle. For the other fuel failures, based on calculations, it was concluded that additional through-coating failure of as-fabricated SiC failed particles, and an excessive increase of internal pressure by the accelerated irradiation were identified to be the failure mechanisms. The release fraction of fission gases increases with burnup due to the irradiation-enhanced diffusion.

Fuel failures may be expected to increase when the fuel reaches a burnup of 7–8% FIMA due to the internal pressure increase and the changes in the SiC layer tangential stresses from compression to tension, depending on particularly the layer thicknesses (i.e., buffer and SiC), and when the fuel is exposed to high temperatures exceeding 1600°C (which will be discussed in the next section) due to the degradation of the SiC layer by CO attack or increased fission product attack.

7.8. FUEL FAILURES UNDER ACCIDENT CONDITIONS

Four classes of accident conditions were considered in [14]:

- Core heat up due to the loss of helium pressure and flow, resulting in the fuel achieving a peak temperature of roughly 1600°C. Note that a temperature of 1800°C was considered to envelope phenomena of interest.
- Reactivity insertion due to rod ejection accident (that was considered to be the basis for the accident conditions). A sudden reactivity addition could result in locally high fuel temperatures and/or fuel damage.
- Air ingress, resulting from the complete severing of the horizontal vessel between the reactor vessel and the power conversion system vessel, is followed by depressurization and air diffusion into the core. To mitigate the depressurization accident with air ingress, it is crucial to control or limit the oxidation rate and exposure duration. This can be achieved by delaying the entry of air into the core until the core temperatures have sufficiently reduced and by restricting the air flow rate through the core. It is essential to keep fuel temperatures and reactant flow rates within limits to prevent fuel particle damage. Core matrix degradation occurs rapidly at temperatures exceeding 1600°C. The loss of protective matrix material may expose fuel particles, leading to chemical attack and the release of retained fission products into the coolant. Sustained air ingress with high carbon reaction rates is only possible if a self-sustaining flow of air is established while the fuel is still at a relatively high temperature. Once the OPyC is breached, SiO_2 is

produced as SiC oxidizes, forming a layer on the particle that impedes further reaction. However, in an oxygen-limited atmosphere above 1300°C, SiO_2 may convert to volatile SiO, allowing the SiC to SiO reaction to continue unabated.

— Water ingress due to failure of a heat exchanger tube with reactor depressurization: the failure of the pressure boundary and depressurizing the core with a heat exchanger failure would allow water entry under low driving pressure.

Table 68 lists the behaviour of kernel and coating layers during the four postulated accidents: depressurization – core heat up, reactivity insertion, air and water ingress accidents. Table 69 through Table 74 list the parameters of interest identified for the kernel and the coating layers under the four postulated accidents. The parameters with high importance were marked with 'X' in the tables.

TABLE 68. BEHAVIOUR OF KERNEL AND COATING LAYERS DURING ACCIDENTS

Layer	Depressurization (core heat-up)	Reactivity insertion	Air ingress	Water ingress
Kernel	The temperatures in the fuel are expected to be below the melting point of the kernel material. The diffusion of fission products from the kernel will be increased at high temperatures during a heat up accident, having an impact on the integrity of the SiC, which is the primary barrier to the release of fission products. In particular, noble metals are of interest as they are known to attack the SiC layer. If the time at temperature is long enough and if SiC is corroded during normal operation, the SiC layer may fail by the additional diffusion of the fission products at higher temperatures. In addition, CO that migrates through cracks in IPyC can attack SiC and increase the stresses in the layers. As the temperature increases, the oxygen potential of the kernel may change, shifting the equilibrium. The vapour pressure of the volatile species is expected to increase.	In a reactivity insertion accident, the abrupt deposition of energy leads to a rapid increase in the kernel temperature, causing the sudden release of fission products that were trapped within the grains and crystal structure. Additionally, there is a possibility of kernel melting during this event. Sudden release of fission products will result in a sudden pressure increase or pressure pulse, over-pressurizing and breaking the coating layers, and further increasing the temperature of the coatings. Based on experiments, the energy deposition rate for fuel damage is in the range of 1000–2000 J/g.	It is expected that the kernel will behave similarly as in the case of a core heat up accident, if the fuel particle is intact; however, it can be oxidized by oxygen or CO (the oxygen is converted to CO by the large amount of carbon in the reactor system) if the kernel is exposed. In the course of oxidation, the structure of the kernel changes and it releases much of its stored fission product inventory relatively quickly. It is possible that a catalyst can increase reaction rates under certain conditions.	If the fuel particle remains intact during a water ingress accident, the behaviour of the kernel is similar to that during a core heat up event. However, when the kernel is exposed, it can undergo oxidation by water. This oxidation process alters the structure of the kernel, leading to a rapid release of a significant portion of its stored fission product inventory. The extent of this effect seems to be influenced by the burnup level. Exposed kernels are particularly vulnerable to water vapour, reacting swiftly and releasing fission products at a faster rate, while intact particles experience minimal impact. Therefore, under water ingress conditions, the predominant fuel releases are governed by the releases from exposed kernels,

TABLE 68. BEHAVIOUR OF KERNEL AND COATING LAYERS DURING ACCIDENTS

Layer	Depressurization (core heat-up)	Reactivity insertion	Air ingress	Water ingress
				especially at moderate temperatures. The behaviour is contingent on the partial pressure of water, and the potential influence of catalysts on reaction rates under specific conditions remains uncertain.
Buffer	Similar to the normal operating conditions, during a core heat up accident, the buffer layer is to provide void volume for gases in order to control the gas pressure in the fuel particle. It also provides little retention of fission products.	Since the buffer layer provides the expansion space for any gases suddenly released from the kernel during the accident, it affects the response of the coated fuel particle to a reactivity event, which could result in a short pressure pulse that could damage the coating.	For an intact particle, the buffer will behave similarly as in the case of a core heat up accident; however, once exposed, it will be oxidized, and the oxidizer will be quickly transported to the kernel. The buffer in an exposed particle provides little resistance to fission product transport. A catalyst can increase reaction rates under certain conditions.	The buffer layer functions similarly to its role in a core heat up accident for an intact particle. However, when exposed, it provides minimal resistance to the transport of fission products and may undergo slow oxidation upon exposure to water. Additionally, the buffer layer facilitates the rapid transport of water vapour to the kernel.
IPyC	Generally, the IPyC layer maintains the SiC layer in compression. Nevertheless, if the IPyC layer breaks or cracks, it can lead to elevated localized stresses in the SiC layer. Moreover, a crack in the IPyC layer can expose the SiC layer to CO, potentially causing corrosion at elevated temperatures during accidents. Full debonding of the IPyC may result in a change in the pressure loading. Decomposition or chemical reactions of the IPyC layer are not of importance for a core heat up accident. Similar to the normal operating conditions, dimensional	During the pulse, the particle pressure may increase the stresses in the IPyC layer. The particle integrity can be maintained by accommodating the particle pressure.	For an intact particle, the IPyC will behave similarly as in the case of a core heat up accident; however, once exposed (i.e., the OPyC and SiC layers are damaged), the IPyC undergoes a chemical reaction. This layer serves as the final barrier for gases (metals would have initiated diffusion through the layer), and thus, the failure of this layer results in the release of any contained	During water ingress, the considerations for the IPyC are akin to those during a core heat up event, with the additional factor of chemical attack. If the OPyC and SiC layers remain intact, the IPyC will not be exposed to a new environment. However, if the other layers become damaged or cracked, the IPyC would be exposed to water and susceptible to a chemical reaction. Given that this

TABLE 68. BEHAVIOUR OF KERNEL AND COATING LAYERS DURING ACCIDENTS

Layer	Depressurization (core heat-up)	Reactivity insertion	Air ingress	Water ingress
	stability is important for a core heat up accident. Metallic fission products will diffuse through the layer at a greater rate, but the gaseous diffusion should still be small for an intact layer.		gases and exposes the kernel to the oxidizing environment, triggering subsequent reactions within the kernel.	layer serves as the final barrier for gases (metals will have started diffusing through the layer), its failure would result in the release of any stored gases and expose the kernel to the oxidizing environment, initiating subsequent reactions within the kernel.
SiC	The integrity of the layer can be compromised by pressure, corrosion, and decomposition. Pressure can be managed through particle design, corrosion by sequestering most fission products, and controlling operating temperature to mitigate Pd attack (and CO attack in case of IPyC cracking). Decomposition can be addressed by restricting the maximum accident temperature. Above 1600°C, porosities start developing in SiC, indicating that thermal decomposition mechanisms are active, and above 2000°C, thermal decomposition of SiC is a dominate failure mechanism. Corrosion of SiC layer during normal operation can weaken the SiC layer and can cause the failure under the greater pressure at accident temperatures. Silver and radiological important fission products diffuse through the SiC, as the temperatures increase and the time at temperature becomes significant. Accident releases tend to increase with peak accident	The SiC layer has to remain undamaged despite the induced stresses. In the event of a substantial internal pressure pulse, the fracture toughness of SiC becomes crucial. The temperature of the layer during and after the incident is a significant factor, as elevated pressures and layer temperatures around or above 1600°C could lead to failure.	Under air ingress conditions, the SiC layer behaves similarly to a core heat up accident until the OPyC is subjected to chemical attack. If the OPyC weakens and fails, the stress state of the SiC may change, potentially leading to failure if the SiC has been weakened by other factors. Following OPyC loss, the SiC is exposed to air and CO. Depending on the conditions, either SiO_2 or SiO (lower oxygen content) can form. SiO_2 is non-volatile and impedes the reaction rate, as the reactant has to diffuse through it. On the other hand, SiO is gaseous and does not provide this protection. The specific details of the reacting environment determine the damage and rate of damage to the SiC layer. Thinning of	Under water ingress conditions, the SiC layer behaves similarly to a core heat up accident until the OPyC is subjected to chemical attack. As the OPyC weakens and fails, the stress state of the SiC may change, potentially leading to failure if the SiC has been weakened by other factors. The SiC would then be exposed to water and possibly some generated CO. Thinning of this layer can result in particle failure due to SiC failure as pressure vessel, and penetration of the layer will lead to the release of metallic fission products and expose the IPyC to oxidation.

TABLE 68. BEHAVIOUR OF KERNEL AND COATING LAYERS DURING ACCIDENTS

Layer	Depressurization (core heat-up)	Reactivity insertion	Air ingress	Water ingress
	temperature, and larger releases occur above 1600°C. SiC can be attacked by both fission products and fabrication impurities.		this layer can result in particle failure due to SiC failure as the pressure vessel, and penetration of the layer will lead to the release of metallic fission products and expose the IPyC to oxidation.	
OPyC	The OPyC maintains the SiC layer in compression, and the rupture or cracking of the OPyC can result in heightened local stresses in the SiC, increasing the risk of its failure. For a core heat up accident, the decomposition or chemical reactions of the OPyC layer are not critical, though dimensional stability remains important. During a core heat up accident, metallic fission products may diffuse through the layer at an accelerated rate, but the gaseous diffusion is anticipated to be minimal for an intact layer. There is a possibility that fission products trapped within the layer (intercalation) could be released at higher temperatures. Additional interaction between the OPyC and the matrix is not expected.	The OPyC layer is to support the SiC during a power pulse to keep it from exceeding its ultimate stress	In the presence of air ingress, the OPyC initially behaves similarly to its role in a core heat up scenario. However, over time, it experiences erosion. The loss of OPyC due to chemical attack has a comparable impact, with the added consequence that the SiC layer is now vulnerable to attack.	In the presence of water ingress, the OPyC initially functions similarly to its role in a core heat up scenario, followed by erosion. The loss of OPyC due to chemical attack has a comparable effect, with the additional consequence that the SiC layer is now exposed to attack.

TABLE 69. PARAMETERS, CHARACTERISTICS AND PHENOMENA OF HIGH IMPORTANCE TO ACCIDENT CONDITIONS FOR KERNEL

	Depressurization (core heat-up)	Reactivity insertion	Air ingress	Water ingress
ALL ACCIDENTS				
Maximum fuel temperature: Temperature affects chemical reactions and fission product diffusion.	X	X	X	X
Time at temperature:	X	X	X	X

TABLE 69. PARAMETERS, CHARACTERISTICS AND PHENOMENA OF HIGH IMPORTANCE TO ACCIDENT CONDITIONS FOR KERNEL

	Depressurization (core heat-up)	**Reactivity insertion**	**Air ingress**	**Water ingress**
Diffusive releases are determined based on time at temperature. Time at temperature has an impact on the amount of material reacted and fission product release.				
Thermodynamic state of fission products: Chemical form of fission products affects their diffusion/mobility. The chemical state of fission products, which is a function of temperature for oxidizing elements, determines the vapour pressure and how they migrate. The goal is to tie rare earth elements and limit CO production.	X	X	X	X
Energy transport: Determines kernel temperature and fission product diffusion from the kernel.				
Oxygen flux: Oxygen diffusion can increase at high temperatures. A substantial repositioning of oxygen has the potential to alter the oxygen potential and, consequently, the chemical species of fission products within the kernel.				
Grain growth: Grain growth is expected to occur[19] at high temperatures, affecting the fission product diffusion from kernel to the kernel-buffer interface: "Grain growth could release fission products from the grain to the grain boundary region, thus enhancing transport."				
Buffer carbon-kernel interaction: Some interaction between the kernel and the buffer carbon is possible at high temperatures: the kernel periphery may react with the buffer carbon.				
DEPRESSURIZATION AND REACTIVITY ACCIDENTS				
Condensed phase diffusion: Higher temperatures increase the diffusion rate. Diffusion of fission products is affected by how the accident phenomena progress (e.g. grain growth, occurrence of kernel melting).	X	X		
Gas phase diffusion Higher temperatures increase the diffusion rate.	X			
REACTIVITY ACCIDENTS				
Energy deposition (total):		X		

[19] In Ref. [14], it was indicated "grain growth may increase fission product transport", though the reason was not explained. In the conventional UO_2 pellets, larger grains would provide longer mean free path to fission product diffusion to the grain boundaries and hence delays the fission product release, as grain boundaries provides an easy path for fission product/gas release. However, in TRISO, if the grain size becomes as large as the kernel size, the diffusion of fission products to the kernel-buffer interface will be facilitated, and hence fission product transport is increased.

TABLE 69. PARAMETERS, CHARACTERISTICS AND PHENOMENA OF HIGH IMPORTANCE TO ACCIDENT CONDITIONS FOR KERNEL

	Depressurization (core heat-up)	Reactivity insertion	Air ingress	Water ingress
Energy deposition determines the fuel and core temperature.				
Energy deposition rate: Energy deposition rate determines the likelihood of impulsive fuel damage and system response time.				
AIR/WATER INGRESS ACCIDENTS				
The kinetics of chemical attack by air or water will govern the rates at which fission products are released.				
Chemical attack by air / water – catalysis: Some fission products or impurities can increase the reaction rate.				
Chemical attack by air / water – changes in chemical form of fission products: The availability of oxygen may elevate the oxygen potential of the kernel, leading to alterations in the chemical form of fission products, thereby influencing transport properties.				
AIR INGRESS ACCIDENTS				
Chemical attack by air – changes in graphite properties: Modifications in the characteristics of graphite (or PyC) have the potential to impact the transport of fission products or the liberation of entrapped fission products.				
Chemical attack by air – holdup reversal: Structural and chemical changes to the kernel may release stored fission products.				
Chemical attack by air – temperature distribution: The temperature of the kernel and surrounding material will affect reaction rates and the transport of fission products.				
WATER INGRESS ACCIDENTS				
Chemical attack by water – changes in kernel properties: Structural and chemical changes to the kernel may release stored fission products.				

TABLE 70. PARAMETERS, CHARACTERISTICS AND PHENOMENA OF HIGH IMPORTANCE TO ACCIDENT CONDITIONS FOR THE BUFFER LAYER

Parameters, characteristics and phenomena	Depressurization (core heat-up)	Reactivity	Air ingress	Water ingress
ALL ACCIDENTS				

TABLE 70. PARAMETERS, CHARACTERISTICS AND PHENOMENA OF HIGH IMPORTANCE TO ACCIDENT CONDITIONS FOR THE BUFFER LAYER

Parameters, characteristics and phenomena	Depressurization (core heat-up)	Reactivity	Air ingress	Water ingress
Gas phase diffusion The fission product transport through the buffer layer is expected to be high at accident temperatures. This layer offers little impedance to the transport of fission products.	X	X		
Response to kernel swelling: The buffer layer must be weak enough that it will deform to accommodate the kernel.		X		
Condensed phase diffusion The fission product transport through the buffer layer is expected to be high at accident temperatures. This layer offers little impedance to the transport of fission products.				
Maximum fuel gaseous fission product uptake: The buffer layer is designed to have sufficient free volume to accommodate the fission products, and to withstand the pressure.				
Layer oxidation: Any oxygen released from the kernel will oxidize a small portion of the buffer. In addition, in the case of water ingress, the buffer will be oxidized from the outside if the particle is cracked or broken.				
Thermal gradient: The gradient may influence the transport of fission products to the other layers. 1) During core heat up conditions, the gradient across the buffer is much less than normal operation. Thermal gradient can cause carbon transfer from side to other side of the kernel. 2) During reactivity insertion accident, gradients can be high for a high energy pulse.				
Irradiation and thermal shrinkage: Irradiation and thermal shrinkage can cause cracks in the buffer layer. Due to these cracks, fission products can locally accumulate (at higher temperatures) on the IPyC.				
AIR/WATER INGRESS ACCIDENTS				
Chemical attack by air/water- kinetics: The reaction rates will control the fission product release rate.				

TABLE 70. PARAMETERS, CHARACTERISTICS AND PHENOMENA OF HIGH IMPORTANCE TO ACCIDENT CONDITIONS FOR THE BUFFER LAYER

Parameters, characteristics and phenomena	**Depressurization (core heat-up)**	**Reactivity**	**Air ingress**	**Water ingress**
Chemical attack by air/water – changes in chemical form of fission products: The variation in chemical form of the fission products may change their transport.				
Chemical attack by air/water – changes in graphite properties: Changes in the graphite (or PyC) properties may influence the fission product transport and release.				
Chemical attack by air/water – holdup reversal: Fission products stored in the buffer may be released quickly if the buffer structure is attacked.				
Chemical attack by air/water – temperature distribution: The temperature determines reaction rates and transport.				
AIR INGRESS ACCIDENTS				
Chemical attack by air- catalysis: Some impurities or fission products may increase the reaction rate.				

TABLE 71. PARAMETERS, CHARACTERISTICS AND PHENOMENA OF HIGH IMPORTANCE TO ACCIDENT CONDITIONS FOR THE IPYC LAYER

Parameters, characteristics and phenomena	**Depressurization (core heat-up)**	**Reactivity**	**Air ingress**	**Water ingress**
ALL ACCIDENTS				
Gas phase diffusion Gaseous fission products are generally maintained by IPyC. High local accident temperatures and chemical attack can increase the diffusion rate.	X	X	X	
Cracking: Cracks can change the stress distribution. In the case of air/water ingress, if other layers are broken, cracking will allow the oxidizer to the kernel.	X		X	X
Pressure loading (CO): High CO production will result in high pressures. High pressures can challenge IPyC layer. The IPyC can help keep the SiC in compression if the bonding between layers is strong.	X	X		X

TABLE 71. PARAMETERS, CHARACTERISTICS AND PHENOMENA OF HIGH IMPORTANCE TO ACCIDENT CONDITIONS FOR THE IPYC LAYER

Parameters, characteristics and phenomena	**Depressurization (core heat-up)**	**Reactivity**	**Air ingress**	**Water ingress**
Pressure loading (fission products): High pressure could challenge this layer. The IPyC products can help keep the SiC in tension if the bonding between layers is strong.		X		
Layer oxidation: In some cases, internal oxidation of the layer can be significant. IPyC cracks can allow CO to the SiC, resulting in corrosion of SiC (inside out). In the case of air/water ingress accident, cracks in the OPyC and SiC will allow oxidation of the IPyC (outside in).			X	X
Condensed phase diffusion Metallic fission products are not well retained. In addition, high local accident temperatures and chemical attack can increase the diffusion rate.				
Stress state (compression/tension): Breaking or debonding of IPyC results in changes in stress distribution of the particle, and hence failures.				
Intercalation/trapping: The IPyC can accumulate fission products during normal operation that can be released during an accident. Chemical attack of the IPyC may allow release of trapped fission products.				
AIR/WATER INGRESS ACCIDENTS				
Chemical attack by air/water- kinetics: The reaction rates will control the fission product release rate.			X	X
Chemical attack by air/water – changes in chemical form of fission products: The changes in chemical form of the fission products may change transport properties.			X	X
Chemical attack by air/water – temperature distribution: The temperature of the IPyC and surrounding materials determines reaction rates and transport.			X	X
Chemical attack by air/water - catalysis: Some impurities or fission products may increase the reaction rate.				

TABLE 71. PARAMETERS, CHARACTERISTICS AND PHENOMENA OF HIGH IMPORTANCE TO ACCIDENT CONDITIONS FOR THE IPYC LAYER

Parameters, characteristics and phenomena	**Depressurization (core heat-up)**	**Reactivity**	**Air ingress**	**Water ingress**
Chemical attack by air/water – changes in graphite properties: Changes in the graphite (or PyC) properties may influence the transport or release of fission products.				
Chemical attack by air/water – holdup reversal: Serious damage to the IPyC will allow access to the buffer and its relatively large inventory of fission products.				

TABLE 72. PARAMETERS, CHARACTERISTICS AND PHENOMENA OF HIGH IMPORTANCE TO ACCIDENT CONDITIONS FOR THE SIC LAYER

Parameters, characteristics and phenomena	**Depressurization (core heat-up)**	**Reactivity insertion**	**Air ingress**	**Water ingress**
ALL ACCIDENTS				
Gas phase diffusion Fission product diffusion is low through the SiC even at accident temperatures.	X		X	X
Condensed phase diffusion: Silver can diffuse through SiC at temperatures above 1100°C.				
Thermal deterioration or decomposition: Above 1600-1800°C, SiC begins to decompose and its ability to retain fission products is greatly reduced.				
Fission product corrosion: Some fission products may migrate into the SiC and corrode it. This corrosion process is a function of temperature and can begin during normal operation. Pd is of main concern, though other fission products can also cause corrosion of SiC layer.				
Heavy metal diffusion: The diffusion of fissile material can change the core configuration. A concern is major particle cracking and the expulsion of the kernel material during the pulse.				
Layer Oxidation: Attack of the SiC layer by CO can be due to a failed OPyC at high temperatures or due to IPyC failure (e.g. cracks at high temperatures) (inside out). The formation of either SiO or SiO_2 can determine the rate at which the SiC is eroded.				

TABLE 72. PARAMETERS, CHARACTERISTICS AND PHENOMENA OF HIGH IMPORTANCE TO ACCIDENT CONDITIONS FOR THE SIC LAYER

Parameters, characteristics and phenomena	Depressurization (core heat-up)	Reactivity insertion	Air ingress	Water ingress
Fission gas release through undetected defects: In order to release gases, both PyCs have to be cracked in addition to a failed SiC. The SiC can be attacked by fission products and fabrication impurities. Undetected defective SiC could have poor accident behaviour not apparent during normal operations.	X			
Fission gas release through failures (e.g. cracking): Failure of the SiC will allow the release of metallic fission products even with intact PyCs.	X	X	X	X
Thermodynamics of the SiC fission product system: At the higher temperatures of interest, SiC is just stable to oxidation in its intact particle environment, though SiC is known to be attacked by some noble metals. At higher burnups and temperatures, other chemical concerns may arise.	X			
Sintering: Accident temperatures can result in microstructural changes: change the crystal or grain structure of the SiC layer.				
AIR/WATER INGRESS ACCIDENTS				
Chemical attack by air/water- kinetics: The reaction rates will determine the rate of fission product release.			X	
Chemical attack by air/water – changes in chemical form of fission products: The variation in chemical form of the fission products may change their transport.			X	X
Chemical attack by air/water – temperature distribution: The temperature of the SiC and surrounding materials determines reaction rates and transport.			X	X
Chemical attack by air/water - catalysis: Some impurities or fission products may increase the reaction rate.				
Chemical attack by air/water – changes in SiC properties: Changes in the SiC properties may influence the transport or release of fission products.				

TABLE 72. PARAMETERS, CHARACTERISTICS AND PHENOMENA OF HIGH IMPORTANCE TO ACCIDENT CONDITIONS FOR THE SIC LAYER

Parameters, characteristics and phenomena	Depressurization (core heat-up)	Reactivity insertion	Air ingress	Water ingress
Chemical attack by air/water – holdup reversal: Significant harm to the SiC layer will permit entry to the IPyC layer, which exhibits relatively poor retention of metallic fission products.				

TABLE 73. PARAMETERS, CHARACTERISTICS AND PHENOMENA OF HIGH IMPORTANCE TO ACCIDENT CONDITIONS FOR THE OPYC LAYER

Parameters, characteristics and phenomena	Depressurization (core heat-up)	Reactivity insertion	Air ingress	Water ingress
ALL ACCIDENTS				
Gas phase diffusion: Gaseous fission product diffusion is low.	X		X	
Condensed phase diffusion: Metallic fission product diffusion is high.				
Layer Oxidation: The OPyC is the most exposed layer. It would be the first layer subjected to oxidation (sound particle). Cracked layers may allow CO to the OPyC or coolant impurities may attack it.			X	X
Stress state (compression/tension): Debondig of OPyC from SiC results in changes in stress distribution but the effect is not as significant as that of IPyC debonding from SiC, though failure will increase the likelihood of SiC failure.				
Cracking: Cracks can lead to higher probability of SiC failure. Gaseous will be released if the other layers fail.	X			
Trapping: Chemical attack of the OPyC may allow release of trapped fission products.				
Intercalation: Chemical attack of the OPyC may allow release of trapped fission products.				
Fission gas release through failures (e.g. cracking): Failure of the SiC will allow the release of metallic fission products even with intact PyCs.	X		X	

TABLE 73. PARAMETERS, CHARACTERISTICS AND PHENOMENA OF HIGH IMPORTANCE TO ACCIDENT CONDITIONS FOR THE OPYC LAYER

Parameters, characteristics and phenomena	Depressurization (core heat-up)	Reactivity insertion	Air ingress	Water ingress
Thermodynamics of the SiC fission product system: At the higher temperatures of interest, SiC is just stable to oxidation in its intact particle environment, though SiC can be attacked by some noble metals.	X			
Sintering: Accident temperatures can result in changes of the crystal or grain structure of the SiC layer.				
AIR/WATER INGRESS ACCIDENTS				
Chemical attack by air/water- kinetics: The reaction rates will control the fission product release rate.			X	X
Chemical attack by air/water – changes in chemical form of fission products: Changes in chemical form of the fission products can change transport properties.			X	X
Chemical attack by air/water – temperature distribution: The temperature of the SiC and surrounding materials determines reaction rates and transport.			X	X
Chemical attack by air/water - catalysis: Some impurities or fission products may increase the reaction rate.				
Chemical attack by air/water – changes in graphite properties: Changes in the graphite properties may influence the transport or release of fission products.				
Chemical attack by air/water – holdup reversal: Attack of the OPyC may result in the release of any stored fission products.				

TABLE 74. PARAMETERS, CHARACTERISTICS AND PHENOMENA OF HIGH IMPORTANCE TO ACCIDENT CONDITIONS FOR THE FUEL ELEMENT

Parameters, characteristics and phenomena	Depressurization (core heat-up)	Reactivity	Air ingress	Water ingress
ALL ACCIDENTS				
Irradiation history: Matrix behaviour (shrinkage) is related to time temperature history, and particularly temperatures and fast fluence, changing its chemical reactivity.	X			

TABLE 74. PARAMETERS, CHARACTERISTICS AND PHENOMENA OF HIGH IMPORTANCE TO ACCIDENT CONDITIONS FOR THE FUEL ELEMENT

Parameters, characteristics and phenomena	Depressurization (core heat-up)	Reactivity	Air ingress	Water ingress
Gas phase diffusion: In the matrix, gas diffusion is high. Chemical attack can increase the transport of gases outside the fuel element.		X	X	X
Condensed phase diffusion: Metallic diffusion varies, with less volatile metals exhibiting a high degree of retention in the matrix material.				
Transport of metallic fission products through fuel element-chemical form: The chemical form of fission product controls its transport behaviour. Significant retention of metallic fission products has been observed.	X		X	X
AIR/WATER INGRESS ACCIDENTS				
Chemical attack by air/water- kinetics: The reaction rates will determine the rate of fission product release.			X	X
Chemical attack by air/water – changes in chemical form of fission products: Changes in chemical form of the fission products can change transport properties.			X	X
Chemical attack by air/water – temperature distribution: The temperature of the SiC and surrounding materials determines reaction rates and transport.			X	X
Chemical attack by air/water - catalysis: Some impurities or fission products may increase the reaction rate.				
Chemical attack by air/water – changes in graphite properties: Changes in the graphite properties can affect the transport and release of fission products.			X	
Chemical attack by air/water – holdup reversal: Damage of the matrix material can enhance the release of fission products.				

7.8.1. Summary of parameters of high importance identified for accident conditions

Parameters, characteristics and phenomena were identified to be of high importance during depressurization/ core heat up accident for kernel, coatings and the fuel element, as listed in Table 75. In addition, several parameters were identified by the majority (but not by all the experts) as of high importance, including [14]:

- For the kernel, buffer carbon-kernel interaction below and above 1600°C;
- For the buffer layer, layer oxidation below and above 1600°C;

— For the SiC layer, thermal deterioration or decomposition of SiC layer and fission product corrosion of the SiC layer above 1600°C.

A consensus on their importance of several other parameters for the depressurization/ core heat up accident was not reached; these parameters include [14]:

— For the kernel, oxygen flux;
— For the buffer layer, condensed phase diffusion and thermal gradient.

A total of 12 parameters, characteristics and phenomena were identified to be of high importance during reactivity initiated accident for kernel, coatings and the fuel element, as listed in Table 76. Several parameters were identified by the majority (but not by all the experts) as of high importance, including [14]:

— For the kernel, gas phase diffusion, energy transport – conduction within the kernel;
— For the buffer layer, maximum fuel gaseous fission product uptake;
— For the fuel element, condensed phase diffusion.

A consensus on the importance of several other parameters for the reactivity initiated accident was not reached; these parameters include [14]:

— For the kernel, buffer carbon-kernel interaction;
— For the buffer layer, condensed phase diffusion and thermal gradient;
— For the IPyC layer, stress state (compression/tension);
— For the OPyC layer, gas phase diffusion.

TABLE 75. DEPRESSURIZATION/HEATUP ACCIDENT PARAMETERS OF HIGH IMPORTANCE

Layer	Parameter
Kernel	Condensed phase diffusion
	Gas phase diffusion
	Maximum fuel temperature
	Time at temperature
	Energy transport
	Thermodynamic state of fission products
Buffer	Gas phase diffusion
IPyC	Cracking
	Gas phase diffusion
	Pressure loading
SiC	Fission product release through failures (e.g. cracking)
	Fission product release through undetected failures
	Gas phase diffusion
	Thermodynamics of the SiC fission product system
OPyC	Cracking
	Gas phase diffusion
Fuel element	Irradiation history
	Transport of metallic fission products – chemical form

TABLE 76. REACTIVITY INSERTION ACCIDENT PARAMETERS OF HIGH IMPORTANCE

Layer	Parameter
Kernel	Condensed phase diffusion Maximum fuel temperature Time at temperature Energy deposition (total) Thermodynamic state of fission products
Buffer	Gas phase diffusion Response to kernel swelling
IPyC	Gas phase diffusion Pressure loading (fission products) Pressure loading (CO)
SiC	Fission product release through failures (e.g. cracking)
OPyC	None
Fuel element	Gas phase diffusion

Parameters, characteristics and phenomena were identified to be of high importance during air ingress accident for kernel, coatings and the fuel element, as listed in Table 77. Several parameters were identified by the majority (but not by all the experts) as of high importance, including [14]:

— For the kernel, buffer carbon-kernel interaction;
— For the buffer layer, layer oxidation;
— For the SiC layer, layer oxidation;
— For the OPyC layer, cracking;
— For the fuel element, condensed phase diffusion, chemical attack by air – catalysis.

A consensus on their importance of several other parameters for the air ingress accident was not reached; these parameters include [14]:

— For the kernel, chemical attack by air – kinetics, chemical attack by air – change in chemical form of fission products;
— For the buffer layer, gap phase diffusion, condensed phase diffusion, maximum fuel gaseous fission product uptake, thermal gradient, chemical attack by air – change in chemical form of fission products;
— For the IPyC layer, pressure loading (fission products), pressure loading (CO), stress state (compression/tension);
— For the SiC layer, fission product corrosion;
— For the fuel element, chemical attack by air – holdup reversals.

TABLE 77. AIR INGRESS ACCIDENT PARAMETERS OF HIGH IMPORTANCE

Layer	Parameter
Kernel	Maximum fuel temperature Time at temperature Thermodynamic state of fission products
Buffer	None
IPyC	Gas phase diffusion Layer oxidation Cracking Chemical attack by air: kinetics Chemical attack by air: changes in chemical form of fission products Chemical attack by air: temperature distribution
SiC	Gas phase diffusion Fission product release through failures (e.g. cracking) Chemical attack by air: temperature distribution Chemical attack by air: kinetics
OPyC	Gas phase diffusion Layer oxidation Chemical attack by air: kinetics Chemical attack by air: temperature distribution Chemical attack by air: changes in chemical form of fission products
Fuel element	Gas phase diffusion Transport of metallic fission products through fuel element – chemical form Chemical attack by air: kinetics Chemical attack by air: temperature distribution Chemical attack by air: changes in chemical form of fission products Chemical attack by air: changes in graphite properties

A total of twenty-two parameters, characteristics and phenomena were identified to be of high importance during water ingress accident for kernel, coatings and the fuel element, as listed in Table 78. Several parameters were identified by the majority (but not by all the experts) as of high importance, including [14]:

— For the kernel, buffer carbon-kernel interaction;
— For the buffer layer, layer oxidation;
— For the IPyC layer, gas phase diffusion;
— For the SiC layer, chemical attack by water – kinetics;
— For the OPyC layer, cracking;
— For the fuel element, condensed phase diffusion, chemical attack by water – changes in graphite properties.

A consensus on their importance of several other parameters for the water ingress accident was not reached; these parameters include [14]:

— For the kernel, chemical attack by water – kinetics, chemical attack by water – change in chemical form of fission products;
— For the buffer layer, gas phase diffusion, condensed phase diffusion, maximum fuel gaseous fission product uptake, thermal gradient;
— For the IPyC layer, pressure loading (fission products), stress state (compression/tension), chemical attack by water – changes in graphite properties;
— For the SiC layer, layer oxidation;
— For the OPyC layer, chemical attack by water – changes in graphite properties;
— For the fuel element, chemical attack by air – holdup reversals.

TABLE 78. WATER INGRESS ACCIDENT PARAMETERS OF HIGH IMPORTANCE

Layer	Parameter
Kernel	Maximum fuel temperature Time at temperature Thermodynamic state of fission products
Buffer	None
IPyC	Pressure loading (CO) Layer oxidation Cracking Chemical attack by water: kinetics Chemical attack by water: changes in chemical form of fission products Chemical attack by water: temperature distribution
SiC	Gas phase diffusion Fission product release through failures (e.g. cracking) Chemical attack by water: temperature distribution Chemical attack by air: changes in chemical form of fission products
OPyC	Layer oxidation Chemical attack by water: kinetics Chemical attack by water: temperature distribution Chemical attack by water: changes in chemical form of fission products
Fuel element	Gas phase diffusion Transport of metallic fission products through fuel element – chemical form Chemical attack by water: kinetics Chemical attack by water: temperature distribution Chemical attack by water: changes in chemical form of fission products

7.8.2. Fuel failures under accident conditions (corresponding to core heat up)

Figure 121 [6] shows the fractional release of ^{85}Kr of German UO_2 fuel at temperatures up to 2100°C [6]. No TRISO failures were observed at the temperature of 1600°C, however, TRISO failures occurred after short periods at 1800°C. At the test temperature of 1600°C (see Fig. 122 [6]), no TRISO failures were observed at burnups lower than 10% FIMA, but failures were observed at burnups at ~14% FIMA. Cs release as a function of burnup and temperature is shown in Fig. 123: at 1600°C and burnup <10% FIMA, Cs release remained relatively low, however, increasing burnup and temperature increases SiC layer degradation and Cs release.

Prior to 2000 in US, selected TRISO fuels irradiated in the HFIR at ORNL were subjected to heating tests. HRB-15B capsule included various fuel types of UO_2, UC_2, UCO, UO_2*(1), and UO_2*(2) fuel particles. The irradiation was performed for 169 EFPDs at an accelerated condition up to fuel burnup of 21−25 % FIMA. The PIE of irradiated fuel particles revealed some SiC layer was cracked in each TRISO coated fuel type but mostly in the UCO particles [3,4,10]. Subsequent heating tests were conducted at temperatures up to 1500°C. Ten particles of each fuel type were annealed for 11,866 h at 1500°C [10]. Caesium and silver were released from the UO_2 and UC_2 fuel particles as shown in Figure 124 (a) and (b). These releases exhibited a maximum for the particles with SiC of a columnar grain structure as shown in Fig. 104. The fashion of releasing caesium and silver from the UO_2 fuel particles supported a diffusional release mechanism through the SiC layer.

Individual UO_2 TRISO fuel particles deconsolidated from a compact of Japanese manufacture were heated at 1700°C for 270 h and at 1800°C for 222 h at ORNL [31]. The fuel underwent irradiation at HFIR in the HRB-22 capsule, reaching a burnup of 4.8% FIMA and a fast neutron fluence of 2.1×10^{25} n/m^2 over 89 EFPDs, with a time-averaged maximum temperature of 1100°C. Releases of silver, caesium, europium, and krypton were monitored over time, as illustrated in Figures 125 and 126 for batches of 25 particles at each temperature [10] The krypton release in Fig. 126 showed that one particle failed early in the heating with a release of antimony.

As shown in Table 67, the design value of TRISO failure fraction for fuels developed for NGNP during accidents is $\leq 6.0 \times 10^{-4}$ [6]. Based on annealing testing of US AGR-1 and AGR-2 fuels in the 1600–1800°C range, no TRISO failures in a fuel with UCO kernel were observed in any of the 1600°C annealing tests. Combining the results gives a total TRISO failure fraction in a fuel with UCO kernel of $\leq 6.6 \times 10^{-5}$ [108]. This is a factor of 9 lower than typical reactor design specifications for allowable failures during 1600°C accidents (6×10^{-4}). Fractional release of ^{134}Cs from compacts containing only intact particles at 1600°C was $< 6 \times 10^{-5}$. When a SiC layer in a particle failed, some of the Cs from that particle was released. TRISO failure fraction in a fuel with UCO kernel for the tests at 1800°C after the AGR-1 and AGR-2 irradiations was $\leq 3.0 \times 10^{-4}$. Conversely, the combined (AGR-1 + AGR-2) SiC failure fractions in a fuel with UCO kernel were $\leq 1.7 \times 10^{-4}$ and $\leq 1.3 \times 10^{-3}$ during annealing testing at 1600°C and 1800°C, respectively [108]. In spite of the much higher frequency of SiC failure observed for the irradiated AGR-2 UO_2 fuel compacts during annealing tests relative to UCO, no TRISO failures were observed in the three annealing tests (i.e., tests at 1600, 1700 and 1800°C) [108]. Releases of Ag, Sr, and Eu observed at 1600 and 1700°C is ascribed to the diffusion of these fission products in the fuel matrix during irradiation. Subsequently, these elements are released from the matrix upon exposure to high-temperature heating [108]. In summary, the findings suggest a minimal incremental release of annealing-critical fission products under accident conditions.

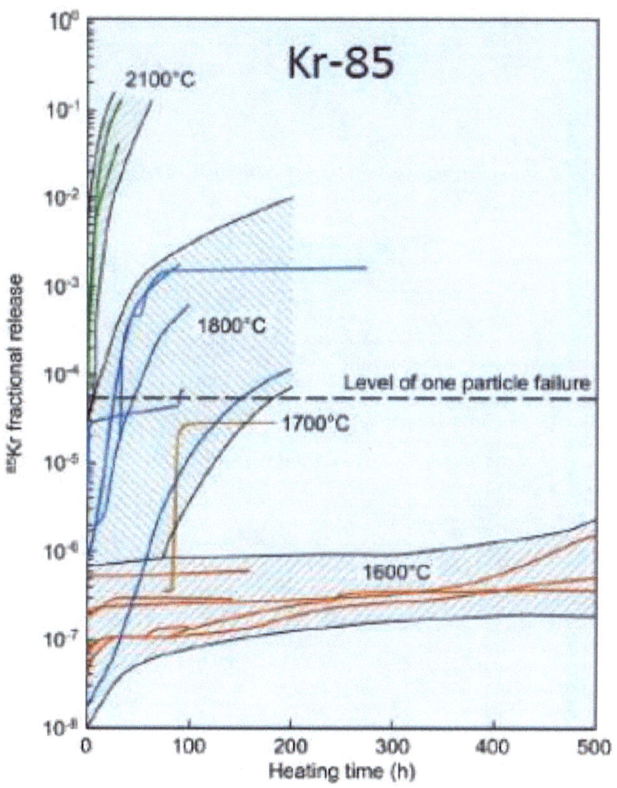

FIG. 121. Temperature effect on ^{85}Kr fractional release of German UO_2 fuel (courtesy of INL).

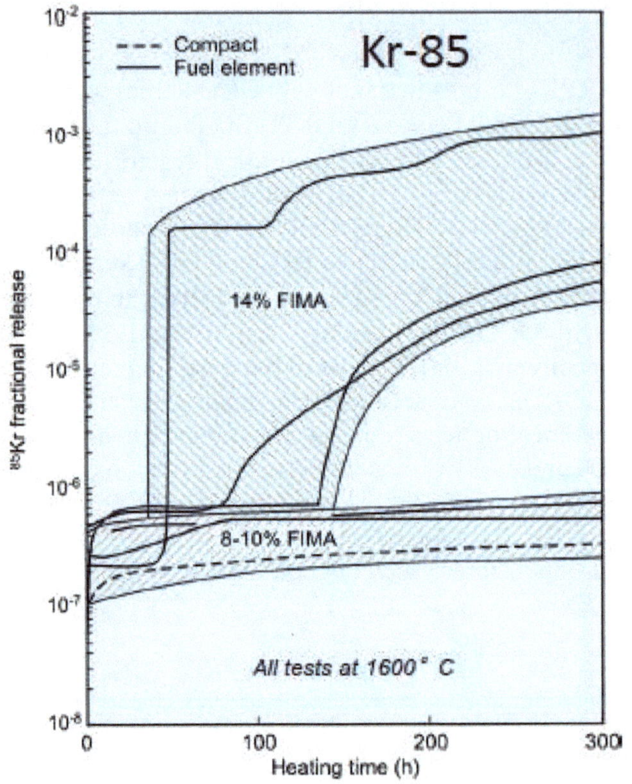

FIG. 122. Burnup effect on ^{85}Kr fractional release of the German UO_2 fuel at the temperature of 1600°C (courtesy of INL).

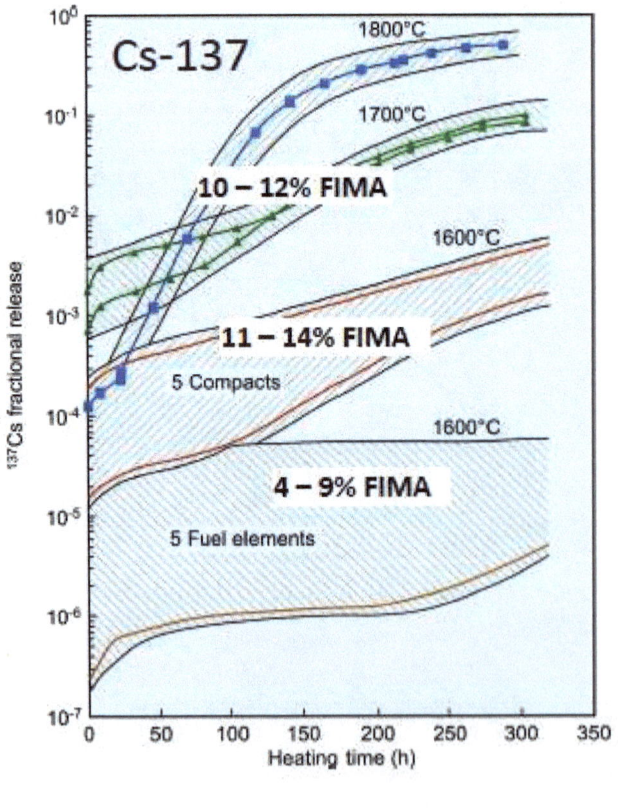

FIG. 123. Burnup and temperature effects on ^{137}Cs fractional release of German UO_2 fuel (courtesy of INL).

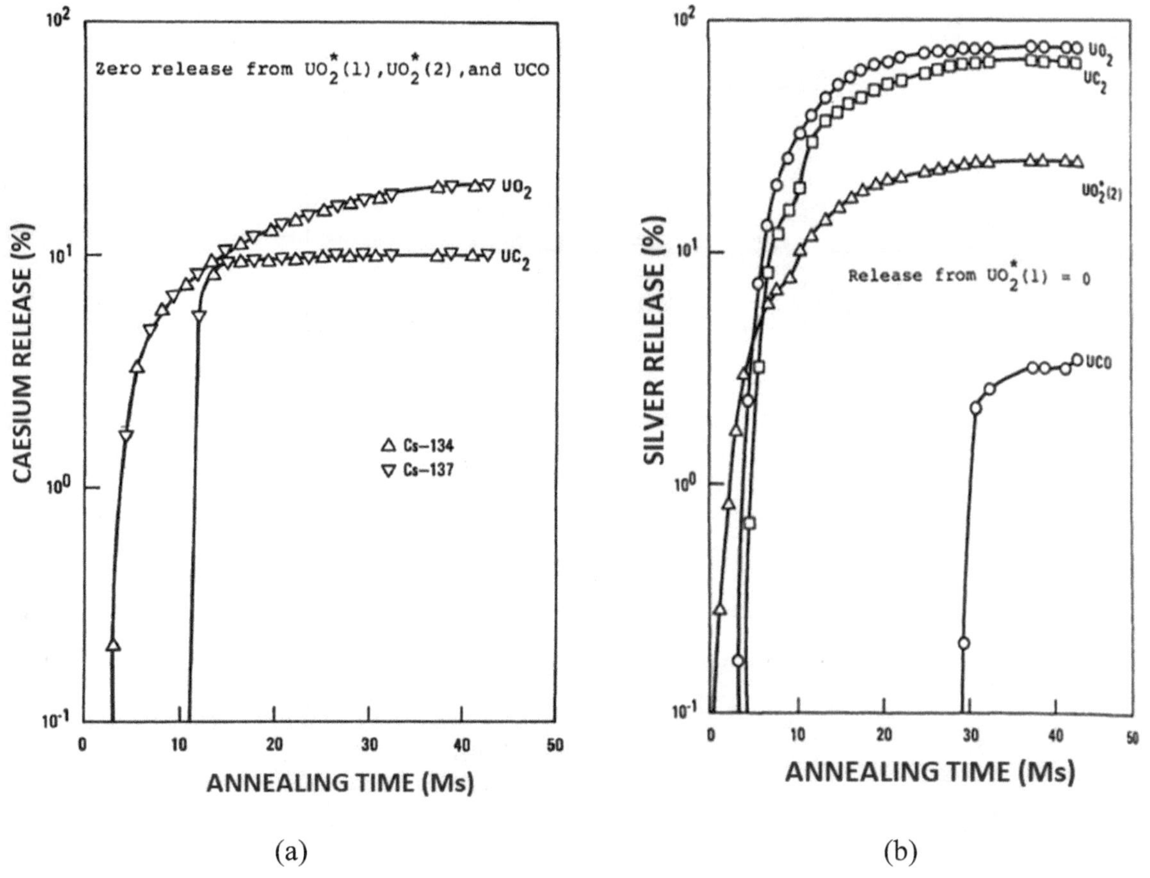

FIG. 124. Release of Cs (a) and Ag (b) from various types of TRISO-coated fuel particles at 1500 ℃ (courtesy of INL).

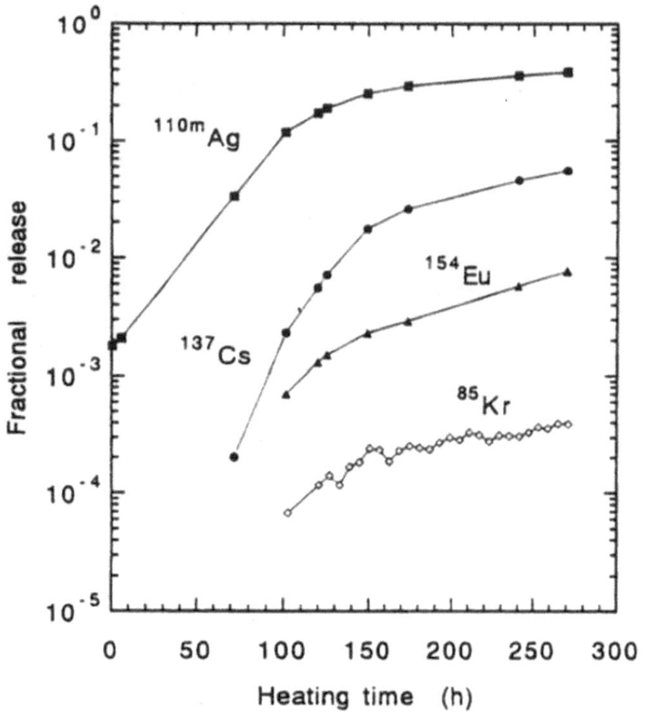

FIG.125. Time-dependent fractional releases of fission products during the ACT3 annealing test at 1700 ℃ for 270 h, obtained by the online measurements of fission gas release and intermittent measurements of metallic fission product release (courtesy of INL).

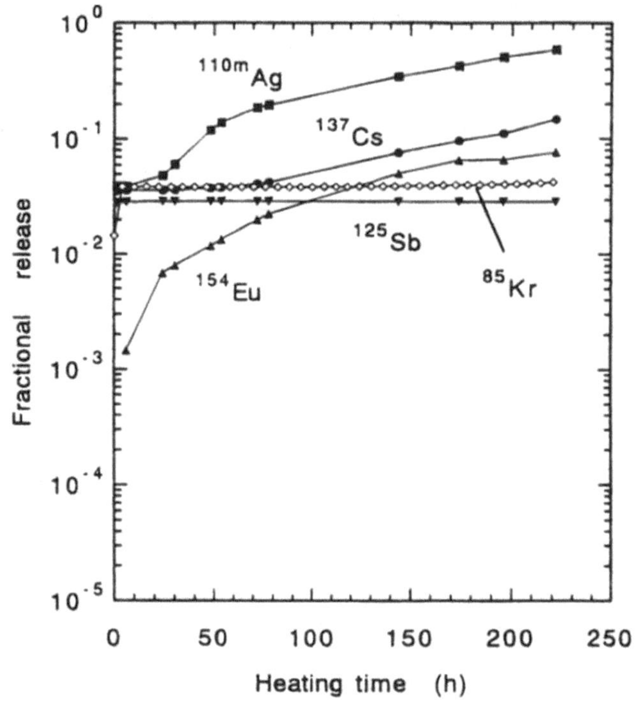

FIG. 126. Time-dependent fractional releases of fission products during the ACT4 heating test at 1800 °C for 222 h, obtained by the online measurements of fission gas release and intermittent measurements of metallic fission product release (courtesy of INL).

7.8.3. Fuel failures under accident conditions (corresponding to reactivity insertion accident)

A rapid escalation in reactor local or core-wide power resulting from a surge in reactivity can stem from malfunctions in the control system, inadvertent control rod ejection, or sudden alterations in the core's internal configuration, such as pebble compaction in a pebble bed reactor [20]. Notably, a pebble bed reactor exhibits low excess reactivity due to continual fuel cycling, while the prismatic core design initially features excess reactivity and burnable poisons that diminish as the core approaches the end of its operational life. In the event of a sudden power pulse, there is a risk of damaging the fuel element and fuel particles, potentially leading to a substantial release of fission products from the fuel. For pulse durations on the order of seconds, it is assumed that a portion of the fuel may fail through the cracking of PyC and SiC layers, while the kernel and buffer layers remain intact [20].

The UO_2 kernel melting and central vaporization for coated particles with an energy deposition of 2300 J/g UO_2 were observed for the Japanese fuel compacts after pulse testing (10–30 ms): uranium vapour penetrated the cracks in the coating layers, though at lower energies, the particle remained almost unchanged, but layer cracking was observed [20].

- Two sets of pulse tests (i.e., short pulses and long pulses) were performed by in a Russian research reactor [20]: a pulse of duration (half width peak) of 0.7 seconds with an energy deposition of 26,000 J/g UO_2 into spherical fuel elements resulted in no failure (i.e., intact) of fuel elements.
- A pulse lasting 30 s, with an energy deposition of 90,000 J/g UO_2, led to the failure of the coated particles and the fragmentation of the fuel element spheres. In addition, Russian scientists performed a set of short pulse (1–2 ms) tests with energy deposition of 100 to 1700 J/g UO_2: they observed coating cracking at < 1050 J/g UO_2 and kernel restructuring at > 1300 J/g UO_2.

A distinct investigation aimed at identifying the necessity for in-pile testing data to support the licensing of TRISO fuel in non-HTGR applications (e.g. LWRs, microreactors, nuclear thermal propulsion, and salt-cooled reactors) revealed that temperature transients might occur much faster than in HTGR DLOFC accidents (e.g. 0.05°C/s in a HTGR DLOFC versus up to 1000°C/s in a non-high temperature reactor using TRISO fuel) [156]. Notably, significant fuel particle failures were observed at energy depositions exceeding 1400 J/g fuel due to kernel melting, while fuel compact graphite matrix failures were observed above 2300 J/g fuel [156]. It is essential to recognize that failure thresholds and mechanisms may differ for particles with fuel materials distinct from UO_2 (e.g. UCO and UN), particles with different irradiation histories, and particles with different structural configurations (layers and relative thickness). Although historical tests have predominantly identified kernel melting as the primary failure mechanism, it is acknowledged that other mechanisms, such as thermal stresses between coating layers during rapid temperature increases, leading to crack propagation and delamination of coating layers, may also be possible.

The failure fraction of particles was generally observed to be ~0.01 at an energy deposition of 1000 J/g UO_2 and 1.0 at 1700 J/g UO_2 (the trend of failure fraction as a function of energy deposition is shown in Fig. 127) (refer to Ref. [156] for details). At the highest energy deposition, ~2300 J/g UO_2, the fuel compact graphite matrix exhibited cracking on the surface and failed due to stresses within the compact. Recent in-pile tests using fresh UO_2 kernels (7.26% ^{235}U enrichment) demonstrated consistent failure of coated fuel particles above 1400 J/g UO_2, coinciding with the point at which the kernels melted.

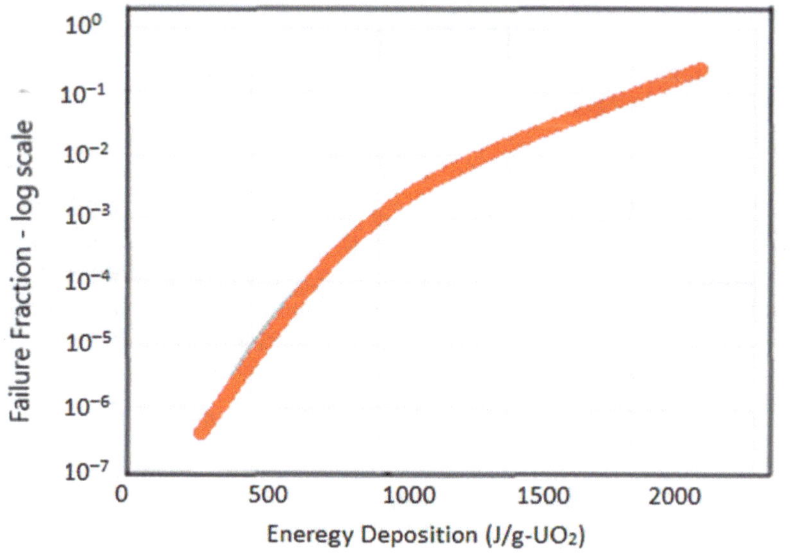

FIG. 127. Failure fraction of fuel particles as a function of energy deposition.

8. FUEL MODELLING AND CODE DEVELOPMENTS

8.1. MODELLING OF COATED PARTICLE FUEL PERFORMANCE

This section describes important fuel models that are necessary in evaluating the integrity and performance of coated particle fuel for use in high-temperature gas cooled SMRs.

8.1.1. Fuel burnup and depletion

Fuel burnup and depletion analysis for a reactor core provides information on the time-dependent variation of fuel burnup, flux and fluence, and the production and depletion of fission product nuclei. The fuel burnup and depletion can be evaluated using various reactor core analysis computer codes such as MCNP-MOCUP-ORIGEN2 [177–179] and MCCARD [180]. The calculated fuel burnup and fast fluence are used as input to material property correlations of a fuel material in order to determine the specific values of the material properties during the operation of the reactor. The nuclide inventory is used as a fission product source term for design and annealing analyses. A chemical equilibrium calculation using the nuclide inventory gives the fission yields of gas species over time that are used in the evaluation of gas pressure build-up in a coated particle.

8.1.2. Pressure build-up in a coated particle

Excess oxygen generated in a kernel is a major contributor to the build-up of the internal gas pressure of a particle. There exist some models which predicts the free oxygen release from an oxide kernel [181–188].

A thermochemical equilibrium calculation using the nuclide inventory generated in the kernel and buffer can produce the yields of gas species in a particle. This method is applicable to any type of kernel i.e., nitride, oxy-carbide, and even oxide kernel. The Booth model is capable of estimating the release quantity of gas atoms from the kernel into the void volume within the kernel and the buffer [189]. The void volume, constituting the open-pore space in the kernel and the buffer, undergoes changes with burnup due to solid and gaseous swelling of the kernel. This swelling leads to densification of the buffer, resulting in a reduction of the void volume. The gas pressure within the void volume can be approximated using the ideal gas law or other relevant equations of state (e.g. [190]). Figure 128 shows an example of the calculation process of the internal gas pressure (that employs the HSC Chemistry™ Software [191] and the Microsoft Excel).

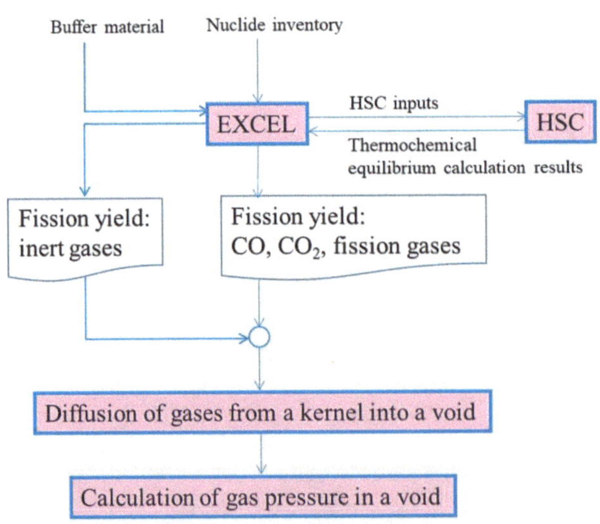

FIG. 128. Calculation process of gas pressure in a particle.

8.1.3. Failure of pressure-bearing layers

Gas pressure built up in a buffer of a particle develops tangential stresses on the coating layers of the particle. The calculated stresses are used to determine the integrity of pressure-bearing layers of the particle. A coating layer is considered failed if the tangential stress acting on the layer exceeds the ultimate tensile strength (UTS) of the material. During irradiation, a coating layer experiences elastic, plastic, thermal, creep deformation together with irradiation-induced dimensional change [192,193]. Gas pressure due to fission gases, CO, and CO_2 applies to the inner surface of the innermost layer, whilst an ambient pressure acts on the outer surface of the outermost layer by a matrix material surrounding the particle.

Three mathematical formulas are used in expressing the mechanical state of coating layers: a stress-strain relation, a strain-displacement relationship, and an equilibrium equation.

The stress-strain relation for a shell-like coating layer at fluence f can be represented in the following matrix equation:

$$[C]\{\sigma\} = \{\varepsilon\} - \{\varepsilon^{pl}\} - \{\varepsilon^{th}\} - \{\varepsilon^{sw}\} - \{\varepsilon^{cr}\} \tag{22}$$

where

$$\{\sigma\}^T = \{\sigma_r \quad \sigma_\theta\}$$

$$\{\varepsilon\}^T = \{\varepsilon_r \quad \varepsilon_\theta\}$$

$$\{\varepsilon^{pl}\}^T = \{\varepsilon_r^{pl} \quad \varepsilon_\theta^{pl}\}$$

$$\{\varepsilon^{th}\}^T = \{\varepsilon_r^{th} \quad \varepsilon_\theta^{th}\}$$

$$\{\varepsilon^{sw}\}^T = \{\varepsilon_r^{sw} \quad \varepsilon_\theta^{sw}\}$$

$$\{\varepsilon^{cr}\}^T = \{\varepsilon_r^{cr} \quad \varepsilon_\theta^{cr}\}$$

$$[C] = \frac{1}{E}\begin{bmatrix} 1 & -2\nu \\ -\nu & 1-\nu \end{bmatrix} \tag{23}$$

σ is the stress (MPa),

ε is the total strain (dimensionless),

ε^{pl} is the plastic strain (dimensionless),

ε^{th} is the thermal strain (dimensionless),

ε^{sw} is the irradiation-induced dimensional change (dimensionless),

ε^{cr} is the irradiation creep strain (dimensionless),

E is the modulus of elasticity (MPa),

ν is the Poisson's ratio (dimensionless),

the subscripts r and θ indicate the radial and circumferential directions, respectively.

The irradiation creep strain vector is given as follows [194]:

$$\{\varepsilon^{cr}\} = \int_0^{\phi} [A]\{\sigma\}\, d\tau \tag{24}$$

where

$$[A] = K\begin{bmatrix} 1 & -2\mu \\ -\mu & 1-\mu \end{bmatrix} \quad (25)$$

K is the irradiation creep coefficient [MPa (10^{25} n/m^2; $E_n > 0.18$ MeV)]$^{-1}$,

μ is the Poisson's ratio in irradiation creep,

ϕ is the neutron flux.

The PyC layer experiences elastic deformation, thermal expansion, irradiation-induced creep and irradiation-induced dimensional change under fast neutron irradiation, however, does not experience plastic deformation [192]. At fluence up to 5×10^{26} n/m^2, β-SiC swells in small amount which is insignificant compared with PyC dimensional change [195]. The SiC layer has been assumed to experience only elastic deformation and thermal expansion under the fast neutron irradiation condition of a conventional HTGR. The new coating layer candidate material, ZrC, has an additional plastic deformation property unlike PyC and SiC [193].

For a spherical shell layer, the strain-displacement relationship and the equilibrium equation are expressed by the following two equations [196]:

$$\begin{Bmatrix} \varepsilon_r \\ \varepsilon_\theta \end{Bmatrix} = \begin{Bmatrix} \frac{\partial u}{\partial r} \\ \frac{u}{r} \end{Bmatrix} \quad (26)$$

$$\frac{\partial \sigma_r}{\partial r} + \frac{2}{r}(\sigma_r - \sigma_\theta) = 0 \quad (27)$$

where u is the radial displacement (μm) and r is the radial coordinate (μm).

In an HTGR compact whose matrix material is graphite, a very small amount of ambient pressure, for example 0.1 MPa, is applied to the outer surface of the outermost layer of a particle. A stiffer compact matrix material like dense SiC put more ambient pressure on the outer surface of the outermost layer of the particle. The existing models for predicting stresses in coated particles were extended in [197] to treat stresses induced by the matrix graphite and neighbouring particles in a pebble.

8.1.4. Kernel migration

In oxide, carbide or oxy-carbide fuel, the kernel of a particle can move in the direction of increasing temperature. This phenomenon, known as the amoeba effect, results from the transport of carbon from the hot side of the kernel to the cool side in the presence of a temperature gradient. The failure of the coating layers of a coated fuel particle is assumed to occur when the kernel touches the coating layers. The following equation expresses the kernel migration within a particle [198]:

$$x_{KM}(t) = \int_0^t KMC(t)\frac{1}{[T(\tau)]^2}\frac{dT(\tau)}{dx}d\tau \quad (28)$$

where

x_{KM} is the kernel migration distance (m),

$KMC(T)$ is the kernel migration coefficient (m^2 K/s),

dT/dx is the temperature gradient across the particle (K/m),

T is the average particle temperature at time t (K), and

t is the total time (s).

Table 79 shows Arrhenius-type kernel migration coefficients, taken from Ref. [198].

TABLE 79. KERNEL MIGRATION COEFFICIENTS

	$KMC = KMC_0 e^{-\frac{Q}{RT}}$,	R = 8.3145 J/(mol K), T = temperature (K)
	KMC_0 (m^2 K/s)	Q (J/mol)
UO_2 (Dragon)	8.106×10^{-7}	24 737
UO_2 (FZJ)	1.053×10^{-6}	34 742
UC_2 and UCO	4.811×10^{-1}	299 217

8.1.5. Fission product attack

High concentrations of silver will attack SiC, however, at the low concentrations typical of operating fuel particles, silver can escape from a particle by diffusing through an intact SiC [199,200]. It has been known that, in all fuel kernel types, palladium is released from the kernel and moves to the inner surface of the SiC layer [201,202]. Rare earths such as lanthanum, cerium, praseodymium, and neodymium mostly form immobile, refractory oxide compounds in oxide-based kernels, such as ThO_2, UCO, and UO_2, whose O/U ratios are 1.1 or more [201,202]. They are released and migrate down the thermal gradient in UC_2 kernel or kernels whose O/U ratios are less than 1.1. In UC_2 kernel or kernels whose O/U ratios are less than 1.1, palladium corrodes the SiC layer at rates faster than the REs at temperatures below 1400°C. But the reverse is true at temperatures above 1400°C [201]. The normal operation temperature of an HTGR is much less than 1400°C. Therefore, palladium is the main corroder of the SiC layer.

The fission product palladium is generated in a kernel during irradiation. Palladium isotopes are produced more in low-enriched uranium kernels than in high-enriched uranium kernels because the fission yield of palladium from plutonium is much larger than that from uranium. The palladium atoms in a kernel diffuse out of the kernel, and then, they transport to the SiC layer through the buffer and IPyC layers. The Pd-SiC interaction starts to occur at the inner surface of the SiC layer. Very small precipitates or nodules of a noble metal compound occur along SiC grain boundaries and inside the SiC grains [200,203]. The nodules move SiC by dissociating SiC at the leading edge of a nodule through the following possible chemical interaction [204]:

$$2Pd + SiC \rightarrow Pd_2Si + C$$

and by re-forming SiC at the trailing edge. The presence of other fission product elements in an operating particle could inhibit the repair process. This leads to local corrosion and thinning of the SiC layer [26].

The rate of SiC corrosion by palladium typically depends on irradiation temperature, kernel composition, palladium concentration, other fission product compositions, and SiC properties. A correlation describing the Pd-SiC reaction has to include the effects of the above factors. The threshold amount of palladium for the Pd-SiC reaction has not been clearly observed [201,204]. It can be assumed that the SiC begins to corrode once a small amount of palladium reaches the inner surface of the SiC layer. It was suggested in [204] that, in a UO_2 kernel, the release from a kernel limits the Pd-SiC interaction rate. They concluded qualitatively that the SiC corroded even in the small number of palladium atoms, and the corrosion depth of SiC is proportional to the one third power of the number of palladium atoms.

The following corrosion rate of SiC through a data base of SiC attack by fission products compiled from irradiation and out-of-pile heating tests was introduced in [205]:

$$\dot{v} = 5.872 \times 10^{-7} e^{-\frac{179500}{RT}} \quad (29)$$

where $\dot{v}$ is the corrosion rate (m/s), T is the temperature (K), R is the gas constant (8.3145 J/mol·K).

Equation (29) was made by using SiC attack data which had been obtained by covering in-pile and out-of-pile tests, REs and Pd attack, accelerated and real time irradiations, HEU and LEU fuel [205].

Equation (29) describes only the temperature dependence of the corrosion rate, but not the enrichment or heavy metal burnup dependence of the reaction rate.

No reaction was found on the ZrC TRISO particles heated in the palladium vapour as a result of out-of-pile heating tests of 1830–2150 K [206]. No fission product attack was found [206] in (1) the ZrC TRISO $UC_{4.6}O_{1.1}$ particles irradiated at 1523K to 86%, (2) the ZrC-TRISO UO_2 and UC_2 particles irradiated at 1328–1403 K to 24.9–28.8% FIMA, (3) the ZrC-TRISO UO_2 particles irradiated at 1673–1923 K to 4.5% FIMA.

8.1.6. Thermal decomposition

At very high temperatures above 2000°C, the SiC layer decomposes into its constituent elements: the silicon vaporizes, leaving a porous carbon structure [207]. The SiC thinning rate due to the thermal decomposition is given as follows [198]:

$$k(T) = \frac{375}{d_0} e^{-\frac{556000}{RT}} \tag{30}$$

where k is the SiC thinning rate (s^{-1}), T is the temperature (K), R is the gas constant (8.3145 J/mol-K).

Equation (30) is useful to evaluate how much the SiC layer of individual particles is thermally decomposed.

Out-of-pile high-temperature heating experiments conducted on unirradiated ZrC-TRISO particles demonstrated the stability of ZrC itself up to the eutectic point of ZrC-C at approximately 3123 K. However, the ZrC-TRISO particles could not endure heating above 2773 K, leading to failure induced by substantial internal pressures [206]. In the post-irradiation heating test involving 101 ZrC-TRISO UO_2 particles sampled from an irradiated fuel compact at 1373 K to 4% FIMA, no failure was detected during the initial heat-up stage, with one particle failure occurring after maintaining the particles at 2673 K for 100 minutes [206]. No ZrC failure was observed in isothermal post-irradiation heating tests on ZrC-TRISO UO_2 particles carried out at 1873 K for 4500 h and at 2073 K for 3000 h. The IPyC and ZrC coating layers were damaged in isothermal post-irradiation heating tests carried out at 2273 K for 100 h. The ZrC damage was attributed to the reaction of ZrC with CO gas caused by the failure of the IPyC layers [206].

8.1.7. Failure analysis for a batch of particles

The current particle manufacturing methods such as gelation of kernel and chemical vapour deposition of coating layers do not produce uniform particles. The sizes and material properties of the prepared particle are statistically indicated. The particle quantities with Gaussian statistical distribution are the diameter of the kernel, the thickness and density of the buffer, and the thicknesses, densities, and BAFs of coating layers. The particle quantities with Weibull statistical distribution are the coating layer strengths. They are stochastically independent of each other.

Failure analysis is applied statistically to each individual particle. The failure of a particle occurs due to mechanical failure, fission product attack, and thermal decomposition during a reactor operation. Many models for mechanical failure mechanisms have been developed [70]: SiC failure, kernel migration, pyrocarbon crack, layer debonding, matrix-OPyC interaction. Some researchers [185,208] used the Monte Carlo method to analyse the SiC failure of TRISOs. The Monte Carlo failure analysis consists of random particle generation and failure judgement using criteria for various failure mechanisms.

The particle quantities can be generated using uniform deviates whose range is from 0 to 1, where the uniform deviate is just a random number that comes from a uniform distribution. Two independent standard normal distribution deviates are produced from two independent uniform distribution deviates using the Box-Muller method [209]:

$$y_1 = \sqrt{-2\ln R^2/R^2}\,(2x_1 - 1) \quad (31)$$

$$y_2 = \sqrt{-2\ln R^2/R^2}\,(2x_2 - 1) \quad (32)$$

$$R^2 = (2x_1 - 1)^2 + (2x_2 - 1)^2 \quad (33)$$

where

y is the standard normal distribution deviate Î (-∞, ∞),

x is the uniform distribution deviate Î (0, 1).

A usual random number generator produces x_1 and x_2. Then, the particle quantities with normal distribution can be calculated using a standard normal distribution deviate y:

$$t = \bar{t} + y \cdot STD_t \quad (34)$$

where $\bar{t}$ and STD_t is the mean value and standard deviation of *t*, respectively, that are experimentally given.

The probability distribution for the strength of a particle coating layer is defined by a Weibull distribution [210]:

$$F(s) = 1 - e^{-ln2\cdot(s/s_{med})^m} \quad (35)$$

where

$F(s)$ is the Weibull cumulative distribution function of s,

s is the strength of a coating layer (MPa),

s_{med} is the median strength of a coating layer (MPa), and

m is the Weibull modulus.

Equation (35) gives the following strength of a coating layer:

$$s = s_{med}\left[-\frac{ln(1-F(s))}{ln2}\right]^{1/m} \quad (36)$$

A usual random number generator produces F in (0, 1).

It is judged that the coating layer is broken if the tangential stress calculated using the particle quantities with normal distribution is greater than the strength calculated using Eq. (36).

8.1.8. Fission product releases from a coated particle

Within a particle embedded in a matrix material, fission products are generated through nuclear fission in the kernel material and heavy metal contamination in the coating layers, with some being transferred to adjacent layers through recoil. In the kernel, fission products diffuse through grains to the grain boundaries and then move through interconnected open pores to reach the kernel surface. Released fission products from the kernel swiftly migrate to the IPyC layer through the buffer's pores and subsequently diffuse through the various coating layers. The fission product transport within each layer of a particle is influenced by various complex mechanisms. While the actual transport behaviour is not fully understood, a simplified approach using classic Fickian diffusion has been employed due to incomplete knowledge. The Fickian diffusivity can be correlated to temperature, burnup, or fluence, as well as the material microstructure.

The fission product transport within a coated fuel particle can be described by the following Fickian diffusion equation:

$$\frac{\partial C_{i,j}(r,t)}{\partial t} = \dot{B}_{i,j}(r,t) + \lambda_{j-1}C_{i,j-1}(r,t)\lambda - \lambda_j C_{i,j}(r,t) + \frac{1}{r^2}\frac{\partial}{\partial r}\left[r^2 D_{i,j}\frac{\partial C_{i,j}(r,t)}{\partial r}\right] \tag{37}$$

where

C is the concentration (mol/m^3),

$\dot{B}$ is the volumetric birth rate (mol/m^3·s),

λ is the decay constant (s^{-1}),

D is the diffusion coefficient (m^2/s),

r is the radial coordinate (m) – see Fig. 129,

t is the time (s),

the subscript i is the layer of a particle (K for kernel, B for buffer, I for IPyC, S for SiC, O for OPyC),

the subscript j is the j-th nuclide.

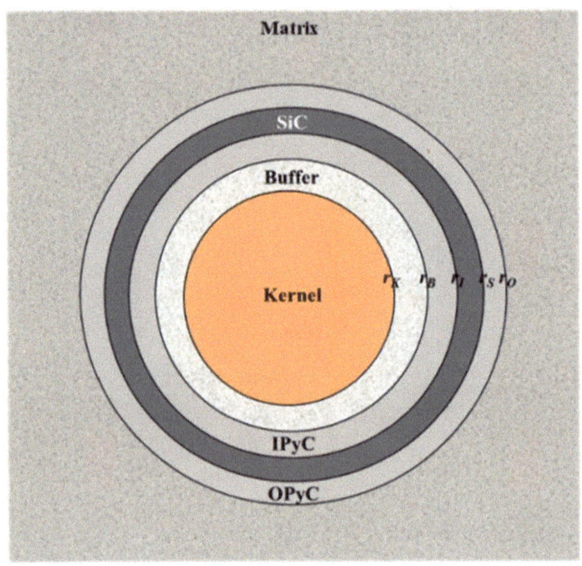

FIG. 129. A TRISO particle in a matrix material.

The initial concentration distribution within a particle is described as a function of the radial coordinate, assuming radial symmetry with a zero diffusion current at the kernel centre. At the particle surface, the concentration is assumed to be zero. Additionally, at the interfaces between two adjacent layers, it is assumed that the mass current flowing out from the left layer surface is equal to that entering the right layer surface. The volumetric birth rate of the fission product in a layer i can be expressed by equation:

$$\dot{B}_i = b_i \dot{B}_{fission,i} + \Delta\dot{B}_{recoil,i} \tag{38}$$

where

b is the heavy metal contamination fraction,

$\dot{B}_{fission} = \xi\dot{F}/6.023 \times 10^{23}$ is the volumetric birth rate due to nuclear fission ($mol/m^3 \cdot s$),

ξ is the fission yield (atom/fission),

$\dot{F}$ is the volumetric fission rate ($fissions/m^3 \cdot s$),

$\dot{B}_{recoil} = \pi(hr^2 - h^3/12)\dot{B}_{fission})$ is the volumetric birth rate due to recoil ($mol/m^3 \cdot s$) [211],

h is the recoil length (m) and r is the radial position of the outer surface of a layer (m).

The heat within a particle originates from the nuclear fission of fissile materials and gamma heating of the particle materials. This heat is conducted through the particle materials into a fuel element. For a conservative calculation, it can be assumed that the temperature distribution within a particle reaches an equilibrium state instantly. The temperature distribution is then described by the following steady-state heat transfer equation:

$$\dot{q}_i(r,t) + \frac{1}{r^2}\frac{\partial}{\partial r}\left(k_i r^2 \frac{\partial T_i}{\partial r}\right) = 0 \quad (39)$$

where T is the temperature (K), $\dot{q}$ is the volumetric heat generation rate (W/m^3), k is the thermal conductivity (W/m·K).

The thermal current at the centre of the kernel is assumed to be zero due to radial symmetry. The temperature at the surface of the particle is considered equal to the temperature of the corresponding part of the fuel element in which the particle is located.

8.1.9. Fission product releases into the coolant

In a fuel region, the fission products are released from the particles and generated from nuclear fissions in heavy metal contamination in the fuel matrix. Metallic fission products migrate through the fuel region, traverse the gap in the case of a fuel block, and pass through the graphite region before ultimately being released into the helium coolant. Gaseous fission products are released from the fuel region into a gap, flowing through the gap into the mainstream of the coolant. Alternatively, they may enter the graphite and diffuse into a coolant hole. This migration is primarily governed by the concentration and temperature within the fuel matrix and structural graphite. Figure 130 represents fission product concentration within a fuel compact.

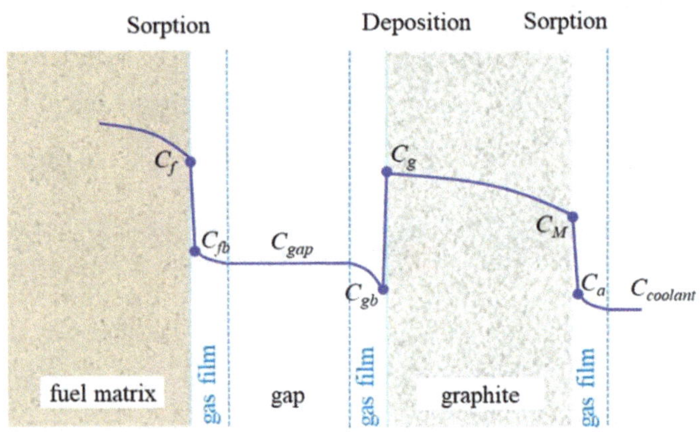

FIG. 130. Concentration profile inside and between a fuel compact and a structural graphite.

The fission product transport within a fuel element can be described by the following Fickian diffusion equation:

$$\frac{\partial C_{i,j}(r,t)}{\partial t} = \dot{S}_{i,j}(r,t) + \lambda_{j-1} C_{i,j-1}(r,t) - \lambda_j C_{i,j}(r,t) + \frac{1}{r^z}\frac{\partial}{\partial r}\left[r^z D_{i,j}\frac{\partial C_{i,j}(r,t)}{\partial r}\right] \qquad (40)$$

where

C is the concentration (mol/m^3),

$\dot{S}$ is the volumetric generation rate (mol/m^3·s),

λ is the decay constant (s^{-1}),

D is the diffusion coefficient (m^2/s),

r is the radial coordinate (m),

t is the time (s),

the subscript i means the layer of a fuel element,

the subscript j means the j-th nuclide,

the superscript z is 0 for a slab, 1 for a cylinder, 2 for a sphere.

The initial concentration distribution is expressed by a function of r coordinate. The generation term is the fission product release rates from the intact, defective and broken particles and the volumetric birth rate due to nuclear fissions in heavy metal contamination:

$$\dot{S}(r,t) = n\sum_{i=0}^{M} f_i \dot{R}_{CFP,i} + f_{HM}\dot{B}_{fission} \qquad (41)$$

where

n is the particle density (particles/m^3),

i is the failure mode of a particle,

M is the total number of failure modes of a particle,

f is the failure fraction,

$\dot{R}_{CFP}$ is the release rate from a particle (mol/s),

f_{HM} is the heavy metal contamination fraction,

$\dot{B}_{fission} = \xi\dot{F}/6.023 \times 10^{23}$ is the volumetric birth rate due to nuclear fission (mol/(m³·s),

ξ is the fission yield (atom/fission) and $\dot{F}$ is the volumetric fission rate (fissions/m³·s).

At the centre of a fuel region, the current is assumed to be zero due to radial coordinate symmetry. Concerning metallic fission products, two boundary layers emerge on both surfaces of a fuel region and within the graphite gap separating them. A sorption isotherm is established between the vapour phases [212], facilitating mass transfers across these two boundary layers. If the mass transfers are rapid, the gas pressures on both sides of a gap equalize. Subsequently, the gas pressures can be described by the sorption isotherms in the following manner:

$$e^{x_f}\left(C_{m,f}^{y_f} + C_{m,t,f}^{y_f-1}C_{m,f}\right) = e^{x_g}\left(C_{m,g}^{y_g} + C_{m,t,g}^{y_g-1}C_{m,g}\right) \tag{42}$$

where

C_m is the mass concentration on the graphite surface (mmol/g),

$C_{m,t}$ is the transition mass concentration (mmol/g) = $\exp(d_1\text{-}d_2T)$,

$X = A + B/T$, $y = G + E/T$, T is the temperature (K),

A, G, d_1 are the sorption isotherm parameters (dimensionless),

B, E, d_2 are the sorption isotherm parameters (K),

the subscripts f and g indicate the fuel and graphite, respectively.

For gaseous fission products in the gap, it can be assumed that the rate of flowing out from the fuel is equal to the rate of entering into the graphite, and the concentrations at the fuel surface and the inner surface of the graphite are the same.

The concentration of a gaseous fission product is zero at the graphite surface facing the coolant. Metallic fission products undergo evaporation on the graphite surface adjacent to the coolant. The concentration on the graphite surface facing the coolant is assumed to achieve sorption isotherm equilibrium with the vapour pressure on the graphite side of the boundary layer, formed between the graphite surface and the bulk coolant. The equilibrium relationships can be expressed using a sorption isotherm [213]:

$$P = e^{x_g}\left(C_{m,g}^{y_g} + C_{m,t,g}^{y_g-1}C_{m,g}\right) \tag{43}$$

where P is the vapour pressure in the graphite side of the boundary layer (Pa).

For metallic fission products, a mass transfer occurs through the boundary layer into the bulk coolant:

$$J(r_c, t) = h(C_a - C_\infty) \tag{44}$$

where

J is the current (mol/m²·s),

C_∞ is the mixed mean concentration of a fission product in the coolant and is usually assumed to be zero (mol/m³),

C_a is the concentration of a fission product in the graphite side of the boundary layer and is calculated by an ideal gas law and the corresponding vapour pressure (mol/m³) = $P/(RT)$,

R is the gas constant (8.3145 J/mol·K),

T is the temperature (K), h is the mass transfer coefficient of a fission product in helium coolant (m/s),

r_c is the r coordinate of graphite surface facing a coolant (m).

It is known that the fission product sorptivity of the graphite increases with exposure to fast neutron fluence, while for the compact matrix material, the sorptivity does not change [213]. For an irradiated graphite, the right-hand side in Equations (42) and (43) is replaced with a modified expression [213]:

$$e^{x_g}\left[\left(\frac{C_{m,g,irr}}{1+g_F C_{m,g,irr}^{\beta_F}}\right)^{y_g} + C_{m,t,g}^{y_g-1}\frac{C_{m,g,irr}}{1+g_H}\right] \tag{45}$$

where

$C_{m,g,irr}$ is the mass concentration on the irradiated graphite surface (mmol/g),

$g_F = \gamma_F \varphi e^{\alpha_F}$,

$g_H = \gamma_H \varphi (10^4/T)^{\alpha_H} e^{-\beta_H (10^4/T)}$,

γ_i = the annealing factor (dimensionless) = $\frac{\delta_i}{1+e^{\varepsilon(T-1423)}}$,

$i = H, F$ denote the Henrian and Freundlich isotherm regions, respectively,

$\alpha_F = \frac{\xi}{T} - \frac{\theta}{T^2}$,

$\beta_H = \kappa - \frac{\lambda}{T} + \frac{\mu}{T^2}$,

φ is the fast neutron fluence (n/m^2; $E_n > 0.18$ MeV),

α_H, δ_H, δ_F, κ are constants (dimensionless),

ε is constant (K^{-1}),

ξ, λ are constants (K),

θ, μ are constants (K^2).

In a fuel element, the heat is generated from nuclear fissions in nuclear materials and heavy metal contamination, and from gamma heating. It is conducted through the fuel region and the structural graphite and removed by the coolant. The temperature distribution in a fuel element is described by the following steady state heat transfer equation:

$$\dot{q}_i(r,t) + \frac{1}{r^z}\frac{\partial}{\partial r}\left(k_i r^z \frac{\partial T_i}{\partial r}\right) = 0 \tag{46}$$

where

T is the temperature (K),

$\dot{q}$ is the volumetric heat generation rate (W/m^3),

k is the thermal conductivity (W/(m K)),

r is the r-coordinate (m),

z is 0 for a slab, 1 for a cylinder, 2 for a sphere.

The thermal current at the centre of the fuel region is assumed to be zero due to radial symmetry. The thermal current at the fuel surface is calculated as the product of the heat transfer coefficient of the gap and the temperature difference between the fuel surface and the gap. Similarly, the thermal current at the graphite surface facing a gap is determined as the product of the heat transfer coefficient of the gap and the temperature difference between the graphite surface and the gap. Heat at the graphite surface facing the coolant is transferred to the helium coolant through convection.

8.2. DEVELOPMENT OF COATED PARTICLE FUEL PERFORMANCE ANALYSIS CODES

Computer codes for predicting the performance of coated particle fuels are described in various references, such as [3,207,212,214]. Table 80 shows some examples of these computer codes.

TABLE 80. AN INTEGRATED SUITE OF COMPUTER CODES FOR PREDICTING THE AMOUNT AND DISTRIBUTION OF RADIONUCLIDES IN A HTGR

Code	Developer	Function	Reference
GARGOYLE ORIGEN-2	GA, USA ORNL, USA	- Radionuclide inventories	[215] [179]
RADC RANDI TRITGO	GA, USA	- Overall radionuclide mass balance for a plant - Advanced plant mass balance - Overall mass balance for tritium	[216] [217] [218]
PISA SOLGASMIX-PV	GA, USA	- Mechanical analysis for a coated particle fuel - Thermochemical equilibrium calculation	[219] [220]
SURVEY	GA, USA	- Full-core coated particle fuel failure and fission gas release rate - Corrosion of fuel element graphite and hydrolysis of failed fuel particles by coolant impurities	[221]
COPAR-FD TRAFIC-FD	GA, USA	- Transient fission product release from coated particle fuels - Full-core release of metallic fission products and actinides	[222] [223]
SORS/NP1	GA, USA	- Transient releases of gaseous and metallic fission products	[224]
OXIDE-4	GA, USA	- Transient response following in leakage of steam and/or air into the primary system	[225]
PADLOC	GA, USA	- Plate out activity distributions of condensable fission products	[226,227]
POLO	GA, USA	- Fractional re-entrainment of plate out activity in the primary coolant circuit during depressurization accidents	[228]
MACCS	SNL, USA	- Radionuclide transport from the primary coolant circuit to the site boundary and the attendant site-boundary doses during postulated accidents	[229]

IAEA's Coordinated Research Project (CRP), CRP-6 [5] was conducted to assess some codes for predicting the performance of coated particle fuels in HTGR through benchmark exercises. Participating codes are listed in Table 81.

TABLE 81. COATED PARTICLE FUEL PERFORMANCE ANALYSIS CODES PARTICIPATED IN THE IAEA CRP-6 BENCHMARK EXCRCISES

Code	Institution	Function	Reference
PANAMA	FZJ, Germany	- Mechanical and failure analysis for TRISOs	[230]
FRESCO II		- Fission product release analysis	[211]
GETTER[a]	PBMR, South Africa	- Fission product inventory and release	–
STRESS3 STAPLE	D.G. Martin, UK	- Mechanical analysis for failure analysis of a batch of TRISOs	–
TRFUEL	Turkey	- Mechanical and failure analysis for TRISO	–
B-2[b]	JAEA, Japan	- Stress calculation for a SiC TRISO	–
FORNAX-A[b]		- Fission product release analysis	–
GOLT	VNIINM, Russia	- Mechanical and failure analysis for TRISO	–
PISA	GA, USA	- Mechanical analysis for a TRISO	[219]
CAPPER		- Analysis of failure and fission gas release	[231]
SORS		- Analysis of release of fission products, transuranics, and other radionuclides	[232]
NRCDIF	NRC, USA	- Fission product release analysis	
MELDIF	SNL, USA	- Fission product release analysis	[233]
PARFUME	INL, USA	- Integrated code	[234]
ATLAS	CEA Cadarache, France	- Integrated code	[235]
COPA	KAERI, Republic of Korea	- Integrated code	[236]

[a] GETTER was developed by the German utility consortium HRB (Hochtemperatur Reaktorbau).
[b] There were no code names at the time of participation.

The normal operation benchmark problems of the IAEA CRP-6 [5] consisted of five parts:

— The first part deals with simple analytical cases to test simple thermo-mechanical behaviours (cases 1 to 3);
— The second part tests a pyrocarbon layer behaviour (cases 4a to 4d);
— The third part treats a single particle with more complexity added with each subsequent case (cases 5 to 8);
— The fourth part treats more complicated benchmarks of actual experiments that have been completed (cases 9 to 12);
— The fifth part encompasses planned experiments (case 13).

Table 82 shows fuel characteristics [5] for the normal operation benchmark, case 8. Table 83 shows the material properties for case 8 [5]. Table 84 shows the internal gas pressure for case 8 [5]. It was assumed that both burnup and fast fluence accumulated linearly with time. The assumed scenario involved subjecting the particle to ten temperature cycles, with the temperature initially held at 873 K, then linearly increasing to 1273 K, followed by an immediate decrease back to 873 K. Each cycle had a period equal to one-tenth of the total irradiation time, corresponding to 100 days.

TABLE 82. FUEL CHARACTERISTICS FOR THE NORMAL OPERATION BENCHMARK, CASE 8

Parameter	Units	Case 8: TRISO cyclic temp.
Oxygen to uranium ratio	Atom ratio	2
Carbon to uranium ratio	Atom ratio	0
U-235 enrichment	Weight %	10
Kernel diameter	mm	500
Buffer thickness	mm	100
IPyC thickness	mm	40
SiC thickness	mm	35
OPyC thickness	mm	40
Kernel density	Mg/m^3	10.8
Buffer density	Mg/m^3	0.95
IPyC density	Mg/m^3	1.9
SiC density	Mg/m^3	3.20
OPyC density	Mg/m^3	1.9
IPyC BAF		1.03
OPyC BAF		1.03
Irradiation conditions		
Irradiation duration	EFPD	1000
End of life burnup	% FIMA	10
End of life fluence	10^{25} n/m^2; $E_n > 0.18$ MeV	3
Constant irradiation temperature[a]	K	873 to 1273 (10 cycles)
End of life internal pressure	MPa	see Table 83
Ambient pressure	MPa	0.1

[a] Temperature increases linearly from 873 to 1273 K for each 100-day cycle and is constant throughout the particle.

TABLE 83. MATERIAL PROPERTIES FOR THE NORMAL OPERATION BENCHMARK CASE 8

Parameter	Units	Case 8: TRISO cyclic temp.
PyC modulus of elasticity	MPa	3.96×10^4
PyC Poisson's ratio		0.33
PyC Poisson's ratio in creep		0.50
PyC coefficient of thermal expansion	K^{-1}	5.35×10^{-6}
PyC creep coefficient	(MPa$\times10^{25}$ n/m^2; $E_n > 0.18$ MeV)$^{-1}$	Creep coefficient [(MPa$\times10^{25}$ n/m^2)$^{-1}$, $E_n > 0.18$ MeV] $= 4.386\times10^{-4} - 9.70\times10^{-7}\,T + 8.0294\times10^{-10}\,T^2$, T in °C.
PyC swelling strain rate	(DL/L)/(10^{25} n/m^2; $E_n > 0.18$ MeV)	Radial swelling/shrinkage rate [(DL/L)/10^{25} n/m^2] $= 4.03266\times10^{-4} j^3 - 2.25937\times10^{-3} j^2 + 9.82884\times10^{-3} j - 1.80613\times10^{-2}$ Tangential swelling/shrinkage rate [(DL/L)/10^{25} n/m^2] $= -4.91648\times10^{-4} j^3 + 2.32979'10^{-3} j^2 + 1.71315\times10^{-3} j - 1.78392\times10^{-2}$ where j = (fast neutron fluence)/10^{25} n/m^2 for $E_n > 0.18$ MeV. Note: divide fast neutron fluences for $E_n > 0.10$ MeV by a factor of 1.10 to obtain fast neutron fluences for $E_n > 0.18$ MeV.
SiC modulus of elasticity	MPa	3.70×10^5
SiC Poisson's ratio		0.13
SiC coefficient of thermal expansion	K^{-1}	4.90×10^{-6}

TABLE 84. INTERNAL PRESSURE FOR THE NORMAL OPERATION BENCHMARK CASE 8

Fast fluence (10^{25} n/m^2; E_n > 0.18 MeV)	**Irradiation duration (EFPD)**	**Internal pressure (MPa)**
0.00	0.00	0.00
0.29	96.67	0.14
0.30	100.00	0.02
0.59	196.67	0.94
0.60	200.00	0.04
0.89	296.67	2.59
0.90	300.00	0.07
1.19	396.67	4.87
1.20	400.00	0.10
1.49	496.67	7.64
1.50	500.00	0.14
1.79	596.67	10.79
1.80	600.00	0.20
2.09	696.67	14.26
2.10	700.00	0.26
2.39	796.67	17.99
2.40	800.00	0.33
2.69	896.67	21.96
2.70	900.00	0.41
2.99	996.67	26.13
3.00	1000.00	0.50

The following observations were made from the normal operation benchmark exercises under the IAEA CRP-6 [5]:

- As for cases 1 to 7, codes' predictions were reasonably in agreement with the analytical solutions, and between codes;
- As for case 8 with cyclic temperature history in a pebble bed fuel, the predicted stresses were in good agreement between codes;
- As for cases 9 to 13 with a real irradiation history, codes' predictions showed discrepancies owing to the uncertainties that could largely be attributed to the physical-chemical models employed. In the case of German-type particles with low gas pressures, the stress in the SiC layer was primarily influenced by the shrinkage of the PyC layer rather than the internal gas pressure within the particle.

The accident condition benchmark conducted in the IAEA CRP-6 [5] consisted of three parts:

- A sensitivity study was conducted to investigate the release of fission products from a fuel particle, commencing with a bare kernel and concluding with an irradiated TRISO particle (cases 1a, 1b, 2a, 2b, 3a, 3b, 3c, 3d, 3e, 4a, 4b, 4c, 4d, 5a, 5b);
- The post-calculation of some well documented irradiation and heating experiments (cases 6a, 6b, 7a, 7b, 8a, 8b, 9);
- The prediction of heating tests which are planned in future (cases 10, 11).

A total of 24 cases have been suggested.

Tables 85 and 86 show the characteristics of the fuel element and the particles in an accident benchmark, case 8b [5], which is the heating test HFR-K3/3. Table 87 describes the irradiation time, temperature, burnup, fast fluence, heating time and temperature information on the irradiation and heating phases of HFR-K3/3 [5]. Diffusion coefficients used for accident analysis are provided in Table 88. Uranium inventory fractions (U/U_{total}) outside the fuel kernel are 1.0×10^{-3} in the buffer, 1.0×10^{-4} in the IPyC layer, 1.0×10^{-6} in the SiC layer, 1.0×10^{-6} in the OPyC layer, and 1.0×10^{-7} in the matrix graphite, respectively.

TABLE 85. CHARACTERISTICS OF THE FUEL ELEMENT FOR THE ACCIDENT BENCHMARK CASE 8b

Parameters	**HFR-K3**
Fuel	Sphere
Fuel element type	GLE-3 LEU phase 1
Matrix graphite grade	A3-27
Matrix density (kg/m^3)	1750
Total fuel element diameter (mm)	59.98
Fuel zone diameter (mm)	47
No of coated particle fuels per sphere/compact	16,350
Packing fraction (%)	-
Heavy metal loading (g/fuel element)	10.22
U-235 content (g/fuel element)	1.004
Fraction of free uranium	3.5×10^{-5}

TABLE 86. CHARACTERISTICS OF THE COATED FUEL PARTICLES FOR THE ACCIDENT BENCHMARK CASE 8b

Parameters	**HFR-K3**
Coated particle batch	EUO 2308
Kernel composition	LEU UO_2
Enrichment (U-235 wt.%)	9.82
Kernel diameter (μm)	497 ± 14.1
Buffer layer thickness (μm)	94 ± 10.3
IPyC layer thickness (μm)	41 ± 4.0
SiC layer thickness (μm)	36 ± 1.7
OPyC layer thickness (μm)	40 ±2.2
Kernel density (g/cm^3)	10.81
Buffer density (g/cm^3)	1.00
IPyC density (g/cm^3)	~ 1.9
SiC density (g/cm^3)	3.20
OPyC density (g/cm^3)	1.88
IPyC Anisotropy BAF	
OPyC Anisotropy BAF	
Fraction of defective SiC	4×10^{-5}

TABLE 87. POST-CALCULATION OF A HEATING TEST HFR-K3/3

Case	Irradiation phase				Heating phase		
	Time (EFPD)	Temperature	Burnup (% FIMA)	Fast neutron fluence (10^{25} n/m^2, $E_n > 0.1$ MeV)	Temperature T (°C)	Time to reach T (h)	Time at T (h)
8b HFR-K3/3	359 (8616 h)	700(surface)-983(centre)	10.6	5.9	300	-	0.5
					1050	1.5	5.5
					1250	0.5	13.5
					1800	12	25.5
					300	1	-
					1050	1.5	19.5
					1250	0.5	19
					1800	12	74.5
							Total: 187

TABLE 88. DIFFUSION COEFFICIENTS FOR ACCIDENT CONDITIONS

	[a] $D = \Sigma_i D_{0,i} Q_i/(RT)$			
	$D_{0,1}$ (m²/s)	**Q_1 (kJ/mol)**	**$D_{0,2}$ (m²/s)**	**Q_2 (kJ/mol)**
Caesium				
in UO_2	5.6×10^{-8}	209	5.2×10^{-4}	362
in buffer	1×10^{-8}	0		
in PyC	6.3×10^{-8}	222		
in SiC	[b] $5.5\times10^{-14}\ e^{\varphi/5}$	125	1.6×10^{-2}	514
in matrix A3-3	3.6×10^{-4}			
in matrix A3-27	3.6×10^{-3}			
Strontium				
in UO_2	2.2×10^{-3}	488		
in buffer	1×10^{-8}	0		
in PyC	2.3×10^{-6}	197		
in SiC	1.2×10^{-9}	205	1.8×10^{6}	791
in matrix	1.0×10^{-2}			
Silver				
in UO_2	6.7×10^{-9}	165		
in buffer	1×10^{-8}	0		
in PyC	5.3×10^{-9}	154		
in SiC	3.6×10^{-9}	215		
in irradiated matrix A3-3	1.6			
Krypton (Iodine)				
in UO_2	8.8×10^{-15}	54	6.0×10^{-1}	480
in buffer	1×10^{-8}	0		
in PyC	1×10^{-30}	0		
in SiC	1×10^{-30}	0		
in matrix	6.0×10^{-6}			

[a] D_0 = the pre-exponent factor (m²/s), Q = the activation energy (J/mol), R = the gas constant = 8.3145 (J K^{-1} mol^{-1}), T = the temperature (K).

[b] φ = the fast neutron fluence (10^{25} n/m²; $E_n > 0.1$ MeV), which is reactor-specific.

Figures 131 and 132 shows the fractional releases of ^{110m}Ag and ^{137}Cs from HFR-K3/3 [5], respectively. Detailed discussion on measured and predicted releases are referred in Ref. [5].

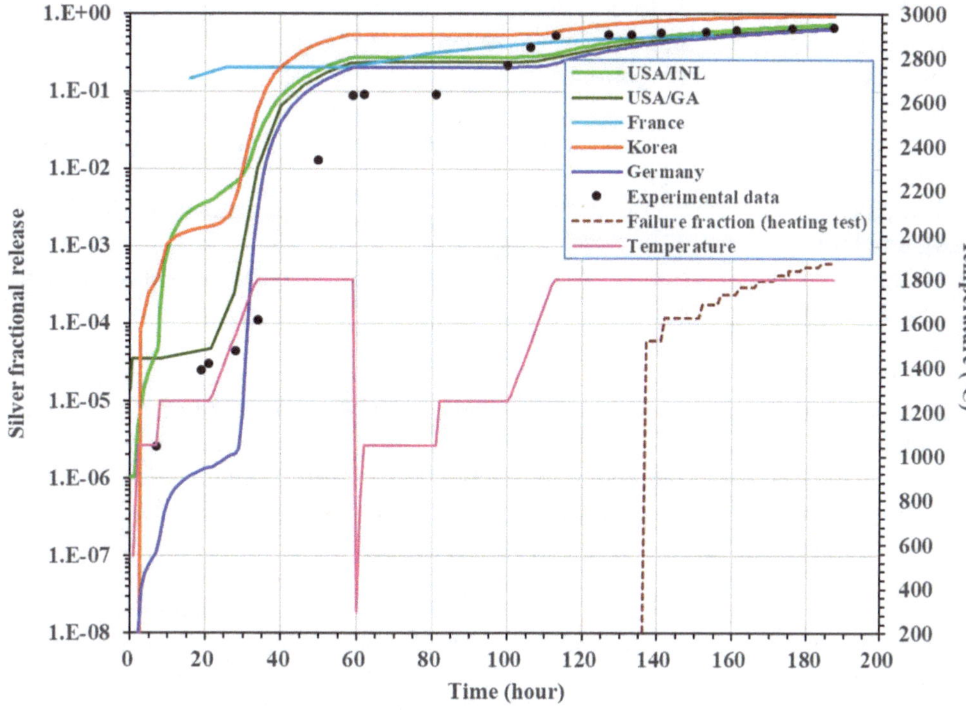

Fig. 131. Fractional release of ^{110m}Ag from HFR-K3/3 (courtesy of IAEA).

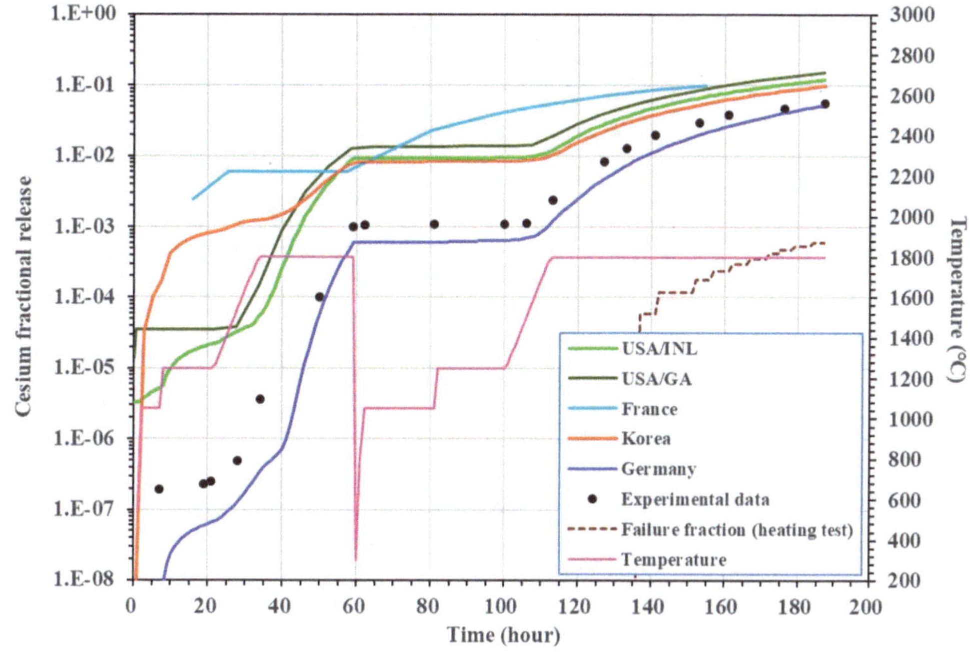

Fig. 132. Fractional release of ^{137}Cs from HFR-K3/3 (courtesy of IAEA).

The Generation IV International Forum (GIF) is a collaborative international initiative involving fourteen members, aimed at conducting research and development to assess the feasibility and performance capabilities of the next generation of nuclear energy systems. In late 2015, GIF initiated a three-year program for benchmarking TRISO fuel performance models under accident conditions, with INL, USA steering the effort. Other participating institutions included the Japan Atomic Energy Agency

(JAEA, Japan) and the Korea Atomic Energy Research Institute (KAERI, Republic of Korea), utilizing the PARFUME code (INL), B-2 and FORNAX-A codes (JAEA), and COPA code (KAERI). The benchmark focused on three main components:

- Modelling a simplified benchmark problem to assess potential numerical calculation issues at low levels of fission product release;
- Modelling fission product release during annealing testing experiments for AGR 1 [237], AGR-2 [118], and HFR EU1bis [69];
- Comparing all modelling results of AGR 1, AGR-2, and HFR EU1bis with corresponding experimental data.

The simplified benchmark case, referred to as NCC (numerical calculation case), aligned with case 5 of the IAEA CRP-6 benchmark on fission product release behaviour models under accident conditions. This case was intended to evaluate the low levels of fission product release. Case 5 of the IAEA CRP-6 showed large discrepancy between the codes' predictions of fission products release, mainly owing to the "effects of the numerical calculation method rather than the physical model" [5]. The NCC was therefore intended to examine if these numerical effects subsisted.

Tables 89–92 list the modelling parameters, irradiation characteristics and temperatures, and annealing test heating plan for the NCC. Selected annealing tests for modelling encompass the initial and subsequent experiments from the AGR program (AGR-1 and AGR-2) and the HFR EU1bis experiment. The radionuclides chosen for the benchmark calculations are a long-lived activation product (^{110m}Ag) and three fission products (^{90}Sr, ^{137}Cs, and ^{85}Kr). The AGR-1 fuel, particle, and compact properties are referred to Section 6.2.2.1.

TABLE 89. NUMERICAL CALCULATION CASE FUEL MODELLING PARAMETERS

Category	Parameter	Mean value
Fuel properties	U-235 enrichment (wt.%)	10
	Oxygen/uranium (atomic ratio)	2
	Carbon/uranium (atomic ratio)	0
	Uranium contamination fraction	0
Particle properties	Kernel diameter (μm)	350
	Buffer thickness (μm)	100
	IPyC thickness (μm)	40
	SiC thickness (μm)	35
	OPyC thickness (μm)	40
	Kernel density (g/cm^3)	10.8
	Kernel theoretical density (g/cm^3)	10.96
	Buffer density (g/cm^3)	0.95
	Buffer theoretical density (g/cm^3)	2.25
	IPyC density (g/cm^3)	1.9
	SiC density (g/cm^3)	3.20
	OPyC density (g/cm^3)	1.9
	IPyC anisotropy (BAF)	1.03
	OPyC anisotropy (BAF)	1.03
	Particle asphericity (SiC level)	1.0
Boundary conditions	Ambient pressure (MPa)	0.1

TABLE 90. NUMERICAL CALCULATION CASE IRRADIATION CHARACTERISTICS

Case	Burnup (%FIMA)	Fast fluence (10^{25} n/m^2 *E* > 0.18 MeV)	Irradiation length (EFPD)
NCC	10[(a)]	2[(a,b)]	1000

a. Burnup and fast fluence assumed to follow linear evolution throughout irradiation.
b. Fast fluence (*E* > 0.18 MeV) = 0.91 × Fast fluence (*E* > 0.1 MeV)

TABLE 91. NUMERICAL CALCULATION CASE IRRADIATION CHARACTERISTICS

Cycle number	Cycle EFPD	Surface temperature (°C)
1	100	Ramp 600 → 1000
2	100	Ramp 600 → 1000
3	100	Ramp 600 → 1000
4	100	Ramp 600 → 1000
5	100	Ramp 600 → 1000
6	100	Ramp 600 → 1000
7	100	Ramp 600 → 1000
8	100	Ramp 600 → 1000
9	100	Ramp 600 → 1000
10	100	Ramp 600 → 1000

TABLE 92. NUMERICAL CALCULATION CASE ANNEALING TEST HEATING PLAN

Time, (h:min)	Temperature, °C
00:00	1000
00:01	1600
200:01	1600

Table 93 lists the irradiation characteristics and annealing test temperatures of the AGR-1 compacts [118]. Table 94 details the timeline of the AGR-1 annealing test heating plans [118].

TABLE 93. AGR-1 COMPACT SELECTION AND IRRADIATION CHARACTERISTICS

Annealing test temperature (°C)	**Compact**	**Number of particles with failed SiC during annealing test**	**Burnup[a] (%FIMA)**	**Fast fluence[a,b] (10^{25} n/m² E > 0.18 MeV)**	**Average temperature (°C)**	**Irradiation length (EFPD)**	**Variant**
1600	6-4-1	1	13.22	2.43	1041	620.2	Baseline
	4-3-3	0	18.52	4.16	1094	620.2	3
1700	4-4-3	0	18.83	4.06	1059	620.2	3
	3-3-1	4	19.00	4.23	1051	620.2	Baseline
1800	5-1-3	7	18.17	3.82	1042	620.2	1
	4-4-1	2	18.84	3.99	1057	620.2	3
	4-3-2	3[©]	16.24	3.68	1057	620.2	3
	3-2-3	11	19.03	4.28	1053	620.2	Baseline
Transient	1-4-2	0	14.83	3.01	1045	620.2	3
	1-1-3	0	15.21	2.86	1018	620.2	3
	1-1-1	0	15.05	2.81	1017	620.2	3

a. Burnup and fast fluence are assumed to follow linear evolution throughout irradiation.
b. Fast fluence (E > 0.18 MeV) = 0.91 × Fast fluence (E > 0.1 MeV)
c. Compact 4-3-2 modelled with 2 particles with exposed kernel and 1 particle with failed IPyC and SiC.

TABLE 94. AGR-1 ANNEALING TEST HEATING PLANS

Transient		**1600°C**		**1700°C**		**1800°C**	
Time (h:min)	Temperature (°C)	Time (h:min)	Temperature (°C)	Time (h:min)	Temperature (°C)	Time (h:min)	Temperature (°C)
00:00	30	00:00	30	0:00	30	0:00	30
00:30	300	03:05	400	3:05	400	3:05	400
22:30	300	05:05	400	5:05	400	5:05	400
24:00	857	12:10	1250	12:10	1250	12:10	1250
94:00	857	24:10	1250	24:10	1250	24:10	1250
97:48	1300	31:10	1600	33:10	1700	35:10	1800
104:30	1585	331:10	1600	333:10	1700	335:10	1800
112:00	1670	333:47	30	335:57	30	338:07	30
124:00	1695						
136:00	1680						
164:00	1620						
214:00	1508						
294:00	1342						
394:00	1200						
396:00	20						

The AGR-2 fuel, particle, and compact properties are described in Section 6.2.2.2. Table 95 lists the AGR-2 irradiation characteristics and annealing test temperatures. Table 96 details the timeline of the AGR-2 annealing test heating plans.

TABLE 95. AGR-2 COMPACT SELECTION AND IRRADIATION CHARACTERISTICS

Annealing test temperature (°C)	Compact	Number of particles with failed SiC during annealing test	Burnup[(a)] (%FIMA)	Fast fluence[(a,b)] (10^{25} n/m^2 $E > 0.18$ MeV)	Average temperature (°C)	Irradiation length (EFPD)	Fuel type
1600	5-2-2	0	12.34	3.39	1141	559.2	UCO
	3-3-2	0[I]	10.54	3.53	1062	559.2	UO_2
1700	3-4-1	0[I]	10.62	3.47	1013	559.2	UO_2
1800	5-4-1	1	12.05	3.12	1071	559.2	UCO

a. Burnup and fast fluence are assumed to follow linear evolution throughout irradiation.
b. Fast fluence ($E > 0.18$ MeV) = 0.91 × Fast fluence ($E > 0.1$ MeV)
c. UO_2 compacts were modelled with zero SiC failures because their exact number could not be determined by post-heating examination.

TABLE 96. AGR-2 ANNEALING TEST HEATING PLANS

1600 °C		1700 °C		1800 °C	
Time (h:min)	Temperature	Time (h:min)	Temperature (em	Time (h:min)	Temperature (em
00:00	30	0:00	30	0:00	30
03:05	400	3:05	400	3:05	400
05:05	400	5:05	400	5:05	400
12:10	1250	12:10	1250	12:10	1250
24:10	1250	24:10	1250	24:10	1250
31:10	1600	33:10	1700	35:10	1800
331:10	1600	196:10	1700[(a)]	37:00	1800
333:47	30	201:50	30	41:00	30[(b)]
				51:00	1200
				63:00	1800
				363:00	1800
				366:00	30

a. The test was terminated after ~163 h at 1700 °C.
b. The furnace shut down after ~37 h.

The HFR-EU1bis fuel, particle, and pebble properties are detailed in Table 97 [68,238]. Tables 98 and 99 list the HFR-EU1bis irradiation characteristics and irradiation temperatures, respectively. Table 100 details the timeline of the HFR-EU1bis annealing test heating plans.

TABLE 97. HFR-EU1BIS FUEL, PARTICLE, AND PEBBLE PROPERTIES

Category	Parameter	Mean value ± standard deviation
Fuel properties	U-235 enrichment (wt%)	16.76
	Oxygen/uranium (atomic ratio)[(a)]	2
	Carbon/uranium (atomic ratio)[(a)]	0
	Uranium contamination fraction	7.8×10^{-6}
Particle properties	Kernel diameter (mm)	502.2 ± 10.6
	Buffer thickness (mm)	94.3 ± 13.0
	IPyC thickness (mm)	40.6 ± 3.7
	SiC thickness (mm)	35.9 ± 2.2
	OPyC thickness (mm)	39.8 ± 3.3
	Kernel density (g/cm^3)	10.86
	Kernel theoretical density (g/cm^3)	10.96
	Buffer density (g/cm^3)	1.012
	Buffer theoretical density (g/cm^3)	2.25
	IPyC density (g/cm^3)	1.87
	SiC density (g/cm^3)	3.20
	OPyC density (g/cm^3)	1.87
	IPyC anisotropy (BAF)	1.02
	OPyC anisotropy (BAF)	1.02
	Particle asphericity (SiC level)	1.04
Pebble properties	Sphere diameter (mm)[(a)]	60
	Fuel zone diameter (mm)	50
	U-235 content (g/pebble)	1.00 ± 0.01
	Heavy metal loading (g/pebble)	6.0
	Number of particles per sphere	9560
	Volume packing fraction (%)	6.2
	A3-3 matrix density (g/cm^3)	1.75
Boundary conditions	Ambient pressure (MPa)	0.1

TABLE 98. HFR-EU1BIS IRRADIATION CHARACTERISTICS

Sphere	Burnup[(a)] (%FIMA)	Fast fluence[(a,b)] (10^{25} n/m^2 $E > 0.1$ MeV)	Irradiation length (EFPD)
1	9.34	2.41	249.55
3	11.07	2.86	249.55
4	11.07	2.86	249.55
5	9.70	2.51	249.55

a. Burnup and fast fluence are assumed to follow linear evolution throughout irradiation.
b. Fast fluence ($E > 0.18$ MeV) = 0.91 × Fast fluence ($E > 0.1$ MeV)

TABLE 99. HFR-EU1BIS IRRADIATION TEMPERATURES

Cycle number	Cycle name	Cycle EFPD	Time (EFPD)	Surface Temperature (°C)				Central temperature (°C)			
				\1	\3	\4	\5	\1	\3	\4	\5
1	04-08	24.97	24.97	1014	1009	1015	1006	1216	1250	1247	1227
2	04-09	24.72	49.69	1026	1024	1030	1020	1215	1251	1248	1227
3	05-01	25.99	75.68	1036	1038	1043	1032	1215	1252	1249	1228
4	05-02	25.67	101.35	1042	1047	1052	1040	1211	1249	1246	1224
5	05-03	25.29	126.64	1053	1062	1066	1052	1211	1252	1249	1226
6	05-04	25.67	152.31	1058	1072	1075	1060	1208	1251	1248	1224
7	05-06	24.26	176.57	1065	1082	1086	1069	1207	1252	1249	1224
8	05-07	25.19	201.76	1067	1088	1091	1073	1202	1248	1245	1220
9	05-08	22.19	223.95	1076	1101	1103	1083	1203	1253	1250	1223
10	05-09	25.60	249.55	1079	1108	1109	1088	1199	1252	1248	1220

TABLE 100. HFR-EU1BIS ANNEALING TEST HEATING PLANS

HFR-EU1bis/1		HFR-EU1bis/3		HFR-EU1bis/4		HFR-EU1bis/5			
Time (h:min)	Temp.	Time (h:min)	Temp. (em	Time (h:min)	Temp. (em	Time (h:min)	Temp. (em	Time (h:min)	Temp. (em
00:00	20	00:00	20	00:00	20	00:00	300	848:00	300
00:30	300	01:00	20	00:30	300	16:00	300	850:00	1800
06:30	300	01:30	300	03:30	300	23:00	950	852:00	300
09:30	1250	04:30	300	05:30	800	126:00	950	854:00	1800
209:30	1250	06:30	1250	53:30	800	136:00	300	856:00	300
215:30	20	96:00	1250	56:30	1250	157:00	300	858:00	1800
216:30	20	98:00	20	66:30	1250	161:00	950	860:00	300
217:00	300	99:00	20	71:30	1320	185:00	950	862:00	1800
220:00	300	99:30	300	76:30	1390	186:00	1050	864:00	300
222:00	1250	102:30	300	86:30	1500	207:00	1050	866:00	1800
232:00	1250	104:30	1250	91:30	1535	210:00	1250	868:00	300
233:30	1320	114:30	1250	96:30	1570	280:00	1250	870:00	1800
239:30	1600	118:00	1412	106:30	1630	284:00	1500	872:00	300
439:30	1600	122:00	1600	115:30	1666	374:00	1500	874:00	1800
455:00	20	322:00	1600	116:30	1670	378:00	1250	876:00	300
456:00	20	324:00	20	126:30	1695	455:00	1250	878:00	1800
456:30	300			136:30	1710	457:00	300	880:00	300
459:30	300			140:00	1711.5	461:00	300	900:00	300
461:30	1250			160:30	1720	481:00	1250	906:00	1800
471:30	1250			280:30	1720	505:00	1250	980:00	1800
473:00	1321			282:00	20	511:00	1600		
479:00	1605					624:00	1600		
481:00	1700					631:00	1800		
631:00	1700					821:00	1800		
646:00	20					844:00	300		

The NCC consists of a 200-h heating phase at 1600°C following a 1000-day irradiation phase of 10 temperature cycles in the range 600–1000 °C. Calculated fractional release data for Ag, Cs, Sr, and Kr after irradiation and the subsequent heating phase are presented in Table 101.

TABLE 101. CALCULATED AG, CS, SR, AND KR RELEASE FRACTIONS FOR THE NCC

	After irradiation				**After 200 h heating**			
	Ag	Cs	Sr	Kr	Ag	Cs	Sr	Kr
INL	6.7×10^{-5}	1.8×10^{-12}	4.3×10^{-11}	2.1×10^{-89}	5.0×10^{-1}	6.7×10^{-4}	3.1×10^{-2}	4.0×10^{-14}
JAEA	2.8×10^{-5}	1.2×10^{-14}	4.2×10^{-6}	7×10^{-122}	4.6×10^{-1}	2.9×10^{-4}	5.5×10^{-2}	1.2×10^{-17}
KAERI	1.6×10^{-5}	9.8×10^{-14}	2.2×10^{-6}	2.6×10^{-90}	4.6×10^{-1}	4.2×10^{-4}	3.3×10^{-2}	1.3×10^{-16}

8.3. SUMMARY AND DISCUSSION

Many countries are developing their own HTGR fuel performance analysis codes. Two benchmarking exercises were conducted with observations:

— The IAEA CRP-6 provided a platform to exchange information on the verification and validation of the codes under normal operating and accident conditions. For the benchmark cases on simple problems most codes provided good agreement with each other, whilest for the benchmark cases on simulating real irradiation or heating experiments, a large discrepancy in predictions among codes were recognized;
— Under Generation IV programme, "Benchmarking of TRISO Fuel Performance Models under Accident Conditions" was carried out after the IAEA CRP-6. The code-to-code comparison showed reasonable agreement for the release of Ag, Cs and Sr; however, it also showed a large discrepancy for Kr. In comparison with measurements, overall codes overpredicted the fission gas releases.

As next steps, the following are considered:

— Fuel codes need to be extended to model UCO;
— Simplified, fast running codes are used for quick design and safety evaluations. On the other hand, multiscale fuel performance analysis codes are also necessary for better understanding of changes in the microstructures and material properties of nuclear fuels. Separate effects tests are necessary to support modelling parameters taken account of the multiscale codes.

REFERENCES

[1] BREY, H.L., "The evolution and future development of the high temperature gas cooled reactor", Presented at GENES4/ANP2003, Paper 1194, Kyoto, Japan, 15-19 September 2003.

[2] INTERNATIONAL ATOMIC ENERGY AGENCY, Fuel Performance and Fission product Behaviour in Gas Cooled Reactors, IAEA-TECDOC-978, IAEA, Vienna (1997).

[3] INTERNATIONAL ATOMIC ENERGY AGENCY, High Temperature Gas Cooled Reactor Fuels and Materials, IAEA-TECDOC-CD-1645, IAEA, Vienna (2010).

[4] INTERNATIONAL ATOMIC ENERGY AGENCY, Advances in High Temperature Gas Cooled Reactor Fuel Technology, IAEA-TECDOC-CD-1674, IAEA, Vienna (2012).

[5] INTERNATIONAL ATOMIC ENERGY AGENCY, Performance Analysis Review of Thorium TRISO Coated Particles during Manufacture, Irradiation and Accident Condition Heating Tests, IAEA-TECDOC -1761, IAEA, Vienna (2015).

[6] DEMKOWICZ, P.A., TRISO Fuel: Design, Manufacturing, and Performance, INL/MIS-19-52869, INL, Idaho Falls (2019).

[7] INTERNATIONAL ATOMIC ENERGY AGENCY, Safety of Nuclear Power Plants: Design, Specific Safety Requirements No. SSR-2/1 (Rev 1), IAEA, Vienna (2016).

[8] INTERNATIONAL ATOMIC ENERGY AGENCY, IAEA Safety Glossary – Terminology Used in Nuclear Safety and Radiation Protection, 2018 Edition, IAEA, Vienna (2019).

[9] BELLES, R., POORE, W., BROWN, N., FLANAGAN, G., HOLBROOK, M., MOE, W., SOFU, T., Proposed Adaptation of the Standard Review Plan NUREG-0800, Chapter 4 (Reactor) for Sodium-Cooled Fast Reactors and Modular High-Temperature Gas-Cooled Reactors, ORNL/TM-2017/151, Oak Ridge National Laboratory (2017).

[10] PETTI, D.A., MAKI, J.T., BUONGIORNO, J., HOBBINS, R.R., MILLER, G.K., Key Differences in the Fabrication, Irradiation and Safety Testing of US and German TRISO-coated Particle Fuel and Their Implications on Fuel Performance, INEEL/EXT-02-00300, INL, Idaho Falls (2002).

[11] POWERS, J., WORRALL, A., TERRANI, K.A., GEHIN, J.C., SNEAD, L.L., "Fully ceramic microencapsulated fuels: characteristics and potential LWR applications", Presented to PHYSOR 2014, Kyoto, Japan, September 28 – October 3, 2014.

[12] DEMKOWICZ, P.A., LIU, B., HUNN, J.D., Coated particle fuel: historical perspectives and current progress, J. Nucl. Mater. **515** (2019) 434-450.

[13] DEMKOWICZ, P.A., HUNN, J.D., Two decades of DOE investment lays foundation for TRISO-fuelled reactors in the US, Nuclear News **63**, August 2020.

[14] MORRIS, R.N., PETTI, D.A., POWERS, D.A., BOYACK, B.E., RUBIN, M.B., TRISO-Coated Particle Fuel Phenomenon Identification and Ranking Tables (PIRTs) for Fission Product Transport Due to Manufacturing, Operations, and Accidents, NUREG/CR-6844, Vol. 1 (2004).

[15] MAKI, J.T., PETTI, D.A., KNUDSON, D.L., MILLER, G.K., The challenges associated with high burnup, high temperature and accelerated irradiation for TRISO-coated particle fuel, J. Nucl. Mater. **371** (2007) 270-280.

[16] SKERJANC, W.F., MAKI, J. T., COLLIN, B. P., PETTI, D. A., Evaluation of Design Parameters for TRISO-coated Fuel Particles to Establish Manufacturing Critical Limits Using PARFUME, INL/JOU-15-34392, INL, Idaho Falls (2016); also see J. Nucl. Mater. **469** (2016) 99-105.

[17] MARSDEN, B.J., HAVERTY, M., BODEL, W., HALL, G.N., JONES, A.N., MUMMERY, P.M., TREIFI, M., Dimensional change, irradiation creep and thermal/mechanical property changes in nuclear graphite, International Materials Reviews **61** (2016) 155-182.

[18] HAVERTY, M., The Microstructure, Texture and Thermal Expansion of Nuclear Graphite, PhD Thesis, Faculty of Engineering and Physical Sciences, University of Manchester (2015).

[19] BOER, B., OUGOUAG, A.M., Final Report on Utilization of TRU TRISO Fuel as Applied to HTR Systems Part I: Pebble Bed Reactors, INL/EXT-11-21436, INL, Idaho Falls (2011).

[20] LOWE, T., BRADLEY, R.S., YUE, S., BARII, K., GELB, J., ROHBECK, N., TURNER, J., WITHERS, P.J., Microstructural analysis of TRISO particles using multi-scale X-ray computed tomography, J. Nucl. Mater. **461** (2015) pp. 29-36.

[21] KOVACS, W.J., YOUNG, C.A., HARMON, D.P., Pre-Irradiation Report of TRISO and BISO Coated ThO_2 Particles for Irradiation in Capsules HT-31 and HT-33, GA-A13923 / UC-77 (1976).

[22] HALES, J.D., JIANG, W., TOPTAN, A., GAMBLE, K., BISON TRISO Modelling Advancements and Validation to AGR-1 Data, INL/EXT-20-59368, INL, Idaho Falls (2020).

[23] SCHAPPEL, D.P., Improvements to the Predictive Capability of FCM Fuel Performance Modelling, PhD thesis, University of Tennessee (2017).

[24] MARTIN, D.G., Considerations pertaining to the achievement of high burn-ups in HTR fuel Nucl. Eng. Des. **213** (2002) 241–258.

[25] INTERNATIONAL ATOMIC ENERGY AGENCY, Irradiation Damage in Graphite Due to Fast Neutrons in Fission and Fusion Systems, IAEA-TECDOC-1154, IAEA, Vienna (2000).

[26] TERRANI, K.A., KIGGANS, J.O., KATOH, Y., SHIMODA, K., MONTGOMERY, F.C., ARMSTRONG, B.L., PARISH, C.M., HINOKI, T., HUNN, J.D., SNEAD, L.L., Fabrication and characterization of fully ceramic microencapsulated fuels, J. Nucl. Mater. **426** (2012) 268–276.

[27] HUSCHKA, H., VENET, P., Préparation et propriétés des particules de combustible nucléaire pour réacteurs à gaz à haute temperature, Bulletin de la Société Francaise Céramique **73** (1966) 51–63 (in French).

[28] HUSCHKA, H., VYGEN, P, Coated fuel particles: requirements and status of fabrication technology, Nucl. Techn. **35** (1977) 238–245.

[29] PODO, L.A., HORSLEY, G. W, Studies on the Preparation of Th/U Dicarbide and Thoria/Urania Fuel Kernels by the Sol-Gel Technique in the Dragon Fuel Element Development Laboratory, Dragon Project Report DP-431, Winfrith (1966).

[30] COGLIATI, G., et al., Proc. Symposium on Sol-Gel-Processes for the Production of Ceramic Nuclear Fuels, Turin/Italy (1968).

[31] FERGUSON, D.E., DEAN, O.C., HASS, P.A, ORNL Report 542, Oak Ridge (1968).

[32] WYMER, R.G., Laboratory and Engineering Studies of Sol-Gel Processes at Oak Ridge National Laboratory, ORNL-TM-2205, Oak Ridge (1968).

[33] FOERTHMANN, R., NAOUMIDIS, A., NICKEL, H., Untersuchungen zur Herstellung Kugelförmiger Kernbrennstoffteilchen Nach Einem Sol-Gel-Verfahren, KFA Report, Jül 583 RW (1969) (in German).

[34] BOND, W.D., FINNEY, B.C, HAAS, P.A., HAWS, C.C., PARROTT, J.R., Proc. Symposium on Sol-Gel-Processes and Reactor Fuel Cycles, CONF 700502 Gatlinburg/USA, (1970) 123.

[35] COGLIATI, G., FACCHINI, A., Proc. Symposium on Sol-Gel Processes and Reactor Fuel Cycles, CONF 700502, Gatlinburg/USA, May (1970) 210.

[36] KADNER, M., BAIER J., Production of Fuel Kernels for High Temperature Reactor Fuel Elements, Kerntechnik 18 (1977) 413–420.

[37] NAEFE, P., ZIMMER, E. Preparation of uranium kernels by an external gelation process, Nucl. Techn. **42** (1979) 163–171.

[38] KADNER, M., BAIER J., Production of fuel kernels for high temperature reactor fuel elements, Kerntechnik **18** (1977) 413–420.

[39] MINATO, K., FUKUDA, K., Structure of chemically vapour deposited silicon carbide for coated fuel particles, J. Mater Sci. **23** (1988) 699.

[40] CHAROLLAIS F., et al., CEA and AREVA R&D on HTR fuel fabrication and presentation of the CAPRI experimental manufacturing line, Nucl. Eng. Des. **236** (2006) 534–542.

[41] HAASE, V., et al., Handbook of Inorganic Chemistry, Silicon, 8th Ed., vol. B3, Springer-Verlag, New York (1986) pp.31.

[42] KRAUTWASSER, P., et al., Raman spectral characterization of silicon carbide nuclear fuel coatings, Journal of American Ceramic Society **66** (1983) 424–434.

[43] LÓPEZ-HONORATO, E., MEADOWS, P.J., TAN, J., XIAO, P., Control of stoichiometry, microstructure, and mechanical properties in SiC coatings produced by fluidized bed chemical vapour deposition, Journal of Material Research **23** (2008) 1785.

[44] CHIN, J., et al., The structure of chemical vapour deposited silicon carbide, Thin Solid Films **40** (1977) 57–72.

[45] XU, S.J., ZHOU, J G., YANG, B., ZHANG, B Z., Effect of deposition temperature on the properties of pyrolytic SiC, Journal of Nuclear Materials **224** (1995) 12–16.

[46] VOICE, E.H., Silicon Carbide as a Fission Product Barrier in Nuclear Fuels, in Silicon Carbide (Eds. HENISCH, H & ROY, R.), Pergamon Press, New York (1969).

[47] VOICE, E.H., LAMB, D.N., The Deposition and Structure of Pyrolytic Silicon Carbide, DRAGON Project Report DP-677, Winfrith (1969).

[48] ABLITZER, C., CHAROLLAIS, F., DUGNE, O., MENDES, A., MOULINIER, D., PEREZ, M., in Euro Course on Coated Particle Fuel, IAEA-EU-RAPHAEL-PUMA, held in NRG, Petten, The Netherlands (2007).

[49] OXLEY, J.H., BROWNING, M. F., VEIGEL, N. D., J. M. BLOCHER, J. M., Microminiaturized Fuel elements by vapour-deposition techniques, Industrial and Engineering Chemistry **1** (1962) 102–107.

[50] OXLEY, J.H., in Vapour Deposition, (Edited by. POWELL, C.F., OXLEY, J.H., BLOCHER, J.M), Wiley, New York (1966).

[51] HROVAT, M., NICKEL, H., KOIZLIK, K., Über die Entwicklung eines Matrixmaterials zu Herstellung gepresster Brennelemente für Hochtemperatur Reaktoren, Forschungszentrum Jülich Report Jül-969 (1973) (in German).

[52] SCHULZE, R.E., SCHULZE, H.A., RIND, W., Graphite Matrix Materials for Spherical HTR Fuel Elements, Forschungsanlage Jülich report Jül-Spec-167 (1982).

[53] LIANG, T. X., "Recent R&D of HTR fuel in INET", Presented at the IAEA Technical Meeting on HTR Fuel and Fuel Cycle, Vienna, 6–9 September 2010.

[54] DELLE, W.W., et al., Quality Control Procedures for HTR Fuel Element Components. Report Jül-1333, Forschungszentrum Jülich, Germany (1976).

[55] NELSON, A T., HUNN, J D, KARNOWSKI, T P, Current Development Status of a Particle Size Analyzer for Coated Particle Fuel, Oak Ridge National Laboratory, <http://info.ornl.gov/sites/publications/files/Pub6657.pdf >, (2004).

[56] HUNN, J.D., JELLISON, G.E., LOWDEN, R.A., Increase in pyrolytic carbon optical Anisotropy and density during processing of coated particle fuel due to heat treatment, J. Nucl. Mat. **374** (2008) 445–452.

[57] LIU, B., Fabricating Process and Quality Control Diagram of the Fuel Element, Institute of Nuclear and New Energy Technologies, Tsinghua University, Beijing, China (2011).

[58] KIM, K.W., LEE, Y.W., CHO, M.S., PARK, J.Y., RA, S.W and PARK, J.B., Non-destructive measurement of the coating thickness for simulated TRISO coated fuel particles using phase contrast X ray radiography, Nuclear Eng. And Des. **238** (2008) 3285–3291.

[59] HITTNER, D., LOMMERS, L., SHAHROKHI, F., "R&D needs for near-term HTRs", Proc. of HTR-2010, Prague (2010).

[60] BANCHET, J., TISSEUR, D., GUILLERMIER, P., "X-ray radiography: A alternative method for determining HTR fuel particle coating layer thickness and density", Proc. HTR-2008, Washington D.C., U.S.A. (2008).

[61] KASHYAP, Y.S., YADAV, P.S., SARKAR, P.S., AGGARWAL, A., ROY, T., SINHA, A., DASGUPTA., K., SATHIYAMOORTHY, D., Application of X ray phase contrast imaging

technique in the study of pyrocarbon-coated zirconia kernels, Nondestructive Testing and Evaluation International **42** (2009) 388–388.

[62] RAGOSS, H., Experimentelle Absicherung des Brennelementverhaltens für Normalbetriebs- und Störfallbedingungen des HTR-Moduls, Siemens Report No. 76.00526.9, Bensberg (1987) (in German).

[63] RAGOSS, H., NABLIEK, H., Fuel Specifications for the Siemens MODUL, Interatom-Forschungszentrum Jülich Joint Technical Note, Bensberg (1989).

[64] VENTER, J.H., and NABIELEK, H., "Performance Envelope of Modern HTR TRISO Fuel", Proc. HTR2006, Johannesburg (2006).

[65] HACKSTEIN, K.G., et al., Stand der brennelementtechnologie für hochtemperaturreaktor-kugelhaufenreaktoren. Atomkernenergie/Atomtechnik **47** (1985)163–169) (in German).

[66] MEHNER, A. W., HEIT, W., RÖLLIG, K., RAGOSS, H., MÜLLER, H., Spherical fuel Elements for HTR manufacture and qualification by irradiation testing, J. Nucl. Mat. **171** (1990) 9–18.

[67] INTERNATIONAL ATOMIC ENERGY AGENCY, Safety Related Design and Economic Aspects for HTGRs, IAEA-TECDOC-1210, IAEA, Vienna (1999).

[68] NABIELEK, H., et al., Selection of Benchmark Cases for Mechanical Failure Prediction, Technical Note in EU HTR-F Project No. HTR-F1-04/08-D-3.1.1, Forschungszentrum, Jülich, Germany (2005).

[69] FÜTTERER, M., BERG, G., MARMIE, A., TOSCANO, E.H., FRIES, D., BAKKER, K., De GROOT, S., Results of AVR fuel pebble irradiation at increased temperature and burnup in the HFR Petten, Nucl. Eng. Des. **238** (2008) 2877–2885.

[70] PETTI, D.A., DEMKOWICZ, P.A., MAKI, J.T., HOBBINS, R.R., "TRISO-coated particle fuel performance", Section 3.07, Comprehensive Nuclear materials, ed. by Rudy M. KONINGS and Roger E. STOLLER, Elsevier Inc. (2012).

[71] AHLF, J., CONRAD, R., CUNDY, M., SCHEURER, H., Irradiation experiments on high temperature gas-cooled reactor fuels and graphite at the High Flux Reactor Petten, J. Nucl. Mater. **171** (1990) 31-36.

[72] MAKI, J.T., et al., NP-MHTGR Fuel Development Program Results, INEEL-EXT-02-01268, INL, Idaho Falls (2002).

[73] HAYASHI, K., et al., Irradiation Experiments of the 13th – 15th OGL-1 Fuel Assemblies, JAERI-Research 2000-001, JAERI (2000).

[74] MONTGOMERY, F.C., et al., HRB-22 Irradiation Test Data Report, ORNL/M-4080, Rev. 0, ORNL, Oak Ridge (1995).

[75] HAYASHI, K., et al., Distribution of fission and activation products in the graphite sleeves of HTGR fuel rods: First and second OGL-1 fuels, J. Nucl. Mater. **116** (1983) 233-243.

[76] FUKUDA, K., et al., Irradiation experiments of VHTR reference fuels by in-pile gas loop (OGL-1), J. Atomic Energy Society of Japan, **26 [1]** (1984) 57-74. (1984).

[77] HAYASHI, K., et al., Distribution of fission products in irradiated graphite materials of HTGR fuel assemblies: Third and fourth OGL-1 fuels, J. Nucl. Mater. **136** (1985) 207-217.

[78] HAYASHI, K., et al., Irradiation Experiments of the 6th – 12th OGL-1 Fuel Assemblies, JAERI-Research 94-017, JAERI (1994).

[79] INTERNATIONAL ATOMIC ENERGY AGENCY, Fuel Technology and Performance of Non-Water Cooled Reactors, IAEA-TECDOC-850 (1995).

[80] MINATO, K., et al., HRB-22 Capsule Irradiation Test for HTGR Fuel (JAERI/US DOE Collaborative Irradiation Test), JAERI-Research 98-021, JAERI (1998).

[81] ACHARYA, R., SAWA, K., HRB-22 Pre-irradiation Thermal Analysis, ORNL/M-3563, ORNL, Oak Ridge (1995).

[82] TOBITA, T., et al., Metallic Fission Product Release from Failed Coated Fuel Particles (ICF-51H Capsule Irradiation Test), JAERI-Research 96-014, JAERI (1996).

[83] MORRIS, R.N., et al., The Core Conduction Cooldown Test Facility: Current Status and Issues, ORNL/NPR-91/7, ORNL, Oak Ridge (1992).

[84] BULLOCK, R.E., Fission product release during post-irradiation annealing of several types of coated fuel particles, J. Nucl. Mater. **125** (1984) 304-319.

[85] SAWA, K., et al., Development of ceramics-coated particle fuel for very high-temperature gas-cooled reactors, J. Atomic Energy Society of Japan **6** [2] (2007) 113-125.

[86] DE GROOT, S., et al., Modelling of the HFR-EU1bis experiment and thermos-mechanical evaluation, Nucl. Eng. Des. **238** (2008) 3114-3120.

[87] TANG, C., et al., Fuel irradiation of the first batches produced for the Chinese HTR-10, Nucl. Eng. Des. **236** (2006) 107-113.

[88] LAURIE, M., et al., Results of the HTR-EU1 fuel irradiation of INET and AVR pebbles in the HFR Petten, Nucl. Eng. Des. **251** (2012) 117-123.

[89] TANG, C., et al., Irradiation testing of matrix material for spherical HTR-10 fuel elements, Nucl. Eng. Des. **238** (2008) 2886-2892.

[90] DONG, Y., Design, safety features & progress of HTR-PM, 24 January 24 2018, gif_webinar_htr-pm_dong_201801v2_final.

[91] KNOL, S., et al., HTR-PM Fuel pebble irradiation qualification in the high flux reactor in Petten, Nucl. Eng. Des. **329** (2018) 82-88.

[92] FREIS, D., et al., Burn-up determination and accident testing of HTR-PM fuel elements irradiated in the HFR Petten, Nucl. Eng. Des. **357** (2020) 110414.

[93] DE GROOT, S., et. Al, Fission-Product Behaviour during Irradiation of TRISO-Coated Particles in the HFREU1bis Experiment, HTR2008-58125 (2008).

[94] BARRACHIN, M., et al., Fission-product behaviour in irradiated TRISO-coated particles: Results of the HFR-EU1bis experiment and their interpretation, J. Nucl. Mater. **415** (2011) 104-116.

[95] RAY, I.L.E., THIELE, H., MATZKE, H.J., Transmission electron microscopy study of fission product behaviour in high burnup UO_2, J. Nucl. Mater. **188** (1992) 90-95.

[96] MINATO, K., et al., Fission Product behaviour in TRISO-coated UO_2 fuel particles, J. Nucl. Mater. **208** (1994) 266-281.

[97] KLEYKAMP, H., The chemical state of the fission product in oxide fuels, J. Nucl. Mater. **131** (1985) 221-246.

[98] FREIS, D., Störfallsimulationen und Nachbestrahlungsuntersuchungen an kugelförmigen Brennelementen für Hochtemperaturreaktoren, Ph.D. Thesis (2010) (in German)

[99] FREIS, D., et al., Accident testing of high-temperature reactor fuel elements from the HFR-EU1bis irradiation, Nucl. Eng. Des. **241** (2011) 2813-2821.

[100] LAURIE, M., et al., JRC's on-line fission gas release monitoring system in the high flux reactor Petten, Nucl. Eng. Des. **251** (2012) 325-329.

[101] DE GROOT, S., et al.," RAPHAEL-FT & generation IV PYCASSO-I irradiation", Proc. 4th Int. Top. Meeting on High Temperature Reactor Technology HTR2008, Washington, DC (2008).

[102] DEMKOWICZ, P.A., TRISO Fuel Part II: Accident Performance and NRC Engagement, INL/MIS-21-62430, INL, Idaho Falls (2021).

[103] PETTI, D., COLLIN, B., MARSHALL, D., A Summary of the Results from the DOE AGR Fuel Development and Qualification Program, INL-EXT-16-40784 Rev 0, INL (2017).

[104] DEMKOWICZ, P.A., AGR Program Highlights – 2020 ART-GCR Status Meeting, INL/MIS-20-60355 Rev. 0, INL (2020).

[105] MITCHELL, T.R., Technical Program Plan for the Next Generation Nuclear Plant Advanced Gas Reactor Fuel Development and Qualification Program, PLN-3636 Rev. 9, INL/MIS-10-20662, INL (2020).

[106] DEMKOWICZ, P., et al., "TRISO fuel experience and capabilities in the DOE advanced gas, reactor program", GAIN-EPRI-NEI Advanced Fuels Workshop, March 5-6, 2019, Boise State University, USA.

[107] DEMKOWICZ, P.A., et al., AGR-1 Post Irradiation Examination Final Report, INL/EXT-15-36407 Rev.0, INL (2015).

[108] SCHILTHELM, S., MCINTYRE, B., SLOAN, S., Uranium Oxycarbide (UCO) Tristructural Isotropic (TRISO) Coated Particle Fuel Performance, EPRI-AR-1 (NP), 3002015750, Electric Power Research Institute (EPRI) (2019).

[109] KIM, B.G., et al., The first irradiation testing and PIE or TRISO-coated particle fuel in Korea, Nucl. Eng. Des. **329** (2018) 34-45.

[110] MAKI, J.T., AGR-1 Irradiation Experiment Test Plan, INL/EXT-05-00593 Rev. 3, INL (2009).

[111] HUNN, J.D., LOWDEN, R.A., Data Compilation for AGR-1 Baseline Coated Particle Composite LEU01-46T, Rev. 1, ORNL/TM-2006/019, ORNL (2006).

[112] JELLISON, G.E. JR. and HUNN, J.D., Optical anisotropy measurements of TRISO nuclear fuel particle cross sections: the method. J. Nucl. Mater. **372** (2008) 36–44.

[113] HUNN, J.D., SAVAGE, T.W., KEHN, J.S., AGR-1 Fuel Compact Pre-Irradiation Characterization Summary Report, ORNL/TM-2012/295, ORNL (2012).

[114] MAKI, J.T., PARFUME User's Guide, INL/EXT-10-18317, INL (2010).

[115] PLOGER, S.A., et al., Microscopic analysis of irradiated AGR-1 coated particle fuel compacts, Nucl. Eng. Des. **271** (2014) 221-230.

[116] DEMKOWICZ, P.A., et al., Key results from irradiation and post-irradiation examination of AGR-1 UCO TRISO fuel, Nucl. Eng. Des. **329** (2018) 102-109.

[117] COLLIN, B.P., AGR-2 Irradiation Experiment Test Plan, PLN-3798 Rev. 2 (2018).

[118] COLLIN, B.P., AGR-2 Irradiation Test Final As-Run Report. INL/EXT-14-32277, Rev. 4 (2018).

[119] DEMKOWICZ, P.A., AGR-2 Post-Irradiation Examination Plan, PLN-4616 Rev. 0, INL/MIS-13-30534, INL (2013).

[120] DEMKOWICZ, P.A., et al., Results of the AGR-2 TRISO fuel performance demonstration irradiation experiment in the advanced test reactor, Annals of Nuclear Energy **150** (2021) 107833.

[121] STEMPIEN, J.D., et. al, AGR-2 TRISO Fuel Post-Irradiation Examination Final Report, INL/EXT-21-64279, INL (2021).

[122] GROVER, S.B., et al., Completion of the First NGNP Advanced Gas Reactor Fuel Irradiation Experiment, AGR-1, in the Advanced Test Reactor, INL/CON-10-18032, INL (2010); also Proceedings of HTR 2010 Prague, Czech Republic, October 18-20, 2010, Paper 104.

[123] HUNN, J.D., et al., Safety Testing and Destructive Examination of AGR-2 UCO Compacts 2-3-2 and AGR-2 UO_2 Compact 3-4-1, ORNL/TM-2018/956, ORNL (2018).

[124] HUNN, J.D., et al., Safety Testing and Destructive Examination of AGR-2 UCO Compacts 2-1-2, ORNL/TM-2019/1201, ORNL (2019).

[125] HUNN, J.D., et al., Safety Testing and Destructive Examination of AGR-2 UO_2 Compacts 3-1-1, ORNL/TM-2020/1451, ORNL (2020).

[126] GERCZAK, T.J., et al., AGR-2 Loose Particle Heating Tests in the Furnace for Irradiated TRISO Testing, ORNL/TM-2020/1715, ORNL (2020).

[127] HUNN, J.D., et al., Safety Testing and Destructive Examination of AGR-2 UCO Compacts 6-4-3, ORNL/TM-2019/1200, ORNL (2019).

[128] Advanced Reactor Technologies: Gas-Cooled Reactor Research and Development Quarterly Report, July, August, and September 2020, INL/EXT-21-61295, INL (2021).

[129] Final Safety Evaluation; Uranium Oxycarbide (UCO) Tristructural Isotropic (TRISO) Coated

Particle Fuel Performance: EPRI-AR-1(NP), EPRI; ML20216A453, (visited www.nrc.gov on Aug. 10, 2021).

[130] Advanced Reactor Technologies: Very High Temperature Reactor Research and Development Quarterly Report (Jan., Feb., Mar. 2020), INL/EXT-20-59634, Rev. 0, INL (2020).

[131] GERCZAK, T.J., HUNN, J.D., MORRIS, R.N., MONTGOMERY, F.C., SKITT, D.J., BALDWIN, C.A., DYER, J.A., ECKHART, B.D., analysis of fission product distribution and composition in the TRISO layers of AGR-2 fuel, Nucl. Eng. Des. **364** (2020) 110656.

[132] HUNN, J.D., BALDWIN, C.A., MONTGOMERY, F.C., GERCZAK, T.J., MORRIS, R.N., HELMREICH, G.W., DEMKOWICZ, P.A., HARP, J.M., STEMPIEN, J.D., Initial examination of fuel compacts and TRISO particles from the US AGR-2 irradiation test, Nuclear Engineering and Design **329** (2018) 89-101.

[133] RICE, F.J., STEMPIEN, J.D., DEMKOWICZ, P.A., Ceramography of irradiated TRISO fuel from the AGR-2 experiment, Nuclear Engineering and Design **329** (2018) 73-81.

[134] MORRIS, R.N., HUNN, J.D., BALDWIN, C.A., MONTGOMERY, F.C., GERCZAK, T., DEMKOWICZ, P.A., Initial results from safety testing of US AGR-2 irradiation test fuel, Nuclear Engineering and Design **329** (2018) 124-133.

[135] RICE, F.J., STEMPIEN, J.D., DEMKOWICZ, P.A., Ceramography of Irradiated TRISO Fuel from the AGR-2 Experiment, INL/EXT-16-39462, INL (2016).

[136] HUNN, J.D., et al., Destructive PIE and Safety Testing of Six AGR-2 UO2 Capsule 3 Compacts. ORNL/TM-2020/1612, ORNL (2020).

[137] COLLIN, B.P., AGR-3/4 Irradiation Experiment Test Plan, PLN-3867 Rev. 1, INL/MIS-11-22351, INL (2015).

[138] COLLIN, B.P., AGR-3/4 Irradiation Test Final As-Run Report, INL/EXT-15-35550, Rev. 1, INL (2016).

[139] HARP, J.M., et al., Initial Gamma Spectrometry Examination of the AGR-3/4 Irradiation, INL/CON-16-39738, INL (2016); also International Topical Meeting on High Temperature Reactor Technology (HTR 2016).

[140] DEMKOWICZ, P.A., AGR-3/4 Phase 2 Post-Irradiation Examination Plan, PLN-5382 Rev. 0, INL/MIS-17-41954, May 2017.

[141] STEMPIEN, J. et al., AGR-3/4 TRISO Fuel Compact Ceramography, INL/EXT-20-57610, INL (2020).

[142] EINERSON, J.J., et al., Analysis of fission gas release-to-birth ratio data from the AGR irradiations, Nucl. Eng. Des. **306** (2016) 14-23.

[143] PHAM, B.T., SCATES, D.M., AGR-1, AGR-2 and AGR-3/4 Release-to-Birth Ratio Data Analysis, INL/EXT-14-32970-Rev002, INL (2019).

[144] STEMPIEN, J.D., et al., Measurement of Fission product Concentration Profiles in AGR-3/4 TRISO Fuel Graphite Matrix and Nuclear Graphites, INL/EXT-21-62863, INL (2021).

[145] STEMPIEN, J.D., et al., AGR-3/4 Irradiation Test Train Disassembly and Component Metrology First Look Report, INL/EXT-16-38005, INL (2016).

[146] COLLIN, B.P., AGR-5/6/7 Irradiation Experiment Test Plan, PLN-5245, Rev. 1, INL (2018).

[147] STEMPIEN, J.D., AGR-5/6/7 Post-Irradiation Examination Plan, PLN-6110 Rev. 0, INL/EXT-20-59111, INL (2020).

[148] PHAM, B.T. et al., AGR-5/6/7 Irradiation Test Final As-Run Report, INL/EXT-21-64221, INL (2021).

[149] HELMREICH, G.W., HUNN, J.D., SKITT, D.J., DYER, J.A., SCHUMACHER, A.T. Acceptance Test Data for the AGR-5/6/7 Irradiation Test Fuel TRISO Composite. ORNL/TM-2017/037, ORNL (2017).

[150] STEMPIEN, J.D. et al., Initial Observations from Advanced Gas Reactor (AGR)-5/6/7 Capsule 1, INL/RPT-22-66720, Idaho National Laboratory (2022).

[151] KIM, B.G., et al., Irradiation Test Plan of Coated Particle Fuel at HANARO, KAERI/TR-4875/2013. KAERI (2013).

[152] DE GROOT, S., et al., HTR fuel coating separate effect test PYCASSO, Nucl. Eng. Des. **240** (2010) 2392-2400.

[153] KNOL, S., et al., PYCASSO: irradiation of HTR coated particles at high temperatures, Nucl. Eng. Des. **251** (2012) 150-156.

[154] KNOL, S., et al., PYCASSO: X-ray tomography on HTR coated particles, Nucl. Eng. Des. **271** (2014) 206-208.

[155] LIU, D., et al., X-ray tomography study on the crushing strength and irradiation behaviour of dedicated tristructural isotropic nuclear fuel particles at 1000°C, Mater. Des. **187** (2020) 108382.

[156] BROWN, N.R., A review of in-pile fuel safety tests of TRISO fuel forms and future testing opportunities in non-HTGR applications, J. Nucl. Mater. **534** (2020) 152139.

[157] FUKUDA, K., HAYASHI, K., SHIBA, K., "Fuel Behaviour and Fission Product Release under HTGR Accident Conditions-Fission Product Transport Processes in Reactor Accidents", Fission Product Transport Processes in Reactor Accidents ed. by J.T. Rogers, Hemisphere Publishing, New York, NY (USA), 1990, ISBN 0-89116-876-1, pp. 197-204. Hemisphere Publishing, 79 Madison Avenue, New York, NY 10016 (USA).

[158] UMEDA, M., SUGIYAMA, T., NAGASE, F., FUKETA, T., UETA, S., SAWA, K., Behaviour of coated fuel particle of high-temperature gas-cooled reactor under reactivity initiated accident conditions, J. Nucl. Sci. Technol. **47**[11] (2010) 991-997.

[159] OLANDER, D.R., Fundamental Aspects of Nuclear Reactor Fuel Elements, TID-26711-P1, Technical Information Centre, US Department of Energy (1976).

[160] LEWIS, B.J., THOMPSON, W.T., IGLESIAS, F.C., "Fission Product Chemistry in Oxide Fuels", Comprehensive Nuclear Materials, ed. by R.J.M. Konings, Volume 2, pp. 515-546, Elsevier.

[161] MINATO, K., OGAWA. T., FUKUDA, K., SHIMIZU, M., TAYAMA, Y., TAKAHASHI, I., Fission product behaviour in TRISO-coated UO_2 fuel particles, Journal of Nuclear Materials **208** (1994) 266-281.

[162] VAN ROOYEN, I.J., YANG, Y., HOLESINGER, T., BACHHAV, M., FU, Z., Microstructure and Fission Product Distribution Examination in the UCO kernel of TRISO Fuel Particles, INL/CON-18-44479-Revision-0, INL (2018); Also in the Proceedings of High Temperature Reactor Technology (HTR) 2018, Paper HTR 2018-3012, Warsaw, Poland, October 8-10, 2018.

[163] WRIGHT, K.E., VAN ROOYEN, I.J., "EPMA of irradiated AGR-1 TRISO fuel particles", Proceedings of 8th International Topical Meeting on High Temperature Reactor Technology, HTR2016, Las Vegas, 2016.

[164] VAN ROOYEN I.J., OLIVIER, E. J., NEETHLING, J.H., "Fission products silver, palladium and cadmium identification in neutron irradiated SiC using a Cs-corrected HRTEM", Proceedings of the 7th International Topical Meeting on High Temperature Reactor Technology, Weihai, China, October 27-31, 2014, (HTR2014), Paper HTR2014-31255.

[165] VAN ROOYEN, I.J., OLIVIER, E. J., NEETHLING, J.H., Fission products silver, palladium and cadmium identification in neutron irradiated SiC using a Cs-corrected HRTEM, Journal of Nuclear Materials **476** (2016) pp. 93-101.

[166] WEN, H., VAN ROOYEN, I.J., Distribution of fission products palladium, silver, cerium and caesium in the uncorroded areas of the locally corroded SiC layer of a neutron irradiated TRISO fuel particle, Journal of the European Ceramic Society **37** (2017) pp. 380051-3284.

[167] WEN, H., VAN ROOYEN, I.J., HUNN, J.D., GERCZAK, T.J., Electron microscopy study of Pd, Ag, and Cs in carbon areas in the locally corroded SiC layer in a neutron-irradiated TRISO fuel particle, Journal of the European Ceramic Society **38** (2018) pp. 978-4188.

[168] BOOTH, A.H., A Method of Calculating Fission Gas Release from UO_2 Fuel and its Application to the X-2-f Loop Test, AECL- 496 (1957).

[169] VAN DER VEN, A., Fission Product Transport in TRISO Particle Layers under Operating and Off-Normal Conditions, Nuclear Energy University Program, Project No. 10-924 (2014).

[170] JIANG, W., HALES, J.D., SPENCER, B.W., COLLIN, B.P., SLAUGHTER, A.E., NOVASCONE, S.R., TOPTAN, A., GAMBLE, K.A., GARDNER, R., TRISO particle fuel performance and failure analysis with BISON, Journal of Nuclear Materials **548** (2021) 152795.

[171] MILLER, G.K., PETTI, D.A., VARACALLE, D.J., MAKI, J.T., Consideration of the effects on fuel particle behaviour from shrinkage cracks in the inner pyrocarbon layer, Journal of Nuclear Materials **295** (2001) pp. 205-212.

[172] HOMAN, F.J., LONG, E.L., Jr., "Irradiation Performance of HTGR Recycle Fissile Fuel", ORNL/TM-5502 (1976).

[173] MONTGOMETY, F., "Fission-Product SiC Reaction in HTGR Fuel", GA 1485(Rev. 11/80), PED-81-2200 (Rev/1), DOE/ET/35300—T3, 1981.

[174] SAWA, K., TOBITA, T., Investigation of irradiation behaviour of sic-coated fuel particle at extended burnup, Nuclear Technology **142** (2003) pp. 250-259.

[175] MARTIN, D.G., SAWA, K., UETA, S., SUMITA, J., "A Study of Fuel Failure Behaviour in High Burnup HTGR Fuel – Analysis by STRESS3 and STAPLE Codes", JAERI-Research 2001-033 (2001).

[176] KOVACS, W.J., BONGARTZ, K., GOODIN, D., TRISO-Coated HTGR Fuel Pressure-Vessel Performance Models, GA-A16807 UC-77 (1983).

[177] WERNER, C.J., MCNP® User's Manual Code Version 6.2. LA-UR-17-29981 (2017).

[178] MOORE, R.L, SCHNITZLER, B.G., WEMPLE, C.A., BABCOCK, R.S., WESSOL, D.E., MOCUP: MCNP-ORIGEN2 Coupled Utility Program. INEL-95/0523 (1995).

[179] CROFF, G., A User's Manual for the ORIGEN2 Computer Code. ORNL/TM-7175 (1980).

[180] SHIM, H.J., HAN B.S., JUNG J.S., PARK, H.J., KIM, C.H., MCCARD: Monte Carlo Code for Advanced Reactor Design and Analysis. Nuclear Engineering and Technology **44** (2012) pp. 161-176.

[181] PROKSCH, E., STRIGL, A., NABIELEK, H., Production of Carbon Monoxide during Burn-up of UO_2 Kerneled HTR Fuel Particles, J. Nucl. Mater. **107** (1982) 280-285.

[182] HORSLEY, G.W. et al., Influence of irradiation temperature, burnup, and fuel composition on gas pressure (Xe, Kr, CO, CO2) in coated particle fuels, J. Am/. Ceram. Soc. **59** [1-2] (1976) 1-4.

[183] LINDEMER, T.B., Measurement and interpretation of CO and Kr + Xe in irradiated ThO_2-Containing HTGR fuel particles, Journal of American Ceramic Society **60** (1977) pp.409-416.

[184] HOMAN, F., NABIELEK, H., YANG, L., Low Enriched Fuel Particle Performance Review. GA-A14759 (1978).

[185] KOVACS, W.J., BONGARTZ, K., GOODIN, D.T., High-temperature gas-cooled reactor fuel pressure vessel performance models, Nuclear Technology **68** (1985) pp. 344-354.

[186] Deep-Burn Modular Helium Reactor Fuel Development Plan, ORNL/TM-2002/135, GA-224-0-TRT-000167. P. 3-17 (2002).

[187] RICHARDS, M.B., SHENOY, A.S., BROWN, L.C., BUCKINGHAM, R.T., HARVEGO, E.A., PEDDICORD, K.L., REZA, S.M.M., COUPEY, J.P., "H2-MHR pre-conceptual design report: SI-based plant", GA-A25401, P. 3-40 (2002).

[188] Development of Improved Models and Designs For Coated-Particle Gas Reactor Fuels, INEEL/EXT-05-02615 (2004).

[189] BECK, S.D., The Diffusion of Radioactive Fission Products from Porous Fuel Elements, BMI-1433 (1960).

[190] REDLICH, O., KWONG, J.N.S., On the thermodynamics of solutions; an equation of state; fugacity of gaseous solutions, Chemical Reviews **44** [1] (1949) pp. 233-244.

[191] Outotec Oy, Outotec HSC Chemistry® 7.1. Outotec Oy, Finland, 2009.

[192] NP-MHTGR Material Models of Pyrocarbon and Pyrolytic Silicon Carbide. CEGA-002820, Rev. 1 (1993).

[193] JACKSON, H.F., LEE, W.E., "2.13 properties and characteristics of ZrC", Comprehensive Nuclear Materials, ed. by J.M. Konings, Elsevier Ltd. (2012).

[194] KAAE, J.L., On Irradiation-induced creep of pyrolytic carbon in a general state of stress, Journal of Nuclear Materials **34** (1970) pp. 206-208.

[195] PRICE, R.J., Properties of silicon carbide for nuclear fuel particle coatings, Nuclear Technology **35** (1977) pp. 320-336.

[196] TIMOSHENKO, S.P., GOODIER, J.N., *Theory of Elasticity*, 3rd ed. McGraw-Hill, Inc. (1970).

[197] BOER, B., OUGOUAG, A.M., KLOOSTERMAN, J.L., MILLER, G.K., Stress analysis of coated particle fuel in graphite of high-temperature reactors, Nuclear Technology **162** (2008) pp. 276-292.

[198] VERFONDERN, K., "3.24 TRISO fuel performance modelling and simulation", Comprehensive Nuclear Materials, Ed. by J.M. Konings, Elsevier Ltd. (2012).

[199] PEARSON, R.L., LINDEMER, T.B., BEAHM, E.C., Simulated Fission Product-SiC Interaction in TRISO-Coated LEU or MEU HTGR Fuel Particles. ORNL/TM-6991 (1980).

[200] PEARSON, R.L., LAUF, R.J., LINDEMER, T.B., The Interaction of Palladium, the Rare Earths, and Silver with Silicon Carbide in HTGR Fuel Particles, ORNL/TM-8059 (1982).

[201] TIEGS, T.N., Fission product Pd-SiC interaction in irradiated coated-particle fuels, Nuclear Technology **57** (1981) pp. 389-398.

[202] STANSFIELD, O.M., HOMAN, F.J., SIMON, W.A., TURNER, R.F., Interaction of Fission Products and SiC in TRISO Fuel Particles: A Limiting Design Parameter, GA-A17183 (1983).

[203] VAN ROOYEN, I.J., LILLO, T.M., WU, Y.Q., 2014. Identification of silver and palladium in irradiated TRISO coated particles of the AGR-1 experiment. Journal of Nuclear Materials **446** (2014) pp. 178-186.

[204] MINATO, K., et al., Fission product palladium-silicon carbide interaction in HTGR fuel particles, J. Nucl. Mater. **172** (1990) 184-196.

[205] MONTGOMERY, F., Fission product SiC Reaction in HTGR Fuel. GA-905837 (1981).

[206] MINATO, K., OGAWA, T., "3.08. Advanced concepts in TRISO fuel", Comprehensive Nuclear materials, 1st Ed., Editors-in-Chief; Rudy M. Konings and Roger E. Stoller, Vol. 3; Advanced Fuels/Fuel Cladding/Nuclear Fuel Performance Modelling and Simulation, Elsevier Inc. (2012).

[207] HANSON, D.L., A Review of Radionuclide Release from HTGR Cores during Normal Operation. EPRI Report 1009382 (2004).

[208] MILLER, G.K., MAKI, J.T., KNUDSON, D.L., PETTI, D.A., Calculating failure probabilities for TRISO-coated fuel particles using an integral formulation, Journal of Nuclear Materials **399** (2010) pp. 154-161.

[209] BOX, G.E.P., MULLER, M.E., A note on the generation of random normal deviates, The Annals of Mathematical Statistics **29** (1958) 610-611.

[210] GREEN, A.E., BOURNE, A.J., *Reliability Technology*, John Wiley & Sons Ltd. (1958).

[211] KROHN, H., FINKEN, R., FRESCO-II: Ein Rechenprogramm zur Berechnung der Spaltproduktfreisetzung aus kugelförmigen HTR-Brennelementen in Bestrahlungsund Ausheizexperimenten, Rep. Jül-Spez-212. Research Centre Jülich (1983) (in German).

[212] MARTIN, R.C., Compilation of Fuel Performance and Fission Product Transport Models and Database for MHTGR Design (1993).

[213] MYERS, B.F., BELL, W.E., Caesium Transport Data for HTGR Systems, GA-A13990 (1979).

[214] HANSON, D.L., Software Requirement for Predicting Radionuclide Transport, PC-000537, Rev. 0. General Atomics (2007).

[215] SHIRLEY, G., User's Manual and Description for GARGOYLE: A Fuel Cycle Analysis Code with a Decay Heat Calculation Capability. CEGA-002922, Rev. 0. CEGA Corporation (1993).

[216] EICHENBERG, T.W., RADC Users Manual, CEGA-002814 (1993).

[217] HUDRITSCH, W.W., RANDI, A Zero Dimensional Computer Program for Calculating HTGR Activities, GA-A14091 (1977).

[218] HANSON, D.L., Review of Tritium Behaviour in High Temperature Gas-Cooled Reactors, PC-000535, Rev. 0. General Atomics (2006).

[219] PELESSONE, D., PISA: A One-Dimensional Spherically Symmetric Computer Program to Perform Thermal and Stress Analysis of Irradiated Fuel Particles, General Atomics Unpublished report CEGA-M-92-2052 (1992).

[220] BESMANN, T.M., SOLGASMIX-PV, a Computer Program to Calculate Equilibrium Relationships in Complex Chemical Systems, ORNL/TM-5775 (1977).

[221] PFREMMER, R.D., Software Description and User's Manual for GT MHR Fuel Performance Code, SURVEY. GA Document 911009/0 (2002).

[222] TZUNG, F., COPAR-FD, A Finite Difference Program to Compute Release of Metallic Fission Products from Coated Fuel Particles: Code Theory and Users Manual, CEGA-002098 (1992).

[223] TZUNG, F., TRAFIC-FD, A Finite Difference Program to Compute Release of Metallic Fission Products from an HTGR Core: Code Theory and Users Manual, CEGA-001904 (1992).

[224] CADWALLADER, G.J., SORS/NP1 Code Description and User's Manual, CEGA-002092 (1993).

[225] TANGIRALA, V., BETTS. W., OXIDE-4 Software Design Description and User's Manual, CEGA-001871, Rev. 0 (1993).

[226] HUDRITSCH, W.W., SMITH, P.D., PADLOC, A One Dimensional Computer Program for Calculating Coolant and Plateout Fission Product Concentrations, USDOE Report GA-A14401 (1977).

[227] HUDRITSCH, W.W., PADLOC, A One Dimensional Computer Program for Calculating Coolant and Plateout Fission Product Concentrations, Part II. GA-A14401 (1981).

[228] BOLIN, J.M., POLO Users Manual, DOE-HTGR-88332, Rev. 0 (1989).

[229] CHANIN, D., YOUNG, M.L., Code manual for MACCS2: Volume 1, User's Guide, NUREG/CR-6613 SAND97-0594 Vol. 1 (1998).

[230] PELLETIER, M., NABIELEK, H., ABRAM, T., MARTIN, D., HTR-F Project: Selection of properties and models for the coated particle – Fuel kernel. Commissariat a l'Energie Atomique et aux Energies Alternatives, Technical Note SESC/LSC 03-028, August 2003.

[231] BRADLEY, J.S., RICHARDS, M.B., Software Design Description and User's Manual for CAPPER Irradiation Capsule Performance Computer Code, CEGA-002309. General Atomics, San Diego, CA (1992).

[232] SCHWARTZ, M.H., SEDGELEY, D.B., MENDONCA, M.M., SORS: Computer Programs for Analyzing Fission Product Release from HTGR Cores During Transient Temperature Excursions, GA-A12462 (1974).

[233] HUMPHRIES, L.L., BEENY, B.A., GELBARD, F., LOUIE, D.L., PHILLIPS, J., Computer Code Manuals, Vol. 1: Primer and Users' Guide, Version 2.2.9541 2017. SAND2017-0455 O (2017)

[234] MILLER, G.K., PETTI, D.A., MAKI, J.T., KNUDSON, D.L., PARFUME Theory and Model Basis Report, INL/EXT-08-14497 (2009).

[235] PHÉLIP, M., MICHEL, F., PELLETIER, M., DEGENEVE, G., GUILLERMIER, P., "The ATLAS HTR fuel simulation code objectives, description and first results", #Paper B03. In: 2nd International Topical Meeting on High Temperature Reactor Technology, Beijing, China, September 22-24, 2004.

[236] KIM, Y.M., JO, C.K., COPA Ver. 1.0: Theory Report. KAERI/TR-7945/2019 (2019).

[237] COLLIN, B.P., AGR-1 Irradiation Test Final As-Run Report, INL/EXT-10-18097 Rev. 3 (2015).

[238] VERFONDERN, K., NABIELEK, H., PANAMA Ein Rechenprogramm zur Vorhersage des Partikelbruchanteils von TRISO-Partikeln unter Störfallbedingungen. Report Jül-Spez-298. Research Centre Jülich, Germany (1985) (in German).

ANNEX I.

LIST OF MODULAR HTGRS AND ADVANCED HTGR-TYPE REACTORS

Table I-1 provides a list of modular HTGRs and advanced HTGR-type reactors, taken from Ref. [I-1].

REFERENCES

[I-1] INTERNATIONAL ATOMIC ENERGY AGENCY, Advances in Small Modular Reactor Technology Developments, A Supplement to: IAEA Advanced Reactors Information System (ARIS), 2022 Edition, IAEA, Vienna (2022).

Table I-1 LIST OF MODULAR HTGRS AND ANDVANCED HTGR-TYPE REACTORS

Reactor	Type	Output MW(th)/MW(e)	Fuel type	Enrichment (%)	Core discharge burnup (GWd/t)	Temp. (ºC) (Inlet/outlet)	Designers	Country	Status
HIGH TEMPERATURE GAS COOLED SMALL MODULAR REACTORS									
HTR-PM	HTGR	2×250/210	Sphere (TRISO coated particle fuel)	8.5	90	250/750	INET, Tsinghua University	China	In operation
StarCore	HTGR	Block One: 35/14 Block Two: 50/20 Block Three: 150/60 One to six modules /plant	TRISO prismatic	15	60	280/750	StarCore Nuclear	Canada/UK/USA	Pre-conceptual design
GTHTR300	HTGR	<600/100–300	UO_2 TRISO ceramic coated particle	14	120	587–633/ 850-950	JAEA	Japan	Basic Design
GT-MHR	HTGR	600/288	Coated particle fuel in compacts, hexagonal prism graphite blocks of 0.36 m	14–18 LEU or wt.Pu	100–720 (depends on fuel type)	490/850	JSC Afrikantov OKBM	Russian Federation	Preliminary design completed
MHR-T	HTGR	4×600/4×205.5	Coated particle fuel in compacts, hexagonal prism graphite block of 0.36 m	< 20 LEU	125	578/950	JSC Afrikantov OKBM	Russian Federation	Conceptual design
MHR-100	HTGR	215/25–87 (depends on configuration)	Hexagonal prism graphite blocks with coated particle fuel	LEU < 20	-	490–553/795–950 (depends on configuration)	JSC Afrikantov OKBM	Russian Federation	Conceptual design
PBMR-400	HTGR	400/165	Pebble bed with coated particle fuel	9.6 LEU or wt.Pu	-	500/900	PBMR SOC Ltd	South Africa	Preliminary design
A-HTR-100	HTGR	100/50	Pebble bed with coated particle fuel	LEU or WPu	86	406/1200	Eskom Holdings SOC Ltd	South Africa	Conceptual design completed
HTMR-100	HTGR	100/35 single module plant	TRISO particles in pebbles: LEU, Th/LEU, Th/HEU or Th/Pu	Various*	80–90	250/750	Steenkampskraal Thorium Limited	South Africa	Basic design
Xe-100	HTGR	200/82.5	UCO TRISO/pebbles	15.5	165	260/750	X-Energy LLC	USA	Basic design
SC-HTGR	HTGR	625/272	UCO TRISO particle fuel in hexagonal graphite blocks	14.5 ave. / 18.5 max.	165	325/750	Framatome, Inc.	USA	Preliminary design

Table I-1 LIST OF MODULAR HTGRS AND ANDVANCED HTGR-TYPE REACTORS

Reactor	Type	Output MW(th)/MW(e)	Fuel type	Enrichment (%)	Core discharge burnup (GWd/t)	Temp. (°C) (Inlet/outlet)	Designers	Country	Status
HTR-10	HTGR	10/2.5	Spherical elements with TRISO particles fuel (UO_2 kernel)	17	80	250/700	INET, Tsinghua University	China	Operable
HTTR-30	HTGR	30/	UO_2 TRISO ceramic coated particle	3 – 10 (6 avg.)	22 (33 max.) Ave. fuel discharged burnup, GWd/tHM	395/850 (950 max.)	JAEA	Japan	In operation
RDE/Micro-PeLUIt	HTGR	10/3	Spherical elements with coated particle fuel	17	80	250/750	BATAN	Indonesia	Conceptual design
MOLTEN SALT SMALL MODULAR REACTORS									
KP-FHR	Pebble-bed salt cooled Reactor	320/140	TRISO particles in graphite pebble matrix / pebble bed	19.75	-	550/650	KAIROS Power, LLC.	USA	Conceptual design
Mk1 PB-FHR	FHR	236/100	TRISO particles in graphite pebble matrix / pebble bed	19.9	180 (Fuel burnup)	600/700	University of California at Berkeley	USA	Pre-conceptual design
MICRO-SIZED SMALL MODULAR REACTORS									
Energy Well	FHTR	20/8	TRISO	15	70	650/700	Centrum výzkumu Řež	Czech Republic	Pre-conceptual design
U-Battery	HTGR	10/4	TRISO / Hexagonal	< 20	~80 (ave.)	-	Urenco	UK	Conceptual design
Westinghouse eVinci (MMR)	Heat pipe	7-12/2-3.5	TRISO or another encapsulation	5 – 19.75	Not Disclosed	NA800 (Core temperature)	Westinghouse Electric Company, LLC.	USA	Conceptual Design Completed
MMR	HTGR	15/>5	FCM or TRISO graphite/Hexagonal	HALEU 19.75	> 60	Helium 300/630, solar salt stored at 560	USNC (Ultra Safe Nuclear Corporation)	USA	Preliminary design
TCR*	HTGR	~3/	UCO TRISO or UN TRISO FCM	< 19.75 HALEU <20			ORNL	USA	Conceptual design

ANNEX II.

EXAMPLE OF FUEL SPECIFICATIONS

Specifications of some coated particle fuel designs are described in this Annex.

TABLE II-1. SPECIFICATION OF US NEW PRODUCTION REACTOR (NPR) FUEL COMPACT

FUEL KERNEL	
Material	UCO
Enrichment (wt.%)	$93.15^{+0.15}_{-1.00}$
Impurities (wt.-ppm)	5000
C/U atomic ratio	0.5
O/U atomic ratio	1.4 – 1.7
Diameter (μm)	195^{+15}_{-50}
Density (g/cm^3)	10.3
COATINGS	
Buffer layer thickness (μm)	90 – 110
Buffer layer density (g/cm^3)	0.80 – 1.10
IPyC thickness (μm)	40 – 60
IPyC density (g/cm^3)	1.85 – 1.95
IPyC BAF	1.20
SiC thickness (μm)	35 – 40
SiC density (g/cm^3)	3.18
OPyC thickness (μm)	30 – 50
OPyC density (g/cm^3)	1.80 – 1.95
OPyC BAF	Not specified
PPyC[20] thickness (μm)	40 – 60
PPyC density (g/cm^3)	0.80 – 1.10
FUEL COMPACT	
Length (mm)	49.3 ± 0.5
Diameter (mm)	12.37 – 12.62
Burnable impurities (B, Cd, Eu, Gd, Li, Sm) (ppm EBC[21])	5
Non-burnable impurities (Na, S, Ca, Yb, Ti, V, Cr, Lu, Mn, Fe, Co, Al, In, Ta, Cs, La, Ce, W, Pr, Nd, Tb, Ho, Er, Tm) (ppm EBC)	1
Total free uranium fraction	6×10^{-5}

* Data taken from [II-1].

[20] PPyC stands for protective (or overcoating) pyro carbon.
[21] EBC stands for Equivalent Boron Concentration.

TABLE II-2. SPECIFICATION OF US AGR FUEL COMPACT

	Specification of fuels for irradiation tests AGR-5/6/7	**Specification of fuels for irradiation test AGR-1**
FUEL KERNEL		
Material	UCO	UCO
Enrichment (%)	15.5 ±0.1	19.8±0.1
Diameter (mm)	425 ± 10, also a critical region[22] was defined as < 375 and > 475	350 ± 10, also a critical region was defined as < 300 and > 400
Density (g/cm³)	10.4	10.4
Uranium fraction (gU/gU-CO)	0.885	0.87
C/U atomic ratio	0.4 ±0.1	0.5 ±0.2
O/U atomic ratio	1.5 ± 0.2	1.5 ± 0.2
(C + O) / U atomic ratio	2.0	2.0
Individual impurities (wt.-ppm): Li, Na, Al, Cl, Ca, V, Cr, Mn, Fe ©, Co, Ni, Cu, and Zn	100 each	100 each
Process impurities (wt.-ppm): P, S	1500 each	1500 each
Aspect ratio or sphericity (ellipticity)	Not specified, but a critical region was defined as 1.05, it was also indicated the fraction of composite kernels in the critical region to be ≤ 0.10	Not specified, but a critical region was defined as 1.05, it was also indicated the fraction of composite kernels in the critical region to be ≤ 0.10
COATINGS		
Buffer layer thickness (mm)	100 ± 15	100 ± 15
Buffer layer density (g/cm³)	1.05 ± 0.10	1.03 ± 0.15
IPyC thickness (mm)	40 ± 4	40 ± 4.
IPyC density (g/cm³)	1.90 ± 0.05	1.90 ± 0.05
IPyC BAF	Not provided.	Not provided.
SiC thickness (mm)	35 ± 3	35 ± 3
SiC density (g/cm³)	3.19	3.19
SiC aspect ratio	Not specified, but a critical region was defined as 1.14.	Not provided.

22 How to determine the critical region is not reported.

TABLE II-2. SPECIFICATION OF US AGR FUEL COMPACT

	Specification of fuels for irradiation tests AGR-5/6/7	**Specification of fuels for irradiation test AGR-1**
OPyC thickness (mm)	40 ± 4	40 ± 4
OPyC density (g/cm^3)	1.90 ± 0.05	1.90 ± 0.05
OPyC BAF	Not provided.	≤Not provided.
Defective IPyC coating fraction	1.0×10^{-4}	2.0×10^{-4}
Defective SiC coating fraction	1.0×10^{-4}	1.0×10^{-4} also another value of 2.0×10^{-4} was defined
Defective OPyC defect fraction	1.0×10^{-4}	0.01
Missing OPyC layer defect fraction		3.0×10^{-4}
FUEL COMPACT		
Mean uranium loading (gU/compact) Nominally 40% packing fraction Nominally 25% packing fraction	1.36 ± 0.10 0.90 ± 0.08	0.905 ± 0.04
Length (mm)	Not specified, but critical limits were set as 24.4 and 25.3.	Not specified, but a lower critical limit of 25.02 and an upper limit 25.40 were set.
Diameter (mm)	Not specified, but critical limits were set as 12.20 and 12.44.	Not specified, but a lower critical limit of 12.22 and an upper limit of 12.46 were set.
Matrix density ($g \times cm^{-3}$)	1.65	Not provided.
Iron (Fe) outside of SiC per compact (mg)	25, also a critical limit of ³ 100 was set.	25, also an upper critical limit of 100 was set.
Transition metals (Cr, Mn, Co, and Ni) outside SiC per compact (mg)	50 each element.	≤ 75 each element
Calcium (Ca) outside SiC per compact (mg)	50	90
Aluminium (Al) Al outside SiC per compact (mg)	50	45
Titanium (Ti) and Vanadium (V) outside SiC per compact (mg)	240	400
Chlorine (Cl) content outside SiC in compact (wt.- ppm)	Not provided	30
Dispersed uranium fraction (DUF)[23]; (g $U_{leached}$/g U_{sample})	1.0×10^{-5}	1.0×10^{-4}

---

[23] In Refs. [II-2, II-3], a term “Heavy metal contamination fraction, g exposed U/g U in compact” is used, and it is believed that it is equivalent to DUF. In addition, these terms somewhat are equivalent to total free uranium fraction, which includes both terms: matrix contamination and particles with defective SiC.

TABLE II-3. SPECIFICATION FOR GERMAN FUEL PEBBLES

FUEL KERNEL	
Material	UO_2
Enrichment (wt%)	8.0 ± 0.1
Diameter (mm)	480 – 520
Density (g/cm^3)	10.4
Sphericity	
Fraction of odd-shaped kernels	
COATING LAYERS	
Buffer layer thickness (mm)	72 – 108
Buffer layer density (g/cm^3)	1.05
IPyC thickness (mm)	30 – 50
IPyC density (g/m^3)	1.91 ± 0.1
IPyC BAF	1.1
IPyC OPTAF	
SiC thickness (mm)	31 – 39
SiC density (g/cm^3)	3.18
OPyC thickness (mm)	25 – 45
OPyC density (g/cm^3)	1.91 ± 0.1
OPyC BAF	1.1
OPyC OAF	
FUEL ELEMENT MATRIX	
Density (g/cm^3)	1.75±0.02
Ash content (ppm)	
Thermal conductivity, at 1000°C (W/cm·K)	≥ 25
Standard specific corrosion rate (mg/cm^2·h)	≤ 1000°C
Standard abrasion rate (mg/cm^2·h)	
Crushing strength (kN)	≥ 18
Total free uranium fraction	6×10^{-5}

* Data taken from Ref. [II-1].

TABLE II-4. SPECIFICATION FOR JAPANESE HTTR FUEL COMPACTS

FUEL KERNEL	
Material	UO_2
Enrichment (wt%)	6
Diameter (mm)	600 ± 55
Density (g/cm^3)	10.63 ± 0.26
Impurity (ppm) – EBC	< 3.0
COATING LAYERS	
Buffer layer thickness (mm)	60 ± 12
Buffer layer density (g/cm^3)	1.1 ± 0.1
IPyC thickness (mm)	30 ± 6
IPyC density (g/cm^3)	$1.85^{+0.10}_{-0.05}$
SiC thickness (mm)	25^{+12}_{0}
SiC density (g/cm^3)	3.2
OPyC thickness (mm)	45 ± 6

TABLE II-4. SPECIFICATION FOR JAPANESE HTTR FUEL COMPACTS

OPyC density (g/cm^3)	$1.85^{+0.10}_{-0.05}$
IPyC/OPyC OAF	1.03
COATED FUEL PARTICLE	
Diameter (mm)	920^{+50}_{-30}
Sphericity	Up to 1.2
ANNULAR FUEL COMPACT	
Particle packing factor (vol%)	30 ± 3
Impurity (ppm) – equivalent boron content	5
Outer diameter (mm)	26.0 ± 0.1
Inner diameter (mm)	10.0 ± 0.1
Height (mm)	39.0 ± 0.5
Matrix density (g/cm^3)	1.70 ± 0.05
Compressive strength (N)	4900
Total free uranium fraction	2×10^{-3}

* Data taken from Ref. [II-1].

TABLE II-5. SPECIFICATION FOR CHINESE HTR-10 FUEL BEBBLES

FUEL KERNEL	
Material	UO_2
Diameter (mm)	500 ± 50
Density (g/cm^3)	10.4
Sphericity	< 1.2
O/U ratio	2.01
Fraction of odd-shaped kernels	5×10^{-4}
COATING LAYERS	
Buffer layer thickness (mm)	90 ± 36
Buffer layer density (g/cm^3)	1.10
IPyC thickness (mm)	40 ± 20
IPyC density (g/cm^3)	1.9 ± 0.1
SiC thickness (mm)	35 ± 10
SiC density (g/cm^3)	3.18
OPyC thickness (mm)	40 ± 20
OPyC density (g/cm^3)	1.9 ± 0.1
IPyC/OPyC OAF	1.03
GRAPHITE MATRIX	
Density (g/cm^3)	1.70
Total ash (ppm)	300
Li content (ppm)	0.3
Impurity (ppm) – equivalent boron content	3.0
Thermal conductivity (W/cm·K)	0.25 at 1000°C
Corrosion rate (mg/cm^2·h)	1.3 at 1000°C in He + 1 vol% H_2O
Erosion rate (mg/h)	6 per fuel element
Breaking loading (kN)	18
FUEL ELEMENT	
Diameter (mm)	59.6 – 60.2

TABLE II-5. SPECIFICATION FOR CHINESE HTR-10 FUEL BEBBLES

Thickness of fuel free shell (mm)	4.0 – 6.0
Total free uranium fraction	3×10^{-4}

* Data taken from Ref. [II-1].

TABLE II-6. SPECIFICATION FOR KOREAN KAERI FUEL PEBBLES

Properties	Design value	Measured value	Remarks
Kernel			
- Diameter (μm)	480 ± 30	483.50 ± 15	300–350 particles/batch (PSA*)
- Density (g/cm³)	10.65 ± 0.25	10.68 ± 0.023	6g of sample weight (Pyconometer)
- U-235 enrich. (wt%)	4.5 ± 0.10	4.504 ± 0.004	chemical analysis (TIMS)
- O/U ratio	2.00 ± 0.01	2.003 ± 0.002	30 mg of sample weight (TG/DTA)
- Total uranium (wt%)	≥ 87.0	88.13	cal. Value
- Sphericity (aspect ratio)	<1.2	≤ 1.04	average
Coated fuel particle			
- Buffer thickness (μm)	95 ± 45	102.91 ± 30	CG* (10 particles/batch)
- Buffer density (g/cm³)	1.00 ± 0.10	1.052 ± 0.002	X-ray (10 particles/batch)
- IPyC thickness (μm)	40 ± 20	40.55 ± 0.53	CG* (10 particles/batch)
- IPyC density (g/cm³)	1.85 ± 0.20	1.91 ± 0.01	DC* (10 particles/batch)
- Anisotropic index of IPyC	≤ 1.03	1.0165 ± 0.003	OM* (10 particles/batch)
- SiC thickness (μm)	35 ± 10	36.08 ± 0.15	CG* (10 particles/batch)
- SiC density (g/cm³)	≥ 3.18	3.182 ± 0.004	DC* (10 particles/batch)
- OPyC thickness (μm)	40 ± 20	46.3 ± 3.05	CG* (10 particles/batch)
- OPyC density (g/cm³)	1.85 ± 0.20	1.88 ± 0.07	DC* (10 particles/batch)
- Anisotropic index of OPyC	≤ 1.03	1.019 ± 0.004	OM* (10 particles/batch)
- Particle dia. (mm)	0.90 ± 0.10	0.95 ± 0.011	PSA* (10 particles/batch)
- Average wt. (g)	0.001392	0.001392	
Fuel compact			
- Compact mass (g)	1.050	1.028 ± 0.024	Average of 14 compacts
- Mean U loading (gU)	0.14	0.135	Average/compact
- Diameter (mm)	8.0 ± 0.2	7.981 ± 0.006	Average of 14 compacts
- Length (mm)	10.0 ± 0.5	9.995 ± 0.270	Average of 14 compacts
- No. of compact (ea)	9(rod1)/5(rod2)	9(rod1)/5(rod2)	
- Packing (volume) fraction (%)	20	19.703 ± 0.44	263 particles/compact
Matrix graphite (M) & structural graphite (G) specimen			
- Diameter (mm)	8.0 ± 0.2	7.976 ± 0.004	Average of 8 specimens
- Length (mm)	5 ± 0.2	5.028 ± 0.008	Average of 8 specimens
- No. of specimens(ea)	8(rod 2)	8(rod 2)	
- Density (g/cm³)	1.7 ± 0.1	1.770 ± 0.004	Average of 8 specimens

*CG: Ceramography; DC: Density column; PSA: Particle size analysis; OM: Optical microscopy

** Data provided by KAERI.

REFERENCES

[II-1] PETTI, D.A., MAKI, J.T., BUONGIORNO, J., HOBBINS, R.R., MILLER, G.K., Key Differences in the Fabrication, Irradiation and Safety Testing of US and German TRISO-coated Particle Fuel and Their Implications on Fuel Performance, INEEL/EXT-02-00300, INL, Idaho Falls (2002).

[II-2] MARSHALL, D.W., "AGR-5/6/7 Fuel Specification", SPC-1352, INL/MIS-11-21423, INL, Idaho Falls (2017).

[II-3] BARNES, C.M., AGR-1 Fuel Product Specification and Characterization Guidance, INL/MIS-05-00238-Revision-1, INL, Idaho Falls (2006)

LIST OF ABBREVIATIONS

ADU	Ammonium diuranate
ADUN	Acid-deficient uranyl nitrate
AGR	Advanced Gas Cooled Reactor
ALARA	As low as reasonably achievable
APT	Atom probe tomography
AVR	Arbeitsgemeinschaft Versuchs Reaktor in Germany
BAF	Bacon Anisotropy Factor
BEC	Backscattered electron composition
BISO	Bi-structure isotopic
BOL	Beginning of life
CNL	Canadian Nuclear Laboratories
CTE	Coefficient of thermal expansion
CVD	Chemical vapour deposition
DFT	Density functional theory
DTF	Design to fail
DLOFC	Depressurized loss-of-forced cooling
EPMA	Electron probe microanalyzer
FB-CVD	Fluidized bed-chemical vapor deposition
FCM fuel	Fully ceramic microencapsulated fuel
FHR	Fluoride Salt-Cooled High-Temperature Reactor
FIMA	Fission per initial metal atom
FP	Fission product
FPD	Full power day
FSV	Fort St. Vrain
GA	General Atom
GCR	Gas Cooled Reactor

Gen IV	Generation IV
GFR	Gas-cooled fast reactor
GT-MHR	Gas Turbine-Modular Helium Reactor
HALEU	High Assay Low Enriched Uranium
HEU	High enriched uranium
HFIR	High Flux Isotope Reactor
HTGR	High temperature gas-cooled reactor
HTTR	High Temperature Test Reactor in Japan
INET	Institute of Nuclear and New Energy Technology
INL	Idaho National Laboratory
IPyC	Inner isotropic pyrocarbon.
JAEA	Japan Atomic Energy Agency (former Japan Atomic Energy Research Institute, JAERI)
JMTR	Japan Materials Testing Reactor
KAERI	Korea Atomic Energy Research Institute
KCMI	Kernel-coating mechanical interaction
LASL	Los Alamos Scientific Laboratory
LEU	Low enriched uranium
LOCA	Loss-of-coolant accident
LWR	Light water reactor
MHTGR	Modular HTGR
MMR	Micro modular reactor
NCC	Numerical calculation case
NGNP	Next Generation Nuclear Plant
NPR	New production Reactor
NSRR	Nuclear Safety Research Reactor
OAF	Optical Anisotropy Factor
OPyC	Outer isotropic pyrocarbon
ORNL	Oak Ridge Nuclear Laboratory

PBMR	Pebble Bed Modular Reactor in South Africa
PIE	Post-irradiation examination
PyC	Isotropic pyrocarbon
PSA	Particl size analyzer
RE	Rare earth
R/B	Release to birth
R&D	Research and development
SEM	Scanning electron microscophy
SiC	Silicon carbide
SMR	Small modular reactor
STEM	Scanning transmission electron microscopy
TD	Theoretical density
TEM	Transmission electron microscopy
THTR	Thorium High Temperature Reactor
TRISO	Tri-structural isotropic
USNC	Ultra Safe Nuclear Corporation
UUNG	Uranium Naturel Graphite Gaz reactor in France
VHTR	Very High Temperature Reactor
XRD	X-Ray Diffraction analysis

CONTRIBUTORS TO DRAFTING AND REVIEW

Cheon, J-S.	Korea Atomic Energy Research Institute (KAERI), Republic of Korea
Demkowicz, P.	Idaho National Laboratory (INL), United States of America
Fomichenko, P.	National Research Centre "Kurchatov Institute", Russian Federation
Kim, B.G.	Korea Atomic Energy Research Institute (KAERI), Republic of Korea
Kim, Y.M.	Korea Atomic Energy Research Institute (KAERI), Republic of Korea
Liu, B.	Tsinghua University, China
Nabielek, H.	Research Center Jülich (FZJ), Germany
Onder, N.	Canada Nuclear Laboratories (CNL), Canada
Sim, K.S.	International Atomic Energy Agency (IAEA)

Consultants Meetings (virtual)

27–28 April 2021,

4–17 March 2022.

IAEA
International Atomic Energy Agency

CONTACT IAEA PUBLISHING

Feedback on IAEA publications may be given via the on-line form available at:
www.iaea.org/publications/feedback

This form may also be used to report safety issues or environmental queries concerning IAEA publications.

Alternatively, contact IAEA Publishing:

Publishing Section
International Atomic Energy Agency
Vienna International Centre, PO Box 100, 1400 Vienna, Austria
Telephone: +43 1 2600 22529 or 22530
Email: sales.publications@iaea.org
www.iaea.org/publications

Priced and unpriced IAEA publications may be ordered directly from the IAEA.

ORDERING LOCALLY

Priced IAEA publications may be purchased from regional distributors and from major local booksellers.

25-01175E

FSC
www.fsc.org
MIX
Papier | Fördert
gute Waldnutzung
FSC® C083411

Zeitfracht Medien GmbH
Ferdinand-Jühlke-Straße 7
99095 Erfurt, Deutschland
produktsicherheit@kolibri360.de